Biochemical Spectroscopy

Biochemical Spectroscopy

by

Richard Alan Morton, Ph.D., D.Sc., Sc.D., F.R.I.C., F.R.S.
Emeritus Professor of Biochemistry
University of Liverpool

Volume 1

A company now owned by
The Institute of Physics
Techno House
Redcliffe Way
Bristol BS1 6NX
England

First Published February 1975

ISBN 0 85274 286 X

A company now owned by
The Institute of Physics
Techno House
Redcliffe Way
Bristol BS1 6NX
England

Contents

VOLUME 2

13. Vitamins and Coenzymes

Preface

The first edition of my book on *The application of absorption spectra to the study of vitamins and hormones* (1939) included in its Introduction the following paragraphs:

> Studies in absorption spectra interest chemists from many different points of view. Applied to diatomic molecules and the simpler polyatomic molecules, they yield valuable information on the intimate structure of the molecules, so that moments of inertia, internuclear distances, valency vibrations, valence angles and characteristic frequencies may be deduced. For the large polyatomic molecules with which this booklet is concerned information of such precision is at present out of the question.
>
> Spectroscopic research in this field is thus much more empirical, and its value lies in its use *as an adjunct* to biochemical and organic methods of study. In spite, however, of this serious limitation, the recent developments in knowledge concerning physiologically active substances owe much to spectral absorption curves in the two directions of elucidating photochemical changes and in providing characteristic 'labels' for substances the existence or importance of which rests on biological methods of experimentation. Vitamin research must always accept the animal test as the first and last court of appeal; the main service of absorption spectra lies in the possibility (which may not, of course, always eventuate) of supplementing the physiological description of an X-substance by means of a physical criterion capable of aiding in identification and analysis.

The Preface to the second edition (1942) acknowledged my debt to E. C. C. Baly and I. M. Heilbron, and to the Medical Research Council which, through Edward Mellanby, supported my investigations on vitamins A, D and E and, incidentally, introduced me to inter-laboratory cooperation and the committee-round, from which escape is so difficult. In this Preface I wrote:

> In the last twenty years the subject of Absorption Spectra has ceased to be a minor and auxiliary speciality within chemistry and has become a very versatile tool, almost indispensable in many fields of research. The subject thus affords a good vantage point. More and more workers, having other main interests, find it necessary to make use of absorption spectra, and they need a certain minimum acquaintance with the methods of experimentation and interpretation. Similarly, the spectroscopists (of whom there are several varieties) need to see their subject against a wider background. This book is a contribution to the work of liaison. It reviews many brilliant papers, distinguished by patient work, great skill and insight, but it is much more a record of small advances, the cumulative effect of which is prodigious in its implications. The great days of line spectra are over, with the developments of quantum theory in the first third of the century, sub-atomics has become neat

and tidy. Similarly, quantum mechanics is providing an adequate theory of the absorption spectra of simple molecules. A great clarification of the theory of large molecules has begun to take shape and in the next few years the study of absorption spectra will play its part in a process destined to affect the outlook of all chemists.

Much of the quotation needs no apology today, although the naive reference to sub-atomics must raise a smile. (My first war-time work concerned gas-warfare, which, to the leading chemists of the day, was a more immediate threat than atomic energy.)

At the head of the first chapter of the second edition was the following quotation:

> 'Les données numériques qui caracterisent l'absorption inégale des diverses lumières conduiront peut-être prochainement à une méthode d'analyse chimique universelle.'
> (BERTHELOT, *Science et philosophie.*)

In 1941 nobody could have foreseen how today ultraviolet and infrared spectroscopy, nuclear magnetic resonance and mass spectroscopy together with optical rotatory dispersion, circular dichroism and other techniques have lightened the labours of chemists and biochemists. Berthelot's hopes have come true.

I have often been asked by the Publishers and by many scientific friends to write a third edition. My failure to do so was due first to pre-occupation with the work of a long succession of research students to whom I owe much. In the second place it was due to the burden of service on committees academic, official and semi-official, and in particular to the task of reading and understanding the relevant literature. Such work is doubtless necessary and I have felt it to be part of the price the academic scientist has to pay for the privileges he enjoys.

A good deal of water has flowed under the bridges since 1942 and it seems that an introduction to many aspects of spectroscopy has become necessary in undergraduate courses for chemists, biochemists and, indeed, many biologists. Thanks to advances in electronics and in the design of instruments generally, users are now more concerned with what an instrument can do rather than how it does it. The making of measurements has become easier, quicker and more accurate than was formerly the case. Moreover, there is no lack of introductory textbooks dealing with the theory and practice of the main spectroscopic methods.

I have therefore in the present volumes used a new title and have aimed at displaying various spectroscopic approaches in action, sometimes concerted and at other times working selectively, in the solution of biochemical problems. The modern biochemist needs a very wide background knowledge in both biology and chemistry, but he cannot be a master of every trade. The designers of scientific instruments have relieved him of many tasks and specialists in physical methods have assembled a body of classified information which he can put to good use.

Those of us whose interest in biochemistry began some fifty years ago have seen immense advances and have enjoyed participating in them and in learning about them. I have tried in this book to *interest* the reader in both the biochemistry and the spectroscopy, but I have also borne in mind the worker in a particular field who needs rather full references to the literature.

In the past twenty-five years new Schools of Biochemistry have multiplied all over the world and the number of students has increased greatly. The prevailing pattern of biochemical research is a phase of surging advance leading to rapid and large-scale consolidation, followed by diminishing returns and redirection of effort. In preparing this book I found plenty of 'applied' problems requiring attention and holding out promise. A mature science has responsibility to undertake research that is necessary, timely and practicable. It is encouraging to know that very many biochemists are doing that kind of work and gaining satisfaction from it.

I am indebted to the Leverhulme Trust for the award of an Emeritus Fellowship which permitted me to continue writing after retirement. Mrs M. Hilditch acted as part-time secretary under this arrangement and her assistance has been invaluable. The late Professor R. J. Pumphrey kindly offered me a room in the Zoology Department and Professor A. J. Cain extended the hospitality. I have throughout had the use of the University Library and have often visited the Library of the Chemical Society at Burlington House. I wish also to thank the various societies and authors whose diagrams I have used and to apologize for any inadvertent failure to contact authors for permission. The sources of most diagrams are given in the figure captions, with a complete reference at the end of the chapter, but the following should also be included: Fig. 2.4, Golterman & Clymo (1967); Figs. 4.1–4.7, Morton *et al.* (1934); Fig. 4.10, Morton & Stubbs (1939); Figs. 4.11–4.23, Morton & Stubbs

(1940); Figs. 6.6–6.8, 6.11–6.12, Mayneord & Roe (1935); Fig. 8.9, Wetlaufer *et al.* (1959); Fig. 8.23, Bendit (1967); Figs. 8.30–8.31, Shifrin (1965); Figs. 9.6–9.11, Mason (1959); Fig. 9.14, Hearn *et al.* (1951); Figs. 13.24–13.25, Siegel *et al.* (1959); Fig. 15.18, Crawford & Jensen (1971); Fig. 16.10, Dowling (1960); Fig. 17.1, White *et al.* (1963); Figs. 17.3–17.4, Kishi *et al.* (1966).

Finally I must thank my wife for her tolerance and patience as well as her encouragement over the years.

The University of Liverpool
October 1974

1 Introduction

1.1 Introductory

In recent years the task of establishing the structures of most new natural products has been made far easier than once seemed possible. High resolution mass spectroscopy leads to accurate molecular weights and sets definite limits to possible empirical formulae. Fragmentation patterns, nuclear magnetic resonance, infrared absorption, Raman spectra, ultraviolet absorption, optical rotatory dispersion and circular dichroism all provide evidence such that chemists today enjoy a flying start. X-ray crystallography too makes its own superb contribution.

The designers and makers of elaborate scientific instruments have greatly facilitated the gathering of reliable information. Most of the techniques are now explained, at least in principle, in undergraduate courses. Each physical method has, however, specialists thrusting into new territory and revealing more and more sophisticated aspects of structure and function. An important aspect of the new situation is that a good part of the effort once needed for recognizing and characterizing the molecules participating in complex biological processes is now deflected to elucidating modes of action. Biochemists should be able to interpret the information derived from a wide range of spectroscopic investigations. They will then be more free, and more competent, to clarify the roles of individual compounds in living organisms.

In the period immediately after the 1914–1918 War experimental work on infrared absorption spectra of organic compounds was a task for specialists. Coblentz had laid down firm foundations but serious technical difficulties persisted and even as late as 1938 it could be said that 'the technique for accurate work is neither easily nor cheaply acquired'. The advances in instrumentation which came later gradually reduced to simple routine much of the gathering of data.

The pioneers of absorption spectroscopy in the ultraviolet and visible regions (Hartley, Dobbie, Hantzsch, Baly and others) managed to obtain only semi-quantitative curves. They had to use photographic methods for recording spectral transmission by solutions of organic substances, and although the blackening of the developed photographic plate could be measured accurately, both theoretical and practical difficulties remained. In particular when the product It was constant (I = light intensity, t = time of exposure) the blackening was not constant. In fact a relation It^n = constant held, but *n* (the Schwarzschild exponent) showed considerable variation between different kinds of photographic plates. The advent of suitable photometers permitted reasonable accuracy to be achieved in the measurement of absorption spectra in the visible and ultraviolet regions. Twyman's sector photometer (invented just before the 1914–1918 War) came into wide use among specialists. The light source was a high tension condensed electric spark between metallic electrodes. A long-focus sector photometer was made specially for Baly's group at Liverpool so as to allow an intense arc between iron and nickel rods to be used instead of a spark. Various improved photometers were devised (Judd-Lewis, Spekker) but the Twyman instrument proved easier to use. It was not theoretically impeccable (because of the Schwarzschild factor) but there was a compensatory intermittency effect on the blackening of the plate. The measurements remained tedious and some skill and experience was needed for $\pm 2\%$ accuracy in measuring extinction coefficients. The light sources moreover emitted line spectra, a fact which militated against the measurement of absorption displaying fine structure, particularly in vapour spectra. Light sources with continuous spectra were available for work in the visible and near-ultraviolet regions but there was still need for a stable light source having a spectrum continuous down to 200 nm. An early attempt in that direction was a condensed spark between aluminium electrodes under water. The spectrum was continuous but the spark was erratic and very noisy. An underwater spark between tungsten electrodes (Fulweiler & Barnes, 1922) was perfected at the Bureau of Standards by Brode. It involved a rotatory auxiliary spark gap in air and a spark under water actuated by a Tesla discharge at a very high voltage. This light source, in conjunction with a short-focus Twyman sector photometer, proved highly satisfactory apart from its noise and the fact that it was a source of electrical interference.

A hydrogen discharge tube associated with the names of Bay and Steiner was a further development. Different versions of hydrogen 'lamps' were all at first quite sizable and had to be connected to a source of (purified) hydrogen; they were however noiseless and provided a continuous spectrum down to 210 nm.

Considerable advances in electronics led to the use of small and highly reliable hydrogen 'lamps' which today form part of standard photoelectric spectrophotometers. They are now taken very much for granted but they were however the result of a prolonged effort to find the best light source for a new generation of spectrophotometers.

Visual spectrophotometery over the range 440–760 nm continues to have some advantages and the combination of the Hilger constant deviation spectrometer and the Nutting (or König–Martens) photometer has a fine record of work done. Nevertheless this technique is obsolescent and for many purposes recording spectrophotometers meet the needs better.

Photoelectric spectrophotometery began with very 'individual' assemblies using a mercury vapour lamp as light source and the work of von Halban (von Halban & Siedentopf, 1922) overcame numerous difficulties. Historians of the subject will find many other significant developments (cf. Suhrmann & Kollath, 1928; Warburg *et al.*, 1929). The work of Michaelson & Liebhafsky (1936) assisted Hardy in his important studies which led to recording ultraviolet spectrophotometers.

The appearance of the Cary & Beckman spectrophotometers, followed by many rather similar instruments, led to an enormous widening of the field of users of ultraviolet absorption. The manually operated photoelectric spectrophotometers were capable of an accuracy of $\pm 0{\cdot}2\%$ and this allowed a fresh approach to analytical work particularly in respect of correction for irrelevant absorptions. On the technical side Edisbury's book (1966) is of great value to new entrants to this field.

With respect to infrared spectroscopy applied to organic substances, here too remarkable advances in instrumentation have been made with parallel developments in the interpretation of spectra. Rock salt prisms remain widely used although prisms of KBr, LiF, CaF_2 and CsBr are also used. Prism-grating spectrophotometers and precision-grating spectrometers are now available. Selection of an instrument depends on the nature of the problem to be tackled. Some relatively cheap recording spectrophotometers are easy to use and maintain and give excellent service for semi-quantitative work, while the better performance of more expensive instruments meets special needs. The literature distributed by manufacturers is very helpful and most centres of organic or biochemical research now have workers experienced in the field. Infrared instrumentation is discussed briefly by Rao and more fully by Conn & Avery (1960), Goddu (1960) and West (1960), and many others.

Nuclear magnetic resonance was first observed in 1946 but its advantages in the study of organic structures only began to emerge in 1953 (Meyer *et al.*). In twenty years a vast amount of work has been done and the technique has undergone progressive improvements. The main reasons for this effort are that the information provided is different from, but complementary to, that obtained by ultraviolet and infrared investigations. Nuclear magnetic resonance (n.m.r.) has added a new dimension to structural studies. In many problems it can be used easily but the subject has complications demanding subtlety and expertise.

At the time that Aston carried out his early work on mass spectra, the notion that a molecular weight could for example be observed as 346·177 and calculated to be 346·178 for $C_{20}H_{26}O_5$ would have been barely conceivable. High resolution mass spectroscopy and fragmentation patterns have provided powerful new aids for natural-product chemists.

Optical rotatory dispersion and circular dichroism too contribute greatly to the solution of selected problems. Isotopic labelling has made success possible in many difficult investigations. X-ray crystallography has become for some problems the most powerful single approach.

At present instruction in the principles of the various physical methods is an integral part of training in chemistry and biochemistry and elementary expositions abound in the literature. Instrumentation has an increasingly special niche in scientific activity and the plurality of highly sophisticated techniques forces many scientists (feeling the weight of exponential growth in their own fields) to be rather unquestioning 'users'. Nobody can hope to take in everything and division of labour becomes inevitable. It remains necessary to be able to check performance but to what degree the user should try to master the details of instrumentation will depend on his judgment and his aptitude and perhaps also on the accessibility of service agencies to his laboratory.

The study of natural products tends to proceed from isolation, characterization and synthesis to biosynthesis and then to biological modes of action. As the attainable goals of biochemistry emerge more clearly a narrow view of the subject must be eschewed. Each generation of biochemists needs both chemical and biological insight. The plan of the present work is to show biochemical spectroscopy in action and to illustrate something of its diversity.

1.2 Notation

The different radiations are specified in terms of wavelength (λ), wave numbers ($1/\lambda$), and frequencies (ν), e.g. the sodium line D_2, expressed in different units, has a wavelength or frequency:

5889·965 Å, 588·9 nm, 0·589 μm, $5{\cdot}889 \times 10^{-5}$ cm

$1/\lambda = 16\,973{\cdot}52\ \text{cm}^{-1}$ $\nu = 509{\cdot}2 \times 10^{12}$ cm

$$\text{where } \nu \text{ (true frequency)} = \frac{\text{velocity of light}}{\lambda}$$

Spectrophotometry is mainly concerned with the following regions of the spectrum:

Ultraviolet	185–400 nm	(nanometres or millimicrons)
Visible	400–760 nm	
Infrared	0·76–15 μm (microns)	

When light falls on any material it may be reflected, transmitted, or absorbed. In absorption spectrophotometry reflection losses are usually irrelevant because the use of a control (e.g. a comparison cell containing pure solvent) eliminates such complications.

If I_λ is the intensity of incident light of wavelength λ, the weakening of the light on its passage through a small thickness dl of medium is proportional to I_λ and dl

$$-\mathrm{d}I_\lambda = a\ I_\lambda\ \mathrm{d}l \quad \text{or} \quad \frac{-\mathrm{d}I_\lambda}{I_\lambda} = a\ \mathrm{d}l$$

By integration, putting $I_{0\lambda}$ for $l=0$, $\log_e I_0/I = al$.

Stated in other terms, this is Lambert's law, $I = I_0 e^{-al}$ (where I_0 is the intensity of the incident light, and I that of the emergent light, l is the thickness of the absorbing layer in cm, and a is the absorption coefficient), indicating that each successive layer absorbs an equal fraction of the light. By replacing e (natural base) by 10, $I = I_0 . 10^{-Kl}$ and K is the extinction coefficient of Bunsen and Roscoe, defined as the reciprocal of the thickness (l cm) needed to weaken the light to 1/10 its incident intensity, i.e.

$$\log I_0/I = Kl = \mathrm{E}\ \text{(extinction)}$$

Beer's law states that absorption is proportional to the number of molecules in the light path. Thus a 10 cm layer of **M**/1000 solution exerts the same effect as a

1 cm layer of **M**/100 solution or a 1 mm layer of **M**/10 solution, so that

$$I = I_0 \; 10^{-\epsilon cl}$$

where c is concentration (Mol/l), l is thickness in cms and ϵ is the molecular extinction coefficient, ($I = I_0 \, e^{-\epsilon' cl}$ is sometimes used, ϵ' is the molar absorption coefficient). Physicists prefer on occasion to use an absorption index k given by

$$I = I_0 \, e^{-4\pi kl/\lambda}$$

$$\mathrm{k} = \epsilon c \; 0{\cdot}1832\lambda \quad (\lambda \text{ in cm})$$

An absorption *band*, i.e. an approximately symmetrical absorption curve extending over a range on the wavelength scale and rising to a peak at $\lambda_{max.}$ and $\epsilon_{max.}$, may be described conveniently as follows. Two wavelengths λ_1 and λ_2 are found such that $\epsilon = \frac{1}{2}\epsilon_{max.}$. The halfwidth of the band is then $\lambda_2 - \lambda_1$ and the band strength will be:

$$\int_{\nu_2}^{\nu_1} \epsilon \, d\nu$$

where ν_1 and ν_2 are the frequencies corresponding to λ_1 and λ_2.

If the (effective) cross-sectional area of a molecule is put as a and dl is a layer of solution so thin that there is no superposition of molecules in the light path

$$-\mathrm{d}I/I = Nac \, \mathrm{d}l/1000$$

where c is concentration of solute (gram-molecules per litre) and $cN/1000$ is the number of molecules per ml and A is the Avogadro number. Then

$$\log_{10} I_0/I = 0{\cdot}43Nacl \, . \, 10^{-3}$$

and since $\log_{10} I_0/I = \epsilon \mathrm{cl}$

$$\epsilon = 0{\cdot}00043Na = 2{\cdot}6 \times 10^{20}a$$

Not all the molecules will be so oriented that they exert maximal absorption and a factor $\frac{1}{3}$ may be introduced to allow for random orientation. For many organic molecules the cross-sectional area will be about 10 Å^2 or 10^{-15} cm^2. On this basis a representative value for ϵ will be:

$$\epsilon = \tfrac{1}{3}(2{\cdot}6 \, . \, 10^{20} \times 10^{-15})$$

or

$$\epsilon = 8{\cdot}7 \times 10^4$$

The *order* of the extinction coefficient will then be 100 000. Such a value is shown by carotenoids and for vitamin A, $\epsilon = ca.$ 50 000. A probability of 1 is assumed here and for a vast range of substances in which $\epsilon_{max.}$ is 5000–25 000 the probability of the electronic transition is between say 0·05 and 0·5.

The optical density of a medium, d, is equal to ϵcl and is thus identical with E.

The expression E_l^c stands for $\log_{10} I_0/I$ for an l cm layer of a solution of concentration c and is useful as the neatest way of describing an actual measurement, since most spectrophotometers are graduated on a scale corresponding with $\log I_0/I$ (E values 0·1, 0·2, . . .). Modified a little, this notation is applicable to the specification of absorption for natural products whether mixtures or substances of uncertain molecular weight, where c is expressed in percentage (wt/vol) of solute. Thus, if a 0·5 cm layer of 0·1% solution transmits 1/10 of the light at, say 325 nm, E = 1·0 and $\mathrm{E}_{0{\cdot}5\,\mathrm{cm.}}^{0{\cdot}1\%}$ 325nm = 1·0 specifies the measurement, whilst

$$\mathrm{E}_{1\,\mathrm{cm.}}^{1\%} \; 325 \; \mathrm{nm} = \frac{1{\cdot}0 \times 1 \times 1}{0{\cdot}5 \times 0{\cdot}1} = 20$$

reduces it to a basis convenient for comparison with similar solutions.

Absorption spectra are usually set down in the form of graphs of the function $\epsilon(\lambda)$. As ordinates either ϵ or $\log \epsilon$ may be used, the latter having advantages when ϵ varies (with λ) over a wide range. (E and log E are also used). The wavelengths are sometimes* marked off in Å or nm (or μm in the infrared), and an alternative method of plotting ν or $1/\lambda$ has the advantage of giving a truer picture of the relative widths of bands in different parts of the spectrum.

Absorption spectrophotometry consists of measuring the intensities of absorption exerted at different wavelengths, by molecules in the path of the light. The data are plotted in the form of spectral absorption curves, the shape of which will depend upon the nature of the absorbing molecule and its environment. The mechanism of absorption varies with the wavelength of the radiation, or in other words, with the magnitude of the energy quantum. When a quantum of light is absorbed it may (*a*) effect dissociation, (*b*) raise an electron to a higher energy state, (*c*) bring about a change in interatomic vibrations, or (*d*) cause a change in molecular rotation. Near-infrared radiation is made up of quanta which as a rule are too small for electronic activation, and selective absorption in this part of the spectrum corresponds with an increase in vibrational energy. It is often possible to observe fine structure, i.e. narrow absorption maxima indicative of increased rotational energy. Ultraviolet and visible radiations provide larger quanta which are absorbed by effecting electronic transitions within the molecule. Absorption bands which are truly continuous (i.e. do not possess fine structure) correspond with the process of molecular dissociation. Discontinuous bands may have a series of fairly wide regions of selective absorption

* mμ and μ in the older literature.

within which the abscrption curve changes continuously; this type of curve corresponds with electronic activation and concurrent changes in vibrational energy. With some molecules, particularly in the vapour state, the absorption revealed by using a continuous-spectrum light source consists of a large number of narrow bands split up into hundreds of 'lines'. In such cases the main electronic changes (each corresponding with a line in atomic emission spectra) are associated with simultaneous changes in vibration and rotation.

Briefly then, infrared spectrophotometry is interpreted in terms of vibrations within molecules, and the *location* of visible and ultraviolet bands depends upon electronic transitions, whilst the finer structure of such bands depends again on molecular vibrations. The following expressions are typical for the analysis of well-resolved spectra, shown by vapours of relatively simple substances.

Benzene (absorption—*n*, *p*, and *q* integral)

$$\nu_{(1/\lambda)} = \begin{cases} 38\,612{\cdot}4 + 921{\cdot}7n - 162p + 121q \\ 37\,483{\cdot}5 + 924n - 164{\cdot}5p + 121q \text{ (Henri)} \end{cases}$$

Benzene (fluorescence)

$$\nu = \begin{cases} 37\,473 \\ 39\,535 - 990{\cdot}4n - 160{\cdot}8p \text{ (Ingold and Wilson)} \\ 38\,607 \end{cases}$$

The location of the spectra is fixed by the first (electronic) terms, and the narrow bands of small persistence are reduced to order by postulating a very limited number of vibrational constants and integral values for *n*, *p* and *q*.

This book is concerned with solution spectra, and in the present state of knowledge interpretation must often be tentative and empirical, because only very simple molecules are amenable to more rigorous treatment. Selective absorption in the ultraviolet and visible is associated with unsaturated groupings like —C=C—, >C=O, —N=N—, C_6H_5, etc., and the general methods of interpretation, using a concept of chromophoric groups, are dealt with elsewhere.

One of the most frequent difficulties encountered on the experimental side is that of testing the adjustment of spectrophotometric apparatus. For calibration a simple substance should be chosen, it should be readily available in a high state of purity and its absorption spectrum in solution should be accurately known, preferably as the result of concordant photoelectric determinations from several different laboratories. In the ultraviolet region these requirements are met by a 0·003N solution of potassium chromate in 0·05N KOH. This solution has been studied by many workers and the curve given in Fig. 1.1 is not far from the truth.

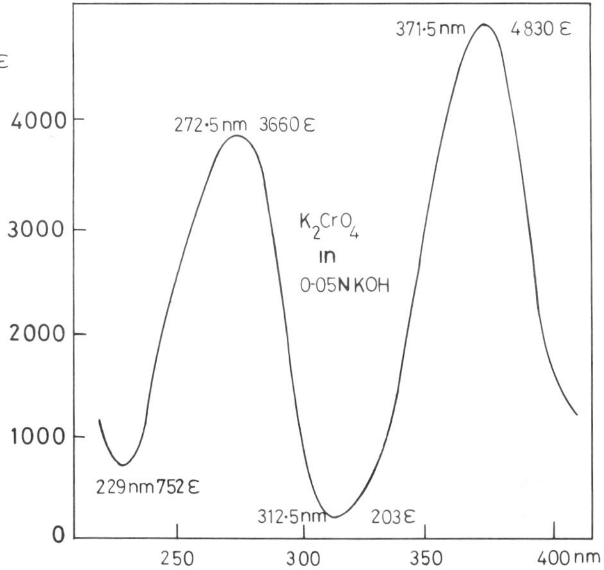

Figure 1.1 Absorption spectrum of 0·003 **N** potassium chromate solution in 0·05 **N** potassium hydroxide

1.3 Literature

1.3.1 Collections of Data

Organic Electronic Spectral Data, Vol. I (1960), covering 1946–1952, Ed. M. J. Kamlet; *Vol. II*, covering 1953–1955, Ed. H. E. Ungnade; *Vol. III* (1966), covering 1956–1957, Eds. L. A. Kaplan and O. H. Wheeler; *Vol. IV* (1963), covering 1958–1959, Eds. J. P. Phillips and F. C. Nachod; *Vol. V* (1969), covering 1960–61, Eds. J. P. Phillips, R. E. Lyle and P. R. Jones; *Vol. VI* (1970), covering 1962–1963, Eds. J. P. Phillips, L. D. Freedman and J. C. Craig. Organic Electron Spectral Data Inc., Wiley-Interscience, New York, London, Sydney, Toronto.

Each volume contains organic compounds assembled under empirical formulae in order of increasing number of carbon atoms. For each substance recorded the information includes name, solvent used, $\lambda_{max.}$nm, log $\epsilon_{max.}$ reference. Each journal has a code number, and the reference includes names of authors. The system is convenient and effective.

Absorption Spectra in the Visible and Ultraviolet Region. Ed. L. Láng, Akadémiai Kiadó, Vols. I–XVIII appearing between 1963 and 1973, some volumes in revised editions. Full data on 3403 compounds to Vol. XVIII, absorption curves and numerical data. Also cumulative index IV, VI–X and XI–XV.

Ultraviolet and Visible Absorption Spectra Index for 1930–1954 (1956); *Index for 1955–1959* (1961); *Index for 1958–1962* (1964). H. M. Hershenon. Academic Press, New York, London.

These volumes enable published spectra to be located. The indexing is easy to follow and the system is effective.

Ultraviolet Spectra of Aromatic Compounds (1951). R. A. Friedel and M. Orchin, J. Wiley, New York (579 compounds). An early but very useful collection of graphs.

U.V. Atlas of Organic Compounds (1966–1971), 5 volumes. Butterworths, London, and Verlag Chemie Weinheim. 1200 graphs. Published in collaboration with the Photoelectric Spectrophotometry Group and Institute of Spectrochemistry and Applied Spectroscopy, Dortmund.

Each sheet contains the name and structure of compound, molecular weight, solvent used, cell-length, purity and some information concerning infrared spectra. Absorption curves are plotted (cm^{-1}) against log ϵ but there is also a wavelength scale. Each maximum is given numerical values $\nu_{max.}$, $\lambda_{max.}$nm, $\epsilon_{max.}$ Each volume has a formula index in English and in German. When using this collection for the first time the method of classifying organic compounds in the work should be mastered. There are valuable commentaries and data on related compounds are sometimes assembled on extra sheets.

Sadtler Ultraviolet Spectra, 28 000 spectra with special sections for biochemist, pharmaceutical compounds and dyestuffs, etc. Sadtler Research Laboratories, Philadelphia, U.S.A.

Subscribers receive reproductions of spectra.

Ultraviolet spectral data. American Petroleum Institute Research Project 44. Carnegie Institution Pittsburgh, and U.S. Bureau of Standards.

Spectral Data of Natural Products, Vol. I (1970). K. Yamaguchi. Elsevier, Amsterdam, London, New York.

A valuable compilation on compounds the structures of which were known before 1963. Compounds are classified into 20 main groups. Structures and spectroscopic data (u.v., i.r., n.m.r., m.s. and c.d.) are recorded.

Ultraviolet and Visible Spectra listed and indexed (1963), American Society for Testing Materials (ASTM). Punched Cards for spectra of 10 000+ compounds. IBM sorter needed.

Handbook of Ultraviolet and Visible Absorption Spectra of Organic Compounds (1967). K. Hirayama. Plenum Press, New York. Also in *Jekken Kagaku Koza, Vol. 1.* Chemical Society of Japan, Maruzen, Tokyo.

Atlas of Spectral Data and Physical Constants of Organic Compounds (1973). Ed. J. G. Grasselli. C.R.C. Press, A Division of the Chemical Rubber Company, Cleveland, Ohio.

This work, which was published too late to be used here, is clearly an important collection.

Physical Properties of the Steroid Hormones (1963). Ed. L. L. Engel. Pergamon Press, Oxford, London.

Atlas of Steroid Spectra (1965). W. Neudert & H. Röepke. Springer-Verlag, Berlin.

Infrared Absorption Spectra of Steroids (1953). K. Dobriner, E. Katzenellenbagen & R. N. Jones. *Vol. II* (1958), G. Roberts, B. Gallagher & R. N. Jones. Interscience Publishers, New York.

Atlas of Protein Structure in the Ultraviolet and Visible Regions (1972). Ed. D. M. Kirschenbaum. Adam Hilger, London.

Sadtler Infrared Spectra (1959). 20 000 compounds indexed by name, formula, strongest maxima, etc. Sadtler Research Laboratories, Philadelphia, U.S.A.

An Index of Published Infrared Spectra (1960). Eds. M. B. B. Thomas & E. R. Adams. Vols. I and II. Ministry of Aviation. Her Majesty's Stationery Office, London. Valuable for locating published spectra.

Infrared Absorption Spectra for 1945–1957 (1959), H. M. Hershenon; *for 1958–1962* (1964). Academic Press, New York.

Nuclear Magnetic Resonance Spectral Data. American Petroleum Institute Project 44. Agricultural and Mechanical College of Texas.

High Resolution n.m.r. Spectra Catalogue, Vol. I (1962). *Vol. II* (1963), Varian Associates.

Nuclear Magnetic Resonance Spectra and Chemical Structure (1967) W. Brügel. Academic Press, New York, London.

1.3.2 Some Earlier References to Literature

Reports to the British Association for the Advancement of Science 1899–1903, 1916, 1920, 1922. Compounds are listed alphabetically and references to the literature are given. Now of historical interest.

International Critical Tables (1929), Vol. V, p. 359. Section by V. Henri. Contains early quantitative data.

Tables Annuelles de Constantes et données numeriques (1934). Gauthier-Villars, Paris. Contains further quantitative data collected by Henri.

Absorption Spectra of Natural Products (1937). F. Ellinger Tabulae Biologicae, XII, p. 291. Junk, Den Haag.

Bibliography and Abstracts on Absorption Spectrophotometry (1939) O. J. Walker. Adam Hilger, London.

Landolt Börnstein Zahlenwerte (1951), 6th Edition. Vol. I. Part III. A. Eucken & K. H. Hallwege. Springer-Verlag, Berlin. Contains over 1000 spectra and classified references to many others.

1.3.3 Early General Treatises

Etudes de Photochimie (1919). V. Henri. Gautier-Villars, Paris.

Hundbuch der Spectroscopie, (1900–1934). 8 Volumes. H. G. J. Kayser & H. Konen. Herzel, Leipzig. (Vols. III and IV deal with absorption spectra).

Spectroscopy (1927–1929). 3 Volumes. E. C. C. Baly. Longmans, London.

Optische Methoden der Chemie (1927). F. Weigert. Akademische Verlag, Leipzig.

Chemical Spectroscopy (1943). W. R. Brode. J. Wiley, New York.

Infrared Determination of Organic Structures (1949). H. M. Randall, N. Fuson, R. G. Fowler & J. R. Dangl. Van Nostrand, London, New York.

1.3.4 Techniques and General References

The Practice of Absorption Spectrophotometry (1934). F. Twyman & C. B. Allsop. Adam Hilger, London.

Practical Aspects of Absorption Spectrophotometry (1938). R. A. Morton. Royal Institute of Chemistry, London.

Optical Methods of Chemical Analysis (1942). T. R. P. Gibb. McGraw-Hill, New York.

Absorption Spectrophotometry (1949). 3rd Edition 1968. G. F. Lothian. Adam Hilger, London.

Analytical Absorption Spectroscopy (1950). Ed. M. G. Mellon. J. Wiley, New York.

Manual of Standardized Procedures for Spectrophotometric Chemistry (1951). Fister. Standard Scientific Supply Corporation, New York.

Modern Methods of Plant Analysis (1955–1956). 4 Volumes. Eds. K. Paech and M. V. Tracey. Springer-Verlag, Berlin.

Practical Hints on Absorption Spectrophotometry (Ultraviolet and Visible) (1966). 3rd Impression 1969. J. R. Edisbury. Adam Hilger, London.

Techniques of Organic Chemistry. Chemical Applications of Spectroscopy (1956). Ed. W. West. Interscience Publishers, New York.

Electronic Absorption Spectra and Geometry of Organic Molecules. An Application of Molecular Orbital Theory (1967). H. Suzuki. Academic Press, New York, London.

The Theory of the Electronic Spectra of Organic Molecules (1963). J. N. Murrell. Methuen, London.

Molecular Spectra and Molecular Structure (1950). G. Herzberg. Van Nostrand, New York.

Theory and Applications of Ultraviolet Spectroscopy (1962). H. H. Jaffé & M. Orchin. J. Wiley, New York, London.

Absorption Spectroscopy (1962). R. P. Bauman. John Wiley, New York.

Applications of Absorption Spectroscopy of Organic Compounds (1965). J. R. Dyer. Prentice Hall, Inglewood Cliffs, New Jersey.

Ultraviolet and Visible Light Absorption (1955). E. A. Braude in *Determination of Organic Structures by Physical Methods*. Eds. E. A. Braude & F. C. Nachod. Academic Press, New York.

Gillam & Stern's Introduction to Electronic Absorption Spectroscopy in Organic Chemistry (1970). 3rd Edition. Eds. E. S. Stern & C. J. Timmons. Edward Arnold (Publishers) Ltd., London.

Absorptions-spektroscopie im ultravioletten und sichtbaren Spectralbereich (1962). B. Hampel. Vieweg, Braunschweig.

Absorption Spectra in the Visible and Ultraviolet Region (1961). L. Lang. Academic Press, New York.

Thorpe's Dictionary of Chemistry (1946). Vol. 8. Molecular Spectra Visible and Ultraviolet. R. A. Morton. Longmans, London.

The Application of Absorption Spectra to the study of Vitamins, Hormones and Coenzymes (1942). R. A. Morton. 2nd Edition. Adam Hilger, London.

Spectrophotometry in the Ultraviolet and Visible Regions. (1962). R. A. Morton in *Comprehensive Biochemistry*, Vol. 3, p. 67. Eds. M. Florkin & E. H. Stotz.

Ultraviolet and Visible Spectroscopy Chemical Applications (1967). 2nd Edition. C. N. Rao. Butterworths, London.

Interpretation of the Ultraviolet Spectra of Natural Products (1964). A. I. Scott. Pergamon Press, Oxford, London, New York.

Electronic Absorption Spectroscopy, 'Jikken Kagaku Koza' (1957). *Chem. Soc. Japan*, Vol. **3**, Chapter 7. H. Baba. Maruzen, Tokyo. (See also *ibid.* 1965, **11**.)

The Acridines (1951). 2nd Ed. 1966, A. Albert, Edward Arnold, London. Chapter 11 (2nd Edition) reviews in considerable detail (54 pages) ultraviolet and visible spectra, infrared spectra and fluorescence and phosphorescence of acridines.

Fluorescence and Phosphorescence (1949). P. Pringsheim. Interscience Publishers, New York.

Fluorescence of Solutions (1953). E. J. Bowen & F. Wokes. Longmans, London.

Fluorescence (1962). A. Ehrenberg & H. Theorell *Comprehensive Biochemistry*, *Vol. 3*. Eds. M. Florkin & E. H. Stotz. Elsevier, Amsterdam, New York.

Fluorescenz Organische Verbindungen (1951). T. Förster. Vanderhoeck and Ruprecht, Göttingen.

Light Absorption and Photochemistry (1947). E. J. Bowen. *Quarterly Reviews*, Chemical Society, London.

Fluorescence Assay in Biology and Medicine (1962). 1st Edition. 2nd Edition, 1964. S. Udenfriend. Academic Press, London, New York.

Guide to the NMR Empirical Method: A Workbook (1967). R. H. Bible Jr. Plenum Press, New York.

Applications of Nuclear Magnetic Resonance Spectroscopy in Organic Chemistry (1969). 1st Edition. L. M. Jackman. 1969. 2nd Edition. L. M. Jackman & S. Sternhell. Pergamon Press, Oxford.

Nuclear Magnetic Resonance Applications to Organic Chemistry (1959). J. D. Roberts. McGraw-Hill, New York.

Nuclear Magnetic Resonance Spectroscopy (1969). F. A. Bovey. Academic Press, New York, London.

The Interpretation of n.m.r. Spectra (1962). K. B. Wiberg & B. J. Nist. Benjamin, New York.

Magnetic Resonances in Biological Research (1971). Ed. C. Franconi. Gordon & Breach, London, New York.

High Resolution Nuclear Magnetic Resonance (1959). J. A. Pople, W. G. Schneider and H. J. Bernstein. McGraw-Hill, New York.

Electronic Paramagnetic Resonance (1962). S. I. Weissman. Vol. 3 in *Comprehensive Biochemistry*, Eds. M. F Florkin & E. H. Stotz. Elsevier, Amsterdam, New York.

Nuclear Magnetic Resonance (1962). C. D. Jardetzky and O. Jardetzky. Vol. 3, p. 208, in *Comprehensive Biochemistry*, Eds. M. Florkin & E. H. Stotz. Elsevier, Amsterdam, New York.

Massenspektrometrie organischer Verbindungen (1969). W. Benz. Akademische Verdagsgesellschaft, Frankfurt a-M.

Mass Spectrometry of Organic Compounds (1967). H. Budzikiewicz, C. Djerassi & D. H. Williams. Holden-Day, San Francisco.

Mass Spectrometry (1962). K. Biemann. McGraw-Hill, New York.

A Programmed Introduction to Infrared Spectroscopy (1972). B. W. Cook & K. Jones. Heyden, London, New York.

Chemical Applications of Infrared Spectroscopy (1963). C. N. Rao. Academic Press, New York, London.

Symposium on Biological Applications of Infrared Spectroscopy (1957–8). *Ann. N.Y. Acad. Sci.*, **69**, 1–254.

Applications of Infra-red Spectroscopy in Biochemistry, Biology and Medicine (1971). F. S. Parker. Adam Hilger, London.

Infrared Light Absorption (1955). R. C. Gore & E. S. Waight in Braude & Nachod, *loc. cit.*, 195.

Practical Hints on Infra-red Spectrometry from a Forensic Analyst (1971). M. J. de Faubert Maunder. Adam Hilger, London.

Steroid Sapogenenins (1953). R. N. Jones, E. Katzenellenbogen & K. Dobriner. *N.R.C. Bull.*, No. 2929. National Research Council, Canada.

Investigations of Infrared Spectra. Part 1. Infrared absorption spectra. (1905). W. W. Coblentz. Carnegie Institution. Publ. 35. Washington, D.C.

Chemical Applications of Infrared Spectroscopy (1956). A Chemical Application of Spectroscopy. Ed. W. West. Also R. N. Jones & C. Sandorfy, same volume.

The Raman Effect and its Chemical Applications (1939). J. H. Hibben. Reinhold, New York.

Infrared determination of organic structures (1950). H. M. Randall, R. B. Fowler, N. Fuson & J. R. Dangl. Van Nostrand, New York.

The Infrared Spectra of Complex Molecules (1954). L. J. Bellamy. Methuen, London. 2nd Edition 1958.

Application of Infrared and Raman Spectroscopy to the Elucidation of Molecular Structure (1956). In Technique of Organic Chemistry. Ed. A. Weissburger. Vol. 9. Interscience, New York.

Introduction to Infrared Spectroscopy (1962). W. Brügel. Methuen, London.

Infrared Spectroscopy Applications to Organic Chemistry (1972). M. Avram & Gh. D. Mateescu (translated by L. Bîrlădeanu). Wiley-Interscience, New York, London.

Infrared and Raman Spectroscopy (1962). In *Determination of Organic Structures by Physical Methods.* Ed. F. C. Nachod and W. F. Phillips. Academic Press, New York.

Introduction to Practical Infrared Spectroscopy (1960). A. D. Cross. Butterworths, London.

Laboratory Methods in Infrared Spectroscopy (1972). Eds. R. G. J. Miller & B. C. Stace. Heyden, London, New York.

Spectroscopic methods. W. Vetter, G. Englert, N. Rigassi, & U. Schwieter. In *Carotenoids*, Ed. O. Isler, Birkhauser, Basel. (1971).

Fat Soluble Vitamins (1970). Ed. R. A. Morton. Vol. 9. In *International Encyclopaedia of Food and Nutrition.* Pergamon Press, Oxford, London.

Biochemistry of Quinones (1965). Ed. R. A. Morton. Academic Press, London, New York.

Carotenoids (1971). Ed. O. Isler. Birkhauser Verlag, Basel, Stuttgart (in English).

Spectrophotometry in Medicine (1943). By L. Heilmeyer. Trans. A. Jordan and T. L. Tippel. Adam Hilger, London.

The Bile Pigments (1953). C. H. Gray. Methuen, London.

The Bile Pigments in Health and Disease (1961). C. H. Gray and C. C. Thomas. Springfield, Illinois.

Leaf Analysis (1951). H. Lundegårdh (translated by R. L. Mitchell). Adam Hilger, London.

Photosynthesis and Related Process E. I. Rabinowitch. Vol. 1 (1945); Vol. 2, Part I (1951); Vol. 2, Part 2 (1956). Interscience Publishers, New York.

Haematin Compounds and Bile Pigments (1949). R. Lemberg & J. W. Legge. Interscience, London, New York.

Porphyrins and Metalloporphyrins (1964). J. E. Falk. Elsevier, Amsterdam, London, New York. (Contains spectroscopic data.)

Haematin Enzymes (1961). Eds. J. F. Falk, R. Lemberg & R. K. Morton. Pergamon Press, Oxford.

Comparative Biochemistry of Photoreactive Systems (1960). Ed. M. B. Allen. Academic Press, New York.

The Nature of Animal Colours (1960). H. M. Fox and G. M. Vevers. Sidgwick & Jackson, London.

Carotenoids other than Vitamin A. IUPAC. Butterworths, London (1969). Also in *Pure and Applied Chemistry*, 1969, **20**, No. 4.

The Systematic Identification of Flavonoids (1970). T. J. Mabry, K. R. Markham and M. B. Thomas. Springer Verlag, Berlin, Heidelberg, New York. xi + 354 pp. 325 figs.

Cis-trans *isomeric carotenoids, Vitamins A and Arylpolyenes* (1962). L. Zechmeister. Springer Verlag, Vienna.

Methods of Biochemical Analysis (1951–1972). Ed. D. Glick. Interscience, John Wiley, New York, London.
(Volume numbers shown in bold numbers)

Spectrophotometry of translucent materials. Opal glass Methods. Shibata, K., **9**.

Spectrophotometry of opaque biological materials. Reflection methods. Shibata, K., 9.

Automated analysis of absorbent and fluorescent substances separated on paper strips. Boulton, A. A., **16**.

Mass Spectrometry in the determination of structure of certain natural products containing sugars. Hanessian S., **19**.

Biochemical Applications of Magnetic Resonance. Jardetzky, C. and Jardetzky, O., **9**.

Use of infrared analysis on determination of carbohydrate structure. Baker, S. A., Bourne, E. J. & Whiffen, D. H., **3**.

Infrared analysis of vitamins, hormones and coenzymes. Rosenkrantz, H., **5**.

Analysis of steroids by infrared spectrometry. Rosenkrantz, H., **2**.

Fluorimetric analysis of corticoids. Silber, R. H., **14**.

Fluorimetric assay of enzymes. Roth, M., **17**.

Assay of enzymes related to serotonin. Udenfriend, S., Weissbach, H. & Brodie, B. B., **6**.

Determination of ATP by firefly luminescence. Stochler, B. L. & Totter, J. R., **1**.

Measurement of luciferin and luciferase. Chase, A. M., **8**.

Principles and practice of bioluminescence assay. Strehler, B. L., **16**.

Application of optical rotatory dispersion and circular dichroism to the study of biopolymers. Tinoco, I. Jr., **18**.

Determination of carotene. Bickoff, E. M., **4**.

Determination of Vitamin A and carotenoids. (*a*) Embree, N. D., Ames, S. R., Lehman, R. W. & Harris, P. L., **4**. (b) McLaren, D. S., Read, W. W. C., Auden, Z. L. & Tchalian, M., **15**.

Determination of coenzyme Q (ubiquinone). Crane, F. L. & Dilley, R. A., **11**.

Chemical determination of flavins, Yagi, K., **10**.

New color reactions for determination of sugars in polysaccharides. Dische, Z., **2**.

Methods for the determination of thiamine. Mickelsen, O. & Yamamoto, R. S., **6**.

Enzymic microdeterminations of uric acid, hypoxanthine, xanthine, adenine and xanthopterin by ultraviolet spectrophotometry. Plesner, P. & Kalchar, H. M., **3**.

Nucleic acids. Estimation. Volkin, E. & Cohn, W. E., **1**; Webb, J. M. & Leuy, H. B., **6**; Muhro, H. N. & Fleck, A., **14**; Use of Ethidium bromide. Le Pecq, J. B., **20**.

Separation, identification and estimation of prostaglandins. Shaw, J. E. & Ramwell, P. W., **17**.

Spectrophotometric assay of cytochrome c oxidase. Smith, L., **2**.

Separation and determination of bile pigments. Brodersen, S. & Jacobsen, J., **17**.

Separation and determination of bile acids. Sjovall, J., **12**.

Determination of prophyrins in biological materials. Schwartz, S., Berg, M. H., Bossenmaier, I. & Dinsmore, H., **8**.

Comprehensive Biochemistry. Eds. M. Florkin & E. H. Stotz. Elsevier, Amsterdam.

Vol. **3** (1962). Methods for the study of molecules. Analysis by emission spectroscopy, N. H. Nachtrieb. Spectrophotometry in the ultraviolet and visible regions, R. A. Morton. Infrared spectra of compounds of biological interest, L. J. Bellamy. Fluorescence, A. Ehrenberg and H. Theorell. Electron paramagnetic resonance, S. I. Weissman. Nuclear magnetic resonance, C. D. Jardetzky and O. Jardetzky.

Other volumes contain many incidental references to spectroscopic methods applied to different classes of compounds of biochemical interest.

References

1.1

Baly, E. C. C., Morton, R. A. & Riding, R. W. (1926). *Proc. Roy. Soc.*, A, **113**, 709.

Cary, H. H. (1946). *Rev. Sci. Inst.*, **17**, 558.

Cary, H. H. & Beckman, A. O. (1941). *J. opt. Soc. Amer.*, **31**, 682.

Conn, G. K. T. & Avery, D. G. (1960). *Infrared Methods.* Academic Press, New York.

Edisbury, J. R. (1966). *Practical Hints on Absorption Spectrometry.* 3rd Impression 1969. Adam Hilger, London.

Fulweiler, W. H. & Barnes, J. (1922). *J. Franklin Inst.*, **194**, 83.

Goddu, R. F. (1960). In *Advances in Analytical Chemistry and Instrumentation.* Ed. C. N. Reiley. Vol. I, 347. J. Wiley-Interscience, New York.

Hardy, A. C. (1935). *J. opt. Soc. Amer.*, **25**, 305.

Meyer, L. H., Saika, A. & Gutowsky, H. S. (1953). *J. Amer. Chem. Soc.*, **75**, 4567.

Michaelson, J. L. & Liebhafsky, H. A. (1936). *Gen. Elect. Rev.*, **39**, 445.

Rao, C. N. R. (1963). *Chemical Applications of Infrared Spectroscopy.* Academic Press, New York.

Suhrmann, R. & Kollath, W. (1927). *Biochem. Z.*, **184**, 216; (1928), *Strahlentherape*, **32**, 389.

Warburg, O., Negelein, E. & Christian, W. (1929). *Biochem. Z.*, **214**, 26, 34, 38.

West, W. (1960). *Physical Methods of Organic Chemistry* in *Techniques of Organic Chemistry.* Ed. A. Weissberger Vol. 1. Part II. Interscience Publishers, New York.

Von Halban, H. & Ebert, L. (1924). *Z. Physikal Chem.*, **112**, 329.

Von Halban, H. & Eisenbrand, J. (1927). *Proc. Roy. Soc.* A. **116**, 153.

Von Halban, H. & Siedentopf, K. (1922). *Z. Physikal Chem.*, **100**, 208.

2 Water and Sunlight in a Biological Context

2.1 Water

The optical properties of water are clearly important in oceanography and in limnology, in particular with regard to the maintenance of aquatic forms of life. Jerlov (1963) has discussed theory, terminology and instrumentation in the field of optical oceanography.

The main facts about the absorption spectrum of liquid water can be stated briefly. Water is very transparent towards visible light and to ultraviolet light as far as 210 nm. On the other hand, water absorbs infrared radiation very strongly. The earth's atmosphere transmits little infrared beyond 2 μm and on the short wave side the spectrum of sunlight does not extend below about 295–300 nm so that in nature radiations between 200 and 295 nm are not significant. The salts dissolved in sea water increase the absorption to only a very minor extent (Lenoble, 1956). The ultraviolet absorption of 'average' sea water is said to be about twice that of simulated or artificial sea water, the difference being attributed to minute amounts of organic matter (Armstrong & Boalch, 1961). Light scattered by small particles is linearly polarized down to considerable depths. The blue colour of the sea arises in part from the fact that absorption in the visible is selective with a maximum transmission around 460–480 nm.

In oceanography the absorption coefficient of water κ_λ is defined by $dI_\lambda = \kappa_\lambda I_\lambda \, dx$ where the thickness is defined by the limits $x = h$ and $x = h + L$ so that

$$\kappa_\lambda = \frac{2{\cdot}30}{L}(\log I_{\lambda h} - \log I_{\lambda(h+L)}).$$

In other contexts the unit in which L is expressed is 1 cm but here it is usually 1 metre. At the wavelengths where the absorption is very small, quantitative agreement between different investigators is not very good partly because of differences between water samples and partly because of differences in technique.

Water transmits particularly well between 400 and 600 nm. From 600 to 900 nm the transmission falls by a factor of about 50. At 1·3 μm it reaches about 3·1% for a 1 cm layer. Beyond this point the absorption increases rapidly. A layer of circulating water a few cm thick effectively absorbs infrared radiation. Collins (see Dietrich, 1939) also showed that the dissolved salts of sea water had comparatively little effect on the absorption and this was confirmed by Clark & James (1939). Utterback (1936) and Jorgensen & Utterback (1939) found that in the visible region extinction coefficients for the clearest oceanic water were only twice those for pure water, the increases at 460 nm being greater than those at 600–660 nm. At 480 nm 97·5% of the radiation is transmitted by a 1 m layer of clear oceanic water and 63·5% by 1 m of a slightly more turbid coastal water. Different effects are seen at different depths in different localities.

The increased extinction coefficients for many samples of sea water compared with highly purified water are attributed to the presence of tiny particles that scatter, reflect and absorb light. When the particles are small compared with the wave-length of the radiation the scattering is proportional to λ^{-4} (Rayleigh) so that at 460 nm it will be nearly 3 times that at 600 nm. In coastal waters there is also yellow material including perhaps stable metabolites from phytoplankton (Kalle, 1938). Similar material may also be present (albeit in considerably smaller amounts) in the open sea. Kalle considers that when the sea is green, light absorption by the yellow substance may be superimposed on Rayleigh scattering. In the short-wave ultraviolet region absorption is practically complete below 187 nm for a 1 mm layer, below 191 nm for a 1 cm layer, 193 nm for a 2 cm layer and 195 nm for a 4 cm layer. Table 2.1 shows the absorption intensities for pure water in the ultraviolet, visible and infrared

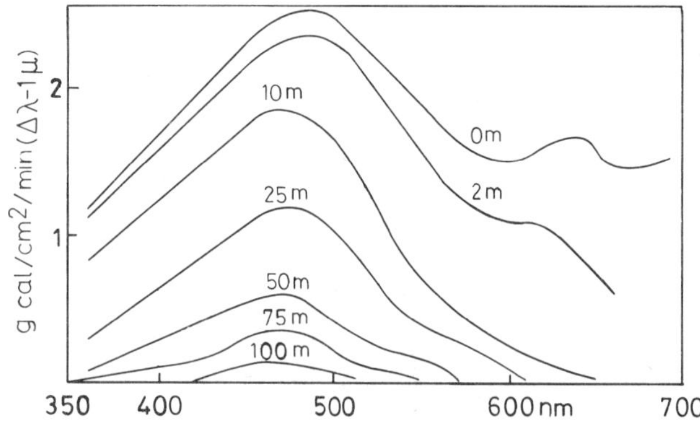

Figure 2.1 Spectral energy of the downward directed light at various depths (after Jerlov, 1951)

Table 2.1 Absorption coefficients per meter of pure water

Ultraviolet Wavelength nm	(*a*) κ/m	(*b*)	(*c*)	Visible* Wavelength nm	κ	Infrared Wavelength μm	κ
220	1·14	7·3	—	410	0·041	0·8	2·40
230	0·98	—	—	440	0·023	0·85	4·12
						0·90	6·55
240	0·93	1·35	—	460	0·015	0·90	28·9
250	0·75	—	—	480	0·015	1·00	39·7
260	0·62	—	—	500	0·016	1·05	17·7
270	0·55	—	—	520	0·019	1·10	20·03
280	0·39	0·77	—	540	0·024	1·20	123·2
290	0·34	—	—	560	0·030	1·30	115·0
300	0·28	0·64	—	580	0·055	1·40	160·0
310	0·27	—	—	600	0·125	1·50	194·0
320	0·23	0·43	0·58	620	0·78	1·60	80·0
330	0·22	—	—	650	0·210	1·70	73·0
340	0·20	0·28	0·38	700	0·84	1·80	170·0
350	0·18	—	—	750	2·72	1·90	73·00
360	0·16	0·19	0·28	—	—	2·00	85·00
370	0·15	—	—	—	—	2·10	39·00
380	0·14	0·13	0·148	—	—	2·20	21·00
390	0·13	—	—	—	—	2·30	24·00
400	0·12	0·08	0·072	—	—	2·40	42·00
						2·50	85·00

(*a*) Lenoble & Saint-Guilly (1955); (*b*) Dawson & Hilbert (1934); (*c*) W. R. Sawyer (1931)
* 350–650 nm due to Sawyer and > 650 nm to J. R. Collins quoted by Dietrich (1939)

regions. The high transparency of water is maintained from 600 nm to 300 nm, but even between 300 and 200 nm water transmits well in 1 m layers (Atkins *et al.*, 1938).

From Table 2.1 it will be seen that the absorbancy of water in the visible is least between 460 and 480 nm. This can be illustrated in other ways (Fig. 2.1, 2.2, 2.3) based on the results of Jerlov (1951) [also given in Friedrich's *Marine Biology* (1969)].

The most prominent infrared band has its peak at 3490 cm^{-1} in absorption (3446 cm^{-1}, Raman spectrum). It is due to O—H stretching vibrations of molecules and in the vapour it occurs at 3657 and 3756 cm^{-1} with 3200 cm^{-1} in ice. The HOH bending modes occur at 1645 cm^{-1} (vapour 1595 cm^{-1}, ice ~1650 cm^{-1}). There is also a broad very weak band with its maximum near 2125 cm^{-1} due probably to intermolecular modes or combinations. There is also a very broad band from 600 to 900 cm^{-1} with its flat maximum near 750 cm^{-1}; it arises from intermolecular modes. Table 2.2 shows the effects due to deuterium.

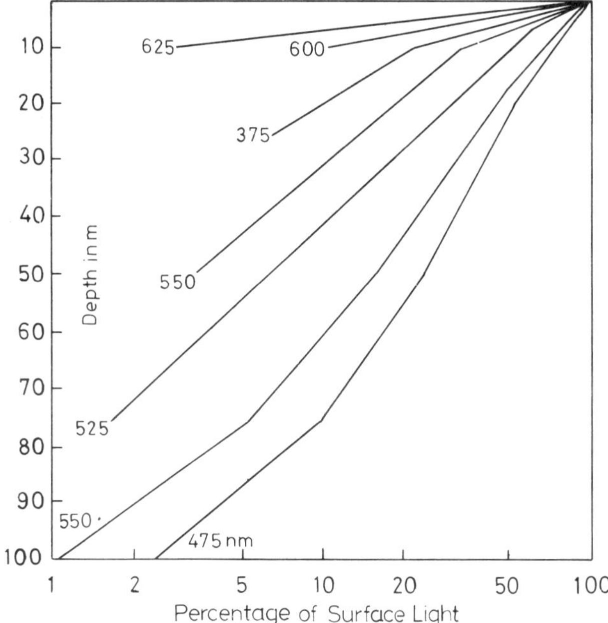

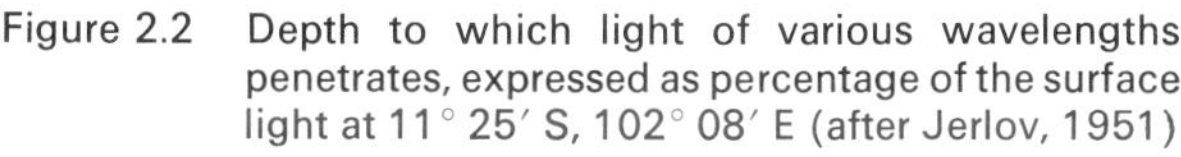
Figure 2.2 Depth to which light of various wavelengths penetrates, expressed as percentage of the surface light at 11° 25′ S, 102° 08′ E (after Jerlov, 1951)

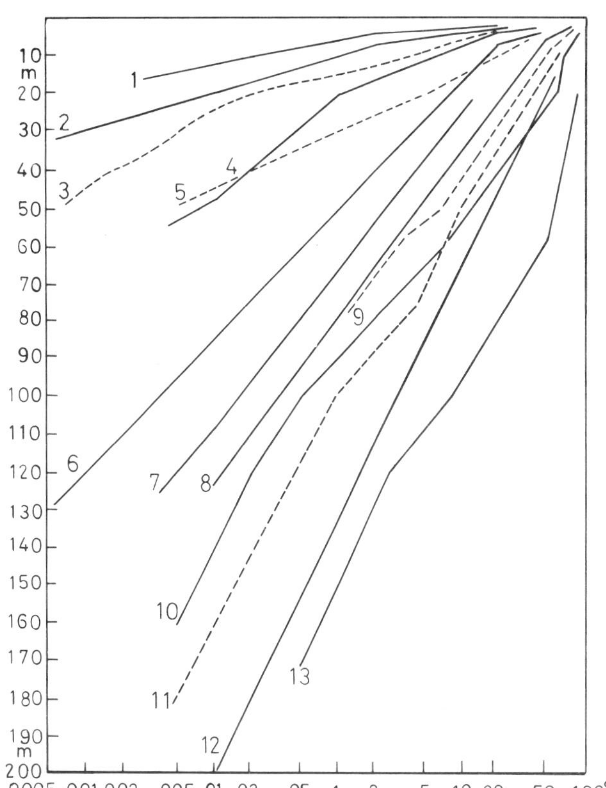

Figure 2.3 Depths to which light penetrates, expressed as percentage of the surface light, in various sea areas (after Clark, 1939 and Jerlov, 1951). 1, Woods Hole Harbour; 2, Mean curve for various coastal waters; 3, 8, 10, 11, 13, Sargasso Sea; 4, Gulf of Mexico near mouth of the Mississippi; 5, Gulf of Maine; 6, Coastal water; 7, Caribbean Sea; 9, Gulf of Mexico (from Friedrich, 1969)

Table 2.2 Vibrational bands

	H_2O		D_2O		HDO in H_2O		HDO in D_2O	
	cm^{-1}	ϵ	cm^{-1}	ϵ	cm^{-1}	ϵ	cm^{-1}	ϵ
	3280	54·5	2450	55·2	2500	42 ± 5	3400	64 ± 5
	3490	62·7	2540	59·8	—	—	—	—
	3920	0·83	2900	0·598	—	—	—	—
Raman	3439	—	2532	—	2516	—	3439	—
	3300	—	2400	—	—	—	—	—

ϵ = extinction coefficient $\times 10^{-3}$ cm^2 mol^{-1}
cf. Walrafen (1967); Eisenberg & Kauzmann (1969)

Sea water and lake or river water contain many solutes and the variations, qualitative and quantitative, can be wide. Analytical methods also range widely. Organic substances in sea water account for 0·4–4·6 mg carbon/litre; 0·03–0·38 mg nitrogen/litre; and 0·009–0·029 mg phosphorus/litre. Although these amounts seem small it has been calculated that perhaps 5 to 6 times as much organic matter is present in the sea in a soluble or colloidal form as exists in the form of plankton. The 'protein' of the sea amounts to about 1·5 g/m^3 and the carbohydrate to 3·9 g/m^3. All sea water probably contains free aminoacids and a whole constellation of products derived from bacteria, yeasts, flagellates, miscellaneous excreta and dead plants and animals.

The importance of photosynthesis in the food chains of the sea is obvious but it is perhaps desirable to lay stress on the aqueous medium. Some flagellates and diatoms are known to take up nutrients such as vitamins and growth factors and many workers (cf. Lucas,

1955) have explored the possibility that nutritionally-exacting organisms require and find in the aqueous environment traces of essential materials.

The analysis of sea water or lake water for organic solutes has not gone far enough despite much effort. In recent years, however, the methods for separating and identifying small amounts of organic substances have been so developed that more rapid progress is now possible if and when the task is given priority. Friedrich (1969) has documented the occurrence of the following in sea water as solutes or in suspensions: uracil, isoleucine, glycine; acetic, formic, glycollic, lactic, malic and citric acids; various fatty acids including lauric, myristic, palmitic, stearic, palmitoleic and linoleic; thiamin and cobalamines, ascorbic acid; carotenoids and chlorophyll derivatives. Jeffrey & Hood (1958) discussed experimental techniques in considerable detail.

Hood (1963) listed investigations on glycerides, sterol esters, free and combined aminoacids and carbohydrates in sea water. It is probable that very small but none-the-less significant amounts of some micronutrients like biotin are best identified by microbiological methods.

Most sea water specimens contain the yellow material already referred to. Its nature has not been settled but one view is that it comes from melanoids. Shapiro (1967), who at one time regarded the average molecular weight of the yellow substance from lake water as 200–300, now puts it near 10 000 and concludes that simple colour measurements have no precise meaning. Substances exhibiting a marked blue fluorescence also occur in extracts from sea water but they have not been fully identified. Special problems arise when particular types of organisms flourish greatly in restricted areas of the sea or of lakes. Excretory products or post-mortem artefacts may have deleterious consequences on other species.

It is now becoming clear that 'chemical' and 'biochemical' oceanography and limnology offer considerable challenges. Inorganic variables such as oxygen content, nitrate–nitrite parameters, orthophosphate, trace metals such as zinc, copper, molybdenum and cobalt afford scope for spectroscopic methods of analysis. In the field of possible organic solutes analytical methods should be directed towards well-defined goals rather than to 'complete analysis'. This means that the analyst should have at call many delicate methods of separation and all the standard spectroscopic techniques so that the many different problems can each be tackled by the best methods.

A few examples may now be offered from the inorganic and the organic fields.

One important parameter in water is the content of dissolved oxygen. De Graaf, Bierbrauwer and Golterman (1967) found that cerium hydroxide [$Ce(OH)_3$] is rapidly oxidized quantitatively by the oxygen in the solution:

$$2Ce^{3+} + \tfrac{1}{2}O_2 + H_2O \rightarrow 2Ce^{4+} + 2OH^-$$

A cerous salt is added to the sample of water and the solution is then made alkaline and a mixture of $Ce(OH)_3$ and $Ce(OH)_4$ is precipitated. To 100 ml of lake water is added 1 ml of 50% $CeCl_3 . 7\,H_2O$ (or an equivalent amount of cerous nitrate) in 0·05N H_2SO_4; this is followed by 0·6 ml of 50% NaOH solution. The container is stoppered and shaken. When the precipitate has settled 4 ml H_2SO_4 (S.G. 1·84) are added. Ceric salts show $\lambda_{max.}$ 320 nm whereas cerous salts have negligible absorption at that wavelength (see Fig. 2.4).

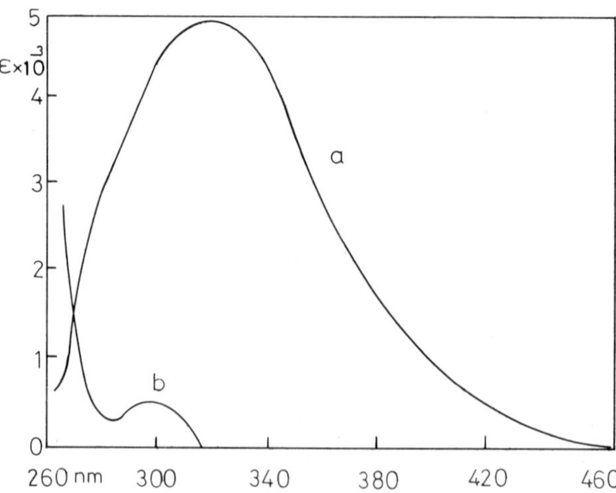

Figure 2.4 Absorption spectra of (*a*) $Ce(SO_4)_2$, 0·01 **N** in H_2SO_4, and (*b*) $CeCl_3$ (10·3%) in 0·05 **N** H_2SO_4

Beer's law is obeyed ($\epsilon_{320\,nm} = 5000$) i.e. E = 5·0 at 320 nm for a 1 cm layer of 0·001 N, ceric sulphate $Ce(SO_4)_2$; this corresponds with 8 mg O_2/l. The ceric ion may also be measured using a filter 405 ($\epsilon_{405\,nm} = 752$).

The classical Winkler method also continues to be used (Golterman, 1969). It is based on the fact that oxygen combines with $Mn(OH)_2$ to form higher hydroxides which on subsequent treatment with iodide in the presence of acid liberate free iodine almost stoichiometrically. Riley (1965) has discussed the method critically in the light of studies by Grasshoff (1962*a*, *b*) and Rotschi (1963).

Biochemical oxygen demand (B.O.D.) is measured by the rate of removal of oxygen by organisms growing at the expense of organic matter in water. This constitutes a rough index of contamination by organic matter such as that present in sewage. The need is to measure oxygen (O_2) concentration before and after incubation in darkness at 20° for 5 days. Not all the oxygen should be used during the test and to that end the sample may need to be diluted and aerated. It may be necessary to introduce a suitable bacterial culture. Various methods of determining dissolved oxygen are available. There is much to be said for consulting *Standard Methods* (1965).

Another parameter used for lake or river water is chemical oxygen demand (C.O.D.) which depends on the concentration of oxidizable organic compounds. Potassium bichromate is used as the reagent:

$$Cr_2O_7^{2-} + 14H^+ + 6e^- \rightarrow 2Cr^{3+} + 7H_2O$$

where a molar solution of bichromate is equivalent to 6 N Cr^{3+}. De Graaf & Golterman (1967) measured the Cr^{3+} produced by its absorption at 585 nm. The unused bichromate, however, makes a small contribution to absorption at that wavelength. At 500 nm where Cr^{3+} shows no absorption the absorption will be due mainly to bichromate. The corrected reading is:

$$E^{585}_{Cr^{3+}} = E^{585}_{\text{sample}} - \frac{E^{500}_{\text{sample}}}{E^{500}_{\text{blank}}} \times E^{585}_{\text{blank}}$$

The procedure requires that after standing for 3 hours (to allow the $H_2SO_4/K_2Cr_2O_7$ mixture to complete its work) reagent blanks, standards and samples should all be measured at 585 and 500 nm in suitable cells.

The determination of orthophosphate and total phosphate in water depends on the fact that in strongly acid solutions orthophosphate forms a yellow complex with molybdate ions. This can be reduced to give a highly coloured blue complex. When ascorbic acid is the reducing agent, the formation of blue colour is assisted by including sodium antimony tartrate ($NaSbOC_4H_4O_6$) in the reagent (Murphy & Riley, 1958, 1962). The blue coloured complex can be extracted by a suitable solvent immiscible with water.

[The reagent consists of $(NH_4)_6Mo_7O_{24}$, $4H_2O$, 4·8 g, 0·1 g sodium antimony tartrate in 4 N H_2SO_4, 400 ml, made up to 500 ml with 4 N H_2SO_4. The ascorbic acid solution, 2 g/100 ml water, is approximately 0·1 **M**. For standardization, 0·1757 g KH_2PO_4 (dry A.R.) is dissolved in water to 1 l, (40 μg P/ml) and a 1/40 dilution, i.e. 25 ml/l is prepared. Before making up to volume 2·5 ml of H_2SO_4 (1+1) and a few drops of chloroform are added. To 40 ml of sample are added molybdate reagent (5 ml) and ascorbic acid (2 ml) with good mixing. The solution is made up to 50 ml].

Spectroscopic measurements are made at 882 nm, 10 minutes after mixing. When the blue colour is weak the solution can be extracted with *n*-hexanol (10 ml). This is separated and made up to 10 ml with isopropanol to clear the solution. Readings are made at 690 nm. Extraction is 90–95% complete but much depends on careful calibration. An alternative procedure involves extraction of the yellow phosphomolybdate into isobutanol followed by reduction by stannous chloride (Proctor & Hood, 1954; Golterman & Würtz, 1961; Golterman, 1969). Extinctions are measured at 720 nm and *n*-hexanol can be substituted for isobutanol.

Carbohydrates are determined in two steps (i) furan derivatives are formed by dehydration with strong sulphuric acid; (ii) the products react with anthrone to form coloured products. Yemm & Willis (1954) found that different sugars react at different rates but under standardized conditions the extinction-coefficients attributable to the different sugars are closest with 15 minutes' heating. Calibration is conducted with the aid of a solution of glucose 10 mg/100 ml. The anthrone reagent is made by dissolving 0·1 g of anthrone in 0·5 l H_2SO_4 (S.G.1.84) and adding the solution carefully to 0·2 l water with 1 g thiourea as antioxidant. The procedure is to place 25 ml of anthrone solution in a tall tube and to introduce cautiously a layer of 5 ml sugar solution (*ca.* 100 μg glucose or equivalent). After mixing, the tube is stoppered loosely and heated in a boiling water bath for exactly 15 minutes. On cooling, all solutions (samples, reagent blanks, standards) are measured in suitable cells against water at 620–630 nm, the absorption peak of the coloured product. (See Golterman, 1969; Hewitt, 1958.)

In the living cell the chlorophyll of the algal chloroplasts (at a concentration which may reach 10%), will be present in complexes containing lipid and protein. When the newer spectrophotometric techniques are used on live algae the main chlorophyll peak (664 nm) is seen to be displaced some 30 nm in the direction of longer wavelengths. There is evidence of several 'complexed forms' of chlorophyll a with slightly different values for $\lambda_{max.}$, but only a single chlorophyll a is seen in solvent extracts (see p. 321).

The enzyme chlorophyllase is highly active in some species; it catalyses the hydrolytic removal of the phytyl groups to give chlorophyllide. Phaeophytin a is obtained when acid removes magnesium and the spectrum changes markedly. Phaeophorbide a is obtained when phytol is removed from phaeophytin (see Patterson & Parsons, 1963).

It is sometimes useful to determine chlorophyll and phaeophytin in water. The pigments will occur in particulate matter which must be collected on magnesium carbonate in a filter tower. The filter plus solid matter is comminuted (in a suitable high speed grinder or even by hand) and extracted by means of a small volume of 90% acetone. After centrifuging, the liquid is introduced into a suitable cell and E values are measured at 663 nm and 750 nm. The solution is transferred to a tube containing a small measured volume (e.g. 0·1 ml) of 4 N HCl, centrifuged again and measurements are again made at 663 and 750 nm. Both chlorophyll and phaeophytin have absorption peaks at 663 nm but the molecular extinction coefficient for phaeophytin is much lower than that of chlorophyll. From the decrease in E at 663 nm which occurs when chlorophyll is converted to phaeophytin on acidification, the chlorophyll content can be calculated. At 750 nm neither pure chlorophyll nor pure phaeophytin exerts significant absorption. The absorption coefficient K for chlorophyll a in 90% acetone is 89–90 at $\lambda_{max.}$ (*ca.* 664 nm) according to Parsons & Strickland (1963) and for phaeophytin it is about 56 (Golterman, 1969). A corrected ratio of extinctions before and after acidification shows a linear relationship for percentage of chlorophyll to total 663 nm pigment. This is perhaps a problem in which the investigator must make his own decisions concerning the information sought and the appropriate programme of calibration.

Yentsch (1967) discussed the analysis of water for chloroplast pigments and the interpretation of the results. Richards & Thompson (1952) had previously used a method for the analysis of chlorophylls and carotenoids in which measurements at different wavelengths were used to solve simultaneous equations aiming at the determination of three chlorophylls plus the plant and animal carotenoids. The method obviously depended on assuming that those were the predominant pigments present. In most aerobic waters chlorophylls a and b are present together with chlorophyll c from higher brown algae, diatoms, dinoflagellates and coccolithophores. Fucoxanthin is the most plentiful carotenoid of brown algae and diatoms. Yentsch & Ryther (1959) separated particulate matter by filtering large volumes of sea water and examined the absorption spectra of extracts in 90% acetone. They also found magnesium-free degradation products. Maximal phaeophytin occurred in sea water at 20 m greater depth than maximal chlorophyll. It appeared that herbivorous zooplankton possess a gut sufficiently acid to transform chlorophyll to phaeophytin and/or phaeophorbide.

The determination of chlorophylls and carotenoids in algal extracts by 'multichromatic' spectrophotometry on 90% aqueous acetone solutions has been moderately successful (Richards & Thompson, 1952; Creitz & Richards, 1955). Parsons & Strickland (1963) recommended the following (provisional) equations:

(i) $C_a = 11{\cdot}6\ E_{665} - 0{\cdot}14\ E_{630} - 1{\cdot}3\ \ E_{645}$
$C_b = 20{\cdot}7\ E_{645} - 4{\cdot}34\ E_{665} - 4{\cdot}42\ E_{630}$
$C_c = 55{\cdot}0\ \ E_{630} - 16{\cdot}3\ E_{645} - 4{\cdot}64\ E_{665}$
C_a, C_b, C_c = concentrations of chlorophylls in mg/l of 90% aqueous acetone.

(ii) Carotenoids for most marine crops MSPU/l = 4·0 E_{480} (MSPU = modified (arbitrary) specific pigment unit = approx. 1 mg for marine algae).

(iii) Carotenoids for chlorophyceae and myxophyceae (MPSU/l 4·0 E_{480}). (E extinctions for 1 cm cell at the chosen wavelengths.)

Individual pigments can be separated. Thus if a 90% aqueous acetone extract (10 ml) is mixed with aqueous sodium chloride (3·5 ml 0·05%) and extracted with

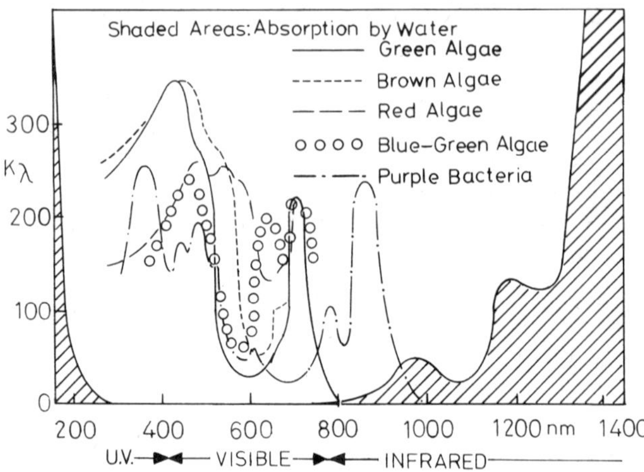

Figure 2.5 Absorption spectra of algae and bacteria in water (from Yentsch, 1967)

hexane (13·5 ml), 90% of the chlorophyll c remains in the hypophasic layer and can be measured directly (or by the change) in E_{450} when on acidification chlorophyll c is converted to phaeophytin c. This gives more trustworthy results than the 'multichromatic' calculation. If then the hexane layer is separated and shaken with 90% aqueous methanol most of the fucoxanthin, peridinin and phaeophorbide are found in the methanol layer. Spectrophotometric readings at 665, 510 and 480 nm then prove serviceable. The hexane layer can be chromatographed to separate carotenoids, chlorophyll a and phaeophytin a (for full details see Goodwin, 1955; Purcell, 1958; Jeffrey, 1961).

Thin-layer chromatography followed by absorption spectroscopy or fluorimetric analysis can be used to explore further the decomposition of chloroplast pigments and the possibilities of biochemical recycling. Progress is being made but the researches are in a quite early phase.

The determination of nitrite and nitrate is often necessary both in freshwater and in sea water. Nitrite is used to diazotize an aromatic amine and the diazonium compound is then coupled with another amine to give an azo dye, the depth of colour being proportional to the nitrite content. Many variations on the Greiss–Ilosvay method have been described. Among the more favoured ones sulphanilamide has been used for diazotization and N-(1-naphthyl)ethylenediamine for coupling. Both reactions proceed well at pH 3–4. Details of the method applied to sea water are given by Bendschneider & Robinson (1952) and Strickland & Parsons (1960). At 14 μg NO_2^- N/l the sensitivity is 0·14 μg NO_2^- N/l and the precision ±0·45 μg NO_2^- N/l. The azo dye can be extracted by means of isobutanol (Zeller 1955), and concentrated so as to permit spectrophotometric determination.

Table 2.3 Pigments of phytoplankton

	Solvent			
	Ether		90% aqueous acetone	
	$\lambda_{max.}$ nm	$E^{1\%}_{1cm}$	$\lambda_{max.}$ nm	$E^{1\%}_{1cm}$
Substance				
Chlorophyll a	662	1010	664	910
(Chlorophyllide a			615	190
almost the same)	430	1320	580	104
			433	1015
Chlorophyll b	644	620	645	540
			593	118
	455	1750	460	1480
Chlorophyll c	628	158	630	195
	580	157	580	115
	450	1690	446	1400
Phaeophytin a	667	637	666·5	566
			610	119
(Phaeophorbide a	408·5	1320	536	130
almost the same)	—	—	505	150
	—	—	409	1310
Fucoxanthin	—	—	480	725
	—	—	450	880
	—	—	430	731
	—	—	480	225
β-Carotene	—	—	455	2500
	—	—	430	1780
Astaxanthin	—	—	480	2500
Peridinin	—	—	480	840
(Sulcatoxanthin)				

(After J. D. H. Strickland, in Riley & Skirrow, 1965).

Nitrate can be reduced to nitrite and Riley & Sinhaseni (1957) used Raney nickel at pH 10·5 as perhaps the best method. The nitrate ion shows an absorption band at 303 nm but the ϵ value is too low to be useful in determining nitrate in water. There is a very much stronger band at 202 nm, $\epsilon_{max.}$ *ca.* 9900 but this is subject to interference except possibly in some fresh waters (cf. Goldman & Jacobs, 1961). A method applicable to sea water (Armstrong, 1963) depends on the appearance of well-marked selective absorption near 230 nm when an aqueous solution containing sodium chloride and nitrate is mixed with an equal volume of concentrated H_2SO_4. The 230 nm absorption conforms to Beer's Law at nitrate concentrations as high as 3 mg NO_3^- N/l and is probably due to nitrosyl chloride. Reduction with hydrazine destroys the nitrosyl chloride and any irrelevant absorption at 230 nm stands revealed.

It seems however that many authorities prefer to determine nitrate via nitrite. The reduction can be effected by hydrazine at pH 9·6 in the presence of catalytic amounts of copper (Mullin & Riley, 1955) but the yield is only 70%. Strickland & Parsons (1960) standardized the conditions and Henrikson (1966) adapted the method and also phosphate determinations to a Technicon Auto Analyser. This reduction procedure has disadvantages (Riley, 1965) and Mottir & Riley (1963) preferred to use a column filled with amalgamated cadmium filings, the nitrite formed being measured according to the method of Bendschneider & Robinson (1952).

It should be noted that when spectrophotometric determinations based on colour tests are done on sea water the intensity of absorption per unit weight of trace substance or element is frequently lower than would be expected from reference determinations in pure water. This reduction in absorption is called with-salt error and can sometimes reach 20% of the control value. The high ionic strength of sea water seems to influence the colour-forming process.

Bauman (1962) has discussed the determination of selectively absorbing constituents in mixtures by measurements at carefully chosen wavelengths. Separate studies on the individual substance allow simultaneous equations to be set up and solved by matrix methods.

Few organic compounds present in sea water can be directly assayed by spectrophotometric methods. Carbohydrates, however, react with N-ethylcarbazole in strong H_2SO_4 to give violet-red colours (Atkins, 1957; Zein–Eldin & May, 1958). Specificity is moderately good and a rather similar procedure involving anthrone (Lewis & Rakestraw, 1955) can be used instead or to give confirmation.

The examination of potable water has a long history of expertise (Institution of Water Engineers, 1960; *Standard Methods*, 1965; Klein, 1968). Some problems however give rise to increasing concern. The newer detergents can give rise to persistent 'pollution' and sewage entering lakes can cause eutrophication as effluents progressively fertilize seasonal growth of algae and weeds. Lake Erie is the major example. The analytical methods in use are not unsatisfactory but the specialists are constantly seeking improvements.

Phenols can be present in water and very small quantities of a few can greatly affect the taste. Larger quantities may poison fish. One well-tried method (Fox & Gauge, 1920) involved coupling a phenol with freshly diazotized sulphanilic acid to form a yellow or orange azo dye; e.g.

$$[SO_3H.C_6H_4{-}N_2]SO_4 + 2C_6H_5OH \rightarrow H_2SO_4 + 2SO_3H.C_6H_4{-}N{=}N{-}C_6H_4OH$$

For calibration a mixture of cresols (*o*-35%, *m*-40%, *p*-25%) is used and today determination should be made spectrophotometrically at $\lambda_{max.}$. Instead of sulphanilic acid *p*-nitraniline can be used (Beier, 1957). Gibbs (1927) allowed the phenol or phenolic mixture to react with 2:6-dibromoquinonechloroimide at pH 9·4 to produce an indophenol dye:

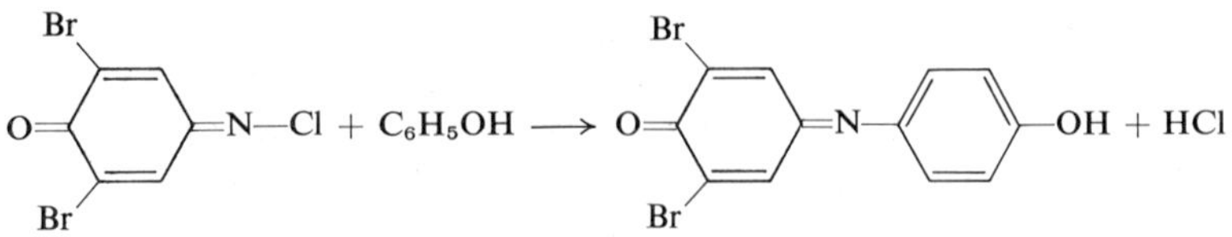

$$O{=}C_6H_2Br_2{=}N{-}Cl + C_6H_5OH \longrightarrow O{=}C_6H_2Br_2{=}N{-}C_6H_4{-}OH + HCl$$

The dye(s) can be extracted into *n*-butyl alcohol (Ettinger & Ruchhoft, 1948). There are technical difficulties since different phenols give differently coloured dyes but reinvestigation using modern separation methods with spectrophotometric monitoring might well be worth while provided dye production can be made quantitative or at least very reproducible.

Standard Methods (1965) recommends the use of 4-aminoantipyrine (4-aminophenazone) in the presence of potassium ferricyanide to react with phenol(s) to give a red colour

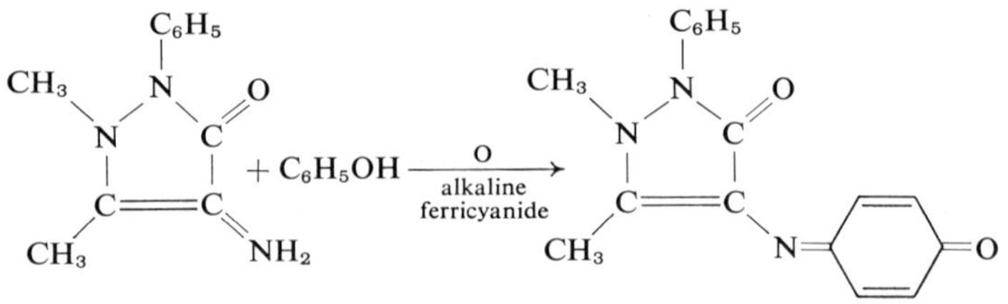

$$\text{(4-aminoantipyrine: } C_6H_5, CH_3, CH_3, O, NH_2) + C_6H_5OH \xrightarrow[\text{alkaline ferricyanide}]{O} \text{(antipyrine: } C_6H_5, CH_3, CH_3, O)\,{-}N{=}C_6H_4{=}O$$

This procedure is very sensitive when it is applicable but its range is limited and it must be used with caution.

At the present time any serious problem of contamination of water by phenols might well be treated as an investigation rather than a routine application of accepted methods. Large volumes of water could be filtered through adsorbents (such as activated carbon or other suitable solid) and the retained organic products eluted. The eluate could be subjected to chromatography, perhaps initially to thin layer chromatography. Spots could be visualized by spraying to form dyes which could be extracted and examined in a spectrophotometer. Larger scale separations could be monitored by ultraviolet absorption (phenolic absorption in acid and in alkali). The important decision concerns how detailed the required information should be; to obtain it is now largely a matter of time and trouble.

A major problem is the determination of synthetic detergents in water. The main types are (*a*) anionic detergents yielding in water, e.g.

$$Na^+ + R.OSO_3^-$$

where R is a long chain (Teepol, Dreft), or

$$Na^+ + R{-}C_6H_4{-}SO_3^-$$

(Tide, Daz); (*b*) non-ionic detergents such as polyglycol ethers of alkyl phenols; e.g.

$$R{-}C_6H_4{-}(C_2H_4O)_nOH$$

(Stergene) and (*c*) cationic detergents such as cetylpyridinium bromide

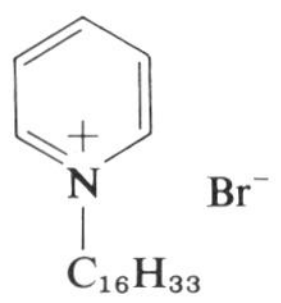

$$\text{C}_5\text{H}_5\text{N}^+\text{–C}_{16}\text{H}_{33}\ \ \text{Br}^-$$

or quarternary ammonium compounds

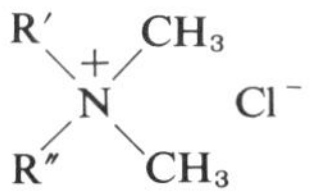

$$\text{R}'\text{R}''\text{N}^+(\text{CH}_3)_2\ \ \text{Cl}^-$$

where R′ and R″ = $CH_3(CH_2)_n$ are long chains. A standard procedure for determining anionic detergents in water is based on the formation of a chloroform-soluble blue complex of methylene blue and the detergent (ABCM–SAC Joint Committee, 1957; see also Longwell & Maniece, 1955). Technical details are given in *Approved Methods* (1960). The optical density of the final chloroform extract is measured at 650 nm. The detergent Monoxol O.T. (dioctylsulphosuccinate) can be bought in a satisfactory state of purity and used as a reference standard.

Degens *et al.* (1953) determined combined detergents (sulphate plus sulphonate) and then hydrolysed sulphate detergents in a second sample by boiling with sulphuric acid. The dye produced on this solution was due to the sulphonate detergent. A quantitative spectrophotometric procedure for very low concentrations of alkyl benzenesulphonate detergents is described by Fairing & Short (1956). Non-ionic detergents present formidable analytical problems (see Klein, 1968).

Schechter *et al.* (1945) determined the insecticide DDT by extracting it from water (ether plus hexane) followed by nitration to a polynitro derivative. This dissolves in benzene and gives a strong blue colour with methanolic sodium methoxide. Berck (1953) adapted this to the examination of river water and/or suspended solids containing very small traces of DDT. Schechter & Hornstein (1952) and Hancock & Laws (1955) described the extraction of gammexane (γ-benzenehexachloride) from water and its indirect determination photometrically by forming a coloured reaction product.

2.2 Sunlight and Ultraviolet light

Exposure of the human body to sunlight results in a rapid reddening (erythema) caused by the visible and infrared radiations, but the effect is transient. Ultraviolet rays produce erythema after a delay of some hours and if exposure has been intense or prolonged the effect may be severe and blisters can be formed.

Exposure of the eyes to ultraviolet radiation from unshielded arcs or the light from a quartz-mercury vapour lamp can have delayed and very painful effects and vision can be irreversibly impaired in extreme cases if goggles are not worn. Less drastic exposure results in pigmentation slowly following erythema. Pioneer work was done by Finsen, who used a carbon arc as source of light (Bang, 1905).

The absorption of light by skin is a complicated process in which the different layers behave differently in ways which vary with the spectral region. Moreover skin exhibits fluorescence, $\lambda_{max.}$ for human skin being near 460 nm. Separate layers of skin were studied by Bachem (1929), Bachem & Reed (1929) and Lucas (1931). Strong selective absorption occurs at 250–300 nm ($\lambda_{max.}$ 280 nm). Bachem found for the horny layer $\lambda_{max.}$ 280 nm with a minimum at 260 nm followed by steeply rising end-absorption. The stratum granulosum absorbs visible light rather more strongly and there is an inflection near 350 nm and the usual 280 nm peak. It is clear however that scattering gives rise to technical difficulties, but these were to some extent overcome by Lucas (1931) when he studied the skin from a blister and found an absorption spectrum consistent with contributions by tyrosine and tryptophan.

The action spectrum for erythema was measured by Hausser & Vahle (1928), by Luckiesh *et al.* (1930) and by Coblentz *et al.* (1931, 1932), see Fig. 2.6. The effective radiation lay mainly between 290 and 307 nm

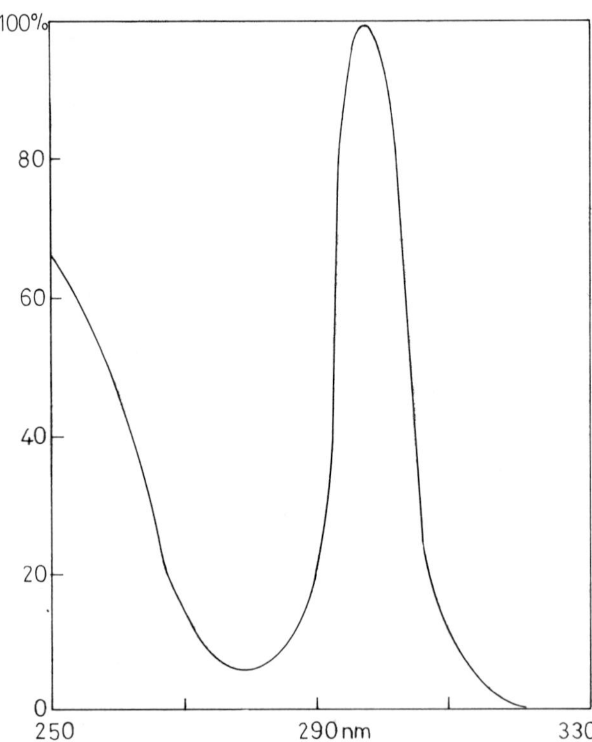

Figure 2.6 Spectral erythema production as percent of maximal effect (based on data by Hausser & Vahle, 1927, and Luckiesh *et al.*, 1930)

(λ_{max}. 297 nm) while between 270 and 290 nm the radiation was less effective and maximal at 280 nm. This action spectrum seems to be less than fully understood.

Creams intended to be used on skin to prevent sunburn are popular but users seem to expect them to favour sun-tan as well. Many contain agents absorbing in the region 290–320 nm. It should be remembered that the spectral intensity of sunlight falls off very rapidly on the short-wave side of about 298 nm as a result of absorption by ozone in the upper atmosphere. The problems of combining in unguents safety in use with protection against sunburn and the safe acceleration of normal pigmentation have not been adequately analysed.

2.2.1 Lethality

Downes and Blunt (1877) recorded a bactericidal action of ultraviolet light applicable equally to moist and dry materials. It was at first thought that free oxygen was necessary but Ward (1893) disproved the idea. He found that although sunlight was slightly bactericidal, only the ultraviolet rays were effective. This conclusion was supported by the observation that interposing a quartz plate between a carbon arc and the bacterial preparation did not diminish the photosterilization whereas a glass plate transmitting little or no light of wavelength shorter than 340 nm reduced it practically to zero. The early experiments in which oxygen was excluded were carried out in glass vessels!

Bernard & Morgan (1903) projected the spectrum of a carbon arc on to an agar plate superficially inoculated with *Bacillus coli.* After exposure for a definite time the plate was incubated and the positions and wavelengths at which the bacteria had been killed were noted. The effective radiation lay between 226 and 329 nm. The authors concluded that the bactericidal rays had too little penetrating power to sterilize body tissues. Browning & Russ (1917) carried out a similar experiment on *Staphylococcus aureus* using a tungsten arc. They found a sharp demarcation at 296 nm, longer wavelengths having no effect. Shorter waves were highly lethal especially between 254 and 280 nm. Browning and Russ distinguished between two regions (*a*) 296–380 nm able to penetrate skin to a considerable extent and (*b*) 210–296 nm strongly bactericidal but penetrating no more than 0·1 nm of human skin.

Many authors were inclined to attribute the lethal effect *in the first instance* to absorption by the tyrosine and tryptophan present in proteins. That this was too simple a view was perhaps implicit in the variability of different bacterial species (see Ellis & Wells); indeed partial photosterilization at one time had a certain vogue inasmuch as a number of disease-causing organisms were among the more photosensitive species. In the period 1909–1915 many patents were taken out for photosterilization of water, wine, meat and even vaccines, but few of them had much permanent technological success.

There was indeed considerable interest in photosterilization but practically no progress was made in understanding the mode of action. Gates (1928) broke new ground when he claimed that the action spectrum (relative lethality of different wavelengths) agreed rather well with the absorption spectrum of nucleic acid. The time was not however ripe for appreciating this since the consequences of absorption by nucleic acids were no better understood than the consequences of absorption by aromatic amino acid residues.

Beukers & Berends (1964) discovered that ultraviolet light could effect dimerization of thymine in DNA and although this is far from being the only photochemical effect, it is a highly significant one. The peak absorption by purine and pyrimidine residues in nucleic acids occurs at 260–265 nm and it is pertinent to consider what such wavelengths can in fact accomplish: (*a*) in *Escherichia coli* (strain B) the first noticeable effect of radiation at 265 nm at low intensity is impaired ability of the cells to divide, (*b*) at a higher intensity colony-forming ability declines and DNA and RNA biosynthesis is impaired, (*c*) at still higher intensities the lethal effect is marked and specific mutation may occur. Bacterial spores are much more resistant and enzymes often require considerable exposure before being fully inactivated. It is convenient to use as a parameter the dose of radiation needed to inactivate bacteria to 37% survival.

$$[e=2{\cdot}71828, \quad e^{-1}=0{\cdot}3679=ca.\ 37\%]$$

By expressing the kinetics of inactivation as an exponential function of dose, 37% survival means an average of one successful 'hit' per sensitive 'target' since

$$N/N_0=e^{-\sigma D}$$

where N_0 is the number of such targets present initially, N is the number surviving, D=dose (ergs/mm^2) and σ is an 'inactivation' cross section (not necessarily related to a physical cross section; σ is better considered as a product of quantum yield and an 'absorption' cross section).

The kinetic aspect is that there is an exponential loss of active units with increasing dose, i.e. of photons absorbed. The 'inactivation' cross section becomes accessible from the slope of the straight line obtained when the logarithm of the surviving fraction N/N_0 is plotted against D, or it can be obtained from the reciprocal of the dose at which N/N_0 (survival) is 0·37.

Quantum yields may be quite small in large molecules like DNA and RNA. Using the definition

$$\phi\ \text{(quantum yield)}=\frac{\text{Number of moles (units) changed}}{\text{Number of Einsteins absorbed}}$$

$$(1\ \text{Einstein}=\text{N}h\nu)$$

ϕ can be 10^{-6} for a macromolecule and 10^6 for a chain reaction. In nucleic acids a low ϕ indicates that damage occurs at 'spots' or very specific loci and most of the radiation absorbed by purines and pyrimidines is dissipated without effecting chemical change. Smith & Hanawalt (1969) state that 99% *E. coli* B/r is killed by 1800 ergs/mm^2 and that 0·1% of the total thymine (itself about one-quarter of all the purine and pyrimidines) is dimerized. At least half the inactivating action of ultraviolet light is due to thymine dimerization so that only 0·025% of the DNA molecule has been changed by a dose that kills 99% of the organisms. They point out that if the small quantum yield is now applied to the molecular weight of a postulated 'active centre' of DNA it raises the value of ϕ to about unity. 'Quantum yield' must therefore be put into appropriate perspective.

The dose rate can be important because at very high intensity and very short duration the Bunsen–Roscoe reciprocity law is not obeyed. In flash photolysis the special technique allows triplet–triplet absorption.

If now the mechanisms of photochemical damage in cells are reconsidered, the possibility that DNA is the major target and that within DNA a limited number of thymine units are specially important, at once raises the question whether the damage can be repaired or not. This will have a bearing on ultraviolet mutagenesis. If the radiation affects the active site of an enzyme, any resultant damage may be insignificant if the genetic outfit for replacing the protein is intact. On the other hand damage localized in the genome may demand repairing processes.

Action spectra for the lethal effect on bacteria have been measured and a classical curve obtained by Gates (1930) on *E. coli* shows a rounded peak near 265 nm and a minimum at 235 nm. Beyond 295–300 nm the effect is vanishingly small. Action spectra for mutagenesis are very similar.

The probability that DNA is the significant absorbing substance in the damaging effect of irradiation has led to *in vitro* studies on the photochemistry of nucleic acids, nucleotides and purine or pyrimidine bases. The sugar moieties are practically transparent in the region 230–300 nm. The photosensitivity of pyrimidines is much greater than that of the purines, the quantum yields being of the order 10^{-3} and 10^{-4} respectively. Although energy transfer is not excluded, the probability is that pyrimidine absorption is the major cause of photochemical damage to DNA.

Irradiation of cytosine or uracil in aqueous solution results in the disappearance of their selective absorption but heating or treatment with acid or alkali causes its reappearance. The chromophoric groups ($\lambda_{max.}$ 260 nm) require a double bond in uracil between carbon atoms 5 and 6 and the irradiation product is 6-hydroxy-5-hydrouracil

```
        O                              O
        ‖                              ‖
  H     C                        H     C
   \  /   \                       \  /   \
    N      CH        hν            N      CH2
    |      ‖       ⇌              |      |
    C      CH       heat           C      CHOH
  //  \   /                      //  \   /
 O     N                        O     N
       |                              |
       H                              H
```

The hydration of cytosine also explains the reversible effect of irradiation. Irradiation of cytidine or cytidylic acid also results in a change of absorption at 265 nm which is reversible. With thymine (2,6-dioxy-5-methylpyrimidine)

```
        O
        ‖
  H     C
   \  /   \
    N      C—CH3
    |      ‖
    C      CH
  //  \   /
 O     N
       |
       H
```

irradiation of aqueous solutions again results in the disappearance of definite selective absorption but the change is not heat-reversible. The quantum yield, $\phi = 0{\cdot}4 \times 10^{-3}$, is not high but if the aqueous solution is frozen the quantum yield rises to about 0·2. The product is a dimer and freezing seems in favour of 'bimolecular interaction'. The dimeric form contains a cyclobutane ring. Radiation in which λ is < 250 nm favours splitting of the dimer to regenerate the monomeric form whereas radiation on the long wave wide of 270 nm favours dimerization.

The cyclobutane ring structure is also invoked for dimers of uracil and cytosine and for the double compounds uracil-thymine, uracil-cytosine and thymine-cytosine. Cytosine dimers can be converted to uracil dimers by deamination. They are formed more slowly but more rapidly broken down (photochemically, 240 nm) than uracil or thymine dimers. If in the DNA of a cell deamination of a cytosine dimer were to occur, and the uracil dimer were to be broken down, a mutation could occur. *Escherichia coli* strains differ in respect of their resistance to irradiation. The resistant strains excise the thymine dimers and the DNA is repaired whilst the sensitive strains fail to repair the DNA. The dimerization of thymine is obviously an important aspect of irradiation damage but it would be easy to over simplify the situation. Thymine residues may on photochemical excitation enter into combination with amino acid residues in proteins, e.g. through sulphydryl or hydroxyl groups. Cytosine residues may undergo similar combinations.

In general, frozen solutions containing DNA exhibit greater photochemical sensitivity than aqueous solutions at room temperature, whereas in dry preparations cyclobutane dimers of thymine are but little formed. Photochemical cross linking of DNA and protein occurs and perhaps the simplest analogue is the formation of 5 S-cysteine-6-hydrouracil.

```
        O
        ‖                    NH2
  H     C                     |
   \  /   \
    N      CH—S—CH2CH—COOH
    |      |
    C      CH2
  //  \   /
 O     N
       |
       H
```

Photochemical changes in DNA are sometimes apparently spontaneously reversible. This applies particularly to some hydration products of pyrimidines but the phenomenon is not fully understood. Direct reversal of dimer formation can occur with larger photons (235 nm to 280 nm) but an even more significant effect is a photochemical reversal catalysed by an enzyme. The action spectrum for photoreactivation of *E. coli* B/r previously exposed to damaging ultraviolet irradiation shows peaks at 360 and 390 nm (and a third peak at 324 nm) and the envelope extends from 300 to 500 nm. Absorption peaks at 360 and 390 nm are also shown in the action spectrum for photoreactivation of ultraviolet-irradiated transforming DNA from *Haemophilus influenzae* with yeast extract. The enzyme has been greatly concentrated but the absorbing prosthetic group has yet to be identified. The photoreactivation applies particularly to the thymine dimer lesions in DNA but it also reactivates other polynucleotide strands in which combinations of pyrimidines form 'dimers'.

A very general function of carotenoids is the protection of cells against harmful effects of light. This photo-oxidation (Griffiths *et al.*, 1955) was made clearer by studies on a blue-green mutant of *Rhodopseudomonas spheroides* which though nearly normal in respect of chlorophylls failed to carry the biosynthesis of carotenoids beyond the phytoene stage. Growth in the mutant proceeded quite normally under anaerobic photosynthetic conditions, but the organism was quickly killed on simultaneous exposure to light and oxygen, with concomitant destruction of bacteriochlorophyll. In darkness oxygen was not harmful. The wild type of *R. spheroides* with its complement of coloured carotenoid was not sensitive to light and oxygen. The destruction of cells was ascribed to a photodynamic action in which bacteriochlorophyll acted as the photosensitizer. The lethal effect is largely independent of temperature and destruction of bacteriochlorophyll is secondary, the true cause being the absence of coloured carotenoid.

Completion of the synthesis of carotenoids is delayed or inhibited in *Rhodospirillum rubrum* by adding diphenylamine to the medium (Goodwin & Osman, 1954) and phytoene accumulates. This is a physiological rather than a genetic effect, since coloured carotenoids are again formed on removing the inhibitor. The cells from which coloured carotenoids are absent are susceptible to destruction by light and oxygen and bacteriochlorophyll is also degraded (Cohen-Bazire & Stanier, 1959).

A deficiency of coloured carotenoids thus increases susceptibility to photodestruction and the role of carotenoids is protective. The lethal effect is perhaps due to a failure in photophosphorylation. In the case of *Chromatium* (strain D) carotenoid biosynthesis is halted by diphenylamine and the isolated chromatophores, though capable of photophosphorylation in the absence of oxygen, lost that capacity on exposure to light and oxygen (Fuller & Anderson, 1958).

It is natural to consider whether the protective effect of coloured carotenoids (as against the absence of protective effect by the closely related phytoene) is merely a light-filter action. This simple idea has been rejected in favour of more sophisticated hypotheses (Stanier, 1960; Calvin, 1955) involving reversible oxidation of carotenoids and the three dimensional structure of photosynthetic entities.

Photo-oxidation is not limited to chlorophyll-containing organisms and photosensitization can be effected by other coloured substances (flavins, some porphyrins). Thus non-pigmented cells of a *Corynebacterium poinsettiae* mutant displayed the photodynamic effect when a synthetic dye (toluidine blue) was used as a photosensitizer (Kunisawa & Stanier, 1958).

The hydrocarbon [^{3}H] benzpyrene sensitizes the killing of *Paramecium* by binding to DNA (Epstein & Burroughs, 1962). The effective light is that absorbed by the hydrocarbon and the addition occurs at the base residues.

The furocoumarin 8-methoxypsoralen acts similarly (Oginsky *et al.*, 1959; Matthews, 1962) and irradiation of the complex results in mutations. The adduct is formed with pyrimidines. Photosensitization is discussed by Simon (1967) and the effects of ultraviolet irradiation and photoreactivations by Setlow (1967). This is an area from which much may be expected.

References

2.1

ABCM–SAC Joint Committee (1957). *Analyst*, **82**, 836.

Armstrong, F. A. J. (1963). *Analyt. Chem.*, **35**, 1292.

Armstrong, F. A. J. & Boalch, G. T. (1961). *U.G.G.I. Monogr.* **10**, 63; and *J. Mar. Biol. Assoc. U.K.*, 1961, **41**, 591.

Atkins, W. R. G. (1957). *J. Cons. int. Explor. Mer.*, **22**, 271.

Atkins, W. R. G., Clarke, G. L., Pettersson, H., Poole, H. H., Utterback, C. L. & Angstron, A. (1938). *J. Cons. int. Explor. Mer.*, **13**, 37.

Bauman, R. P. (1962). In *Absorption Spectroscopy*, J. Wiley, New York.

Beier, E. (1957). *Gas-u Wasserfach*, **98**, 262.

Bendschneider, K. & Robinson, R. J. (1952). *J. Mar. Res.*, **11**, 87.

Berck, B. (1953). *Anal. Chem.*, **25**, 1253.

Clarke, G. L. & James, H. R. (1939). *J. Opt. Soc. Amer.*, **29**, 43.

Creitz, G. I. & Richards, F. A. (1958). *J. Mar. Res.*, **14**, 211.

Dawson, J. H. & Hulbert, E. O. (1934). *J. Opt. Soc. Amer.*, **24**, 175.

Degens, P. N., Evans, H. C., Kommer, J. D. & Winsor, P. A. (1953). *J. Appl. Chem. Lond.*, **3**, 54.

Dietrich, G. (1939). *Ann. d. Hydrogr. u. Mar. Meteor.*, **67**, 411 (quotes observations by W. R. Sawyer and J. R. Collins).

Dorsey, N. E. (1940). *Amer. Chem. Soc. Monograph Series* No. 81, Reinhold, New York.

Eisenberg, D. & Kauzmann W. (1969). *The Structure and properties of water*. Clarendon Press, Oxford.

Ettinger, M. B. & Ruchhoft, C. C. (1948). *Analyt. Chem*,. **20**, 1191.

Fairing, J. D. & Short, F. R. (1956). *Analyt. Chem.*, **28**. 1827.

Fox, J. J. & Gauge, A. J. H. (1920). *J. Soc. Chem. Ind., Lond.*, **39**, 260T; *ibid.*, 1922, **41**, 173T.

Friedrich, H. (1969). In *Marine Biology*. Sedgwick & Jackson, London.

Gibbs, H. D. (1927). *J. biol. Chem.*, **72**, 649.

Goldman, E. & Jacobs, R. (1961). *J. Amer. Wat. Wrks, Ass.*, **53**, 187.

Golterman, H. L. & Clymo, R. S. (1967). In *Chemical Environment in the Aquatic Habitat*. Roy. Netherlands Acad. Sciences, Amsterdam. (North-Holland Publishers.)

Goltermann, H. L. (with R. S. Clymo) (1969). In *Methods for Chemical Analysis of Fresh Waters*. I.B.P., Blackwell. Oxford.

Golterman, H. L. & Wurtz, I. M. (1961). *Analyt. Chim. Acta.*, **25**, 295.

Goodwin, T. W. (1955). In *Modern Methods of Plant Analysis*. Ed. K. Paech and M. V. Tracey, **III**, p. 272. Springer-Verlag, Berlin.

de Graaf Bierbrauwer, I. M. & Golterman, H. L. (1967). In *Chemical Environment in the aquatic environment*. Ed. H. L. Golterman and R. S. Clymo. North-Holland Publishers, Amsterdam.

Grasshoff, K. (1962 *a*, *b*). *Kiel Meeresforsch.*, **18**, 42, 151.

Hancock, W. & Laws, E. Q. (1955). *Analyst*, **80**, 665.

Henricksen, A. (1966). *Analyst*, **91**, 1087.

Hewitt, B. R. (1958). *Nature* (*Lond.*), **182**, 246.

Hood, D. W. (1963). In *Ann. Rev. Oceanography and Marine Biology*. Ed. H. Baines. p. 144.

Jeffrey, L. M. (1961). *Biochem. J.*, **80**, 336.

Jeffrey, L. M. & Hood, D. W. (1958). *J. Mar. Res.*, **17**, 247.

Jerlov, N. G. (1948) *Optical studies of ocean waters*. Rep. Swedish Deep-Sea Exped. 1947–1948, **3**, 1.

Jerlov, N. G. (1963). In *Ann. Rev. Oceanography and Marine Biology*, 1.

Jorgensen, W. & Utterbach, C. L. (1939). *J. Marine Res.*, **2**, 30.

Kalle, K. (1938). *Ann. d. Hydrogr. u. Mar. Meteor.*, **66**, 1.

Klein, L. (1968). In *River Pollution*. **1**. *Chemical Analysis*. Butterworths, London.

Lenoble, J. (1956). *C. R. Acad. Sci., Paris*, **242**, 806.

Lenoble, J. & Saint-Guilly, B. (1955). *C. R. Acad. Sci., Paris*, **240**, 954.

Lewis, J. L. & Rakestraw, N. W. (1955). *J. Mar. Res.*, **14**, 253.

Longwell, J. & Maniece, W. D. (1955). *Analyst*, **80**, 167.

Lucas, C. E. (1955). In *Deep Sea Research*, Suppl. to Vol. **3**, p. 139.

Morris, A. W. & Riley, J. P. (1963). *Analyt. Chim. Acta.*, **29**, 272.

Mullin, J. B. & Riley, J. P. (1955). *Analyt. Chim. Acta.*, **12**, 464.

Murphy, J. & Riley, J. P. (1958). *J. Mar. Biol. Assoc. U.K.*, **37**, 9; (1962). *Analyt. Chim. Acta.*, **27**, 31.

Parsons, T. R. & Strickland, J. D. H. (1963). *J. Mar. Res.*, **21**, 155.

Patterson, J. & Parsons, T. R. (1963). *Limnol. Oceanogr.*, **8**, 355.

Proctor, C. M. & Hood, D. W. (1954). *J. Mar. Res.*, **13**, 122.

Purcell, A. E. (1958). *Analyt. Chem.*, **30**, 1049.

Richards, F. A. & Thompson, T. G. (1952). *J. Mar. Res.*, **11**, 156.

Riley, J. P. (1965). In *Chemical Oceanography*, Vol. **2**. Ed. J. P. Riley and G. Skirrow. Academic Press, London and New York, pp. 313, 314.

Riley, J. P. & Sinhaseni, P. (1957). *J. Mar. Biol. Assocn. U.K.*, **36**, 161.

Rotschi, H. (1963). *Cahuro Oceanogr.*, **15**, 7.

Sawyer, W. R. (1931). *Conts. Can. Biol. & Fish*, **7**, 75.

Sawyer, W. R. & Collins, J. R. (*see* Dietrich, G., 1939).

Schechter, M. S. & Hornstein, I. (1952). *Analyt. Chem.*, **24**, 544.

Schechter, M. S., Solloway, S. B., Hayes, R. A. & Haller, H. L. (1945). *Ind. Eng. Chem.* (*Anal.*), **17**, 704.

Shapiro, J. (1967). In *Chemical Environment in the aquatic environment*. Ed. H. L. Golterman and R. S. Clymo. North-Holland Publishers, Amsterdam.

Standard Methods for the Examination of Water and Waste Water. (1965). *Amer. Publ. Health Assoc.*, 12th Ed., New York.

Strickland, D. H. & Parsons, T. R. (1960). *Fish. Res. Bd. Can. Bull.*, No. **125**, 185.

Sverdrup, U., Johnson, M. W. & Fleming, R. H. (1946). In *The Oceans*. Prentice Hall, Englewood Cliffs, New Jersey, U.S.A.

Utterback, C. L. (1936). *Conseil Perm. Internat. p. l'Explor. de la Mer Rapp. et Proc. Verb*, **101**, pt. 2, No. 4.

Walrafen, G. E. (1967). *J. Chem. Phys.* **47**, 114 also **48**, 244.

Yemm, E. W. & Willis, A. J. (1954). *Biochem. J.*, **67**, 508.

Yentsch, C. S. (1967). In *Chemical environment in the aquatic habitat*. Ed. H. L. Golterman and R. S. Clymo. North-Holland Publishers, Amsterdam, p. 255. Also *Deep Sea Research*, 1965, **12**, 653.

Yentsch, C. S. & Ryther, J. H. (1959). *Deep Sea Research*, **6**, 72.

Zein-Eldin, Z. P. & May, B. A. (1958). *Analyt. Chem.*, **30**, 1935.

Zeller, H. D. (1955). *Analyst*, **80**, 632.

2.2

Bachem, A. (1929). *Amer. J. Physiol.*, **91**, 58.

Bachem, A. & Reed, C. I. (1929). *Amer. J. Physiol.*, **90**, 600; (1931), *ibid.*, **97**, 86.

Bang, S. (1905). *Mitt. Finsen's Medisins Lysinstitut*, **9**, 164.

Bernard, J. E. & Morgan, H. (1903). *Proc. Roy. Soc.*, **72**, 126.

Beukers, R. & Berends, W. (1964). *Chem. Weekblad.*, **60**, 97.

Browning, C. H. & Russ. (1917). *Proc. Roy. Soc.*, **110**, 33.

Calvin, M. (1955). *Nature (Lond.)*, **176**, 1215.

Coblentz, W. W. & Fulton, H. R. (1924). *Bureau Stand. Sci. Papers*, **19**, 495, 641.

Coblentz, W. W., Stair, R. & Hogue, J. M. (1931), *Bureau Stand. J. Res.*, **7**, 723; (1932). *ibid.*, **8**, 541, 759.

Cohen-Bazire, G. & Stanier, R. Y. (1958). *Nature (Lond.)*, **181**, 250.

Downes, A. & Blunt, T. P. (1877). *Proc. Roy. Soc.*, **26**, 488.

Epstein, S. S. & Burroughs, M. (1962), *Nature (Lond.)*, **193**, 337.

Fuller, R. C. & Anderson, I. C. (1958). *Nature (Lond.)*, **181**, 252.

Gates, F. L. (1928). *Science (N.Y.)*, **58**, 479; (1929). *J. Gen. Physiol.*, **13**, 231, 249; (1930) *ibid.*, **14**, 31.

Goodwin, T. W. & Osman, H. G. (1954). *Biochem. J.*, **56**, 222.

Griffiths, M., Sistrom, W. R., Cohen-Bazire, G. & Stanier, R. Y. (1955). *Nature (Lond.)*, **176**, 1211.

Hausser, K. W. & Vahle, W. (1928). *Strahlentherapie*, **27**, 348.

Henri, V. (1914). *Compt. rend. soc. biol.*, **73**, 323.

Kunisawa, R. & Stanier, R. Y. (1958). *Arch. Mikrobiol.*, **31**, 146.

Luckiesh, M., Holladay, L. L. & Taylor, A. H. (1930). *J. Opt. Soc. Amer.*, **20**, 423.

Lucas, N. S. (1931). *Biochem. J.*, **25**, 57.

Matthews, M. M. (1964). *Photochem. Photobiol.*, **3**, 75.

Oginsky, E. L., Green, G. S., Griffith, S. G. & Fowlkes, W. L. (1954). *J. Bact.*, **78**, 821.

Setlow, J. K. (1967). In *Comprehensive Biochemistry*, Vol. **27**, p. 157. Ed. M. Florkin and E. H. Stotz. Elsevier, Amsterdam.

Simon, M. I. (1967). In *Comprehensive Biochemistry*, Vol. **27**, p. 137. Ed. M. Florkin and E. H. Stotz. Elsevier, Amsterdam.

Smith, K. C. & Hanawalt, P. C. (1969). *Molecular Photobiology inactivation and recovery*. Academic Press, New York.

Stanier, R. Y. (1960). *Harvey Lect.*, **54**, 219.

T'so, P. O. P. & Lu, P. *Proc. Nat. Acad. Sci.*, **51**, 17.

Ward, H. M. (1893) *Proc. Roy. Soc.*, **42**, 393; **53**, 23; **54**, 472.

3 Electronic Absorption Spectra

3.1 Ultraviolet Absorption Spectra

Ultraviolet absorption spectra arise by the excitation of electrons from their ground state. The energy of a transition depends on the strength of the binding between atoms. The wavelength of the selective absorption thus depends on two or more atoms held together and constituting a chromophoric group, the properties of which may be modified by its intramolecular environment or by the polarity of a solvent medium. Penetrating theoretical treatments have been advanced in respect of the simpler absorbing molecules but the interpretation of the spectra of polyatomic organic substances rests on largely empirical but well-evidenced generalizations.

A first step to interpretation is to distinguish between σ, π and n electrons. Saturated compounds containing only single bonds have σ valency electrons. The transitions give rise to strong absorption bands in the vacuum ultraviolet region (< 170 nm). These are known as N → V and N → R transitions (Price, 1947) or Rydberg transitions corresponding with ionization potentials. In N→R transitions, a σ electron is removed by stages from both atoms concerned until it is in the end ejected (ionization):

$$\text{C—C} \rightarrow \text{C}^{+}\text{—C}.$$

The absorption shows narrow bands and a continuous region. In N→V transitions both atoms of a link acquire a temporary change: C—C→C^{-}—C^{+} or C—H→C^{-}—H^{+} forming a dipole. The lone pairs of electrons present in many atoms (e.g. the oxygen of acetone) are non-bonding and are designated n (or sometimes p) electrons. They are less strongly held than σ electrons and undergo Rydberg (N→R) transitions at higher wavelengths. Smaller quanta are needed to eject non-bonding n electrons. Ionization potentials, measured by electron bombardment, confirm this. There is also a transition between an n electron and an antibonding σ orbital (σ^*) giving rise to much weaker absorption shifted towards the red.

The electrons of multiply bonded atoms, as in the ethylenic and acetylenic linkages are designated π electrons. Rydberg transitions occur and correspond with selective absorption in the far ultraviolet region. In addition a transition $\pi - \pi^*$ (a π electron excited to an anti-bonding orbital) gives rise in olefines to selective absorption at 180–205 nm, $\epsilon_{max.}$ order 10 000. Furthermore, as in ketones, a transition $n - \pi^*$ gives rise to weak absorption near 280 nm, ϵ *ca.* 15–20, and a more intense band $n - \pi^*$ occurs near 190 nm.

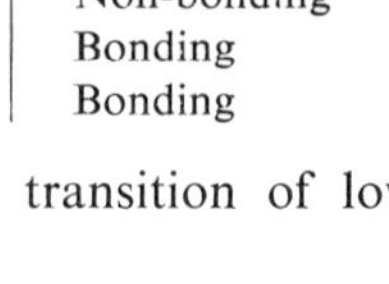

```
    H     H
  σ | σ   |
H—C—C—C—H
    |  ||π |
    H  O  H
       ··
       n
```

The electronic energy levels are:

↑	Anti-bonding	σ	σ^*
	Anti-bonding	π	π^*
Energy	Non-bonding	n	n
	Bonding	π	π
	Bonding	σ	σ

$n \rightarrow \pi^*$ being the transition of lowest energy. The largest transitions N $\rightarrow$ R consist of excitations tending towards ionization. They proceed from a ground state orbital to one of high energy. Other large transitions N $\rightarrow$ V fall into two groups: $\sigma \rightarrow \sigma^*$ from an orbital in the ground state to an orbital of higher energy (e.g. CH_4 at 125 nm) and $\pi \rightarrow \pi^*$ exemplified by $H_2C{=}CH_2$ at 180 nm. Burawoy (1939, 1963) called these K bands. The N $\rightarrow$ Q transitions are from a non-bonding orbital to an anti-bonding π orbital $n \rightarrow \pi^*$. Here the transition is a forbidden one so that the intensity of absorption is very low. The difference between the n and π^* levels may be small so that $\lambda_{max.}$ can be at 280 nm for a ketone or 665 nm for nitroso-*n*-butane. The $n \rightarrow \sigma^*$ transition is to an anti-bonding σ orbital and the energy corresponds to absorption at 190–215 nm. Burawoy (*loc. cit.*) denoted these weak absorption bands as R (Radikal) bands.

The following references carry the discussion further: Sidman (1958); Murrell (1963); Mason (1961); Rao (1967); Scott (1964); Jaffé and Orchin (1962).

3.2 Simple Chromophores

Saturated aliphatic hydrocarbons containing single bonds (σ electrons) exhibit a σ—σ^* transition corresponding with absorption bands in the region of the ultraviolet amenable to study with a vacuum spectrograph. Methane vapour ($\lambda_{max.}$ 125 nm) has the C—H excitation and in ethane ($\lambda_{max.}$ 135 nm) the C—C bond is excited. The ionization potentials are 13·1 and 12·6 eV for methane and ethane. Cyclopropane ($\lambda_{max.}$ 190 nm) displays a big wavelength shift, the strained molecular condition being chromophorically equivalent to unsaturation. The hydrocarbon

$$(CH_3)_3C.CH_2.C(CH_3)_3$$

shows $\lambda_{max.}$ 154 nm $\epsilon_{max.}$ 10 000 in the vapour state. In general, saturated hydrocarbons exhibit little or no selective absorption in the region 190–800 nm. Water has $\lambda_{max.}$ 167 nm and methanol with $\lambda_{max.}$ 177 nm $\epsilon_{max.}$ *ca.* 200, is also transparent in the main ultraviolet region. This applies to most aliphatic saturated alcohols. Dimethyl ether has a peak at 184 nm and diethyl ether when pure is nearly as transparent.

Ammonia vapour has a strong peak at 194 nm while methylamine displays $\lambda_{max.}$ 215 nm and $\epsilon_{max.}$ 600. Trimethylamine vapour has two peaks, at 227 and 199 nm with $\epsilon_{max.}$ 900 and 1950 respectively. Piperidine when pure has little absorption on the long wave side of 200 nm ($\lambda_{max.}$ *ca.* 200 nm $\epsilon_{max.}$ 4000 in vapour state).

3.3 Single Chromophores

The olefine group and the azo group together with the carbonyl group constitute three of the main single chromophores. The theory of the absorption exercised by ethylene has been expounded in detail by Jaffé & Orchin (1962) and a good deal of information has been reduced to order by Scott (1964) and Rao (1967). Every compound with an isolated ethenoid linkage shows strong absorption (ϵ *ca.* 10 000) in the region 190 nm due to a $\pi \rightarrow \pi^*$ transition. Opinions differ concerning the utility of absorption data on alicyclic olefines for diagnostic purposes. Scott (*loc. cit.*) gives a Table recording experience on 27 compounds. The majority of the substances display in ethanol $\lambda_{max.}$ 195–197 nm with $\epsilon_{max.}$ varying mostly between 6000 and 9000. Ellington & Meakins (1960) made it clear that only the tetra-substituted double bond exocyclic to two rings could be identified with certainty, since $\lambda_{max.}$ was moved to 204 nm and $\epsilon_{max.}$ to well over 10 000. Bladon *et al.* (1952) found $\epsilon_{210\,nm}$ a useful diagnostic parameter. It must however be remembered that for work at 190 nm it is necessary to ensure that the performance of the spectrophotometer is good, that the solvents are highly transparent and the compounds very pure. Simple alkynes show only very weak absorption beyond 200 nm but alkyl substituents produce a band with $\lambda_{max.}$ 222 nm ϵ *ca.* 100.

Saturated ketones and aldehydes show bands at 190 nm ($n \rightarrow \sigma^*$ transition) with $\epsilon_{max.}$ under 2000 and a very weak $n \rightarrow \pi^*$ transition with $\lambda_{max.}$ 279 nm, $\epsilon_{max.}$ 13 for acetone in hexane. Data for a large number of saturated ketones have been recorded. The wavelength shifts are quite small and the changes in $\epsilon_{max.}$ even smaller. Thus the compound $(CH_3)_3C.CO.C(CH_3)_3$ has $\lambda_{max.}$ 295 nm and $\epsilon_{max.}$ 20, and cyclohexanone has $\lambda_{max.}$ 285 nm, $\epsilon_{max.}$ 14. In acetaldehyde the weak band is at 290 nm, ϵ *ca.* 15.

Aliphatic azo compounds show weak absorption near 350 nm ($n \rightarrow \pi^*$), but the intensity is low ($\epsilon < 20$) (Wolfe, 1932; Brinton & Volman, 1951; Gillam & Stern, 1958). For azides see Closson *et al.* (1963). Replacement of C=O by C=S has interesting effects. Thus in thiourea ($\lambda_{max.}$ 252 nm, $\epsilon_{max.}$ 15 000) there is a large displacement, with also a weak band near 290 nm. Thiocamphor has a weak ($n \rightarrow \pi^*$) band at 493 nm, thiofenchone at 448 nm and some steroid thioketones are known with $\lambda_{max.}$ 492 nm. In all these cases $\epsilon_{max.}$ is about 12–14. The more intense absorption (ϵ *ca.* 10 000) occurs near 240 nm. In thioacetamide the weak band is at 325–360 nm depending on the solvent, while the strong band is at 265 nm. The displacement can go further in thiobenzophenone ($\lambda_{max.}$ 605 nm, $\epsilon_{max.}$ 66) with a strong (ϵ 17 000) $\pi \rightarrow n^*$ transition at 315 nm (Rao *et al.*, 1961).

The nitroso group is a chromophore in its own right and the green colour of nitroso compounds is due to a weak absorption band at 650–685 nm, $\epsilon_{max.}$ *ca.* 20. In aliphatic compounds the nitro group gives rise to a band at 270–280 nm, again with a low intensity ($\epsilon_{max.}$ 15–30). The absorption closely resembles that of the carbonyl chromophore. Aliphatic nitrates also show $\lambda_{max.}$ *ca.* 270 nm, $\epsilon_{max.}$ 15–40. Aliphatic nitrites show weak absorption in the region 300–400 nm ($\epsilon_{max.}$ < 80). The special characteristic of the aliphatic nitrite spectra is the marked vibrational structure, five or six rather evenly spaced (Δ cm^{-1} *ca.* 970) maxima being recorded. (Ungnade & Smiley, 1956). Differences between *n*-butyl, *iso*butyl, *sec*butyl and *tert*butyl derivatives have diagnostic value. In non-polar solvents nitrosamines show rather similar structure but this disappears in more polar solvents.

3.4 α-β-Unsaturated Ketones

The absorption spectrum of mesityl oxide has been much quoted and studied.

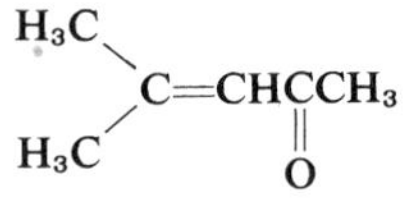

$$(H_3C)_2C{=}CH\overset{\overset{O}{\|}}{C}CH_3$$

Essentially there are two chromophores; there is an electronic transition localized in the ethylenic linkage but influenced by the conjugated carbonyl group and there is a transition localized in the carbonyl group but influenced by the conjugated ethylenic linkage

$$\rangle C\overset{*}{=}CHCOCH_3 \qquad \rangle C{=}CH\overset{*}{C}OCH_3$$

Transition (*a*) is $\pi \to \pi^*$ and (*b*) is $n \to \pi^*$; (*a*) shows $\lambda_{max.}$ 229 nm, $\epsilon_{max.}$ 12 600 and (*b*) 327 nm, $\epsilon_{max.}$ 40 in a hydrocarbon solvent like heptane. The effects of different solvents on the absorption spectra of mesityl oxide are well verified (Table 3.1). The high-intensity band exhibits progressive displacement of $\lambda_{max.}$ in the direction of longer wavelengths as the polarity of the solvent increases. The low-intensity R band is however displaced in the opposite direction and in water becomes merged into the skirt of the high intensity band and is seen only as an inflection.

The bands are also displaced by substituents such as alkyl groups and Woodward (1941, 1942) and Evans & Gillam (1941, 1943) have put forward empirical but thoroughly serviceable generalizations. Thus in unsaturated ketones the π—π^* (K) bands are as shown in Table 3.2. Aldehydes behave similarly.

Table 3.1 Effects of solvents on the absorption bands of mesityl oxide

	(K-band)		(R-band)		
Solvent	$\lambda_{max.}$ nm	$\epsilon_{max.}$	$\lambda_{max.}$ nm	$\epsilon_{max.}$	Reference
n-Hexane	229·5	12 600	327	40	(*a*)
iso-Octane	230·6				(*b*)
Diethyl ether	230	12 600	327	40	(*a*)
Methanol	236·8		308·8		(*b*)
	(238)	10 700	312	55	(*a*)
Ethanol	237	12 600	315	90	(*a*)
Acetonitrile	233·9		314		(*b*)
Chloroform	237·6		315·4		(*b*)
Water	244·5	10 000	305	95	(*a*)
	242·6		masked		(*b*)

(*a*) Scheibe (1925); (*b*) Kosower (1958)

Table 3.2 Positions of $\lambda_{max.}$ in α-β-unsaturated aldehydes and ketones in ethanol

Ketones: $R{-}\underset{\underset{O}{\|}}{C}{-}\underset{\alpha}{C}{=}C\langle^{\beta}_{\beta}$ Aldehydes: R=H

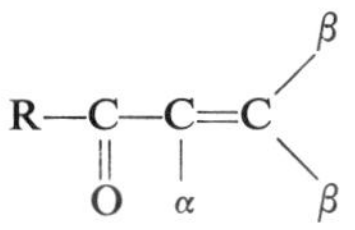

Substitution	$\lambda_{max.}$ nm	Substitution	$\lambda_{max.}$ nm
mono α or β	225 ± 5	mono α or β	220 ± 5
di $\alpha\beta$ or $\beta\beta$	239 ± 5	di $\alpha\beta$ or $\beta\beta$	230 ± 5
tri $\alpha\beta\beta$	254 ± 5	tri $\alpha\beta\beta$	242 ± 5
$\beta\beta$ both H	218		
di (no exocyclic C=C bond)	235 ± 5		
tri (no exocyclic C=C bond)	247 ± 5		

Table 3.3 Absorption spectra of α-β-unsaturated aldehydes and ketones (solvent ethanol)

	Ketones					Aldehydes			
	K-band		R band			K band		R band	
Compound	$\lambda_{max.}$ nm	$\epsilon_{max.}$	$\lambda_{max.}$ nm	$\epsilon_{max.}$	Compound	$\lambda_{max.}$ nm	$\epsilon_{max.}$	$\lambda_{max.}$ nm	$\epsilon_{max.}$
Methylvinylketone	219	3600	324	24	Acraldehyde	208	—	328	13
Methyl isopropenyl	218	8300	319	27	Crotonaldehyde	217	15 650	320·5	19
α-Ionone	228	14 000	305	60	Tiglinaldehyde	228	5000	290	11
Mesityloxide	235	14 000	313·5	60	Citral	238	13 500	324	65
1-Methylcyclohex-1-ene-3-one	235	12 950	310	48	β-*Cyclo*citral	244·5	8300	328	43
3,4-Dimethyl-3-ene-2-one	249	4000	—	—	2-Formylcholest-2-ene	235	12 600	masked	
Cholest-4-en-3-one	240·5	17 000	—	—					

The empirical rules are obeyed well (see Table 3.3). Dannenberg (1939) took Δ^4-cholesten-3-one as a reference substance and accepting the position of the K band in ethanol or methanol, he listed the following corrections for other solvents (in nm):

water −8; chloroform +1; dioxan +5;
ether +7; hexane +11

Ethyl acrylate affords a good example of an αβ-unsaturated ester showing $n \rightarrow \pi^*$ transition. In iso-octane there is a peak at 243 nm, $\epsilon_{max.}$ 70. In acetonitrile the peak is at 239 nm, $\epsilon_{max.}$ 92 and in ethanol there is an inflection near 240 nm. In water there is a strong band ($\lambda_{max.}$ 196 nm, $\epsilon_{max.}$ 11 500) but the selective absorption at 240 nm is masked.

In conjugated systems the individual chromophores interact to produce new absorption with a red (bathochromic) shift. Further in

$$\ddot{X}\text{—C=C—C=C} \rightarrow \overset{\oplus}{X}\text{=C—C=C—}\overset{\ominus}{C}\text{—}$$

with non-bonding electrons at X there is an additional red shift explained in various ways.

The conjugation which occurs in αβ-unsaturated ketones

$$\text{>C=C—C=O} \xrightarrow{h\nu} \text{>}\overset{\oplus}{C}\text{—C=C—O}^{\ominus}$$

gives rise on absorption to transitions in which an electron is moved from an ethenoid π-orbital to the carbonyl π-orbital. Similarly in benzene derivatives a substituent with available *p* electrons (+M donor substituent or an electron withdrawing −M substituent) tends to transfer an electron to or from the aromatic π-orbital system:

Longuet-Higgins & Murrell (1955) described such absorption as E.T. (electron transfer) bands and the $\pi \rightarrow \pi^*$ bands of olefines and of the phenyl group as well as the weak $n \rightarrow \pi^*$ bands of the carbonyl groups as L.E. (local excitation) bands. This notation is accepted by Scott (1964). The intense $\pi \rightarrow \pi^*$ bands are also known as K (Konjugiert) bands and the L.E. bands ($n \rightarrow \pi^*$) bands of aldehydes and ketones as R bands. Many authors have distinguished between the low intensity of R bands and the benzene absorptions at 255 nm and have designated the latter as B bands.

Charge transfer is seen when a permanent charge is introduced, e.g. triphenylmethane ($\lambda_{max.}$ 264 nm) → ion ($\lambda_{max.}$ 282–480 nm), $Ph_3CH \rightarrow Ph_3C^-$. Solvent–solute complexes can give rise to large changes in absorption spectra; thus iodine, which is violet in carbon tetrachloride, forms red or brown solutions in less inert solvents such as ethanol or benzene. The green colour of quinhydrone crystals witnesses to the formation of another type of complex. In the general case of a charge-transfer complex, a new $\lambda_{max.}$ arises well removed towards longer wavelengths from the maxima of the contributing molecules; indeed for the most loosely held electron there can be in the complex an energy decrease 10–20 kcal/mole. The new absorption band is intense ($\epsilon_{max.}$ *ca.* 10 000) and broad. The absence of resolution indicates that the constituents of the complex are not tightly held together in the ground state. This agrees with low heats of formation for the complexes.

Rao (1967) discusses very clearly other aspects of charge-transfer spectra. Data on inter-halogen compounds were obtained by Gillam & Morton (1929, 1931) and discussed by Andrews & Keefer (1950–1952). A good example is provided by Bhaskar *et al.* (1966).

Benzene has weak resolved absorption in the region 230–270 nm, $\epsilon_{max.}$. 60–300 with a main peak at 255 nm.

This is part of a system which (in hexane) consists of 8 narrow bands. Fine structure is revealed in great detail by the absorption spectrum of benzene vapour obtained using a continuous-spectrum light source and photographic recording of the absorption. This low-intensity (B) absorption is attributed to a forbidden electronic transition to a non-ionized (homopolar) state. Benzene also exhibits selective absorption with peaks at 184 nm (log ϵ 4·57) and 202 nm (log ϵ 3·84). These bands, sometimes designated E_1 and E_2, are due to transitions to dipolar excited states which could be represented:

In naphthalene the E_1 and E_2 bands are displaced to 220 nm (log ϵ 5·05) and 285 nm (log ϵ 3·75) while the B band reappears, well resolved, at 310–330 nm. Anthracene shows a main band with $\lambda_{max.}$ 252 nm (log ϵ 5·3) and another near 375 nm (log ϵ 3·9). The very high values for the extinction coefficients of the short-wave peaks of the polycyclic hydrocarbons are notable.

The introduction of substituents into aromatic compounds like benzene tends to have bathochromic effects accompanied by a loss of resolution while the value of $\epsilon_{max.}$ tend to increase. As might be expected, alkyl substituents CH_3, C_2H_5,C_nH_{2n+1}—$CH_2CH{=}CH$ and halogen substituents exert minor effects. Auxochromic substituents, hydroxyl, alkoxy, carboxyl, carboxyester, amino have larger bathochromic effects and quite large increases in $\epsilon_{max.}$ can occur.

As an example, phenol (dissolved in hexane) shows a strong peak at 210 nm ($\epsilon_{max.}$ *ca.* 10 000) and resolved absorption around 272 nm with $\epsilon_{max.}$ 2000. The benzene (B) absorption has undergone a red shift and $\epsilon_{max.}$ has risen almost tenfold. Phenol absorption measured in an alkaline medium shows a further displacement in the direction of longer wavelengths. Anisole (in neutral solution) has an absorption curve resembling that of phenol but alkali has no bathochromic effect.

Dihydroxybenzenes (catechol, resorcinol, quinol, etc.) also show bathochromicity, with the smallest effect in the ortho compounds (cf. also guaiacol and veratrole). Two peaks ($\lambda_{max.}$ 275 and 283 nm) are characteristic with $\epsilon_{max.}$ varying between 2500 and 4500.

Aniline displays two maxima at 234 nm (log ϵ 4·06) and 285 nm (log ϵ 3·24) respectively. In acid the auxochromic effect of the amino group largely diminishes and the new spectrum reverts to the benzene type.

Table 3.4 Principal peaks of benzene and mono-substitution products

	$\lambda_{max.}$ nm	$\epsilon_{max.} \times 10^2$
Benzene	254·3	3·02
Toluene	262	3·02
Monochlorobenzene	264	3·24
Benzoic acid	227	141·3
	267	17·8
Phenol	215	100·0
	273	17·8
Phenylmercaptan	230	100·0
	279	7·94
Anisole	272	22·4
Aniline	234	114·8
	284·5	17·4

Table 3.5

Benzaldehyde				Acetophenone				Benzoic acid			
in Cyclohexane		in Ethanol		in Cyclohexane		in Ethanol		in Cyclohexane		in Ethanol	
$\lambda_{max.}$ nm	log ϵ	$\lambda_{max.}$ nm	log ϵ	$\lambda_{max.}$ nm	log ϵ	$\lambda_{max.}$ nm	log ϵ	$\lambda_{max.}$ nm	log ϵ	$\lambda_{max.}$ nm	log ϵ
241	4·17	244	4·2	238·5	4·10	242	4·08	231·5	4·1	232	4·09
~247	4·10	—	—	—	—	—	—	—	—	—	—
278·5	3·06	280	3·18	278	2·95	278	3·04	274	3·01	275	3·02
287	2·95	—	—	285	2·94	—	—	283·5	2·97	284	2·95
317	1·43	—	—	321	1·64	319	1·90	—	—	—	—
328	2·45	—	—	—	—	—	—	—	—	—	—
338	1·44	—	—	—	—	—	—	—	—	—	—
353	1·30	—	—	—	—	—	—	—	—	—	—

Table 3.6

Aminophenols

Ortho Cyclohexane		pH 3		pH 13		Meta Cyclohexane*		pH 3		0·1 N NaOH		Para Cyclohexane*		pH 1·9	
$\lambda_{max.}$ nm	log ϵ	$\lambda_{max.}$ nm	log ϵ	$\lambda_{max.}$ nm	log ϵ	$\lambda_{max.}$ nm	log ϵ	$\lambda_{max.}$ nm	log ϵ	$\lambda_{max.}$ nm	log ϵ	$\lambda_{max.}$ nm	log ϵ	$\lambda_{max.}$ nm	log ϵ
235	3·85	210	3·81	—	—	234	3·83	~212	3·79	291	3·46	235	3·85	220	3·8
288	3·52	268	3·32	—	—	282·5	3·36	217	3·80	—	—	304	3·40	275	3·2
—	—	273	3·26	296	3·55	—	—	269	3·26	—	—	—	—	—	—
—	—	—	—	—	—	—	—	273	3·23	—	—	—	—	—	—

* +2% ether

Dihydroxybenzenes

Catechol Cyclohexane		Resorcinol Cyclohexane		pH 1·4		pH 11·3		Hydroquinone Cyclohexane		Ethanol		pH 8·9	
$\lambda_{max.}$ nm	log ϵ	$\lambda_{max.}$ nm	log ϵ	$\lambda_{max.}$ nm	log ϵ	$\lambda_{max.}$ nm	log ϵ	$\lambda_{max.}$ nm	log ϵ	$\lambda_{max.}$ nm	log ϵ	$\lambda_{max.}$ nm	log ϵ
214	3·83	217	3·83	~220	3·7	218	4·1	223	3·64	—	—	236	3·8
~270	3·34	266	3·22	—	—	236	3·9	~285	3·40	—	—	285	2·7
374	3·40	273·5	3·35	275	3·3	286	3·5	289	3·46	294	3·47	405	2·0
280	3·35	279·5	3·26	—	—	290	3·5	292	3·46	—	—	430	2·0
								~299	3·34	—	—	—	—

Nitrophenols

Ortho Cyclohexane		pH 12·3		Ethanol		Meta Cyclohexane		0·5 N NaOH		Para Hexane		pH 9·4		Ethanol	
$\lambda_{max.}$ nm	log ϵ	$\lambda_{max.}$ nm	log ϵ	$\lambda_{max.}$ nm	log ϵ	$\lambda_{max.}$ nm	log ϵ	$\lambda_{max.}$ nm	log ϵ	$\lambda_{max.}$ nm	log ϵ	$\lambda_{max.}$ nm	log ϵ	$\lambda_{max.}$ nm	log ϵ
~231	3·6	228	4·16	231	3·55	222	4·01	226	4·28	222	3·91	228	3·78	225	3·85
271	3·87	250	3·60	273	3·78	259	3·78	252	4·16	286	4·01	260	3·00	314	4·11
346	3·57	282	3·61	347	3·51	351	3·12	292	3·78	–290	—	400	4·26	—	—
—	—	418	3·65	—	—	—	—	392	3·30	(295	4·04)	—	—	—	—

Chlorophenols

Ortho Cyclohexane		0·1 N NaOH		Ethanol		Meta Cyclohexane		pH 13		Para Cyclohexane		0·1 N NaOH	
$\lambda_{max.}$ nm	log ϵ	$\lambda_{max.}$ nm	log ϵ	$\lambda_{max.}$ nm	log ϵ	$\lambda_{max.}$ nm	log ϵ	$\lambda_{max.}$ nm	log ϵ	$\lambda_{max.}$ nm	log ϵ	$\lambda_{max.}$ nm	log ϵ
212	3·81	—	—	216	3·81	216	3·81	—	—	224	3·94	—	—
~265	3·19	—	—	274	3·37	266	3·11	—	—	~273	3·15	—	—
271	3·37	—	—	~280	3·30	272	3·31	—	—	279	3·27	298	3·36
278	3·39	293	3·43	—	—	279	3·30	299	3·38	287	3·23	—	—

Cresols

Ortho Cyclohexane		in NaOH		Ethanol		Meta Cyclohexane		Ethanol		NaOH	
$\lambda_{max.}$ nm	log ϵ	$\lambda_{max.}$ nm	log ϵ	$\lambda_{max.}$ nm	log ϵ	$\lambda_{max.}$ nm	log ϵ	$\lambda_{max.}$ nm	log ϵ	$\lambda_{max.}$ nm	log ϵ
212	3·88	236	3·97	273	3·28	215	3·76	—	—	238	3·91
271	3·15	278	3·17	—	—	273	3·18	274·5	3·23	289	3·51
279	3·09	284	3·17	—	—	278	3·18	—	—	—	—
—	—	~296	3·00	—	—	—	—	—	—	—	—

Para Cyclohexane		Ethanol		Alkali	
$\lambda_{max.}$ nm	log ϵ	$\lambda_{max.}$ nm	log ϵ	$\lambda_{max.}$ nm	log ϵ
220	3·71	—	—	236	3·89
278	3·23	281	3·3	295	3·50
283	3·14	288	3·2	—	—

When the phenyl and carbonyl chromophores are conjugated as in benzaldehyde and acetophenone four peaks are seen in suitable solvents:

$\lambda_{max.}$ nm	199	240	278	320
$\epsilon_{max.} \times 10^{-3}$	20	13·2	1·05	0·05

The long wave band is clearly the weak C=O band influenced (i.e. red-shifted) by the proximity of the phenyl group. Table 3.5 shows the spectra in more detail (omitting the absorption near 200 nm). It will be seen that there is a normal substituent effect (*ca.* 240 nm, log ϵ 4, 274–278 nm, log ϵ *ca.* 3) plus the weak carbonyl absorption, resolved only for benzaldehyde in cyclohexane. Such absorption is not seen with benzoic acid.

The spectra of the cresols (Table 3.6) illustrates very clearly the significance of the alkali shift. The importance of pH is also seen in the data for the dihydroxybenzenes and the aminophenols as well as the chlorophenols.

Nitrobenzene in heptane shows peaks at 252, 280 and 330 nm ($\epsilon_{max.} \times 10^{-3}$ 9·5, 1·0 and 0·165 respectively). The nitrophenols show very marked variations depending on *o*, *m*, *p*-isomerism.

From a biochemical standpoint however the most interesting chromophores are those which contain carbonyl groups and hydroxyl groups.

The vapour absorption spectrum of salicylaldehyde was studied by Morton & Stubbs (1940); 35 peaks were recorded between 254·82 and 241·0 nm. They were accounted for by the expressions:

$$1/\lambda \text{ cm}^{-1} = 39\,614{\cdot}9 + 242{\cdot}6p + 57{\cdot}6q \text{ (Series I)}$$
$$p\ 0 \text{ to } 8\ q\ 0 \text{ to } -5$$
$$40\,370{\cdot}1 + 242{\cdot}6p + 57{\cdot}6q \text{ (Series II)}$$
$$p\ 0 \text{ to } 3\ q\ 0 \text{ to } -2$$

(34 614·9 = 252·43 nm
40 370·1 = 247·7 nm;
40 370·1 − 39 614·9 = 755·2 cm^{-1})

In the vapour spectra of benzene, phenol and benzaldehyde the recurring frequency difference are respectively 922, 938 and 954 cm^{-1}. Nothing akin to this occurs in the vapour spectrum of salicylaldehyde. In *o*-hydroxyacetophenone vapour there is less resolution, peaks being seen at 250·3, 249·5, 245·9 and 245·2 nm. The vapour of *o*-methoxyacetophenone shows unresolved absorption except between 296·2 nm and 311·7 nm where the weak absorption shows five subsidiary peaks.

Table 3.7

$COCH_3$, OH, OMe in Ethanol		$COCH_3$, OMe, OH in Ethanol		CHO, OMe in Hexane		$COCH_3$, OMe no solvent given	
$\lambda_{max.}$ nm	log ϵ	$\lambda_{max.}$ nm	log ϵ	$\lambda_{max.}$ nm	log ϵ	$\lambda_{max.}$ nm	log ϵ
270	4·2	228	4·2	246·5	4·08	211	4·3
315	3·8	275	4·0	—	—	242·5	3·90
—	—	303	3·94	—	—	299	3·55
—	—	—	—	310	3·85	330	2·00
						?	3·40

$COCH_3$, OH, Et, OH in Methanol			HO, COBu, OH in Ethanol		HO, $COCH_3$, OH in Cyclohexane		CHO, OMe in Hexane		$COCH_3$, OMe in Cyclohexone	
$\lambda_{max.}$ nm	log ϵ		$\lambda_{max.}$ nm	log ϵ	$\lambda_{max.}$ nm	log ϵ	$\lambda_{max.}$ nm	log ϵ	$\lambda_{max.}$ nm	log ϵ
217	4·48		223	4·10	—	—	249	3·93	214	4·40
231	4·36		270	4·00	263	3·92	309	3·46	244	3·93
323	4·06		342	3·45	334	3·36	—	—	~250	3·79
—	—		(345)	—	—	—	—	—	298	3·45
		Alkali	284	3·88					~308	3·32
			381	3·59						

Table 3.7 *continued*

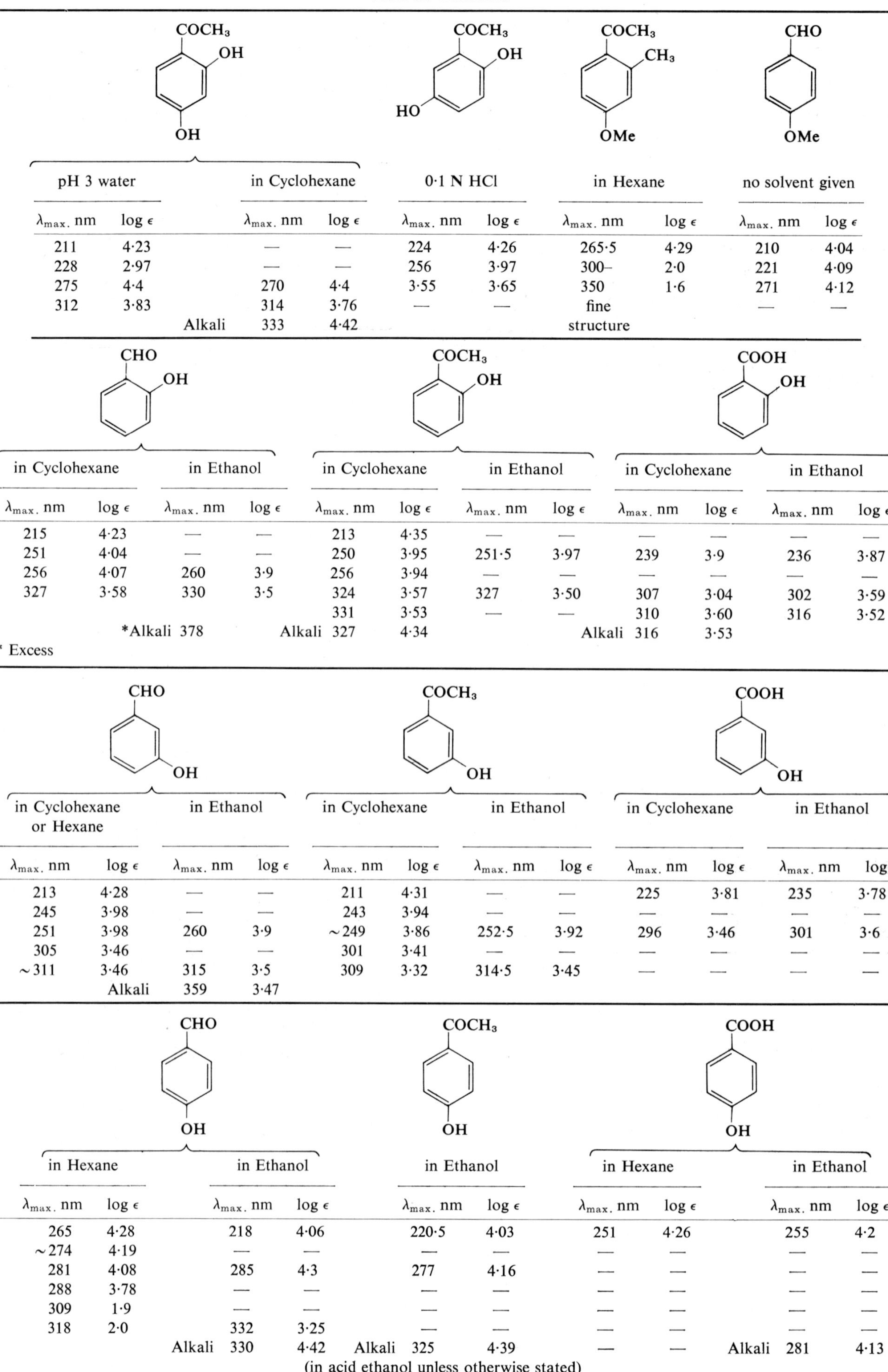

	pH 3 water		in Cyclohexane		0·1 N HCl		in Hexane		no solvent given	
	$\lambda_{max.}$ nm	log ϵ	$\lambda_{max.}$ nm	log ϵ	$\lambda_{max.}$ nm	log ϵ	$\lambda_{max.}$ nm	log ϵ	$\lambda_{max.}$ nm	log ϵ
	211	4·23	—	—	224	4·26	265·5	4·29	210	4·04
	228	2·97	—	—	256	3·97	300–	2·0	221	4·09
	275	4·4	270	4·4	3·55	3·65	350	1·6	271	4·12
	312	3·83	314	3·76	—	—	fine		—	—
Alkali			333	4·42			structure			

in Cyclohexane		in Ethanol		in Cyclohexane		in Ethanol		in Cyclohexane		in Ethanol	
$\lambda_{max.}$ nm	log ϵ	$\lambda_{max.}$ nm	log ϵ	$\lambda_{max.}$ nm	log ϵ	$\lambda_{max.}$ nm	log ϵ	$\lambda_{max.}$ nm	log ϵ	$\lambda_{max.}$ nm	log ϵ
215	4·23	—	—	213	4·35	—	—	—	—	—	—
251	4·04	—	—	250	3·95	251·5	3·97	239	3·9	236	3·87
256	4·07	260	3·9	256	3·94	—	—	—	—	—	—
327	3·58	330	3·5	324	3·57	327	3·50	307	3·04	302	3·59
				331	3·53	—	—	310	3·60	316	3·52
	*Alkali	378	Alkali	327	4·34		Alkali	316	3·53		

* Excess

in Cyclohexane or Hexane		in Ethanol		in Cyclohexane		in Ethanol		in Cyclohexane		in Ethanol	
$\lambda_{max.}$ nm	log ϵ	$\lambda_{max.}$ nm	log ϵ	$\lambda_{max.}$ nm	log ϵ	$\lambda_{max.}$ nm	log ϵ	$\lambda_{max.}$ nm	log ϵ	$\lambda_{max.}$ nm	log ϵ
213	4·28	—	—	211	4·31	—	—	225	3·81	235	3·78
245	3·98	—	—	243	3·94	—	—	—	—	—	—
251	3·98	260	3·9	~249	3·86	252·5	3·92	296	3·46	301	3·6
305	3·46	—	—	301	3·41	—	—	—	—	—	—
~311	3·46	315	3·5	309	3·32	314·5	3·45	—	—	—	—
	Alkali	359	3·47								

	in Hexane			in Ethanol			in Ethanol		in Hexane			in Ethanol	
	$\lambda_{max.}$ nm	log ϵ		$\lambda_{max.}$ nm	log ϵ		$\lambda_{max.}$ nm	log ϵ	$\lambda_{max.}$ nm	log ϵ		$\lambda_{max.}$ nm	log ϵ
	265	4·28		218	4·06		220·5	4·03	251	4·26		255	4·2
	~274	4·19		—	—		—	—	—	—		—	—
	281	4·08		285	4·3		277	4·16	—	—		—	—
	288	3·78		—	—		—	—	—	—		—	—
	309	1·9		—	—		—	—	—	—		—	—
	318	2·0		332	3·25		—	—	—	—		—	—
			Alkali	330	4·42	Alkali	325	4·39	—	—	Alkali	281	4·13

(in acid ethanol unless otherwise stated)

m-Hydroxybenzaldehyde and *m*-methoxybenzaldehyde both show well-resolved absorption in the region 243–252 nm while in *p*-hydroxy and *p*-methoxybenzaldehyde it is the absorption at 273–286 nm which is well resolved. In the vapour state, as in solution using inert hydrocarbon solvents like hexane, intramolecular chelation effects give rise to a shift of the second band (electron transfer) around 325 nm but the differences between hydroxylic solvent maxima and hydrocarbon solvent maxima are small. Conversion of OH to OMe effects a rather larger blue shift for *o* compounds but not for *m*-isomers. Conover (1956) has extended the discussion in terms of solvent-chelation effects first put forward by Morton and Stubbs. In particular Conover emphasizes the role of steric repulsion in accounting for the difference in the effect of methylation of *o*-hydroxybenzaldehyde and *o*-hydroxyacetophenone on the position of the long-wave band. He also uses the steric concept to account for the difference in behaviour with alkali

CHO — O⁻ (benzene ring) COCH₃ — O⁻ (benzene ring)

$\lambda_{max.}$ 383 nm 365 nm

(in the presence of excess of alkali).

Other data on the spectra of hydroxyketones have been obtained by Doub & Vandenbelt (1947), Lemon (1947) and Cram & Cranz (1950).

3.5 Tetracyclines

The tetracyclines are an important group of antibiotics obtained from *Streptomyces* species in the order Actinomycetales. As the name implies they contain a four-ring system, the parent substance being tetracycline or achromycin, but only ring D is aromatic.

R′ R″ OH R‴ H H NMe₂ OH CONH₂ OH O OH O (rings D, C, B, A; positions 1–12, 5a, 12a)

(I)	R′	R″	R‴	
a	H	CH_3	H	Tetracycline
b	Cl	CH_3	H	Aureomycin
c	H	CH_3	OH	Terramycin
d	H	H	H	6-Desmethyl-tetracycline

The absorption spectrum displays peaks at 220, 268 and 355 nm with maxima near 235 and 300 nm for the hydrochloride in 0·1 N HCl (Boothe *et al.*, 1953).

A chlorine substituent at position 7 gives chlortetracycline (aureomycin, the first antibiotic in the group to be discovered: Duggan, 1948; Broschard *et al.*, 1949). On the other hand a hydroxyl group at position 5 gives terramycin or oxytetracycline:

	$\lambda_{max.}$ nm	$\epsilon_{max.} \times 10^{-3}$	
Aureomycin ($C_{22}H_{23}N_2O_8Cl$)	267	14·8	(0·1 N HCl in ethanol)
	375	11·8	
Terramycin ($C_{22}H_{24}N_2O_9$)	267	21·2	(in 0·1 N HCl)
	357	12·3	
	264	18·6	(in 0·1 N NaOH and methanol)
	375	16·2	

(Stephens *et al.*, 1954; Regna *et al.*, 1951; Finlay *et al.*, 1950).

The spectra of the hydroxyaldehydes and ketones have proved to be of considerable value in the study of the more complicated structures of the tetracyclines. Oxytetracycline (Ic) can be dehydrated to anhydrooxytetracycline (II) which can be isomerized to *apo*-oxytetracycline (III) which appears to occur in two

Isochlortetracycline
(blue fluorophore formed from
chlorotetracycline by base)

(II) Anhydroterramycin

(III) Apoterramycin

stereoisomeric forms. Chlortetracycline and tetracycline yield stable anhydro derivatives. The *apo*-tetracyclines (in acid plus an oxidizing agent like $FeCl_3$) yield terrinolide (IV) while a more drastic attack by boiling 12 N H_2SO_4 eliminates dimethylamine and aromatizes ring A, hydrolyses the amide

(IV) Terrinolide

(V) Decarboxamido terrinolide

group and decarboxylates to give decarboxamidoterrinolide (V). *Apo*-tetracycline also yields the 1-methyl-4,8-dihydroxy-2-naphthoic acid (VIa) and the 2,5-dihydroxy-1,4-benzoquinone (VII). The naphthaline chromophore is clearly to be seen in the spectra of terranaphthoic acid (VIa), terranaphthol (VIb) and in the model substance 1,8-dihydroxy-naphthoic acid with a carboxyl group ortho to the hydroxyl (VId). Another very useful reference substance is 5,9,10-trihydroxy-1-oxo-1,2,3,4-tetrahydroanthracene (VIc) the spectrum of which closely agrees with that of anhydroterramycin.

(VIa)

Terranaphthoic acid

(VIb)

Terrenaphthol

(VIc)

5,9,10-trihydroxy-1-oxo
1,2,3,4-tetrahydroanthracene

(VId)

1,8-Dihydroxy-2-naphthoic acid

Alkali attacks oxytetracycline in the presence of zinc to give terracinoic acid (VIII) in good yield and small amounts of isodecarboxyterracinoic acid (IX). When the NMe_2 group of oxytetracycline is replaced by H the product is isomerized by weak alkali to a phthalide.

Conover (1956) compared the spectra of 8-hydroxytetralone (X), 7-hydroxy- and 5-hydroxy-3-methylin-

(VII) 2,5-Dihydroxy-1,4-benzoquinone

(VIII)

Terracinoic acid

(IX)

Isodecarboxyterracinoic acid

(X)

8-Hydroxytetralone

(XII)

5-Hydroxy-3-methyl indanone

(XI)

7-hydroxyindanone

danones (XI and XII) with those of their corresponding methyl esters and related the findings with those of Morton & Stubbs (1940) on the benzaldehyde and acetophenone derivatives. Conover (*loc. cit.*) noted (i)

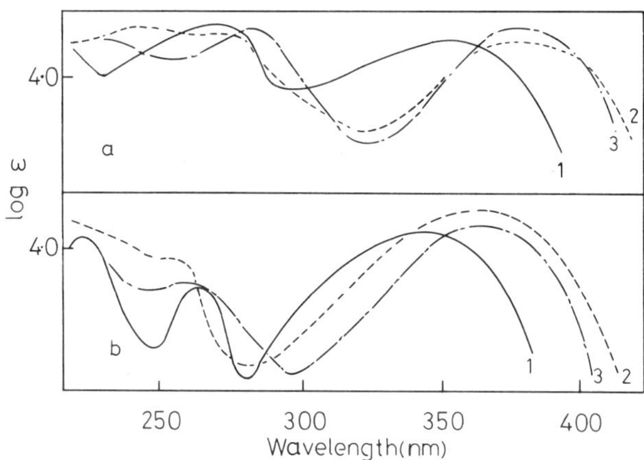

Figure 3.1 Ultraviolet absorption spectra of (*a*) oxytetracycline and (*b*) 2-acetyl-8-hydroxytetralone in (1) 0·01 M-methanolic HCl, (2) 0·01 M-methalonic NaOH, and (3) 0·01 M-methanolic $NiCl_2,6H_2O$ (from Conover, 1956)

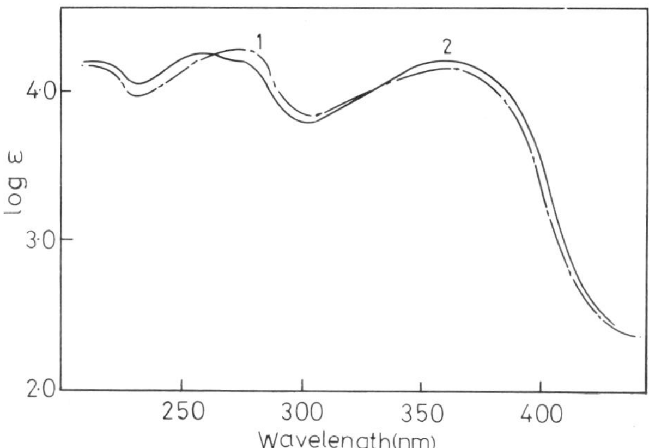

Figure 3.2 Ultraviolet absorption spectra of (1) tetracycline and (2) *epi*tetracycline in 0·01 M-methanolic HCl (from Conover, 1956)

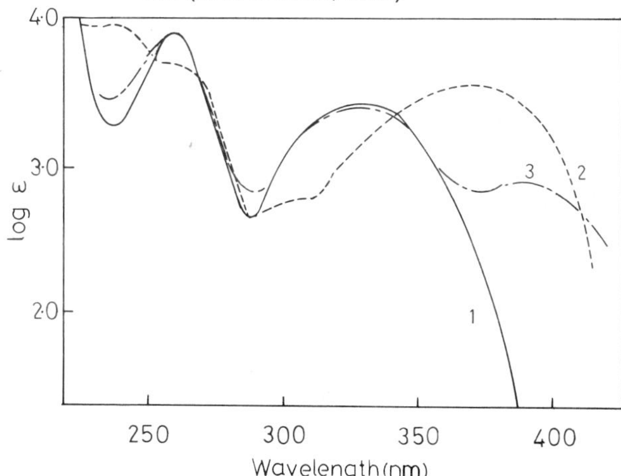

Figure 3.3 Ultraviolet absorption spectra of 8-hydroxytetralone in (1) 95% ethanolic 0·01 M HCl, (2) 95% ethanolic 0·01 M NaOH, and (3) methanolic 0·01 M $NiCl_2,6H_2O$ (from Conover, 1956)

Table 3.8 Long wavelength absorption maxima of *o*-hydroxy- and *o*-alkoxy-aldehydes and ketones

Solvent	Hydrocarbon			Hydroxylic		Alkaline (excess)	
Parent Substance	*o*-OH nm	*o*-Alk nm	λ	*o*-OH nm	*o*-O Alk nm	*o*-OH nm	$\Delta\lambda$ on ionization nm
Benzaldehyde	329	310	19	325	320	383	58
Acetophenone	329	300	29	327	305	365	38
Tetralone	331	307	24	335	317	372	37 (55)*
Indanone	314	306	8	316	312	364	47 (52)*

* OH compound in HCl-Ethanol and in parentheses the figure for the ether in HCl-Ethanol.

that etherification of 7-hydroxy-indanone gave rise to a small hypsochromic (blue) shift (see Table 3·8) quite unlike that shown for the other three *o*-hydroxy-carbonyl compounds; (ii) the red shift is greater for *o*-acetophenone than for *o*-hydroxybenzaldehyde; (iii) the position of $\lambda_{max.}$ in an inert solvent is anomalous for *o*-hydroxyindanone; (iv) none of the four *o*-hydroxy compounds displays much displacement when a hydroxylic solvent is compared with a hydrocarbon solvent; (v) the red shift on ionization is significantly different for *o*-hydroxyacetophenone and *o*-hydroxyaldehyde.

Morton & Stubbs (1940) interpreted their own observations in terms of intramolecular hydrogen bonding and solvent solute interaction.

Conover emphasized coplanarity in the chelated forms achieved by virtue of the hydrogen bonding. In the case of *o*-methoxybenzaldehyde the dipole moment is consistent with the idea that both substituents can rotate through 180° and retain coplanarity in the molecule but in the case of *o*-methoxyacetophenone the bulkiness of the acetyl methyl group prevents a rotation of similar magnitude. This can be interpreted as a steric inhibition of resonance reflected in the position of $\lambda_{max.}$ in *o*-methoxyacetophenone. The difference in $\lambda_{max.}$ in the presence of excess alkali between *o*-hydroxyacetophenone and the *o*-hydroxybenzaldehyde is consistent with greater resonance stability in the latter.

Conover also cites the fact that compared with *o*-hydroxyacetophenone, salicylaldehyde is much the stronger acid; this also is attributed to steric inhibition. In 7-hydroxyindanone the distance between the carbonyl and hydroxyl groups is greater than in the other three compounds and the hydrogen bond is 'longer and weaker'. Chelation does less to stabilize the excited state and the electronic transition requires more energy. That chelation does, however, occur follows from the fact that there is no marked solvent shift. Infrared absorption shows a good peak at 3·02 μ (3310 cm^{-1}) arising from a chelated hydroxyl group. The compound 8-hydroxytetralone (X) shows a chelated hydroxyl band at 3·4 μ and it seems likely that the 7-hydroxyindanone intramolecular hydrogen bond is very weak.

When *o*-methoxyacetophenone and 7-methoxy-3-methylindanone are compared the rigidity of the 5-membered ring makes difficult any twisting of the carbonyl group but the methoxy group assumes an *S-trans*-configuration. This is however not important compared with the difference 329 to 314 nm seen in the second column of Table 3.8. Clearly the weak chelation in 7-hydroxy-3-methylindanone is the main variable. Geissman & Harborne (1956) found for 4-hydroxy- and 4-methoxy-aurones in cyclohexane maxima at 384 and 376 nm respectively and in 95% ethanol 389 and 387 nm. This is very similar to the effect noted for the indanone derivative.

Conover (*loc. cit.*) made effective use of the spectra of metal chelates of both tetracyclines and of model substances. It is of course a matter of considerable biochemical interest to know which of the several possible sites of metal binding actually operates. The 11,12-β-carbonyl system (I, II) is more important than the alternatives.

When a metal chelate is formed in the tetracyclines, a big red shift is seen in solutions of the purified complexes and also very often when the spectra are examined in the presence of excess of di- or tri-valent metal ions. Thus when 2-acetyl-8-hydroxytetralone is examined in methanolic-HCl there is a very broad absorption band 290–390 nm, with $\lambda_{max.}$ *ca.* 350 nm, log ϵ 4·0. In the presence of excess of alkali the whole envelope undergoes a red shift $\lambda_{max.}$ *ca.* 366 nm and a very similar absorption curve is seen when solvent is 0·01 **M** methanolic $NiCl_2 . 6H_2O$. Oxytetracycline shows a rather flat peak at 355 nm (log ε 4·0) displaced to 377 nm in either excess of alkali or in the presence of excess of nickel chloride.

It is interesting that 8-hydroxytetralone shows $\lambda_{max.}$ 335 nm (log ε 3·4) displaced to 375 nm (log ε 3·5), while in the presence of nickel chloride the 340 peak persists with a marked inflection consistent with a band at 390 nm, log ε 2·8. Other model substances were studied similarly and it became clear that the site of metal combination in the antibiotics is the β-dicarbonyl portion of the 2-acyl-8-hydroxytetralone chromophoric unit, since the changes brought about by the introduction of metals into the antibiotics fit this model more closely than any of the others. It seems that a hydroxyl group in position 10 (ring D) contributes to the ready formation of a chelate complex but is not indispensable. Inclusion of both C=O groups in the fused ring system favours chelation.

The structures of the tetracyclines have been discussed in relation to ultraviolet spectra by Scott (1964) and by Conover (*loc. cit.*) whose paper is of great interest but has here been dealt with from one standpoint only.

The derivative terranaphthol ($C_{12}H_{12}O_3$), obtained from terramycin by the action of alkali in the presence of zinc, showed in ethanol a strong absorption maximum at 229 nm (log ε 4·8) with three peaks at 311, 326 and 339 nm (log ε *ca.* 3·9). This spectrum is similar to that of 1,8-dihydroxynaphthalene (see Table 3.9) (Allport & Bu'Lock, 1960; Cowell, King & Morgan, 1961), and different from that of 1,4-dihydroxy-naphthalene: $\lambda_{max.}$ 250, 340 nm, log ε 4·3,

Table 3.9 Absorption peaks of terramycin and related substances (in acid ethanol unless otherwise stated)

Terramycin		Anhydroterramycin		8,9,10-Trihydroxy-1-oxo-1,2,3,4-tetrahydroanthracene	
$\lambda_{max.}$ nm	log ε	$\lambda_{max.}$ nm	log ε	$\lambda_{max.}$ nm	log ε
265	4·27	275	4·55	270	4·55
362	4·13	314	3·4	300	3·5
high unresolved		333	3·2	314	3·4
absorption 280–350 nm		420	3·8	325	3·27
β-Apoterramycin		**Terrinolide**		**Decarboxamidoterrinolide**	
$\lambda_{max.}$ nm	log ε	$\lambda_{max.}$ nm	log ε	$\lambda_{max.}$ nm	log ε
250	4·75	248	4·73	248	4·70
304	3·75	308	3·66	310	3·93
322	3·75	322	3·82	330	3·85
360	3·92	360	4·1	360	4·00
375	3·92	375	4·06	372	4·03
Terranaphthol		**1,8-Dihydroxynaphthol**		**1,8-Dihydroxynaphthoic acid**	
$\lambda_{max.}$ nm	log ε	$\lambda_{max.}$ nm	log ε	$\lambda_{max.}$ nm	log ε
229	4·77	229	4·47	247·5	4·65
311	3·89	302	3·84	290	3·5
326	3·89	320·5	3·84	301	3·5
339	3·89	334	3·90	360	3·8
				371	3·8
Terranaphthoic acid		**Indanone**		**7-Hydroxyindanone**	
$\lambda_{max.}$ nm	log ε	$\lambda_{max.}$ nm	log ε	$\lambda_{max.}$ nm	log ε
335	4·25	243	4·09	255	3·91
311	3·8	(ethanol)		(ethanol)	
342	3·67				

5-Hydroxyindanone shows $\lambda_{max.}$ 273 nm

3·9 (Arbusov, Polezhaeva & Vinogradova, 1960). Distillation of terranaphthol with zinc dust gave 1-methylnaphthalene so that a 4-methyl-1,8-dihydroxy naphthalene moiety became highly plausible. Another product obtained by the action of alkali on terramycin was terracinoic acid:

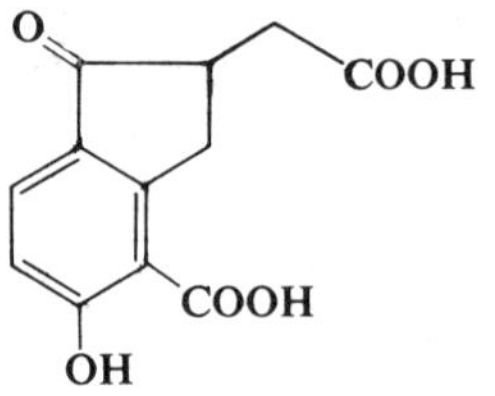

the absorption spectrum of which resembles that of 5-hydroxyindanone. It is accompanied by a small amount of *iso*decarboxyterracinoic acid which has spectrum like that of 7-hydroxyindanone:

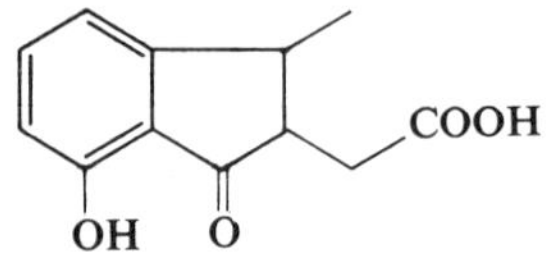

Treatment of terramycin with acid gave as a first product anhydroterramycin (Hochstein *et al.*, 1953) with an absorption spectrum (in acid ethanol) closely akin to that of the anthracene derivative 5,9,10-trihydroxy-1-oxo-1,2,3,4-tetrahydroanthracene.

A further degradation product is apoterramycin which exists in α and β forms showing a marked spectroscopic resemblance to 1,8-dihydroxy-2-naphthoic acid. Further degradation with acid led to β-apoterramycin and two isomeric lactones (with infrared peaks at 1739 cm^{-1}).

Apoterramycin was allotted the partial structure

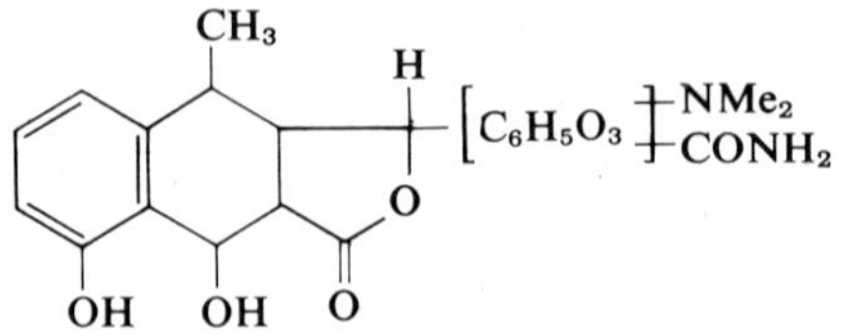

Further treatment with acid yielded terrinolide and decarboxanidoterrinolide (loss of NMe_2 and $CONH_2$), the latter having three main regions of absorption (Hochstein *et al.*, 1953). This compound formed a pentamethoxy derivative which was reduced to a glycol.

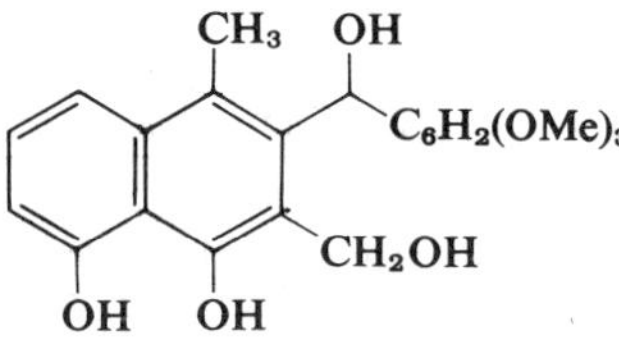

This structure contains two chromophores insulated from one another. The spectra of terranaphthol and a trimethoxybenzene should on addition give a curve close to that of the glycol. Comparison of the curves for trimethoxybenzenes led to a decision in favour of the 1,2,4-isomer. Chemical evidence supported the further approach to structure:

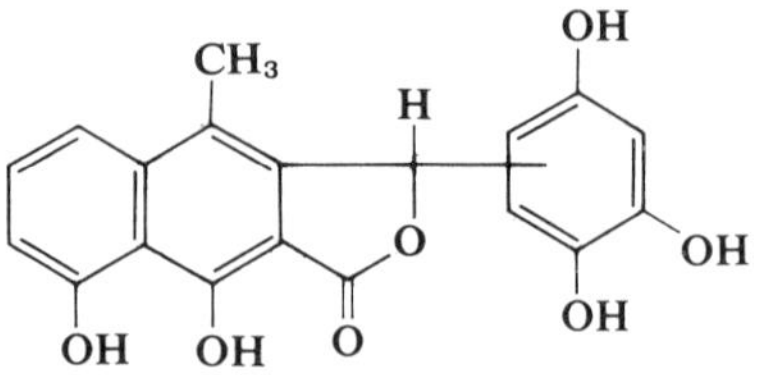

When terramycin was treated with zinc and acetic acid, oxygen and $—NMe_2$ were eliminated to give desoxydesdimethylamineterramycin ($C_{20}H_{19}NO_3$). The 360 nm peak of terramycin disappeared but a strong band occurred with $\lambda_{max.}$ 320 nm. The 360 nm band was provisionally attributed to an enol moiety:

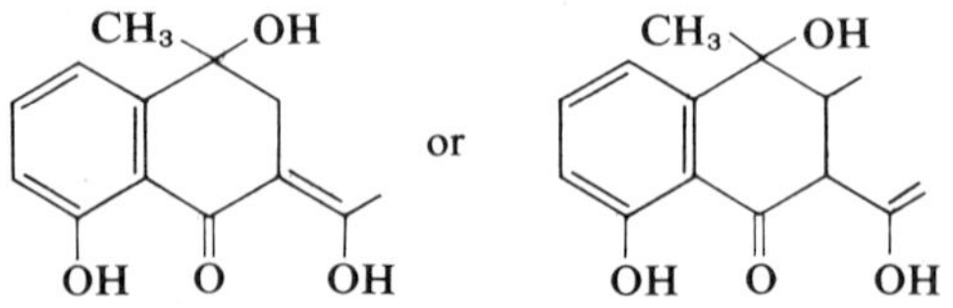

A neat contribution to the problem was obtained by constructing a summation curve for 8-hydroxytetralone and diosphenol, a steroid with a very relevant chromophore.

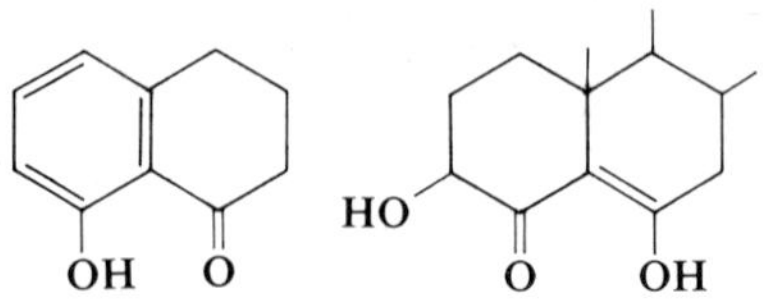

The result closely simulated the curve for desoxydesdimethylaminoterramycin. When the action of zinc and acetic acid on terramycin was stopped at the stage of removing NMe_2 there was little change in ultraviolet absorption. The structure which emerges leaves open

Me OH OH NMe$_2$ OH OH CONH$_2$ O O O O M H

Me OH OH NMe$_2$ O OH O HO O O O NH$_2$ H M

Me OH HO NMe$_2$ OH OH CONH$_2$ HO O O O M

Me OH OH NMe$_2$ O M OH O HO O O O NH$_2$ H

Terramycin

different possibilities for chelation. The problem may be looked at further as follows. The major chromophore can be written:

R O OH

Major chromophore

where R = H or Cl. Hydrogenation of aureomycin by means of palladized charcoal eliminates chlorine and deschloroaureomycin gives the same absorption spectrum as terramycin. The minor chromophore may be written:

OH CONH$_2$ O

Minor chromophore

and is the same for the two antibiotics. Treatment with zinc and acetic acid removes NMe_2 to give desdimethylaminotetracyclines with $\lambda_{max.}$ 265 and 324 nm (ϵ *ca.* 20 000) which lose water to give anhydroterramycin (or aureomycin). Now among the degradation products are isodesoxydesdimethylaminoaureomycin (or terramycin) which yields phthalides:

R CH_3OH OH OH CONH$_2$ O OH O

Desoxydesdimethylaminotetracyclines

$\lambda_{max.}$ nm	265,	324
$\epsilon_{max.} \times 10^{-3}$	20,	17·8

CH_3 OH O C O CONH$_2$ O OH O

Isodesoxydesdimethylaminoterramycin

	CH_3, OH, C=O, OCH_3	Cl, CH_3, C—OH, C=O, OCH_3	$COCH_3$, $COOCH_3$, OCH_3	CH_3, C—OCH_3, C=O, OCH_3	Cl, CH_3, C—OCH_3, C=O, OCH_3
$\lambda_{max.}$ nm	232, 300	235, 302	245, 312	237, 300	236, 312
$\epsilon_{max.} \times 10^{-3}$	6·94, 4·14	8·61, 3·17	9·01, 3·16	6·19, 5·09	9·09, 5·24

The tetracyclines can undergo partial isomerization to quatrimycins (Doershuk *et al.*, 1955) otherwise known as epitetracyclines (Stephens *et al.*, 1956) when held at 25° for 24 hours in methanol–water (2:1 v/v) buffered to pH 4·6 by $MNaH_2PO_4$. The process is reversible but the ratio tetracycline/quatrimycin reaches equilibrium at 1·5 to 1. The isomerization leads to decreased biological activity, since the intrinsic antibiotic activity of quatrimycin is <5% that of tetracycline. The change is believed to be a simple epimerization.

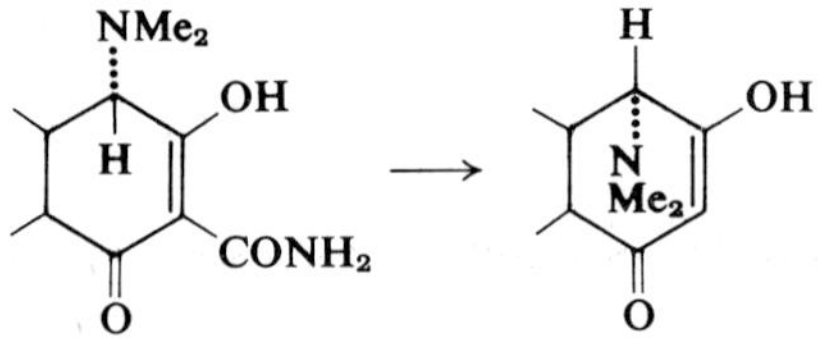

The ultraviolet absorption differ significantly in the region 250–300 nm; tetracycline hydrochloride has a peak at a 275 nm, whereas quatrimycin has weaker absorption with $\lambda_{max.}$ at 260 nm and an inflection at 280 nm. The 355 nm peak of tetracycline is displaced to 360 nm but the spectra are virtually the same as 250 nm and at 375 nm (Doerschuk *et al.*, 1955). The epimers can be separated by chromatography and then determined (Selzer & Wright 1957). This is important if the drugs are to be subjected to prolonged storage when epimerization could involve considerable deterioration.

When the absorption spectrum obtained by subtracting the curve for 7-hydroxy-3-methylphthalide from that of isodesoxydesdimethylaminoterramycin the result is the absorption of a chromophore:

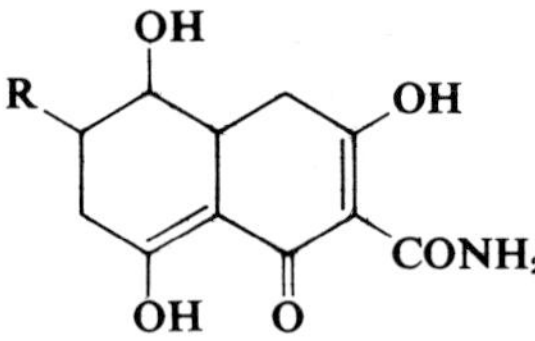

$\lambda_{max.}$ nm	259	320	$\lambda_{min.}$ nm	280
$\epsilon_{max.} \times 10^{-3}$	12·6	17·8	$\epsilon_{min.} \times 10^{-3}$	5·6

Exactly the same curve is obtained by subtracting the absorption of 4-chloro-7-hydroxy-3-methylphthalide from that of isodesoxydesdimethylaminoaureomycin.

The original paper (Hochstein *et al.*, 1953) has a full account. Blackwood *et al.* (1961) record the properties of 6-methylene-tetracyclines which are of very high potency.

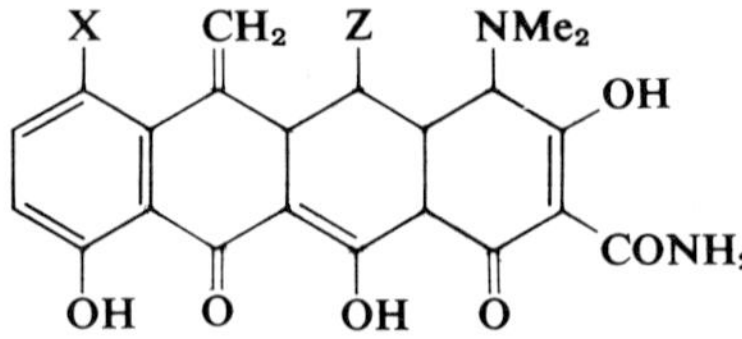

6-Methylenetetracyclines

Z=X=H	$\lambda_{max.}$ nm	254	353
0·1N HCl (MeOH)	$\epsilon_{max.} \times 10^{-3}$	22·4	15·5
Z=OH	$\lambda_{max.}$ nm	245	347
X=Cl	$\epsilon_{max.} \times 10^{-3}$	2·2	15·5

The structures in the tetracycline field were of course arrived at by taking into account all the rather complicated evidence, but there is no doubt that ultraviolet absorption spectra were very useful. The original literature illustrates this better than any summary can do.

3.6 Dyes

When two chromophores are separated by saturated groups (CH_2, $CH_2.CH_2$, CH_2O) they absorb additively. Many dyes with isolated chromophores show such behaviour almost quantitatively. Classical studies on this were done by Brode and his colleagues (Brode & Piper, 1941; Piper & Brode, 1935; Funkhouser & Brode, 1934; Morris & Brode, 1948).

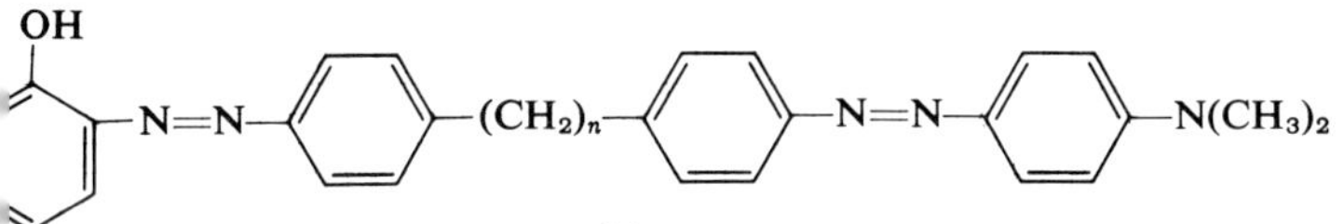

(a)

Thus the absorption spectrum of a disazo dye (*a*) agrees with that of an equimolecular mixture of (*b*) and (*c*).

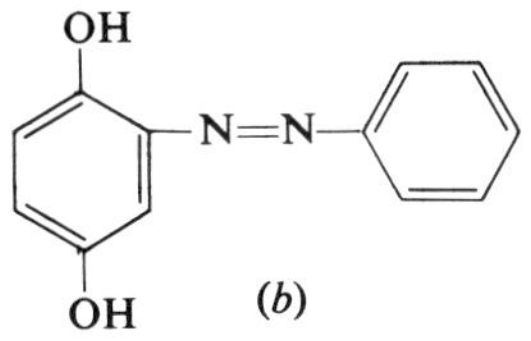

(*b*)

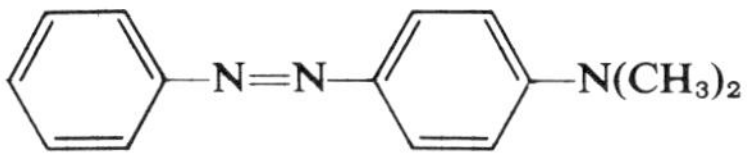

(*c*)

while a symmetrical dye (*d*) shows absorption the same as that of twice the molar concentration of (*e*)

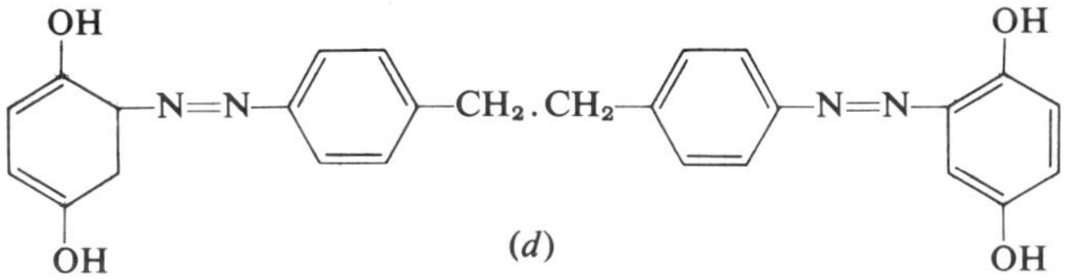

(*d*)

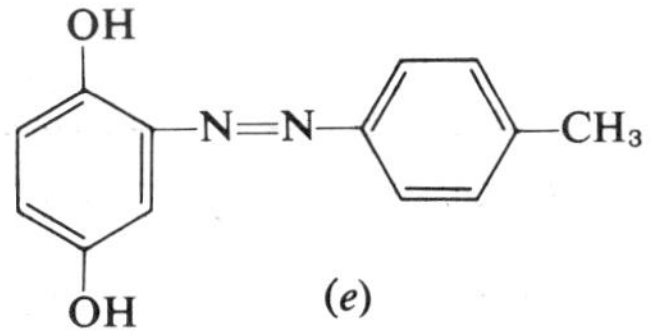

(*e*)

Funkhouser & Brode (1934) examined a bisazo dye (*f*) with linkages through the meta positions in a benzene ring:

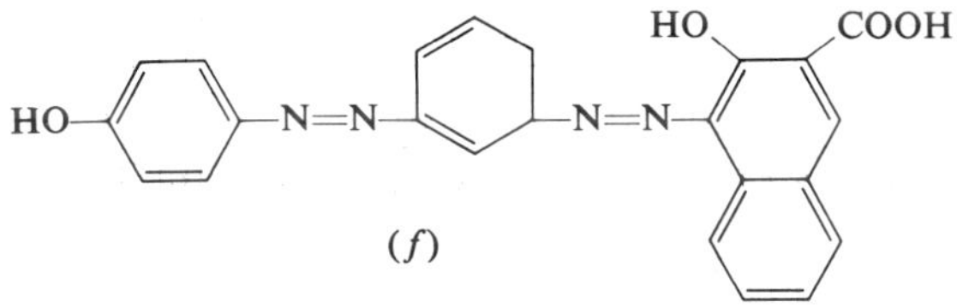

(*f*)

and found its absorption spectrum to correspond with $g+h$:

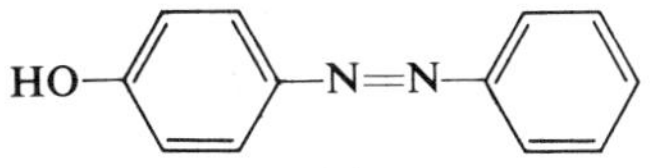

(*g*)

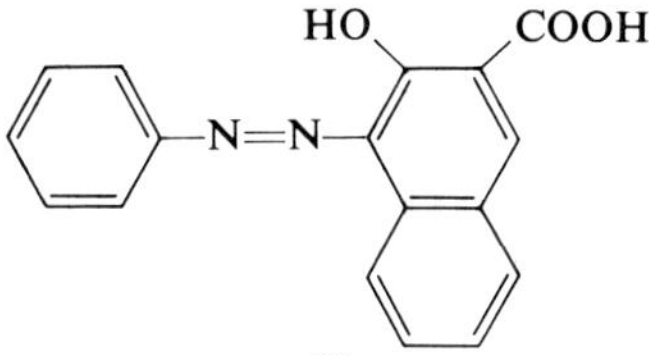

(*h*)

From this it seems that the meta positioning prevents full conjugation, just as the *m*-polyphenyls differ from the *p*-polyphenols (p. 131).

Another example of isolation despite conjugation occurs in the substituted biphenyl dye:

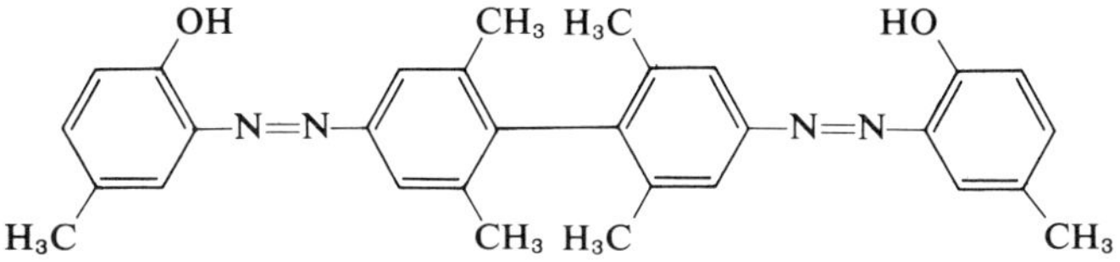

(*i*)

where steric hindrance is clearly a factor.

Bis-dyes of other classes (cyanines and merocyanines) also exhibit deepening of colour when there is conjugation.

When two arylazo groups are linked through a *p*-phenylene group (*j*) there is considerable red shift and an increase in molecular extinction for the main band, but if the two groupings are separated by CH_2 as in (*k*) there is isolation as evidenced by the fact that the absorption is twice that of (*l*):

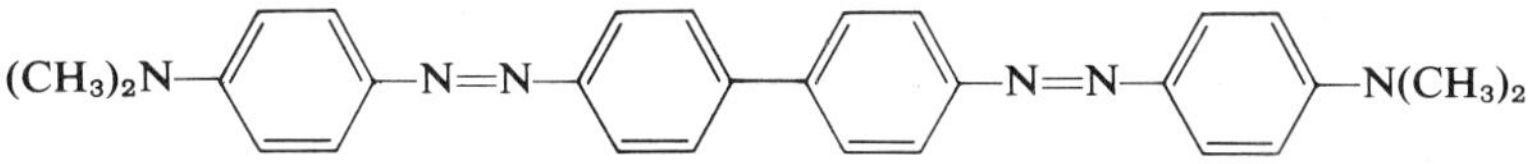

(*j*)

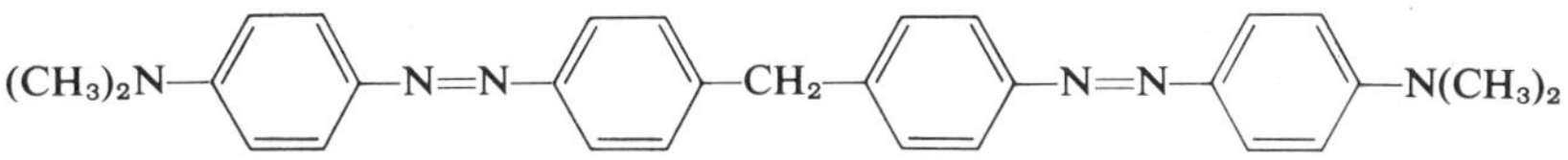

(*k*)

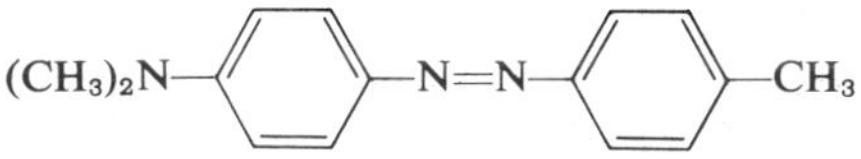

(*l*)

Krasovitskii & Mal'tseva (1966) found that in the bisazomethines of type (*m*) the bathochromic effect depended on the nature of X.

	$\lambda_{max.}$	$\epsilon_{max.}$
X=CH_2	345 nm	28 600
X=O	350	38 000
X=S	358	3700
X=NH	413	4830

Kiprianov (1971) recalled how it had been generally held that two conjugated chromophores in a dye molecule lost their individuality and behaved as one system like a single chromophore with a longer chain, so that $\lambda_{max.}$ underwent a red shift and the absorption became more intense. With his colleagues, Kiprianov showed that this was true only (*i*) when the two moieties absorb quanta of about the same size and (*ii*) when the separate chromophores lie on the same straight line. If the two chromophores are very different they do not interact. When they are not co-linear the two absorption peaks do not merge into one band but move apart. The intensities of the new displaced bands will vary with the angle between the two chromophores.

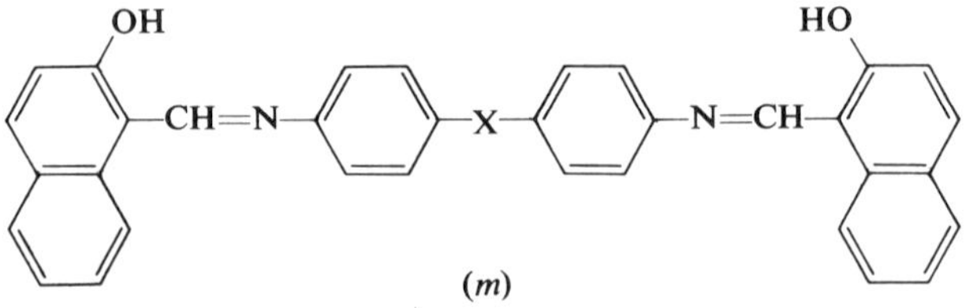

(*m*)

Kiprianov & Mushkalo (1965) found the above to be the case for biscyanine dyes. Kiprianov & Buryak (1968) compared

(*n*)

with

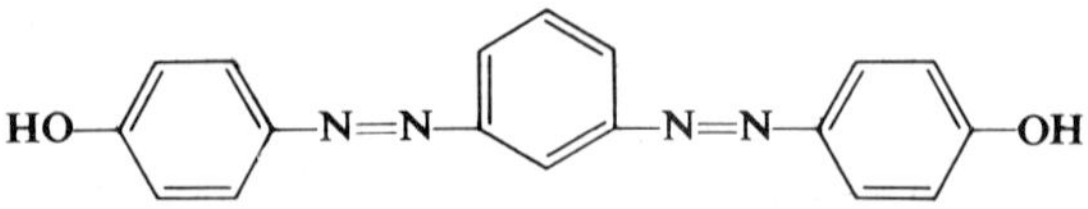

(*o*)

When (*n*) was compared with *p*-hydroxyazobenzene the bathochromic effect was clear. With compound (*o*), despite the *meta* orientation, the conjugation effect is definite but smaller, due probably to non-linearity of the chromophores.

	$\lambda_{max.}$ nm	$\epsilon_{max.} \times 10^{-5}$	
p-hydroxyazobenzene	345	0·2	(ethanol)
	415	0·25	(alkali)
bis-dye (*n*)	400	0·45	(ethanol)
	525	0·52	(alkali)
bis-dye (*o*)	370	0·45	(ethanol)
	~415	0·44	(alkali)
	455	0·52	

The work of Kiprianov and his group needs to be seen in perspective. Förster (1944) noted that the absorption peaks of some dimeric cyanine dyes were at shorter wavelengths than in the monomers and attributed this to the fact that the chromophores were parallel to one another in the dimers. Kuhn (1967) deposited monomolecular layers of simple cyanine dyes on a glass plate so that the chromophoric chains of two neighbouring components A and B were contiguous. The chromophores interacted as coupled oscillators and the absorption bands of dyes A and B were displaced in opposite directions and so moved apart.

Kiprianov & Buryak (1968) determined the absorption curves of three isomeric bis-hemicyanines (*p*), (*q*) and (*r*) and also that of the component hemi-

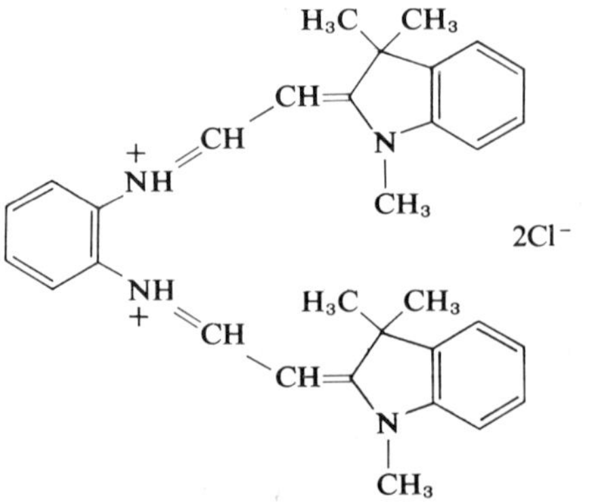

(*p*)

$\lambda_{max.}$ nm	$\epsilon \times 10^{-5}$
381	0·39
435	0·45

(*q*)

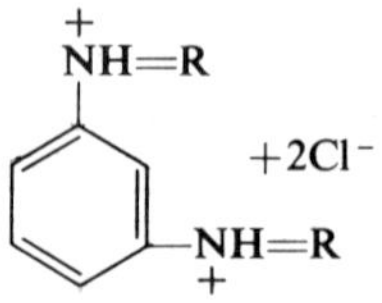

$\lambda_{max.}$ nm	$\epsilon \times 10^{-5}$
~411	0·50
435	0·68

(*r*)

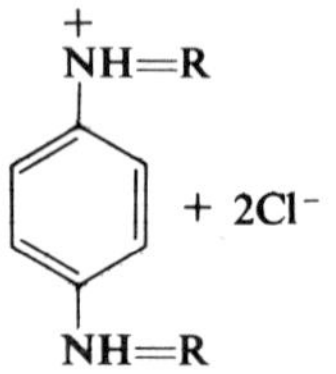

$\lambda_{max.}$ nm	$\epsilon \times 10^{-5}$
395	0·20
490	0·76

(*s*)

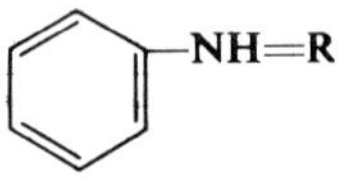

$\lambda_{max.}$ nm	$\epsilon \times 10^{-5}$
413	0·46

H_3C CH_3

R = —CH—CH=

N

CH_3

cyanine (*s*). It will be seen that splitting occurs in all three; (*r*) shows the greatest bathochromic effect while (*p*) and (*q*) clearly display the angular effect.

The implications of this work are considerable and care must be taken in interpreting the spectra of non-linear conjugated systems.

3.7 Conjugated Poly-enes

As the number of ethenoid linkages increases, the principal absorption peak of conjugated poly-enes is shifted in the direction of longer wavelengths but intensities of absorption, though tending to increase significantly, are less predictable. Ethylene shows intense absorption ($\lambda_{max.}$ *ca.* 180 nm, $\epsilon_{max.}$ *ca.* 10^4). The dienes show a peak near 220 nm due to a π—π^* transition.

	In hexane	
	$\lambda_{max.}$ nm	$\epsilon_{max.}$ 10^{-4}
CH_2=CH—CH=CH_2	217	2·1
CH_2=CMe—CH=CH_2	220	2·4
CH_2=CMe—CMe=CH_2	225	2·0
CH_3—CH=CH—CH=CH—CH_3	227	2·3
$(CH_3)_3$C—CH=CH—CH=CH—C$(CH_3)_3$	234	2·3

It is interesting that

$$(CH_3)_2C{=}CH{-}\underset{\displaystyle CH_3}{\underset{|}{C}}{=}CH_2$$

shows $\lambda_{max.}$ 229 nm with $\epsilon_{max.}$ much lower at 8500. The triene H(CH=CH)$_3$H has peaks at 248, 257 and 268 nm ($\epsilon \times 10^{-4}$ 30·5, 42·7 and 34·6 respectively in iso-octane; Sondheimer *et al.*, 1961). The dimethyl analogue octatriene has very similar absorption (Nayler & Whiting, 1955). The tetraene $CH_3(CH{=}CH_4)CH_3$ shows resolved absorption with peaks at 272, 284, 296 and 320 nm, $\epsilon_{max.} \times 10^{-3}$ 4·2, 8·4, 12 and 12 respectively, while the hexaene $CH_3(CH{=}CH)_6CH_3$ shows

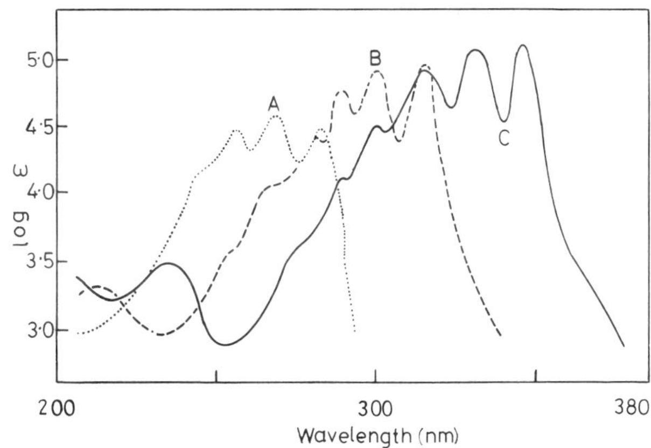

Figure 3.4 Absorption spectra of dimethyl poly-enes (I) in hexane. A, n=3; B, n=4; C, n=5 (from Nayler & Whiting, 1955)

peaks of rather similar intensities at 330, 340, 360 and 372 nm (in chloroform).

The alcohols $CH_3(CH{=}CH)_nCH_2OH$ show peaks at 232, 268 and 310 nm (in ethanol) when n = 2, 3 or 4 and the intensities ($\epsilon_{max.}$) for the two last-mentioned are of the order 50×10^3. The corresponding acids where n = 2, 3 and 4 show peaks (in hexane) at 261, 302 and 330 nm, $\epsilon_{max.} \times 10^{-4}$ being 2·52, 3·65 and 4·92 respectively. Introduction of a methyl group as in methylsorbic acid

$$CH_3CH{=}CH{-}\underset{\displaystyle CH_3}{\underset{|}{C}}{=}CH{-}COOH$$

moves the peak from 261 to 266 nm. The conjugated

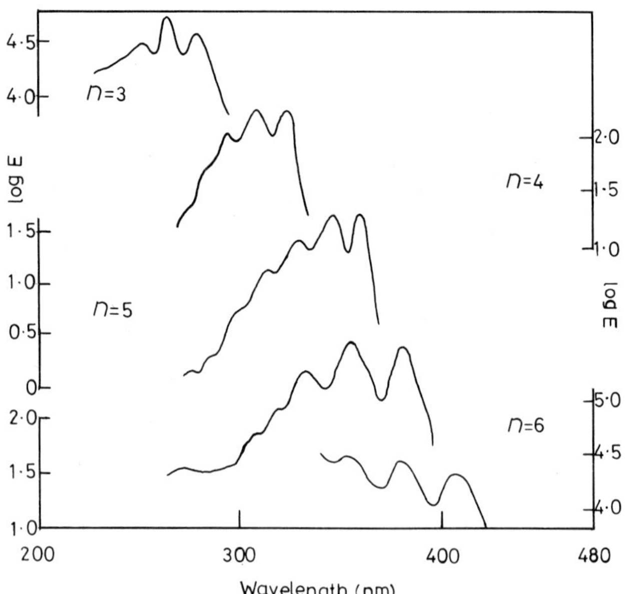

Figure 3.5 Absorption spectra of dimethyl poly-enes in chloroform. Note that except for n=3 and n=6, log E values are instrument readings, obtained with solutions of unknown concentration (from Nayler & Whiting, 1955)

Bohlmann & Mannhardt (*Chem. Ber.* 1956, **89**, 1307) succeeded in preparing in a state of purity all-*trans* isomers of $CH_3.(CH{=}CH)_n.CH_3$ where n=8, 9 and 10. Absorption spectra were obtained using benzene as solvent. It transpired that Nayler & Whiting (1955) in their investigation of dimethyl poly-enes had obtained *cis*-isomers, the all-*trans* forms having been eliminated by relative insolubility. Absorption maxima for the all-*trans* form occur at appreciably longer wavelengths than for *cis*-forms. Marshall & Whiting (*J. Chem. Soc.*, 1956, 4082) agree that mixtures of *cis*-forms were studied in earlier work on the higher dimethyl poly-enes and probably also for Kuhn's diphenyl poly-enes with n=11 and n=15 (*Angew. Chem.* 1937, **50**, 703)

carboxyl group displaces the 268 nm peak ($\epsilon_{max.}$ $5{\cdot}3\times10^4$) for $CH_3(CH{=}CH)_3CH_2OH$ to *ca.* 300 nm, $\epsilon_{max.}$ *ca.* $3{\cdot}6\times10^4$) for the corresponding acid.

Similarly the tetraenol with $\lambda_{max.}$ 310 nm, $\epsilon_{max.}$ $4{\cdot}47\times10^4$ compares with the tetraenoic acid which has $\lambda_{max.}$ 330 nm, $\epsilon_{max.}$ $4{\cdot}92\times10^4$. The carboxyl group behaves (at a first approximation) like an additional conjugated double bond. Thus comparison of the octadienoic acids: $CH_3CH_2CH_2(CH{=}CH)_2COOH$ $\lambda_{max.}$ 260 nm, $CH_3CH_2(CH{=}CH)_2CH_2COOH$ $\lambda_{max.}$ 228 nm, shows clearly the effect of a CH_2 in insulating the ethylenic bonds from the carboxyl group. (See Braude, 1955.)

Introduction of a phenyl group as in cinnamylacrylic acid $C_6H_5(CH{=}CH)_2COOH$ moves the peak for $CH_3(CH{=}CH)_2COOH$ from 261 nm ($\epsilon_{max.}$ $2{\cdot}52\times10^4$) to 314 nm ($\epsilon_{max.}$ $3{\cdot}48\times10^4$) and that for $CH_3(CH{=}CH)_3COOH$ from 302 nm ($\epsilon_{max.}$ $3{\cdot}65\times10^4$) to 340 nm ($\epsilon_{max.}$ $5{\cdot}3\times10^4$) for $C_6H_5(CH{=}CH)_3COOH$. The furyl group also have the effect of extending the conjugation in

(furyl)–$(CH{=}CH)_nCOOH$

n = 0-3

e.g. when n=2 the 260 nm peak is displaced to 351 nm. The family $C_6H_5(CH{=}CH)_nC_6H_5$ shows the phenyl groups extending the conjugation and in, for instance, diphenyldodecahexaene and diphenyltetradecaheptaene (in benzene) the resolved absorption extends into the visible region and the absorption is very strong:

	$\lambda_{max.}$ nm			
hexaene	377	400	420*	445
heptaene	394	413	435†	465

* $\epsilon_{max.}\times10^{-5}=1{\cdot}13$ † 1·30

In this family molecular extinction coefficients rise steadily as n increases. There is also a linear relationship between $\epsilon_{max.}$ and n for the $CH_3(CH{=}CH)_nCHO$ and $CH_3(CH{=}CH)_nCOOH$ families.

The broad picture in this field owes much to a valuable series of papers by Hausser, Kuhn, Smakula and others published in 1935. Despite the lapse of time this work remains classical. More recent studies on simple *all-trans* poly-enes (Woods & Schwartzmann, 1948; Mebane, 1952; Nayler & Whiting, 1955; Bohlmann, 1952; Bohlmann & Mannhardt, 1956) display good resolution (in hexane) into four bands.

Mason (1961) reviewed the literature for various families of poly-enes. The data on *all-trans* dimethyl poly-enes show that the absorption intensity of the first (long wavelength) band conforms with Braude's theory (1955) that the oscillator strength f should increase linearly where* $f=4{\cdot}32\times10^{-9}\int\epsilon\,d\nu$. For the family $CH_3(CH{=}CH)_nCH_3$ where n=2–10 (inclusive) the calculated integrated intensities (oscillator strengths f) agreed reasonably well with observed values.

The infrared absorption spectra of poly-enes were studied by Blout, Fields & Karplus (1948). The strong peak at 1445–1455 cm^{-1} is highly characteristic of the

* f=number of oscillators each of mass m and charge $(10^3mc^2\log_e 10/\pi e^2N)=4{\cdot}32\times10^{-9}$. For a smooth symmetrical band, $f=2{\cdot}2\times10^{-9}\epsilon_{max.}\,\Delta\nu$ where $\Delta\nu$ is band width in cm^{-1} when $\epsilon=\frac{1}{2}\epsilon_{max.}$.

Table 3.10 Principal peaks for some simple poly-enes

$H(CH{=}CH)_nH$	$n=$ 2	3	4	5	6
$\lambda_{max.}$ nm	217	268	304	334	364*
$\epsilon \times 10^{-3}$	21	34·6	—	121	138
$CH_3(CH+CH)_nCH_3$					
$\lambda_{max.}$ nm	227	263	299	326	352†
$\dot{\lambda}_{max.}$ nm	226	274	314	343	370‡
	(0·32)	(0·98)	(1·11)	(1·37)	(1·78)
	—	279	316·5	349	380§
	—	(0·74)	(1·11)	(1·66)	(2·17)

f values in parentheses

* Woods & Schwartzman (1949); † Bohlmann & Mannhardt (1956) in petrol or ether; ‡ *Ibid.* in methanol; § Nayler & Whiting (1955) in chloroform.

Table 3.11 Absorption spectra of some poly-enes

Solvent	Hexane		Ethanol	
	$\lambda_{max.}$ nm	$\epsilon \times 10^{-3}$	$\lambda_{max.}$ nm	$\epsilon \times 10^{-3}$
(*a*) Aldehydes				
Crotonaldehyde	212	16·4	217	15·7
Sorbicaldehyde	263*	27*	270	26·5
Octatrienal	306	40·8	316	37
(*b*) Acids				
Crotonic acid	208†	12·6 (12)	204	11·7
Sorbic acid	261	25·7	254	24·8 (24·5)
2,4,6-Octatrienoic acid	303	36·4 (35·9)	294	36·5
2,4,6,8-Decatetraenoic acid	332	—	327	48·7 (49·0)
(*c*) Acids at −196° in ether-alcohol	$\lambda_{max.}$ nm		$\epsilon_{max.} \times 10^{-3}$	
2,4,6-octatrienoic	304·5 291·5 278·3		46	
2,4,6,8-decatetraenoic	344·5 328 312·4 299		45·5	
2,4,6,8,10-dodecapentaenoic	384 363 344·3 326·4		44·5	

* Hausser *et al.* (1935 *b*); † P. Y. Blanc (1961)
Values in parentheses by G. Cherabim

Table 3.12 Absorption peaks of vapours

	Stilbene	Diphenyl-butadiene	Diphenyl-hexatriene	Diphenyl-octatetraene
I	979 (308)	936 (320)	883 (340)	839 (357·6)
II	1017 (295)	976 (308)	929 (323)	884 (339)
III	—	1025 (293)	971 (309)	931 (322)
IV	—	—	1020 (294)	975 (308·6)
	$\Delta\nu$ 38	40 49	46 42 49	45 47 44

ethylene group. Other quite weak absorption bands occur in the region 1580–1670 cm^{-1} which are not unequivocally assigned. (See also Fenton & Crisler, 1959.)

Hausser, Kuhn & Seitz (1935) examined the spectra of the diphenyl poly-enes in the vapour state. In Table 3.12 the maxima are expressed in frequency ($\nu \times 10^{12}$ sec^{-1}) and in nm in parentheses.

The α,ω-diphenylpoly-enes (all-*trans*), from stilbene to the diphenylheptaene, were all examined at −196° in a mixture of ether and ethanol. The long wave band for each member of the series is denoted ν_0 and in every

Table 3.13 Absorption maxima of diphenylpoly-enes

	$C_6H_5(CH{=}CH)_nC_6H_5$											
	in Ethanol						in Benzene					
n	$\lambda_{max.}$ nm			$\epsilon \times 10^{-3}$			$\lambda_{max.}$ nm			$\epsilon \times 10^{-3}$		
1	314	297	—	21·7	27	—	$\underline{319}$	306	$\underline{294}$	—	24·3	—
2	345	328	—	26	40·8	36·5	$\underline{352}$	334	$\underline{316}$	—	40	—
3	370	349	335	55·6	67·4	54	377	358	343	52	75	54·3
4	396	375	353	71	85	54·5	404	384	363	77	86	58·3
5	—	—	—	—	—	—	424	403	387	88·7	94	61
6	—	—	—	—	—	—	445	420	400	109	113	74
7	—	—	—	—	—	—	465	435	413	121	135	87

Inflections underlined

case there is notable fine structure, the peaks being given by the equation

$$\nu = (\nu_0 + l.37{\cdot}0 + m.47{\cdot}1)10^{12}\ \text{sec}^{-1}$$

where l and $m = 0, 1, 2, 3 \ldots$

Stilbene, for example, shows peaks at 917·4 (327 nm), 961·6 (312), 1003·3 (299) and 1048·3 (286). $\Delta\nu$ being 44·2, 42 and 45×10^{12}. The heptaene has its long wave peak at $\nu = 637{\cdot}10^{12}$ (471 nm). The vibrational structure consistently shows frequency differences of 37 and 47·1 corresponding with 1230 and 1570 cm^{-1} and the intermediate members of the family give similar results.

The poly-ene acids were also examined at $-196°$ in ether-alcohol and the spacings between maxima were approximately $\Delta\nu = 45 \times 10^{12}$ or 1500 cm^{-1}.

In the region 220 to 250 nm diphenyl shows at $-196°$ (in ether-alcohol) peaks at 222 and 228 nm, stilbene has 3 peaks at 225, 232 and 239 nm, the diene at 233, 241 and 240 nm, the triene at 231·7, 238·5, 246 and 253 nm and the tetraene at 243, 250·6 and 259 nm. These bands are due to the phenyl groups, influenced by the conjugated olefinic linkages. Essentially then, the diphenyl poly-enes show additively the poly-ene bands steadily rising in intensity and displaced in the direction of longer wavelengths as n increases and the phenyl bands are somewhat changed in position as n increases, but with unresponsive molecular extinction coefficients.

Hausser, Kuhn & Kuhn (1935) studied the fluorescence spectra of α,ω-diphenylpoly-enes in a paper of great illustrative value. At $-190°$ the fluorescence bands followed equations in which $1/\lambda$ [$\nu_0(cm^{-1})$] represented the position of the fluorescence peak with the shortest wavelength and values of $\triangle\nu$ (cm^{-1}) of 1550 and 1160 recurred. At 20° some fine structure appeared, but was less marked than at low temperature (see Figs. 3.6 *a*, *b*).

It will be seen that band I moves steadily from $\nu = 892$ (336 nm) to 535 (560 nm) and the spacing is on the average 40×10^{12} or 1200 cm^{-1}.

The fluorescence spectra of the entire family are covered well by the expression

$$\nu\ (cm^{-1}) = \nu_0 - a\nu' - b\nu''$$

where a and b are integers and ν' and ν'' are 1550 and

Table 3.14 Fluorescence maxima (ν) of diphenyl poly-enes $Ph(CH{=}CH)_nPh$ with spacings $\Delta\nu$ in xylol at 20°

	$n=1$		$n=2$		$n=3$		$n=4$		$n=5$		$n=6$	
Maximum	ν	$\Delta\nu$	ν	$\Delta\nu$	ν	$\Delta\nu$	ν	$\Delta\nu$	ν	$\Delta\nu$	ν	$\Delta\nu$
I	892		828		730		655		590		535	
		41		41		31		42		40		40
II	851		787		699		613		550		495	
		39		40		40		39		39		39
III	812		747		659		574		511		456	
		41		38		39		41		41		
IV	771		709		620		533		470			
						42		39				
					578		494					

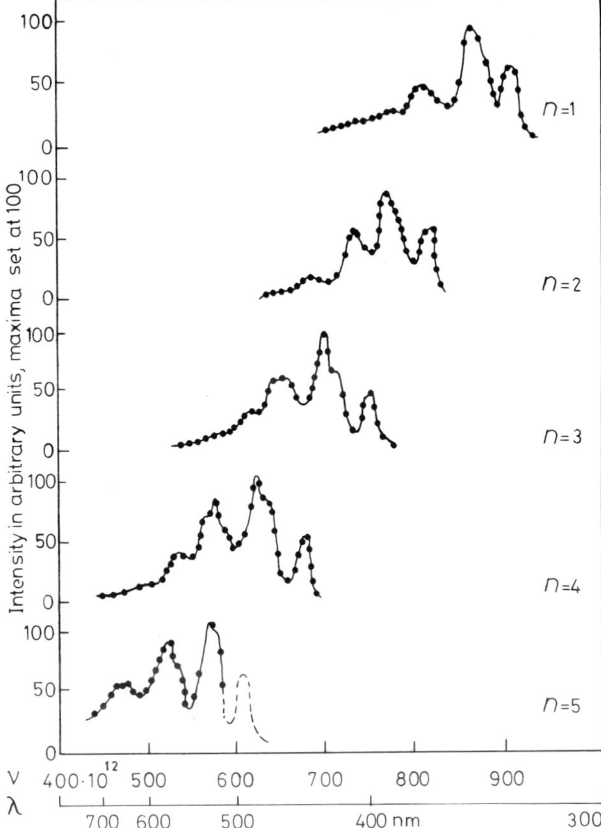

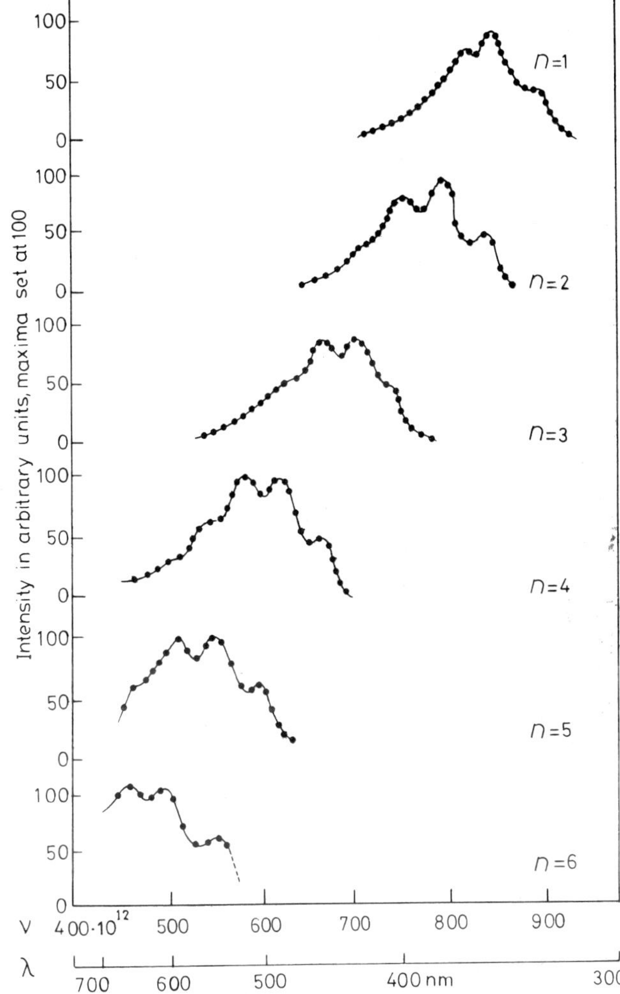

Figure 3.6 (*a*) Fluorescence spectra in xylene of solutions of diphenylpoly-enes at −196°, (*b*) right; Fluorescence spectra in xylene of solutions of diphenylpoly-enes at room temperature (20°) (from Hausser, Kuhn & Kuhn, 1935)

1160 cm^{-1} respectively and ν_0 values are in cm^{-1}, stilbene 29 730, diphenylbutadiene 27 150, diphenylhexatriene 24 900, diphenylhexatriene 24 900, diphenyloctatetraene 22 135 and diphenyldecapentaene 19 870.

The fluorescence spectra did not vary with the wavelength of the exciting light. In general the fluorescence spectra were mirror images of the absorption spectra (see Fig. 3.7), fairly accurately so in respect of the position of maxima on the wavelength scale, but less accurately in respect of intensities of absorption or emission.

The effects of varying the solvent and the temperature were studied in detail. In general, emission intensi-

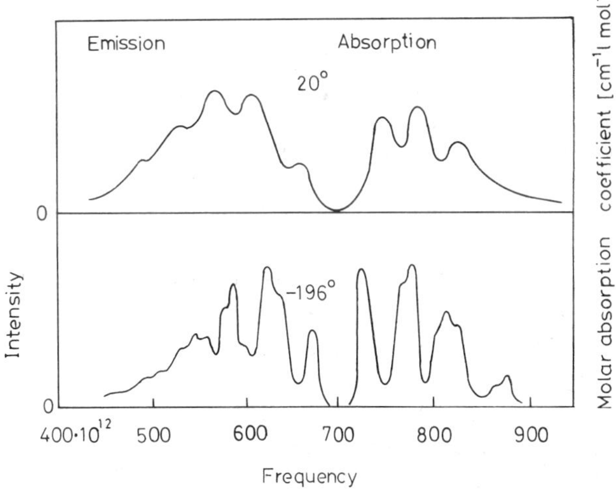

Figure 3.7 The mirror image symmetry of fluorescence and absorption spectra for diphenyloctatetra-ene in xylene at 20° and −196° (from Hausser, Kuhn & Kuhn, 1935)

ties fell with increasing temperature by about 50% for a 100 degree rise depending partly on the concentration. When allowance is made for change of volume the true temperature coefficient is nearer −0·4 for 100 deg.

The first absorption band of diphenyloctatetraene at 26° occurred at 403·5 nm and was displaced linearly as the temperature rose to 136° (396·5 nm). The emission was more gradually displaced towards longer wavelengths. Dodecapentaenoic acid is of somewhat greater biological interest than the diphenylpolyenes. It shows a very well resolved fluorescence spectrum at −196° with 15 maxima between $\nu = 698 \times 10^{12}$ and 480×10^{12} (429–625 nm) with well marked spacings $\Delta\nu = 46$ and 34×10^{12} (1535 and 1140 cm^{-1}).

Cyclic conjugated dienes may be homoannular or heteroannular; e.g.,

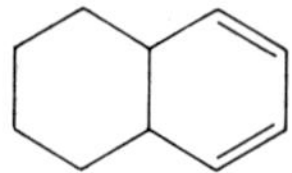 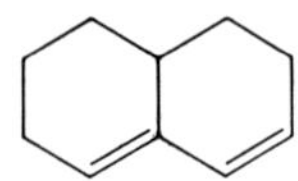 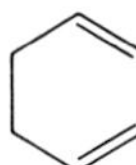

and a good deal of experience has been expressed as empirical rules (Dannenberg, 1939; Booker *et al.*, 1840; Woodward, 1942; Fieser & Fieser, 1959).

Booker *et al.* summarized the then available data (see below) and Woodward's initial statements were:

(i) That each alkyl substituent or ring residue attached to the diene chromophore shifts $\lambda_{max.}$ by 5 nm to the red;

(ii) Each exocyclic double bond causes a further displacement of 5 nm in the same direction, the effect being repeated if the bond is exocyclic to two rings. Thus for: butadiene ($\lambda_{max.}$ taken as 217 nm)

2,4-hexadiene

$\lambda_{max.}$ 217+2×5=227 nm (obs. 227 nm)

2,3-dimethylbutadiene

$\lambda_{max.}$ 217+2×5=227 nm (obs. 227 nm)

2,4-tetramethylbutadiene

$\lambda_{max.}$ 217+4×5=237 nm (obs. 234 nm)

Now the homoannular conjugated dienes like cyclohexadiene (see above) are necessarily in a *cis* form and $\lambda_{max.}$ is moved to 256 nm and $\epsilon_{max.}$ reduced to 8000. This contrasts with the heteroannular cholesta-4,6-diene with $\lambda_{max.}$ 238 nm.

The following data (cf. Morton, 1962) illustrate the argument:

Absorption peaks of dienes

	$\lambda_{max.}$ nm (range)
Acyclic dienes	
(*a*) only acyclic substituents	217–228
(*b*) one cyclic substituent	235·5–236·5
(*c*) two cyclic substituents	246–248
Semicyclic dienes	230–242
Cyclohexadienes	256–265
Bicyclic dienes	236
Polycyclic dienes	
(*a*) homoannular	260–282
(*b*) heteroannular	235–248

Rules for diene maxima in steroids

	$\lambda_{max.}$ nm
Parent heteroannular diene	214
Parent homoannular diene	253
Increments:	
Double bond extending conjugation	30
Alkyl substituent or ring residue	5
Exocyclic double bond	5
Polar grouping OAc	0
OAlkyl	6

(iii) *Conjugated trienes* are not common in the steroids but a few are known. Ergost-4,7,9,22-tetra-en-3-one enol acetate has $\lambda_{max.}$ 339, 356 and 375 nm, $\epsilon_{max.}$ 15 000, 17 400 and 13 000 respectively. Cholest-4,6,8,11-tetra-en-3β-ol has a maximum at 335 nm, $\epsilon_{max.}$ 13 500 (in ether) with no vibrational structure.

Cooke *et al.* (1952) and Bohlmann (1955) recorded the spectra of dimethyl poly-ynes

$$CH_3—(C{\equiv}C)_n—CH_3$$

where $n=2$ to 6.

These spectra are notable in having very high intensities and well-resolved absorption on the short wave side and equally well-resolved but quite weak absorption on the long-wave side (see Figs. 3.8 and 3.9).

The molecular extinction coefficients for the higher poly-ynes are very great. Indeed, Jones *et al.* (1951) drew attention to the $E^{1\%}_{1\,cm}$ values of 25 500 ($\epsilon_{max.}$ 445 000) recorded for the hexa-yne as uniquely great at that time and for this series the $\epsilon_{max.}$ values are roughly $n \times 72\,000$. Taking $\lambda_{max.}$ as the wavelength of the first band on each series in the dimethyl poly-yne spectra

Table 3.15 Principal maxima of conjugated poly-ynes $CH_3—(C{\equiv}C)_n—CH_3$ in ethanol
(First, i.e. long wavelength peaks)

	Intense series		Weak series	
n	$\lambda_{max.}$ nm	$\epsilon_{max.} \times 10^{-3}$	$\lambda_{max.}$ nm	$\epsilon_{max.} \times 10^{-3}$
2	—	—	250	0·16
3	207	135	306	0·12
4	234	281	354	0·105
5	260·5	352	384	0·12
6	284	445	430	0·10
7	310	530	452	0·30

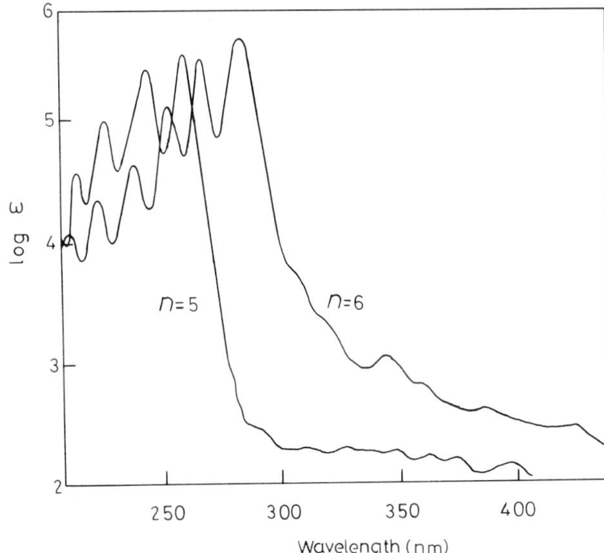

Figure 3.8 Absorption spectra of dimethylpenta- and dimethylhexa-acetylenes in alcohol (from Cook *et al.*, 1952)

Figure 3.9 Absorption spectrum of deca 2,4,6,8-tetrayne (from Armitage, Jones & Whiting, 1952)

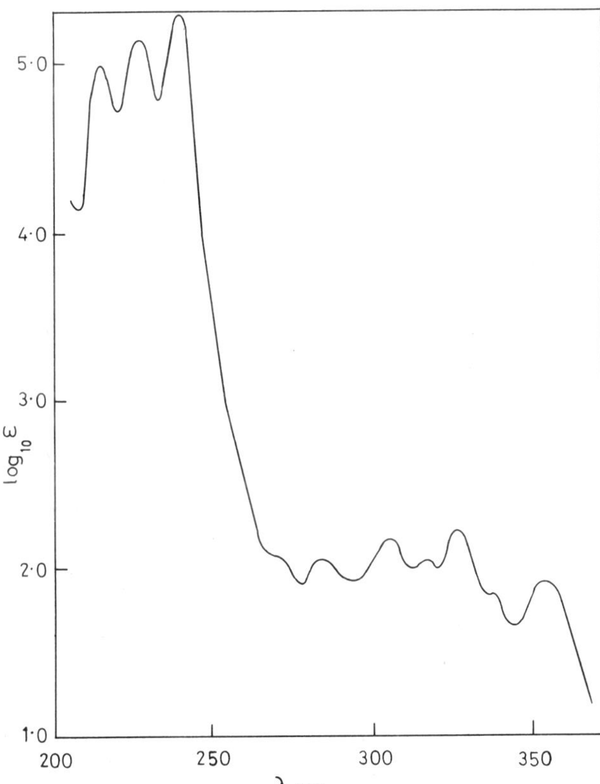

of $\lambda^2_{max.}$ plotted against n gives linear relationships (see Fig. 3.10).

In a well-insulated family of di-ynes:

$$XCH_2—CH_2—(C{\equiv}C)_2—CH_2CH_2X$$

where X=H or ethyl or hydroxyl, the absorption is weak in the region 228, 239, 253·5 nm (ϵ 230–450). and the substituent has little effect except that Cl or Br causes small shifts. The separation between peaks is again *ca.* 2000 cm^{-1}.

The fine structures are illustrated by the spectra of the two penta-ynes (Armitage *et al.*, 1952) and some hexa-ynes (Cook *et al.*, 1952) in Tables 3.17 and 3.18.

In the strong series as well as the weak series the separation between the sub-peaks is between 2000 and 2300 cm^{-1}, whereas the typical spacing for poly-enes

Figure 3.10 Plot of $\lambda_{max}{}^2$ (nm^2) against n for dimethylpolyene spectra (from Cook *et al.*, 1952)

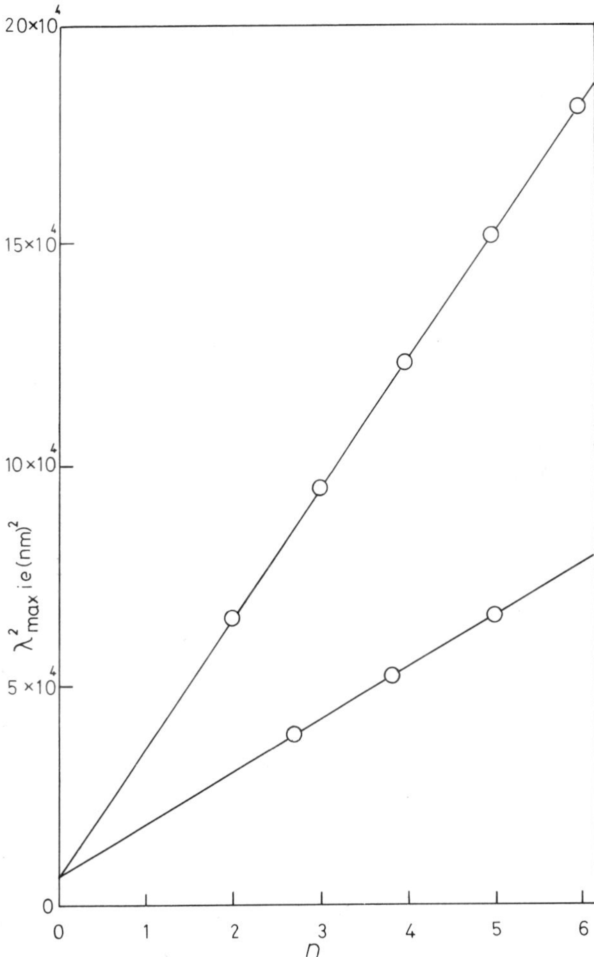

Table 3.16 Absorption maxima of diacetylenes (nm)

(*a*) R′—C≡C—C≡C—R″		D		C		B		A	
R′	R″	λ nm	ε	nm	ε	nm	ε	nm	ε
Me	H	—	—	227	370	236·5	390	249	210
Me	Me	218·5	300	226·5	360	236	330	250	160
Bu	Bu	—	—	228	440	239·5	390	254	240
CH_2OH	H	—	—	228	415	239	415	253	230
CH_2OH	Me	—	—	230·5	395	241·5	385	255	245
CH_2CH_2OH	CH_2CH_2OH	215·5	330	228	450	239·5	390	254	330
(*b*) $HOC(R')(R'')-(C{\equiv}C)_2H$									
H	CH(Me)=CH			232·5	830	245	790	250	459
H	C_6H_5			—	—	247·5	1300	260·5	920
C_6H_5	C_6H_5			—	—	248·5	2240	261·0	1660

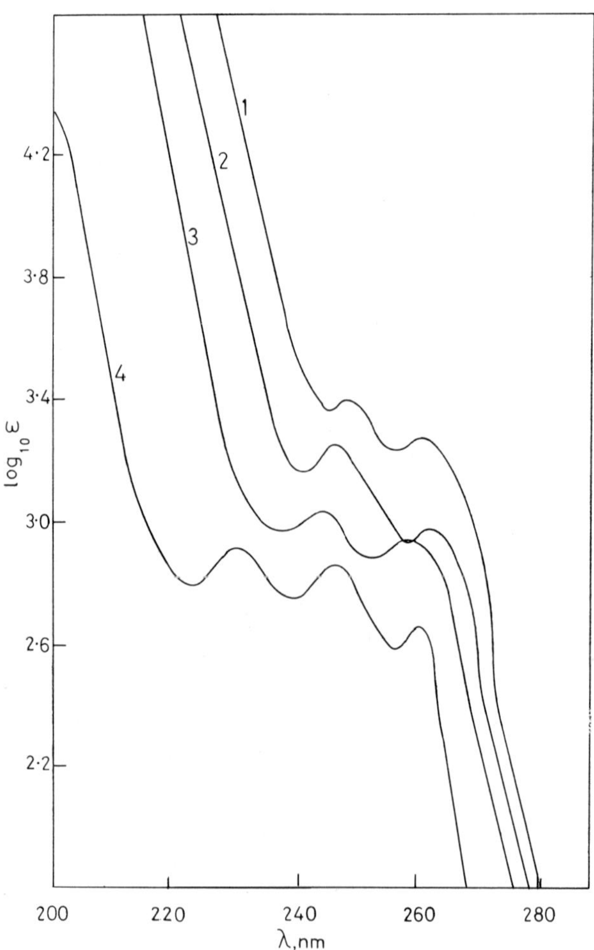

is *ca.* 1600 cm^{-1}. This fine structure makes it easy to recognize polyacetylenic grouping. (See also Nakagawa, 1950.)

The very intense bands in the poly-ynes are ascribed by Scott to a Rydberg transition.

The experimental results can be interpreted in terms of oscillator strengths (Table 3.19).

Braude's (1950) single chromophore-area theory requires *f* to increase linearly with *n* (neglecting end-of-chain anomalies likely to occur in the lower polyenes). This theory is roughly valid.

The spectra of poly-ene-ynes are collected in Table 3.20

Figure 3.11 Absorption spectra of conjugated diynes:
(1) HO—CPh_2—C≡C—C≡C—CPh_2OH,
(2) HO—CHPh—C≡C—C≡C—CHPhOH,
(3) HO—CPhMe—C≡C—C≡C—CPhMeOH,
(4) CH_3CH=CH—CH(OH)—C≡C—C≡C—CH(OH)—CH=$CHCH_3$
(from Armitage *et al.* 1952)

Table 3.17 Absorption maxima of two penta-ynes

(*a*)	CH_3—$(C{\equiv}C)_5$—CH_3								
$\lambda_{max.}$ nm	215	224	233·5	247	260·5	218	228	254	265
$\epsilon_{max.}$ 10^{-3}	10	29	95	243	352	15	44	240	355
(*b*)	$HOCMe_2(C{\equiv}C)_5CMe_2OH$								
$\lambda_{max.}$ nm	324·5	347·5	373·5	394		314	337·5	363	393
$\epsilon_{max.} \times 10^{-3}$	0·23	0·21	0·15	0·12		0·785	0·67	0·64	0·535

Table 3.18 Absorption spectra of hexa-ynes

	Intense maxima ($\epsilon \times 10^{-4}$) of hexa-ynes $R(C{\equiv}C)_6R$											
R	nm	ϵ	nm	ϵ	nm	ϵ	nm	ϵ	nm	ϵ	nm	ϵ
CH_3	221	1·1	231	1·75	242	4·7	255	12·8	268·5	31·7	284	44·5
C_2H_3	222	0·9	230	1·45	244	2·2	258	5·0	270	7·4	285	9·0
$(CH_3)_2COH$			235	2·8	245	3·0	258	3·4	272	4·0	289	4·8

(After Cook, Jones & Whiting, 1952.)

Table 3.19 Oscillator strengths

	Dimethylpoly-enes		Di*tert*butylpoly-ynes	
n	$f_{calc.}$	$f_{exp.}$	$f_{exp.}$	Ratio*
3	0·94	0·74	2·08	2·81
4	1·21	1·11	2·99	2·69
5	1·47	1·66	4·10	2·48
6	1·74	2·17	5·15	2·37

* Column 4/column 3.

Table 3.20 Absorption maxima of poly-ene-ynes

[$\lambda_{max.}$ nm $\epsilon_{max.} \times 10^{-3}$, ϵ values in parentheses]
R—$(CH{=}CH)_{n'}$—$(C{\equiv}C)_n$—$(CH{=}CH)_{n''}$—R

n'	n	n''	First band group				Second band group					
1	1	0	—	—	—	235(15)	—	—	—	—	—	—
1	1	1	—	—	266(20)	275(16)	—	—	—	—	—	—
2	1	0	—	—	—	—	230(2)	238(5)	251(10)	264(14)	280(11)	—
2	1	1	—	230(31)	235(30)	244(23)	261(8)	276(14)	290(20)	311(17)	—	—
2	2	1	—	213(17)	252(33)	267(29)	281(17)	296(30)	316(40)	338(29)	—	—
2	2	2	—	254(28)	276(34)	290(33)	312(32)	334(45)	359(43)	—	—	—
2	3	0	—	—	242(13)	252(22)	287(17)	303(39)	319(63)	336(65)	—	—
2	3	1	—	—	264(32)	274(27)	310(36)	330(57)	354(42)	—	—	—
2	3	3	—	355(51)	374(60)	405(55)	—	—	—	—	—	—
3	1	0	—	—	231(81)	241(110)	258(4)	273(7)	290(13)	308(17)	330(11)	—
3	1	1	—	245(25)	259(45)	269(75)	289(11)	305(18)	—	—	—	—
3	2	0	—	246(27)	258(60)	268(110)	289(16)	305(28)	325(42)	348(34)	—	—
3	2	1	235(49)	242(66)	273(96)	289(81)	300(24)	309(20)	320(30)	330(20)	342(39)	369(29)
4	1	1	245(50)	260(70)	274(100)	294(75)	318(8)	341(12)	366(14)	396(7)		—
5	1	0	—	265	286	327	348	377	410	—	—	—
5	1	1	270(105)	280(123)	297(145)	312(94)	342(10)	367(13)	398(15)	433(8)	—	—

From Bohlmann (1955).

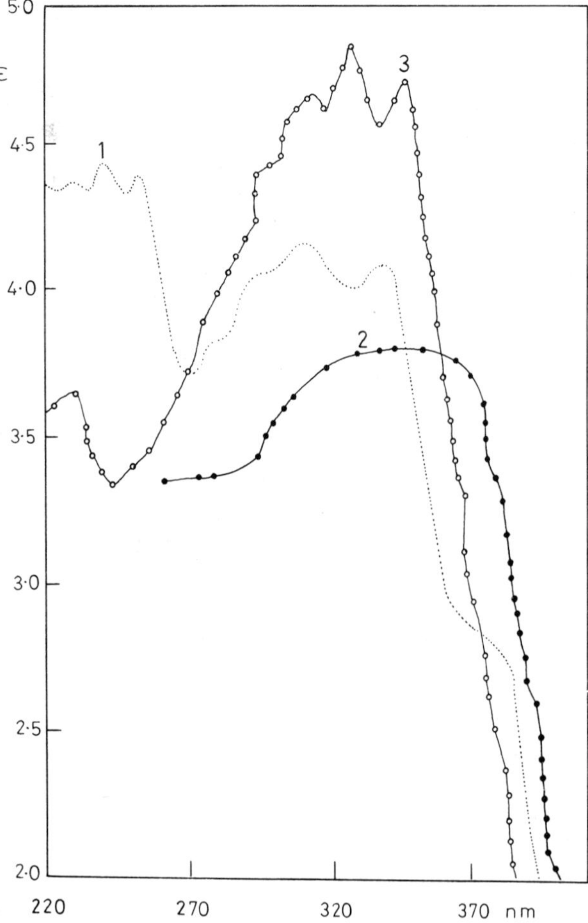

Figure 3.12 Ultraviolet absorption of hexane solutions of: (1) matricaria ester, (2) compositcumulene I, and (3) methyl decatetraenoate (from Sörenson and Stavholt, 1950 *a*)

Figure 3.13 Ultraviolet absorption of dehydromatricaria ester in hexane, —●●●●. Dehydromatricaria ester from *Erigeron canadense* L. -o-o-o-o. The curve for the ester obtained from *Artenmisia vulgaris* agrees almost perfectly (from Sörenson & Stavholt, 1950 *a*)

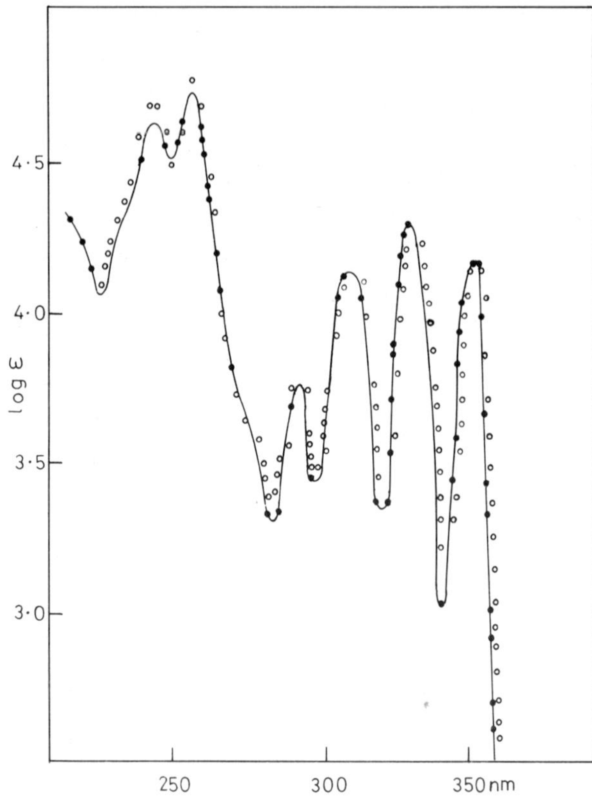

Bell *et al.* recorded data on an ene-diyne-aldehyde in which the conjugated carbonyl group has a marked effect

$$CH_3CH_2CH_2—(C{\equiv}C)_2—CH\overset{cis}{=}CHCHO$$

$\lambda_{max.}$ nm	220·5	229·5	248·5	278	283	295	314
$\epsilon_{max.} \times 10^{-3}$	15	20	19	10	8·9	21	23

while the dienediyne

$$CH_3CH{=}CH—(C{\equiv}C)_2—CH{=}CHCH_2OH$$

shows the effect of a further olefine group:

$\lambda_{max.}$ nm	218	237	246·5	262·5	277·5	294·4	314
$\epsilon_{max.} \times 10^{-3}$	26·5	30	24	7·4	14	21·5	18

Sörensen & Stavholt (1950 *a*) studied the essential oil obtained from the scentless mayweed (*Matricaria indora* L) and found that it contained a very unsaturated derivative of methyl caprate. The compound, designated matricaria ester ($C_{11}H_{10}O_2$), had been obtained earlier by Stene & Stavholt (1941). Its constitution is

$$CH_3CH{=}CH—(C{\equiv}C)_2—CH{=}CHCOOCH_3$$

$\lambda_{max.}$ nm	232	246	259	314	336
$\epsilon_{max.} \times 10^{-3}$	22·2	25·4	23	15·4	12·9

(Ligthelm & Schwartz, 1950) and the effect of the carbomethoxy group is again evident. The compound was accompanied by small amounts of hexahydromatricaria ester which on hydrogenation took up three molecules of hydrogen to give methyl caprate. This compound is designated 'compositcumulene I' and it displays a very broad absorption band with $\lambda_{max.}$ near 345 nm (see Fig. 3.12). By analogy with

$$(C_6H_5)_2—C{=}C{=}C{=}C—(C_6H_5)$$

it probably has the cumulene chromophore

$$>C{=}C{=}C{=}C<$$

(See Bohlmann & Kieslich, 1954.)

Sörensen & Stavholt (1950 *b*) studied essential oils from *Erigeron* species such as *Erigeron canadense* L (marestail, horseweed) which contained 0·2% of matricaria ester (see Fig. 3.13). *Erigeron boreal* (Vierh) oil contained 51% and *E. polunum* nearly 90% in the oil. Other species contained a dihydromatricaria ester $C_{11}H_{12}O_2$; this was similar to a compound obtained by Williams, Smirnov & Gol'mov (1935) from *Lachnophyllum gossypinum* Bge. The ester *cis*-lachnophyllum was obtained by Sörensen and Stavholt (1950 *b*) from *Erigeron acre* L

$CH_3CH_2CH_2—(C≡C)_2—CH=CHCOOCH_3$

in good yield

$\lambda_{max.}$ nm	224·5	276	291·5	309·5
$\epsilon_{max.} \times 10^{-3}$	35·4	9·5	14·4	15·8

The *trans* form synthesized by Bruun, Haug & Sörensen (1950) had rather better fine structure and the peaks were displaced by about 4 nm from the *cis* peaks in the direction of shorter wavelengths (see Fig. 3.14). The essential oil from the carline thistle (*Carline acaulis* L) contains the highly unsaturated ester 1-acetoxy-n-trideca-2,10,12-triene-4,6,8-triyne

$CH_2=CHCH=CH—(C≡C)_3—CH=CHCH_2COOCH_3$

$\lambda_{max.}$ nm
231 241·5 277·5 289 300 308·5 319·5 330 340 368·5
Δ cm^{-1} 913 1120 1000 890 2280
2033 2120 1890

$\epsilon_{max.} \times 10^{-3}$
49 66·4 95 81·4 24 19·5 30·6 19·5 39 28·9

(Sörensen & Sörensen 1954).
The essential oil from mugwort yielded dehydromatricaria ester

$CH_3(C≡C)_3CH\overset{trans}{=}CHCOOCH_3$

$\lambda_{max.}$ nm	244·7	255	268	283·5	300·8	320·4	343·5
$\epsilon_{max.} \times 10^{-3}$	71·8	106·2	8·6	15	32	45	6

Artemisia vulgaris L contains in the root an unsaturated component, the structure being either the above or

$CH_3CH=CH—(C≡C)_3—COOCH_3$

(Bohlmann *et al.*, 1957).

Sörensen & Sörensen (1954) investigated the essential oils from various species of garden Coreopsis. Among the products present the following were identified:

(A) Dodeca-1,11-diene-3,5,7,9-tetrayne
(roots of all coreopsis species)

$CH_2=CH—(C≡C)_4—CH=CH_2$

(B) 1-Phenyl-n-hept-5-ene-1,3-diyn-7-yl acetate
(all parts of plant)

$C_6H_5—(C≡C)_3CH\overset{trans}{=}CH—CH_2COOCH_3$

(C) Trideca-1,3,5,11-tetraene-7,9-diyne
(leaves)

$CH_3CH=CH—(C≡C)_2—(CH=CH)_2—CH=CH_2$

(D) 1-Phenyl-n-undeca-7,9-diene-1,3,5-triyne

$C_6H_5—(C≡C)_3—(CH=CH)_2CH_3$

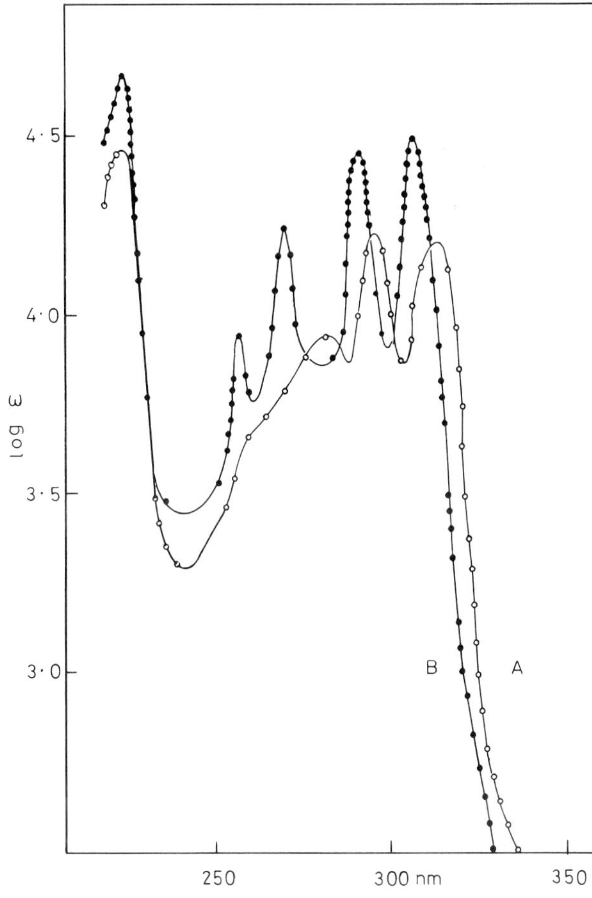

Figure 3.14 Ultraviolet absorption of lachnophyllum ester in hexane, (A) —o-o-o-o *cis*-lachnophyllum ester from *Erigeron acre* L., (B) —●●●● *trans*-lachnophyllum ester (Bruun, Haug & Sörensen, 1950) (from Sörensen & Stavholt, 1950 *a*)

The following are the principal peaks:

(A)

$\lambda_{max.}$ nm	$\epsilon_{max.} \times 10^{-3}$
390·5	13·4
361	20·9
337	17·2
315·5	11·65
287·5	143
271	189
258	143

(B)

$\lambda_{max.}$ nm	$\epsilon_{max.} \times 10^{-3}$
317·5	27·6
298	32·4
281	25·7
263	15·2
253	40·8
245	37·2
222	32·0

(C)

$\lambda_{max.}$ nm	$\epsilon_{max.} \times 10^{-3}$
353·5	42·2
330	57·2
310	36
—	—
254	32
~247	26·7
—	—
—	—

(D)

$\lambda_{max.}$ nm	$\epsilon_{max.} \times 10^{-3}$
367	17·1
~3525	8·8
3405	12
3285	18·4
3186	10·8
3095	12·4
2990	41·2
2875	45·2

The spacings of the main long wave bands are around 2000–2100 cm^{-1}. Infrared spectra in this series show peaks for C≡C at 2130 (2100–2200) COOR 1754, 1225, CH=CH 944 and C_6H_5 1600, 1490, 1445, 917, 687 cm^{-1}.

3.8 Natural Poly-ynes

The diyne chromophore is seen in benzyldiacetylene

$$C_6H_5CH_2—C≡C—C≡CH$$

(*a*) obtained from *Artemisia fruitescens* L and in capillene, (*b*) present in several species (Harada, 1957; Bohlmann *et al.*, 1962):

$\lambda_{max.}$ nm,
(*a*) 196 206·5 226 283·5 247 252 257 263·5 267;
(*b*) 239 253.

$\epsilon_{max.} \times 10^{-3}$,
(*a*) 19·0 12·0 0·66 0·56 0·31 0·10 0·27 0·18 11·5;
(*b*) 0·54 0·43.

Replacement of the methylene group by carbonyl in capillin (*c*) has a marked bathochromic effect

(*b*) capillene $C_6H_5CH_2—C≡C—C≡C—CH_3$

(*c*) capillin $C_6H_5CO—C≡C—C≡C—CH_3$

$\lambda_{max.}$ nm,
(*b*) 227 240·5 253·5;
(*c*) 268 283;
(*d*) 262·5 273 287.

$\epsilon_{max.} \times 10^{-3}$,
(*b*) 2·30 3·2 6·1;
(*c*) 8·7 6·7;
(*d*) 14·1 18·2 18·5.

while phenylpentadiyne (2,4) one (1) (d) obtained from *Chrysanthemum segetum* L

$$C_6H_5CO—C≡C—C≡CH$$

(Bohlmann *et al.*, 1961), though similar to (*c*), shows surprising differences in absorption.

An interesting chromophore is found in products from *Aethusa cynapium* L. Aethusin (*e*) is a hydrocarbon

$$\overset{13}{C}H_3—CH=CH—(C≡C)_2—(CH=CH)_2—CH_2—CH_3$$

and introduction of a hydroxyl group at position 13 gives a primary alcohol aethusin-13-ol (aethusinol B) while aethusanol A has the structure (*f*)

$$CH_3—CH=CH—(C≡C)_2—CH=CH—\underset{\displaystyle OH}{\underset{|}{CH}}—CH_2—CH_2—CH_3$$

(Bohlmann *et al.*, 1960).
Aethusin and aethusin-13-ol are spectroscopically very similar showing the following maxima

$\lambda_{max.}$ nm	249	267	279	296	313	337
$\epsilon \times 10^{-4}$	3·16	2·7	1·66	2·92	3·96	2·88

while aethusanol A shows

$\lambda_{max.}$ nm	230	240	248	265	~270	279	295	313
$\epsilon \times 10^{-4}$	3·98	4·0	2·51	1·0	0·79	1·86	2·95	2·29

the loss of one conjugated ethylenic linkage having a definite effect, e.g. in eliminating the 337 nm peak. The same chromophore as that of (*e*) appears in a hydrocarbon centaur X_4 from *Centaurea ruthenica* L (*f*)

$$CH_3—CH=CH—(C≡C)_2—(CH=CH)_2—(CH_2)_4—CH=CH_2$$

$\lambda_{max.}$ nm	250·5	266	279·5	296	314	336 (in ether)
$\epsilon \times 10^{-4}$	4·1	3·28	1·965	3·20	4·34	3·63

An alcohol (*g*) from *Matricaria chamomilla* L

$$CH_3—(C≡C)_2—(CH=CH)_3—CH_2—CH_2OH$$

is chromophorically related to aethusin

$\lambda_{max.}$ nm	242	252	~302	316	332·5
$\epsilon \times 10^{-4}$	1·12	2·0	3·45	6·02	6·25

Further extension of the chromophore occurs in a hydrocarbon (*h*) from *Achillea* and *Anthemis* spp.

$CH_3—(C≡C)_3—(CH=CH)_2—CH=CH_2$

$\lambda_{max.}$ nm 283 311 331 356

and a tetrayne diene (*i*) occurs in *Centaura* spp:

(*i*) $\lambda_{max.}$ nm 258 271 287 314·5 336 360·5 389·5

A tridecatetraynedienol was obtained from *Bidens leucanthus* L.

$CH_2=CH—(C≡C)_4—CH=CH—CH_2OH$

It has four intense peaks in the region 250–290 nm (ϵ $5–11 \times 10^{-4}$) and four very well defined peaks near 341, 336, 361 and 392 nm, ϵ *ca.* 10 000. When the terminal CH_2OH is replaced by CHO, the conjugation is further extended and there is good resolution from 230 to 305 nm (ϵ $5–8 \times 10^{-4}$) and four deep bands with peaks near 332, 348, 373 and 404 nm with ϵ values $1–2 \times 10^{-4}$. The infrared spectra of these compounds show absorption peaks:—

	cm^{-1}
—C≡C—	2190–2000, 2120
$CH=CH_2$	1830–1850, 965, 910
CH=CH	1615–1630, 945–950, 980–987
$(CH=CH_2)$ *trans*	980, 946
$≡C—CH_3$	1385
$(CH=CH)_2$	1460, 1423

It should be noticed that compounds possessing other chromophores may fail to display evidence of the presence of acetylenic linkages. Thus Bohlmann and his colleagues found that comparison of the spectra of butin-(2)-yl-isocoumarin and isocoumarin itself showed the latter to be dominant. Similarly frutescin and desmethylfrutescin displaced the dominance of the aromatic chromophore. A single CH_2 group provides very considerable isolation of the chromophores.

$CH_2—C≡C—CH_3$

O

O

butin-(2)yl-isocoumarin

$\lambda_{max.}$ nm	228	240	255	266	280	318
$\epsilon_{max.} \times 10^{-3}$	30	20	8·5	10·5	9·0	4·0

$CH_2—C≡C—C≡C—R$

$COOCH_3$

OCH_3

$R=CH_3$ (frutescin) R=H (desmethylfrutescin)

280	280	(in ether)
2·9	2·7	

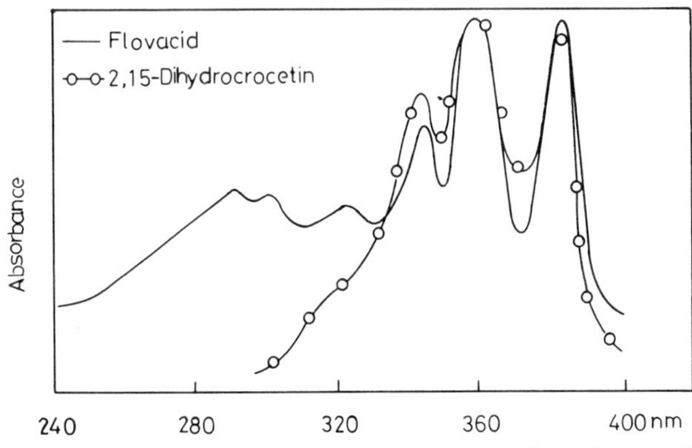

Figure 3.15 Ultraviolet absorption spectra in ethanol of flavacid and 2,15-dihydrocrocetin, HOOC—CH(Me)—CH:CH—CH:C(Me)—CH:CH—CH:CH—C(Me):CH—CH:CH—CH(Me)COOH. The absorbance of the antibiotic is arbitrarily shown as equal to that of the reference compound (after Oroshnik *et al.*, 1955)

A number of antibiotic substances are poly-ynes, e.g. mycomycin:

$HC≡C—C≡C—CH=CH=CH—CH=CH—CH_2COOH$

agrocybin:

$CH_3—C≡C—C≡C—C≡C—CONH_2$

diatretyne:

$HOOC—CH=CH—C≡C—C≡C—CONH_2$

and fumigauin yields on mild hydrolysis

HOOC—CH=OH—CH=CH—CH=CHCOOH

Nystatin, flavacid and candidin are conjugated polyenes rather than polyenynes (see page 61):

	$\lambda_{max.}$ nm			chromophore
Nystatin	292	3045	318	(tetraene)
Flavacid	341	358	379	(hexa-ene)
Candidin	363	383	406	(hepta-ene)
Fumigauin	322	336	357	(tetraendioic acid)

Figs. 3.15 and 3.16 show how the spectrum of flavacid resembles that of 2,15-dihydrocrocetin and nystatin that of methyl α-parinarate.

Figure 3.16 Ultraviolet absorption spectra in ethanol of nystatin and methyl α-parinarate (after Oroshnik *et al.*, 1955)

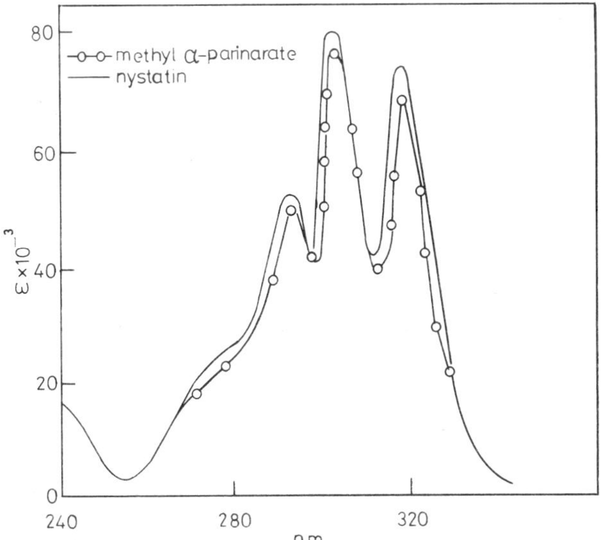

3.9 Ultraviolet Spectrophotometry of Fatty Acids

The carboxyl group of fatty acids gives rise to weak short wavelength absorption ($\epsilon_{max.}$ *ca.* 50); myristic acid has for example $\lambda_{max.}$ 205 nm (Rusoff *et al.*, 1945) but in some branched-chain fatty acids this selective absorption is stronger (Cason & Sumbell, 1951). For most saturated fatty acids, absorption on the long wave side of 215–220 nm is however negligible. Oleic and other monoethenoid fatty acids absorb strongly at about 180–185 nm. Dienoic and trienoic fatty acids in which the double bonds are separated by at least one CH_2 group (e.g. linoleic and linolenic acids) display absorption spectra in which the double bonds function additively. The data on the common polyunsaturated acids in which the double bonds are not conjugated sometimes show departures from additivity in the region 205–215 nm, but tentative generalizations (Barnes *et al.*, 1945) have been questioned (Silk & Hahn, 1954; Whitcutt & Sutton, 1956). There does seem to be a fairly regular increase in end-absorption with increasing unsaturation, but the necessary purification of the acids is not easy.

Conjugated poly-ene acids show much more marked effects. Table 3.21 shows how conjugated diene acids have $\lambda_{max.}$ near 230 nm, triene acids near 270 nm and tetraenoic acids around 300 nm (see also Fig. 3.17).

Some comments may serve to fill in the background. Stillingic acid (deca-2:4-dienoic acid) has the chromophoric group:

—CH=CH—CH=CH.COOH

and the 230 nm band of the isolated diene chromophore —CH=CH—CH=CH is displaced to 260 nm. The value of $\epsilon_{max.}$ (24 000) is low compared with that shown by the dodeca homologue containing the same chromophore (33 400). Synthetic α-γ-dienoic acids (Crossley & Hilditch, 1950; Crombie, 1955; Crombie & Manzoor-I-Khuda, 1956; Allan *et al.*, 1955) also show a high value of $\epsilon_{max.}$. The spectra of the conjugated triethenoid acids all show a degree of resolution not displayed by stillingic and related acids. The spectra moreover depend to a marked degree on *cis-trans* isomerism. The all-*trans* forms have the highest molecular extinction coefficients while on the wavelength axis the peaks for *cis* isomers show displacements towards longer wavelengths.

The first acetylenic fatty acid to be isolated was octadec-6-ynoic (tariric) but selective absorption in the region >220 nm appears only when the triple bond is conjugated with another unsaturated centre.

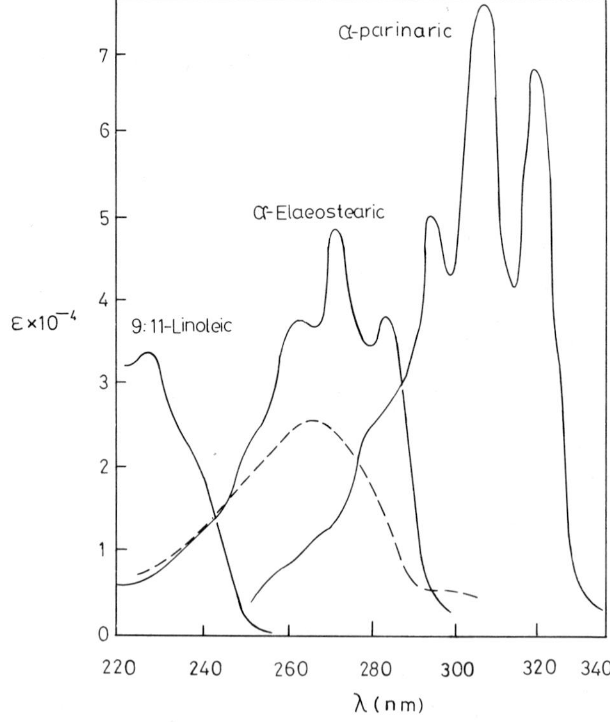

Figure 3.17 Absorption spectra of *trans-trans* 9:11-linoleic, α-elaeostearic and α-parinaric acids in ethanol (from Research Association of British Paint, Colour and Varnish Manufacturers). -----, methyl deca-2:4-dienoate (from Crossley & Hilditch, 1950)

Boleka nuts (*Onguekoe gore* Engler) contain octadec-17-ene-9:11-diynoic acid (erythrogenic or isanic acid). The insulated 17-ene group has no effect on the ultraviolet absorption but the chromophore

—C≡C—C≡C—

is responsible for sharp peaks at 227, 238 and 254 nm in which the values of $\epsilon_{max.}$ (Table 3.21) are quite low (Black & Weedon, 1953; Seher, 1954). An ene-diyne impurity shows maxima at 227, 240, 253·5, 267 and 283·5 nm with a peak intensity at 267 nm of 17 400. Ximenynic acid, apparently identical with santalbic acid, had the

—C=C—C≡C—

chromophore with $\lambda_{max.}$ 229 nm, $\epsilon_{max.}$ 16 200 (Ligthelm *et al.*, 1952; Gunstone & McGee, 1954; Gunstone & Russell, 1954; Grigor *et al.*, 1955).

The molecular extinction coefficients given in different publications do not always agree closely. There are difficulties in attaining purity, especially in respect of separated *cis-trans* isomers and the sharpness of narrow bands means that recorded values of $\epsilon_{max.}$ can vary with slit width. The principal poly-unsaturated fatty acids occurring naturally are however those which, like linoleic, linolenic and arachidonic acids, have their double bonds 'insulated' by CH_2 groups. When the double bonds undergo rearrangement to form a conjugated system, marked selective absorption appears. In fact poly-unsaturated acids can be converted to conjugated isomers by heating in alkali and the intensity of the resulting ultraviolet absorption reflects the rather variable efficiency of the isomerization. For analytical purposes the phenomenon can however only give meaningful and reproducible results under very strictly standardized conditions. (See Polgár & Jungnickel, 1956).

Alkali isomerization of unsaturated fatty acids has been known for some time (Morton *et al.*, 1931; Gillam *et al.*, 1931; Dann & Moore, 1933) and Moore, (1937) established the fact that the acids of linseed oil gave rise to a conjugated triethenoid acid resembling ψ-elaeostearic acid (octadec-10:12:14-trienoic acid). The process was incomplete and a conjugated dienoic acid was formed at the same time (see also Kass & Burr, 1939). It appears that some poly-unsaturated acids exist that are 'non-conjugatable'. The easily

Figure 3.18 Absorption spectra in cyclohexane of geometrical isomers of octadeca-9:11:13-trienoic acid. α-elaeostearic, – – – – –; β-elaeostearic, ———; punicic acid, ... (from Research Association of British Paint, Colour and Varnish Manufacturers)

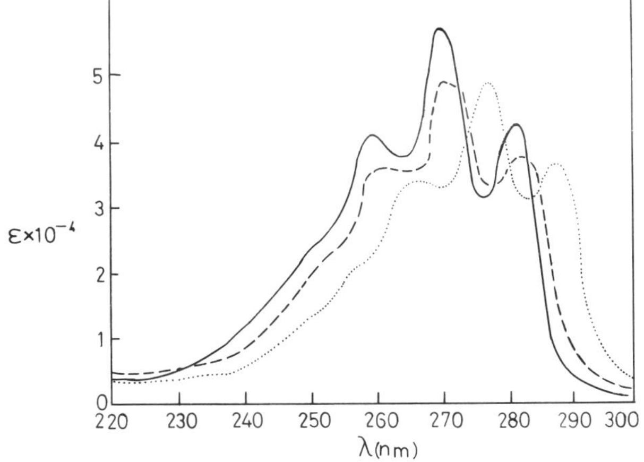

conjugatable acids all contain the grouping: —CH=CH—CH_2—CH=CH—, i.e. they are 1:4-dienes, but Japanese workers have argued that 1,5-dienoic acids also undergo isomerization to give conjugated dienes (Toyama & Tsucheya, 1935 *a*, *b*, *c*, 1936; Herb, 1955). Either 1:5-dienoic acids can yield conjugated acids on alkali isomerization or the structures attributed to the acids studied by the Japanese workers are erroneous.

Edisbury *et al.* (1935) found that alkali isomerization proceeded more quickly at higher temperatures and Kass & Burr (1939) substituted ethylene glycol for ethanol and by so doing raised the working temperature and thereby speeded up practical analytical work. A detailed procedure was described by Mitchell *et al.* (1943) for the determination of linoleic and linolenic acids in mixtures. Ethylene glycol containing potassium hydroxide (6·5% w/v) was heated in a glass tube to 180°C ± 1° under standard conditions. A small vial containing 0·1 g of fat or fatty acids was then dropped in, and after mixing, heating at 180°C was continued for 25 min. The tube was removed, cooled quickly and ethanol was added to a suitable volume, and the intensities of absorption were measured at 268 and 234 nm. Purified linoleic and linolenic acids were subjected to the same treatment and absorbances were determined under exactly the same conditions. The absorption at 268 nm was attributed wholly to the isomerized linolenic acid and its contribution to absorption at 234 nm was noted from the results with the linolenic acid standard (Fig. 3.19). This allowed a correction of the reading at 234 nm obtained on the sample, so that the contribution at that wavelength of the isomerized linoleic acid could be correctly evaluated. Mitchell *et al.* (1943) agreed with Moore (1937) that isomerization to give conjugated trienoic acid was very incomplete. In fact $E^{1\%}_{1cm}$ 268 nm for purified ψ-elaeostearic acid is about 2000 whereas the isomerized linolenic acid gave no more than 550. This meant that the analytical procedure was empirical and non-stoichiometric and that therefore experimental conditions needed to be rigidly specified and adhered to.

Arachidonic acid yields on isomerization a conjugated tetraenoic acid (Mowry *et al.*, 1942). From the isomerization mixture a crystalline material ($\lambda_{max.}$ 286, 300, 315 nm) was isolated (see Beadle & Kraybill, 1:44). Pitt & Morton (1957) discussed the different procedures in some detail. The variables include time of heating, temperature, choice of solvent, methods of

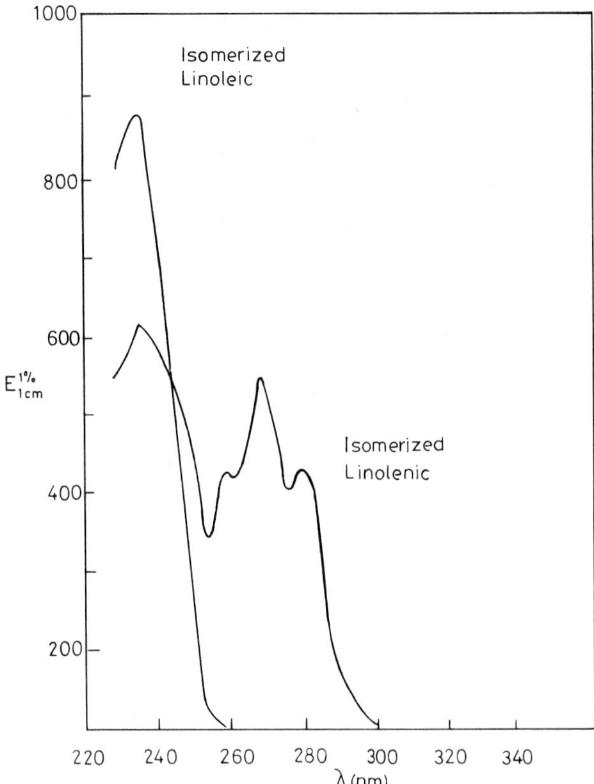

Figure 3.19 Absorption spectra of linoleic and linolenic acids after isomerization in glycol containing 6·5% KOH (from Mitchell *et al.*, 1943)

calibration, differences between *cis-trans* isomerides and correction procedures. Glycol or glycerol are the preferred solvents. Linolenic acid is isomerized optimally in 15 minutes at 170°C but linoleic acid takes 1 hour to isomerize to the maximum extent at 180°C (Hilditch *et al.*, 1945). Two parallel experiments, one at 170° and the other at 180°, were advocated and later improved (Hilditch *et al.*, 1951) but many workers have preferred the simpler isomerization at one selected temperature. Better preparations of all-*cis* acids allowed more satisfactory calibrations (Riemenschneider *et al.*, 1949; Herb *et al.*, 1951).

Corrections for irrelevant absorption in the isomerized products were introduced by Brice & Swain (1945). β-Elaeostearic acid is a typical conjugated triene with a peak at 268 nm and a minimum at 262 nm. Using the three wavelengths 262, 268, 274 nm (i.e. choosing two wavelengths equidistant from $\lambda_{max.}$) and assuming that from 262 to 274 nm any irrelevant absorption is linear the following equations holds (Fig. 3.20)

$$d-g=k_{268}-\tfrac{1}{2}(k_{262}+k_{274})=\Delta k$$

where k_x represents the specific extinction coefficient at x nm. The extinction at $\lambda_{max.}$ divided by Δk is here 2·8 for the pure substance so that $k_{268corr.}=2{\cdot}8\ (k_{268}-\tfrac{1}{2}(k_{262}+k_{274})$, the three k values being observed. Irrelevant absorption will elevate k_{268} but it will not increase Δk, a true measure of β-elaeostearic acid content. Refinements were introduced after

Figure 3.20 Correction for irrelevant absorption: β-elaeostearic acid (from Brice & Swain, 1945)

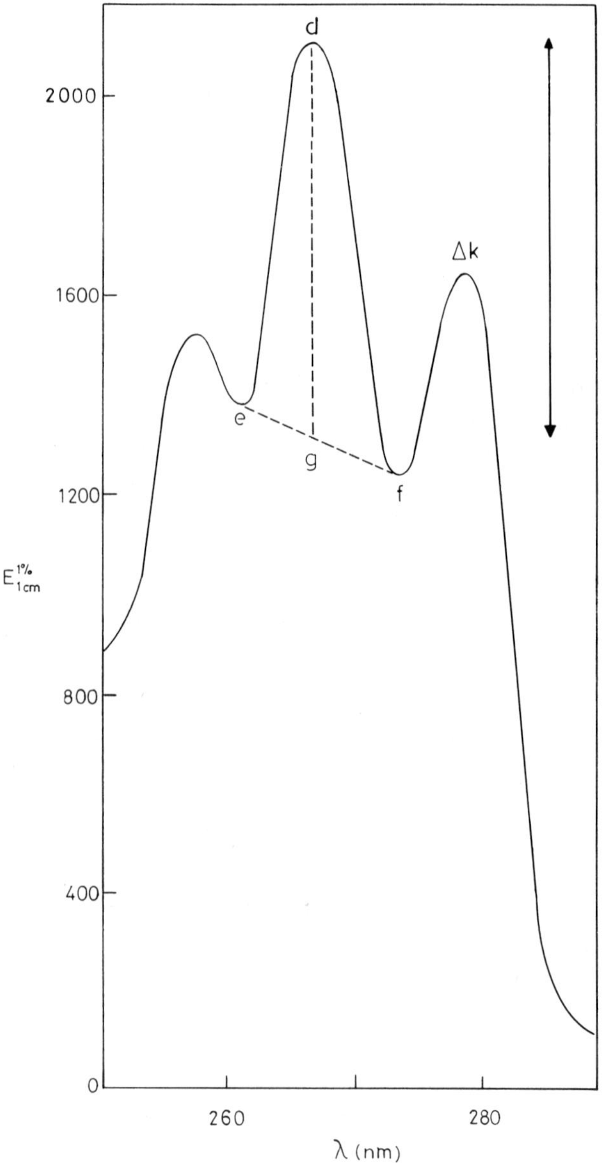

collaborative studies (Spectroscopy Committee, American Oil Chemists' Society, 1949, 1951, 1953, 1955; Brice *et al.*, 1952; O'Connor *et al.*, 1952).

Unsaturated acids up to hexa-enes have been isomerized by heating for 15 minutes in 21% potassium hydroxide (Herb & Riemenschneider, 1952, 1953) to give maximum selective absorption (see Fig. 3.21). Tables 3.21 and 3.22 from Pitt & Morton (1957) summarize the experience gained.

When the methods were developed, alkali isomeriza-

Table 3.21 Ultraviolet absorption data for fatty acids

	Configuration	Solvent*	$\lambda_{max.}$ nm	$\epsilon_{max.}$
Octadeca-9:11-dienoic	*trans-trans*	P	231·5	33 300
(9:11-linoleic)		E	231	33 300
	cis-trans	E	234	24 600
Octadeca-10:12-dienoic	*trans-trans*	P	231	32 200
(10:12-linoleic)		E	231	32 800
(as methyl ester)	*trans-cis*		232	28 700
	cis-cis	M	235	27 900
Octadeca-9:11:13-trienoic	*cis-trans-trans*	C	262	36 200
(α-elaeostearic)			271·5	48 100
			283	37 000
		E	261	36 000
			271	47 000
			281	38 000
(β-elaeostearic)	*trans-trans-trans*	C	259	49 300
			269	54 900
			280	41 700
		E	259	47 000
			268	61 000
			279	49 000
18-Hydroxyoctadeca-9:11:13-trienoic acid	*cis-trans-trans*	E	261	40 500
			271	52 000
(α-kamlolenic)			282	42 000
(β-kamlolenic)	*trans-trans-trans*	E	258	45 000
			268	60 000
			279	47 000
Ocatdeca-9:11:13:15-tetraenoic	2 or 3 *trans*	E	292	59 500
	1 or 2 *cis*		305	76 700
(α-parinaric)			319	69 200
(β-parinaric)	All-*trans*	E	288	—
			301	81 700
			316	—
2:4:6-Trimethyltetracos-2-enoic (mycolipenic)		M	214	13 200
Deca-2:4-dienoic (stillingic)	*trans-cis*	E	200	24 000
Octadeca-11-en-9-ynoic (ximenynic)	*trans*	C	229	16 200
(santalbic)		E	228	16 600
Octadec-17-en-9:11-diynoic		E	227	370
(erythrogenic, isanic)			238	344
			253	120
8-Hydroxyoctadeca-14-en-10:12-diynoic			230	*ca.* 6120
			240	6250
(isanolic)			252·5	6380
			265	8240
			279	6520

Abbreviated from Pitt & Morton (1957).

* P = light petroleum; E = ethanol; M = methanol; C = cyclohexane.

Table 3.22 $E^{1\%}_{1cm}$ values obtained from poly-unsaturated acids by standard alkali-isomerization procedures

Acid	$\lambda_{max.}$ nm	$E^{1\%}_{1cm}$ in glycol containing 6·5 KOH 25 minutes	21 KOH 15 minutes
Linoleic	233	921	916
Linolenic	233	616	475
	268	507	905
Arachidonic	233	569	397
	268	528	482
	315	215	606
Pentaenoic	233	—	415
(50% C_{20} 50% C_{22})	268	—	436
	315	—	697
	346	—	690
Hexaenoic	233	—	417
	268	—	522
	315	—	296
	346	—	277
	375	—	293

From Pitt & Morton (1957) based on Herb & Riemenschneider (1952, 1953) and Riemenschneider (1954)

Figure 3.21 Absorption spectra of eicosapentaenoate after isomerization by (A) 21% KOH in glycol and (B) a weak-alkali procedure (from Herb & Riemenschneider, 1952)

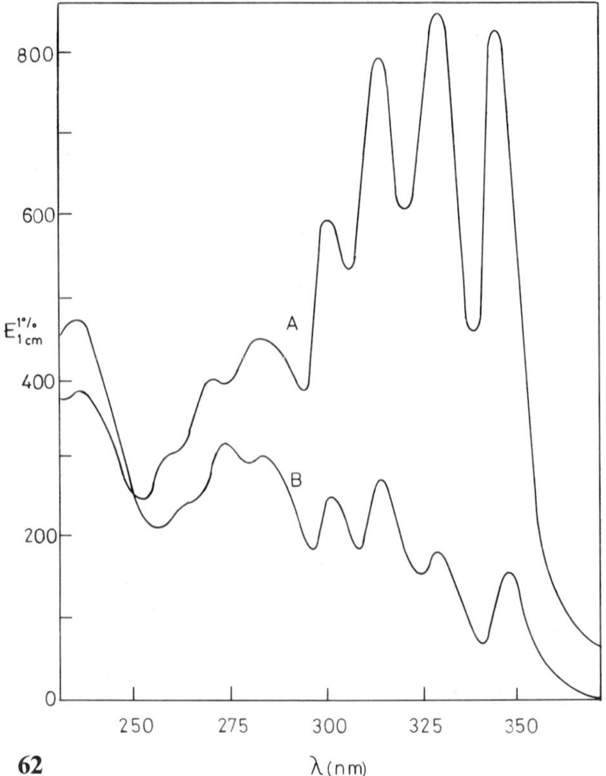

tion was admittedly empirical, but very serviceable procedures emerged. Moreover there was no better method available. Gas-liquid chromatography and other techniques have modified the position. It is probably true that for oils containing di- and trienoic acids, but little tetraenoic or pentaenoic acids, the spectrophotometric method can be made substantially accurate, but for highly processed oils and fish oils the analyst must be prepared to proceed warily and to use more than one approach.

The aim of the present account has been to present the basic spectroscopic considerations rather than to pursue each analytical problem to a conclusion. For this the original literature is helpful.

A biochemical problem of importance was broached by Dann *et al.* (1935) who fed cod liver oil, sardine oil, rape oil and linseed oil to cows under defined experimental conditions. It was observed that the butter fats produced showed much greater absorption at 230 nm and that there was a marked conversion of 'pro-absorptive' into 'absorptive' acids. A molecular rearrangement (rather than an oxidation) was postulated, The investigation, necessarily exploratory at the time, illustrates biochemical issues, some of which have been followed up and others remain to be clarified.

3.9.1 Other Spectroscopic Aspects of Fatty Acid Research

In studying fatty acids it is obviously necessary to establish molecular weights, the numbers and positions of olefinic linkages, *cis* and *trans* configurations, presence of branched chains and of hydroxyl groups and their location in the molecule. Thanks to the use of spectroscopic methods the task is now much less formidable than formerly.

In the infrared region fatty acids show strong absorption near 3030 cm^{-1}. In methyl esters and the corresponding alcohols this is ascribed to the C—H vibration but in free fatty acids there is a contribution due to hydrogen bonding with the OH of the carboxyl group. A band at 1449 cm^{-1} (single or double) due to C—H bending occurs regularly (1449, 1429 approx.). The CH_3 group gives rise to a deformation vibration at 1389–1351 cm^{-1}. When there are four or more directly linked CH_2 groups a peak near 719 cm^{-1} appears (CH_2 wagging). Long chain fatty acids have bands at 2703 and 934 cm^{-1} due to association through OH; they disappear on esterification. A doublet at 1282 and 1250 cm^{-1} is ascribed to the C—O link in carboxyl and a strong band at 1717–1711 cm^{-1} is due to the C=O bond in carboxyl. In methyl esters this is displaced to 1748–1739 cm^{-1} and a triplet 1250, 1205, 1176 cm^{-1} appears regularly in the methyl esters of long chain fatty acids. In triglycerides a strong band near 1163 cm^{-1} is flanked by weaker bands at 1250 and 1111 cm^{-1}. Fatty alcohols show a peak at 1053 cm^{-1}.

The characteristic olefinic infrared absorption occurs near 6·0–6·1 μm (1650 cm^{-1}) but only when there is a change in dipole moment. It is therefore absent from or very weak in infrared absorption of symmetrical compounds like ethylene, but it occurs in all Raman spectra. The chromophore

—C=C—C=C—

shows an unsymmetrical stretching vibration at 1600 cm^{-1} in the Raman spectrum, but for a number of reasons the 1600–1650 cm^{-1} absorption is not very serviceable in fatty acid studies. The grouping R CH=CH_2 gives rise to peaks at 995 cm^{-1} and 913–909 cm^{-1} whereas R′R″CH_2 has a sharp peak at 889–890 cm^{-1}. The *trans* configuration

```
R       H
 \     /
  C=C
 /     \
H       R
```

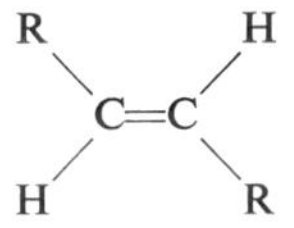

shows a peak at 963–976 cm^{-1} consistently. The *cis* configuration seems to give rise to a more variable peak at 714 to 680 cm^{-1}, which is less useful. The grouping R′R″C=CHR‴ has again variable absorption at 780–840 cm^{-1}.

The measurement of molecular extinction coefficients in the infrared region is not as yet very reproducible from one laboratory to another, but the 10·36 μm (965 cm^{-1}) absorption affords a good example of the utility of intensity measurements. This band was shown (Rasmussen *et al.*, 1947) to be due to *trans* forms, and Shreve *et al.* (1950) verified this with some accuracy for acids, methyl esters and alcohols containing isolated *trans* ethylenic linkages. Using a thin layer (0·1054 cm) of liquid the estimated absorption (ϵ) at 965 cm^{-1} for oleic, petroselinic, palmitic and stearic acids was $35{\cdot}5 \pm 2$, whereas for elaidic and petroselaidic acids it was 157 ± 1 (difference 122). Similarly for methyl oleate and stearate, ϵ was about 10 and for methyl elaidate or petroselaidate about 131–134 (difference 121–124). For oleyl and stearyl alcohol ϵ was about 17 as against 140 for elaidyl alcohol (difference 123). Triolein showed 74, trimyristin 62 and trielaidin 420·6 (difference 115–119 per acid residue). These results indicate that there is weak unselective absorption around 965 cm^{-1} for the *cis* forms and strong selective absorption for the *trans* forms. Shreve *et al.* (1950) have arrived at simple equations for the analysis of mixtures. It should however be emphasized that the method needs to be calibrated by the actual user.

Holman & Edmondson (1956) studied the absorption spectra of fatty acids in the near infrared region and obtained very interesting results. In this part of the spectrum (0·9–3 μm) the spectra exhibit overtones and combinations of more familiar fundamental vibrations. Thus the fundamental acidic OH (bonded) frequency is about 3030 cm^{-1} and the first overtone is 6250 cm^{-1}. Most of the absorption peaks can be interpreted as overtones and combinations but the utility is empirical. The conjugated unsaturation of β-elaeostearic acid gives rise to a triplet 4348, 4310, 4290. A more general result is that a band at 2·14–2·15 μm (4650–4670 cm^{-1}) is characteristic of the *cis* configuration and increases in intensity with increasing *cis* unsaturation. In that region the *trans* configuration gives rise to no selective absorption (see also Goddu, 1957). Terminal unsaturation is detectable by peaks at 6170 and 4760 cm^{-1}. In some respects quantitative analytical work is easier with absorption bands in the region 0·8–2·8 μm than 2·8 to 15 μm. (See also p. 209.)

Mass spectrometry in fat analysis has been reviewed by Stenhagen (1968). The normal fatty acid will give the molecule ion, particularly well if the methyl ester is used. Very strong readings are obtained at m/e = 74 and 87. The former depends on 2–3 cleavage accompanied by rearrangement of one H atom from the rest of the molecule

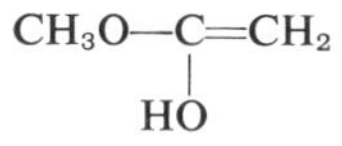

```
CH3O—C=CH2   m/e 74
     |
     HO
```

and a series of ions

$[CH_3OCO(CH_2)_n]^+$ (*n* > 1 is formed)
m/e 87, 101, etc.

Clearly unsaturated acids will have their own cleavage patterns. Thus methyl linoleate with 5,6-cleavage can give a fragment with m/e 111.

Nuclear magnetic resonance spectra of fatty acids have been summarized by Hopkins (1965, 1968). Table 3.23 shows some of the main signals. Gunstone & Ismail (1967) examined all of the sixteen C-18 monoenoic acids and were able to show that n.m.r. spectra could be used to identify the four isomers with

Table 3.23 Nuclear magnetic resonance of common fatty acid methyl esters

	τ	(a)	(b)	(c)	(d)	(e)
Methyl oleate		4·72	6·38	7·83	8·72	9·12
Methyl linoleate		4·68	6·38	—	—	9·12
Methyl linolenate	(3·5)	4·62	7·2	7·5	8·67	9·03
Methyl ricinoleate		4·62	6·39	7·93	8·69	9·12

(a) olefinic H, (b) OCH_3, (c) CH_2 CO, (d) CH_2, (e) CH_3—C (*non-allylic*)

unsaturation closest to the carboxyl group and the four with unsaturation closest to the methyl group. 3-Hexadecenoic was recognized in *Helenium* seed oil and 4-hexadecenoic acid in *Lindera* seed oil. In 16-octadecenoic acid the 9.1 shift for terminal methyl is displaced to 8·38 as a result of a deshielding by the ethylenic grouping. Christie & Holman (1967) studied the C-18 dienoic acids with the grouping

$$—CH{=}CH—CH_2—CH{=}CH—$$

There are altogether 13 isomers, and of these five with unsaturation near the middle of the carbon chain (6,9 to 10,13) cannot be distinguished by nuclear magnetic resonance, but the remaining eight can all be identified.

Bagby *et al.* (1964) isolated 5,6-octadecadienoic acid from *Leonotis* seed oil and observed a band at 4·9–5·0 τ for the allene protons CH=CH=CH. An acid (octadeca-5,6-16-trienoic) isolated from *Lamium* seed oil by Mikolajczak *et al.* (1967) also shows the allenic signal.

A fatty acid with the following structure

$$HC{\equiv}C—(CH_2)_n—\underset{\diagdown\ \diagup \atop CH_2}{C{=}C}—(CH_2)_nCOOCH_3$$

was described by Jevans & Hopkins (1968). The terminal proton A gave $\tau=8{\cdot}2$ (infrared 3320 cm^{-1}) and $\tau=9{\cdot}2$ for the cyclopropene protons. Hopkins (1968) records that autoxidation of soyabean oil produces *inter alia* dec-1-yne (τ triplet 8·2, α-proton 7·9 [cf. Smouse *et al.*, 1965]).

Conjugated dienes exhibit a multiplet in the region $\tau=4{\cdot}2$ (Tallent *et al*, 1966). *Cis-trans* isomers have been studied by Schaumberg & Bernstein (1968) (see also p. 154) and Tulloch (1966) has reported on hydroxystearic acids.

Mono-, di- and triglycerides can readily be distinguished (Serdarevich & Carroll, 1968) (see Table 3.24) and glyceryl ethers have also been studied (Wood & Snyder, 1967). The assignments for protons in the cyclopropane ring have been established (cf. Hopkins, 1968; Minnikin, 1966). Goldfine (1964) examined the aldehydes from plasmalogens and found that some of the aldehydes (obtained as dimethylacetals)

Table 3.24 Glyceryl proton peaks

$$\begin{array}{l} H_2C—O\ \ X \\ \ \ \ | \\ \ HC—O\ \ Y \\ \ \ \ | \\ H_2C—O\ \ Z \end{array}$$

X	Y	Z				
COR	H	H	—	5·7	6·05	6·47
H	COR	H	4·85	—	—	6·4
COR_1	COR_2	H	4·85	5·7	—	6·4
COR_1	H	COR_2	—	5·7	—	—
COR_1	COR_2	COR_3	4·85	5·7	—	—

$$C\begin{array}{l}\diagup OCH_3 \\ \diagdown OCH_3\end{array}\quad 6{\cdot}78\tau \qquad CH\begin{array}{l}\diagup O \\ \diagdown O\end{array}\quad 5{\cdot}74\tau$$

had a cyclopropane ring (*cis* ring 9·4, 10·37 τ)

Slotboom *et al.* (1967) synthesized a plasmalogen and found that while the natural compound was *cis* (olefinic proton a doublet 4·1 τ, J 6·4 Hz) the synthetic substance was *trans*, 3·7 τ, J 12·5 Hz).

These considerations can be brought to a focus by considering the work of Hopkins *et al.* (1969) on a new trienoic acid present in pollen and attractive to the honey bee.

Lepage & Boch (1968) had proved that an attractant present in some pollens was a free fatty acid. Hopkins *et al.* fractionated the lipid extract by adsorption chromatography on silicic acid. The mixed acids were esterified with methanol using boron trifluoride etherate as catalyst. The infrared absorption spectra showed the expected peaks for unsaturated acids with additions at 980 and 1660 cm^{-1} and the n.m.r. spectrum had uncommon signals at 2·8–3·3, 4·1–4·4 and 6·28 τ. The esters were further fractionated by gas-liquid chromatography on polar and non-polar columns and the presence of an unidentified C_{18} acid was indicated. Further fractionation by countercurrent (monitored by g.l.c.) gave a small amount of the unknown methyl ester, and this on hydrogenation gave methyl stearate. The infrared spectrum of the

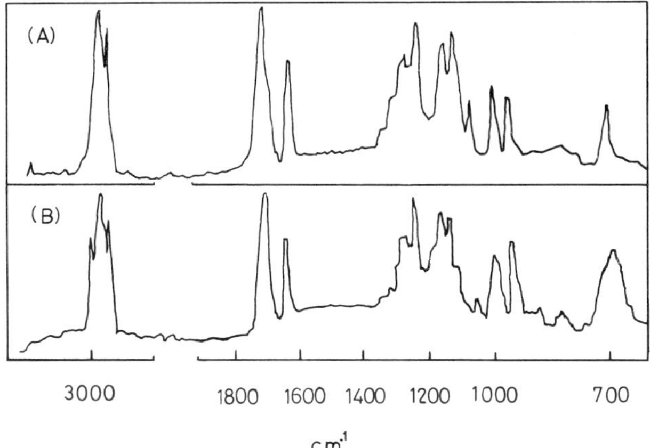

Figure 3.22 The infrared spectra of (A) methyl *trans*-2-octadeconoate and (B) methyl ester of the pollen acid, $CH_3(CH_2)_4CH{=}CHCH_2CH{=}CH(CH_2)_5CH{=}CHCO_2H$, octadeca-*trans*-2, *cis*-12-trienoic acid ($C_{18}H_{30}O_2$) (from Hopkins *et al.*, 1969)

unknown methyl ester showed a strong peak at 1040 cm^{-1}. Since this also occurred in methyl *trans*-2 decenoate (Fig. 3.22) the peak was assigned to the *trans* $\alpha\beta$-unsaturated ester group. The carbonyl (ester) peak was displaced from 1735 to 1725 cm^{-1} by the $\alpha\beta$-olefinic linkage. In addition peaks at 7100 cm^{-1} and 3000 cm^{-1} were present (*cis* HC═CH—). There was no peak at 965 cm^{-1}, so that the isolated double bonds were *cis* rather than *trans*. The mass spectrum of the methyl ester had a molecular ion peak at mass 292 corresponding with a methyl octadecatrienoate.

The n.m.r. spectrum (Fig. 3.23) had signals around 3·0 and 4·2 τ characteristic of two *trans* $\alpha\beta$-olefinic protons (Hopkins, 1965; Purcell *et al.*, 1966). The

Figure 3.23 The n.m.r. spectrum at 60 mHz in $CDCl_3$ (D=deuterium) of the methyl ester of the pollen acid (from Hopkins *et al.*, 1969)

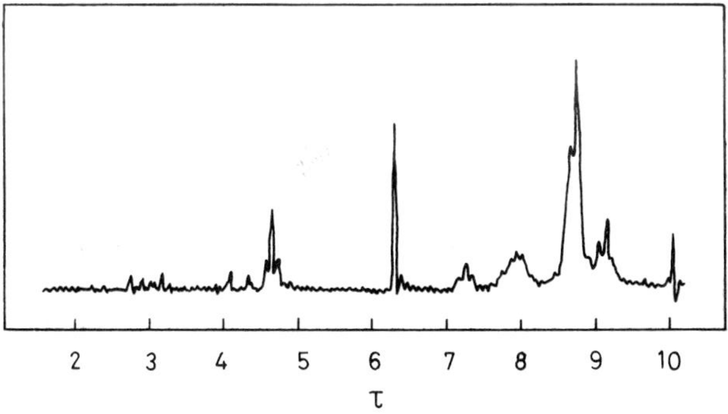

signal at 4·63 τ corresponded with 4 protons on non-conjugated olefinic groups with the *cis* configuration (Schaumberg & Bernstein, 1968). The signal at 7·22 τ (2H) corresponded with ═CH.CH_2—CH═ (diallylic methylene). The terminal methyl group had an important triplet near 9·1 τ, The ester peak was displaced (from 6.38) to 6·28 τ because of the α-olefinic group.

Oxidative cleavage of the new acid gave hexanoic as the sole monobasic acid on degradation, so that the ethylenic linkage furthest from the carboxyl group must be at C_{12}. The absence of a peak at 6·60 τ excluded a 2,5 grouping for the diene (Christie & Holman, 1967) and a 12,15 grouping in a C_{18} chain would have given rise to a well marked triplet at 9·0 τ. Since neither signal was present the only alternative structure for the 1,4-diene group was at positions 9 and 12, so that the acid must be octadeca-*trans*-2,*cis*-9,*cis*-12 trienoic acid.

$$CH_3(CH_2)_4CH\overset{cis}{=}CHCH_2CH\overset{cis}{=}CH(CH_2)_5CH\overset{trans}{=}CH.CO_2H$$

Biological tests showed that very small amounts of the acid were strikingly attractive to foraging bees. Other insect attractants possess a *trans*-2 double bond (Jacobson, 1965; Shearer & Boch, 1966).

A surprising development has recently occurred. New hormones from some reproductive parts of plants have been reported by Mitchell *et al.* (1970, 1971). Mandava & Mitchell (1972) isolated and characterized hormones from the pollen of rape (*Brassica naphus* L). They have been given the name *brassins*.

The pollen was extracted by means of ether and fractionation was effected by chromatography on silica gel to yield a mixture of biologically active (Mitchell & Livingstone, 1968) products. Column chromatography, using Florisil (activated magnesium silicate) resulted in fractions varying in biological activity. The most active fractions each gave a single spot in thin layer chromatography with several solvent systems. The most plentiful brassins showed strong end absorption below 200 nm in methanol, consistent with the presence of the isolated ethylenic chromophore. In alkali this absorption was replaced by a peak at 224 nm due probably to a conjugated diene chromophore brought into being by alkali isomerization (see p. 62). The infrared absorption pointed to the presence of OH (strong 3400 cm^{-1}) and to an ester linkage (1720 cm^{-1}) while CH stretching vibrations (2700, 2900 and 3050 cm^{-1}) indicated an aliphatic chain. The n.m.r. spectrum showed τ=8·74 (CH_2 protons) 9·1 (CH_3) and 7·1–8·5 (olefinic protons attached to carbons near oxygen atoms or double bonds). Mass spectroscopy on the trimethylsilyl derivative showed m/e 728 for the major component with m/e 638 and 548 after the loss of TMS OH groups. The fragmentation pattern indicated the presence of a TMS sugar residue. The evidence showed that the hormone contained equimolar proportions of glucose and linolenic acid (18·3) linked in the form of an ester. The minor components of the brassin mixture were glucose esters derived from palmitic, stearic, oleic and linoleic acids. Saponification of brassins followed by methylation of the acids allowed unequivocal detection of acids by gas chromatography. The glucose was set free by mild acid hydrolysis and characterized by processes leading to the formation of sorbital hexa-acetate.

Brassins are thus made up of fatty acids linked by an ester bond to glucose at carbon 1 in the β-form. They have a role in both cell elongation and cell division in intact plants. Milborrow & Price (1973) have severely criticized this work. They regard the published evidence as inadequate to justify the view that the brassins are a novel complex of fatty plant hormones. Further they say that biological activity has not been correlated with physical measurement to attribute safely the 'brassin activity' to any specific substance.

Streptomyces species synthesize a number of poly-ene antibiotics that inhibit the growth of some fungi and have certain clinical uses.

The organism *Streptomyces noursei* produces nystatin (fungicidin) $C_{46}H_{77}O_{19}N$ while amphotericin B $C_{46}H_{73}O_2N$ is produced by another species. Each consists of a C_{40} moiety linked to mycosamine, an amino sugar. The ultraviolet absorption of nystatin shows that it contains a tetra-ene grouping, while that of amphotericin B indicates a heptaene fragment (Dutcher, Walters & Wintersteiner, 1955; Donovick *et al.*, 1956). The peaks for nystatin occur at 290, 304 and 319 nm with ϵ values $\times 10^{-3}$ of 60, 80 and 65 respectively. Very similar absorption peaks are shown for antimycin A, rimocidin and chromin (Oroshnik *et al.*, 1955).

Pimaracin ($C_{34}H_{49}O_{14}$) contains mycosamine and a large lactone ring structure with a conjugated tetraene portion (Patrick, Williams & Webb, 1958; Zondag, Posthuma & Berends, 1960). The ultraviolet absorption in water shows:

$\lambda_{max.}$ nm	215	~281	292	305	319
$\epsilon_{max.} \times 10^{-3}$	1·585	2·51	39·8	6·31	6·31

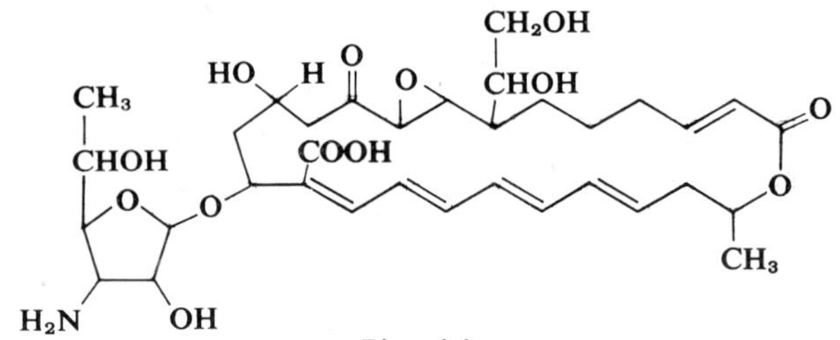

CH₂OH CHOH HO H O O CH₃ CHOH COOH O O O CH₃ H₂N OH

Pimaricin

$OHC—(CH=CH)_5—\overset{12}{C}H_2\overset{13}{C}(OH)CH_3$ $HOCH_2—(CH=CH)_5C(OH)CH_3$

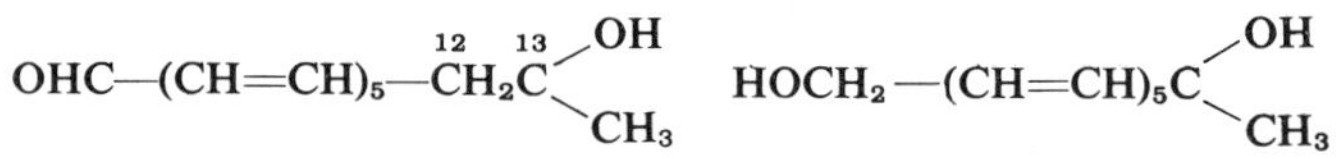

At about 222 nm the value of ϵ is near 22 400 and the =C—C=O chromophore clearly contributes to this. Under the action of alkali, pimaracin yields 13-hydroxytetradecapenta-en-al with $\lambda_{max.}$ nm in 375 in methanol. Dodecapenta-en-al shows peaks at 263 and 377 nm, ϵ 9500, 51 000 in ethanol (Hausser, Kuhn, Smakula & Hoffer, 1935; Blout & Fields, 1948). The product from pimaricin is reduced to borohydride to the 1,13-hydroxy derivative with absorption maxima at 313, 328 and 346 nm.

Fungichromin, obtained from a strain of *Streptomyces cellulosae* has the formula $C_{35}H_{60}O_{13}$ and shows

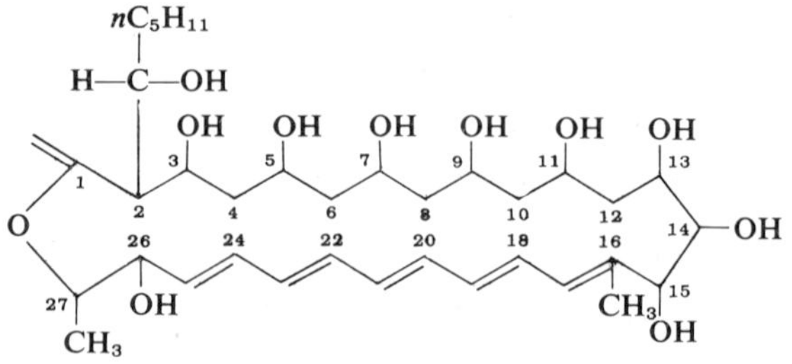

nC_5H_{11} H—C—OH OH OH OH OH OH OH 3 5 7 9 11 13 1 O 2 26 4 24 6 22 8 20 10 18 12 16 14 OH 27 OH CH₃ 15 CH₃ OH

Fungichromin

ultraviolet absorption characteristic of the conjugated penta-ene chromophore. When treated with sodium carbonate and oxidized by sodium periodate, fungichromin gives a product identified as 2-methyl-2,4,6,8,10-dodecapenta-enedial (*a*) which on oxidation by means of silver oxide yields the corresponding dicarboxylic acid (*b*) or on reduction by sodium borohydride gives the diol (*c*).

There is a close resemblance between the selective absorption of fungichromin and that of the diol. The dicarboxylic acid shows the shift characteristic of the extended chromophore and the dial illustrates how this is carried further (Cope *et al.*, 1958, 1962).

Lagosin (Ball *et al.*, 1957) $C_{35}H_{58}O_{12}$ like fungichromin has a macrocyclic ring and the same chromophore (Dhar *et al.*, 1964) which also occurs in filipin (deoxylagosine). Pentamycin (Umezawa & Tanaka, 1958) is a similar antibiotic showing the characteristic spectrum and several other less fully characterized penta-ene antibiotics are known (Korzybsky *et al.*, 1967).

(*c*) $HOCH_2—C(CH_3)=CH—(CH=CH)_4CH_2OH$

$\lambda_{max.}$ nm	$\epsilon_{max.} \times 10^{-3}$	in ethanol
240	*ca.* 3·5	
319	65·3	
334	107	
351	108·5	

(*a*) $O=C(H)—C(CH_3)=CH—(CH=CH)_4C(H)=O$

$\lambda_{max.}$ nm	$\epsilon_{max.} \times 10^{-3}$	in ethanol
222	7·64	
284	6·2	
384	77	
402	75	

(*b*) $O=C(HO)—C(CH_3)=CH—(CH=CH)_4C(OH)=O$

$\lambda_{max.}$ nm	$\epsilon_{max.} \times 10^{-3}$	in ethanol
211	8·0	
270	4·4	
350	46	
364·5	86	
382·5	81	

Fungichromin

$\lambda_{max.}$ nm	$\epsilon_{max.} \times 10^{-3}$	in ethanol
311		
322·5		
338·5	97·7–99·7	
357	($E^{1\%}_{1cm}$ 1420–1450)	

Pentamycin

$\lambda_{max.}$ nm	$E^{1\%}_{1cm}$	in ethanol
322	899	
338	1450	
356	1500	
~308–313		

Several conjugated hexa-ene antibiotics are known. They all show three absorption peaks at 337–341, 356–358 and 376–380 nm. Mycelin is anomalous and may have two chromophores.

			$\lambda_{max.}$ nm		
Crytocidin	290	305	341	358	380
Flavacid			341	358	379
Fradicin			340	356	376
Mediocin			337·5	356	377
Mycelin	243	294	320	335	373

Hepta-ene antibiotics include:

		$\lambda_{max.}$ nm		
Hanycin	345	363	383	406
Candidin		363	383	406
Amphotericin B		360	380	409
[Amphotericin A		295	305	320]
Candicidin		359·5	379·5	401·5
Ascosin		358	377	399
Trichomycin		358	377	399
Perimycin		361	383	406

Candidin ($C_{46}H_{76}O_{16}N$) forms a sodium salt soluble in water and the solution shows absorption peaks at 234, 282, 345, 360, 380 and 405 nm. On acidification the long wave peaks increase in intensity and 345 nm peak disappears. A second chromophoric group must be postulated but the structure is not fully known.

Trichomycin has a chromophore of the type

$CH_3CONH—C_6H_4—COCH_2^-$

which would be expected to produce absorption at 230–300 nm, compare

	$\lambda_{max.}$ nm	$\epsilon_{max.} \times 10^{-2}$
p-aminoacetophenone	234	6·7
(in ethanol)	318	20·3
Aniline (in ethanol)	234	9·37
	286	1·50

but this does not seem to have been observed (Korzybski *et al.*, 1967). Levorin and Candicidin also contain the *p*-aminoacetophenone chromophore and the expected bands are in fact seen. On the other hand perimycin contains a *p*-aminophenylacetone group:

$CH_3COCH_2—C_6H_4—NH_2$

and a careful comparison of the spectra of purified specimens of trichomycin, candicidin and perimycin would be interesting.

Numerous poly-ene and poly-yne groupings occur in a variety of antibiotics. The work of Celmer & Solomons (1952, 1953) on mycomycin and isomycomycin (derived from it by alkali isomerization) is of considerable spectroscopic interest.

Mycomycin ($C_{13}H_{10}O_2$) despite its evident unsaturation, shows a rather simple absorption spectrum with an inflection at 256 nm and two well-defined peaks:

$$\overset{12}{CH}{\equiv}\overset{11}{C}{-}\overset{10}{C}{\equiv}\overset{9}{C}{-}\overset{8}{CH}{=}\overset{7}{C}{=}CH{-}\overset{6}{CH}{=}\overset{5}{CH}{-}\overset{4}{CH}{=}\overset{3}{CH}{-}CH_2COOH$$

Mycomycin
(in diethyl ether)

$\lambda_{max.}$ nm	~256	267	281	$\lambda_{min.}$ nm	225	276
$\epsilon_{max.}\times 10^{-4}$	3·5	6·1	6·7		low	4·9

This spectrum would fit a triene or a triyne. The infrared absorption (in dioxan) gave the following peaks $\equiv$CH—H 3180, —C$\equiv$C— (disubstituted) 2200, CH=C=C= (allene) 1930 and —COOH 1730 cm^{-1},

The isomeric substance (in diethyl ether) exhibits

$$\overset{12}{CH_3}{-}\overset{11}{C}{\equiv}\overset{10}{C}{-}\overset{9}{C}{\equiv}\overset{8}{C}{-}\overset{7}{C}{\equiv}C{-}\overset{6}{CH}{=}\overset{5}{CH}{-}\overset{4}{CH}{=}\overset{3}{CH}{-}CH_2COOH$$

Isomycomycin

a much more complex ultraviolet absorption spectrum:

$\lambda_{max.}$ nm	246	257·5	267	287·5	305·5	324	347
$\epsilon_{max.}\times 10^{-4}$	2·4	5·8	11	1·4	2·7	4·1	3·4

The 2200 cm^{-1} peak for disubstituted —C$\equiv$C— has disappeared but the carboxyl peak at 1730 cm^{-1} persists. There is then a *prima facie* case for a tri-yne structure following the very high $\epsilon_{max.}$ (110 000) at 267 nm in the isomycomycin or its methyl ester.

Celmer & Solomons prepared dodeca-2,10-diene-4,6,8-tri-yne and obtained the following spectrum:

$$\overset{12}{CH_3}{-}\overset{11}{CH}{=}\overset{10}{CH}{-}\overset{9}{C}{\equiv}\overset{8}{C}{-}\overset{7}{C}{\equiv}\overset{6}{C}{-}\overset{5}{C}{\equiv}\overset{4}{C}{-}\overset{3}{CH}{=}\overset{2}{CH}{-}\overset{1}{CH_3}$$

2,10-dodecadiene-4,6,8-tri-yne

$\lambda_{max.}$ nm	245	259	269	280	305·5	325	348
$\epsilon_{max.}\times 10^{-4}$	25·4	43·1	76·4	11·4	18·4	25·8	20·1
				Δ cm^{-1}	1900	2000	2000

Model substances indicated that the chromophores:

—$(C{\equiv}C)_2$—CH=CH—
—CH=CH—C$\equiv$C—C$\equiv$CH—
—$(CH{=}CH)_3$

were each very plausible for mycomycin itself, but the existence of an allene group evidenced by a strong infrared peak at 1930 cm^{-1} had to be fitted into the interpretation of chromophores. *Trans* dehydromatricara ester $CH_3(C{\equiv}C)_3\overset{trans}{CH{=}CH}.COOCH_3$ showed absorption similar to that of isomycomycin

$\lambda_{max.}$ nm	244·7	255	268	283·5	300	324	343·5
$\epsilon_{max.}\times 10^{-4}$	7·18	10·62	8·6	1·5	3·2	4·5	6·0

The allene group (Celmer & Solomons, 1953) does not function like a conjugated diene:

	$\gamma_{max.}$ nm
R—CH=C=CH_2	170
R—C(—CO_2R)=C=CH_2	210–215
$(C_6H_5)_2C{=}C{=}C{-}(C_6H_5)_2$	267
$C_{10}H_7CH{=}C{=}C{=}CHC_6H_5$	298
R—CH=CH—R	185
R—C(—CO_2R)=CH_2	212–214
$(C_6H_5)_2C{=}CH_2$	250
$C_{10}H_7CH{=}CH_2$	295

If we now look at the structures of mycomycin (3,5,7,8-tridecatetra-ene-10,12-diynoic acid) $C_{13}H_{10}O_2$ and isomycomycin $C_{13}H_{10}$ it will be seen that mycomycin contains two chromophores, an ene-diyne and a tri-yne. The central carbon atom of the allene group acts as insulator. The isomerization to isomycomycin results in sustained conjugation of the unsaturated centres.

Bu'Lock, Jones & Leeming (1955) investigated the spectra of nemotin and its derivatives.

The lactone yields nemotinic acid and on treatment with alkali gives nemotin A, an ene-tiyne. Catalytic hydrogenation of nemotin gives undecanoic acid.

Nemotin	$\lambda_{max.}$ nm	208·5	236·5	249	278
	$\epsilon_{max.}\times 10^{-3}$	44·7	6·17	10·5	12·3

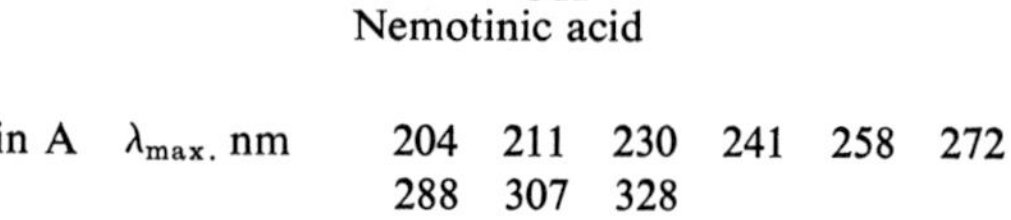

HC$\equiv$C—C$\equiv$C—CH=C=CH—CH—CH_2—CH_2
(lactone ring: CH—O———C=O—CH_2)

Nemotin

Nemotinic acid	$\lambda_{max.}$ nm	209	237	249·5	278·5
	$\epsilon_{max.}\times 10^{-3}$	44·7	5·62	10·0	12·3

HC$\equiv$C—C$\equiv$C—CH=C=CH—CH(OH)—CH_2CH_2COOH

Nemotinic acid

Nemotin A	$\lambda_{max.}$ nm	204	211	230	241	258	272
		288	307	328			
	$\epsilon_{max.}\times 10^{-3}$	26	23	59	85	2·8	6
		12	15·5	11			

HC$\equiv$C—$(C{\equiv}C)_2CH{=}CHCH_2CH_2COOH$

Nemotin A

Seher (1954) studied 8-hydroxyoctadec-14-ene-10, 12-diynoic acid (isanolic acid) and found

$\lambda_{max.}$ nm	230	240	252·5	265	279
$\epsilon_{max.} \times 10^{-3}$	6·12	6·25	6·38	8·24	8·52

The spectrum of nemotin is consistent with an ene-diyne chromophore well removed from the carboxyl group. The infrared spectrum of nemotin displays the 2130 cm^{-1} C≡C stretching vibration and the 1960 cm^{-1} allene C=C=C stretch. The terminal H—C≡C is indicated by a band at 3280 cm^{-1}, the γ-lactone C=O by 1790 cm^{-1} and the 1150 cm^{-1} C—O stretching vibration.

The change in ultraviolet absorption spectrum on formation of nemotin A indicates that the isomerization produces an ene-tri-yne chromophore.

Elaiomycin (Stevens, Gillis & French, 1958) has the chromophoric group

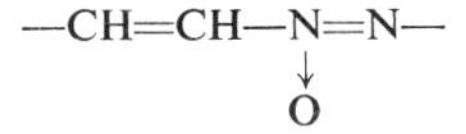

—CH=CH—N=N—
↓
O

with an ultraviolet peak for the monoacetate at 237·5 nm, $\epsilon_{max.}$ 11 000. The infrared absorption shows maxima at 1658 cm^{-1} weak (C=C) and 3425 cm^{-1} due probably to hydroxyl. The structure arrived at by degradation is

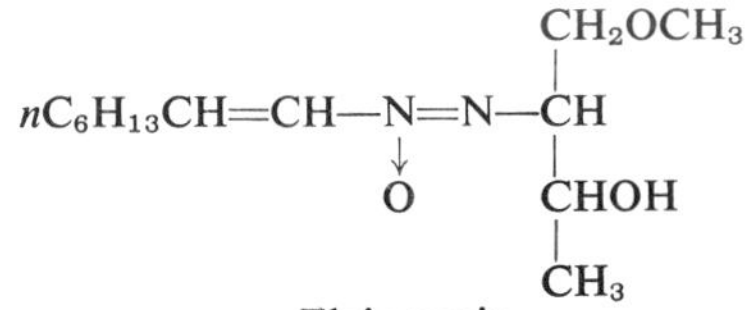

$nC_6H_{13}CH=CH—N(\rightarrow O)=N—CH(CH_2OCH_3)—CHOH—CH_3$

Elaiomycin

Sondheimer & Wolovsky (1959) studied a number of synthetic macrocyclic substances derived from a cyclotetracosa-octayne. Under the action of butoxide the molecule undergoes rearrangement to give cyclotetracosa-1,3,7,9,13,15,19,21-octaene-5,11,17,23-tetrayne.

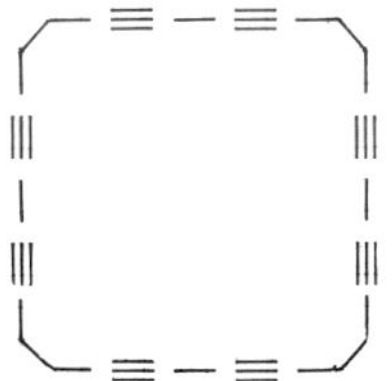

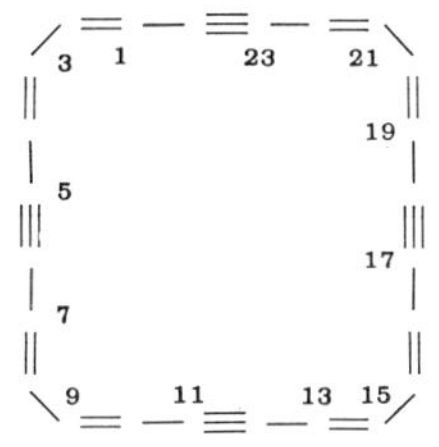

Here conjugation is fully sustained:

	Chloroform			Benzene	
$\lambda_{max.}$ nm	240	330	346	333	350
$\epsilon_{max.} \times 10^{-3}$	28	125	206	120	208

with weak absorption extending as far as 600 nm. Infrared peaks (KBr disc) occur at 3020, 2160, 1590, 1415, 1295, 1178, 1098, *972*, *930*, 846, 763, 755 cm^{-1}.

Partial hydrogenation to the dodecaene $C_{24}H_{24}$ gives

$\lambda_{max.}$ nm	350	363	512 to 750
$\epsilon_{max.} \times 10^{-3}$	195	201	1·74

Sondheimer & Gaoni (1960) prepared cyclotetradecaheptaene ($C_{14}H_{14}$) which does not exist in a planar form and is unstable, lacking aromaticity. In iso-octane it shows absorption peaks at 314 and 374 nm, with $\epsilon_{max.}$ values of 69 000 and 5700. There is no infrared peak at 2100 cm^{-1}. This can be contrasted with cyclotetradec-hexane-13-diyne ($C_{14}H_{12}$) which has peaks at

$\lambda_{max.}$ nm	312	365	392
$\epsilon_{max.} \times 10^{-3}$	92	7·1	4·7

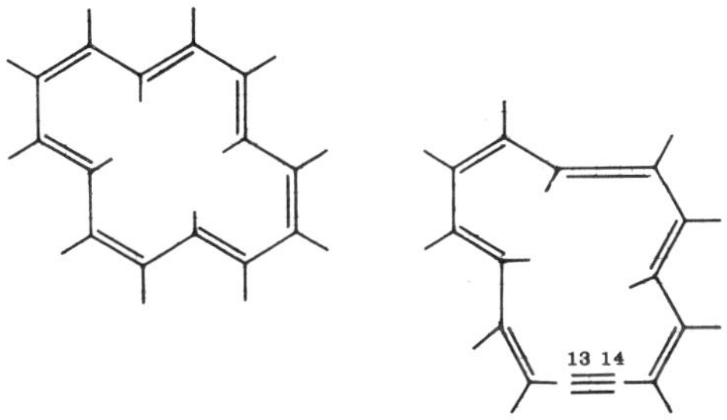

and an infrared peak at 2100 cm^{-1} (C≡C).

An isomeric form shows:

$\lambda_{max.}$ nm	309	402	410	424	586
$\epsilon_{max.} \times 10^{-3}$	175	6·9	9·5	23·6	2·9

with an infrared peak at 2125 cm^{-1}.

References

3.1

Burawoy, A. (1939). *J. chem. Soc.*, 1177; (1941). *J. Chem. Soc.*, 20.

Burawoy, A. (1930). *Ber. dtsch. chem. Ges.*, **63**, 3155.

Jaffé, H. H. & Orchin, M. (1962). In *Theory and Applications of Ultraviolet Spectroscopy.* John Wiley, New York, London.

Mason, S. F. (1961). *Quart. Revs. Chem. Soc.*, **15**, 287. (*Electronic Absorption*). (1963) *Ibid.*, **17**, 36. (*Spectra of Molecules*).

Murrell, J. N. (1963). In *The Theory of the Electronic Spectra of Organic Molecules.* Methuen, London.

Price, W. C. (1947). *Chem. Revs.*, **41**, 257.

Rao, C. N. R. (1967). In *Ultraviolet and Visible Spectroscopy Chemical Applications.* Butterworths, London.

Scott, A. I. (1964). In *Interpretation of the Ultraviolet Spectra of Natural Products.* Pergamon Press, Oxford, London, New York.

Sidman, J. W. (1958). *Chem. Revs.*, **58**, 689.

3.3

Bladon, P., Henblest, H. B. & Wood, G. W. (1952). *J. Chem. Soc.*, 2727.

Brinton, R. K. & Volman, D. H. (1951). *J. Chem. Phys.*, **19**, 1394.

Closson, W. D., Brady, S. F., Kosower, E. M. & Huang, P.-K. C. (1963). *J. Org. Chem.*, **28**, 1161.

Ellington, P. S. & Meakins, G. D. (1960). *J. Chem. Soc.*, 697.

Gillam, A. E. & Stern, E. S. (1958). In *Electronic Absorption Spectroscopy in Organic Chemistry*, 2nd Edition. Arnold, London.

Jaffé, H. H. & Orchin, M. (1962). In *Theory and Applications of Ultraviolet Spectroscopy.* John Wiley, New York, London.

Rao, C. N. R., Balasubramanian, A. & Ramachandran, J. (1962). *J. Sci. Indust. Res.* (*India*), **20 B**, 382.

Rao, C. N. R. (1967). In *Ultraviolet and Visible Spectroscopy Chemical Applications.* 2nd Edition. Butterworths, London.

Scott, A. I. (1964). In *Interpretation of the Ultraviolet Spectra of Natural Products.* Pergamon Press, Oxford, London, New York.

Ungnade, H. E. & Smiley, R. A. (1956). *J. Org. Chem.*, **21**, 993.

Wolfe, H. (1932). *Z. Physiol. Chem.*, **B17**, 46.

3.4

Andrews, L. J. & Keefer, R. M. (1951). *J. Amer. Chem. Soc.*, **73**, 462, 4169. (1952) *Ibid.*, **74**, 4500.

Bhaskar, K. R., Gosovi, R. K. & Rao, C. N. R. (1966). *Trans. Faraday Soc.*, **62**, 29.

Cram, D. J. & Cranz, F. W. (1950). *J. Amer. Chem. Soc.*, **72**, 595.

Dannenberg, H. (1939). *Abhandl. Preuss. Akad. Wiss.*, **21**, 3.

Doub, L. & Vandenbelt, J. M. (1947). *J. Amer. Chem. Soc.*, **69**, 2714.

Evans, L. K. & Gillam, E. A. (1941). *J. Chem. Soc.*, 815; (1943). *J. Chem. Soc.*, 565.

Geissman, T. A. & Harborne, J. B. (1956). *J. Amer. Chem. Soc.*, **78**, 832.

Gillam, A. E. & Morton, R. A. (1929, 1931). *Proc. Roy. Soc.* A **124**, 604 and A **132**, 152.

Kosower, E. M. (1958). *J. Amer. Chem. Soc.*, **80**, 3253.

Lenon, H. W. (1947). *J. Amer. Chem. Soc.*, **69**, 2998.

Longuet-Higgins, H. C. & Murrell, J. N. (1955). *Proc. Phys. Soc.*, **68a**, 60.

Morton, R. A. & Stubbs, A. L. (1940). *J. Chem. Soc.*, 1347.

Murrell, J. N. (1956). *J. Chem. Soc. Spec. Pub.* No. 3.

Scheibe, G. (1925). *Ber. dtsch. chem. Ges.*, **58**, 586.

Scott, A. I. (1964). In *Interpretation of the Ultraviolet Spectra of Natural Products*, Pergamon Press, Oxford.

Woodward, R. B. (1941). *J. Amer. Chem. Soc.*, **63**, 1123; (1942). *J. Amer. Chem. Soc.*, **64**, 72.

3.5

Allport, D. C. & Bu'Lock, J. D. (1960). *J. Chem. Soc.*, 654.

Arbusov, B. A., Polezhaeva, N. A. & Vinogradovna, V. S. (1960). *Isvest. Acad. Nauk. SSSR*, **60**, 1219.

Blackwood, R. K., Beereboon, J. J., Rennhard, H. H., v. Wittenau, M. S. & Stephens, C. R. (1961). *J. Amer. Chem. Soc.*, **83**, 2773.

Boothe, J. H., Wilkinson, R. G., Kushner, S. & Williams, J. H. (1953). *J. Amer. Chem. Soc.*, **75**, 1732.

Broschard, R. W. *et al.* (1949). *Science, N.Y.*, **109**, 199.

Conover, L. H. (1956). *Special Publication No. 5*, 48, Chemical Society, London.

Cowell, E. J., King, F. E. & Morgan, J. W. W. (1961). *J. Chem. Soc.*, 702.

Doerschuk, A. P., Bitler, B. A. & McCormick, J. R. D. (1955). *J. Amer. Chem. Soc.*, **77**, 4687.

Duggar, B. M. (1948). *Ann. N.Y. Acad. Sci.*, **51**, 177.

Finlay, A. C., Hobby, G. L., Pan, S. Y., Regna, P. P., Boutien, J. B., Seeley, D. B., Shull, G. M., Sobin, B. A., Solomons, I. A., Vinson, J. W. & Kane, J. H. (1950). *Science, N.Y.*, **111**, 85.

Hochstein, F. A., Stephens, C. R., Conover, L. H., Regna, P. P., Pasternack, R., Gordon, P. N., Pilgri, F. J., Brunings, K. J. & Woodward, R. B. (1953). *J. Amer. Chem. Soc.*, **75**, 5455.

Regna, P. P. *et al.* (1951). *J. Amer. Chem. Soc.*, **73**, 4211.

Selzer, G. B. & Wright, W. W. (1957). *Antibiotics & Chemotherapy*, **7**, 292.

Stephens, C. R. *et al.* (1956). *J. Amer. Chem. Soc.*, **78**, 1515.

3.6

Brode, W. R. & Piper, J. (1941). *J. Amer. Chem. Soc.*, **63**, 1504.

Förster, T. (1946). *Naturwiss.*, **33**, 166.

Funkhouser, J. & Brode, W. R. (1934). *J. Amer. Chem. Soc.*, **56**, 2172.

Kiprianov, A. L. (1971). *Russian Chemical Reviews* (English trans.), **40**, 594.

Kiprianov, A. L. & Buryak, V. Ya. (1968). *Ukrain. Khim. Zhur.*, **35**, 179.

Kiprianov, A. L. & Mushkalo, I. L. (1965). *Zhur. Org. Khim.*, **1**, 744, 750.

Krasovitskii, B. M. & Mal'tseva, N. L. (1966). *Zhur. Org. Khim.*, **2**, 894.

Kühn, H. (1967). *Naturwiss.*, **54**, 429.

Morris, R. & Brode, W. R. (1948). *J. Amer. Chem. Soc.*, **70**, 2487.

Piper, J. & Brode, W. R. (1935). *J. Amer. Chem. Soc.*, **57**, 137.

3.7 3.8, 3.9

Allan, J. H. L., Jones, E. R. H. & Whiting, M. C. (1955). *J. Chem. Soc.*, 1862.

Armitage, J. B., Cook, C. L., Entwhistle, N. E., Jones, E. R. H. & Whiting, M. C. (1952). *J. Chem. Soc.*, 1998.

Armitage, J. B., Cook, C. L., Jones, E. R. H. & Whiting, M. C. (1952). *J. Chem. Soc.*, 2010.

Armitage, J. B., Jones, E. R. H. & Whiting, M. C. (1952). *J. Chem. Soc.*, 2014.

Armitage, J. B. & Whiting, M. C. (1952). *J. Chem. Soc.*, 2005.

Bagby, M. O., Smith, C. R. Jr. & Wolff, I. A. (1964). *Chem. & Ind.*, 1861.

Ball, S., Bessell, C. J. & Mortimer, A. (1957). *J. Gen. Microbiol.*, **17**, 96.

Barnes, R. H., Rusoff, I. I., Miller, E. S. & Burr, G. O. (1945). *Indust. Eng. Chem.* (*Anal.*), **10**, 385.

Beadle, B. W. & Kraybill, H. R. (1944). *J. Amer. Chem. Soc.*, **66**, 1232.

Bell, I., Jones, E. R. H. & Whiting, M. C. (1958). *J. Chem. Soc.*, 1313.

Black, H. K. and Weedon, B. C. L. (1953). *J. Chem. Soc.*, 1785.

Blanc, P. Y. (1961). *Helv. Chim. Acta*, **44**, 1.

Blout, E. R. & Fields, M. (1948). *J. Amer. Chem. Soc.*, **70**, 189.

Blout, E. R., Fields, M. & Karplus, R. (1948). *J. Amer. Chem. Soc.*, **70**, 194.

Bohlmann, F. (1951). *Ber. dtsch. chem. Ges.*, **84**, 545, 785; (1952). *Ber. dtsch. chem. Ges.*, **85**, 386; (1953). *Ber. dtsch. chem. Ges.*, **86**, 63, 657; (1953). *Angew. Chem.*, **65**, 385; (1955). *Angew Chem.*, **67**, 389.

Bohlmann, F. *et al.* (1960). *Ber. dtsch chem. Ges.*, **93**, 981, 1931. (1961) *Ibid.*, **94**, 958, 3179. (1962) *Ibid.*, **95**, 39, 602, 1315, 1733, 2939. (1963) *Ibid.*, **96**, 226.

Bohlmann, F. & Kieslich, K. (1954). *Ber. dtsch. chem. Ges.*, **87**, 1363.

Bohlmann, F. & Inhoffen, E. (1956). *Ber. dtsch. chem. Ges.*, **89**, 21, 1276.

Bohlmann, F., Inhoffen, E. & Herlst, P. (1957). *Ber. dtsch chem. Ges.*, **90**, 174.

Bohlmann, F. & Mannhardt, H. J. (1955). *Ber. dtsch. chem. Ges.*, **88**, 429.

Bohlmann, F. & Mannhardt, H. J. (1956). *Ber. dtsch. chem. Ges.*, **89**, 1307.

Bohlmann, F. & Mannhardt, H. J. (1957). Acetylenverbindungen im Pflanzenreich, 1–70. In *Progress in the Chemistry of Natural Products XIV*, Ed. L. Zechmeister. Springer, Vienna.

Booker, H., Evans, L. K. & Gillam, A. E. (1940). *J. Chem. Soc.*, 1453.

Braude, E. A. (1950). *J. Chem. Soc.*, 379.

Braude, E. A. (1955). In *Determination of Organic Structures by Physical Methods.* Chapter 4. Ed. E. A. Braude and F. C. Nachod. Academic Press, New York.

Brice, B. A. & Swain, M. L. (1945). *J. Opt. Soc. Amer.*, **35**, 532.

Brice, B. A., Swain, M. L., Herb, S. F., Nichols, P. L. & Riemenschneider, R. W. (1952). *J. Amer. Oil Chem. Soc.*, **29**, 279.

Bruun, T., Haug, O. M. & Sörensen, N. A. (1950). *Acta. Chem. Scand.*, **4**, 850.

Bu'Lock, J. D., Jones, E. R. H. & Leeming, P. R. (1955). *J. Chem. Soc.*, 4270.

Cason, J. & Sumbell, G. (1952). *J. Org. Chem.*, **16**, 1177, 1181.

Celmer, W. D. & Solomons, I. A. (1952). *J. Amer. Chem. Soc.*, **84**, 1870, 3838; (1953). *Ibid.*, **85**, 1372.

Christie, W. W. & Holman, R. T. (1967). *Chem. Phys. Lipids*, **1**, 407.

Coggeshall, N. D. (1950). *J. Amer. Chem. Soc.*, **72**, 2836.

Coggeshall, N. D. (1960). In *Physical Chemistry of Hydrocarbons*, Ed. A. Farkas, Vol. 1. Academic Press, New York, London.

Cook, C. L., Jones, E. R. H. & Whiting, M. C. (1952). *J. Chem. Soc.*, 2883.

Cope, A. C. & Johnson, H. E. (1958). *J. Amer. Chem. Soc.*, **80**, 1504.

Cope, A. C. & Kagan, F. (1958). *J. Amer. Chem. Soc.*, **80**, 5499 (also see (1962). *Ibid.*, **84**, 2170).

Crombie, L. (1955). *J. Chem. Soc.*, 1007.

Crombie, L. & Manzoor-I-Khuda, H. (1956). *Chem. & Ind.*, 409.

Crossley, A. & Hilditch, T. P. (1950). *J. Sci. Food & Agric.*, **1**, 292.

Dann, W. J. & Moore, T. (1933). *Biochem. J.*, **27**, 1166.

Dann, W. J., Moore, T., Booth, R. G., Golding, J. & Kon, S. K. (1935). *Biochem. J.*, **29**, 138.

Dannenberg, H. (1939). *Abhandl. preuss. Akad. Wiss.*, **21**, 3.

Dhar, M. L., Thaller, V. & Whiting, M. C. (1964). *J. Chem. Soc.*, 842.

Donovick, R. (1956). *Antibiotics Annual*, 579. (See also (1956). *Giorn Microbiol.*, **2**, 147).

Dutcher, J. D., Walters, D. R. & Wintersteiner, O. (1955). *Resumé commun. 3rd. Intern. Congr. Biochemistry*, Brussels.

Edisbury, R., Morton, R. A. & Lovern, J. A. (1935). *Biochem. J.*, **29**, 899.
Fenton, A. J. Jr. & Crisler, R. O. (1959). *J. Amer. Oil Chem. Soc.*, **36**, 620.
Fieser, L. F. & Fieser, M. (1959). *Steroids*. P. 15. Reinhold and Chapman & Hall, New York and London.
Gillam, A. E., Heilbron, I. M., Hilditch, T. P. & Morton, R. A. (1931). *Biochem. J.*, **25**, 30.
Goddu, R. F. (1957). *Anal. Chem.*, **29**, 1990.
Goldfine, H. (1964). *J. Biol. Chem.*, **239**, 2130.
Grigor, J., McInnes, D., McLean, J. & Hogg, A. J. P. (1955). *J. Chem. Soc.*, 1069.
Gunstone, F. D. & McGee, M. A. (1954). *Chem. & Ind.*, 1112.
Gunstone, F. D. & Russell, W.-C. (1955). *J. Chem. Soc.*, 3782.
Harada, R. (1957). *J. Chem. Soc. Japan*, **78**, 1031.
Hausser, H. K., Kuhn, R. & Kuhn, E. (1935). *Z. Physikal Chem.*, B **29**, 417.
Hausser, H. K., Kuhn, R., Smakula, A. & Deutsch, A. (1935). *Z. Physikal Chem.*, B **29**, 384.
Hausser, K. W., Kuhn, R., Smakula, A. & Hoffer, M. (1935). *Z. Physikal Chem.*, B **29**, 371.
Hausser, K. W., Kuhn, R., Smakula, A. & Kreuchen, K. H. (1935). *Z. Physikal Chem.*, B **29**, 363.
Hausser, K. W., Kuhn, R. & Seitz, G. (1935). *Z. Physikal Chem.*, B **29**, 391.
Hausser, K. W. & Smakula, A. (1935). *Angew. Chem.*, **47**, 657.
Herb, S. F. (1955). *J. Amer. Oil Chem. Soc.*, **32**, 153.
Herb, S. F. & Riemenschneider, R. W. (1952). *J. Amer. Oil Chem. Soc.*, **29**, 456; (1953). *Analyt. Chem.*, **25**, 953.
Herb, S. F., Riemenschneider, R. W. & Donaldson, J. (1951). *J. Amer. Chem. Soc.*, **28**, 55.
Herb, S. F., Witnauer, L. P. & Riemenschneider, R. W. (1951). *J. Amer. Oil Chem. Soc.*, **28**, 505.
Hilditch, T. P., Morton, R. A. & Riley, J. P. (1945). *Analyst*, **70**, 68.
Hilditch, T. P., Patel, C. B. & Riley, J. P. (1951). *Analyst*. **76**, 81.
Holman, R. T. & Greenberg, S. I. (1954). *Arch. Biochem. Biophys.*, **49**, 49.
Holman, R. T. & Edmondson, P. R. (1956). *Anal. Chem.*, **28**, 1533.
Hopkins, C. Y. (1965). In *Progress in the Chemistry of Fats and Other Lipids*. Vol. 8, p. 213. Ed. R. T. Holman, Pergamon Press, Oxford; (1968). *J. Amer. Oil Chem. Soc.*, **45**, 778.
Hopkins, C. Y., Jeavans, A. W. & Bloch, R. (1969). *Canad. J. Biochem.*, **47**, 433.
Jacobson, M. (1965). *Insect Sex Attractments*. Interscience Publishers, New York.
Jeavans, A. W. & Hopkins, C. Y. (1968). *Tetrahedron Lett.*, 2167.
Jones, E. R. H., Lee, H. H. & Whiting, M. C. (1960). *J. Chem. Soc.*, 3483.
Jones, E. R. H., Whiting, M. C., Armitage, J. B., Cook, C. L. & Entwhistle, W. (1951). *Nature, London*, **168**, 900.

Kass, J. P. & Burr, G. O. (1939). *J. Amer. Chem. Soc.*, **61**, 3292.
Korzybsky, T. & Karylowicz, W. (1961). Antibiotics (trans. from Polish). Fischer Verlag, Jena. See also, *Chem. Abstr.* (1948) **48**, 9018a.
Lepage, M. & Boch, R. (1968). *Lipids*, **3**, 530.
Ligthelm, S. P. & Schwartz, M. H. (1950). *J. Amer. Chem. Soc.*, **72**, 1868.
Ligthelm, S. P., Schwartz, H. M. & von Holdt, M. M. (1952). *J. Chem. Soc.*, 1088.
MacColl, A. (1947). *Quart. Revs. Chem. Soc.*, **1**, 16.
Mandava, N. & Mitchell, J. W. (1972). *Chem. & Ind.*, 930.
Mason, S. F. (1961). *Quart. Revs. Chem. Soc.*, **15**, 287.
Machotra, S. S. & Whiting, M. C. (1960). *J. Chem. Soc.*, 3812.
Mebane, A. D. (1952). *J. Amer. Chem. Soc.*, **74**, 5227.
Mikolajczak, K. L. Rogers, M. F., Smith, C. R., Jr. & Wolff, I. A. (1967). *Biochem. J.*, **105**, 1245.
Milborrow, B. V. & Pryce, R. J. (1973). *Nature, London*, **243**, 46.
Minnikin, D. E. (1966). *Chem. & Ind.*, 2167.
Morton, R. A. (1962). In *Comprehensive Biochemistry*, Vol. 3. Eds. M. Florkin & E. H. Stotz. Elsevier, Amsterdam.
Mitchell, J. H., Kraybill, H. R. & Zscheile, F. P. (1943). *Indust. Eng. Chem. (Anal.)*, **15**, 1.
Mitchell, J. W. & Livington, G. A. (1968). *Agric. Handbook*, No. 336, U.S. Department of Agriculture.
Mitchell, J. W., Mandava, N., Worley, J. F. Plimmer, J. R. & Smith, M. V. (1970). *Nature, London*, **225**, 1065.
Mitchell, J. W., Mandava, N., Worley, J. F. & Drowns, M. E. (1971). *J. Agric. Fd. Chem.*, **19**, 39
Moore, T. (1937). *Biochem. J.*, **31**, 138.
Morton, R. A., Heilbron, I. M. & Thompson, A. (1931). *Biochem. J.*, **25**, 20.
Mowry, D. T., Brode, W. R. & Brown, J. B. (1942). *J. Biol. Chem.*, **142**, 671.
Nakagawa, M. (1950). *Proc. Japan Acad.*, **26**, 38, 43.
Nayler, P. & Whiting, M. C. (1955). *J. Chem. Soc.*, 3037, 3042.
O'Connor, R. T., Stanbury, M. F., Damar, H. G. & Stark, S. M. (1952). *J. Amer. Oil Chem. Soc.*, **29**, 461.
Oroshnik, W., Vining, L. C., McBride, A. D. & Taber, W. A. (1955). *Science, N.Y.*, 121.
Patrick, J. B., Williams, R. P. & Webb, J. S. (1958). *J. Amer. Chem. Soc.*, **80**, 6089.
Pitt, G. A. J. & Morton, R. A. (1957). In *Progress in the Chemistry of Fats*, Vol. 4, p. 228. Pergamon Press, Oxford.
Polgar, A. & Jungnickel, J. L. (1956). In *Olefinic Unsaturation in Organic Analysis*, Vol. 3. Interscience Publishers, New York.
Postuna, J. & Behrends, W. (1960). *Biochem. Biophys. Acta.*, **41**, 538.
Purcell, J. M., Morris, S. G. & Susi, H. (1966). *Anal. Chem.*, **38**, 588.
Rasmussen, R. S., Brittain, R. R. & Zucco, P. S. (1947). *J. Chem. Phys.*, **15**, 135.

Riemenschneider, R. W. (1954). *J. Amer. Oil Chem. Soc.*, **31**, 517.

Riemenschneider, R. W., Herb, S. F. & Nichols, P. L. (1949). *J. Amer. Oil Chem. Soc.*, **26**, 371.

Rusoff, I. I., Platt, J. R., Klevens, H. B. & Burr, G. O. (1945). *J. Amer. Chem. Soc.*, **67**, 673.

Schaumberg, K. & Bernstein, H. J. (1968). *Lipids*, **3**, 193.

Seher, A. (1954). *Liebigs Ann.*, **589**, 222.

Serdarevich, B. & Carroll, K. K. (1966). *Canad. J. Biochem.*, **44**, 743.

Shearer, D. A. & Boch, R. (1966). *Insect Physiol.*, **12**, 1513.

Shreve, O. D., Heether, M. R., Knight, H. B. & Swern, D. (1950). *Anal. Chem.*, **22**, 12, 1261.

Silk, H. M. & Hahn, H. H. (1954). *Biochem. J.*, **577**, 582.

Slotboom, A. J., de Haas, H. G. & van Deenen, L. L. M. (1967). *Chem. Phys. Lipids*, **1**, 192.

Smouse, T. H., Mookerjee, B. D. & Change, S. S. (1965). *Chem. & Ind.*, 1301.

Sondheimer, F. Ben Efraim, D. A. & Wolovsky, R. (1961). *J. Amer. Chem. Soc.*, **83**, 1675.

Sondheimer, F. & Gaoni, Y. (1960). *J. Amer. Chem. Soc.*, **82**, 5765.

Sondheimer, F. & Wolovsky, R. (1959). *J. Amer. Chem. Soc.*, **81**, 4755; (1960). *Ibid.*, **82**, 754; (1959). *Tetrahedron Lett.*, **3**, 3.

Sörensen, J. G., Holme, D., Borlaug, E. T. & Sörensen, N. A. (1954). *Acta. Chem. Scand.*, **8**, 1769.

Sörensen, J. G. & Sörensen, N. A. (1954). *Acta. Chem. Scand.*, **8**, 1763.

Sörensen, N. A. & Stavholt, K. (1950 *a*). *Acta. Chem. Scand.*, **4**, 1080; (1950 *b*). *Acta. Chem. Scand.*, **4**, 1575.

Spectroscopy Committee, American Oil Chemists Society. (1949). *J. Amer. Oil Chem. Soc.*, **26**, 399; (1951). *Ibid.*, **28**, 331; (1953). *Ibid.*, **30**, 352; (1955). *Ibid.*, **32**, 452; (1956). *Ibid.*, **33**, 289.

Stavholt, K. & Sörensen, N. A. (1950). *Acta. Chem. Scand.*, **4**, 1567; (1954). *Ibid.*, **8**, 1741.

Stene, J. & Stavholt, K. (1941). *Liebigs. Ann.*, **549**, 80.

Stenhagen, E. (1968). In *Mass Spectrometry in Fat Analysis* in *Analysis and Characterization of Oils, Fats and Fat Products.* Vol. 2, Ed. H. A. Boekenoogen. Interscience Publishers, New York.

Stevens, C. L., Gillis, B. T., French, J. C. & Haskell, T. H. (1958). *J. Amer. Chem. Soc.*, **80**, 6089.

Tallent, W. H., Harris, H. J. & Wolff, I. A. (1966). *Tetrahedron Lett.*, 349.

Toyama, Y. & Tschuchiya, T. (1935). *Bull. Chem. Soc. Japan.* from *Chemical Abstracts*, **29**, 6208, 8378.

Tulloch, A. F. (1966). *J. Amer. Oil Chem. Soc.*, **43**, 670.

Umezawa, S. & Tanaka, Y. (1958). *J. Antibiotics*, **11**, 26.

Whitcutt, J. M. & Sutton, D. A. (1956). *Biochem. J.*, **63**, 469.

Whitfield, G. B., Brock, I. E., Ammann, A., Gottlieb, D. & Carter, H. E. (1955). *J. Amer. Chem. Soc.*, **77**, 4799.

Williams, W. W., Smirnov, V. S. & Gol'mov, V. P. (1935), *J. Gen. Chem. (USSR)*, **5**, 1195.

Wood, R. & Snyder, F. (1967). *J. Lipid. Res.*, **8**, 494.

Woods, G. F. & Schwartzman, L. H. (1948). *J. Amer. Chem. Soc.*, **70**, 3394; (1949). *Ibid.*, **71**, 1396.

Woodward, R. B. (1942). *J. Amer. Chem. Soc.*, **64**, 72.

Zondag, E., Posthuma, J. & Berends, E. (1960). *Biochim. Biophys. Acta*, **39**, 178.

4 Aldehydes, Ketones and Tautomerism

4.1 Carbonyl Compounds

The carbonyl group has two C—O σ bonding electrons and two electrons in the π bond; there are also two lone pairs, one in the s and the other in the pi orbital. Simple ketones and aldehydes exhibit bands in the far ultraviolet region and a weak band ($\epsilon_{max.}$ *ca.* 20) occurs in the region 275–290 nm. This is due to a singlet symmetry-forbidden transition of an oxygen 2p lone pair electron to the carbonyl antibonding π^* orbital ($n \rightarrow \pi^*$). Formaldehyde and acetone in the vapour state show strong bands in the 'vacuum' spectroscopy region with maxima at 175 nm, ϵ 1800, and 155·5 nm, ϵ 23 000 for the former and 166 nm, ϵ 3000 and 150 nm, ϵ 30 000 for the latter. These are designated $n \rightarrow \sigma^*$ and $\pi \rightarrow \pi^*$ transitions (McMurry, 1941). Formaldehyde also shows very weak ($\epsilon_{max.}$, *ca.* 10^{-3}) $n \rightarrow \pi^*$ triplet absorption at 390 nm. For a theoretical discussion, see Jaffé and Orchin (p. 105). Aliphatic ketones were studied in early classical papers by Rice (1915–1920). The band shown by acetone ($\lambda_{max.}$ 279 nm, $\epsilon_{max.}$ 15) in hexane undergoes small shifts in position and intensity as the alkyl groups are made larger, but even with hexamethylacetone $(CH_3)_3.C.CO.C.(CH_3)_3$ $\lambda_{max.}$ 296 nm, $\epsilon_{max.}$ 20 the effect of substituents is small.

The vapour of acetone shows the same $\lambda_{max.}$ as the hexane solution but $\epsilon_{max.}$ appears to be a little lower. The position of $\lambda_{max.}$ varies a little with change of solvent. If the solvents are placed in order of hydrogen bonding ability, i.e. water, methanol, ethanol, chloroform, hexane, $\lambda_{max.}$ is displaced in that order from 265 to 279 nm. Cyclic compounds and substances like camphor show the normal carbonyl absorption. Compounds like CH_3COCl and CH_3COBr show weak absorption ($\epsilon < 100$) around 240–250 nm. A red shift occurs with CH_3COCH_2Br to 310 nm (ϵ 83).

Table 4.1 Absorption spectra of simple ketones

(*a*) Effect of substituents on $\lambda_{max.}$ nm (in $CHCl_3$)

	$\lambda_{max.}$ nm		$\lambda_{max.}$ nm
Acetone	274		
Methylethylketone	277	Ethylpropyl	280
Methylnonylketone	279	Dipropylketone	282
Methylisopropyl	282	Hexamethylacetone	294
Pinacoline (methyl-tertbutyl)	285		

(*b*) Effect on solvent on $\lambda_{max.}$ of acetone

	$\lambda_{max.}$ nm
Hydrocarbon solvents	278
Methanol	270
Ethanol	272
Diethyl ether	277·4
Water	265
Concentrated hydrochloric acid	259

(*c*) $\epsilon_{max.}$ values (in heptane)

Acetone	17·1	Methylethylketone	19·6
Methylpropyl and higher homologues	21·2		

Benson & Kistiakowsky (1942) examined cyclobutanone in hexane and obtained the spectrum:

$\lambda_{max.}$ nm	274	281	290	303	311
$\epsilon_{max.}$	17	19	19	13·2	8·7

The resolution was not very good and the wave-number differences (Δ cm^{-1} 910, 1110, 1480, 1010) were variable. Cyclopentanone showed maxima as follows:

$\lambda_{max.}$ nm	278	290	300	310	323
$\epsilon_{max.}$	13·2	16·2	17	13·2	6·9

Again the resolution was imperfect and the separations between subpeaks were (Δ cm^{-1} 1490, 1150, 1070, 1300). Cyclohexanone showed a single unresolved peak $\lambda_{max.}$ 290 nm, $\epsilon_{max.}$ 11·2.

The n $\rightharpoonup$ π^* forbidden transition is also seen in acetic acid in ethanol ($\lambda_{max.}$ 204 nm, ϵ 41), and ethylacetate ($\lambda_{max.}$ 204 nm, ϵ 60); in acetamide the peak is at 214 nm for aqueous solution.

$\alpha\beta$-Unsaturated ketones exhibit constitutive effects due to the conjugation of double bonds. Burawoy (1939, 1941) referred to the strong absorption due to conjugation as arising from K-chromophores (German Konjugierte) and to low intensity bands at longer wavelengths as due to R-chromophores (Radikal). Mesityloxide $(CH_3)_2C{=}CH.COCH_3$ and phorone $(CH_3)_2C{=}CHCOCH{=}C(CH_3)_2$ exhibit the characteristic effects of conjugation:

R band, C=O

	$\lambda_{max.}$ nm	$\epsilon_{max.}$
[Acetone	279	15]
Mesityloxide (in hexane)	327	40
Phorone	375	80
$CH_3(CH{=}CH)_2COCH_3$	masked	
HC≡C.CHO	330	17

K band, C=C

	$\lambda_{max.}$ nm	$\epsilon_{max.} \times 10^{-3}$
[Acetone	—	—]
Mesityloxide (in hexane)	229	12·6
Phorone	295	23·5
$CH_3(CH{=}CH)_2COCH_3$	264	20·8
HC≡C.CHO	212	4·6

In the case of mesityloxide, change of solvent from water to hexane causes a red shift for the R band and a blue shift for the K band.

The displacements shown by the K bands have been reduced to order by Booker *et al.* (1940), Evans & Gillam (1942; 1943), Gillam & West (1942), and by Woodward (1941, 1942).

Table 4.2 Absorption peaks for mesityloxide

Solvent	$\lambda_{max.}$ nm	$\epsilon_{max.} \times 10^{-3}$	$\lambda_{max.}$ nm	$\epsilon_{max.}$
Water	244·5	10	305	95
Methanol	238	10·7	312	85
Ethanol	237	12·6	315	90
Diethyl ether	230	12·6	326	40
Hexane	229·5	12·6	327	40

(Scheibe *et al.*, 1925).

Considering $\alpha\beta$-unsaturated ketones generally:

$$\mathrm{R{-}C({=}O){-}\overset{\alpha}{C}{=}C{<}^{\beta}_{\beta}}$$

a very large body of information fits Table 4.3. Cyclopentenones (Gillam & West, 1943) are anomalous.

Evans & Gillam (1941) found that introduction of a methyl group on the β carbon atom resulted in a significantly larger shift of $\lambda_{max.}$ for the K band than a methyl group on the α carbon atom. Moreover in substances with 5 membered rings the bathochromic effect was less than would have been predicted from Woodward's rule, e.g. isothujone showed $\lambda_{max.}$ 237·5 and tetrahydropyrethrolone 232 nm when 247 nm was expected

CH(CH₃)₂, CH₃, CH₂, CH₃, O; C₅H₁₁, CH₃, O, H, CH₂, HO

Isothujone   Tetrahydropyrethrolone

The information collected by Gillam and his group and by Woodward is considerable in amount. It is more-

Table 4.3 Position of short wave absorption maximum for $\alpha\beta$-unsaturated ketones (solvent, ethanol)

Substituents	$\lambda_{max.}$ nm
mono α or β	225±5
di $\alpha\beta$ or $\beta\beta$	239±5
tri $\alpha\beta\beta$	254±5
$\beta\beta$ both H	218
di, no exocyclic C=C bond	235±5
one exocyclic C=C bond	240±5
tri, no exocyclic C=C bond	247±5
one exocyclic C=C bond	252±5

Table 4.4 Absorption maxima for some $\alpha\beta$-unsaturated ketones

	$\lambda_{max.}$ nm	$\epsilon_{max.} \times 10^{-3}$	$\lambda_{max.}$ nm	$\epsilon_{max.}$	Type of substitution
Methylvinylketone	215	3·6	324	24	none
2-Methylpent-1-en-3-one	225	7·9	319·5	27	α
Ethylideneacetone	225	9·75	312·5	38	β
Butylidenacetone	225	10·6	310	39	β
Mesityloxide	235	14·0	313·5	60	$\beta\beta$
1-Methylcyclohex-1-en-3-one	235	12·95	310	48	$\beta\beta$
2-Methyl-pent-3-en-2-one	235	11·1	310	42	$\alpha\beta$
2,3,4,5-Tetrahydroacetophenone	235	9·65	305	44	$\alpha\beta$
3,4-Dimethylpent-3-en-2-one	247	>4·0	—	—	$\alpha\beta\beta$

over correlated with the absorption spectra of conjugated dienes (see p. 45). Good summaries are given by Gillam & Stern (1958) and by Rao (1967) and Scott (1964).

The absorption spectra of derivatives of ketones possess considerable diagnostic value. Their high molecular extinction coefficients allow measurements to be made on very small quantities.

The semicarbazones of saturated ketones or of unconjugated enones show $\lambda_{max.}$ 225–230 nm (mostly 228·5–229·5 nm) $\epsilon_{max.}$ *ca.* 11 400. For $\alpha\beta$-unsaturated ketones the semicarbazones show $\lambda_{max.}$ 265 nm, $\epsilon_{max.}$ 25 000. The thiosemicarbazones obtained from saturated ketones show a band at 228–230 nm due to the thiosemicarbazide moiety (ϵ *ca.* 7000) and a semicarbazone band at 271–273 nm, $\epsilon_{max.}$ *ca.* 22 400. The derivatives from $\alpha\beta$-unsaturated ketones absorb similarly at 245 nm (ϵ 10 000) and 300 nm (ϵ 30 000). 2,4-Dinitrophenylhydrazones exhibit $\lambda_{max.}$ 360 nm, $\epsilon_{max.}$ 20 000 for derivatives of saturated ketones and 380 nm, $\epsilon_{max.}$ 25 000 for derivatives of $\alpha\beta$-unsaturated ketones. Oximes of $\alpha\beta$-unsaturated ketones absorb at 235 nm, $\epsilon_{max.}$ 15 000. Gillam & Stern (1958) have collected the data (see also Evans & Gillam, 1943; Wolf, 1929; Forster *et al.*, 1947; Benson & Kistiakovsky, 1942; Mariella & Raube, 1952. Jones *et al.* (1942) studied cholestenone, sitostenone and related derivatives ($\lambda_{max.}$ 240–341 nm, $\epsilon_{max.}$ 17–18 000 and 307–312 nm, $\epsilon_{max.}$ 70–100). The oximes absorbed maximally at 240 nm, $\epsilon_{max.}$ 23 000, the semicarbazones at 271 nm, $\epsilon_{max.}$ *ca.* 27 000 and the 2,4-dinitrophenylhydrazones at 302 nm, $\epsilon_{max.}$ 31–32 000.

α-Diketones exhibit two weak absorption bands (Ford & Parry, 1958; Holman *et al.*, 1945; Leonard & Mader, 1950). Biacetyl, $CH_3CO\ COCH_3$ showed $\lambda_{max.}$ 280 and 420 nm, ϵ 20 and 10 respectively and camphorquinone $\lambda_{max.}$ 280 and 466 nm, ϵ 23 and 31 respectively is similar, while the hexamethyl compound $(CH_3)_3C.CO.CO.C(CH_3)_3$ shows peaks at 285 and 365 nm (ϵ 53 and 21). Leonard & Mader studied the family

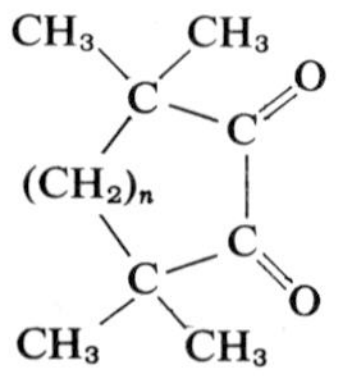

CH₃ CH₃ C C=O (CH₂)ₙ C C=O CH₃ CH₃

where n=2, 3, 4 and 14

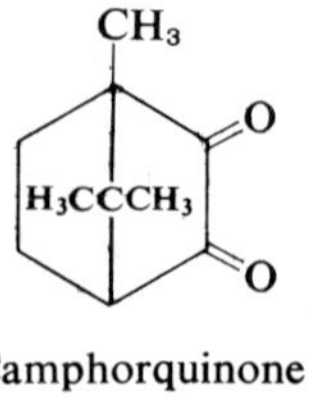

CH₃ H₃CCCH₃ =O =O

Camphorquinone

The band at the longer wavelength (due to an $n' \rightarrow \pi^{*\prime}$ transition) varied in position from 337 nm (n=3), 343 nm (n=4), 380 nm (n=2), 384 nm (n=14), an effect correlated with the angle of twist about the intercarbonyl bond, but the other band varied much less and for n=2, 3 or 4 was *ca.* 298 nm.

In potassium hydroxide solution diacetyl shows three peaks, 252 nm, ϵ 20 000, ~300 nm, ϵ 3550 and 425 nm, ϵ 21·4, presumably due to

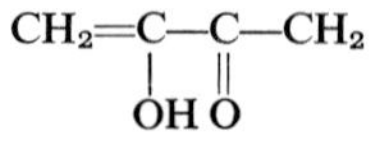

$$CH_2{=}\underset{OH}{C}{-}\underset{O}{\overset{}{C}}{-}CH_2$$

Acetyl propionyl also shows in ethanol two peaks 270 and 420 nm, ϵ 21·2 and 10 respectively. In alkali, peaks appear at 265, 340 and *ca.* 440 nm, ϵ 21 000, 212 and 21 respectively (Holman *et al.*, 1945). 9,10-Diketostearic acid in alkali shows similar bands at 260, 310 and ~400 nm.

4.2 Tautomerism

In an early investigation (Morton & Rosney, 1926) on keto-enol tautomerism, ethylacetoacetate was cooled to $-80°$ to reduce the vapour pressure. It showed in the vapour state a single band $\lambda_{max.}$ 238 nm, ϵ about 1000. Thin layers of the liquid ester showed $\lambda_{max.}$ 243 nm, ϵ 367. The same band appeared in solutions in ethanol (ϵ 2000) and in hexane (ϵ 9100). The ratio $\epsilon_{ethanol}/\epsilon_{hexane}$ agreed with the relative enol contents

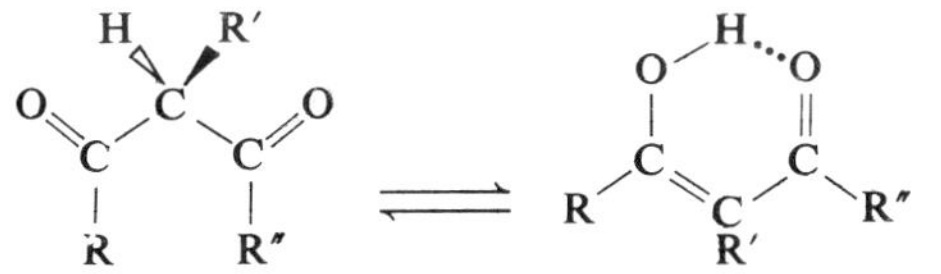

$$RC(=O)-C(H)(R')-C(=O)R'' \rightleftharpoons RC(-O-H\cdots O=)=C(R')-C(=O)R''$$

measured by bromine absorption (Meyer, 1912) so that the 243 nm peak belonged to the enol form. The absorption underwent a red shift to 255 nm with an increased keto content in the ester of mono-ethylaceto-acetate and a blue shift to 235 nm in the ethoxycroto-nate

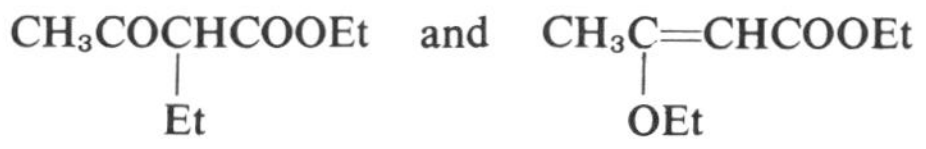

$CH_3COCH(Et)COOEt$ and $CH_3C(OEt){=}CHCOOEt$

In aqueous solution the absorption of ethylaceto-acetate was feeble and showed broad absorption extending from 230 to 320 nm, ϵ being less than 100 throughout. The enol content was negligibly small; the absorption of the keto from (cf. acetone $\lambda_{max.}$ 275 nm, ϵ 17·2) would be expected to be similarly located and weak. The predominantly ketonic liquid ester showed a marked inflection at 270–280 nm. Liquid dialkyl esters of the type:

$CH_3COC(Et_2)COOEt$

showed a well-defined band with $\lambda_{max.}$ near 290 nm, which could only be due to a ketonic form.

Since the hexane solution of ethylacetoacetate contains (by bromine absorption) 80% of enol, the pure enol should have $\lambda_{max.}$ 243 nm, ϵ 11 300.

Acetylacetone exhibits a strong band with $\lambda_{max.}$ 275 nm in both aqueous and ethanolic solutions. The ratio of extinction coefficients (8920/2200=4·05) agrees well with the enol contents measured chemically. The maximum for the enolic form coincides in position with the expected maximum for a ketonic form but is vastly stronger (see also Grossmann, 1924). The effect of alkali on ethylacetoacetate is to give a new and intense peak at 275 nm. This has been attributed to a chelated enol of the type:

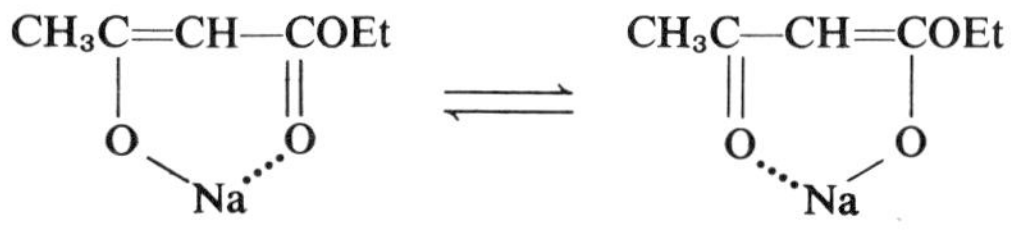

$$CH_3C(-O-)=CH-C(=O\cdots)OEt \;(\text{chelated to Na}) \rightleftharpoons CH_3C(=O\cdots)-CH=C(-O-)OEt \;(\text{chelated to Na})$$

In the case of acetylacetone a displaced peak appears near 290 nm the position varying a little with the alkali metal: $\lambda_{max.}$ nm in aqueous LiOH 291·7, in NaOH 292·6, in KOH 293·5, in RbOH 296·7 and in piperidine 295·5. This band is also ascribed to a chelated form:

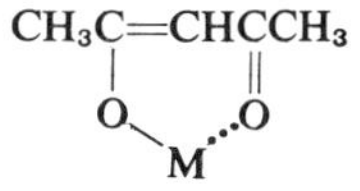

$$CH_3C(-O-)=CHC(=O\cdots)CH_3 \;(\text{chelated to M})$$

Oxalacetic acid exists in two solid forms, both enolic. The form with a lower melting point is the *cis* modification and is often referred to as hydroxymaleic acid.

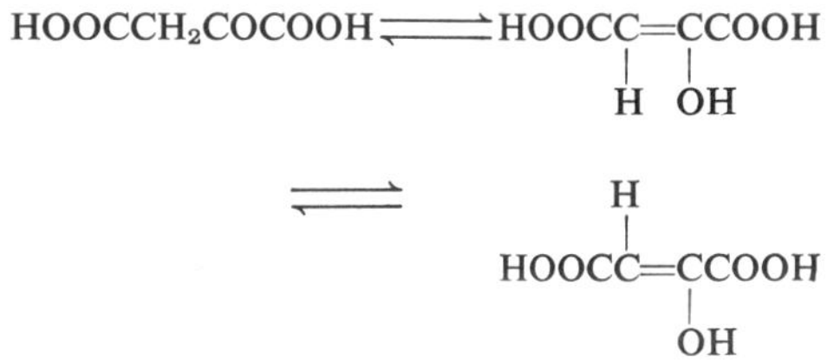

$$HOOCCH_2COCOOH \rightleftharpoons HOOCC(H){=}C(OH)COOH \rightleftharpoons HOOCC(H){=}C(OH)COOH$$

It undergoes ketonization when dissolved in water.

When hydroxymaleic acid is dissolved in dilute sodium hydroxide an enol band ($\lambda_{max.}$ 267 nm) is at once seen. If the solution has a pH between 5 and 9·9 the intensity of absorption at 267 nm falls quickly for 7–15 minutes depending on the temperature. The broad peak 242–262 nm which then appears corresponds to a mixture of keto and enol forms. Gelles & Hay (1957) give at 260 nm $\epsilon_{max.}=8800$. In perchloric acid solution $\epsilon_{260\,nm}=420$, corresponding with some 5% of enol. On standing, further change occurs, probably due to decarboxylation. Diethyloxalacetate examined by nuclear magnetic resonance contained 79% enol (Kumler *et al.*, 1962).

Loewus *et al.* (1955) reduced oxalacetate in the presence of malic dehydrogenase from wheat germ with monodeuterated NAD (prepared by reducing NAD with CH_3CD_2OH in the presence of yeast alcohol dehydrogenase). The malic dehydrogenase transferred H from NAD to the substrate with the stereospecificity for NAD already established for alchohol dehydrogenase. It was shown that oxalacetate in the keto form, and not the enol, is reduced enzymatically because the rate of the reaction can be limited by the rate of ketonization from the enol. Mg^{++} increases the rate at which the tautomeric change proceeds. Pyruvate is also reduced enzymatically in the keto form. An alternative method of using nuclear magnetic resonance spectra (Allen & Divek, 1966) depends on the fact that the protons $—CH_2C(=O)—$ show $\tau = 5{\cdot}5$ and the CH of the enol form $CH{=}C(OH)$ has $\tau = 3-4{\cdot}5$. The $—OH$ band is shown as a broad line at low field.

Ethylmesityloxideoxalate has been obtained in three forms: α, m.p. 21°, β, m.p. 59–60° and a dimeric form, m.p. 175° (Federlin, 1907). The α and β forms come to equilibrium in the dark in about 12 weeks, while diffuse daylight effects the changes α → β → dimer with precipitation of the latter from several solvents. The α-form is enolic (red colour with $FeCl_3$)

$$(CH_3)_2C{=}CHCOCH{=}C(OH)COOEt$$

or

$$(CH_3)_2C{=}CH{-}C(OH){=}CHCOCOOEt$$

while the ketonic nature of the β form was disputed and a cyclic structure advanced instead. The dimer shows negligible selective absorption

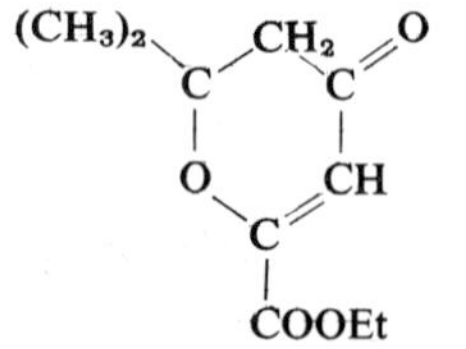

(CH₃)₂C–CH₂–C(=O)–CH=C(COOEt)–O– (ring)

The spectra are noted below in Table 4.5

It may be granted that the α form is enolic in either or both the forms shown.

In phorone

$$(CH_3)_2C{=}CHCOCH{=}C(CH_3)_2$$

the chromophore

$$—C{=}C—C(=O)—C{=}C$$

gives a strong band at 295 nm $\epsilon_{max.}$ 23 500, whereas the dienone chromophore in

$$CH_3(CH{=}CH)_2COCH_3$$

or

$$CH_3(CH{=}CH)_2CHO$$

shows $\lambda_{max.}$ at 264 and 270 nm, $\epsilon_{max.}$ 20 800 and 26 500 respectively . The diketo structure is unlikely to fit either the α or β form, but the problem could well be reopened using modern methods.

Ethylformylphenyl acetate exists in at least two forms is another problem worthy of re-examination inasmuch as it is more likely to prove to be a case of *cis-trans* isomerism than of keto-enol tautomerism.

Ethyldiacetylsuccinate exists in two forms, m.p. 31 and 92°. Each shows $\lambda_{max.}$ 249 nm and the $\epsilon_{max.}$ values are 6250 and 7600 respectively. With alkali both give $\lambda_{max.}$ 275 nm, $\epsilon_{max.}$ depending on the alkali/ester ratio. The isomerism here is probably *cis-trans* for a mono-enol.

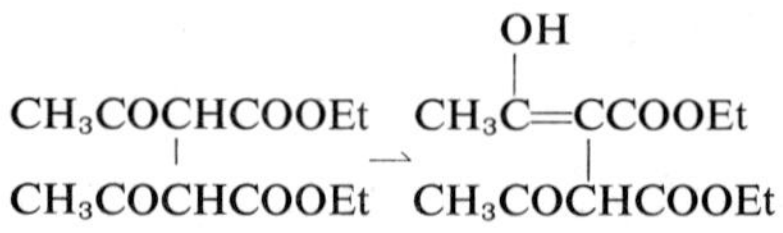

$$\begin{matrix} CH_3COCHCOOEt \\ | \\ CH_3COCHCOOEt \end{matrix} \rightarrow \begin{matrix} CH_3C(OH){=}CCOOEt \\ | \\ CH_3COCHCOOEt \end{matrix}$$

Murthy *et al.* (1962) made a detailed study of solvent effects in keto-enol equilibria for ethyl acetoacetate,

Table 4.5 Absorption spectra of ethylmesityloxideoxalate

	$\lambda_{max.}$ nm	$\epsilon_{max.}$	$\lambda_{max.}$ nm	$\epsilon_{max.}$
α form	312	14 100	242	3 900
(in ethanol, hexane or $CHCl_3$)				
(in ethanolic NaOEt)	351	21 250	287	3 800
in excess	246	9 200	—	—
β form (fresh solutions)				
in ethanol	282·5	12 200	228	3 800
hexane	277·5	11 500	231	3 300
(in ethanolic NaOEt)	351	13 000	—	—
	287	7 500	—	—
	250	7 500	—	—

(Morton & Rogers, 1926).

acetylacetone, ethylcyclopentanone-2-carboxylate and methyl-1,4-methylcyclopentanone-3,4,5-tricarboxylate:

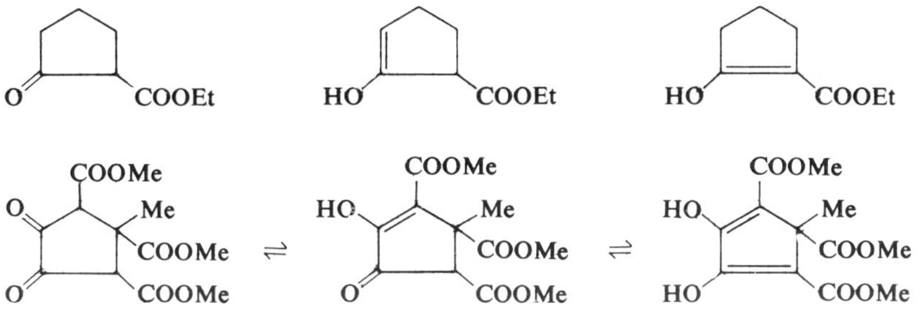

λ_{maxn} nm 270–280 328

The latter shows two peaks attributed respectively to the mono-enol and di-enol. In cyclohexane the dienol accounts for about half the solute and $\epsilon_{max.}$ is about 12 500. Table 4.6 shows the findings for different solvents.

The results for the tricarboxylate are somewhat equivocal. Infrared spectra for the ethylcyclopentanone-2-carboxylate in CCl_4 show (cm^{-1}) 1750 (ring ketone), 1730 (β-keto ester), 1670 (enolized β-keto ester), 1620 (C=C stretch). In acetonitrile the latter band is very weak. The tricarboxylate shows peaks at 1750, 1680 and 1616, 1580 (C=C of the enols).

Enolization of β-diketones, β-ketoacids and β-keto esters has also been followed using infrared absorption by Park *et al.* (1953) and Murthy *et al.* (1962). The ketonic form of acetylatone showed a C=O stretching vibration at 1709 cm^{-1}. The enol had a broad carbonyl band at the 1640–1530 cm^{-1}. If the enol is acetylated

$$CH_3C(OCOCH_3)=CHCOCH_3$$

bands are seen at 1701, 1695 and 1633 due to the acetate carbonyl and the $\alpha\beta$-unsaturated ketone and the olefinic link C=C. Some Raman spectra data were given by Morton *et al.* (1934).

Kasturi *et al.* (1962) studied the system:

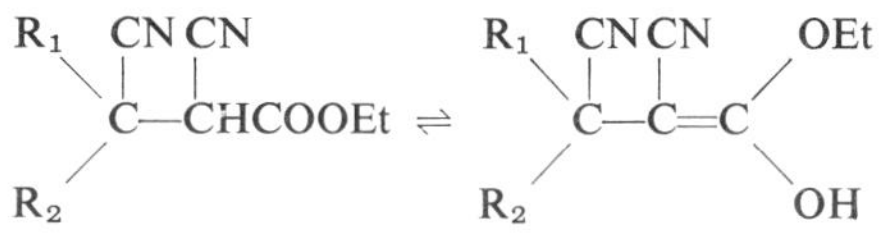

$$R_1R_2C(CN)-CH(CN)COOEt \rightleftharpoons R_1R_2C(CN)-C(CN)=C(OEt)OH$$

where R_1=CH_3 and R_2 is H, CH_3, C_2H_5 or CH_2COOCH_5.

In each case a peak at 247 nm was shown when the solvent was ethanol, but not by solutions in non-polar solvents. With increasing concentrations the intensities decreased. The band was ascribed to association of the enol form.

Benzoylcamphor exists in separable ketonic and enolic forms.

Again the maxima for the tautomeric forms agree in position but not in intensity. (see Table 4.7).

Table 4.6 Solvent effects in keto-enol tautomerism

Solvent	Acetoacetate $\lambda_{max.}$ 244 nm % enol	Acetylacetone $\lambda_{max.}$ 272 nm % enol	Ethylcyclopentanone-2-carboxylate $\lambda_{max.}$ 254 nm % enol	Methyl-4-methyl-cyclopentano-1,2-dione-3,4,5-tri-carboxylate $\lambda_{max.}$ %270 nm	$\lambda_{max.}$ %325nm
Cyclohexane	43	88	22	34	49
Dioxane	8	73	—	—	—
Ether	25	85	6	35	34
Acetonitrile	4	55	1·5	39	27
Water	0·5	16	0·4	44	27
Ethanol	8	73	3	37	39
Methanol	6	70	2·5	38	38

Table 4.7 Absorption maxima for benzoylcamphor

Solvent	Ketonic form		Enolic form	
	$\lambda_{max.}$ nm	ϵ	$\lambda_{max.}$ nm	ϵ
Ethanol	312·5	103	315	10 600
Piperidine	324	123	328	15 250
Freshly dissolved	307·2	1380	—	—
In ethanolic sodium ethoxide	275	1040	—	—
After standing some days	327	12 500	—	—

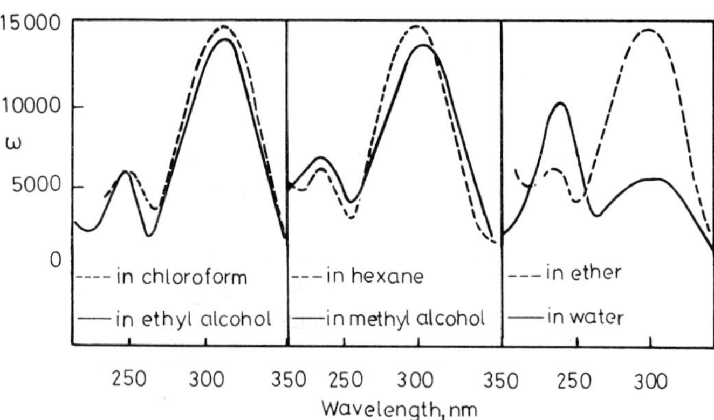

Figure 4.1 Absorption spectra of benzoylacetone in various solvents

Benzoylacetone (Morton *et al.*, 1934) was studied in detail:

$C_6H_5COCH_2COCH_3$ →

→ $C_6H_5C(OH){=}CHC(=O)CH_3$

→ $C_6H_5{-}C(=O){-}CH{=}C(OH)CH_3$

→ $C_6H_5{-}C(OH){=}CH{-}C(OH){=}CH_2$

→ $C_6H_5{-}C(OH){=}C{=}C(OH)CH_3$

This substance has two absorption peaks, the maxima being at 247 and 310 nm (in ethanol ϵ 5620 and 14 800 respectively). The intensities vary with the solvent, thus in water ϵ_{250} nm is 10 200 and $\epsilon_{312\cdot5}$ nm is 5600 (Fig. 4.1). Since in general water is a ketonizing solvent, the 247–250 peaks may be allocated to a keto form and the 310 nm peak to an enol. A diketo form would be expected to resemble acetophenone.

C-Methylbenzoylacetone can be separated into rather stable keto and enol forms. The keto isomeride shows an absorption spectrum very like that of acetophenone or benzaldehyde and the spectrum of the C-dimethylbenzoylacetone

$C_6H_5C(=O)C(CH_3)_2COCH_3$

which cannot enolize, is very similar. The enolic form of C-methylbenzoylacetone shows a strong band with $\lambda_{max.}$ 310 nm, ϵ 12 000.

It seems highly probable that the enolic form (*b*) of benzoylacetone predominates in hydrocarbon solvents with the benzoyl group intact. The C-methyl derivative however is more likely to be:

$C_6H_5C(OH){=}C(CH_3)COCH_3$

3-Methylcyclopentane-1,2-dione (Bredenberg, 1957, 1959) can be written in three ways

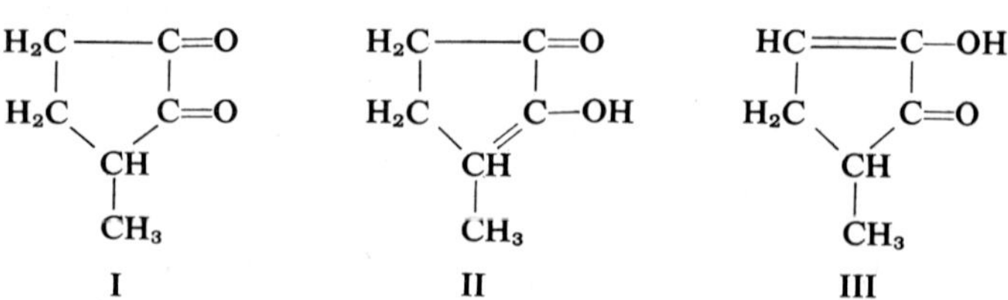

Table 4.8 Diketo forms of benzoylacetone and its C-methyl derivatives compared with benzaldehyde and acetophenone

	$\lambda_{max.}$ nm	ϵ	$\lambda_{max.}$ nm	ϵ	$\lambda_{max.}$ nm	ϵ
Benzoylacetone	247	13 000	280	1000	320	50
C-Methylbenzoylacetone	247	11 000	284	1500	310	200
C-Dimethylbenzoylacetone	237	14 500	(278)	1050	(310)	100
Acetophenone	241·5	16 250	278·5	1150	320	54
Benzaldehyde	244	16 250	281·5	1640	328	20

This substance shows an absorption peak in water at 256 nm (log ε 4·14), in ethanol at 258·5 nm (log ε 4·15) and in cyclohexane at 251 nm (log ε 4·15). Inasmuch as the equilibria between keto and enol forms vary with the solvent the constancy here argues in favour of a single structure, perhaps II rather than III and certainly not I.

Nuclear magnetic resonance spectra in $CDCl_3$ showed CH_3 τ 7·97, CH_2 τ 7·57 nearly equivalent and O—*H* 3·25, areas 3:4:1 which fits II, whereas III should give a more complicated spectrum. In pyridine solution there was no change, again negativing an equilibrium mixture. In the infrared there was no olefinic stretch (CCl_4), again favouring II. Peaks occurred at (cm^{-1}) 1710 in chloroform and 1715 in CCl_4. The double bond showed (cm^{-1}) in the solid state 1650, in $CHCl_3$ 1667 and in CCl_4 1670.

Mühle & Tamm (1962, 1963) made a study of certain cyclic β-diketones and a few points from their work may be noted here. Three ketones were prepared and compared, namely 2,2-dimethyl-4,4-dimethyl- and 2,2,4,4-tetramethylcholestadione-(1,3).

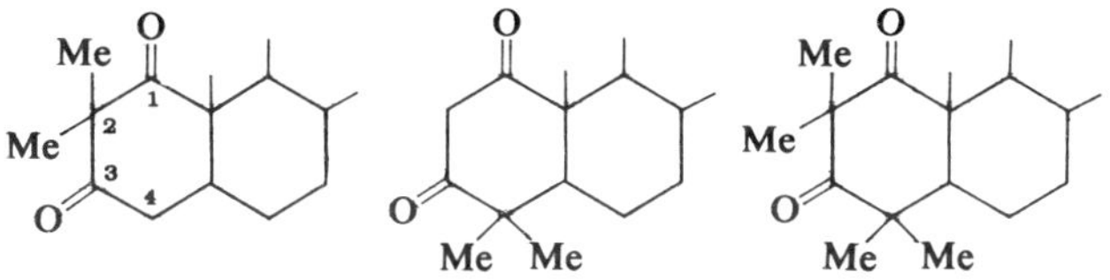

The spectroscopic findings are given in Table 4.9. The tetrasubstituted diketone cannot enolize and the weak absorption at 270–294 nm is characteristic of a ketone. The 2,2-dimethyl-1,3-diketone displays in both ethanol and cyclohexane the typical absorption of the saturated diketone with no enolization despite the presence of CH_2 at position 4. On the other hand the 4,4-dimethyl derivatives in the solid state are like cholestadione-(1,3) wholly ketonic since the principal infrared bands are at 1718 and 1712 cm^{-1}, a typical C=O doublet. In cyclohexane the ultraviolet absorption is still predominantly ketonic but in dioxan an enolic form predominates. The enol could be:

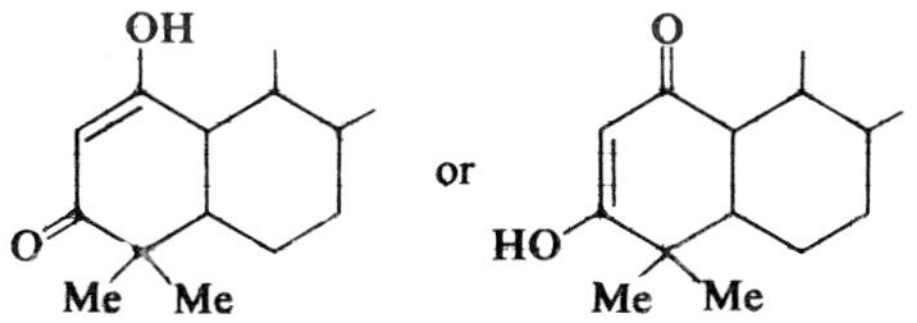

or

In the case of the cholestadione-(1,3) the evidence (Tamm, 1960) of methylation suggested that both

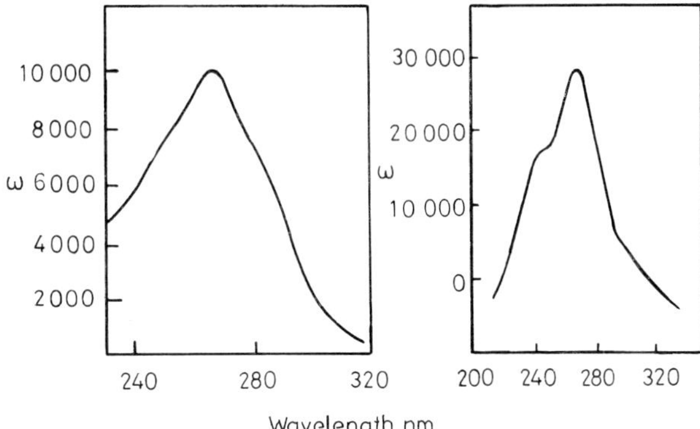

Figure 4.2 Absorption spectra in hexane of benzoylacetone methyl ether, $C_6H_5 \cdot C(OMe):CH \cdot CO \cdot CH_3$ (left) and methylbenzoylacetone ethyl ether, $C_6H_5 \cdot CO \cdot CH:C(OEt) \cdot CH_3$ (right)

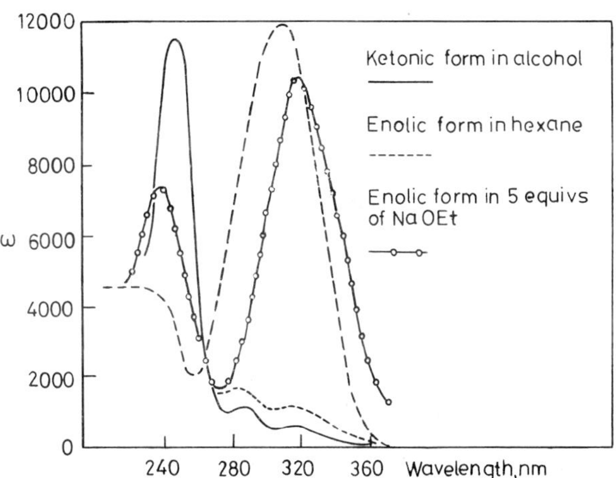

Figure 4.3 Absorption spectra of the ketonic and enolic forms of methylbenzoylacetone in various solvents

Figure 4.4 Absorption spectra of dimethylbenzoylacetone

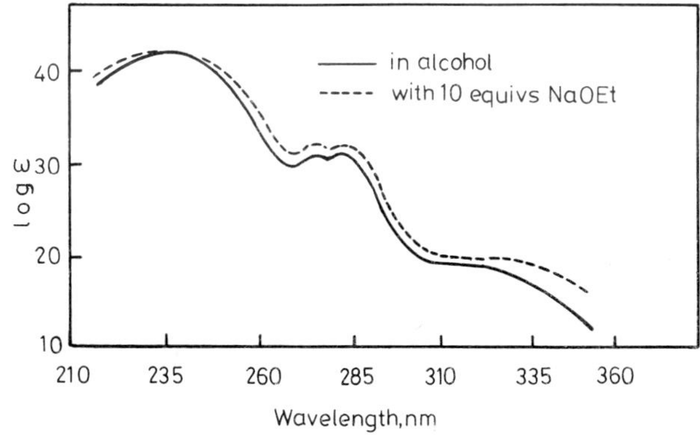

Table 4.9 Absorption spectra of cholestadione-(1,3) derivatives

Compound	Solvent	$\lambda_{max.}$ nm	log ϵ
Cholestadione-(1,3)	Ethanol	255	4·10
	Hexane	299	1·83
	Dioxan	301	1·88
4,4-Dimethyl-	Cyclohexane	199, 262, −295	3·48, 2·26, 2·09
	Dioxan	248, 290	3·39, 2·12
2,2-Dimethyl-	Ethanol	201, 265, −275	3·05, 1·87
	Cyclohexane	192, 215, 292	3·14, 3·13, 1·69
2,2,4,4-Tetramethyl-	Cyclohexane	191, 211, 271, 294	3·0, 3·0, 1·88, 1·82
1-Methoxy-4,4-dimethyl Δ^1 cholestenone-(3)	Cyclohexane	243, 314	4·33, 2·17
	(infrared 1650, 1589 cm^{-1})		
3-Methoxy-4,4-dimethyl Δ^2 cholestenone-(1)		239·5, 317	4·20, 1·90
	(infrared 1661, 1618 cm^{-1})		

enols were formed with the 3-hydroxy form predominating (3:2). In the case of the 4,4-dimethyl derivative it is concluded that the two enols are present in equal amounts.

Three structures can be postulated for benzoylcyclohexanone

	Keto	Enol	Enol
$\lambda_{max.}$ nm	245	245	314
$\epsilon_{max.} \times 10^{-3}$	9·74	7·8	4·4
	(80%)	(20%)	

These results may be compared with those for benzoylacetone where the keto form has $\lambda_{max.}$ 247 nm, $\epsilon_{max.}$ 13 000 and the enol form $\epsilon_{247\,nm}$ = 5600 and $\epsilon_{310\,nm}$ = 15 000. Campbell & Gilow (1960) examined various derivatives of benzoylcyclohexanone with ring substituents. The *p*-fluoro and *p*-chloro derivatives yielded separable keto and enol forms (Table 4.10). The size of the ring system significantly influences the percentage of enol, thus 2-benzoylindanone has 77%, 2-benzoyltetralone 25% and 2-benzoylcyclohexanone 32%.

Table 4.10 Keto-enol tautomerism in benzoylcyclohexanones

	Equilibrium mixture		Keto form		Enol form	
Compound	$\lambda_{max.}$ nm	$\epsilon \times 10^{-3}$	$\lambda_{max.}$ nm	$\epsilon \times 10^{-3}$	$\lambda_{max.}$ nm	$\epsilon \times 10^{-3}$
p-Fluoro	248	12·4	*248	14	*248	5·2
	315	5·0	315	1·7	315	12·4
	(enol. *ca.* 25%)					
m-Chloro	244	4·8	*244	14·8	*244	4·8
	315	11·2	292	2·3	315	11·2
	(enol 100%)					
p-Methoxy	280	14·9	280	16·3	enol *ca.* 9%	
	325	3·3	—	—	—	—
	(enol *ca.* 9%)					
p-Methyl	257	12·7	251	15·3	enol *ca.* 18%	
	319	11·6	—	—	—	—
	(enol *ca.* 18%)					
o-Methyl	230	1·6	negligible		230	1·1
	302	11·6	—	—	302	11·8
	(enol *ca.* 100%)					
o-Chloro	230	1·6	—	—	230	1·1
	302	11·6	—	—	302	11·8
	(enol *ca.* 100%)					

* Isolated.
(After Campbell & Gilow, 1960).

The evidence makes it now necessary to enquire (*a*) whether dienols are formed from β-diketones and (*b*) whether the enol is chelated. In piperidine (0·001 **M** 1 mm light path) benzoylacetone exhibits $\lambda_{max.}$ 333·5 nm, $\epsilon_{max.}$ 14 000 and it appears that more than 90% of the solute exists in an enol form. The bathochromic shift 310–333·5 nm is large but there is no increase in $\epsilon_{max.}$ and dienolization becomes improbable. Benzoylcamphor which cannot dienolize shows in the same solvent a red shift 315–328 nm with no change in $\epsilon_{max.}$.

Anhydrous sodium benzoylacetone (m.p. 225°) is prepared in a mixture of dry ether and hexane by the interaction of sodium ethoxide and benzoylacetone. Sidgwick & Brewer (1925) recrystallized the product from 96% ethanol and obtained a dihydrate (m.p. 112°) which they formulated as:

C_6H_5—C(—O)=CH—C(=O)—CH_3 chelated to Na, with two H_2O coordinated to Na

It was insoluble in hexane but dissolved in pure (alcohol-free) chloroform to give $\lambda_{max.}$ 314 nm, $\epsilon_{max.}$ 14 000 with apparently no 247 nm band. When benzoylacetone was dissolved in absolute ethanol containing 1 equivalent of sodium ethoxide, a single absorption band $\lambda_{max.}$ 320 nm, $\epsilon_{max.}$ 14 500 appeared with no true 247 nm band. With increasing excess of sodium ethoxide new bands appeared with $\lambda_{max.}$ 325 nm, $\epsilon_{max.}$ rising to 18 250, and an additional peak at $\lambda_{max.}$ 238 nm, ϵ 8000–9000. The presence of two isosbestic points

Figure 4.5 Absorption spectra of benzoylacetone in alcoholic sodium ethoxide

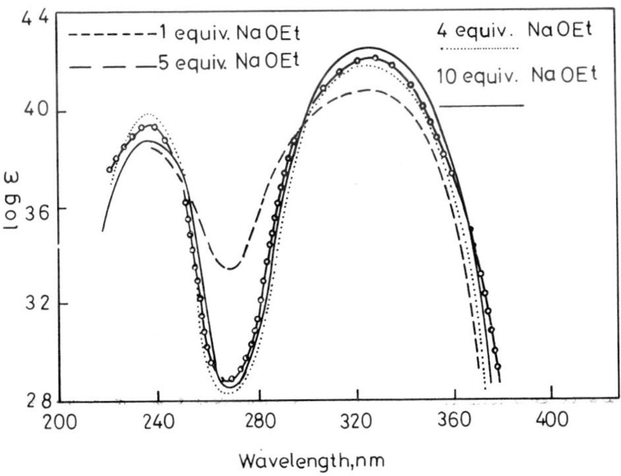

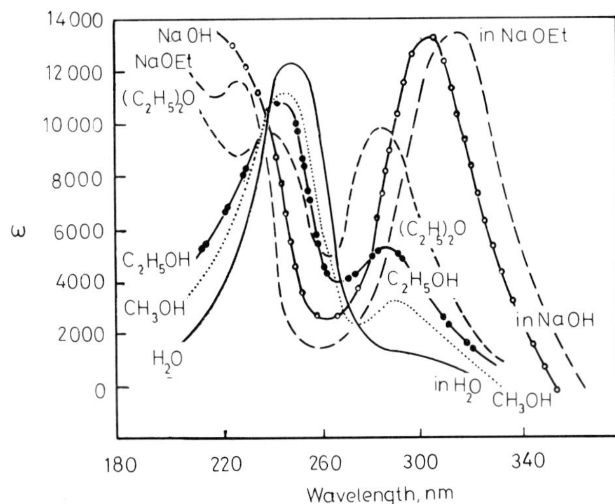

Figure 4.6 Absorption spectra of ethyl benzoylacetate in various solvents

suggested a two-component system (Fig. 4.5). Crystalline lithium, aluminium and cerium derivatives also showed in ethanol peaks at 320–322 and 240–248 nm.

Acetylphenylacetylene when dissolved in concentrated sulphuric acid forms benzoylacetone which is precipitated on dilution (Nef, 1899). The scheme:

$$C_6H_5C{\equiv}CCOCH_3 + H_2O \rightarrow C_6H_5C(OH){=}CHCOCH_3$$

is probably much too simple but this enol must at least be a connecting link. Benzoylacetone in sulphuric acid shows a very broad band (285–380 nm $\lambda_{max.}$ 348 nm, $\epsilon_{max.}$ *ca.* 40 000) with narrow bands at 269 and 277 nm, $\epsilon_{max.}$ 4100 and 4500. The data of Scheibe (1925) on benzophenone and phorone in concentrated sulphuric acid also show displacements which must be attributed to complex formation at the carbonyl group or possibly to charge transfer.

The question arises whether under certain conditions benzoylacetone can form any dienol. C-Methylbenzoyl acetone in the presence of several equivalents of sodium ethoxide shows maxima at 239 and 319 nm ($\epsilon_{max.}$ 7400 and 10 500). Dienolization of the type

$$C_6H_5C(OH){=}C{=}C(OH)CH_3$$

is here impossible but

$$C_6H_5C(OH){=}C(CH_3){-}C(OH){=}CH_2$$

is conceivable. In the case of acetylacetone, bromine

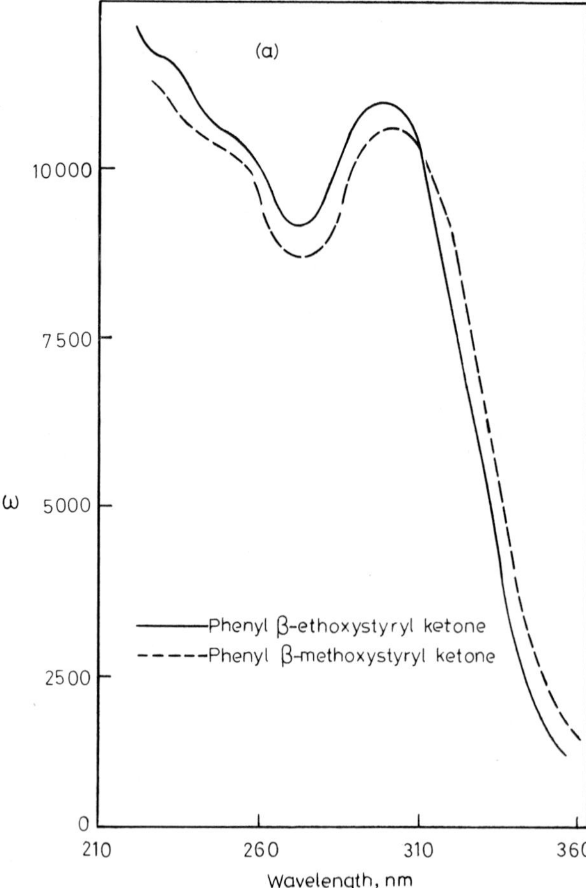

Figure 4.7(a) Absorption spectra of phenyl β-ethoxystyryl ketone and phenyl β-methoxystyrylketone
(b) Absorption spectra of dibenzoylmethane in various solvents

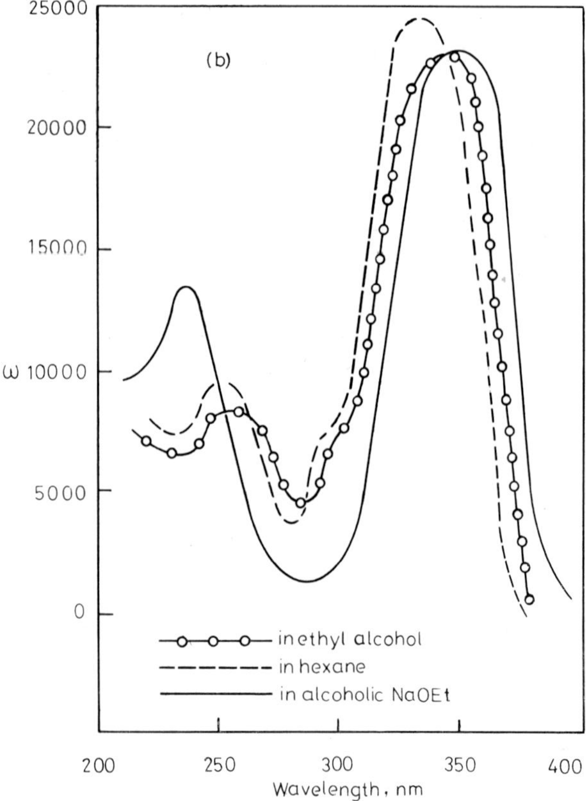

absorption indicated 130% enolization but neither of the following formulations

$$CH_3C{=}C{=}CCH_3 \quad (\text{OH on C-2 and C-4}) \qquad CH_3C{=}CH{-}C{=}CH_2 \quad (\text{OH on C-2 and C-4})$$

would appear to account for the 292 nm band seen in alkaline solutions.

It is possible that at the dilutions needed for spectrophotometry enolic benzoylacetone is not chelated. The evidence is worth scrutiny. Styrylmethylketone (*a*) has one band $\lambda_{max.}$ 285 nm, $\epsilon_{max.}$ 23 000, while phenylstyryl ketone (*b*) shows a similar band at 310 nm, ϵ 26 000. α-Hydroxystyrylmethylketone (*c*) (benzylmethylglyoxalenol *a*) also shows a 310 nm band, $\epsilon_{max.}$ 20 000. When (*a*) is compared with (*c*) there is a red shift, whereas β-methoxystyrylmethylketone (*d*) shows a peak at 274 nm, $\epsilon_{max.}$ 10 300 and phenyl-β-ethoxy or methoxystyrylketone shows $\lambda_{max.}$ 284 nm ($\epsilon_{max.}$ 33 000) with an inflection at 250 nm. Thus there are blue shifts if (*a*) ($\lambda_{max.}$ 285 nm) is compared with (*d*) ($\lambda_{max.}$ 274 nm) and (*b*) ($\lambda_{max.}$ 310 nm) is compared with (*e*).

$$C_6H_5CH{=}CHCOCH_3 \; (a) \qquad C_6H_5CH{=}CHCOC_6H_5 \; (b)$$

$$C_6H_5CH{=}C(OH)COCH_3 \; (c) \qquad C_6H_5C(OMe){=}CHCOCH_3 \; (d)$$

$$C_6H_5C(OMe){=}CHCOC_6H_5 \; (e)$$

Since enolic benzoylacetone and enolic benzylmethylglyoxal both show $\lambda_{max.}$ 310 nm strongly, the six-membered chelate ring, possible for the former but not the latter, cannot be essential for the chromophore. Similarly the 247 nm band of benzoylacetone enol appears to be an undisplaced benzoyl band which does not agree with the 238 nm band seen in the chelated metal derivatives.

Either there is no hydrogen bonding in dilute solution of the enols dealt with here or it must differ in its chromophoric effects from the chelation shown by metallic derivatives. Morton *et al.* (1934) summarized the evidence already available at that time. The acetylphenylacetylene curve shows good resolution in hexane and it is interesting that the envelope almost coincides with that of the methyl ether derived from $\lambda_{max.}$ 274 nm, ϵ 10 300.

Keto-enol tautomerism in β-dicarbonyls has been

studied by nuclear magnetic resonance spectra. Burdett & Rogers (1964) and Allen & Divek (1966) studied acyclic compounds and determined equilibrium constants (enol/keto). Tables 4.11 to 4.13 contain a selection of their findings.

The keto α-protons are deshielded by substitution of electron-withdrawing groups in the α-position. Both keto and enol α protons are deshielded by substitution of electron-withdrawing groups in place of the acetylmethyl group. The enol OH protons show considerable variations in τ but all the values are low, suggesting the presence of intramolecular hydrogen bonds. The infrared carbonyl chelated frequency varies between 1650 and 1605 cm^{-1} and τ varies between −1·2 and −6·6, the relation being roughly linear for β-diketones and β-keto esters.

For most β-keto esters the ketonic form is predominant and there is little steric interaction between R and R″ because OR″ increases the mean RR″ distance. This does not apply to the β-diketones where enolic forms tend to predominate. Substitution of bulky α-substituents causes steric hindrance between the R′ and R (or R″) group protons, especially in enols. The proportion of enol then tends to decrease. In dibenzoylmethane and benzoylacetone (1-phenyl-1,3-butanedione) the presence of the aromatic ring increases the proportion of enol. The aromatic ring could not assume a position parallel to an intramolecular six-

Table 4.11 Nuclear magnetic resonance in keto and enol forms of β-diketones, R′.CO.CHR″.CO.R‴, b=enol, c=keto

				Terminal CH_3 group			3-Proton			
	R′	R″	R‴	CH_3^b	CH_3^c	CH_2^c	CH^c	CH^b	OH	% enol in neat liquid
Compound										
I*	CH_3	H	CH_3	8·0	7·83	6·38	—	4·43	−5·57	79
II	Ph	H	CH_3	8·0	7·88.	6·16	—	4·05	−6·3	98†
III	Ph	H	Ph	—	—	—	—	3·27	−7·0	100†
IV	CF_3	H	CH_3	7·80	7·73	6·05	—	4·00	—	97
V	CF_3	H	CF_3	—	—	5·87	—	3·57	—	100
VI	CH_3	CH_3	CH_3	7·92	7·83	—	6·20	—	−6·5	28 8·75 8·13 α CH_3
VII	CH_3	C_2H_5	CH_3	8·38	8·30	—	6·66	—	—	42
VIII	CH_3	CH_2Ph	CH_3	8·13	8·05	—	5·88	—	—	46
IX	CH_3	CH_3	Ph	7·90	7·83	—	5·40	—	—	4
X	Ph	CH_3	Ph	—	—	—	—	5·02	—	0

* Effect of solvent on enol content (%), no solvent 22, dilute solutions in cyclohexane, carbon tetrachloride 94–95, acetone 25, dimethyl sulphoxide 60.
† In CCl_4.
(from Allen & Divek, 1966).

Table 4.12 Proton chemical shifts for β-keto esters,

RCOCH$_2$COOR′ ⇌ R—C(OH)=CH—C(=O)—OR, k=keto, e=enol

	Ethyl		Acetyl		α-Proton				
	CH_3	CH_2	CH_3^e	CH_3^k	CH_2^k	CH^k	CH^e	OH^e	
Ethylacetoacetate	8·77	5·85	8·05	7·78	6·52	—	5·03	−2·2	—
Methylacetoacetate	—	—	—	7·92	5·33	—	—	—	7·92 OCH_3
Butylacetoacetate	—	—	8·10	7·83	6·59	—	5·17	−2·2	—
Ethyl-α-methylacetoacetate	8·77	5·87	8·02	7·82	—	6·47	—	−2·62	8·75 α CH_3
Ethylbenzoylacetate	8·88[k]	5·9[k]	—	—	6·03	—	4·3	−2·83	—
	8·78[e]	5·78[e]	—	—	—	—	—	—	—
β-Bromoethylacetoacetate	—	5·52	8·02	7·90	6·42	—	4·88	−1·87	—

Table 4.13 Percentage enolic form and equilibrium constants for β-diketones and β-keto esters from p.m.r. data, k=keto, e=enol

Compound	% enol	K_a	Proton signals integrated
Acetylacetone	81	4·3	CH^e/CH_2^k
α-Bromoacetylacetone	46	0·85	CH_3^e/CH_3^k
α-Methylacetylacetone	30	0·43	α CH_3^e/α CH_3^k
Benzoylacetone (in CCl_4)	100	—	—
Dibenzoylmethane (in CCl_4)	100	—	—
1,3-Indanedione (in $CHCl_3$)	0	—	—
2-Phenyl-1,3-indanedione (in dioxane)	0	—	—
Ethylacetoacetate	8	0·09	CH_3^e/CH_3^k
Methylacetoacetate	0	0	—
Butylacetoacetate	15	0·18	CH_3^k/ethyl CH_2
Ethylbenzoylacetate	22	0·28	$CH^e/(CH_3)_3^{e+k}$
Bromoethylacetoacetate	6	0·06	CH_3^e/CH_3^k

membered ring of the enol because of steric interaction with the enol α-proton:

$$C_6H_5-\underset{\substack{|\\O}}{C}=CH-\underset{\substack{\|\\O}}{C}-CH_3 \quad (O\text{—}H\cdots O)$$

Neither 1,3-indandione nor 2-phenyl-1,3-indandione gives evidence of enol formation in any of several solvents.

A substantial investigation of n.m.r. spectra was made by Allen & Divek (1966) with results summarized in Table 4.11.

Gorodetsky *et al.* (1967) studied the tautomerism of benzoylacetone totally labelled with ^{17}O.

$$C_6H_5-\underset{\substack{\|\\^{17}O}}{C}-CH_2-\underset{\substack{\|\\^{17}O}}{C}-CH_3$$

The chemical shifts of the ^{17}O resonances were studied by Christ *et al.* (1961) and they recorded a range extending over 1000 p.p.m. which they divided into intervals characteristic of functional groups containing the oxygenation. The chemical shifts for hydroxyl oxygens are about ± 40 p.p.m., whereas those for a carbonyl group are near -550 p.p.m., both relative to water. Thus monoketones include: acetone (-572), cyclohexanone (559), cyclopentadone (548). Ketonic forms of β-diketones show 500–600 p.p.m. (mainly 540–580) and although two resonances are expected they were not separated in the early work. The enolic form of acetylacetone showed a line at 274 p.p.m. and asymmetric-related compounds showed a line at 258–283 p.p.m. Benzoylacetone showed lines at 233 and 291, indicating the presence of two enols in unequal amounts. Thereupon benzoylacetone was synthesized bearing the ^{17}O label only on the oxygen of the benzoyl group. One line only appeared in the n.m.r. spectrum, namely at 234, and this indicated that the predominant enol was

$$C_6H_5-\underset{\substack{|\\OH}}{C}=CH-COCH_3$$

a conclusion already reached from ultraviolet absorption.

Despite the range of methods deployed and the amount of work done over some 50 years, the subject of keto-enol tautomerism still presents problems due no doubt in part to the differences in conditions required by the various techniques.

In β-diketones and β-keto esters there is in terms of ultraviolet absorption spectra a distinction between enolic forms and chelated metallic derivatives. The evidence that benzoylacetone enolizes at the benzoyl carbonyl and methylbenzoylacetone on the acetyl carbonyl is against the hydrogen bonding to form a six-membered ring. There is chemical evidence for some dienolization but no dienol structure can readily be fitted into a simple chromophore theory.

4.2.1 Infrared Absorption Spectra for Keto-enol Systems

These have been studied by Rasmussen *et al.* (1949), Rasmussen & Brittain (1949), Hunsberger (1950), Park *et al.* (1953) and Bellamy & Beecher (1954). Kohrausch & Pongratz (1934) studied Raman spectra. Several of these groups found it convenient to consider aromatic hydroxyaldehydes and hydroxyketones at the same time. There is wide agreement that conjugation-chelation occurs.

Table 4.14, based on a collection of data made by Bellamy & Beecher (1954), shows how the carbonyl frequency in enol chelates is modified, in the extreme case from *ca.* 1725 to *ca.* 1600 cm^{-1}. Fig. 4.8 shows that when the frequency is plotted against double bond character straight lines are obtained for aldehydes and ketones, acids and esters, the modified carbonyl frequency being lowest in each case for 100% double bond character as defined by Pauling's fractional values (Hunsberger *et al.*, 1952). It will be noted from Table 4.14 that although conjugation chelation

Table 4.14 Assigned infrared carbonyl frequencies

Compound	Frequency (cm^{-1})	Assignment	Concentration effect
Acetylacetone	1724	Keto	(*a*)
	1608	Enolchelate	
Benzoylacetone	1724	Keto	(*a*)
	1600	Enolchelate	
Dibenzoylmethane	1600	Enolchelate	(*a*)
2-Formylcyclohexanone	1706	Ketoaldehyde	(*a*)
	1596	Enolchelate	
Ethylacetoacetate	1733	Estercarbonyl	(*a*)
	1709	Ketocarbonyl	
	1645	Chelate	
Ethyl-2-oxo-*cyclo*-hexane-1-carboxylate	1724	Ketocarbonyl	(*a*)
	1704	Estercarbonyl	
	1634	Chelate	
5,5′-Dimethyl*cyclo*-hexane-1,3-dione	1733	Keto	(*b*)
	1706	Free enol	(*b*)
	1603	Chelate	(*b*)
Salicylaldehyde	1668	Chelate	(*a*)
2,4-Dihydroxybenzaldehyde	1653	Chelate	(*a*)
1,5-Dihydroxyanthraquinone	1631	Chelate	(*a*)
	1626	Chelate	(*a*)
1,2,5,8-Tetrahydroxyanthraquinone	1600	Chelate	(*a*)

certainly occurs in salicylaldehyde and related compounds the double bond character is less than 50%. This also applies to the ethylacetoacetate chelate.

The simple conjugation effect on the carbonyl frequency is seen in phorone and acetophenone 1672 and 1686 cm^{-1}. In diacetone alcohol hydrogen bonding has a small effect (1712 cm^{-1}) and in diacetyl (one band 1718 cm^{-1}) it is negligible compared with benzil (1681 cm^{-1}) which is still small. In acetylacetone there is a big shift and the modified C=O absorption seems to mask the C=C frequency which in acetylacetone enolacetate is revealed at 1633 (1761 and 1695 cm^{-1} for C=O). In *o*-hydroxy-acetophenone and salicylaldehyde a conjugated chelate system is postulated. The unconjugated hydrogen bonded hydroxyl shows 3333 cm^{-1} but in β-keto enols this is moved to 2703 cm^{-1} with C=O 1653 cm^{-1}. Similar values are formed in 2-hydroxy-1-naphthaldehyde and 1-hydroxy-2-naphthaldehyde (*ca.* 1635 cm^{-1}).

Park *et al.* (1953) discuss 'syn' (left) and 'anti' (right) forms of acetylacetone:

H_3C–C(OH)=CH–C(=O)–CH_3 (syn) $\qquad$ HO(H_3C)C=CH–C(=O)–CH_3 (anti)

Hydrogen bonding cannot occur in the 'anti' form.

Meyer (1912) found by bromine absorption that acetylacetone in pyridine contained 179–202% of enol, perhaps due to

$CH_3C(OH)$=CH–C(OH)=CH_2

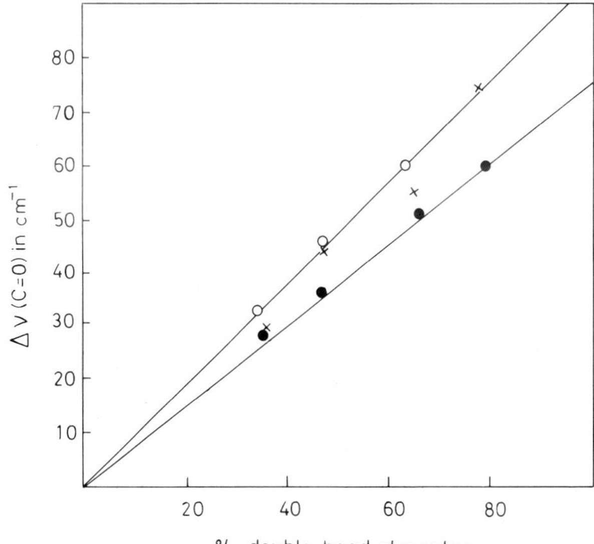

Figure 4.8 Relation of infrared data to aromatic bond character. ○: hydroxymethyl esters; ●: hydroxyaldehydes; × hydroxy methyl ketones (after Hunsberger, Ketcham & Gatowsky, 1952)

For the compound 5,5-dimethyl*cyclo*hexane-1,3-dione conjugated chelation is sterically ruled out, nevertheless the enol shows 2632 and 1605 cm^{-1}. The only explanation offered here is formation of a dimer

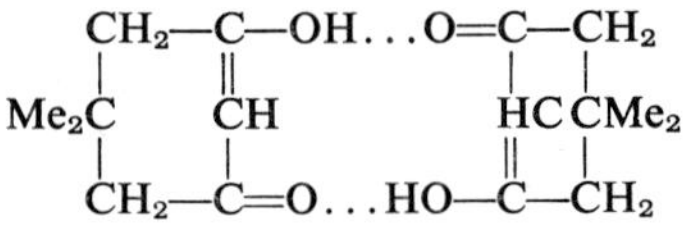

```
     CH2—C—OH...O=C—CH2
     |   ||        | |
Me2C     CH        HC CMe2
     |   |         || |
     CH2—C=O...HO—C—CH2
```

Bellamy & Beecher (1954) consider that in acids and ester the modified carbonyl frequency is a linear function of the double bond character but the OH or OR groups reduce the slope of the line.

Bellamy & Branch (1954) studied the metal chelates but did not find that the results were easy to interpret convincingly (Table 4.15).

α-Keto acids are well-known intermediates in the metabolism of aminoacids and Stephani & Meister (1971) have investigated the dimeric compound derived from asparagine via α-ketosuccinamic acid. The structure attributed to the dimer (4,6-dihydroxy-2-oxo-piperidine-5-carboxamide-4,6-dicarboxylic acid) is shown below.

The sodium salt of the dimer was dissolved in 1 N HCl, placed at 37° and the absorption spectrum of the solution measured at zero time. The curve showed inflections at 250–270 nm ($\epsilon=100$) and at 310–340 nm ($\epsilon=10$). After standing for 1 hour at 37° a maximum appeared at 260 nm ($\epsilon=110$) and the inflection at 310–340 nm became stronger (*ca.* 20). On standing at 37° for 8 hours the 260 nm peak became much stronger ($\epsilon=2230$) while the inflection, now near 325 nm, reached $\epsilon=525$. The 260 nm band is ascribed to the unsaturated amino chromophore in (*a*) and the band at 325 nm to a new absorbing unit. With further incubation (for 21 hours) the intensity of the 260 nm band fell ($\epsilon=$ *ca.* 1500) and maxima appear near 330 nm ($\epsilon=$ *ca.* 1000). The new entity was regarded as (*b*), a dehydration product of (*c*). Further incubation (48 hours) resulted in a very well developed absorption curve with $\lambda_{max.}$ 325 nm, $\epsilon=$ *ca.* 7000. (All the ϵ values given here are based on the concentrations expressed in terms of the original dimer.) An authentic specimen of 2-hydroxypyridine-4,6-dicarboxylic acid (*d*) showed in water $\lambda_{max.}$ 325 nm, $\epsilon=1220$.

The α-keto succinamic dimer was reduced (platinum oxide, hydrogen, acid medium) to yield product (*e*) showing $\lambda_{max.}$ 262 nm, $\epsilon=2100$ which in alkaline medium (1 N NaOH) showed $\lambda_{max.}$ 286 nm ($\epsilon=22\,400$) The effect of alkali was to open the ring and enolize one of the carbonyl groups.

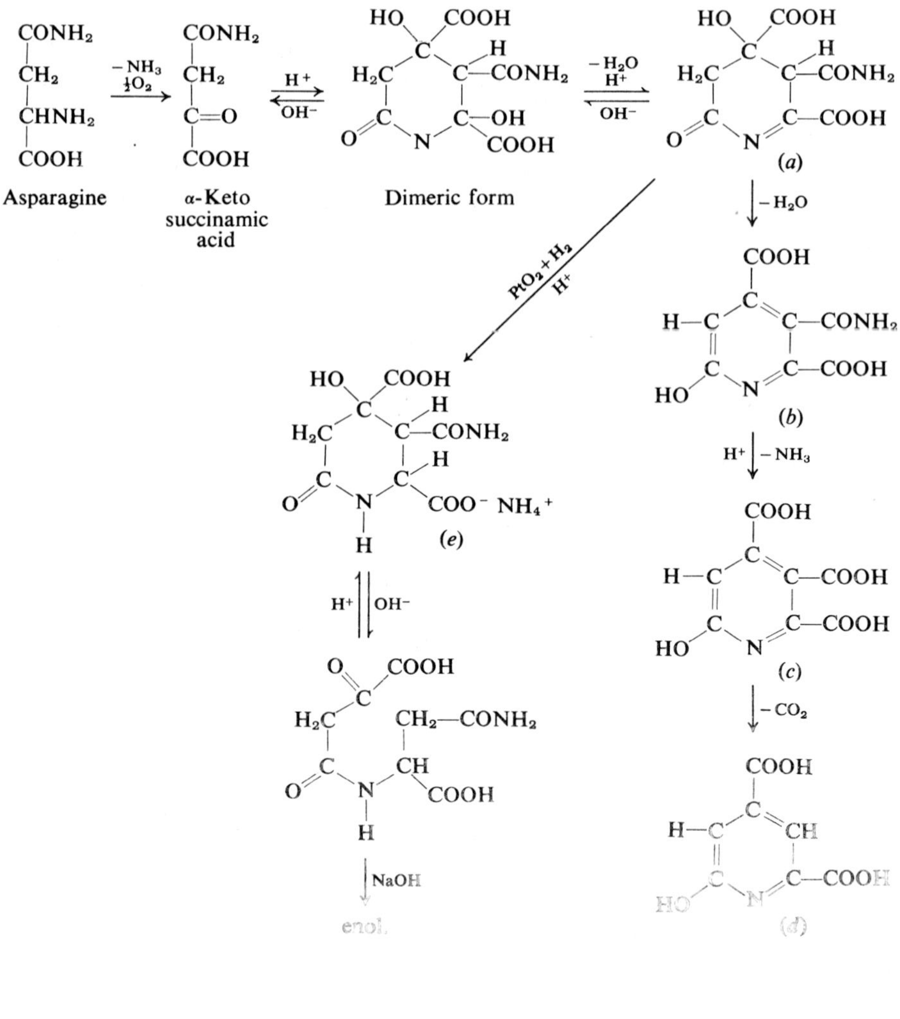

Asparagine → ($-NH_3$, $\frac{1}{2}O_2$) → α-Keto succinamic acid ⇌ (H^+ / OH^-) Dimeric form ⇌ ($-H_2O$, H^+ / OH^-) (*a*)

(*a*) → ($-H_2O$) → (*b*) → (H^+, $-NH_3$) → (*c*) → ($-CO_2$) → (*d*)

Dimeric form → (PtO_2+H_2, H^+) → (*e*) ⇌ (H^+ / OH^-) → (NaOH) → enol

Table 4.15 Copper chelate compounds

C=O frequencies cm^{-1} *in dilute chloroform solutions* Ligand	cm^{-1}	
Acetylacetone	1580	1389
Ethylacetoacetate	1570	1282
Benzoylacetone	1550	1389
Dibenzoylmethane	1524	1391
2-Formyl*cyclo*hexanone	1592	1353
Ethyl-2-oxo-*cyclo*hexanecarboxylate	1590	1290

Carbonyl frequency cm^{-1} *of metal chelate compounds*

	Salicylaldehyde (cm^{-1})	Acetylacetone (cm^{-1})	
Copper	1608	1580	1390
Cobalt	1656	1586	1373
Nickel	1652	1605	1389
Magnesium	1681	1590	1397
Zinc	1658	1577	1373
Cadmium	1585	—	—
Iron	—	1575	1370

Morton & Stubbs (1939) studied thione-thiol tautomerism in 2-mercaptobenzthiazole. In ethanol the thione form is predominant and this appears to be independent of the solvent (Koch, 1949). The substitution product 6-methyl-2-mercaptobenzthiazole behaves similarly (Hasan & Hunter, 1936). In alkaline solutions the 2-mercaptobenzthiazoles form the thiols, the sodium derivative of which shows a spectrum resembling the S-alkyl compounds. Mercaptotetrazoles have also been studied (Lieber *et al.*, 1959).

2-Mercapto-5-aminothiazole in neutral solution shows $\lambda_{max.}$ 228 and 338 nm, (ϵ 4·5 and $10{\cdot}7\times10^{-3}$) due to the thione form. This is also true for the 4-alkyl derivatives but in alkali the thiol structure appears with an absorption spectrum close to that of the N-acetyl-S-methyl derivative ($\lambda_{max.}$ *ca.* 300 nm, $\epsilon_{max.}$ *ca.* 2000). 2-Mercapto-5-amino-4-phenylbenzthiazole appears to exist as the thiol even in neutral solution. Comparisons with related quinolone derivatives given in Table 4.16 are instructive. Here replacement of O by S has a moderate bathochromic effect (309 → 327 nm). The classical case of carbostyril (*h*) was first studied by Hartley & Dobbie (1899) but the technique then in use was inadequate. Morton & Rogers (1925) found that the parent substance had two equally strong bands with peaks at 269 and 327 nm while the N-methyl ether had a very similar spectrum. The *o*-alkyl ethers were different in that the main absorption was split into two peaks at 322 and 308·5 nm, $\epsilon_{max.}$ 4500 and 3700 respectively (see Table 4.16).

The absorption spectra of hydroxyaldehydes and hydroxy ketones containing an aromatic grouping raise many interesting issues.

Benzene (see p. 5) in hexane shows well-resolved absorption between 230 and 270 nm, ϵ 60–300. Neglecting the fine structure, $\lambda_{max.}$ for the envelope is near 255 nm. There is also a band at 190 nm, ϵ 10 000. Compared with cyclohexadiene ($\lambda_{max.}$ 260 nm, ϵ 4550) the phenyl group absorbs much less strongly and must be treated as a chromophore distinct from C=C—C=C. Substitution by certain groups in

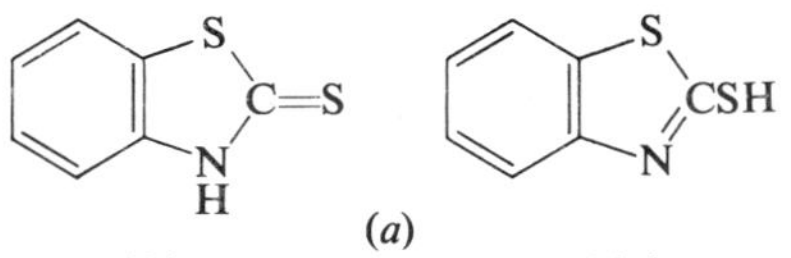

(*a*) Thione Thiol 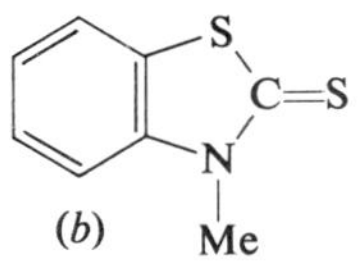 (*b*) 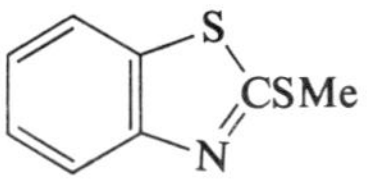 (*c*) 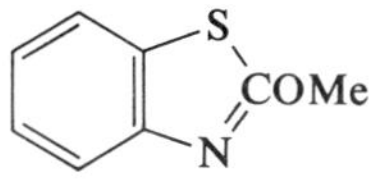 (*d*)

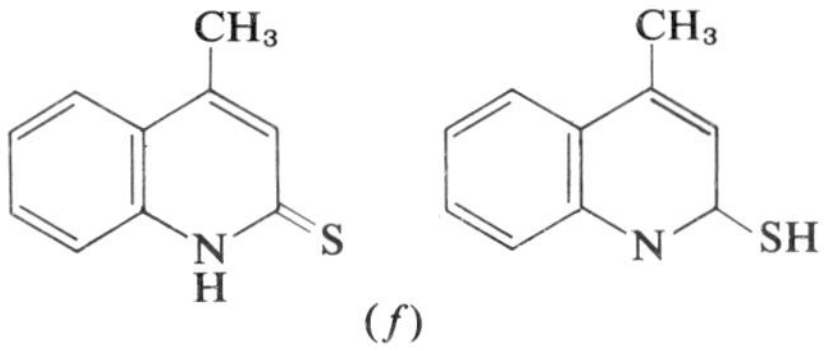

(*f*) 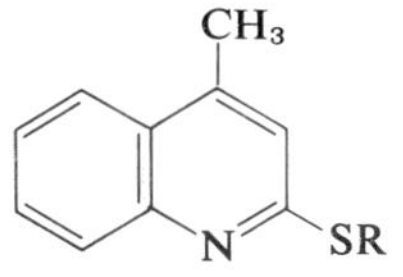 (*g*) 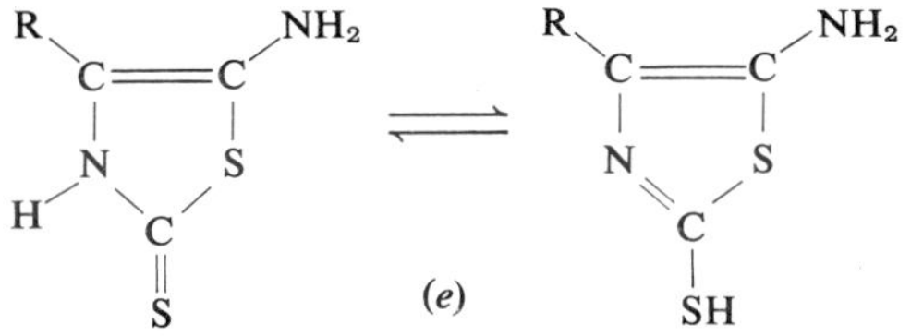 (*e*)

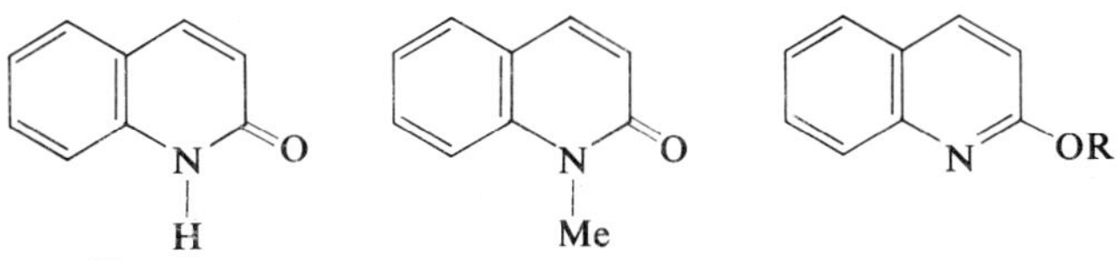

(*h*)

Table 4.16 Absorption maxima of mercaptobenzthiazoles and related substances

Compound	$\lambda_{max.}$ nm	$\epsilon_{max.} \times 10^{-3}$	$\lambda_{max.}$ nm	$\epsilon_{max.} \times 10^{-3}$	$\lambda_{max.}$ nm	$\epsilon_{max.} \times 10^{-3}$	$\lambda_{max.}$ nm	$\epsilon_{max.} \times 10^{-3}$
2-Mercaptobenzthiazole								
(*a*)	235	1·26	~252	6·92	~282	2·19	325	26·9
(*b*)	231	13·5	~255	7·41	~282	2·69	324·5	25·7
	241	13·8	—	—	—	—	—	—
(*c*)	224	22·91	—	—	—	—	—	—
	~244	8·32	280	12·4	~290	10·2	300·5	8·51
(*d*)	—	—	~260	7·94	~282·5	1·12	296·2	10
(1-Hydroxy-5-methyl-benzthiazole)	—	—	—	—	291·2	1·0	—	—
2-Thiol-4-methylquinoline								
(*f*)	241	2·72	278·5	22·9	—	—	383	12·6
(*g*) (R=Me)*	255	27	277	5·89	~327	5·62	339	6·46
* R=C_2H_5 or isopropyl very similar curves.								
Carbostyril								
(*h*)	269	7	327	6·75				
N ether	270·5	6·55	328	6·1				
O ether	308·5	4·5	322·6	4·5				

Figure 4.9 Absorption spectra of 1-thiolbenzthiazole, 1-thio-2-methyl-1:2-dihydrobenzthiazole and 1-methylthiobenzthiazole

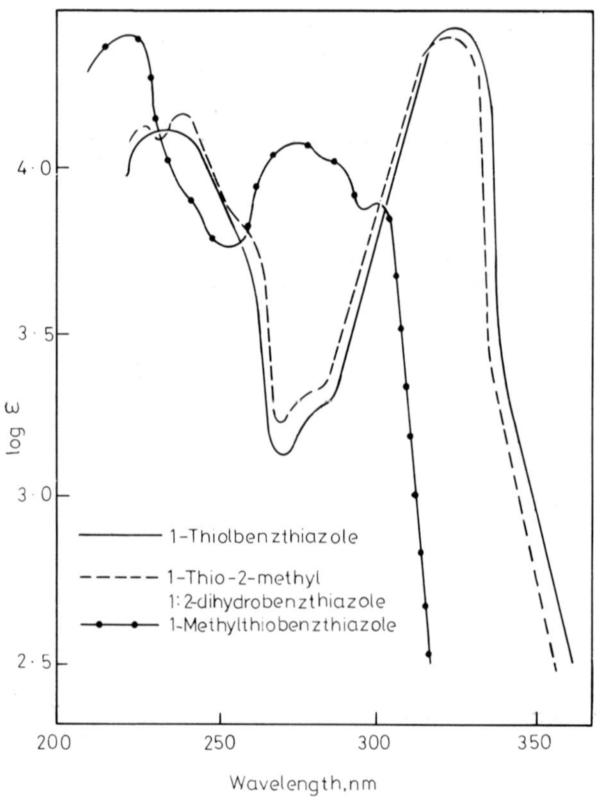

Figure 4.10 Absorption spectra of 2-thiol-4-methylquinoline and 2-*iso*propylthio-4-methylquinoline

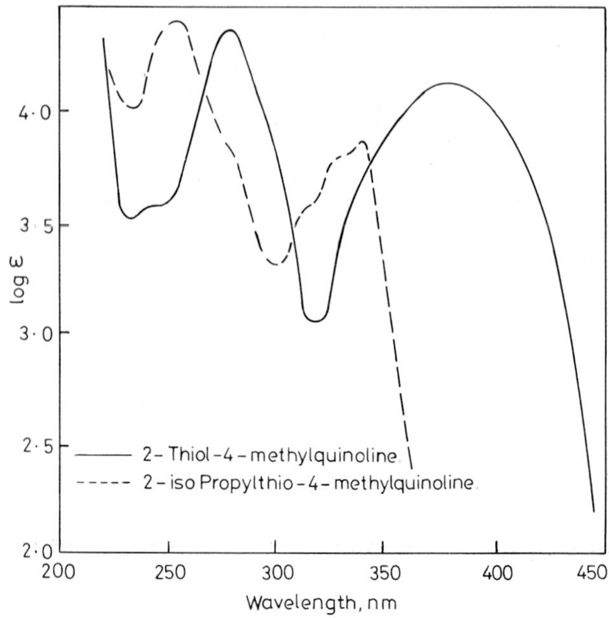

themselves transparent in the region 210–300 nm effects red shifts, thus phenol has peaks at 215 and 273 nm, ϵ 10 000 and 1780 while aniline has peaks at 234 and 284·5 nm, ϵ 11 500 and 1740 respectively. The substituents, besides their bathochromic effect, raise the intensity of the longer wavelength band from 3- to 10-fold. The two bands are benzenoid in origin. Alkyl or halogen substituents have much less marked effects on the benzenoid absorption.

Disubstitution with alkyl or halogen groups results in red shifts that are not very large, and only small changes in $\epsilon_{max.}$. When the substituents are —OH, —COOH or NH_2 the effect of a second substituent is less than that of the first.

If the two substituents separately exert considerable effects new constitutive effects appear and the importance of position isomerism is more marked. The spectra of the hydroxy anilines closely resemble those of the phenylenediamines but considerable displacements occur in the hydroxy- and aminobenzoic acids. The red shift is greatest for the *o*-compounds and the intensity increment is fairly large. The *p*-derivatives exhibit a single band of quite high intensity and this masks the possible presence of a second absorption band. All three *o*-, *m*- and *p*-hydroxybenzoic acids show a band in the far ultraviolet 206, 207, ~210 nm (ϵ 21·4, 21·4, $11·5\times10^{-3}$ respectively) attributable to the carboxyl group.

When the phenyl and carbonyl chromophores occur in the same molecule as in benzaldehyde or acetophenone each modifies the other but the spectrum is an additive one. The ultraviolet absorption spectra of the two substances are very similar and can be summarized:

$\lambda_{max.}$ nm	199	240	278	320
$\epsilon_{max.}$	20 000	13 200	1050	50

The 320 nm band is due to the carbonyl group influenced by the phenyl group and the 240 and 278 nm peaks reflect the phenyl chromophore modified by the conjugated carbonyl group

$$\underset{\displaystyle O^*}{\overset{}{C_6H_5\underset{\|}{C}}} \qquad \underset{\displaystyle O}{C_6H_5\underset{\|}{C}—R}$$

The 278 nm band is the 255 nm benzene band displaced and the 240 nm peak corresponds to the stronger band observed with phenol, benzoic acid and aniline. The conjugated carbonyl group thus behaves towards the benzenoid chromophore like OH, NH_2 and COOH.

The absorption spectrum of *o*-hydroxybenzaldehyde may first be considered. Introduction of the hydroxyl group increases the intensity of the 325 nm band from ϵ 18 to 3020 and the 243 nm peak is displaced to 255 nm. In *o*-methoxybenzaldehyde the bathochromic effects though smaller are essentially similar.

Table 4.17 Aromatic hydroxy-aldehydes, hydroxy-ketones and methylethers

	Hexane		Ethanol		Water		Alkali 1 eq.		NaOH (excess)		Vapour
	$\lambda_{max.}$ nm	log ϵ	$\lambda_{max.}$ nm	log ϵ	$\lambda_{max.}$ nm	log ϵ	$\lambda_{max.}$ nm	log ϵ	$\lambda_{max.}$ nm	log ϵ	$\lambda_{max.}$ nm
OH, CHO (ortho)	215	4·23	—	—	—	—	—	—	—	—	—
	250	4·04	250	4·0	255	4·0	261	3·85	265	3·81	251·5
	257	4·07	—	—	—	—	—	—	—	—	—
	320·5	3·56	—	—	—	—	~327	3·10	—	—	320
	327	3·58	325	3·48	325	3·43	379	3·70	378	3·74	—
OH, $COCH_3$ (ortho)	213	4·25	—	—	—	—	—	—	—	—	—
	250	3·95	251·5	3·97	252·5	4·01	254	3·88	256·5	3·76	243
	256	3·94	—	—	—	—	—	—	—	—	250
	324	3·51	—	—	324·5	3·52	—	—	—	—	321
	331	3·53	327	8·50	—	—	341	3·50	359	3·71	—
$COCH_3$, OH (para)	—	—	220·5	4·03	—	—	225	3·71	235	3·81	—
	—	—	276	4·13	—	—	236	3·70	—	—	—
	—	—	—	—	—	—	~278	3·79	—	—	—
	—	—	—	—	—	—	325	4·18	325	4·39	—

Table 4.17 *continued*

	Hexane		Ethanol		Water		Alkali 1 eq.		NaOH (excess)		Vapour
	$\lambda_{max.}$ nm	log ε	$\lambda_{max.}$ nm	log ε	$\lambda_{max.}$ nm	log ε	$\lambda_{max.}$ nm	log ε	$\lambda_{max.}$ nm	log ε	$\lambda_{max.}$ nm
CHO; OH	265	4·28	221	4·14	220	4·06	240	3·81	239	3·93	252
	281	4·08	284	4·24	284	4·24	~288	3·75	~275	3·35	~298
	288	3·78	~332	2·25	~333	2·78	325	4·36	330	4·42	—
	309	1·9	—	—	—	—	329	—	—	—	—
	318	2·0	—	—	—	—	—	—	—	—	—
CHO; OH	213	4·28	255–260	3·93	253	4·00	239	4·13	237	4·37	239
	247	4·00	—	—	—	—	~257	3·90	~264	3·81	296
	~282	3·17	~293	3·04	—	—	321	3·27	~312	2·78	—
	305	3·5	316	3·46	315	3·42	~352	3·15	359	3·47	329
	~311	3·46	—	3·50	—	—	—	—	—	—	—
$COCH_3$; OH	211	4·31	—	—	—	—	—	—	—	—	—
	243	3·94	—	—	—	—	—	—	—	—	—
	~248	3·86	—	—	—	—	—	—	—	—	—
	301	3·61	—	—	—	—	—	—	—	—	—
	309	3·32	—	—	—	—	—	—	—	—	—
			Acid 0·1 N HCl		pH 11		0·1 N NaOH				
$COCH_3$; OH; OH	—	—	205	3·18	206	4·05	—	—	—	—	—
	—	—	229	4·11	248	4·10	262	4·00	—	—	—
	—	—	274	4·0	~290	3·65	314	3·95	—	—	—
	—	—	305	3·87	342	4·21	380	4·03	—	—	—
$COCH_3$; OH; HO; OH	214	4·12	—	—	—	—	—	—	—	—	—
	241	4·08	—	—	—	—	—	—	—	—	—
	280	4·03	280	4·00	—	—	—	—	—	—	—
	349	3·85	345	3·79	—	—	—	—	—	—	—
$COCH_3$; HO; OH; MeO; OMe; OMe	283	4·10	—	—	—	—	—	—	—	—	—
	356	3·45	—	—	—	—	—	—	—	—	—
	Methanol										
$COCH_3$; OH; Et; OH	217	4·48	—	—	—	—	—	—	—	—	—
	231	4·36	—	—	—	—	—	—	—	—	—
	278	4·34	—	—	—	—	—	—	—	—	—
	323	4·06	—	—	—	—	—	—	—	—	—

Table 4.17 *continued*

	Hexane		Ethanol		Cyclohexane		Alkali 1 eq.		NaOH (excess)		Vapour
	$\lambda_{max.}$ nm	log ε	$\lambda_{max.}$ nm	log ε	$\lambda_{max.}$ nm	log ε	$\lambda_{max.}$ nm	log ε	$\lambda_{max.}$ nm	log ε	$\lambda_{max.}$ nm
$COCH_3$, OCH_3	242·5	3·90	240	4·06	211	4·31	—	—	—	—	288·5
	300	3·58	305	3·58	~238	3·91	—	—	—	—	—
	~330	2·0	—	—	243	3·92	—	—	—	—	—
	—	—	—	—	298	3·53	—	—	—	—	—
	—	—	—	—	~308	3·53	—	—	—	—	—
CHO, OCH_3	246·5	4·08	253	4·07	—	—	—	—	—	—	238·5
	310	3·75	319·5	3·63	—	—	—	—	—	—	299
CHO, OCH_3	249	3·93	252·5	3·92	—	—	—	—	—	—	240
	309	3·46	314·5	3·45	—	—	—	—	—	—	296·5
CHO, OCH_3	265(fs)	4·29	219	4·07	—	—	—	—	—	—	—
	300–	—	277	4·17	—	—	—	—	—	—	—
	350(fs)	—	—	—	—	—	—	—	—	—	—
$COCH_3$, OH, OCH_3	—	—	270	4·2	—	—	—	—	—	—	—
	—	—	315	3·8	—	—	—	—	—	—	—
COBu, HO, OH	—	—	223	4·10	—	—	—	—	—	—	—
	—	—	270	4·03	—	—	—	—	212	4·11	—
	—	—	342	3·45	—	—	—	—	285	3·93	—
	—	—	—	—	—	—	—	—	387	3·67	—
CHO	—	—	244	4·2	241	4·17	—	—	—	—	—
	—	—	—	—	~247	4·10	—	—	—	—	—
	—	—	280	3·18	278·5	3·05	—	—	—	—	—
	—	—	(279	3·25)	287	2·95	—	—	—	—	—
	—	—	—	—	307	1·28	—	—	—	—	—
		—	—	—	317	1·43	—	—	—	—	—
	—	—	325	1·25	328	1·45	—	—	—	—	—
	—	—	—	—	348	1·44	—	—	—	—	—
	—	—	—	—	353	1·30	—	—	—	—	—
$COCH_3$	—	—	242	4·18	238	4·10	—	—	—	—	—
	—	—	278	3·04	278	1·95	—	—	—	—	—
	—	—	—	—	285	2·94	—	—	—	—	—
	—	—	319	1·90	321	1·64	—	—	—	—	—

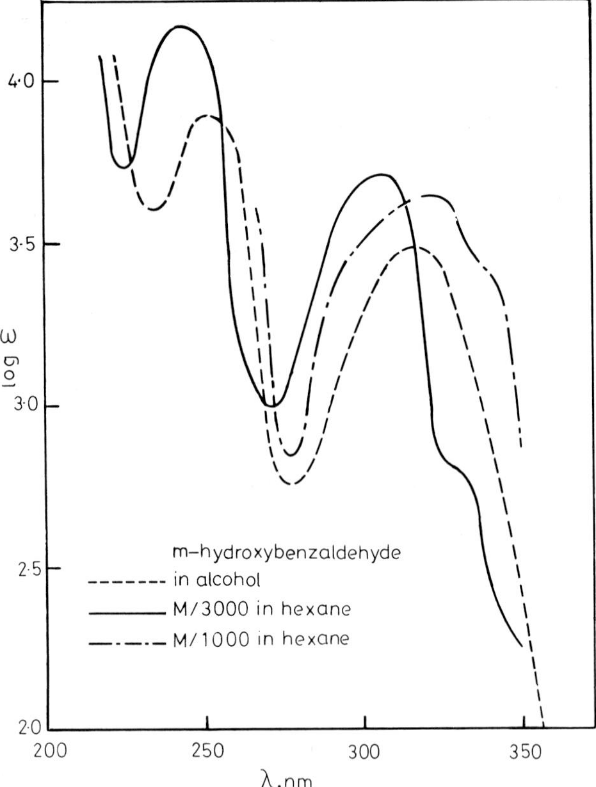

Figure 4.11 Absorption spectra of m-hydroxybenzaldehyde in alcohol, M/3000 in hexane and M/1000 in hexane

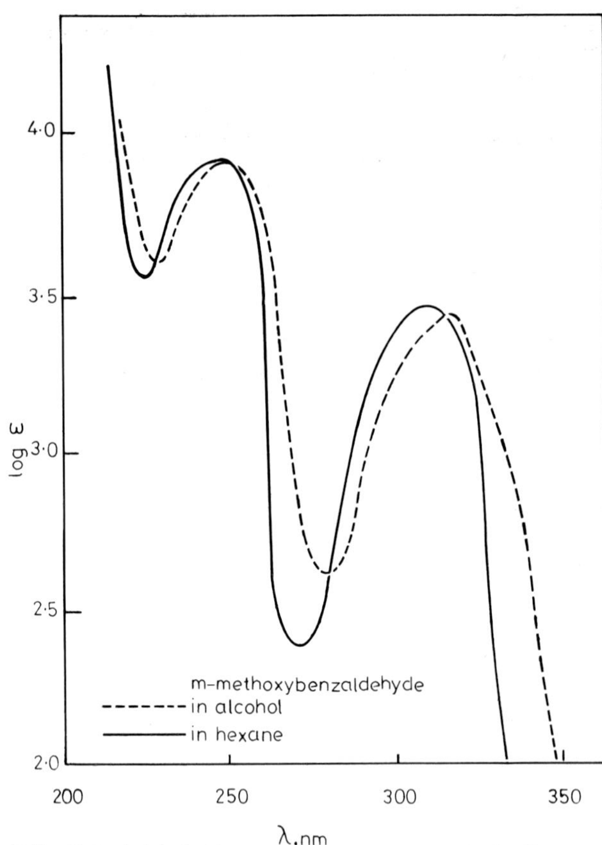

Figure 4.13 Absorption spectra of m-methoxybenzaldehyde in alcohol and hexane

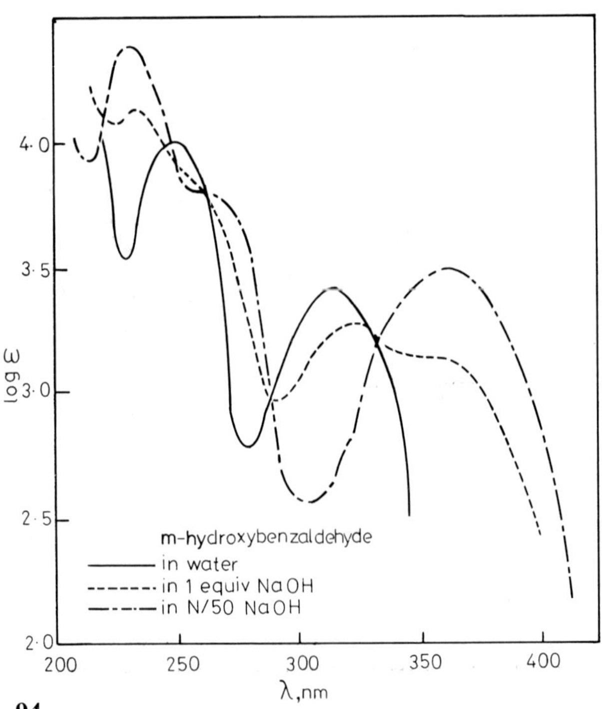

Figure 4.12 Absorption spectra of m-hydroxybenzaldehyde in water and sodium hydroxide

Note on Table 4.17

Benzaldehyde and acetophenone give very similar spectra with three regions $\lambda_{max.}$ *ca.* 240 nm ϵ *ca.* 15 000; 280 nm, ϵ 1000 and 320 nm, ϵ *ca.* 20. The *m*-hydroxy derivatives and their methyl ethers display very similar spectra except that the weak carbonyl absorption is much stronger ϵ 300–310 nm, being of the order 3000.

The *o*-hydroxy compounds show $\lambda_{max.}$ near 250 nm, ϵ *ca.* 10 000 and 325 nm, ϵ *ca.* 4000. The *o*-methyl ethers show little change at 240–250 nm and the long wave band undergoes a blue shift 325–310 nm.

The *p*-hydroxy compounds show bands at 220–230 nm, ϵ *ca.* 10 000, a very strong band *ca.* 280 nm, ϵ *ca.* 15 000 and a very weak 320 nm band which could easily be masked. (See Morton & Stubbs, 1939, 1940; Lemon, 1947.)

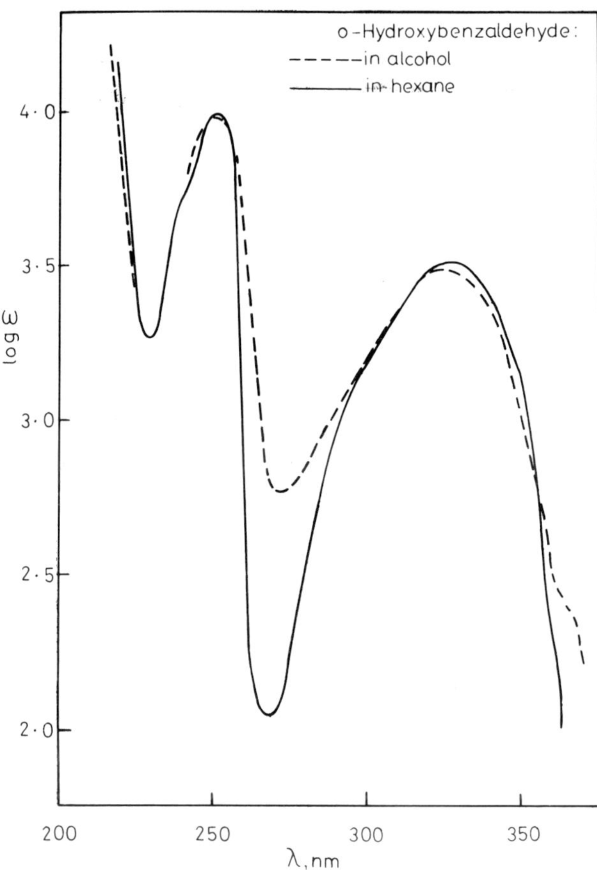

Figure 4.14 Absorption spectra of o-hydroxybenzaldehyde in alcohol and hexane

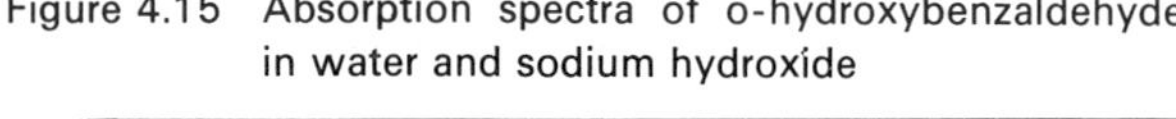
Figure 4.15 Absorption spectra of o-hydroxybenzaldehyde in water and sodium hydroxide

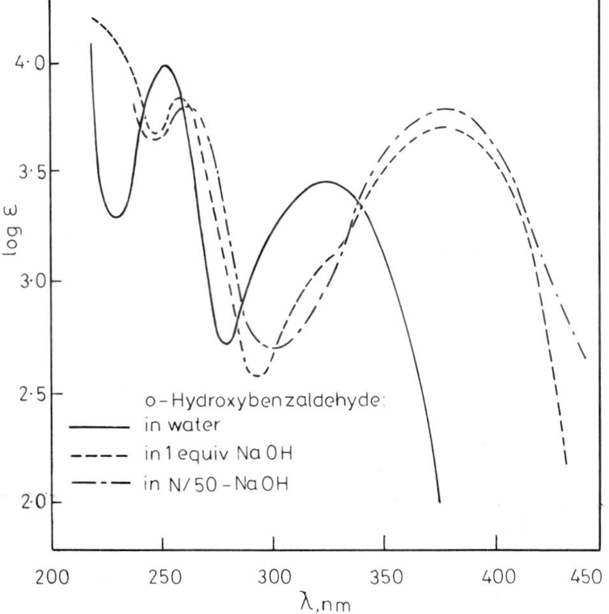

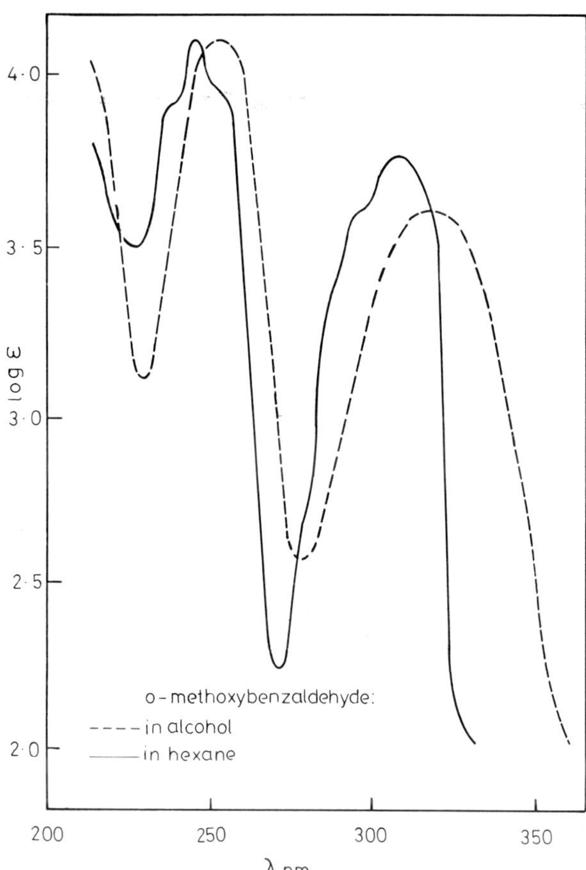

Figure 4.16 Absorption spectra of o-methoxybenzaldehyde in alcohol and hexane

Figure 4.17 Absorption spectra of o-hydroxyacetophonone in alcohol and hexane

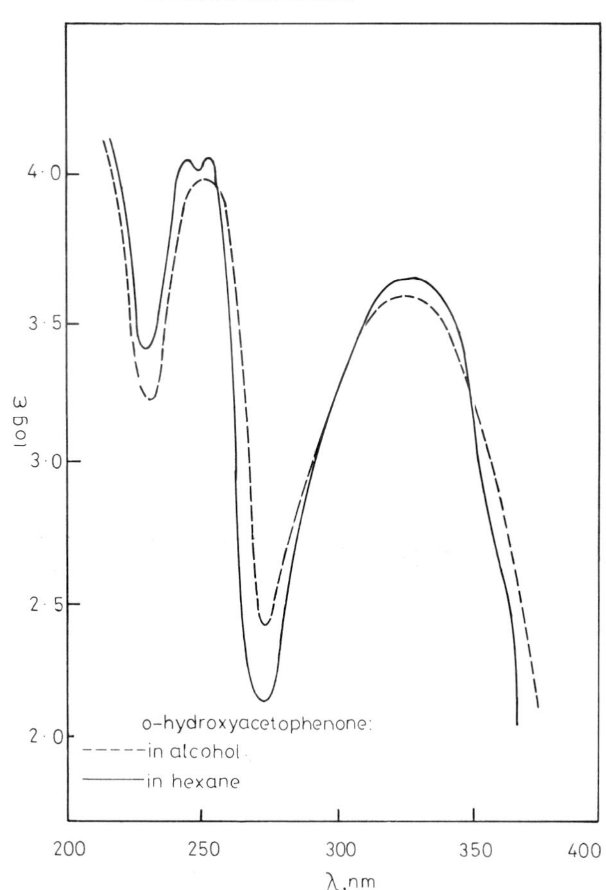

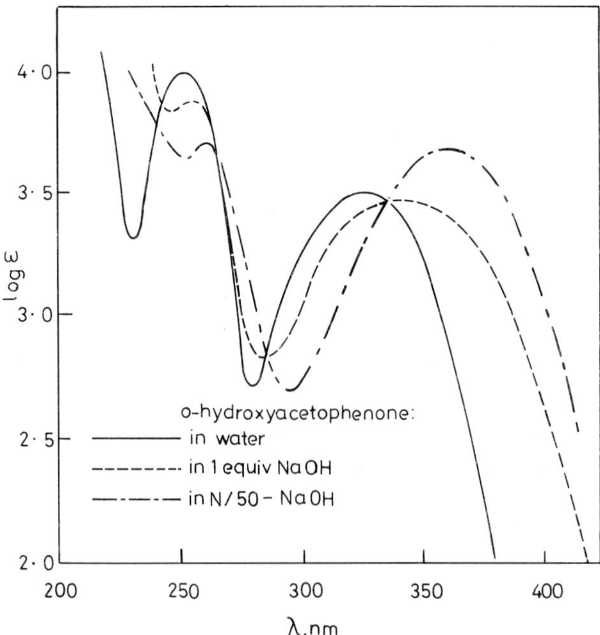

Figure 4.18 Absorption spectra of o-hydroxyacetophenone in water and sodium hydroxide

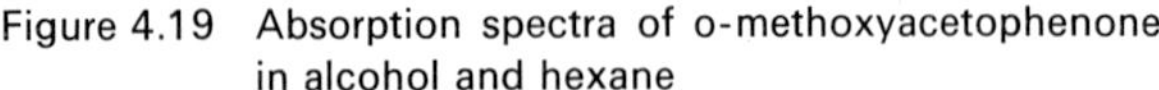

Figure 4.19 Absorption spectra of o-methoxyacetophenone in alcohol and hexane

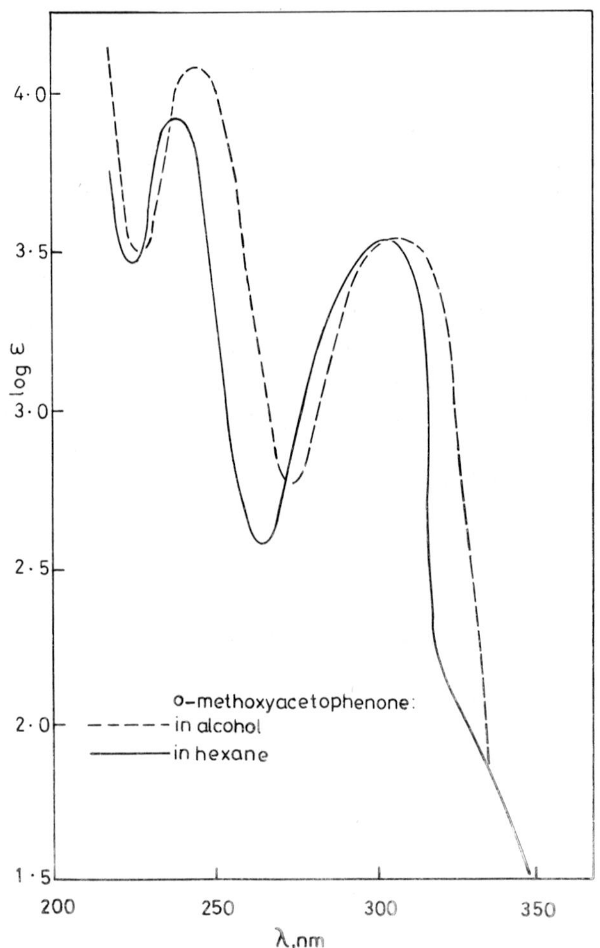

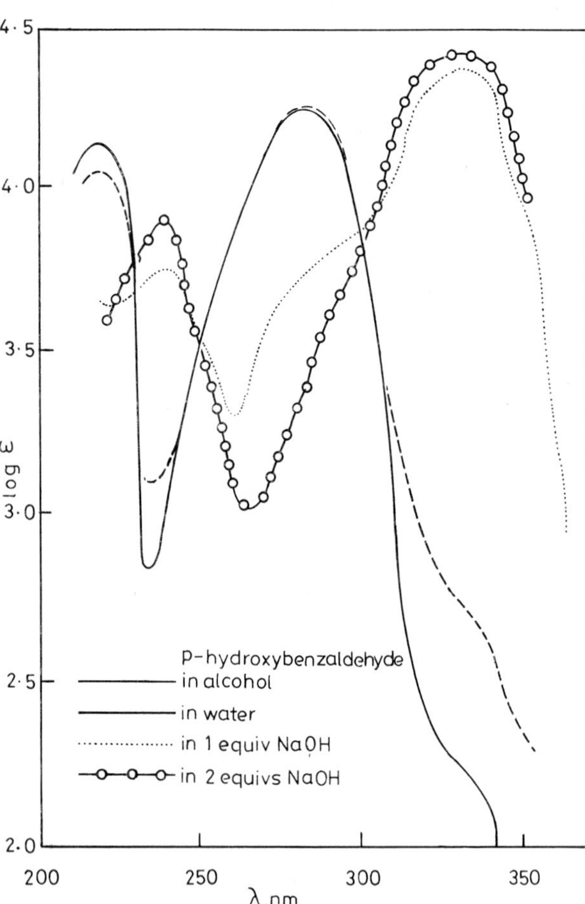

Figure 4.20 Absorption spectra of p-hydroxybenzaldehyde in alcohol, water and sodium hydroxide

The di- and tri-hydroxy derivatives show the *o*-substituent and the *p*-hydroxy exert stronger effects than the *m*-hydroxy group. Apart from loss of fine structure in ethanol or water the solvent has no strong effect. Alkali causes red shifts and increases the intensity of the long wave band.

The shift is greater for *o*-derivatives than for *p* but the intensity increment is much greater for the *p*. Even the *m* compounds show a large red shift with alkali without concomitant rise in ϵ max. In 3,4-dihydroxyacetophenone the *p*-effect is dominant.

The 2,6-dihydroxybutyrophenone spectrum has a large bathochromic shift to 342 nm but no great increase in ϵ max. With alkali the red shift is large—to 387 nm.

Johnson & Critchfield (1961) determined small amounts of 1,4-benzoquinone by the reaction with 2,4-dinitrophenyl-hydrazine followed by treatment of the bis-hydrazone with diethanolamine in pyridine to

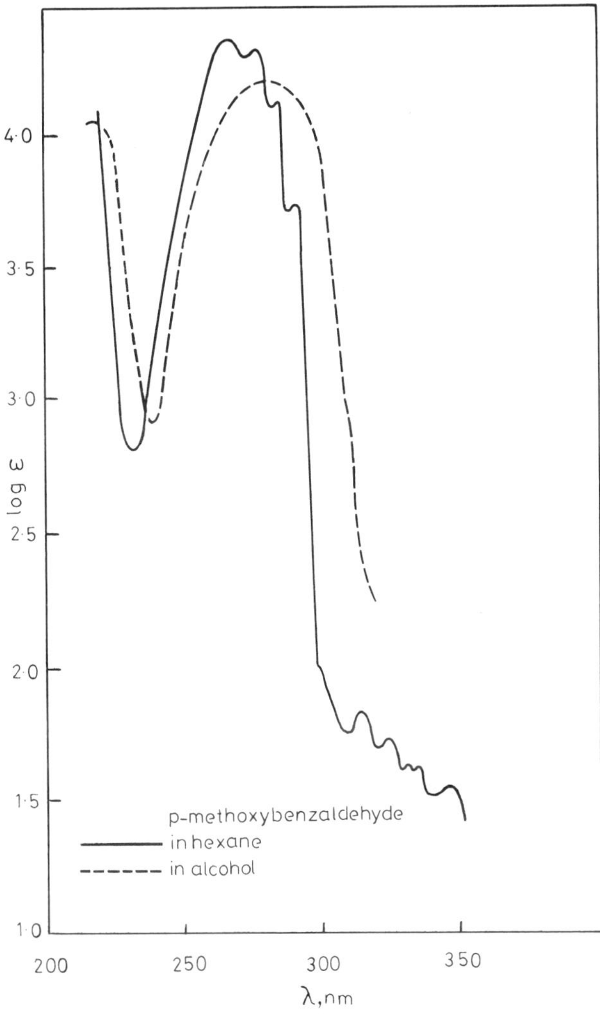

Figure 4.21 Absorption spectra of p-methoxybenzaldehyde in hexane and alcohol

give a blue ion which was measured by spectrophotometry. Johnson *et al.* (1962) extended the method to other multifunctional carbonyl compounds (cf. Neuberg & Strauss, 1945, 1946). The colour due to excess of 2,4-dinitrophenylhydrazine is eliminated by treating the reaction mixture with 2,4-pentanedione which forms a colourless pyrazole.

The procedure is to dissolve the carbonyl compound in an appropriate solvent and add to an aliquot (not more than 3 ml) 1 mg of 10% sulphuric acid and 1 ml of the dinitrophenyl-hydrazine reagent. The mixture is then left for a fixed time at a temperature previously found to be suitable. When the reaction has been completed two drops of pentanedione are added to remove excess of reagent. After cooling, water is added and the product is extracted into 5 ml of methylene chloride. The lower layer is then separated and eva-

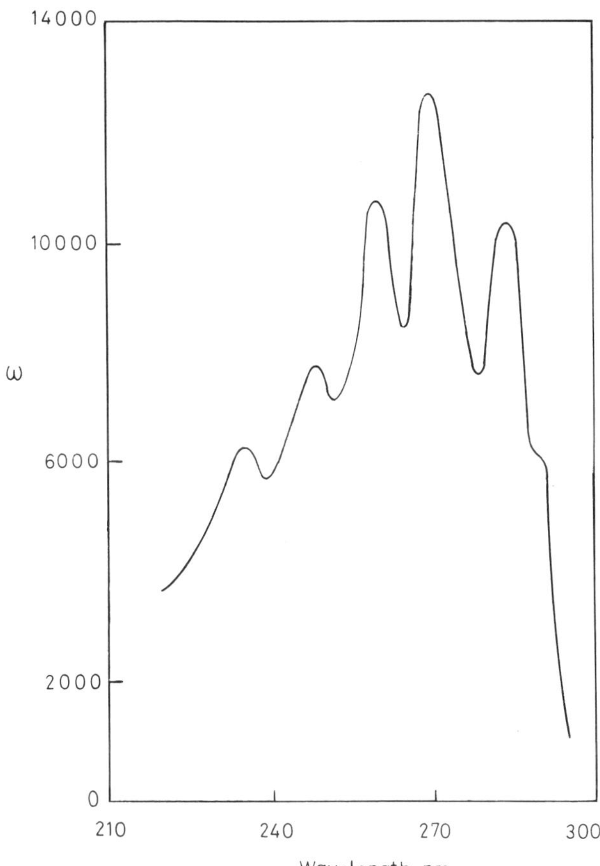

Figure 4.22 Absorption spectrum of acetylphenylacetylene in hexane

Figure 4.23 Absorption spectrum of phenylacetylene in hexane

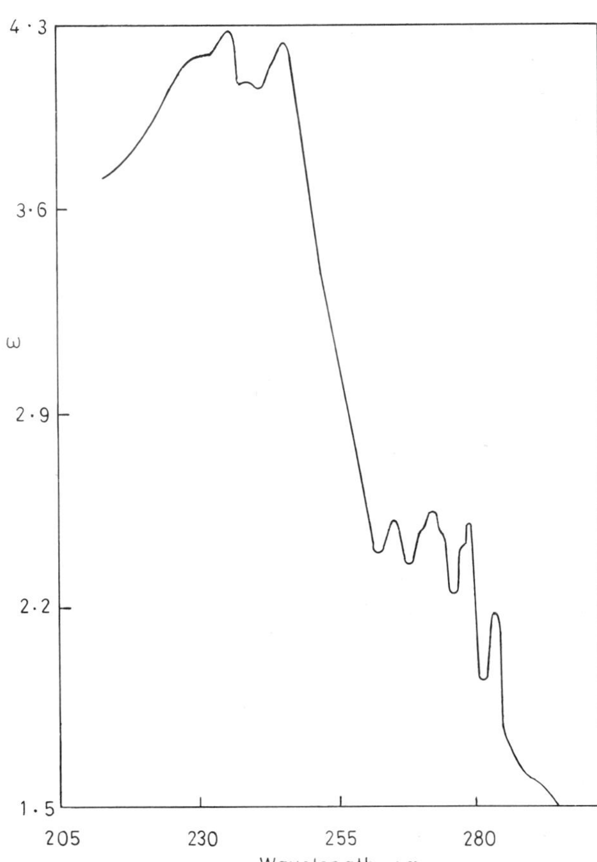

Table 4.18 Bis (or tris) 2,4-dinitrophenylhydrazones examined in a basic medium

Carbonyl compounds	Solvent	Reaction conditions (Temp. °C*)	$\lambda_{max.}$ nm	ϵ/μmole
1,4-Benzoquinone	w	25	615	3·24
1,2-Naphthoquinone	mc	70	640	2·23
1,4-Naphthoquinone	mc	70	645	2·94
Benzil	mc	80	580	1·70
1,2,3-Triketohydrindene	w	50	520	1·33
Glyoxal	w	98	580	3·54
Pyruvic aldehyde	w	98 (15 min.)	575	2·52
Diacetyl	w	98	570	1·49

* Time 30 min. except where shown.

porated under suction at room temperature. The diethanolamine-pyridine reagent (10 ml) is then added and the colour is measured.

The colour obtained when a bis-2,4-dinitrophenylhydrazone is treated with a base is attributed to nitrogen deprotonization to produce a fully conjugated ion which in the case of the product from 1,4-benzoquinone contains three conjugated quinoid moieties.

Vicinal dicarbonyl compounds form red colours with $\lambda_{max.}$ 520–580 nm whereas conjugated (or quinonoid) dicarbonyl compounds give blue colours with $\lambda_{max.}$ 615–645 nm. The tris-2,4-dinitrophenylhydrazone of triketohydrindene shows the colour of both types and indeed a peak occurs at 520 nm with a marked inflection at 620 nm at about half the intensity of the main peak.

4.3 S-acetyl-N-acetyl-cysteamine

The synthetic compound S-acetoacetyl-N-acetyl-cysteamine (or S-acetoacetyl-N-acetylthioethanolamine) $CH_3 . CO . CH_2 . CO—S—CH_2 . CH_2 . NH_2 . COCH_3$ shows a λ peak at 303 nm ϵ 5500 in slightly alkaline solution (pH 8) due to the enolic form. At pH 6·2 the absorption in this region has disappeared and all that remains is a flat inflection near 280 nm (see Fig. 4.24). At both pH values there is a good peak near 235 nm but this varies little. The 303 nm band is attributed to the chromophore $CH_3—C(OH)=CH.COS—R$ (Lynen & Ochoa, 1953). In the presence of magnesium

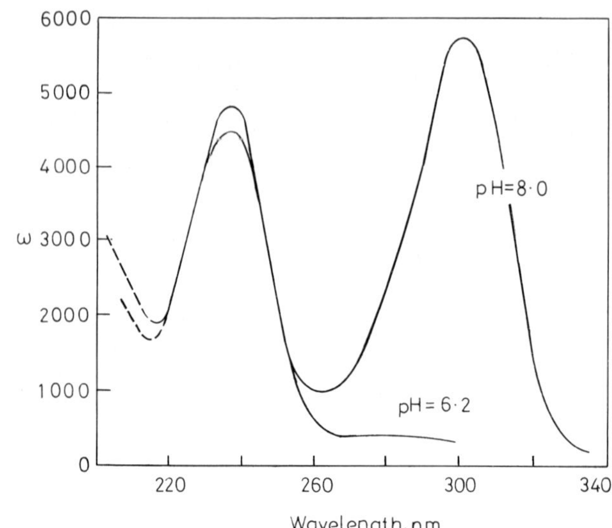

Figure 4.24 Ultraviolet absorption spectrum of S-acetoacetyl-N-acetylthioethanolamine. It is not clear which curve is which below 255 nm (after Lynen & Ochoa, 1953)

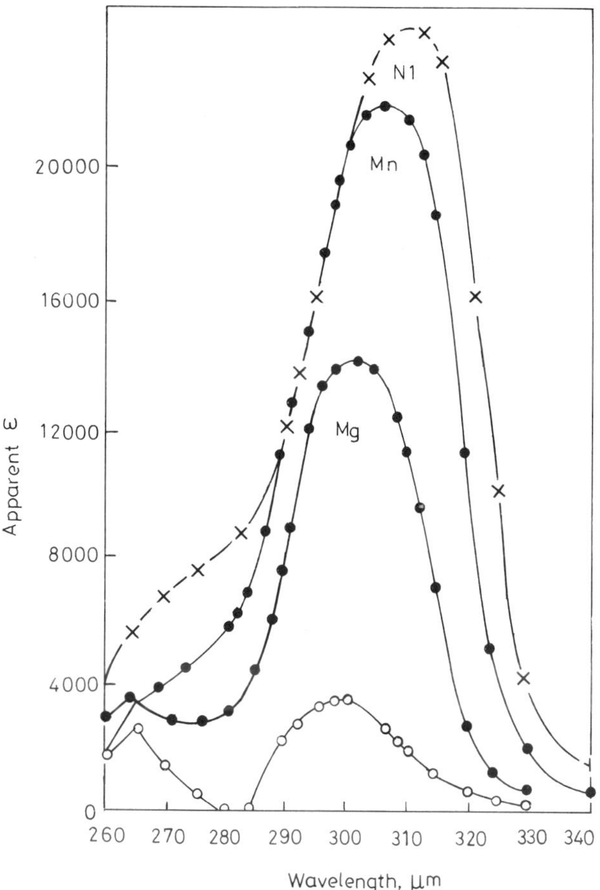

Figure 4.25 Effect of divalent ions on the absorption spectrum of acetoacetyl-S-CoA. These are difference absorption spectra obtained by first measuring the enolate (or enolate+chelate) spectrum at pH 8·1, then adding HCl (final concentration, 0·1 **N**) to obtain the non-enolate absorption spectrum. The latter is subtracted from the former. The difference spectrum between 260 and 280 nm may be attributed in part to changes in absorption of the adenosine moiety of the thio ester. The lowest curve represents acetoacetyl-S-CoA spectrum in the absence of metal. The metal indicated was added as chloride salt, final concentration 5×10^{-3} **M**. Acetoacetyl-S-CoA concentration $0{\cdot}68\times10^{-4}$ **M** (total acetoacetic thio ester concentration $0{\cdot}76\times10^{-4}$ **M**). Standard optical conditions (from Morton, 1959)

a chelate complex is formed and $\epsilon_{max.}$ is doubled. (see Fig. 4.25).

Acetoacetyl-S-pantetheine Mg chelate

Table 4.19 Effect of pH on enolization and chelation*

	Metal ions not added (%)		Magnesium added 5×10^{-3} M Mg^{++}		
pH	Keto	Enol	Keto	Enol	Chelate
		(*a*) of acetoacetyl-S-CoA			
7·05	97·6	2·4†	93·09	0·37	5·7
8·13	87·8	13·2	53·4	2·40	43·2
9·15	66·2	33·8	19·5	9·83	70·7
10·25	12·7	87·3	~0	~0	~100
		(*b*) of acetoacetyl-S-panthetheine			
6·85	98·0	2·0	96·2	0·86	2·91
8·05	82·3	17·7	70·7	11·2	18·1
9·40	24·8	75·2	8·2	29·0	62·8
10·25	6	94	—	—	—

* Abbreviated from Tables by Stern (1956).

† Calculated from the equation % enol = 100 ϵ/25 000 (enol peak at 303 nm).

(See also Lynen *et al.*, 1952; Stern *et al.*, 1956; Morton, 1959.)

Formyl indole (3-hydroxymethylene indolenine)

Acetoacetyl-CoA also shows the 303 nm peak and a 260 nm peak due to the purine moiety. With increasing pH the former increases in intensity while the latter is substantially unchanged. Table 4.19 summarizes the results for acetoacetyl-S-pantotheine and acetoacetyl-CoA.

4.4 Tropone and Tropolones

Tropone (Dauben & Ringold, 1951; Von Doering & Detert, 1951) is a typical non-benzenoid aromatic compound. It is a ketone having three conjugated double bonds in a 7-membered ring. Its ultraviolet absorption (Hosoya *et al.*, 1962) shows peaks around 230 nm (ϵ *ca.* 18 000) and 300 nm (ϵ 4500) with some resolution. The 2,4,6-octatrienal

$$CH_3CH{=}CH{-}CH{=}CH{-}CH{=}CH{-}CHO$$

which has the same number of conjugated double bonds and a carbonyl group also exhibits two ultraviolet absorption peaks (at 222 nm and 308 nm, with ϵ values *ca.* 4000 and 40 000).

Tropone

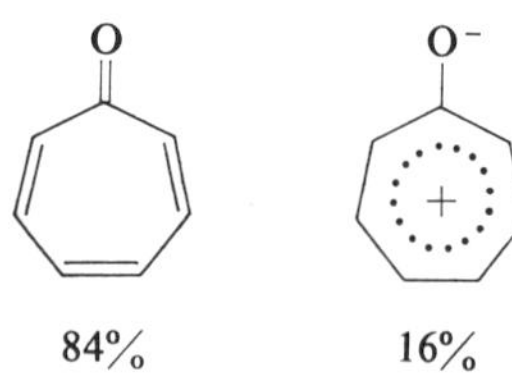

	$\lambda_{max.}$ nm	$\epsilon \times 10^{-3}$
	235	18
	(fine structure)	
	~286	4·1
	298	4·5
	308	4·0
$\lambda_{min.}$	250	0·4

Octatrienal

	$\lambda_{max.}$ nm	$\epsilon \times 10^{-3}$
	222	4·2
	—	—
	—	—
	308	40
$\lambda_{min.}$	250	very low

The two compounds thus resemble one another in respect of band locations but the relative intensities are reversed. Hosoya *et al.* made a theoretical study of the structures of the two molecules and calculated energy levels. It emerged that tropone can be described as a resonance hybrid of two canonical forms:

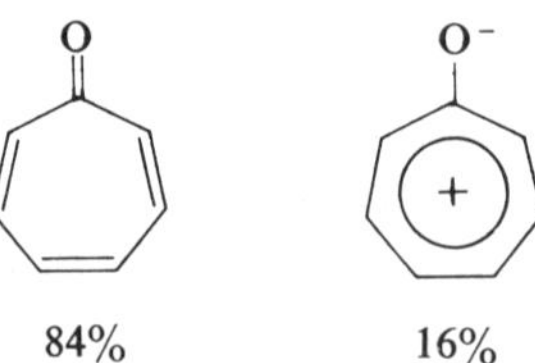

In strong acids tropone shows greatly enhanced absorption. Thus in concentrated sulphuric acid there are peaks near 304–309 nm, ϵ 9000 and 227 nm, ϵ *ca.* 37 000. Tropone has basic properties and protonation occurs with formation of a troponium ion.

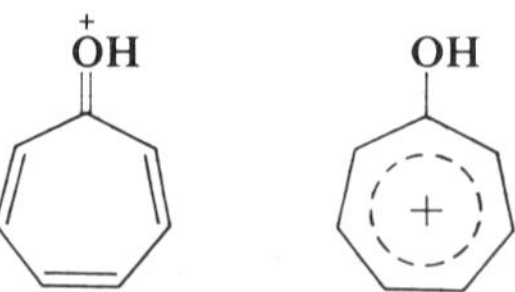

Troponium ion

Tropyliumperchlorate in sulphuric acid on the other hand shows weaker absorption ($\lambda_{max.}$ 273, ϵ 4500), although there is also a strong peak at 218 nm (ϵ 43 500).

Tropolone has a hydroxyl group at a position adjacent to the carbonyl group of tropone.

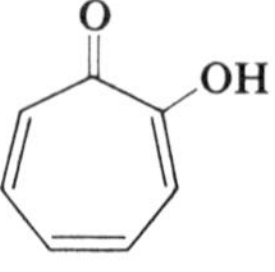

Tropolone

It has a very interesting absorption spectrum (Cook *et al.*, 1951; Hosoya *et al.*, 1962). In cyclohexane there is a very strong absorption (with some resolution at *ca.* 225, 232 and 240 nm, ϵ *ca.* 20 000) and beyond a very low minimum (ϵ *ca.* 550) at 263 nm there is a well-structured and stronger absorption in the region 280–400 nm:

$\lambda_{max.}$ nm	~288	~315	322	340
$\epsilon_{max.}$ (approx.)	2400	5600	7100	4170

$\lambda_{max.}$ nm	357	380	~385
$\epsilon_{max.}$ (approx.)	5000	5000	1500

The 230 nm band seems to correspond with the 220 nm peak of tropone and is probably due to charge transfer from the triene moiety to the carbonyl, while the new absorption may reflect charge transfer between the ring and the hydroxyl group. In water the spectrum is broadly similar but lacks fine structure ($\lambda_{max.}$ 235, 315 and 345 nm, ϵ *ca.* 28 000, 7000 and 5600). In 0·5 N NaOH the 240 nm peak is very strong (*ca.* 170 000) with inflections near 250 and 262 nm and two strong broad bands with peaks at 332 and 392 nm (ϵ *ca.* 41 500 and 34 000 respectively).

The infrared spectrum (Ikegami, 1961) includes peaks at 3208 (OH), 1613s (CO), 1548 v.s. (C=C), 1480 v.s. C—OH, 1425 v.s., 1266 v.s., 1238 v.s., 754, 708, the peaks at 1613 and 1547–1573 cm^{-1} being characteristic for the tropolone ring. The different bands of tropolone are differently influenced by substituents. Thus the very strong absorption near 240 nm is practically unaffected by alkyl substituents but the other bands are more responsive (Mukai, 1958).

Natural tropolones (Nozoe, 1959) are found in a red cedar (*Thuja plicata*) and include the α-, β- and γ-isopropyltropolones which exhibit ultraviolet spectra like that of tropolone. Dolabrinol, pygmaein and α-thujaplicinol are obtained from *Cupressus pygmaea* (Lemm.) Sarg and isopygmaein from *Papuacedrus loricellensis* (Schlechter) L. (Zavarin, 1962; Zavarin & Anderson, 1961).

The substituted tropolones dolabrinol, pygmaein, α-thujaplicinol and isopygmaein exhibit spectra broadly similar to that of tropolone itself; the extra conjugated double bond (in the side-chain) of dolabrinol has little effect other than to *lessen* the intensity of absorption compared with α-thujaplicinol. The methoxyl group in isopygmein also has a hypochromic effect.

Colchicine is obtained from the seeds and corms of the autumn crocus (*Colchicum autumnale*) and colchiceine is derived from it by heating with dilute sulphuric acid. Colchicine has long been used as a specific against gout. It is also important for its mutagenic effects.

It dissolves in sulphuric acid to give a yellow solution and on adding a drop of nitric acid the solution becomes first green then violet and finally red. Ferric chloride gives a deep green colour with colchiceine but not colchicine. These empirical tests are useful. The structures shown below were suggested by Cook and his colleagues (Cook, 1945) and supported by infrared spectra (Scott & Tarbell, 1950; Kemp & Tarbell, 1950).

α-Dolabrinol

OH

HO

O

u.v. absorption

$\lambda_{max.}$ nm	$\epsilon_{max.} \times 10^{-3}$
248·5	20·4
325	4·79
356	3·98
365	4·90
374	6·17

infrared cm^{-1} KBr (S unless otherwise stated)

3250
1613w
1540, 1520, 1455, 1420
1375, 1285, 1265, 1240, 1195
962, 908, 895, 815, 795
(all m)

n.m.r.

α-Thujaplicinol

HO

O

OH

$\lambda_{max.}$ nm	$\epsilon_{max.} \times 10^{-3}$
247·5	42·7
324	5·37
356	6·61
364	7·59
372·5	11·22

3230
1615w
1538, 1520, 1455–1470, 1420
1292, 1262, 1220, 1200, 1180
8·76d (J = 7) $CHMe_2$
6·26 (—$CHMe_2$)

Isopygmaein

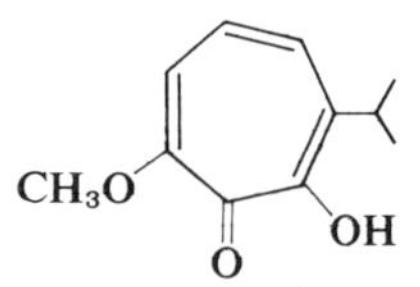

$\lambda_{max.}$ nm	$\epsilon_{max.} \times 10^{-3}$
250	29·5
~313	4·27
326	5·62
358	5·62
373	5·99

3220
1590m
1555, 1475, 1463, 1410
1337, 1290, 1243, 1230, 1210, 1168
1140, 1095
990, 920, 765, 675 (all m)

Pygmaein

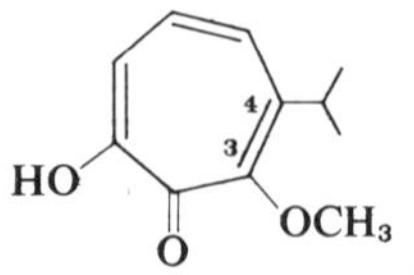

$\lambda_{max.}$ nm	$\epsilon_{max.} \times 10^{-3}$
251	28·2
~310	4·36
325	5·13
358	7

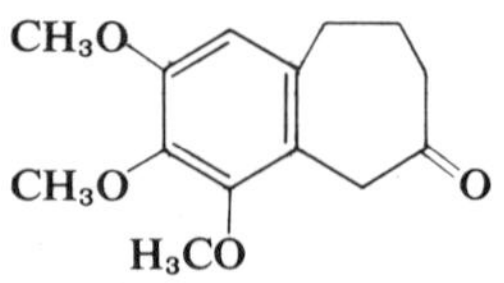

$\lambda_{max.}$ nm	$\epsilon_{max.} \times 10^{-3}$
220	16·6
274	1·07

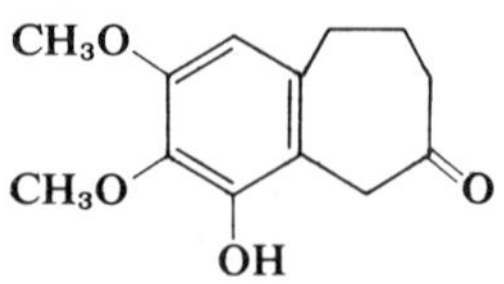

$\lambda_{max.}$ nm	$\epsilon_{max.} \times 10^{-3}$
264	9·12
~300	3·02

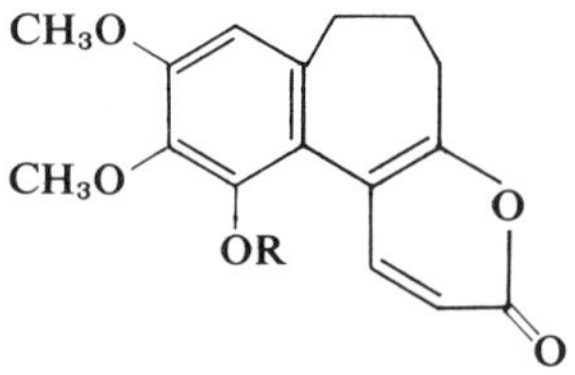

	$\lambda_{max.}$ nm	$\epsilon_{max.} \times 10^{-3}$
R=H	223	21·4
	263	18·2
	330	4·27
R=CH_3	223	22·9
	261	19·5
	325	4·47

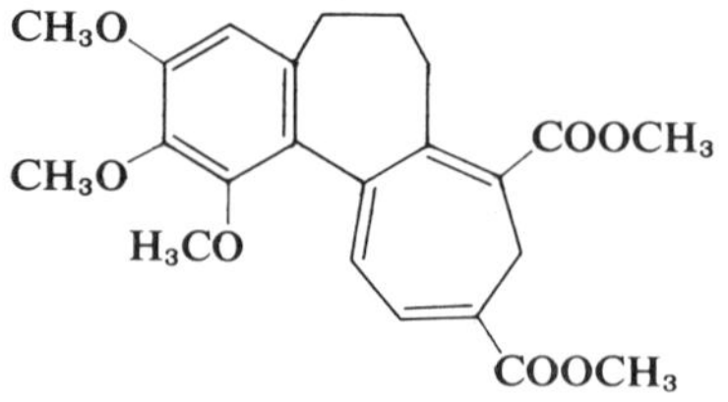

$\lambda_{max.}$ nm	$\epsilon_{max.} \times 10^{-3}$
232	24
~270	11·75
338	9·12

In colchicine the trimethoxybenzene chromophore (270 nm, ϵ 650) is conjugated to the tropolone moiety and the intensity of absorption near 360 nm is considerably increased. It is nevertheless the trienone chromophore in the 7-membered ring system that dominates the spectrum. Desacetylcolchicine shows an essentially similar absorption curve. Desacetylcolchiceine and colchicene (Nakamura, 1962) do not greatly differ, but the switching of relative positions of C=O and —C—OCH_3 in the tropolone ring has noticeable effects.

Colchicine
(synthetic *dl*-)

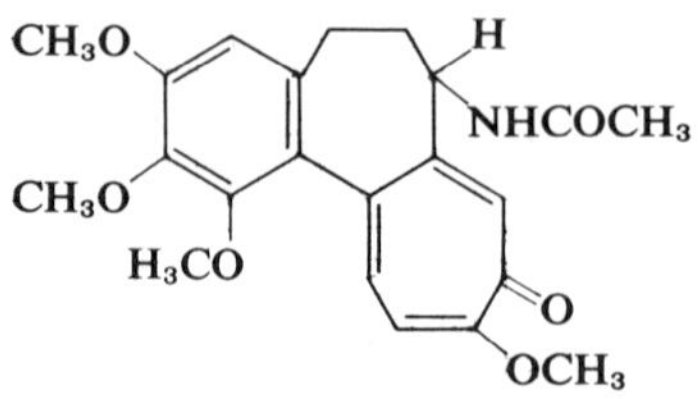

$\lambda_{max.}$ nm	$\epsilon_{max.} \times 10^{-3}$	cm^{-1}
234	28·8	3460
246	30·9	3285 (NH)
357	16·6	1674
		1620
		1597 C=O

Colchiceine
(synthetic *dl*-)

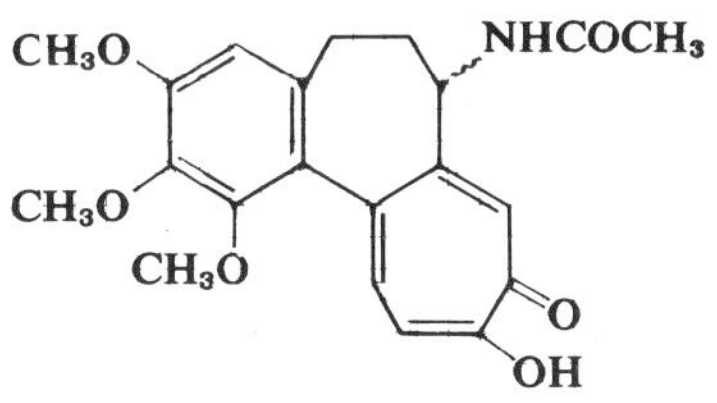

$\lambda_{max.}$ nm	$\epsilon_{max.} \times 10^{-3}$	cm^{-1}
243	35·5	3497
349	19·1	3289
(405	2·09)	1681
		1613
		1553

Desacetylaminocolchiceine

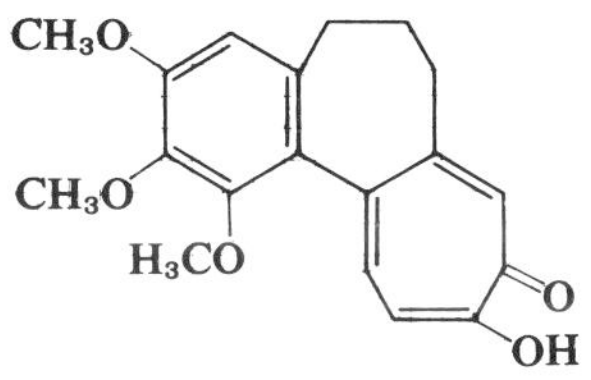

$\lambda_{max.}$ nm	$\epsilon_{max.} \times 10^{-3}$
231	28·8
244	34·7
354	17·8

Desacetylcolchicine

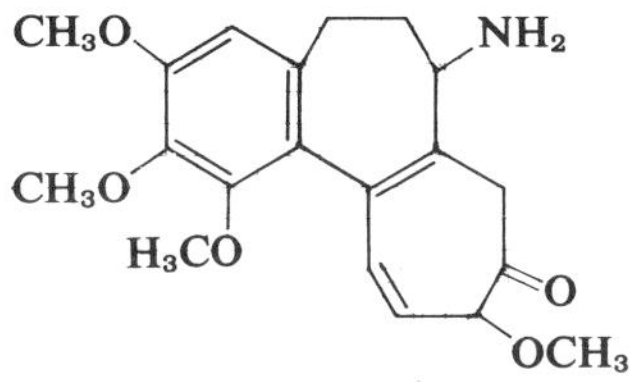

$\lambda_{max.}$ nm	$\epsilon_{max.} \times 10^{-3}$
232	25·1
242	25·7
355	14·45

Desacetylisocolchicine

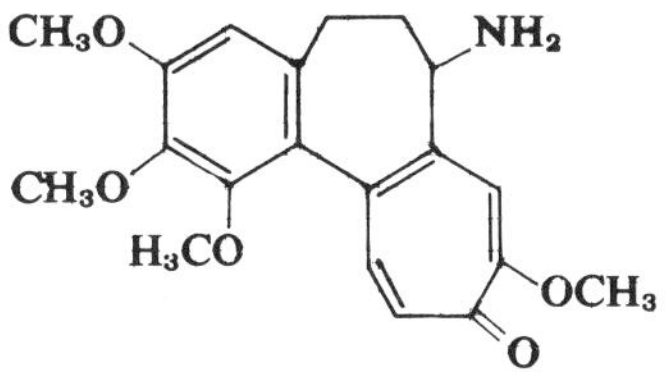

$\lambda_{max.}$ nm	$\epsilon_{max.} \times 10^{-3}$
227	24·0
246	28·2
346	18·20

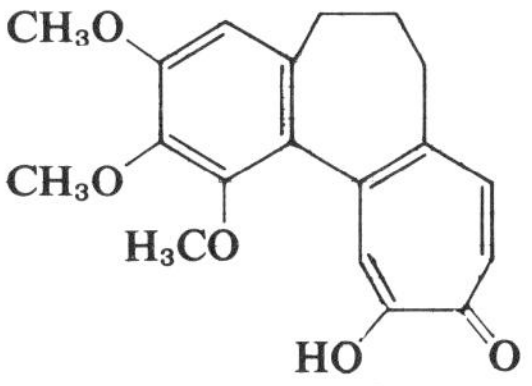

(*a*)

$\lambda_{max.}$ nm	$\epsilon_{max.} \times 10^{-3}$
242	30·2
251	22·9
308	17·8
~352	9·55
372	10·7
385	10·2

(*b*) in 0·1N NaOH+ethanol

$\lambda_{max.}$ nm	$\epsilon_{max.} \times 10^{-3}$
~225	17·4
246	31·6
289	20·9
~335	12·2
348	14·1
415	12·9

There are numerous difficulties in the interpretation of these ultraviolet spectra despite the overall similarities. Nevertheless, an extended study by Schreiber *et al.* (1961), which culminated in the synthesis of racemic colchicine, contains a large amount of valuable spectroscopic work.

4.5 Absorption Spectra of 2,4-Dinitrophenylhydrazones

Valuable work in this field has been done by Braude & Jones (1945), Roberts & Green (1946), Djerassi & Ryan (1949), Johnson (1951, 1953) and Jones *et al.* (1956). The outcome has been that the ultraviolet absorption spectra of 2,4-dinitrophenylhydrazones can provide information concerning the aliphatic (saturated or α-β-unsaturated) or aromatic nature of the parent ketone or aldehyde, and can often allow an estimate of the molecular weight of an unknown substance. These very real advantages are enhanced by the fact that 2,4-dinitrophenylhydrazones are often important in separations.

Table 4.20 Absorption maxima of 2,4-dinitrophenylhydrazones

Substance	Solvent*	$\lambda_{max.}$ nm	$\epsilon_{max.}$	Reference
2,4-Dinitrophenylhydrazine	E	219	12 000	b
		258	9500	
		350	14 500	
		415	6000	
	C	260·5	10 000	b
		343·5	14 500	
		400	6000	
2,4-Dinitro-aniline	E	223	11 500	b
		256	10 000	
		335	14 500	b
		385	7000	
(1) Aliphatic ketones				
Acetone-DNPH	E	360	21 000	a
	E	362	21 500	b
	C	366	22 500	b
	C	364	22 400	c
Cyclopentanone-DNPH	E	363	21 500	a
	E	363	22 500	b
	C	367	24 500	b
Di-isopropylketone DNPH	E	363	22 000	a
Diethylketone DNPH	E	362	22 000	b
	C	366	25 000	b
Methylethylketone DNPH	E	362	23 000	b
	C	365	22 000	b
(2) Aliphatic aldehydes				
Formaldehyde DNPH	E	349	18 200	a
	E	348	23 000	b
	C	348	23 000	b
	C	344	19 000	c
Acetaldehyde DNPH	E	356	21 000	a
	E	360	21 500	b
	C	360	25 000	b
	C	354	22 200	c
Propionaldehyde DNPH	E	359	21 800	a
	E	351	21 500	b
	C	361	21 000	b
	C	356	22 500	c
n-Hexanal DNPH	E	356	20 500	b
	C	358	20 000	b
	C	358	22 100	c
Phenylacetaldehyde DNPH	C	355	22 500	c
Hydrocinnamaldehyde DNPH	C	358	22 800	c

Table 4.20 *continued*

Substance	Solvent*	λ_{max}. nm	ϵ_{max} .	Reference
(3) α-β-Unsaturated aldehydes and ketones				
Crotonaldehyde DNPH	E	377	26 600	a
	E	373	28 500	b
	C	372	29 000	b
	C	373	27 200	c
Tiglinaldehyde DNPH	E	377	27 500	a
CH_3.CH:CMe.CHO	C	376	30 000	c
Acrolein DNPH	E	366	25 500	b
	C	367	26 000	b
	C	368	27 200	c
Mesityloxide DNPH	E	379	23 100	a
(CMe_2:CH.COMe)	E	379	22 500	b
	C	389	25 000	b
	C	385	24 300	c
α-Ionone DNPH	C	377	28 000	b
—CH=CH.$COCH_3$	C	389	25 000	b
	C	387	30 900	d
Acetyl-Δ^1-cyclohexene DNPH	E	377	24 000	b
(—C:C.COMe)	C	387	27 500	b
(4) Aromatic aldehydes and ketones				
Benzaldehyde DNPH	E	378	29 200	a
	E	377	29 500	b
	C	377	28 500	b
	C	378	28 300	c
Acetophenone DNPH	E	377	26 000	b
	C	380	27 500	b
p-Methyl-acetophenone DNPH	E	383	27 600	a
	C	382	27 300	c
Benzophenone DNPH	E	383	28 300	a
	E	379	26 500	b
	C	391	29 000	b
	C	386	30 000	c
Salicylaldehyde DNPH	E	387	29 500	a
	C	381	29 600	c
(5) Unsaturated aromatic aldehydes and ketones				
Cinnamicaldehyde DNPH	E	394	38 000	a
	C	390	38 800	c
Chalcone DNPH	E	395	36 400	a
	C	399	37 400	c
(6) Conjugated dienals, etc.				
CHMe:CMe.CH:CH.CHO DNPH	C	392	42 000	b
Furfuraldehyde DNPH	E	390	27 000	b
	C	388	26 500	b
yellow form	C	379	27 700	c
red form	C	386	26 500	c
ψIonone DNPH				
—CMe:CH.CH:CH.COMe	E	395	32 000	b
	C	407	34 500	b
β-Ionone DNPH	E	385	26 500	b
—CMe:C—CH:CH.COMe	C	388	27 500	b
|	C	392	28 840	d
2-Furanacrylaldehyde DNPH	C	400	37 900	c
	E	396	43 500	b
(—CH:CMe.COEt)	C	400	42 500	b

The reagent 2,4-dinitrophenylhydrazine shows well-marked maxima (see Table 4.20) and the curves are clearly related to those of 2,4-dinitroaniline. The different groups of workers agree that the principal absorption peak of the 2,4-dinitrophenylhydrazones of simple aliphatic aldehydes (in ethanol) occurs near 350–360 nm (average 355 nm) with $\epsilon_{max.}$ at 21 500 approximately. When the solvent is chloroform the, results of Jones *et al.* (1956) indicate $\lambda_{max.}$ *ca.* 355 nm $\epsilon_{max.}$ 22 000. The 2,4-dinitrophenylhydrazones of simple ketones show $\lambda_{max.}$ *ca.* 363 nm, $\epsilon_{max.}$ *ca.* 22 000 (in thanol) $\lambda_{max.}$ 365 nm, $\epsilon_{max.}$ 22 400 in chloroform. Thus for either solvent the intensity of absorption is characteristic of the chromophore

NO_2 / NO_2—C_6H_3—NH.N=C(R')(R'')

where R′ is H or alkyl and R″ is alkyl.

α-β-Unsaturated aldehydes or ketones give appreciably different 2,4-dinitrophenylhydrazones from the above. On the whole agreement is good (see Table 4.20). It is clear that the chromophore:

NO_2 / NO_2—C_6H_3—NH—N=C(R′)—C(R″)=C(R‴)—R″

shows $\lambda_{max.}$ displaced on the direction of longer wave-lengths (> 370 nm) with a rise in $\epsilon_{max.}$ (26 000–30 000). Aromatic aldehydes and ketones show rather larger displacements in the maxima of their 2,4-dinitrophenyhydrazones (Table 4.20). Here the chromophore

NO_2 / NO_2—C_6H_3—NH—N=C(—)—C_6H_5

moves $\lambda_{max.}$ to about 380 nm and $\epsilon_{max.}$ to a little under 30 000, while the chromophore

NO_2 / NO_2—C_6H_3—NH—N=C—C=C—C_6H_5

gives $\lambda_{max.}$ displaced to about 390 nm with $\epsilon_{max.}$ rising to 36 000–39 000.

Djerassi & Ryan (1949) recorded the ultraviolet absorption spectra of a number of steroidal dinitrophenylhydrazones (see Figs. 4.26 and 4.27). This paper shows the usefulness of the method. Jones *et al.* (1956) went further and examined the spectra in an alkaline medium. The original chloroform solution was diluted with 0·25 N ethanolic sodium hydroxide to about 10 μg/ml. The solution was shaken for 10 seconds and the spectrum determined at once on a Cary recording spectrophotometer using the appropriate chloroform–ethanolic alkali in the solvent cell. In order to follow the change in spectrum the recording (350–800 nm) was repeated every 15 minutes.

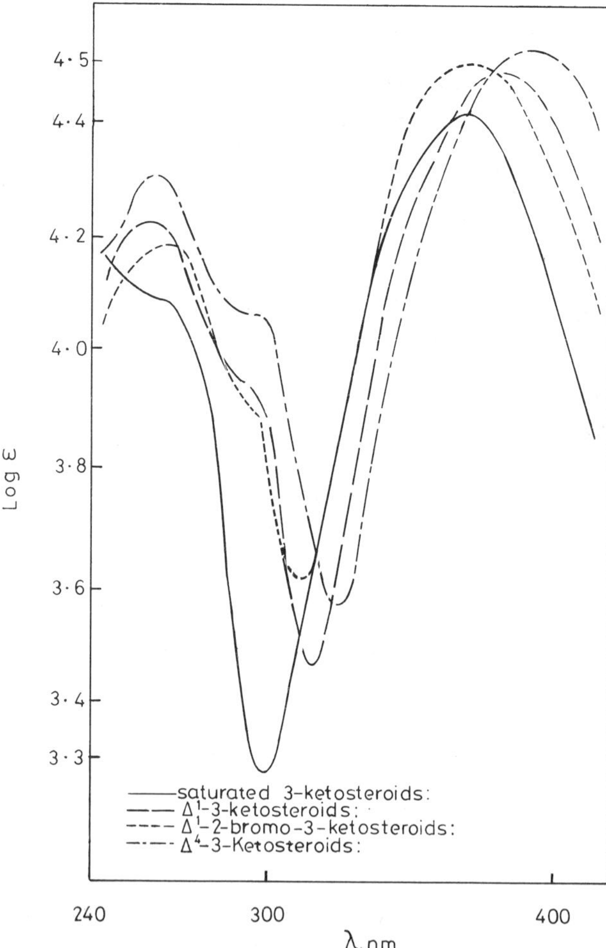

Figure 4.26 Ultraviolet absorption spectra in chloroform solution of dinitrophenylhydrazones of saturated 3-ketosteroids, Δ^1-3-ketosteroids, Δ^1-2-bromo-3-ketosteroids and Δ^4-3-ketosteroids (from Djerassi & Ryan, 1949)

All the DNPH derivatives of simple aldehydes and ketones freshly examined in an alkaline medium showed two absorption peaks at about 432 nm and 515–535 nm respectively. The initial values for $\epsilon_{max.}$ at the 432 peak averaged 22 600. Similarly the olefinic aldehyde and ketone 2,4-dinitrophenylhydrazones

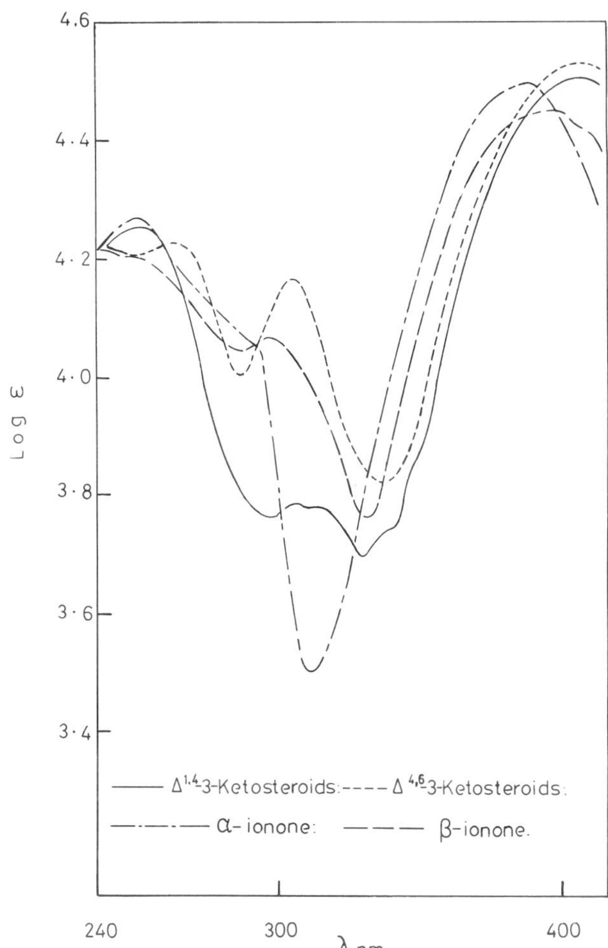

Figure 4.27 Ultraviolet absorption spectra in chloroform solution of dinitrophenylhydrazones of $\Delta^{1,4}$-3-ketosteroids, $\Delta^{4,6}$-3-ketosteroids, α-ionine and β-ionine (from Djerassi & Ryan, 1949)

showed an absorption peak at 452–459 nm, $\epsilon_{max.}$ *ca.* 28 000. The alkaline solutions of the corresponding derivatives of benzaldehyde and acetophenone and derivatives displayed $\lambda_{max.}$ 460–480 nm, $\epsilon_{max.}$ *ca.* 35 000 (with some anomalies). Unsaturated aromatic aldehydes like cinnamaldehyde showed further displacement ($\lambda_{max.}$ 490–510 nm, $\epsilon_{max.}$ up to 42 500).

A variety of furan aldehyde and ketone 2,4-dinitrophenylhydrazones showed broadly consistent displacements on the wavelength scale and on the ε axis depending on the extension of the 2,4-dinitrophenylhydrazine chromophore by conjugated double bonds.

Jones *et al.* point out that it is convenient to use chloroform as the solvent throughout. In general, aliphatic aldehydes showed DNPH peaks at 355–360 and the ketones 364–367 nm, with $\epsilon_{max.}$ values around 22 000. The α-β-unsaturated aldehydes and ketones showed bathochromic effects ($\lambda_{max.}$ up to 390 nm); the molecular extinction coefficients tended to be higher. The aromatic aldehydes and ketones showed rather similar spectral changes except when the aromatic grouping was chromophorically isolated from the DNPH group by a CH_2 group. Jones *et al.* also recorded effects due to substitution in the benzene ring.

In the course of their study of the spectra of alkaline solutions they observed especially rapid fading of colour for aliphatic aldehyde DNPH derivatives. The ketone derivatives also faded but less quickly (see Fig. 4.28). There seems no doubt that sufficient information exists to make it worth while on investigating a new 2,4-dinitrophenylhydrazone to examine its spectrum in neutral solution and in alkaline solution and to use the work of Jones *et al.* as a guide to interpretation.

Figure 4.28 Light absorption of 2:4 dinitrophenylhydrazones in chloroform solution: (1) Acetaldehyde, (2) Crotonaldehyde, (3) 3-Methylsorbaldehyde, (4) Octatrienal. Inset: When the number *n* of double bonds in $[C:C]_n \cdot C:N \cdot NHX$, o–o, is increased, the relation $\lambda^2_{max} = an$ is obeyed, but the constant *a* has a smaller value than in the case of the system $[C:C]_n C:O$, ●●● (from Braude & Jones, 1945)

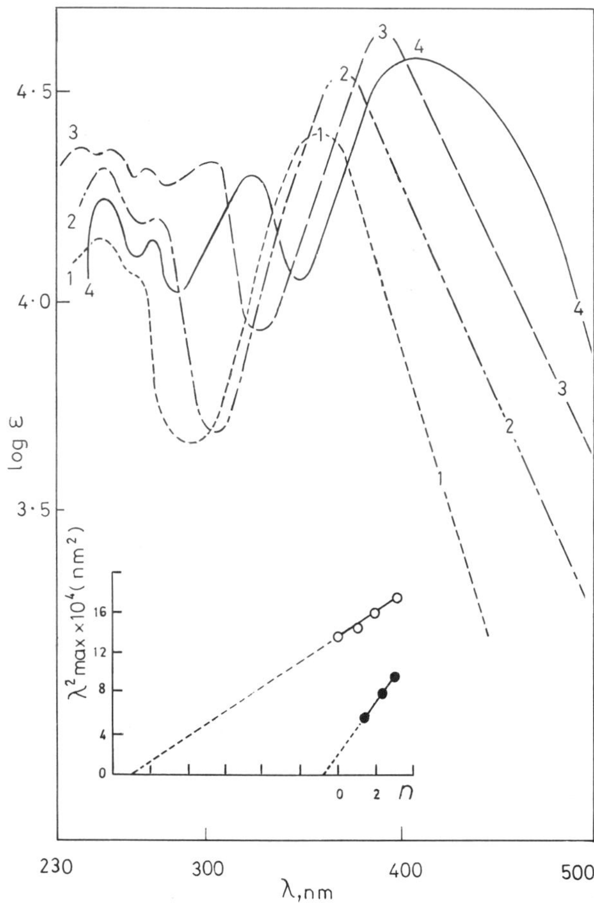

4.6 Infared Spectra

Jones *et al.* examined the infrared spectra of a large number of 2,4-dinitrophenylhydrazones in potassium bromide discs.

The exact position of the N—H stretching absorption (*ca.* 3·05 μm, 3280 cm^{-1}) varied with the structure of the parent aldehyde or ketone and the differences, though not great, distinguish aldehydes and ketones.

In a number of cases the absorption bands of *trans* olefinic 2,4-dinitrophenylhydrazones (*ca.* 1650 cm^{-1} and 966–981 cm^{-1}) were found to be informative. Aromatic 2,4-dinitrophenylhydrazones show absorption in the region 660–770 cm^{-1} and do not display the aliphatic band at about 2880 cm^{-1}.

References

4.1 and 4.2

Allen, G. & Divek, R. A. (1966). *J. Chem. Soc.*, 161.

Bellamy, L. J. & Beecher, L. (1954). *J. Chem. Soc.*, 4487.

Bellamy, L. J. & Branch, R. F. (1954). *J. Chem. Soc.*, 4491.

Benson, S. W. & Kistiakowsky, G. B. (1942). *J. Amer. Chem. Soc.*, **64**, 80.

Booker, A., Evans, L. K. & Gillam, A. E. (1940). *J. Chem. Soc.*, 1453.

Burdett, J. L. & Rogers, M. T. (1964). *J. Amer. Chem. Soc.*, **86**, 2105.

Burawoy, A. (1939). *J. Chem. Soc.*, 1177; (1941). *Ibid.*, 20.

Bredenberg, J. B. (1957). *Acta Chem. Scanda*, **11**, 927. (1959). *Ibid.*, **13**.

Campbell, R. D. & Gilow, A. M. (1960). *J. Amer. Chem. Soc.*, **82**, 5426.

Christ, B., Kuhn, G. & Müller, K. H. (1961). *Arzenei mittel Forsch.*, **11**, 129.

Evans, L. K. & Gillam, A. E. (1941). *J. Chem. Soc.*, 815; (1943). *Ibid.*, 431, 565.

Federlin, W. (1907). *Liebigs Ann.*, **356**, 251.

Ford, C. H. & Parry, F. (1958). *Spectrochim Acta.*, **12**, 78.

Gelles, E. & Hay, R. W. (1957). *J. Chem. Soc.*, 3673.

Gillam, A. E. & Stern, E. S. (1958). In *Electronic Absorption Spectroscopy*. Edward Arnold, London.

Gillam, A. E. & West, T. F. (1942). *J. Chem. Soc.*, 811; (1942), *ibid.*, 483, 486.

Gorodetsky, M., Luz, Z. & Maxur, Y. (1967). *J. Amer. Chem. Soc.*, **89**, 1183.

Grossmann, P. (1924). *Z. Physikal. Chem.*, **109**, 305.

Hartley, W. N. & Dobbie, J. J. (1889). *J. Chem. Soc. Trans.*, **75**, 640.

Hassan, A. Hunter, R. F. & Firdaus, A. A. (1936). *J. Chem. Soc.*, 1672.

Holman, R. T., Lundberg, W. O. & Burr, G. O. (1945). *J. Amer. Chem. Soc.*, **67**, 1669.

Hunsberger, I. M. (1950). *J. Amer. Chem. Soc.*, **72**, 5626.

Hunsberger, I. M., Ketcham, R. & Gutowsky, H. S. (1952). *J. Amer. Chem. Soc.*, **74**, 4839.

Johnson, D. P. & Critchfield, F. E. (1961). *Anal. Chem.*, **33**, 910.

Johnson, D. P. & Ruck, J. E. (1962). *Anal. Chem.*, **34**, 1389.

Jones, E. R. H., Wilkinson, P. A. & Kerlogue, R. H. (1942). *J. Chem. Soc.*, 391.

Kasturi, T., Mylari, B. N., Balasumbramian, A. & Rao, C. N. R. (1962). *Canad. J. Chem.*, **40**, 2272.

Koch, H. P. (1942). *Chem. & Ind.*, **61**, 273.

Koch, H. P. (1949). *J. Chem. Soc.*, 401.

Kohlrausch, K. W. F. & Pongratz, A. (1934). *Ber. dtsch chem. Ges.*, **67B**, 1465.

Kornberg, A., Ochoa, S. & Mehler, A. (1948). *J. Biol. Chem.*, **174**, 159.

Kumler, W., Kun, E. & Shoolery, J. N. (1962). *J. Org. Chem.*, **27**, 1165.

Lemon, H. W. (1947). *J. Amer. Chem. Soc.*, **69**, 2998.

Leonard, N. J. & Mader, P. M. (1950). *J. Amer. Chem. Soc.*, **72**, 5388.

Lieber, E., Rao, C. N. R., Chao, T. S. & Rubinstein, H. (1959). *Canad. J. Chem.*, **36**, 1441.

Loewus, F. A., Chen, T. T. & Vennesland, B. (1955). *J. Biol. Chem.*, **55**, 787.

Mariella, R. P. & Raube, R. R. (1952). *J. Amer. Chem. Soc.*, **74**, 518.
McMurry, H. L. (1941). *J. Chem. Phys.*, **9**, 141, 231.
Meyer, K. H. (1912). *Ber. dtsch chem. Ges.*, **44**, 2718, 2725, 2729, 2767.
Morton, R. A., Hassan, A. & Calloway, T. C. (1934). *J. Chem. Soc.*, 883.
Morton, R. A. & Rogers, E. (1925). *J. Chem. Soc.*, 2698; (1926). *Ibid.*, 713.
Morton, R. A. & Rosney, W. C. V. (1926). *J. Chem. Soc.*, 706.
Morton, R. A. & Stubbs, A. L. (1939). *J. Chem. Soc.*, 1321. (1940). *Ibid.*, 1347.
Mühle, H. & Tamm, C. (1962). *Helv. Chim. Acta.*, **45**, 1473; (1963). *Ibid.*, **46**, 208.
Murthy, A. S. N., Balasubramanian, A., Rao, C. N. R. & Kasturi, T. R. (1962). *Canad. J. Chem.*, **40**, 2267.
Nef, J. U. (1899). *Annalen*, **308**, 279.
Neuberg, C. & Strauss, E. (1945). *Arch. Biochem.*, **7**, 211; (1946). *Ibid.*, **11**, 457.
Park, J. D., Brown, H. A. & Lacher, J. R. (1953). *J. Amer. Chem Soc.*, **75**, 4753.
Rao, C. N. R. (1967). *Ultraviolet and Visible Spectroscopy Chemical Applications.* 2nd Edition. Butterworths, London.
Rasmussen, R. S. & Brattain, R. B. (1949). *J. Amer. Chem. Soc.*, **71**, 1073.
Rasmussen, R. S., Tunnicliff, D. D. & Brattain, R. B. (1949). *J. Amer. Chem. Soc.*, **71**, 1068.
Rice, F. O. (1915). *Proc. Roy. Soc.*, **91**, 76; (1920). *J. Amer. Chem. Soc.*, **43**, 727.
Scott, A. I. (1964). *Ultraviolet Spectra of Natural Products.* Pergamon Press, Oxford, London.
Scheibe, G. (1925). *Ber. dtsch chem. Ges.*, **58**, 586.
Stephani, R. A. & Meister, A. (1971). *J. Biol. Chem.*, **246**, 7115.
Sidgwick, N. V. & Brewer, F. M. (1925). *J. Chem. Soc.*, **127**, 2379.
Tamm, C. (1960). *Helv. Chim. Acta.*, **43**, 1700.
Wolf, K. L. (1929). *Z. Physikal Chem.*, B **2**, 39.
Woodward, R. B. (1941). *J. Amer. Chem. Soc.*, **63**, 1123; (1942). *Ibid.*, **64**, 71, 76.

4.3

Lynen, F. & Ochoa, S. (1953). *Biochem. Biophys. Acta.*, **12**, 299.
Lynen, F., Wessely, L., Wieland, O. & Rueff, L. (1952). *Angew. Chem.*, **64**, 687.
Morton, R. A. (1959). In *International Symposium on Microchemistry.* Pergamon Press, London, p. 355.
Stern, J. R. (1956). *J, Biol. Chem.*, **221**, 33.
Stern, J. R., Coon, M. J., Del Campillo, A. & Schneider, M. C. (1956). *J. biol. Chem.*, **221**, 15.

4.4

Cook, J. W. (1945). *Chem. & Ind.*, 403.
Cook, J. W., Gibb, A. P., Raphael, R. A. & Somerville, A. R. (1951). *J. Chem. Soc.*, 503.
Dauben, H. J. Jr. & Ringold, H. J. (1951). *J. Amer. Chem. Soc.*, **73**, 876.
Hosoya, H., Tanaka, J. & Nagakura, S. (1962). *Tetrahedron Lett.*, **18**, 859.
Ikegami, Y. (1961). *Bull. Chem. Soc. Japan.*, **34**, 94.
Kemp, D. A. & Tarbell, D. S. (1950). *J. Amer. Chem. Soc.*, **72**, 243.
Mukai, T. (1958). *J. Chem. Soc. Japan*, **79**, 1547.
Nakamura, T. (1962). *Chem. Pharm. Bull. Japan*, **10**, 281, 291, 299.
Nozoe, T. (1959). In *Non-benzenoid Aromatic Hydrocarbons.* Chap. 7. Ed. D. Ginsburg. Interscience Publishers, New York.
Schreiber, J., Leimgruber, W., Pesaro, M., Schudel, P., Threlfall, T. & Echenmoser, A. (1961). *Helv. chim. Acta.*, **44**, 540.
Scott, G. P. & Tarbell, D. S. (1950). *J. Amer. Chem. Soc.*, **72**, 240.
Von Doering, W. E. & Detert, F. L. (1951). *J. Amer. Chem. Soc.*, **73**, 876.
Zavarin, E. (1962). *J. Org. Chem.*, **27**, 3368.
Zavarin, E. & Anderson, A. B. (1961). *J. Org. Chem.*, **26**, 173, 1679.

4.5

Braude, E. A. & Jones, E. R. H. (1945). *J. Chem. Soc.*, 498.
Djerassi, C. & Ryan, E. (1949). *J. Amer. Chem. Soc.*, **71**, 1000.
Johnson, G. D. (1951). *J. Amer. Chem. Soc.*, **73**, 5888; (1953). *Ibid.*, **75**, 2720; (1956) *Anal. Chem.*, **28**, 191.
Jones, L. A., Holmes, J. C. & Seligman, R. B. (1956). *Anal. Chem.*, **28**, 191.
Roberts, J. D. & Green, C. (1946). *J. Amer. Chem. Soc.*, **68**, 214.

5 Polyprenols and Glyceryl Ethers

5.1 Polyprenols

Geraniol, farnesol, linalool and nerolidol are well-known unsaturated alcohols, but as the double bonds are not conjugated there is no selective absorption. Solanesol, discovered much later, has nine (unconjugated) prenyl groups ($C_{45}H_{73}OH$) in the *all-trans* configuration. The same nonaprenyl grouping occurs in plastoquinone (PQ–9) from which solanachromene is derived.

Liverpool workers obtained from the unsaponifiable fractions of liver, kidney and spleen lipids a product which they designated dolichol. The best fraction obtained in the early work corresponded with a polyprenol $C_{100}H_{13}OH$ (Pennock *et al.*, 1960). The infrared absorption closely resembled that of solanesol except that a peak at 100 μm characteristic of the allylic group

$$=\underset{|}{C}—CH_2OH$$

hydroxyl was missing whereas a new C—OH band appeared at 9·4 μm. Hydrogenation indicated 19 double bonds and the following formula,

$$H—[CH_2—\underset{\displaystyle CH_3}{\underset{|}{C}}=CH—CH_2]_{19}—CH_2—\underset{\displaystyle CH_3}{\underset{|}{C}}HCH_2CH_2OH$$

which fitted the mass spectra. N.M.R. spectra indicated that 16 internal isoprene units were in the *cis* configuration (Burgos *et al.*, 1964). Dolichol existed in pig liver largely as ester. Improved methods of separation by thin layer chromatography showed that at least six components were present: C_{85}, C_{90}, C_{95}, C_{100}, C_{105} and C_{110} (Dunphy *et al.*, 1967).

Yeast also contains dolichols C_{70}, C_{75}, C_{80}, C_{85} C_{90}, thus partially overlapping with the mammalian sequence.

Aspergillus fumigatus mycelium gave a mixture of seven polyprenols (C_{90}—C_{120}) while the spadix of *Arum maculata* contained polyprenols resembling those of pig tissues and human tissues, the C_{95} and C_{100} isoprenologues being predominant (Hemming *et al.*, 1963).

This work led to an extended search for polyprenols. The leaves of horse chestnut (*Aesculus hippocastanum*) yielded castaprenols-*n* (where *n* = number of isoprene units) beginning at castaprenol-10 (C_{50}) and ending at castaprenol-13 (C_{65}). The rubber plant *Ficus elasticus* yielded ficaprenols-10, -11, -12 and -13.

The spectroscopic interest of these polyprenols lies in the determination of the *cis*- and *trans*- configurations:

$$—CH_2—\overset{\displaystyle H}{\overset{|}{C}}=\overset{\displaystyle CH_3}{\overset{|}{C}}—CH_2— \quad \text{and} \quad —CH_2—\underset{\displaystyle H}{\underset{|}{C}}=\overset{\displaystyle CH_3}{\overset{|}{C}}—CH_2—$$

(Bates *et al.*, 1962, 1963; Feeney & Hemming, 1967).

The latter workers made measurements both in benzene and carbon tetrachloride; the *cis* 'OH-terminal' residue has a peak at $\tau = 8·28$ separated from other peaks but the *trans* 'OH-terminal' peak would be hidden under the *cis* internal residue peak at $\tau = 8.34$. In benzene the *cis* 'OH terminal' peak ($\tau = 8·355$) is hidden under the *trans* internal residue peak ($\tau = 8{\cdot}38$) while the *trans* ('OH-terminal' residue ($\tau = 8{\cdot}51$) is clear of the other peaks. Benzene thus allows a *trans* 'OH-terminal' to be recognized while carbon tetrachloride allows a *cis* 'OH-terminal' residue to be identified.

Lindgren (1964) isolated betulaprenol from birchwood; it is in fact a mixture of C_{35}, C_{40} and C_{45} polyprenols. The leaves are richer in betulaprenol-10 (C_{50}) with smaller amounts of lower isoprenologues. The leaves of the rubber tree *Hevea brasiliensis* contain solanesol and heveaprenols-10, -11, -12 and -13 which from the n.m.r. spectra have *cis* moieties preponderantly. The C_{55} compound has a primary allylic hydroxyl, seven *cis* isoprene units, including that nearest to the hydroxyl. In general the polyprenols of leaves have three internal *trans* residues.

Lactobacilli contain bactoprenol (Thorne & Kodicek, 1962, 1966) made up in the main of a C_{55} alcohol with one double bond filled in so that it resembles dolichol-11. *Lactobacillus plantarum* has a C_{55} polyprenol with two internal *trans*, one *α-cis* and seven internal *cis* isoprene units (Gough *et al.*, 1970).

It has been shown that the C_{55} alcohol and the dolichols have important carrier functions in the biosynthesis of lipopolysaccharides (for a summary see Morton, 1972). This is a new development and the main relevance of spectroscopy in this connection may well turn out to be in respect of establishing the *cis-trans* structures.

Table 5.1 NMR Spectra of Polyprenols

Chemical shifts (τ) ($\pm 0{\cdot}01$) in benzene	in CCl_4	Assignments
8·5 ↓ 8·2	8·5 ↓ 8·2	$R{-}C(\mathit{CH_3}){=}CHR$
7·95w, 7·87~, 7·83, 7·80~	8·00, 7·97	$-\mathit{CH_2}-\overset{\vert}{C}=$
6·06, 5·90 d	6·10, 6·02 d	$=CH-\mathit{CH_2}-OH$
4·66, 4·59, 4·52 t	4·74, 4·66, 4·58 t	$-\overset{\vert}{C}=\mathit{CH}-CH_2OH$

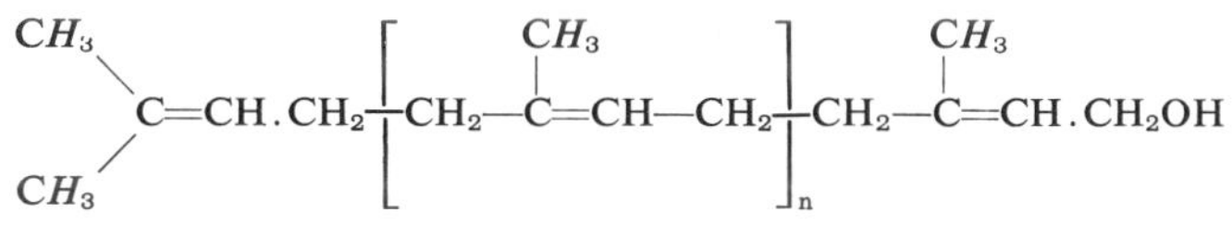

$$(\mathit{CH_3})_2C{=}CH.CH_2{-}\left[CH_2{-}\overset{\mathit{CH_3}}{\overset{\vert}{C}}{=}CH{-}CH_2\right]_n{-}CH_2{-}\overset{\mathit{CH_3}}{\overset{\vert}{C}}{=}CH.CH_2OH$$

(*a*) Residue containing the ω methyl group	(*b*) In benzene *cis* (τ)	*trans*	(*c*) In carbon tetrachloride *cis* (τ)	*trans*
a (ω-terminal)	8·32	8·43	8·34	8·42
b (internal)	8·25	8·38	8·34	8·42
c (OH-terminal)	8·355	8·51	8·28	8·34

w = weak, ~ = shoulder, d = doublet, t = triplet. The resonating proton is italicized.

Table 5.1 *continued*

	Isoprene residue at hydroxylend	Internal residues	ω residue
Leaf polyprenols			
Decaprenol	*cis*	3 *trans* 5 *cis*	*trans*
Undecaprenol	*cis*	3 *trans* 6 *cis*	*trans*
Dodecaprenol	*cis*	3 *trans* 7 *cis*	*trans*
Silver birch wood polyprenols			
Betulaprenol-7	*cis*	2 *trans* 3 *cis*	*trans*
Betulaprenol-8	*cis*	2 *trans* 4 *cis*	*trans*
Lactobacillus plantarum polyprenol			
Undecaprenol	*cis*	2 *trans* 7 *cis*	*trans*
Aspergillus fumigatus fresenius polyprenols			
Hexahydropolyprenols $n=18$–24 *a*, ψ and ω residues saturated			

5.2 Glyceryl Ethers: Batyl, Selachyl and Chimyl Alcohols

The glyceryl ethers, like the polyprenols, do not show selective absorption in the ultraviolet region. The current state of knowledge about them shows but little debt to spectroscopic methods. The following brief account may perhaps suggest that the application of modern methods could well be fruitful.

Lipids in which long-chain alkyl residues are linked to glycerol by ether bonds are widespread in natural products. The alkyl groups in such ether linkages may be saturated, or unsaturated as in the plasmalogens which display the αβ-unsaturated (vinyl) ether link.

The saturated glyceryl ethers include: batyl alcohol (D-α-stearyl glycerol ether), chimyl alcohol (D-α-palmityl glycerol ether) and selachyl alcohol (D-α-oleyl glycerol ether).

The liver oils of elasmobranch fishes are peculiarly rich in the above, together with other glyceryl ethers having alkyl residues from C_{14} to C_{22} (Karnovsky *et al.*, 1946; Hallgren & Larsen, 1962).

It was reported by Marberg & Wiles (1937, 1938) that agranulocytosis, a blood dyscrasia usually brought about by drugs, could be relieved by the administration of bone marrow lipids in large doses and by smaller doses of the unsaponifiable fraction. A. L. Stubbs and the present writer worked up about 50 kg of ox bone marrow and fractionated the unsaponifiable portion. The substances isolated included *inter alia* cholesterol, carotene and batyl alcohol. The late Professor T. B. Davie endeavoured to produce agranulocytosis in experimental animals so that the fractions could be tested, but either there was no agranulocytosis or it proved quickly fatal. Projected clinical trials had to be given up under war-time conditions. The first published evidence that bone marrow contained batyl alcohol came from Holmes *et al.* (1941). More recently Thompson & Hanahan (1963) showed that the native lipids of ox bone marrow included ethers found in the phosphatidylethanolamine fraction. Carter *et al.* (1958) had already obtained from egg yolk phospholipids subjected to mild alkaline hydrolysis an α-alkyl ether of glycerophosphorylethanolamine. Svennerholm & Thorin

(1960) obtained a similar compound from human brain and from calf brain. Hanahan & Watts (1961) isolated an α^1-alkoxy-β-glycerophosphorylethanolmaine from bovine erythrocytes and Hanahan *et al.* (1963), confirming this, obtained a mixture of diacyl and monoalkylmonoacylglycerophosphoryl-ethanolamines, the ether portion accounting for three-quarters of the fraction. All the acids present were unsaturated.

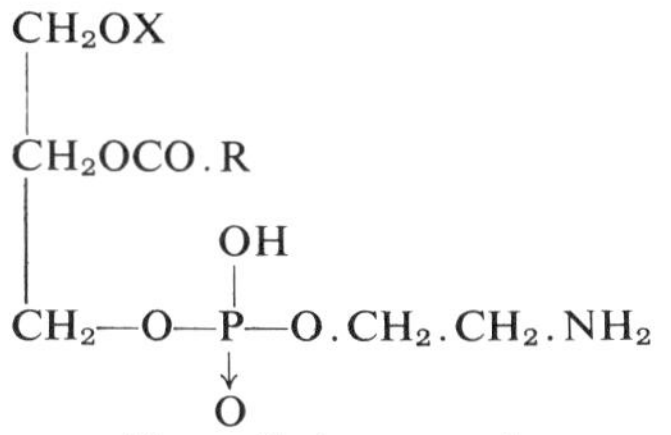

```
CH₂OX
|
CH₂OCO.R
|         OH
|         |
CH₂—O—P—O.CH₂.CH₂.NH₂
          ↓
          O
```

Monoalkylmonoacyl derivative

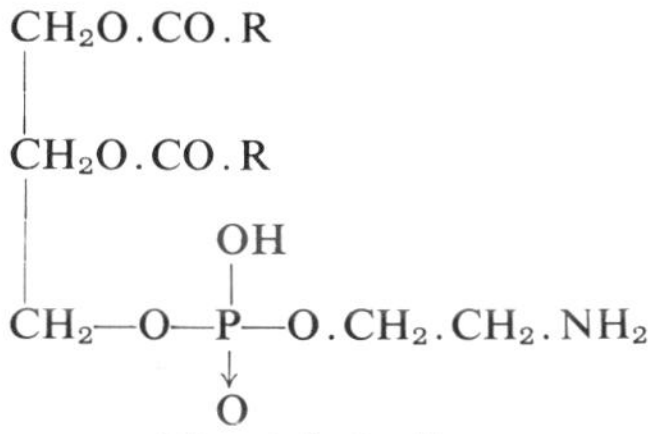

```
CH₂O.CO.R
|
CH₂O.CO.R
|         OH
|         |
CH₂—O—P—O.CH₂.CH₂.NH₂
          ↓
          O
```

Diacyl derivative

R = hexadecyl, octadecyl (C_{16} or C_{18}) or octadecenyl, with one ethylenic linkage.

Bovine bone marrow lipid varies in composition with the age of the animal and the particular bone. Thus femurs and tibiae from calves 2–3 weeks old contain haemopoietic bone marrow in which the phospholipid accounts for one part and the neutral glycerides for 70 parts. As the calf grows older the proportion of phospholipid falls even further, but diacyl derivatives of glycerol ethers are found. In adult bovines the proportion of phospholipid to neutral fat may be between 1/12 and 1/17. A study of normal human bone marrow (Lund *et al.*, 1962) recorded less than 3% of phospholipid in the extractable lipid but glycerol ethers were not mentioned, presumably because the amounts of material obtained were small. Considerable evidence has accumulated to support the idea that batyl alcohol and other glycerol ethers (or compounds containing them) have a definite role in haemopoiesis. Linnan *et al.* (1959) found that when batyl alcohol was fed to normal rats it produced hyperplasia of all myeloid elements. The response was identical with that induced by the heat-stable erythropoietic factor of plasma. Brohult (1958, 1960, 1962) found that the administration of glycerol ethers had a protective effect against the leukopenia produced by irradiation (cf. drug-induced agranulocytosis) and that batyl alcohol stimulated growth. Prelog *et al.* (1943) had found batyl alcohol in pig spleen and erythropoietic activity had been confirmed (Linnan & Bethell, 1960). Brohult claimed a certain value for glycerol ethers in acute leukaemias of childhood.

Marrow tissue is the site of formation of circulating red cells. We now know that the ethanolaminephospholipids of the erythrocyte outer membrane contain a substantial proportion of glycerol ethers. The relative proportions of batyl, chimyl and selachyl alcohols in the phosphatidylethanolamines of ox marrow lipid are in good agreement with the proportions in ox erythrocyte lipid. It is perhaps significant that ox erythrocyte lipid has little or no phosphatidylcholine although this is plentiful in ox bone-marrow lipid.

The bone-marrow cell preparations have the usual mixture of nuclei, microsomes and mitochondria and the glyceryl ethers are actually present in the latter. In red cells, however, the intracellular organelles are lacking and the phospholipids belong exclusively to the erythrocyte membrane. (It would be valuable to know whether this applies to human red cell membranes.)

Thompson & Hanahan (1963) fractionated bone marrow phospholipid by column chromatography on silicic acid and used thin layer chromatography for monitoring. Phosphatidylethanolamines accounted for 28% of the total phosphorus and phosphatidylcholines for 43%, with phosphatidyl-serine and -inositol together accounting for 12% and sphingomyelin for 12%, leaving 6% unidentified. In the phosphatidylethanolamines the fatty acid/phosphorus ratio of 1·42 was consistent with 40% diacylphospholipid and 60% monoalkyl monoacylphospholipid. Plasmalogen (22 moles %) was nearly as plentiful as glycerol ether (28 moles %). The phosphatidylcholines appeared to contain 3 and 12 moles % of plasmalogen and glycerol ether respectively. Table 5.2 illustrates the situation.

Incubation of bone marrow preparations with suitably labelled glucose or palmitic acid showed incorporation into phospholipids. A small but consistent proportion was found in plasmalogens and in glyceryl ethers. The former could well have been precursors of the latter (see Hanahan & Thompson, 1963).

The metabolic fate of glyceryl ethers might throw light on their functions. Blomstrand & Ahrens (1959), using labelled chimyl alcohol, found that when it was

Table 5.2 Distribution of glyceryl ethers in phospholipid fractions

	Side-chains				
	*16·0	16·1	17·1	18·0	18·1
(*a*) Bovine bone-marrow					
Phosphatidylethanolamine	% 33·6	Trace	—	29·0	37
Phosphatidylcholine	% 40·3	1·6	—	15·2	42·9
(*b*) Bovine erythrocytes					
Phosphatidylethanolamine	% 31	7	4	31	26

* Number of carbon atoms first, of double bonds second.
[chimyl alcohol, 1-hexadecylglycerol 16·0; batyl alcohol, 1-octadecylglycerol 18·0; selachyl alcohol, 1-octadecenyglycerol 18·1].

fed to rats or humans, absorption was rapid and labelled palmitic acid was formed. Ether linkages are relatively stable but an enzyme system must be available and efficient. Tietz *et al.* (1964) made an enzyme preparation from liver tissue which affected the oxidation of long-chain alkyl ethers of glycerol to fatty acids and free glycerol. It was found that the process involved free aldehydes as intermediates. The first step required molecular oxygen and the tetrahydro form of an unconjugated pteridine as a cofactor, 2-amino-4-hydroxy-6-methyltetrahydropteridine being very effective, whereas tetrahydrofolate was much less so. The initial product was probably a hemiacetal that spontaneously underwent fission to give the aldehyde and glycerol. The aldehyde then underwent oxidation catalysed by an oxidase requiring a pyridine nucleotide coenzyme:

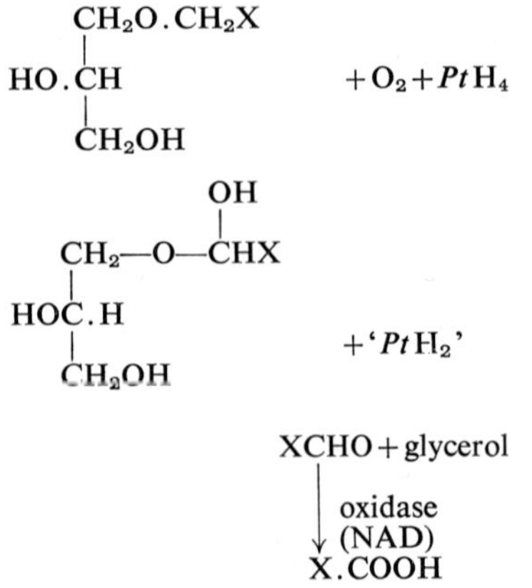

$CH_2O.CH_2X$
HO.CH $\qquad + O_2 + PtH_4$
CH_2OH

OH
CH_2—O—CHX
HOC.H $\qquad$ + 'PtH_2'
CH_2OH

XCHO + glycerol
↓ oxidase (NAD)
X.COOH

Pfleger *et al.* (1967) confirmed and extended these findings. Kaufman (1959, 1961, 1962) has shown how in the hydroxylation of phenylalanine a pteridine cofactor is needed. The glyceryl ether system, like the phenylalanine hydroxylation, responds to NADPH and a pteridine reductase which restores the 'oxidized' pteridine to the tetrahydro state.

It is not clear how glyceryl ethers influence haemopoiesis but control mechanisms are probably at work. The pteridine-requiring metabolic step bears some analogy with the role of tetrahydrofolate in the synthesis of thymidylate, where the tetrahydrofolate acts as a coenzyme for the transfer of a C unit at the level of oxidation of formaldehyde and also as a reducing agent. The catalytic function of the folate cofactor demands that the dihydrofolate be reduced back to the tetrahydrofolate in a reaction catalysed by a dihydrofolate reductase. Aminopterin and amethopterin, which are known to block dihydrofolate reductase, are among the more promising agents for controlling leukaemia.

The role of glyceryl ethers of the type of chimyl, batyl and selachyl alcohols in haemopoiesis is far from being known in detail. That there is such a role is now very probable, but it seems unlikely to be either very simple or very direct.

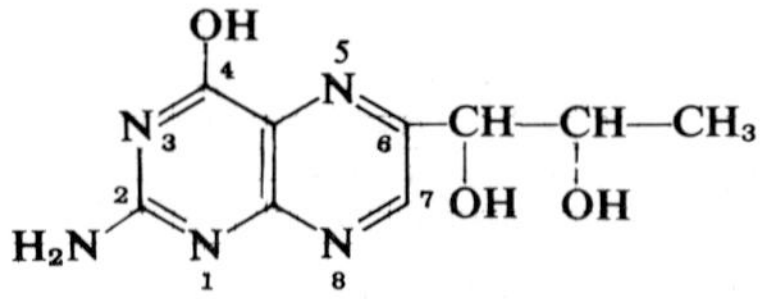

OH, N 3, 4, 5 N, 6, —CH—CH—CH_3, 7, OH OH, 2, H_2N, N 1, N 8

Biopterin (above) is 2-amino-4-hydroxy-6-[1,2-dihydroxypropyl]-(*L-erythro*)-pteridine. Sepiapterin (from the eyes of a sepia *Drosophila* mutant) undergoes reduction with borohydride to give:

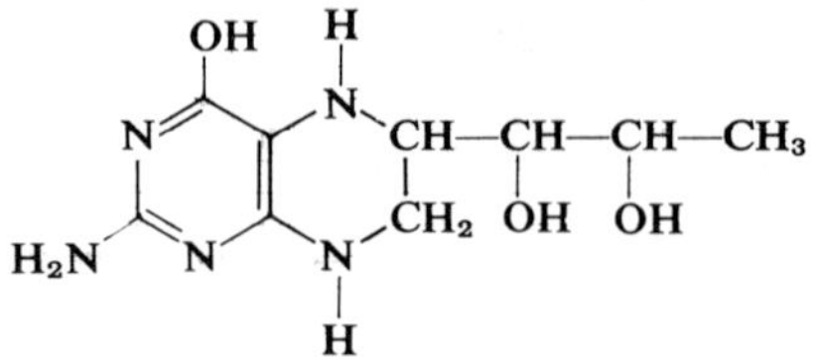

OH, H, N, N, CH—CH—CH—CH_3, CH_2 OH OH, H_2N, N, N, H

which is oxidased to biopterin (or an isomer). The phenylalanine cofactor as isolated is 7,8-dihydrobiopterin which would be reduced to 5,6,7,8-tetrahydrobiopterin, the true active form in hydroxylation. It seems certain that the cofactor in the enzyme system effecting the oxidation of glyceryl ethers is not a folate but a derivative of 2-amino-4-pteridine.

The empirical discovery of a haemopoietic factor in bone-marrow lipid unsaponifiable fraction may yet lead to valuable therapeutic agents. Promising though the outlook seems to be, a good deal of basic biochemistry may have to be clarified before the most effective interventions can be discovered. This is an area where modern spectroscopic techniques could be effectively deployed.

References

5.1

Bates, R. B. *et al.* (1962). *Chem. Ind.*, 1020.

Bates, R. B., Gale, D. M. & Gruner, B. J. (1963). *J. Org. Chem.*, **28**, 1086.

Burgos, J., Butterworth, P. H. W., Hemming, F. W. & Morton, R. A. (1964). *Biochem. J.*, **91**, 22P.

Dunphy, P. J. *et al.* (1967). *Biochim. Biophys. Acta.*, **136**, 136.

Feeney, J. & Hemming, F. W. (1967). *Anal. Biochem.*, **20**, 1.

Gough, D. P. & Hemming, F. W. (1970). *Biochem. J.*, **117**, 309; **118**, 163.

Hemming, F. W., Morton, R. A. & Pennock, J. F. (1963). *Proc. Roy. Soc.*, B **158**, 291.

Lindgren, B. O. (1964). *Acta Chem. Scand.*, **18**, 836.

Morton, R. A. (1972). In *Current Trends in the Biochemistry of Lipids*, p. 203. Ed. J. Ganguly & R. M. S. Smellie (Biochemical Society Symposia No. 35). Academic Press, London, New York.

Pennock, J. F., Hemming, F. W. & Morton, R. A. (1960). *Nature* (*Lond.*), **186**, 470; (1960). *Biochem. J.*, **74**, 38P.

Thorne, K. J. L. & Kodicek, E. (1962). *Biochim. Biophys. Acta*, **59**, 273, 280, 295; (1966). *Biochem. J.*, **99**, 123.

5.2

Blomstrand, R. & Ahrens, E. H. Jr. *Proc. Soc. Exptl. Biol. Med.*, **100**, 802.

Brohult, A. (1958). *Nature* (*Lond.*), **181**, 1484.

Brohult, A. (1960). *Nature* (*Lond.*), **188**, 59.

Brohult, A. (1962). *Nature* (*Lond.*), **193**, 1304.

Carter, H. E., Smith, D. B., & Jones, D. W. (1958). *J. Biol Chem.*, **232**, 681.

Hallgren, B. & Larsson, S. (1962). *J. Lipid Research*, **3**, 31.

Hanahan, D. J., Eckolm, J. & Jackson, C. M. (1963). *Biochemistry*, **2**, 630.

Hanahan, D. J. & Watts, R. (1960). *J. Biol. Chem.*, **136**, PC.59.

Holmes, H. N., Corbet, R. E., Geiger, W. B. & Kornblum, N. (1941). *J. Amer. Chem. Soc.*, **63**, 2607.

Karnovsky, M. L. & Brumm, A. F. (1955). *J. Biol. Chem.*, **215**, 689.

Kaufman, S. (1959). *J. Biol. Chem.*, **234**, 2677.

Kaufman, S. (1961). *J. Biol. Chem.*, **236**, 804.

Kaufman, S. (1963). *Proc. Natl. Acad. Sci. U.S.*, **50**, 1805.

Linnan, J. W. & Bethell, F. (1960). Ciba Foundation Symposium on Erythropoiesis, 369.

Linnan, J. W., Long, M. J., Korst, D. R. & Bethell, F. H. (1959). *J. Lab. Clin. Med.*, **54**, 335.

Lund, P. K., Djahanguir, M. A. & Mathies, J. C. (1962). *J. Lipid Research*, **3**, 95.

Marberg, C. M. & Wiles, H. O. (1937). *J. Amer. Med. Assoc.*, **109**, 1965; (1938). *Arch. Internal. Med.*, **61**, 409.

Pfleger, R. C., Pcantadosi, C. & Snyder, F. (1967). *Biochem. Biophys. Acta.*, **144**, 633.

Prelog, V., Rujicka, L. & Stern, P. (1943). *Helv. Chim. Acta.*, **26**, 2222.

Svennerholme, L. & Thorin, H. (1960). *Biochem. Biophys. Acta.*, **41**, 371.

Thompson, G. A. Jr. & Hanahan, D. J., (1963). *Biochemistry*, **2**, 641.

Tietz, A., Lindberg, M. & Kennedy, E. (1964). *J. Biol. Chem.*, **239**, 4085

6. Aromatic Compounds

6.1 Polycyclic Hydrocarbons

6.1.1 Nomenclature

Clar (1952) drew a distinction between *cata*-condensed and *peri*-condensed polycyclic compounds. The former have empirical formulae $C_{4n+2}H_{2n+4}$ (C_6H_6, $C_{10}H_8$, $C_{14}H_{10}$, etc.) and structures such that no carbon atom belongs to more than two rings. The *cata*-condensed aromatic hydrocarbons are made up of *acenes* (anthracene, naphthacene or tetracene, pentacene, etc.), where the rings are arranged linearly, and *phenes*, where the system of rings is bent, as in phenanthrene (triphene) and tetraphene, etc. The notation illustrated below

Anthracene

Naphthacene

Pentacene

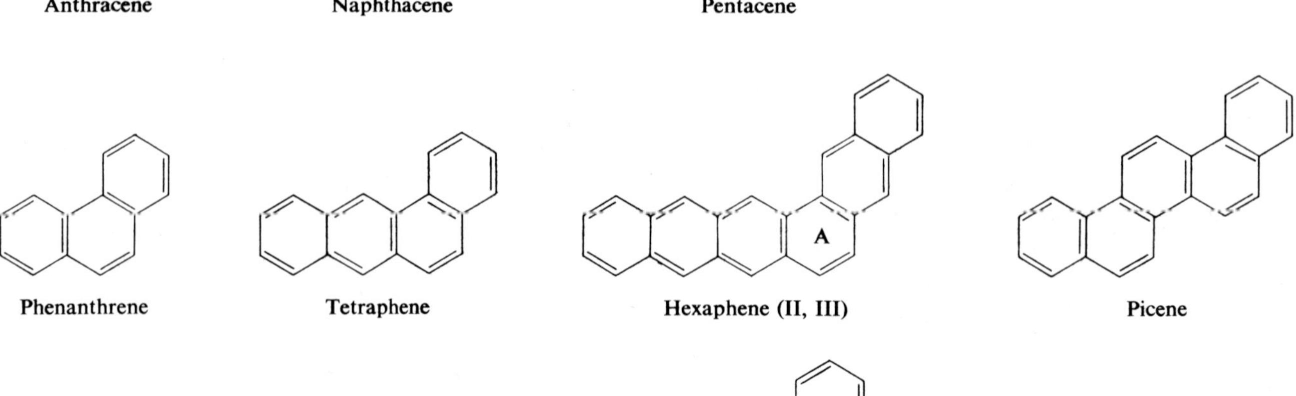

Phenanthrene

Tetraphene

Hexaphene (II, III)

Picene

A

A

Hexaphene (I, IV)

Heptaphene (II, IV)

applies to many hydrocarbons, chrysene and benzo[*c*]-phenanthrene being rather exceptional. The *peri*-condensed polynuclear hydrocarbons are illustrated by pyrene and perylene. The substituted fluorenes are also shown below.

Chrysene

Benzophenanthrene

Fluorene

Perylene

Pyrene

1,2-benzfluorene

2,3-benzfluorene

Perinapthane

Fluoranthene

3,4-benzfluorene

Absorption curves for a large number of aromatic hydrocarbons are readily accessible (Friedel & Orchin, 1951; Clar, 1952; Jaffé & Orchin, 1962) and there is a good earlier collecton by R. N. Jones (1943).

Most polynuclear hydrocarbons display three quite distinct regions of selective absorption, the molecular extinction coefficients rising rapidly as the wavelengths decrease. Clar (1952) notes that the peaks in the short wavelength ultraviolet region have high intensities ($\epsilon_{max.}$ 32 000–160 000), the intermediate peaks have moderate intensities ($\epsilon_{max.}$ 4000 to 13 000) and the long wavelength peaks show $\epsilon_{max.}$ 200 to 1600. Clar, who made many early observations (1929–1940), designated the three systems as β, *para* and α respectively. The intermediate (*para*) absorption suffers large displacements in the direction of longer wavelengths in the polyacenes so that the α absorption of, for instance, anthracene is completely masked. On the other hand, the low intensity absorption of phenanthrene ($\epsilon_{max.}$ *ca.* 200) persists between 310 and 345 nm, and in 1,2-benzanthracene at 370–390 nm.

Benzene (cf. p. 89) has three regions of selective absorption associated with 183 nm ($\epsilon_{max.}$ 46 000), 208 nm ($\epsilon_{max.}$ 6900) and 263 nm ($\epsilon_{max.}$ 220) with good vibrational resolution (Fig. 6.1). In naphthalene,

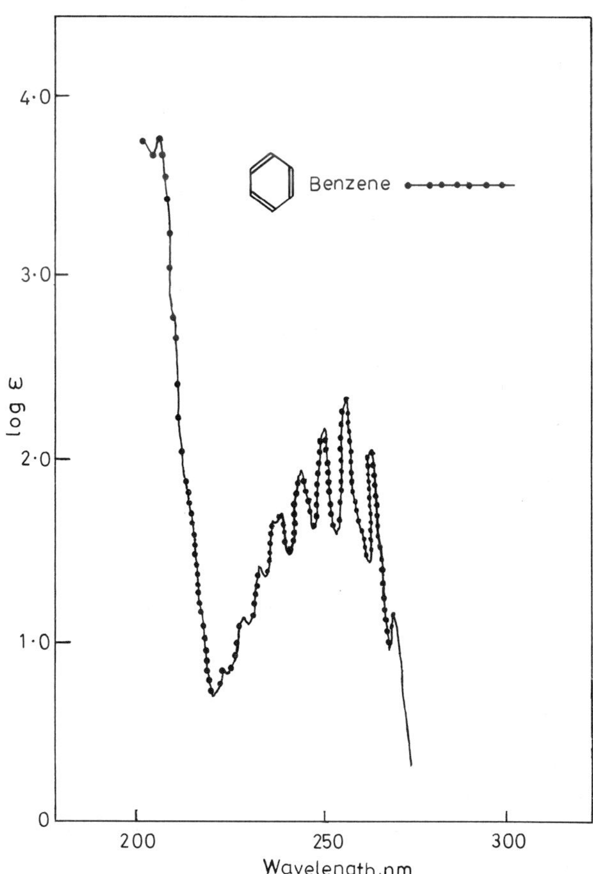

Figure 6.1 Absorption spectrum of benzene (from Mayneord & Roe, 1935)

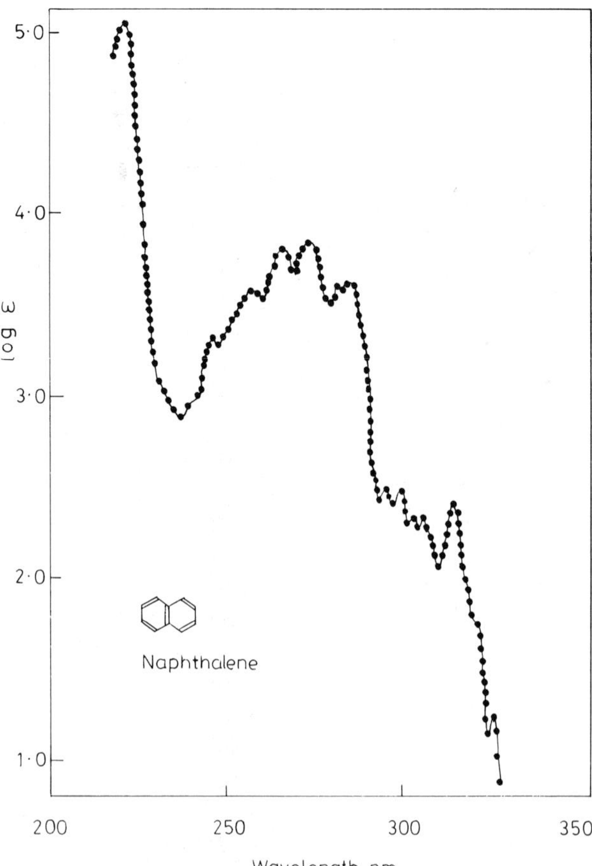

Figure 6.2 Absorption spectrum of naphthalene (from Mayneord & Roe, 1935)

Figure 6.3 Absorption spectrum of anthracene (from Mayneord & Roe, 1935)

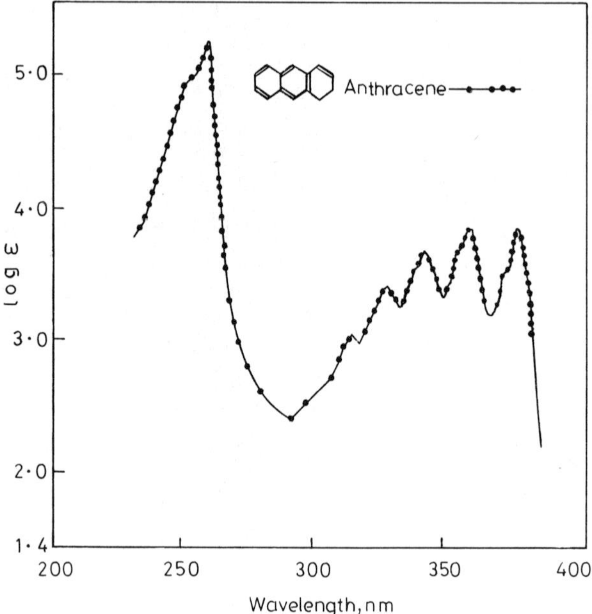

anthracene (Figs. 6.2 and 6.3) and naphthacene, the first band, is displaced to 220 (ϵ 133 000), 256 (ϵ 180 000) and 272 nm (ϵ 180 000) respectively. These transitions have a high probability. The intermediate absorption for naphthalene, anthracene and naphthacene is displaced to 289 (ϵ 9300), 379 (ϵ 9000) and 474 nm (ϵ 12 500) respectively. In each case there are regularly spaced peaks on the short wave side of the initial maximum noted above. The long wave absorption begins at about 310 nm for naphthalene and, like the 263 nm absorption of benzene, it has the low probability of a 'forbidden' transition.

Platt (1949) and Klevens & Platt (1949) have introduced in place of Clar's notation for the β process $'B$, for the *para*, $'L_a$, and for the α absorption, $'L_b$. Their interpretation has been summarized by Jaffé & Orchin (1962) and this classification of the spectra of the *cata*-condensed hydrocarbons has been freely used in interpreting the spectra (cf. Rudenberg & Parr, 1954; Araki & Murai, 1954; Moffit, 1954). There are undoubtedly marked regularities in the spacing of the vibrational band peaks and some success has attended the detailed study of the spectra.

The effect of substitution in benzene is illustrated in Table 6.1.

The naphthalene spectrum (Fig. 6.2) is not greatly changed in 1,5-dimethylnaphthalene, but the long wavelength absorption at 300–330 nm is very well resolved and appreciably strengthened in both 2,6-dimethylnaphthalene and 1,8-dimethylnaphthalene. Perinaphthane on the other hand fails to show resolved absorption in this region, but a peak at 289 nm (ϵ 5000) (with shoulders at 280 and 300 nm) persists, as does a strong peak at 230 nm (ϵ 50 000). The phenylnaphthalenes:

1-phenylnaphthalene		2-phenylnaphthalene	
λ_{max}, nm	ϵ_{max}	λ_{max}, nm	ϵ_{max}
225	50 000	250	52 500
287	7950	285	1100

and the different binaphthyls display broadly similar curves, with the vibrational fine structure diminished and the long wave absorption masked.

Fluorene (Fig. 6.4) has bands at 290 and 300 nm, and a strong peak near 264 nm (Jones, 1945; Friedel & Orchin 1951). The three benzofluorenes show interesting differences; between 230 and 300 nm the spectra of the 1,2- and 2,3-benzofluorenes are essen-

Table 6.1 Absorption peaks of substitution products

1,2,4-Tri-isopropylbenzene		1,3,5-Tri-isopropylbenzene		9,10-Dihydroanthracene	
$\lambda_{max.}$ nm	$\epsilon_{max.}$	$\lambda_{max.}$ nm	$\epsilon_{max.}$	$\lambda_{max.}$ nm	$\epsilon_{max.}$
198	40 740	198	40 740	210	12 600
216	10 230	216	9333	215	14 130
260	275	258	148	252	1260
267	417	263	148	264	1000
275	436	270	148	271	1000

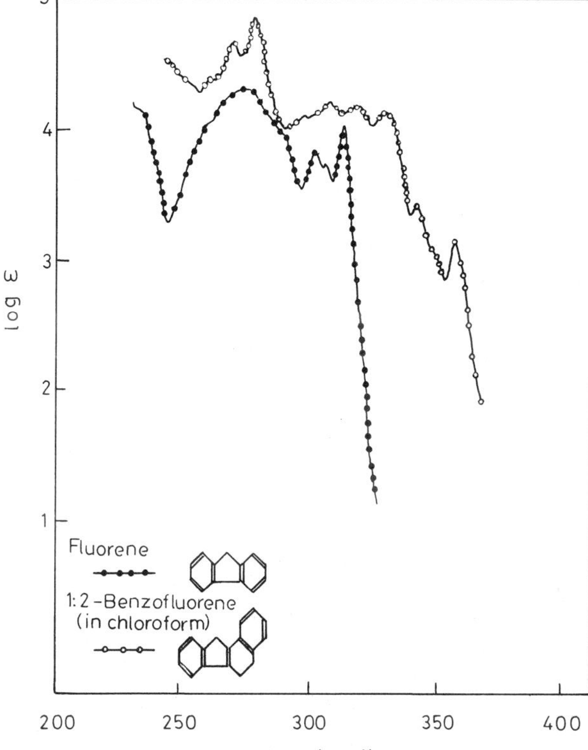

Figure 6.4 Absorption spectra of fluorene and 1,2-benzofluorene in chloroform (from Mayneord & Roe, 1935)

tially similar, while between 300 and 350 nm the latter shows stronger absorption:

1,2-benzofluorene			2,3-benzofluorene		
$\lambda_{max.}$ nm	$\epsilon_{max.}$	$\lambda_{min.}$	$\lambda_{max.}$ nm	$\epsilon_{max.}$	$\lambda_{min.}$
328	5250	340	324	7590	335
335	2510	(ϵ 590)	331	2510	(ϵ 1450)
344	1120	—	340	5750	—

3,4-benzofluorene

$\lambda_{max.}$ nm	$\epsilon_{max.}$	$\lambda_{min.}$
320	13 800	330
328	12 000	(ϵ 7950)
335	15 900	—

On the other hand the 3,4-benzofluorene (Fig. 6.5) shows a marked overlap of the middle ultraviolet and long wave absorption so that the sharp 335 nm peak displays a high value for $\epsilon_{max.}$, namely 15 900.

Anthracene and naphthacene exhibit very large displacements of the middle ultraviolet absorption: Anthracene (in cyclohexane)

$\lambda_{max.}$ nm	~245	252	~295	310
$\epsilon_{max.} \times 10^{-3}$	100	200	0·447	1·12
$\lambda_{max.}$ nm	322	339	355	375
$\epsilon_{max.} \times 10^{-3}$	2·85	6·31	10	7·9

In addition to the above peaks there are several

Figure 6.5 Absorption spectra of 3,4-benzofluorene and 1,2,5,6-dibenzofluorene

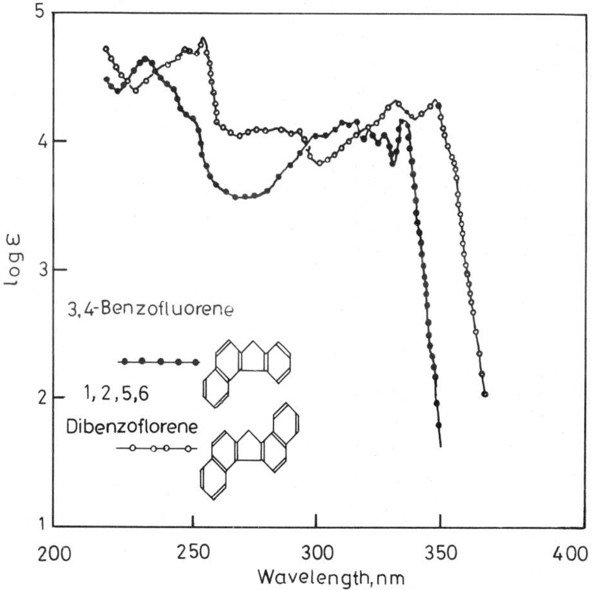

inflections consistent with incomplete resolution. In ethanol the absorption is similar except that much of the structure is smoothed out.

The compound 9,10-dihydroanthracene shows the predictable reversion to a benzenoid type of absorption. The spectra of anthracene derivatives with extra-nuclear conjugated unsaturation show further displacements:

1-Ethynylanthracene (in cyclohexane)

$\lambda_{max.}$ nm	255	275	305	322	334
$\epsilon_{max.} \times 10^{-3}$	112	10·96	0·4	1·07	2·75
$\lambda_{max.}$ nm	352	371	386	391	
$\epsilon_{max.} \times 10^{-3}$	6·03	10·95	4·79	11·5	

1,8-Diethynylanthracene (in cyclohexane)

$\lambda_{max.}$ nm	258	279	291	~312	320
$\epsilon_{max.} \times 10^{-3}$	13·5	12	12	0·74	1·66
$\lambda_{max.}$ nm	345	362	381	402	
$\epsilon_{max.} - 10^{-3}$	0·4	9·55	16·6	17·4	

9-Styrylanthracene (in hexane)

$\lambda_{max.}$ nm	260	287	~300	387
$\epsilon_{max.} \times 10^{-3}$	100	8·7	6·3	10·2

There is little significant change in the positions of maxima when two anthracene units are joined, but there is a very notable increase in intensity between 280 and 320 nm:

2,2′-Dianthryl (in $C_{10}H_5Cl_3$)

$\lambda_{max.}$ nm	285	303	317	355	388	402
$\epsilon_{max.} \times 10^{-3}$	69	51·3	46·8	11·2	15·5	16·2

A break in the conjugation at positions 1 and 7 within the ring system causes a reversion to a naphthalenic absorption:

1,2-Diethyl-1,2-dihydroanthracene (in ethanol)

$\lambda_{max.}$ nm	~229	235	243	252	261	278
$\epsilon_{max.} \times 10^{-3}$	20·9	25·1	28·2	21·9	16·2	7·95
$\lambda_{max.}$ nm	285	298	320	326	336	341
$\epsilon_{max.} \times 10^{-3}$	9·33	8·51	0·23	0·246	0·115	0·159

Extension of the polynuclear system shifts the spectrum greatly:

Naphthacene [tetracene] (in ethanol)

$\lambda_{max.}$ nm	280	295	400	420	450	475	min. 340
$\epsilon_{max.} \times 10^{-3}$	250	31·6	2·00	5·0	1·0	12·5	0·2

1,2-Benzanthracene (in ethanol)

$\lambda_{max.}$ nm	229	245	256	268	280	290	300
$\epsilon_{max.} \times 10^{-3}$	40	20	31·6	50	79·4	100	1·60
$\lambda_{max.}$ nm	314	330	341	360	375	383	
$\epsilon_{max.} \times 10^{-3}$	5·25	7·08	6·5	5	0·5	0·83	

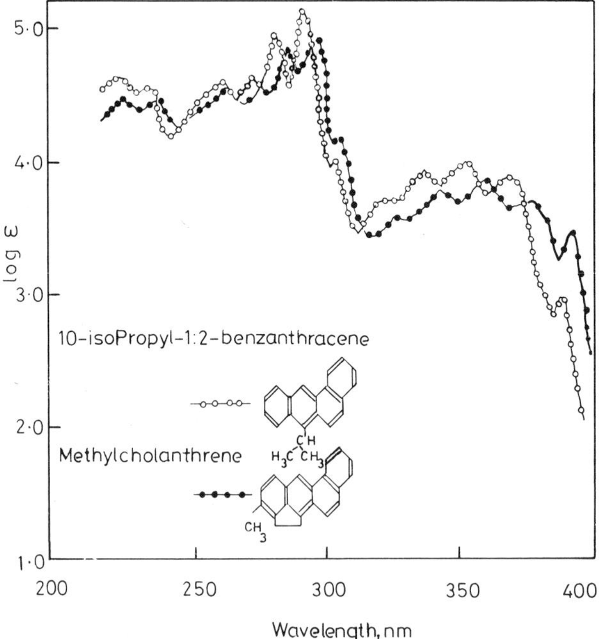

Figure 6.6 Absorption spectra of 10-*iso*propyl-1,2-benzanthracene and methylcholanthrene

Figure 6.7 Absorption spectrum of 1,2,5,6-dibenzanthracene

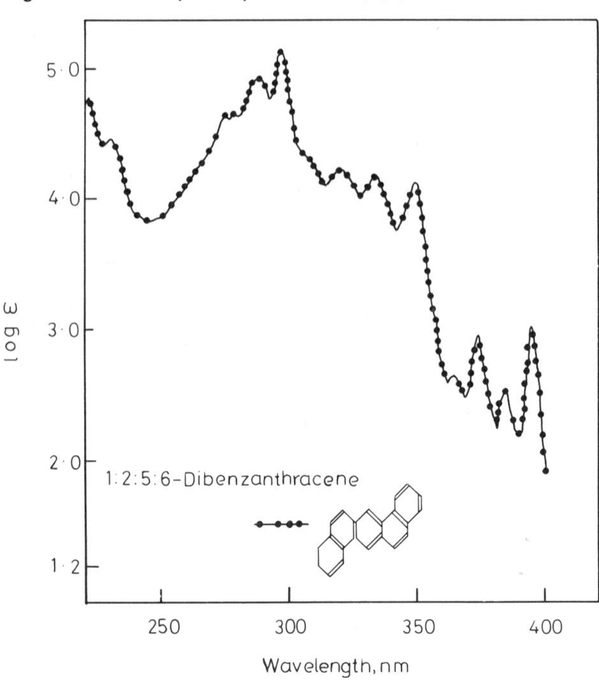

Pentacene

$\lambda_{max.}$ nm	290	302	~330	350	404
$\epsilon_{max.} \times 10^{-3}$	56	148	8·9	6·9	0·68
$\lambda_{max.}$ nm	429	470	505	540	585
$\epsilon_{max.} \times 10^{-3}$	0·71	0·87	2·0	4·3	5·9

1,2,3,4-Dibenzanthracene

$\lambda_{max.}$ nm	338	345	360	375
$\epsilon_{max.} \times 10^{-}$	10	5	7·0	0·4

1,2,5,6-Dibenzanthracene (in benzene)

$\lambda_{max.}$ nm	290	300	322	337	351	373	385	395
$\epsilon_{max.} \times 10^{-3}$	87	125	20·9	16·2	17·4	1·26	0·5	

[6,13-Dihydropentacene reverts to naphthalene type of absorption].

Mayneord & Roe (1935) recorded a spectrum for 9,10-dihydro-1,2,5,6-dibenzanthracene which resembled that of the parent substance quite closely, whereas the *cis* and *trans* forms of 9,10-dimethyl-9,10-dihydro-1,2,5,6-dibenzanthracene showed almost identical spectra of the naphthalenic type. If the findings are correct, the two methylene groups in 9,10-dihydro-1,2,5,6-dibenzanthracene fail to isolate two naphthalene residues whereas the —CH group is fully effective.
|
CH_2

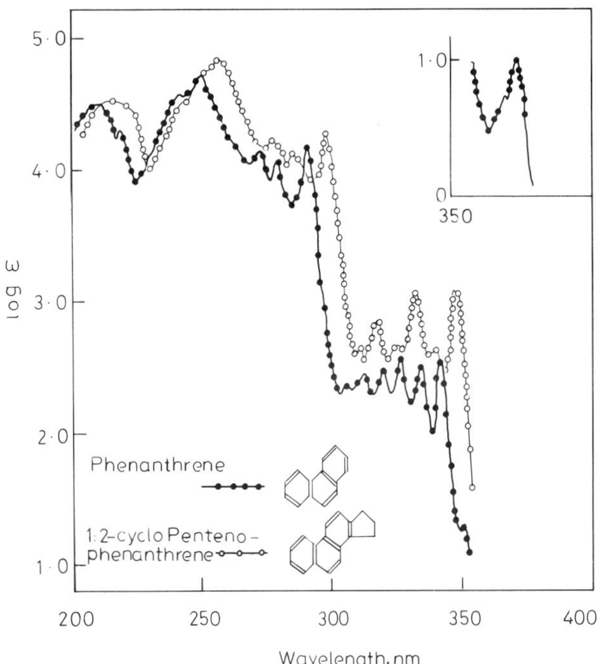

Figure 6.8 Absorption spectra of phenanthrene and 1,2-*cyclo*penteno-phenanthrene

6.1.2 Phenanthrene and its Derivatives

The absorption spectrum of phenanthrene has been recorded a number of times but with only moderate agreement. There are certainly quantitative resemblances between the spectra of anthracene and phenanthrene but intensities for the latter are much lower. The reduction in effective molecular length due to the presence of an angular ring compared with the linear arrangement reduced the absorbance at the 251 nm peak to $\epsilon = 65\ 000$. The displacement of the 270–290 nm absorption is much less in phenanthrene and this allows the weak absorption that is masked in the anthracene spectrum to emerge.

The phenanthrene absorption in hydrocarbon solvents thus consists of four distinct regions (i) two bands *ca.* 216 and 223 nm, $\epsilon_{max.}$ *ca.* 22 500; (ii) very strong absorption at 246 and 252 nm, $\epsilon_{max.}$ *ca.* 50 000 and 65 000 respectively; (iii) peaks at 275, 282 and 294 nm, with $\epsilon_{max.}$ 22 000, 11 000 and 22 000 respectively; (iv) very much weaker absorption made up of five (or six) roughly equally spaced narrow bands between 309 and 348 nm, ϵ values between 200 and 320 (Friedel & Orchin, 1951) (see Fig. 6.8). The spectrum in ethanol

Figure 6.9 Absorption spectra of 3,4-benzphenanthrene and 1′,2′-naphtha-2,3-fluorene. (See also Cook *et al.*, *J. Chem. Soc.*, **1935**, 1319)

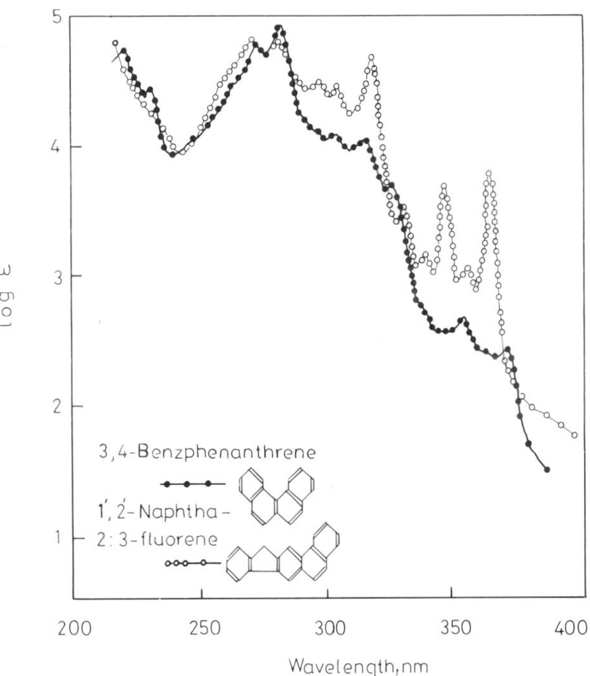

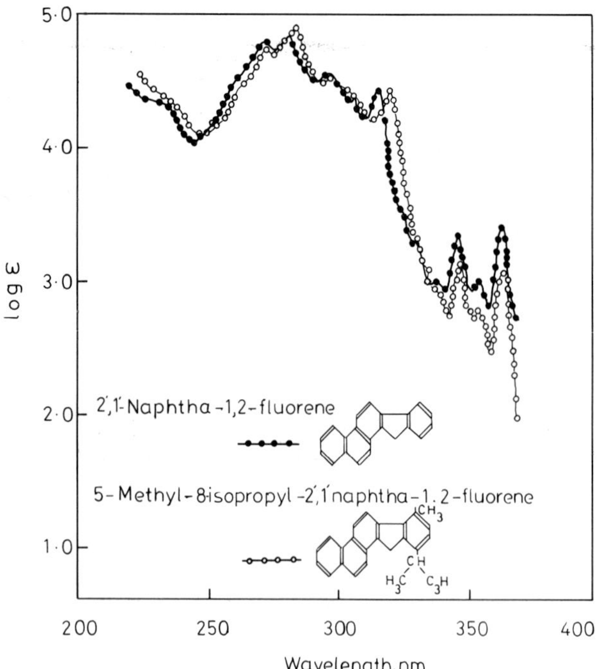

Figure 6.10 Absorption spectra of 2′,1′-naphtha-1,2-fluorene and 5-methyl-8-*iso*propyl-2′,1-naphtha,1,2-fluorene. (See also Cook *et al.*, *J. Chem. Soc.*, **1934**, 1727)

is very similar but the resolution is less marked. The alkylphenanthrenes broadly resemble the parent hydrocarbon in respect of spectra but there is always some loss of vibrational structure on substitution. Tetracyclic compounds derived from phenanthrene have been discussed by Klevens & Platt (1949) and the energy levels have been very well accounted for (see also R. N. Jones, 1945; Friedel & Orchin, 1951; Jaffé & Orchin, 1962).

Several biphenanthrenes ($C_{28}H_{18}$) have been studied. They include 2,2-biphenanthrene, the 9,9′-isomeride and two isomers of m.p. 175 and 211° respectively. All show resolved absorption with peaks having $\epsilon_{max.}$ values of 30 000 to 90 000 in the region below 320 nm. The high intensity masks altogether any weak bands in the region 320 to 360 nm. Phenanthrene-9(1-naphthyl) and phenanthrene-9-(2-naphtyl) also show resolved absorption in the region 220 to 300 nm, with ϵ values for 20 000 to 100 000 and again any weak long wavelength absorption is masked. The compounds 2,3-benzofluoranthene and 8,9-benzofluoranthrene show intense absorption with $\epsilon_{max.}$ *ca.* 50 000 in the 250–260 nm region. The former has moderately strong bands from 305–420 nm (ϵ values all < 5000) while the latter has rather stronger bands, particularly between 280 and 315 nm ($\epsilon_{max.}$ up to 50 000).

17-Ethyl 1,17-dimethyl
17H-cyclopenta[α]phenanthrene

$\lambda_{max.}$ nm	$\epsilon_{max.}$
224	20 100
244	27 500
266	46 800
298	10 960
310	16 600
323	16 000
347	832
365	457

4,17-Dimethyl, 16,17-dihydro,
15H-cyclopentanophenanthrene

$\lambda_{max.}$ nm	$\epsilon_{max.}$
216	38 000
256	50 100
264	60 260
284	13 800
294	13 200
307	16 600
322	525
339	617
353	490

17-Ethyl, 15,16-dihydro, 1,17-dimethyl
cyclopenta[α]phenanthrene

$\lambda_{max.}$ nm	$\epsilon_{max.}$
213	26 920
257	70 800
282	12 900
293	11 200
305	14 800
322	550
337	525
353	380

The interpretation of the spectra of polycyclic hydrocarbons owes much to Klevens & Platt (1949) whose theory has been well summarized by Jaffé & Orchin (1962). The important variables are (*a*) the chain length in the poly-ene series associated with longitudinal polarization and a red shift for each addition to the poly-ene sequence; (*b*) electronic transitions associated with tranvserse vibrations perpendicular to the long axis of the molecule. Naphthalene derivatives show effects that are, for instance, very different for 1-naphthylamine and 2-naphthylamine. The former exhibits a very broad region of unresolved absorption between 270 and 370 nm, $\lambda_{max.}$ *ca.* 321 nm, log ϵ 3·7 and there is a second unresolved peak at 240 nm, log ϵ 4·35. In 2-naphthylamine there are three regions of selective absorption 235 nm, log ϵ 4·8, 285–295 nm, log ϵ 3·3–3·2 and unresolved absorption 310–380 nm, $\lambda_{max.}$ *ca.* 336 nm, log ϵ 3·3; the middle region is but little affected by the introduction of the amino group, whereas the absorption in regions on either side undergoes a red shift and becomes stronger. The spectra for a considerable range of alkyl substituted naphthalenes have been recorded (Mosby, 1953; Bailey *et al.*, 1947; Morton & Gouveia, 1934; Hellbronner *et al.*, 1949; Jones, 1943; Platt, 1951; Friedel & Orchin, 1951) and the overall general agreement with the Klevens and Platt classification is good. The empirical use of hydrocarbon spectra remains important for a number of biochemical problems and the appearance or non-appearance of resolution is a factor. Jones (1945), noting the excellent resolution in the spectra of fluorene derivatives, referred to the fine structure or Fs effect. Introduction of a methylene group increases the rigidity of the molecule and hinders the dissipation of the energy of electronic excitation and preserves the vibrational structure. Alkyl groups and auxochromes generally tend to smooth out the absorption curve by making easier the loss of energy via the vibrational states of the electronically excited molecule.

Comparison of the spectra of anthracene and phenanthrene shows that although there is good resolution in both, the strongest band (associated with longitudinal polarization) is extremely intense in anthracene ($\lambda_{max.}$ 254 nm, ϵ 180 000) and in phenanthrene is decidedly weaker though still strong ($\lambda_{max.}$ 253 nm, ϵ 65 000). The anthracene absorption in the region 290–380 nm shows six well-developed peaks with intensities reaching 7600 and introduction of methyl groups at positions 9 and 10 allows the general pattern to be retained with small red shifts and relatively small increases in intensity of absorption at 360–420 nm. Phenanthrene absorption is much closer to that of some dimethylnaphthalenes in the regions 270–300 nm and 300–350 nm. The group of six well-defined peaks between 305 and 350 nm, ϵ values $\not> 250$ is very characteristic.

The spectra of triphenylene, 3,4-benzphenanthrene, 1,2-benzanthracene and chrysene (see Fig. 6.13) are consonant with a broad pattern based on phenanthrene.

Pyrene (Fig. 6.11) a peri-condensed hydrocarbon, shows a spectrum of considerable complexity and the presence of double bonds internal to the structure is an obstacle to a satisfactory interpretation. Benzpyrene (Fig. 6.12) and picene (Fig. 6.13) raise further issues.

The attempt to reduce to order the spectra of polycyclic aromatic hydrocarbons represents a considerable achievement though a partial one. Biochemists are in the main interested in these hydrocarbon compounds and their spectra because of the need to detect and determine the carcinogenic members of the group. Such work is necessarily empirical but it may well turn out that broadly based spectroscopic research may have a bearing on structure–carcinogenicity relationships.

Figure 6.11 Absorption spectrum of pyrene

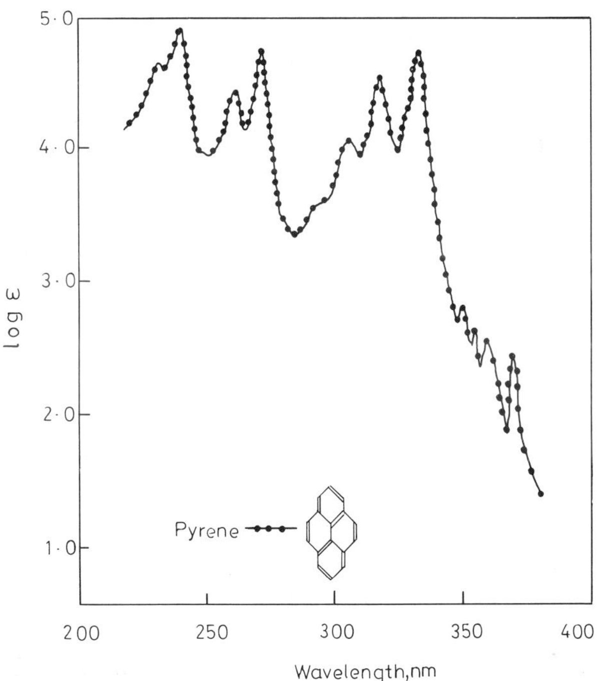

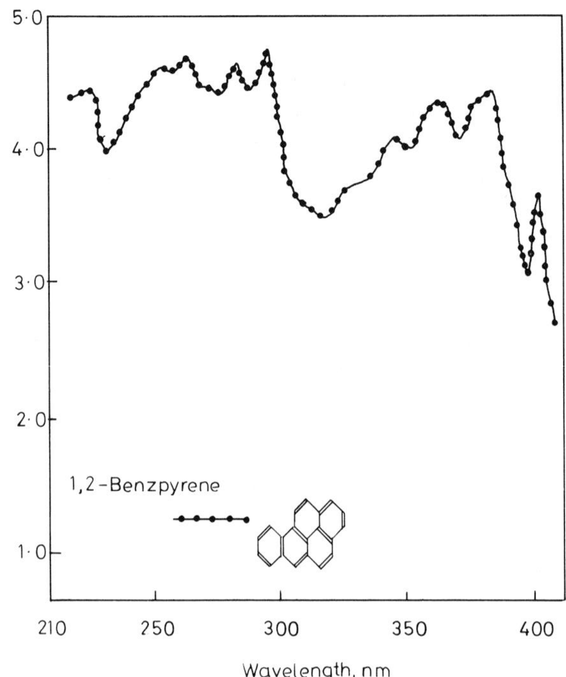

Figure 6.12 Absorption spectrum of 1,2-benzpyrene

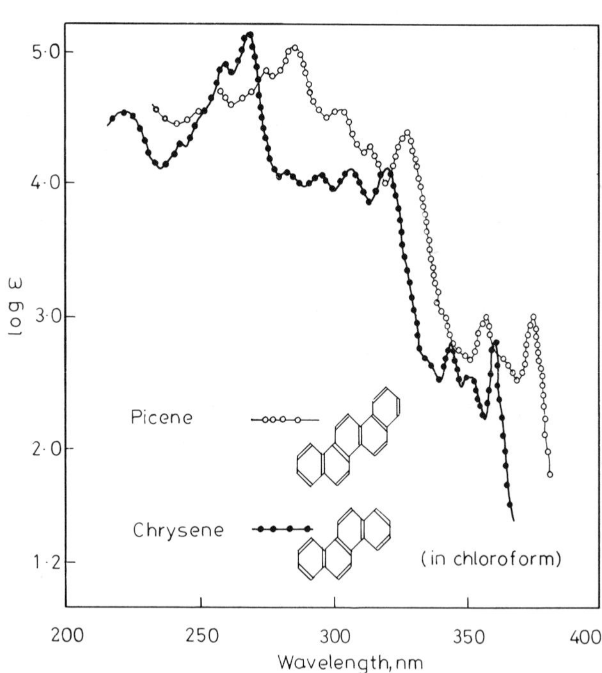

Figure 6.13 Absorption spectra of picene and chrysene in chloroform

Table 6.2 Some polycyclic hydrocarbons

Chrysene in ethanol		Triphenylene in ethanol		3,4-Benzanthracene in ethanol		Perylene in ethanol		Pyrene in ethanol	
$\lambda_{max.}$ nm	$\epsilon_{max.}$	$\lambda_{max.}$ nm	$\epsilon_{max.}$	$\lambda_{max.}$ nm	$\epsilon_{max.}$	$\lambda_{max.}$ nm	$\epsilon_{max.}$	$\lambda_{max.}$ nm	$\epsilon_{max.}$
220	28 180	—	—	215	44 700	208	50 000	233	63 000
241	17 800	—	—	230	31 600	245	33 000	250	100 000
259	63 100	250	100 000	—	—	252	44 700	253	10 000
266	141 300	259	160 000	270	18 200	266	8500	260	25 100
283	12 600	222	20 000	280	23 000	364	3980	277	63 000
295	11 750	284	18 000	—	—	385	11 800	299	6300
307	12 900	290	*ca.* 2000	303	11 500	407	25 800	310	12 600
320	12 600	320	630	314	10 250	434	34 000	325	25 000
343	603	330	398	325	1350	2,8-dimethyl and		339	25 000
352	427	335	795	~340	316	3,9-dimethyl pery-		340	39 800
361	630	340	63	353	316	lenes are very similar		355	1000
—	—	—	—	370	200	to perylene		—	—

6.1.3 Azulene and its Derivatives

Azulene is a good example of a non-alternant hydrocarbon (see Coulson & Rushbrooke, 1940; Coulson and Longuet-Higgins, 1947). In an alternant hydrocarbon like naphthalene the carbon atoms can be divided into two groups, 'starred' and 'unstarred', in which each starred atom is flanked by two unstarred atoms.

(*a*) (*b*) (*c*)

Azulene

In azulene this is not possible. If a molecule has an odd-membered ring it will be non-alternant and every non-alternant compound will contain an odd-membered ring. The structure of naphthalene is such that the π electrons are very evenly distributed on all the carbon atoms whereas azulene has a dipole moment consistent with uneven distribution between the seven-membered ring and the five-membered ring, so that the resonance structures (*b*) and (*c*) contribute substantially to the ground state of the molecule. In some respects azulene resembles a cyclodecapentaene with a trans-annular valency bridge.

Published absorption spectra show certain discrepancies. The following figures are from a recent work of reference (Philips *et al.*, 1969) and are based on recent measurements (Brunke & Poppe, 1960; Scholtz & Treibs, 1961; A.P.I. Project 44; Zimmermann & Joop, 1960; Robertson & King, 1961).

Azulene

In cyclohexane		In pentane		In ethanol	
$\lambda_{max.}$ nm	$\epsilon_{max.}$	$\lambda_{max.}$ nm	$\epsilon_{max.}$	$\lambda_{max.}$ nm	$\epsilon_{max.}$
238	20 000	—	—	240	35 100
271	50 000	—	—	—	—
276	50 000	275	63 000	275	79 400
281	50 000	—	—	280	79 400
296	3980	—	—	—	—
309	1260	—	—	—	—
316	2000	—	—	—	—
322	2510	—	—	—	—
326	3160	—	—	—	—
333	3160	—	—	—	—
337	3980	—	—	—	—
341	5000	340	5000	345	7940
353	1260	—	—	—	—
535	229	—	—	—	—
557	282	—	—	—	—
579	363	575	320	—	—
603	*ca.* 300	—	—	—	—
632	*ca.* 300	—	—	—	—
658	*ca.* 150	650	320	—	—
696	*ca.* 150	—	—	—	—

The major features are: (i) resolved absorption between 250 and 290 nm, high values for absorbance; (ii) well-resolved bands between 300 and 360 nm of moderate intensity and (iii) very sharp narrow bands of low intensity in the visible region.

There is a considerable literature on substituted azulenes (cf. Plattner & Hellbronner, 1948). In the ultraviolet region the spectrum is shifted on methylation, more when the substituent is in the 5-membered ring than when the same entering group is in the 7-membered ring. Even so some bands suffer more displacement than others. In the visible region substitution at positions 1, 3, 5 or 7 displaces the weak bands in the direction of long wavelengths whereas substitution at 2, 4, 6 or 8 shifts the spectrum towards shorter wavelengths. Thus the 696 nm band of azulene moves to 779 nm in 1,3-dimethylazulene and to 666 nm in 4,8-dimethylazulene and to 717 and 721 in 1,2- and 1,4-dimethylazulene respectively. The problem has been discussed theoretically (see Jaffé & Orchin, 1962). Fulvene and fluoranthene are also non-alternant hydrocarbons.

CH_2

Fulvene
(not known)

CH_3
C
CH_3

Dimethylfulvene

$\lambda_{max.}$nm	$\epsilon \times 10^{-3}$
270	16
364	0.2

8 9 7 10 6 1 5 2 4 3

Fluoranthene

$\lambda_{max.}$ nm	$\epsilon \times 10^{-3}$
235	39·8
261	12·0
276	22·4
287	31·6
308	3·55
323	5·75
341	7·76
358	7·95

The study of ultraviolet absorption spectra of aromatic hydrocarbons is of course far from new. Thus Henri and de Laszlo made an elegant experimental analysis of the absorption spectrum of naphthalene vapour as far back as 1924 and Capper & Marsh (1926) published reproduction of spectrograms of condensed nuclear hydrocarbons using a photographic technique. Mayneord & Roe (1935, 1937) published numerous excellent spectra of complex aromatic hydrocarbons and R. N. Jones (1943) reviewed the ultraviolet absorption spectra of aromatic hydrocarbons comprehensively in an article in which 263 references were cited. In more recent times Friedel & Orchin's book (1951) included a fine collection of absorption curves in which log ϵ was plotted against λ. Jaffé and Orchin's book also has many curves and a modern discussion of the work of Clar, Platt and many others.

The ultraviolet absorption spectra of metabolites of polycyclic hydrocarbons are of considerable biochemical relevance. Beale & Roe (1951), building on the metabolic studies of Boyland and his colleagues (1935, 1949, 1950), used absorption spectra to distinguish between

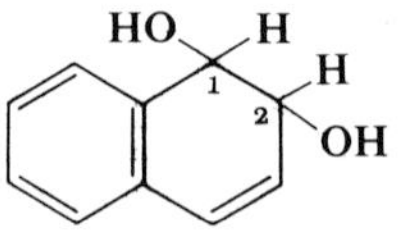

HO H 1 2 H OH and H OH 1 4 HO H

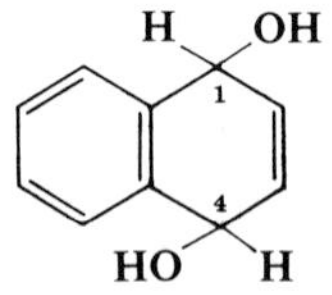

1,4-Dihydronaphthalene, like tetralin, shows benzenoid absorption, inflection 267 nm, ϵ 800 or 740 (Morton & de Gouveia, 1934) whereas 1,2-dihydronaphthalene shows $\lambda_{max.}$ 262 nm, $\epsilon_{max.}$ 10 230. The metabolite from rat and from rabbit obtained either as racemate or optically active shows $\lambda_{max.}$ 263–265 nm, $\epsilon_{max.}$ 7790 to 8410 and is clearly 1,2-dihydro-1,2-dihydroxynaphthalene. The hydrogenated product 1,2-dihydroxy-1,2,3,4-tetrahydronaphthalene shows an inflection at 260 nm and peaks at 273 and 265 nm, $\epsilon_{max.}$ 340 and 320 respectively.

A diol metabolite of anthracene was shown to have the structure

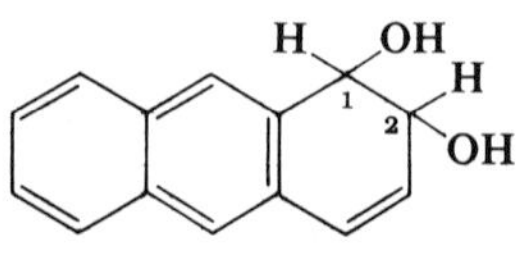

H OH 1 2 H OH

since it exhibits a spectrum very like that of 2-propenylnaphthalene.

Two diols have been isolated as phenanthrene metabolites from rat or rabbit.

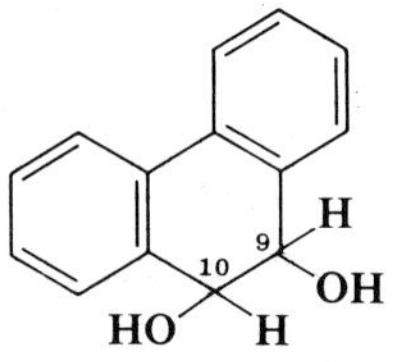

H 9 10 OH HO H

rat

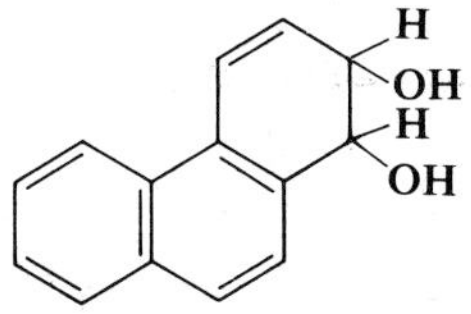

H OH H OH

rabbit

Synthetic *trans*-9,10-dihydro-9,10-dihydroxyphenanthrene with $\lambda_{max.}$ 268, inflection 300 nm, ϵ 15 850 and 2570 proved to be the rat metabolite whereas the synthetic *cis* compound lacked the selective absorption near 300 nm. The other diol, obtained from rabbit, showed a fairly close resemblance with 1-propenylnaphthalene supporting the 1,2-dihydro-1,2-dihydroxyphenanthrene structure. The strongly interacting group at the 1-position in naphthalene causes a large bathochromic shift and the fine structure around 275 nm is lost and the weak bands of naphthalene around 310–320 nm are completely masked.

6.1.4 Carcinogenicity

By about 1962 more than 2000 compounds had been tested for carcinogenicity and about 500 had proved to be tumorigenic. Of these about 150 were polycyclic hydrocarbons (Hartwell, 1951; Shubik & Hartwell, 1957; Oser, 1962). Although many polycyclic hydrocarbons have thus been shown to be carcinogenic under specific conditions, some are clearly more dangerous than others and the problem of assessing risks and taking measures to protect the public from actual or possible hazard is very formidable. There are believed to be thousands of polynuclear hydrocarbons and numerous derivatives are known. Perhaps by now as many as a thousand polynuclears have been screened for carcinogenicity using rodents and up to 200 may be under some suspicion. The work suffers from the defect that few of the suspect compounds were administered orally. On the other hand some substances were tumorigenic at very low dose levels while others were not at higher dosages. It is, however, very difficult to prove that a compound is entirely safe; very few polycyclic hydrocarbons have been shown with any approach to rigour *not* to be carcinogenic, indeed this is deemed to require tests on two species at very high dose levels throughout the normal life span. Neither 'guilt' nor 'innocence' is easy to establish equivocally for the majority of the polynuclears as such, but a few that are both dangerous and widespread have become well recognized. The hazard for man usually manifests itself only after prolonged exposure to polycyclic hydrocarbons at perhaps very low levels of concentrations.

Some possible avenues of exposure may be mentioned. Certain packaging materials for meat contain refined mineral oil consisting in the main of innocuous paraffins but it is on record that up to 200 parts per million of such oil may migrate into meat. Beyond doubt the material so used must be subjected to strict tests.

The tumorigenic hydrocarbons in general contain 4 to 6 rings and display well-resolved selective absorption in the ultraviolet region, but there are no guidelines to suggest that any low level of *intake* of a carcinogen is 'safe'. Moreover the 'more dangerous' polycyclic hydrocarbons have no known specific characteristics which distinguish them spectroscopically from the 'less dangerous' ones. This means that environmental polynuclear hydrocarbon concentrations should be monitored and kept as low as possible. (See Mazee *et al.*).

Haenni & Hall (1960) and Haenni *et al.* (1962) aimed at a specification for white mineral oil that would set an upper limit of 5 p.p.m. for the total polynuclear content. They accepted that nearly all polycyclic hydrocarbons show high absorbances in the region 295 to 299 nm and that absorbances were notably high for all known hydrocarbon carcinogens. They put forward the suggestion that at 275 nm, the extinction (log I_0/I) for a 1 mm layer of white mineral oil should be not more than 0·3 ($E_{1cm}^{100\%}$ 0·3). From 295 to 299 nm a 1 cm layer should show $E < 0{\cdot}225$ and above 299 nm $E < 0{\cdot}18$. This specification would permit only minute amounts of methylbenzanthracenes, benzypyrene, cholanthrene or chrysenes. If 200 parts per million of oil were to migrate into meat the specification would preclude contamination of the food by more than 1 part per billion

$$[200 \times 10^{-6} \times 5 \times 10^{-6} = 1000 \times 10^{-12} = 1 \times 10^{-9}]$$

Lijinsky (1960) and Lijinsky & Keeling (1961) examined 209 samples of paraffin wax and only six had appreciable amounts of selectively absorbing hydrocarbon and 4 out of the 6 had less than 0·01 p.p.m. None of the samples contained a detectable amount of a known carcinogen. Dimethylsulphoxide is almost certainly the best solvent for selectively removing polycyclic hydrocarbons from paraffin wax. The extract can be subjected to paper chromatography (dimethylformamide and iso-octane) or to thin layer chromatography. Mineral oil is allowed in bread by the U.S. Food and Drug Administration up to 1500 p.p.m. subject to conforming to a strict absorbance limit.

The smoking of food and the use of charcoal-broiled meats can introduce small amounts of polycyclic hydrocarbons into a human diet. 'Liquid' smokes are also used; these consist of the water-soluble constituents of smoke and include the so-called pyroligneous acids. Lijinsky & Shubik (1965) found a few parts per billion of pyrene, fluoranthene, chrysene, benzanthracene and other (unidentified) polynuclear compounds. (See also Lijinsky *et al.*, 1962, 1963)

The quantitative isolation of polycyclic hydrocarbons from air or from foods or other materials is far from easy. There are many reasons for this: (*a*) the amounts to be found though often very small are still important; (*b*) reagents and glassware need to be watched carefully lest contaminants should invalidate the analysis; (*c*) several, or possibly many, polynuclear hydrocarbons may be present together in the material under study and the individual substances may be difficult to separate; (*d*) at the low levels of concentration at which scrutiny is frequently necessary, some important polycyclic hydrocarbons can undergo oxidation or photo-oxidation in the course of analytical manipulations to give peroxides or quinones.

Adsorption (column) chromatography, gas-liquid chromatography, paper and thin layer chromatography have all been used (cf. Gunther & Buzzetti, 1965; Sawicki *et al.*, 1964); National Cancer Institute (1962). A useful table of partition coefficients has been published (Hoffman & Wynder, 1962). Identifications should be based on as many tests as possible, but for trace constituents ultraviolet and fluorescence spectra supported by R_F values often provide the only feasible method.

Ultraviolet spectrometry is of course an essential tool. A careful survey of the technical problems of ultraviolet and fluorescence assay methods has been made by Moore *et al.* (1962). Direct measurement of ultraviolet absorption can be very useful for materials that remain apparently homogeneous after paper or thin layer chromatography using two different systems. Even then, plausible 'identifications' should where possible be cross-checked by fluorescence spectra. Sawicki *et al.* (1960 *c*) have published Tables of spectra (mostly for pentane solutions) for 42 hydrocarbons.

Neither ultraviolet absorption nor fluorescence leads to easy or automatic quantitative determinations; the analyst must shrewdly deploy technical skill and experience to deal with individual problems as they arise (cf. Commins, 1962). For instance the activation spectra, fluorescence spectra and absorption spectra lead to identifications that are very likely to be correct if the analyst has made his own selection of the relevant literature and possesses a collection of standard preparations. Fluorescence spectra of the more important polycyclic hydrocarbons have been published by Lyons & Johnston (1957); Lijinsky *et al.* (1961); Sawicki *el at.* (1960 *a*, *b*, *c*) and Van Duuren (1958). It is useful that a pure polycyclic hydrocarbon will give practically the same fluorescence spectrum at all activating wavelengths. If a given preparation shows differences in fluorescence spectra when the incident wavelengths are different, an impurity will be present and a change in the weak long wavelength part of the fluorescence spectrum is specially significant.

Gunther & Buzzetti (1965) placed great emphasis on the use of fluorescence-free solvents, on glassware cleaned by hot concentrated nitric acid (*not* dichromate–sulphuric acid) and on the risk of contaminated spectroscopic cells. They also warn against contamination from rubber, cork and plastic stoppers, stopcock lubricants and plastic tubing or containers.

The study of polycyclic hydrocarbons has become a significant aspect of environmental medicine. Many 'natural' products contain naphthalene or naphthalene derivatives (e.g. clove stem oil, coffee, juniper tar, rice embryo, rice germ); anthracene has been found in beech wood, and azulene derivatives occur fairly widely (e.g. in camphor, chamomile, wormwood and mushrooms). Benzo[*e*]pyrene is found in many smoked foods, roasted coffee and tobacco. Cigarette smoke has been much studied and Gunther & Buzzetti (1965) summarize some of the findings (Table 6.3). Benzopyrene occurs in the atmosphere of cities and Sawicki *et al.* (1960 *b*) quote figures ranging from 1·9 to 29·9 µg/100 m³. Commins (1958) found at three sites in Liverpool: pyrene 4·0–5·0, fluoranthene 2·8–6·7, benzo[*a*]pyrene 4·5–6·8 and benzoperylene 9·8–16·6/100 m³. The institution of smokeless zones reduces the atmospheric pollution but the exhaust gases from petrol- and diesel-powered vehicles contribute very significant quantities of polynuclear hydrocarbons.

Table 6.3 Polycyclic hydrocarbons in cigarette smoke

	µg/100 cigarettes
Benzo[*a*]pyrene	0·5–2·2
Fluoranthene	1·0
3-Methylpyrene	5·0
Pyrene	5·0–11·0

(From Gunther & Buzetti, 1965).

6.1.5 Polynuclear Hydrocarbons in Air-borne Particulates

Methods for collecting air-borne particulates from urban air have been described (Sawicki *et al.*, 1958) and benzene-soluble materials have been extracted and analysed. Column chromatography followed by ultraviolet spectrophotometry (Falk & Steiner, 1952) or fluorometry (Van Duuren, 1958, 1960; Sawicki *et al.*, 1960 *a*) has been used to identify constituents.

Sawicki *et al.* (1960 *b*) describe the preparation of suitably weakened alumina and the conduct of the chromatography to yield some 30 fractions, using pentane followed by pentane-ether (3, 6, 9 and 12% ether) for elution. Air-borne particulates are collected by passing polluted air through a glass fibre filter for 24 hours. In fact mono-, di- and tricyclic aromatic

Figure 6.14 Ultraviolet absorption spectrum of pyrene fraction in pentane (from Sawicki *et al.*, 1960 *a*)

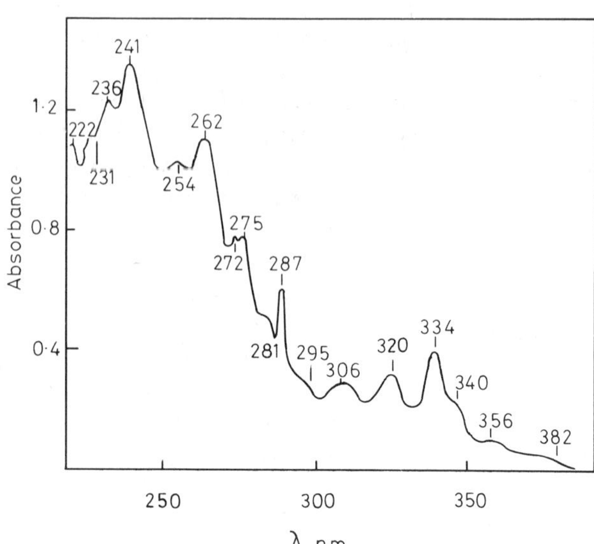

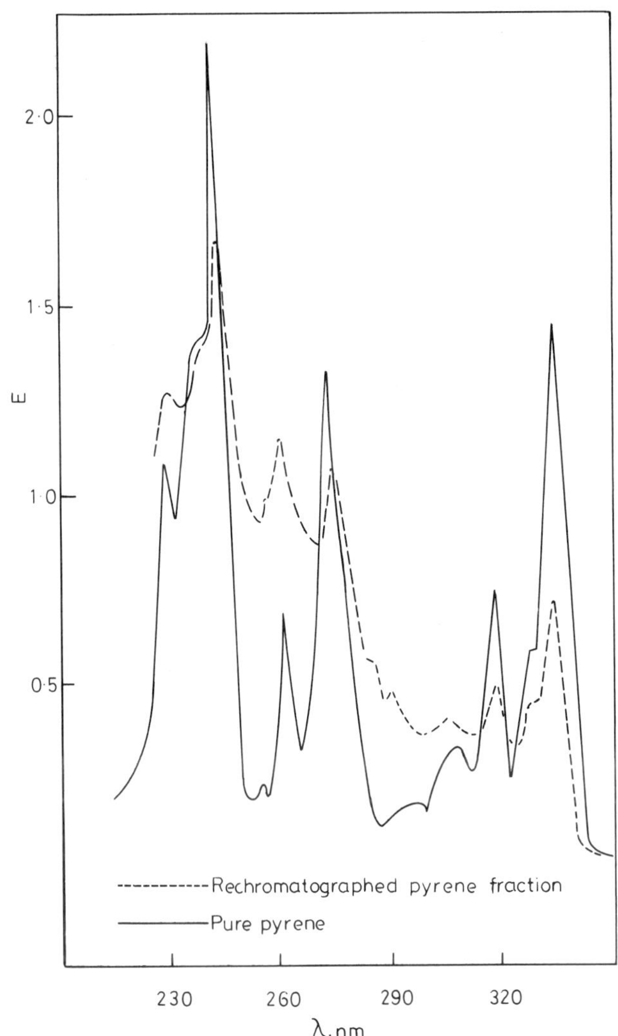

Figure 6.15 Ultraviolet absorption spectra in pentane of rechromatographed pyrene fraction and pure pyrene (from Sawicki *et al.*, 1960 *a*)

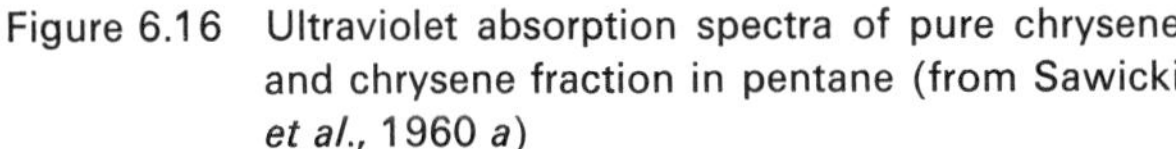

Figure 6.16 Ultraviolet absorption spectra of pure chrysene and chrysene fraction in pentane (from Sawicki *et al.*, 1960 *a*)

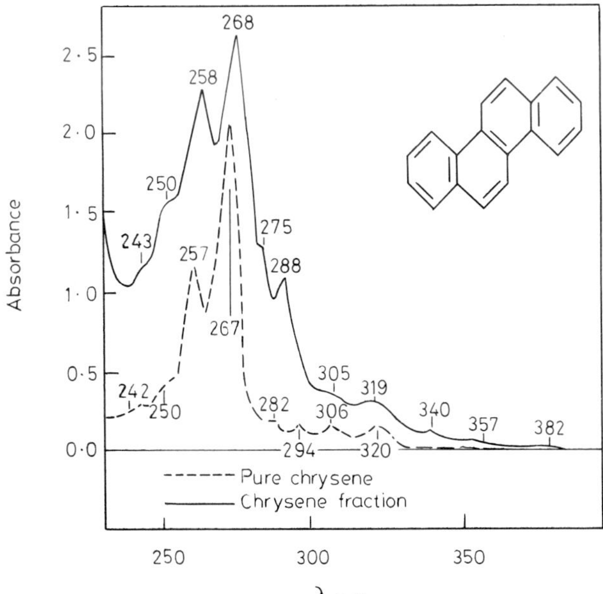

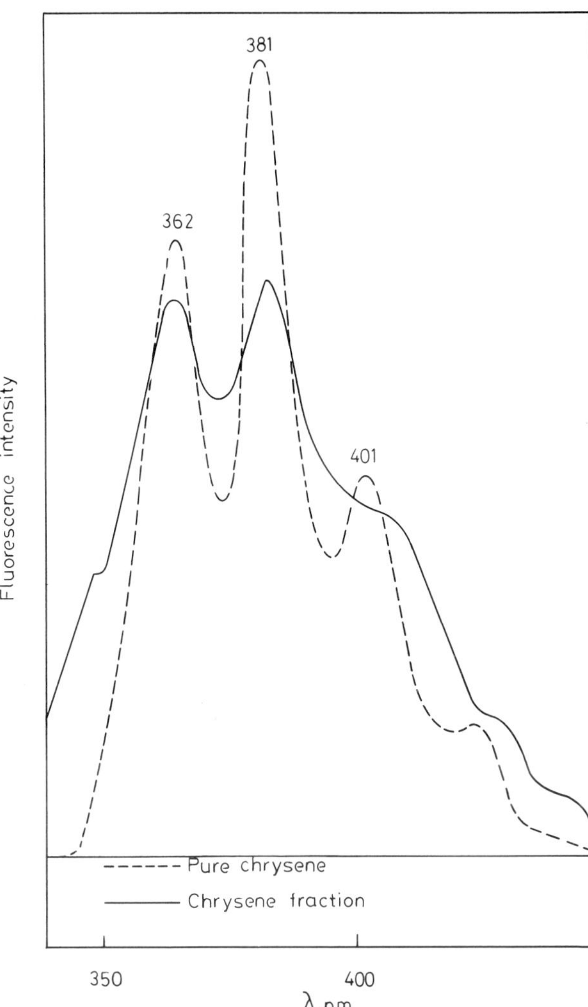

Figure 6.17 Fluorescence spectra at activating wavelength 264 nm of pure chrysene and chrysene fraction in pentane (from Sawicki *et al.*, 1960 *a*)

compounds tend to be lost. Anthracene and phenanthrene are recorded only when a sample is obtained very close to a source of industrial pollution.

Hydrocarbons are eluted in the order: aliphatics, benzene derivatives, naphthalene derivatives, dibenzofuran anthracene, pyrene, benzofluorene, chrysene, benzopyrene, benzoperylene and coronene fractions. Thus with one type of alumina the pyrene fraction was found at the beginning of the 3% ether eluent and the chrysene at the end of it. Ultraviolet absorption spectra were determined in pentane solution (Figs. 6.14 to 6.16). Although most fractions contained two (or more) absorbing entities qualitative analysis, and indeed roughly quantitative analysis, was readily achieved. In addition, activation and fluorescence spectra (Fig. 6.17) in pentane strengthen the identifications (Sawicki *et al.*, 1960 *a*).Finally confirmation can be obtained by a 'destructive' method of analysis such as the measurement of absorption or fluorescence spectra in sulphuric acid or in some instances a suitable colour test can be used.

As an example, one fraction showed absorption peaks at 287, 305, 318, 334 and 354 nm consistent with the presence of pyrene and fluoranthene. The fluorescence spectra of the same solution determined at activating wavelengths of 283 and 300 nm were respectively identical with the fluorescence spectra of fluoranthene and pyrene. The same solution fluorescing at 445 nm gave an activation spectrum identical with that of fluoranthene. The solution was then evaporated and the residue dissolved in a small volume of sulphuric acid. The fluorescence spectrum activated at 351 nm agreed with that of pyrene as also did the activation spectrum for 392 nm fluorescence. Table 6.4 summarizes the methods of identification.

Table 6.4 Criteria for presence of polynuclear hydrocarbons

Method	Solvent	Main bands nm	
Pyrene fraction and pyrene			*Pyrene plus fluoranthene*
U.v. Absorption	pentane	see fig. 6.15	236, 276, 281, 287, 341, 358
Fl. (Ac. 300 nm)	pentane	382, 392	
Ac. (Fl. 392 nm)	H_2SO_4	350, 369, 387, 282	
Fl. (Ac. 351 nm)	H_2SO_4	392, 410, 430	
Benzofluorene fraction and benzo [a]-fluorene			
U.v. Absorption	pentane	254, 262, 273, 303, 316	
Fl. (Ac. 312 nm)	pentane	340, 358	
Ac. (Fl. 340 nm)	pentane	317, 264, 340	
Chrysene fraction and chrysene			*chrysene fraction and benz[α]anthracene*
U.v. Absorption	pentane	257, 267, 320	276, 287, 290, 315, 340, 358, 374, 382
Fl. (Ac. 264 nm)	pentane	381, 361, 401	382, 406, 430
Benzoperylene fraction and		*benzperylene*	*plus anthanthrene fraction*
U.v. Absorption	pentane	287, 290, 312, 320, 344, 361, 377, 382	429, 420, 406, 434
U.v. Absorption	H_2SO_4	405, 513	
Fl. (Ac. 358 nm)	pentane	419, 443, 405, 397 (Ac. 420 nm)	430, 458, 488
Fl. (Ac. 398 nm)	H_2SO_4	442, 421 Fl. (Ac. 563) H_2SO_4	581
Ac. (Fl. 520 nm)	H_2SO_4	no bands	
Ac. (Fl. 442 nm)	H_2SO_4	398 Ac. (Fl. 430 nm) pentane	422, 400, 304, 380, 361
Coronene fraction and		*coronene*	*pure coronene*
U.v. Absorption	pentane	290, 298, 302, 317, 324, 333, 339, 345	
Fl. (Ac. 300 nm)	pentane	410, 419, 433, 447, 454, 473, 464	431, 447, 472
Fl. (Ac. 310 nm)	H_2SO_4	442, 470, 498	442, 470, 498

Fl. (Ac 300 nm) = fluorescence spectrum activated by 300 nm. Ac. (Fl. 392 nm) = activation spectrum for fluorescence at 392 nm. H_2SO_4 = conc. sulphuric acid. (Based on Sawicki *et al.*, 1960)

6.2 Polyphenols

Biphenyl (in light petroleum) shows two strong bands $\lambda_{max.}$ 201 and 247 nm, $\epsilon_{max.}$ 4·65 and $1{\cdot}7 \times 10^{-4}$ respectively. The introduction of substituents produces interesting effects (Klevens & Platt, 1949). In 2,2′-dimethylbiphenyl the short wave band becomes an inflection at *ca.* 220 nm and is weaker, there is an inflection *ca.* 230 nm ϵ *ca.* 5000 and two peaks are seen near 265 and 272 nm at intensities at about 700 and 550. The changes are enhanced in 2,6, 2′,6′-tetramethylbiphenyl and the absorption in the region 255–275 nm is benzenoid in character with resolution into three peaks. The 2,6,2′-trimethylbiphenyl and the 2,6-dimethyl show intermediate changes and all have partially masked selective absorption in the region 220 nm. The 2-methylbiphenyl has a single peak $\lambda_{max.}$ 240 nm, $\epsilon_{max.}$ 10 000. The fully substituted decamethylbiphenyl shows the benzenoid fine structure, a marked inflection near 220 nm. The spectrum closely resembles that of hexamethylbenzene doubled in intensity.

Introduction of methyls at position 3 has relatively little effect on the biphenyl absorption ($\lambda_{max.}$ *ca.* 210 and 252 nm with little change in $\epsilon_{max.}$) while the *p,p′*-derivative (4,4′-trimethylbiphenyl) shows a slightly greater bathochromic effect $\lambda_{max.}$ 258 nm with an increase in $\epsilon_{max.}$.

The characteristic absorption of biphenyl is due to conjugation of the two rings but the most stable configuration requires the two rings to be coplanar. Introduction of substituents in the ortho position has steric effects which produce twists in the molecule with the resonance energy varying as the square of the cosine of the angle of twist (Pickett *et al.*, 1951). The spectra of substituted biphenyls afford an excellent example of steric effects. There are 4 ortho positions and progressive substitution increases the twist so that conjugation across the bond linking the two rings diminishes steadily.

The conjugation effects are seen well in the *p*-polyphenyls (Gillam & Hey, 1939):

Spectra of *p*-polyphenyls

	In hexane		In chloroform	
Compound	$\lambda_{max.}$ nm	$\epsilon_{max.}$	$\lambda_{max.}$ nm	$\epsilon_{max.}$
Diphenyl	246	20 300	251·5	18 300
Terphenyl	276	35 000	280	25 000
Quarterphenyl	292	55 000	300	39 000
Quinquiphenyl	—	—	310	62 500
Sexiphenyl	308	—	317·5	56 600

(Gillam & Hey, 1939; Wenzel, 1953).

but in the *m*-polyphenyls:

Spectra of the *m*-polyphenyls (in chloroform)

Compound	$\lambda_{max.}$ nm	$\epsilon_{max.}$	$\epsilon/n+1$
$n=0$	251·5	18 300	—
1	251·5	44 000	22 000
7	253	184 000	23 000
8	253	213 000	22 500
9	253	215 000	21 500
10	253	233 000	21 200
11	253	252 000	21 000
12	253	283 000	22 800
13	254	309 000	22 700
14	255	320 000	21 300

$\lambda_{max.}$ moves only from 251·5 nm to 253–254 nm, when *n* reaches 15 and $\epsilon_{max.}$ averages 20 000 per C_6 unit. The spectra are consistent with $n/2$ biphenyl units, but inasmuch as Beer's law is not obeyed over a range of concentrations other effects may be superimposed (Pickett *et al.*, 1936).

It must be concluded that in the meta series the phenyl chromophoric units are unconjugated whereas in the para series the effect of one ring on another is cumulative. The narrow bands characteristic of benzene do not appear, but it is has been observed that substituted diphenyl derivatives display structure when the substituents can hinder free rotation about the internuclear bond.

Katritzky & Simmons (1960) and Bridges *et al.* (1965) studied a number of biphenyl derivatives.

The absorption and fluorescence of biphenyl are strong, as would be expected from its resonant structure. The absorption spectrum of 2-hydroxybiphenyl shows two peaks, the 246 nm peak being characteristic of the biphenyl nucleus and the 288 nm peak ϵ 5100 being due to the phenolic moiety. The 4-hydroxyphenyl has a single very strong band $\lambda_{max.}$ 260 nm and probably there is considerable π electron delocalization. The anion is formed at pH > 10 (Kieffer & Rumpf, 1954). 2-Hydroxybiphenyl and 2-methoxybiphenyl illustrate the effects of loss of coplanarity (Braude & Forbes, 1955). The former forms an anion ($\lambda_{max.}$ 309 nm) in alkali. 3-Hydroxybiphenyl also has a less resonant structure than 4-hydroxybiphenyl and an anion is formed in alkali. The 2,2′-dihydroxybiphenyl behaves similarly to the 2-hydroxy compound except that the short wave band tends to be weaker and the *ca.* 280 nm band to be stronger. In alkali the anion absorbs more strongly.

The aminobiphenyls at pH 1 exist as cations. Compared with aniline and toluidine, 2-aminobiphenyl exhibits increased absorption at the short wave end of the spectrum and in 3-amino this tendency is enhanced. 4-Aminobiphenyl bears no relation to the spectra of *p*-toluidine and there is probably little hindrance to coplanarity and the full conjugation effect of biphenyl is preserved.

Table 6.5 Absorption peaks of some substituted biphenyls

	Ethanol		0·01 N HCl		Water		0·1 N NaOH	
Solvent	$\lambda_{max.}$ nm	$\epsilon_{max.} \times 10^{-3}$	$\lambda_{max.}$ nm	$\epsilon_{max.} \times 10^{-3}$	$\lambda_{max.}$ nm	$\epsilon_{max.} \times 10^{-3}$	$\lambda_{max.}$ nm	$\epsilon_{max.} \times 10^{-3}$
Biphenyl	246	16·5	—	—	—	—	—	—
2-Methoxy-	248	10·8	244	10·4	244	10·2	245	9·9
	285	4·6	282	4·1	282	4·1	283	4·0
2-Hydroxy-	246	11·0	243	9·8	243	9·8	252	6·0
	288	5·1	283	4·0	282	4·1	309	5·6
2,2′-Dihydroxy-	242	10·7	236	7·4	234	7·4	250	9·4
	286	7·2	279	6·0	279	5·8	308	8·6
2-Amino-	—	—	233	12·7	220	18·6	217	29·4
	—	—	283–290	1·0	290	2·6	292	3·0
4-Methoxy-	260	20·2	—	—	—	—	—	—
4-Hydroxy-	262	17·5	260	18·0	259	18·2	290	20
4,4′-Dihydroxy-	264	21·0	262	20·4	263	20·4	290	25·6
3-Hydroxy-	251	15·0	251	14·9	250	15·1	~235	25·2
	284	4·2	~280	4·3	280	2·1	308	3·8
3-Amino-	234	22·8	249	14·1	231	22·2	—	—
	300	2·63	—	—	298	3·1	—	—
4-Amino-	280	10·5	247	18·5	272	17·7	273	19·1
[aniline]	234	7·94	254	0·16	230	8·6	—	—
	285	1·39	—	—	280	1·43	—	—
[*o*-toluidine]	234	8·51	—	—	—	—	—	—
	285	2·09	—	—	—	—	—	—
	(hexane)							
[*p*-toluidine]	235	11·75	—	—	—	—	—	—
	290	1·82	—	—	—	—	—	—
	(ethanol)							

Table 6.6 Fluorescence of hydroxybiphenyls

Compound	State	Excitation $\lambda_{max.}$ nm	Fluorescence $\lambda_{max.}$ nm	Relative fluorescence intensity	Maximum fluorescence pH range
Biphenyl	—	270	318	100	0–14
2-Hydroxyl-	un-ionized	262	348	39	0–1
		295	—	83	—
2-Methoxy-	—	262	348	114	0–14
		295	—	229	—
2′,2′-Dimethoxy	un-ionized	256	356	2	−1 to −0·8
		290	—	6	—
	excited state ion	256	400	41	1–6·6
		290	—	385	—
	mono-anion	262	400	266	8·6–1·3
		320	—	1593	—
4-Hydroxy-	un-ionized	288	340	184	1–9
		311	401	136	10–14
4-Methoxy-	molecular	280	338	440	0–14
4,4′-Dihydroxy-	un-ionized	282	354	271	1–7
	mono-anion	300	438	19	10·9–11·3
	di-anion	310	400	7	12·1–14

The work of Bridges *et al.* (1965) on the fluorescence spectra adds a new dimension to the study of the biphenyls. Biphenyl exhibits a strong ultraviolet fluorescence over the whole pH range with an excitation maximum at 270 nm. The fluorescence intensity is used as a standard and given a score of 100 (see Table 6.6). 4-Hydroxybiphenyl at pH 0 shows a phenol fluorescence ($\lambda_{max.}$ 340 nm, $\lambda_{exc.}$ 288 nm) and the maximum intensity is constant between pH 1–8·6. Above this peak the 340 nm fluorescence is gradually replaced by a new peak ($\lambda_{max.}$ 401 nm, $\lambda_{max.}$ ex. 311 nm) with maximum intensity pH 10·4–14. The change corresponds with ionization of the 4-hydroxybiphenyl to the anion. Above pH 14 it is possible that a quinoid structure is formed:

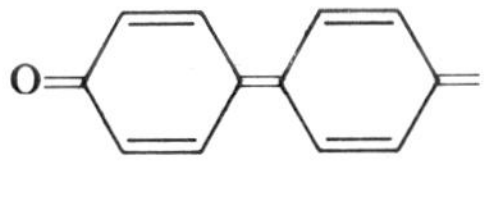

2-Hydroxybiphenyl at pH 1 has little fluorescence but at pH 0 it shows normal phenolic fluorescence ($\lambda_{max.}$ 348 nm excited maximally both at 262 and 295 nm). At pH 2 the fluorescence maxima is at 415 nm due to the anion. At this pH the solute is not ionized but exists in the molecular form and excited state ionization is postulated. At pH 10·4 the excitation peaks are at 272 and 320 nm corresponding to the ionization of 2-hydroxybiphenyl (pK_a 10·01: Kieffer & Rumpf, 1954). The intensity of fluorescence reaches its highest value at pH 12–14.

3-Hydroxybiphenyl also shows excited state ionization. At low pH values there are fluorescence peaks at 345 nm (unionized molecule) and 420 nm (excited state ionization). At pH 0 the fluorescence at 345 nm is 10 times as intense as that at 420 nm. The proportion of excited molecules to undergo excited state ionization increases with increasing pH.

4,4′-Dihydroxybiphenyl at pH values over 12 shows a new fluorescence peak at 400 nm due to excitation (310 nm) of the dianion. At pH values below 8 the 345 nm fluorescence is due to the neutral molecule, but between pH 8 and 11 the mono-anion is responsible for fluorescence with $\lambda_{max.}$ 438 nm.

2,2′-Dihydroxybiphenyl shows excited state ionization at pH 1. The compound ionizes at pH 6–7 and at a pH just over 7 the excitation changes from 290 nm to 320 nm but there is no change in the fluorescence peak. There is no evidence of a di-anion and the mono-anion may be stabilized by hydrogen bonding:

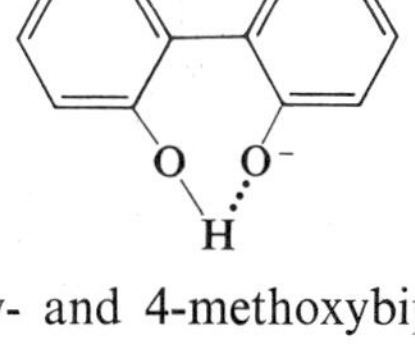

The 2-methoxy- and 4-methoxybiphenyls fluoresce ($\lambda_{max.}$ 348 and 338 nm respectively) over the pH range 0–14 with constant intensity.

The aminobiphenyls exhibit excited state ionization (Table 6.6). At pH −1 they exist as cations and display rather feeble fluorescence, but at pH 1 the wavelength of maximum fluorescence changes to *ca.* 400 nm. The 'neutral' molecule appears at pH 3–3·5 and the fluorescence of this uncharged species can be seen over the pH range 0–14. The most intense fluorescence occurs at pH 5–11. The intensities for the three isomers are 3:2:4-hydroxy-1:2·25:9.

Excited state deprotonation was discovered by Forster (1950) (see also Williams & Bridges, 1964). In the excited state 2- and 3-hydroxybiphenyl form anions at pH 1–2, whereas in the ground state they ionize at pH 9–11. They are therefore 10^8 times stronger acids in the excited state than in the ground state. With 2,2′-dihydroxybiphenyl the increase in acid strength is about 10^6 times. The amino bases tend to become weaker in the excited state, since excited state deprotonation occurs at pH *ca.* 0, whereas ground state deprotonation would be expected at pH about 4. The fluorescent 2- and 4-methoxybiphenyls have no ionizable group and do not exhibit excited state ionization.

Creaven *et al.* (1965) studied the fat of biphenyl incubated with microsomal liver preparations from 11 species of animals. The fluorescences described above permit discrimination between the hydroxylation products 2- and 4-hydroxybiphenyl. All species studied made 4-hydroxybiphenyl and preparations from the liver of mice, hamsters, cats, coypus and frogs produced 2-hydroxybiphenyl also. Only the 4-isomer was produced by the liver preparations from adult rats and rabbits while livers from the young of these species produced both the 2- and 4-isomers.

The determination of the two isomers in the same solution is carried out as follows: (*a*) at pH 13 the mixture shows a single fluorescence peak since the maxima for anionic fluorescence are close together; (*b*) in the pH *ca.* 6 range the 2-isomer absorbs light in the unionized form but fluoresces at 415 nm $\lambda_{exc.}$ 295 nm. The 4-isomer absorbs in the unionized form and emits the unionized state fluorescence at 338 nm (excited by 275 nm). Hence the fluorescence is measured at pH 6 and pH 13.

The two phenols are extracted from the incubation mixture with heptane, transferred to 0·1 N NaOH, centrifuged and an aliquot is brought to pH 5·5 with 0·5 N succinic acid. Fluorescence intensity is then measured at 338 nm with λ 275 nm excitation and at 415 nm with 295 nm excitation. The first reading leads to the 4-hydroxybiphenol concentration and the second to the concentration of the 2-isomer after a correction for a contribution by the 4-isomer.

The fate of biphenyl administered by mouth is more complicated.

6.3 Steric Hindrance

Beaven (1958) and Beaven & Johnson (1959) reported on the properties of methylbiphenyls. The extreme situation is seen in comparing the parent substance I with the fully substituted decamethylbiphenyl II.

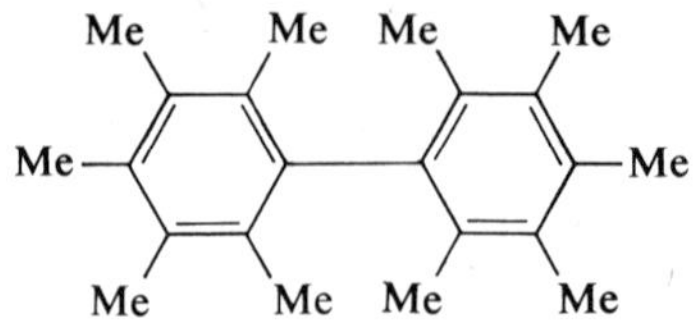

Me Me Me Me
Me — — Me
Me Me Me Me

II

Biphenyl itself shows a very strong broad band with $\lambda_{max.}$ *ca.* 248 nm, $\epsilon_{max.}$ 17 000, whereas decamethylbiphenyl exhibits resolved absorption between 255 and 290 nm, $\lambda_{max.}$ *ca.* 270 nm and $\epsilon_{max.}$ 625. Hexamethylbenzene has very similar absorption between 245 and 286 nm, $\lambda_{max.}$ 271 nm, $\epsilon_{max.}$ 225. There is also a marked inflection at 215–235 nm and a peak at 207 nm. As a first approximation the two pentamethyl halves of the biphenyl derivative each absorb like the hexamethylbenzene, but a closer scrutiny shows that the decamethyl derivative absorbs more strongly than the hexamethylbenzene over the range 207 to 290 nm, the difference being greatest at 240–250 nm. The methyl groups very largely, but not completely, abolish the $\pi \rightarrow \pi^*$ conjugation effect (248 nm).

Steric effects are manifest in the spectrum of 2,2′-dimethylbiphenyl. The substituents, being ortho to the bond linking the two rings, cause a twist and Dewar (1952) found that the resonance energy varied as the cosine of the angle of twist. On the other hand, 3,3′-dimethyl- and 4,4′-dimethylbiphenyl had spectra very like that of biphenyl itself (see Fig. 6.18).

Braude & Forbes (1955) confirmed this view. Attention is drawn to chapters in books edited by Newman (1956) and Gray (1958).

Braude and his colleagues considered that the twisting effects away from coplanarity fell into three spectroscopically distinct types: I, no change in $\lambda_{max.}$ and a small decrease in $\epsilon_{max.}$ characteristic of small twists; II, a blue shift of $\lambda_{max.}$ and a decrease in $\epsilon_{max.}$ as a result of larger twists; and III a near approach to simple additivity, i.e. the two rings absorbing independently, this state of affairs reflecting large twists. Although this classification has not escaped criticism (e.g. Dewar and also Waight & Erskine in Gray, 1958) it remains broadly valid. The whole problem is discussed well by Rao (1967).

Beaven & Johnson (1959) made the following provisional correlations for the infrared spectra of alkyl biphenyls:

cm^{-1}	
1005–1013	always present, but intensity reduced to 1/10 by 3-methyl
ca. 1600	intensity increases with substitution at C-3 or 3′
1165–1170	2,6-dimethyl
1074–1078	one ring unsubstituted
878–895, 1170–1175	3 or 3′ substituent
1518–1572	4 or 4′ substituent
940–955	2-methyl
840–855	3 or 4-methyl
810–835	4 or 2,5-dimethyl
690–705	at least one ring with neither 2 nor 4 methyls.

Introduction of methyl groups at positions 2,6, 2′,6′ results in a resolved benzenoid band at 256–264 nm and 270 nm, ϵ 400–500. The trimethyl derivative 2,6,2′, gives a spectrum intermediate between those of the 2,2′ and 2,6, 2′,6′ derivatives, while a single methyl group at position 2 has a considerable effect, reducing the ϵ value at 265 nm from *ca.* 10 000 to *ca.* 1600 (see Fig. 6.19).

Table 6.7 The effect of position isomerism in dimethylbiphenyls

Dimethylbiphenyls	$\lambda_{max.}$ nm	$\epsilon_{max.}$
2,2′-	265	740
	272	560
	~230	5100
	208	40 000
3,3′-	252	17 000
	207	50 000
4,4′-	258	22 400
	208	40 000

Introduction of larger groups e.g. *tert*-butyl [—$C(CH_3)_3$] in the ortho position reduces the intensity of absorption still further and the absorption spectrum closely resembles that of benzene or toluene. The absorption in the region 208–230 nm undergoes a bathochromic shift, but this is due as much to the

Figure 6.18 Absorption spectra of biphenyl and 2,2′-, 3,3′- and 4,4′-dimethylbiphenyls (from Beaven, 1958)

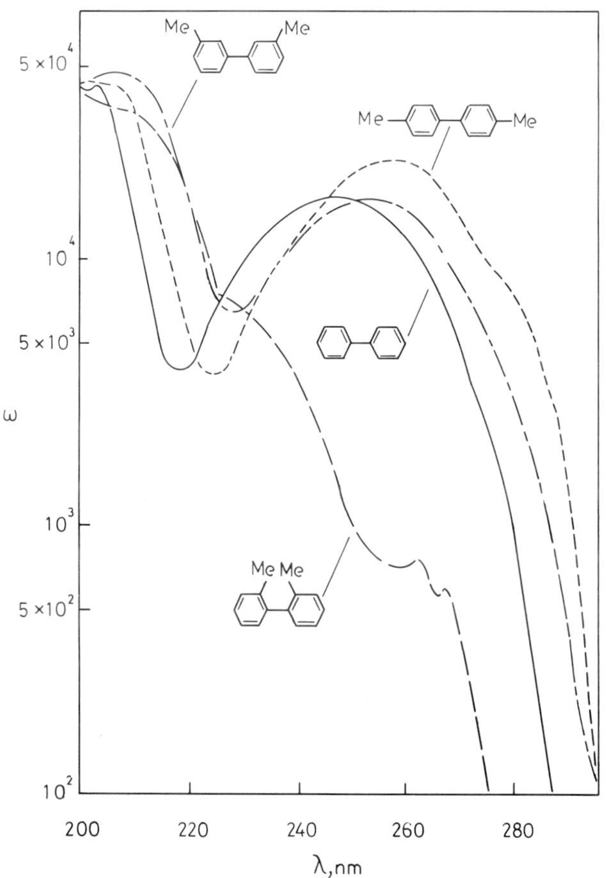

methyl groups themselves as to their location with respect to the bond linking the two rings. The symmetrical 2,4,6, 2′,4′,6′-hexamethylbiphenyl has an absorption spectrum very similar to that of the decamethyl derivative and O'Shaughnessy & Rodebush (1940) using appropriate units found that the curve for bimesityl was nearly the same as twice the curve for mesitylene (1,3,5-trimethyl benzene). Calvin (1939) and Pauling (1939, 1949) argued that substituents could restrict rotation about the 1,1′ bond and could thus influence the degree of conjugation between the phenyl groups. Subsequent studies (Braude & Sondheimer, 1955; Braude & Timmons, 1955; Braude & Forbes, 1955) strengthened the evidence.

The *cis* and *trans* forms of stilbene show (in ethanol) very different spectra (Beale & Roe, 1953):

	$\lambda_{max.}$ nm	$\epsilon_{max.} \times 10^{-3}$
cis	224	24·4
	280	10·5
trans	228	16·4
	295·5	29·0

Calvin & Alter (1951) recorded for the *cis* form $\lambda_{max.}$ 275 nm, ϵ 10 000 and for the *trans* form 295 nm, ϵ 28 000 with inflections at 305 and 325 nm in ethanol (cf. also Smakula & Wasserman, 1931).

cis *trans*

In the *cis* form the chromophore is shorter than in the *trans* form and this is responsible for the 295·5–280 nm difference, while the steric hindrance to coplanarity reduces the intensity of the main 280 nm peak by

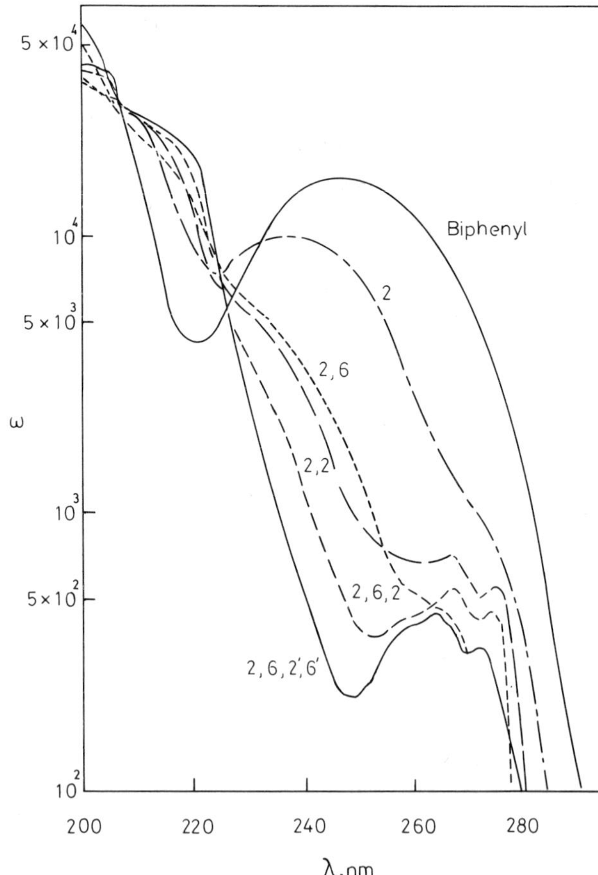

Figure 6.19 Absorption spectra of the 2(6)-methylbiphenyls (from Beaven, 1958)

causing overlapping of the *o* positions. The infrared spectra show for the *cis* form peaks at 1410, 729 and 698 cm^{-1} and for the *trans* form 959 and 691 cm^{-1}. In $\alpha\beta$-dimethyl stilbene the steric hindrance moves the main peak to 240 nm, ϵ 12 000 and absorption intensity near 280 nm is reduced to $\epsilon = 5000$ (Braude, 1949).

Pinckard *et al.* (1948) compared the three steoisomeric 1,4-diphenylbutadienes

all-*trans*
cm^{-1} 985 688

cis-trans
1414 984 945
775 698 690

cis-cis
1368 775 704 695

The *trans-trans* form shows $\lambda_{max.}$ 328 nm, $\epsilon_{max.}$ 56 000 with subsidiary peaks at 314 and 344 nm (see Fig. 6.20). This form is fluorescent. The *cis-cis* form has a

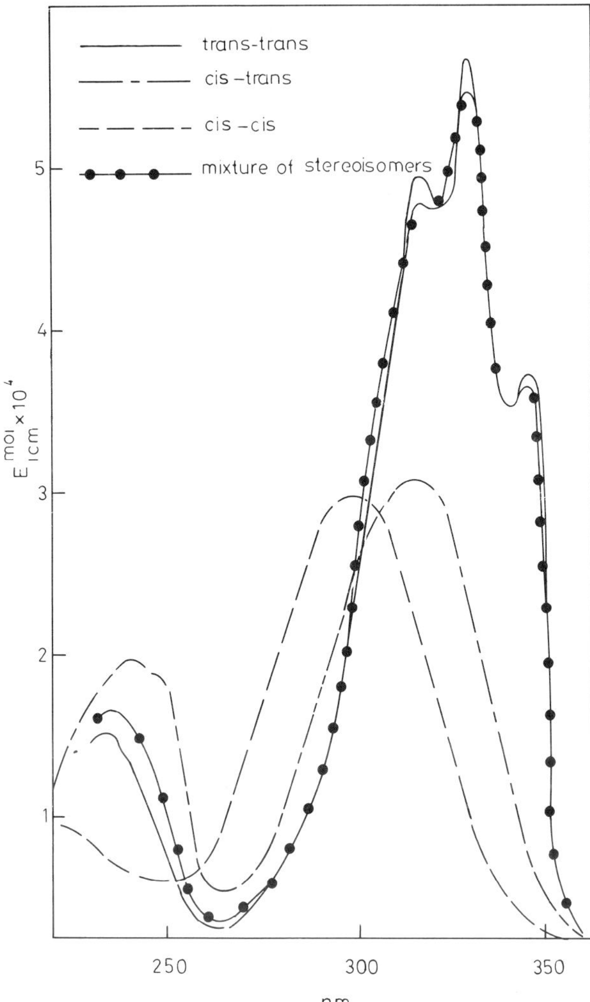

Figure 6.20 Molecular extinction curves of the three stereoisomeric diphenylbutadienes in hexane; *trans-trans, cis-trans, cis-cis* and mixture of steroisomers after iodine catalysis of any of the three forms (from Pinckard *et al.*, 1948)

broad unresolved band with $\lambda_{max.}$ 299 nm, $\epsilon_{max.}$ 28 500, while the *cis-trans* isomer has $\lambda_{max.}$ 313 nm, $\epsilon_{max.}$ 20 200. Neither *cis* form fluoresces visibly. A single change *trans* → *cis* has the major effect on both the location and intensity of absorption. The mono-*cis* form, however, has the highest intensity of absorption at 230–250 nm so that there may well be a masked *cis* peak. *Cis* forms are converted to the all-*trans* form by iodine catalysis or by light.

Beale & Roe (1952) studied methyl derivatives of stilbene and 4-amino stilbene (see Figs. 6.21 and 6.22) and expressed the steric effects by a comparison of oscillator strengths, for the long wavelength band using the expression:

$$f=4{\cdot}315\times 10^{-9}\int \epsilon \, d\bar{\nu}$$

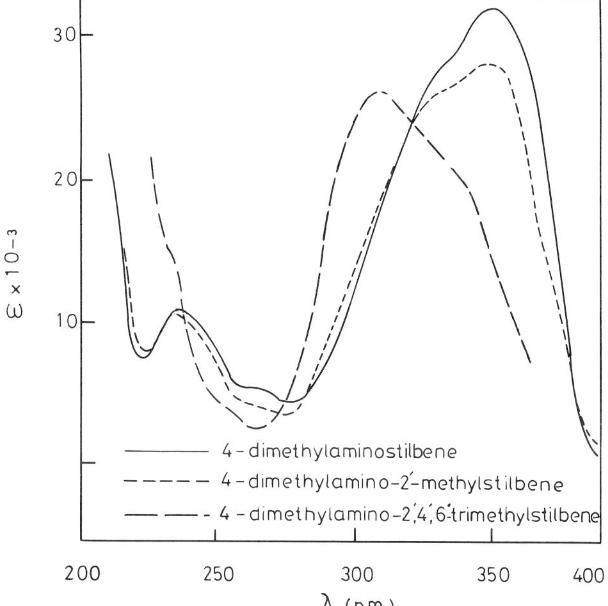

Figure 6.21 Ultraviolet absorption spectra (95% ethanol) of 4-dimethylaminostilbene, 4-dimethylamino-2′-methylstilbene and 4-dimethylamino-2′,4′,6′-trimethylstilbene (from Beale & Roe, 1952)

Figure 6.22 Ultraviolet absorption spectra of: solid line, *trans*-stilbene (in 95% ethanol); short dashes, 2,4,6-trimethylstilbene (in absolute ethanol); and long dashes, 2,4,6,2′,4′,6′-hexamethylstilbene (in absolute ethanol) (from Beale & Roe, 1952)

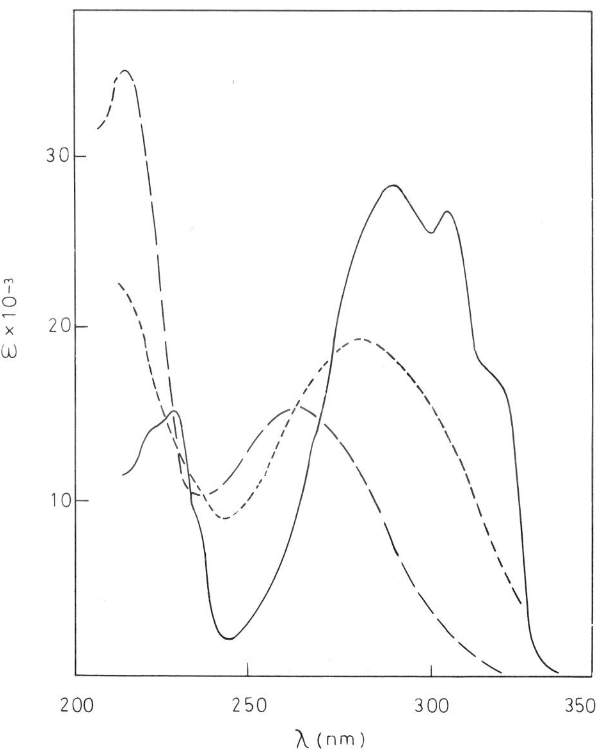

where ϵ is the molar extinction coefficient and $\bar{\nu}$ is the frequency in cm^{-1}. A methyl group in a non-hindering position causes either no change or an increment in f value, i.e. no change with substitution at the 3 or 3′ position but a change in f value of +0·05 with substitution at the 4 or 4′ position. In the 2′ position a methyl group causes a decrement ($\Delta f = 0{\cdot}04$). Introduction of further hindering groups causes larger falls in f and the effects of successive such groups are additive. The $—N(CH_3)_2$ groups when introduced at position 4, cause an increment in f twice as large as that due to CH_3.

6.4 Aspects of Detoxication Processes

6.4.1 Metabolism of Foreign Compounds

The metabolism of foreign compounds is a very large area of biochemistry in which R. T. Williams (1947, 1959) and his collaborators have been responsible for great progress. The work has included studies (*a*) on the fate of simple substances as a basis for work on complex compounds; (*b*) on a variety of medicinal substances varying from sulphonamides to α-methyldopa; (*c*) on a wide range of pesticides; and (*d*) on many toxic substances from thalidomide to carcinogens and alkaloids.

The metabolism of most foreign compounds is made up of at least two enzymatically catalysed reactions. The first stage includes operations which fall into the categories of oxidations, reductions or hydrolyses while the second stage includes conjugation processes synthetic in character. Stage I endows the compound with an active group (such as hydroxyl, carboxyl, amino or thiol) which can enter into a conjugation process. Many compounds already have such reactive groups and may proceed directly to conjugation. Others may be metabolized by first stage processes and elimination from the body may occur without conjugation.

A simple example is benzene, which can be oxidized to phenol (stage I) and then conjugated (stage II) to phenylglucuronic acid or phenylhydrogen sulphate. When phenol is ingested conjugation can occur without further ado. Ethanol is metabolized to carbon dioxide via acetaldehyde and ethanol and there is no conjugation. The main locus of metabolism for foreign substances is the liver, which possesses suitable enzymes present in the endoplasmic reticulum of the hepatic cells.

Detoxication research has been carried out to a considerable extent on rats, rabbits, guinea-pigs and other animals with experiments on larger animals and on man under appropriate circumstances. Valuable work has been done using test substances labelled with radioactive isotopes. The whole range of chromatographic and spectroscopic techniques has been pressed into service.

It can be claimed that this kind of biochemical pharmacology brings together organic chemistry, analytical chemistry, enzymology, genetics and therapeutics. It is important that all the important medicinal substances, food additives and contaminants, as well as pesticides and hazardous chemicals require to be investigated and that the variations between species which are the essence of comparative biochemistry should not be neglected.

It would be possible to find many examples of spectroscopic methods in the service of biochemical pharmacology, but I have chosen one relatively simple area for illustration. Many analytical processes referred to in other sections are however relevant to detoxication studies.

6.4.2 Ultraviolet Absorption Spectra as an Aid to the Study of Detoxication Processes

The metabolic fate of chlorobenzenes affords an example of a biochemical problem in which absorption spectra provide useful information, and contribute significantly to elucidating the problem.

At one end of the scale is the metabolism of monochlorobenzene and at the other end the fate of the insecticide benzene hexachloride (BHC or gammexane). Alkali decomposes the α-, β-, γ- and δ-isomers of BHC to give a mixture of the three isomeric trichlorobenzenes in which the 1,2,4-isomer predominates and accounts for 80% of the total. The insecticide also decomposes when used in the open and harmful effects on wheat seedlings and conifer seedlings are known to be due to trichlorobenzenes, the 1,2,4-isomer being the most damaging towards wheat and the 1,2,5-isomer towards conifer seedlings (Hocking, 1950; Simkover & Shenefelt, 1952). In animals, hexachlorobenzene is only slowly metabolized and indeed may remain in the body for long periods and may cause a distrubance in porphyrin metabolism (Ostrow *et al.*, 1961).

R. T. Williams and his collaborators undertook a

Table 6.8 Absorption spectra of chlorosubstitution products of benzene and related substances

(*a*) Benzene (in heptane)														
$\lambda_{max.}$ nm	183	240	250	255	260									
$\epsilon_{max.}$	45 700	126	251	339	200									
Phenol (in hexane)														
$\lambda_{max.}$ nm	210	265	271	278										
$\epsilon_{max.}$	5250	1590	2510	2510										
Phenol (pH 6·5)			(pH 12·3)											
$\lambda_{max.}$ nm	270		234	286										
$\epsilon_{max.}$	1600		10 000	2·500										
(*b*) Chlorobenzene (in hexane)										(in ethanol)				
$\lambda_{max.}$ nm	211	215	~219	245	251	257	261	264	271	245	251	258	264	272
$\epsilon_{max.}$	7410			69	120	178	170	251	190	90	135	229	280	210
o-Dichlorobenzene (in hexane)										(in ethanol)				
$\lambda_{max.}$ nm	216	219	~224	249	~255	261	~265	270	279	250	256	263	270	277
$\epsilon_{max.}$	8500			76		251		347	324	95	135	230	275	335
m-Dichlorobenzene (in hexane)										(in ethanol)				
$\lambda_{max.}$ nm	271	221	225	250	256	263	270	277		250	256	263	270	278
$\epsilon_{max.}$	8500	7940	4270	70	151	350	372	317		80	140	250	330	270
p-Dichlorobenzene (in hexane)										(in ethanol)				
$\lambda_{max.}$ nm	255	~229		~253	260	265	273	281		258	266	273	281	
$\epsilon_{max.}$	13 500	10 000		129	224	347	501	447		175	290	400	320	
(*c*) 1,2,3-trichlorobenzene (in ethanol)														
$\lambda_{max.}$ nm	265	273	280											
$\epsilon_{max.}$	125	160	125											
1,2,4-trichlorobenzene (in ethanol)														
$\lambda_{max.}$ nm	270	278	287											
$\epsilon_{max.}$	350	565	525											
1,3,5-trichlorobenzene (in ethanol)														
$\lambda_{max.}$ nm	266	273	281											
$\epsilon_{max.}$	180	235	170											
(*d*) 1,2,3,4-tetrachlorobenzene (in hexane)										(in ethanol)				
$\lambda_{max.}$ nm	274	283	292							274	280	291		
$\epsilon_{max.}$	245	350	365							260	330	290		
1,2,3,5-tetrachlorobenzene (in hexane)										(in ethanol)				
$\lambda_{max.}$ nm	274	284	293							274	281	291		
$\epsilon_{max.}$	225	420	515							210	315	275		
1,2,4,5-tetrachlorobenzene (in hexane)										(in ethanol)				
$\lambda_{max.}$ nm	270	278	282	287	296					270	285	294		
$\epsilon_{max.}$	300	620	680	1150	1400					640	1050	1050		

Table 6.8 *continued*

(*e*) Pentachlorobenzene (in hexane)					
$\lambda_{max.}$ nm	261	274	278	289	298
$\epsilon_{max.}$	309	246	251	309	364
(*f*) Hexachlorobenzene (in hexane)					
$\lambda_{max.}$ nm	222	232	291	299	
$\epsilon_{max.}$	26 300	1900	250	229	

systematic study of the fate of halogenobenzenes. It is not proposed here to discuss the whole problem but rather to select certain spectroscopic aspects.

Benzene (Table 6.8) shows weak resolved absorption in the region 240–270 nm $\epsilon_{max.} < 350$. Monochlorobenzene shows the high intensity absorption of benzene below 200 nm displaced to about 215 nm with some structure. The weak absorption at 245–280 nm is only quite slightly displaced towards longer wavelengths and $\epsilon_{max.}$ remains low (<300). The isomeric dichlorobenzenes (Table 6.8 *b*) show the strong absorption (with some resolution) around 220–230 nm and the weak absorption persists but does not extend much beyond 280 nm; *p*-dichlorobenzene has a somewhat higher $\epsilon_{max.}$ (*ca.* 500).

The polychlorobenzenes show a small but measurable shift towards longer wavelengths with main peaks around 290 and 299 nm for penta- and hexa-chlorobenzenes with low values for $\epsilon_{max.}$ maintained.

When administered to rabbits chlorobenzene yields 4-chlorocatechol (30%) and three monochlorophenols. In general, hydroxylation requires an enzyme-generated free hydroxyl radical. The locust, when given chlorobenzene, produces *m*- and *p*-chlorophenols and 4-chlorocatechol each in amounts of about 10% of the dose. The rabbit experiments showed a considerable preponderance of catechols over monophenols excreted (Spencer & Williams, 1950; Azouz *et al.*, 1952, 1953), largely as glucuronides, ethereal sulphates and as mercapturic acid (*p*-chlorophenylmercapturic acid).

m-Dichlorobenzene is oxidized by rabbits mainly to 2,4-dichlorophenol which is excreted conjugated with glucuronic or sulphuric acid. Small amounts of 3,5-dichlorophenol, 3,5-dichlorocatechol and 2,4-dichlorophenylmercapturic acid are also excreted (Parke & Williams, 1955). The compound 3,5-dichlorophenol gives a red colour with a dye (Brentamine Fast Red B salt) while 2,4-dichlorophenol does not. The spectra in acid and alkali are shown in Table 6.9. At 244 nm ϵ_{max} for 2,4-dichlorophenol is 9500 in alkali and ϵ at 244 nm in acid is 300, while for the 3,5-isomer $\epsilon_{max.}$ is 6600 in alkali and 150 for ϵ at 244 nm in acid. If the 3,5-isomer is determined by a colour test, the absorption at 244 nm may be corrected for that component and the 2,4-dichlorophenol estimated from the difference between ϵ at 244 nm in alkali and acid, which for the pure substance is 9200 (9500–300). The sum of the

Figure 6.23 Ultraviolet absorption spectra of the *o*-, *m*- and *p*-chlorophenols; full lines in 0·1 N HCl; broken lines in 0·1 N NaOH (from Smith, Spencer & Williams, 1950)

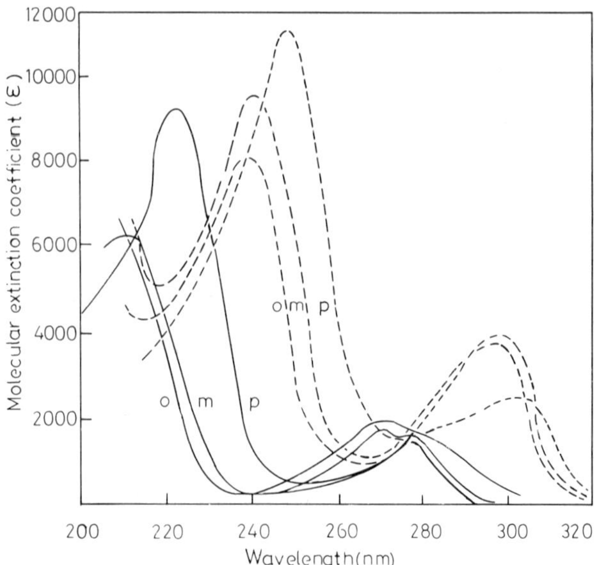

Figure 6.24 Ultraviolet absorption spectrum of 3,4-dihydroxychlorobenzene in water (full line) and its dehydration product, *p*-chlorphenol, in acid (dotted line) and in alkali (from Smith, Spencer & Williams, 1950)

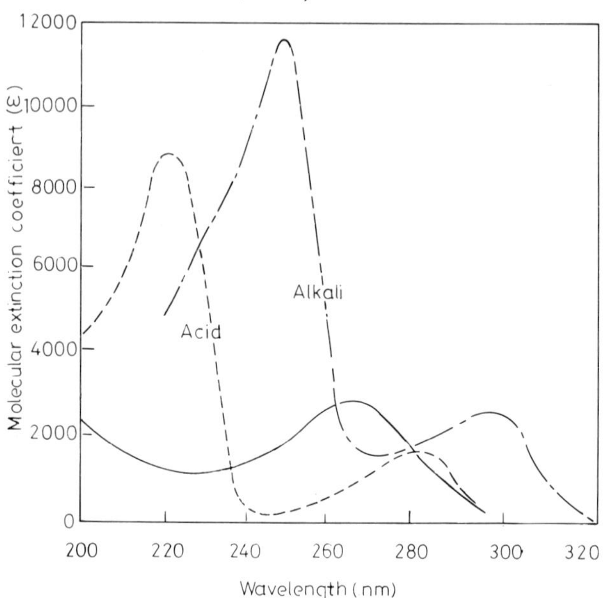

Table 6.9 Ultraviolet absorption maxima of chlorophenols

(*a*)

Chlorophenol	0·1 N NaOH $\lambda_{max.}$ nm	0·1 N NaOH $\epsilon_{max.}$	0·1 N HCl $\lambda_{max.}$ nm	0·1 N HCl $\epsilon_{max.}$	Cyclohexane $\lambda_{max.}$ nm	Cyclohexane $\epsilon_{max.}$
Ortho	236·5	8 300	<210	>6610	212	6460
	294	3 780	284·5	1980	~268	1550
					274	2344
					281	2455
Meta	238·5	9 700	212	6250	216	6460
	292*	3 480	274·5	1720	268	1288
					274	2042
					281	1995
Para	244	11 700	232	8900	224	8710
	298†	2 600	280	1600	~175	1410
					281	1862
					289	1698
Para in ethanol			225	8930		
			284	1775		

In all cases the absorption in acid at 236–244 nm is very low and is almost negligible compared with the absorption in alkali in that region.
At 300 nm the absorption in acid is negligible and is roughly at $\epsilon = 2600$.
* Another report 297, ϵ 2290.
† Another report 299, ϵ 2400.

(*b*)

	0·1 N NaOH $\lambda_{max.}$ nm	0·1 N NaOH $\epsilon_{max.}$	0·1 N HCl $\lambda_{max.}$ nm	0·1 N HCl $\epsilon_{max.}$
OH, Cl, Cl (2,3-)	241	8800	277	2150
	292	5250	280	2000
	300	5000	283	2050
	267 minimum		250 very low minimum	
OH, Cl, Cl (3,4-)	244	13 250	226	6500
	302	3250	282	1950
	275 minimum		285	1900
			250 very low minimum	
OH, Cl, Cl (2,5-)	241	8500	280	2600
	298	4800		
	235 and 267 minima		250 very low minimum	
OH, Cl, Cl (2,4-)	244	9500	285	2000
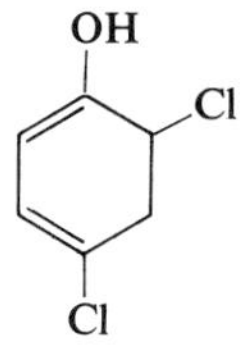	304	3400	245–250 low minimum	

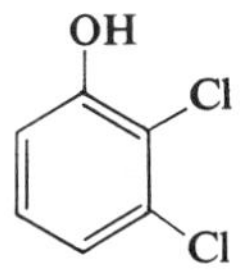

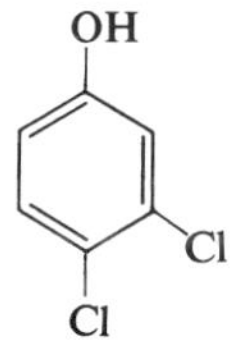

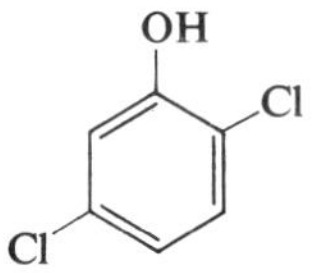

Table 6.9 *continued*

	Solvent			
	0·1 N NaOH		0·1 N HCl	
	$\lambda_{max.}$ nm	$\epsilon_{max.}$	$\lambda_{max.}$ nm	$\epsilon_{max.}$
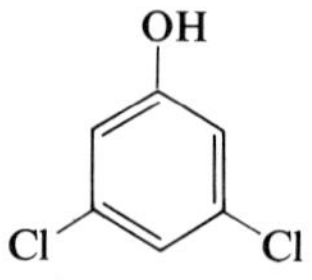	243·5 295 300	6600 4100 4200	223·5 279 287 245–250 low minimum	7000 2300 2100
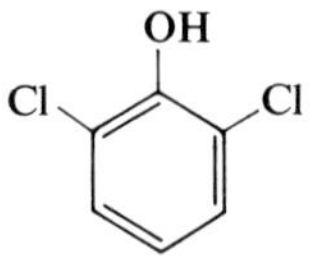	238 301	7100 5250	276 —	2250 —
(*c*)	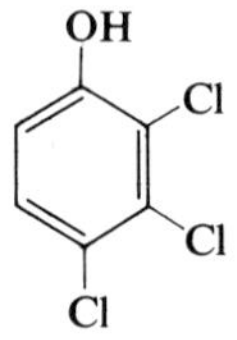 246 309 280 minimum	7100 2800	293 260 low minimum	1950
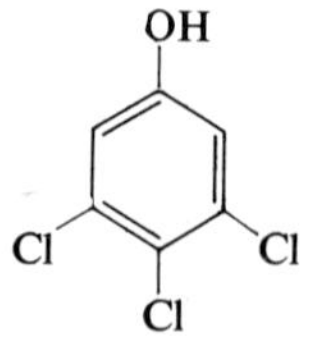	234 307 260	10 800 5050 2000 minimum	290 260 low minimum	1300
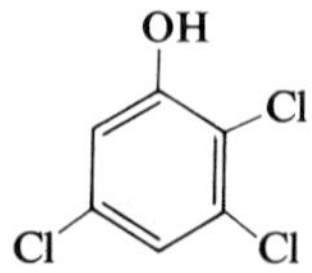	244 298 302	7250 4050 4700	280 290 —	1900 1900 —
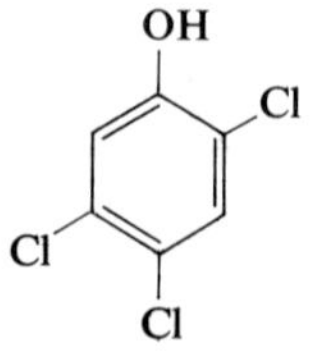	244 310	8900 4200	290 —	2900 —
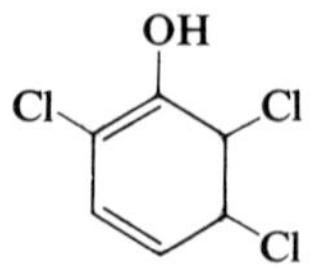	304 —	5650 —	280 289	1900 1950
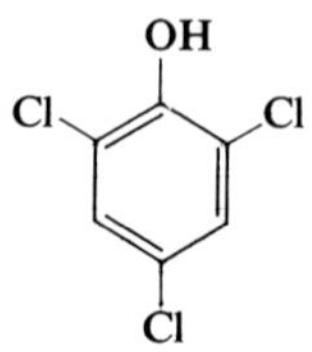	244 311	8950 4900	287 292	2850 2300

Table 6.9 *continued*

(*d*)	Solvent			
	0·1 N NaOH		0·1 N HCl	
	$\lambda_{max.}$ nm	$\epsilon_{max.}$	$\lambda_{max.}$ nm	$\epsilon_{max.}$
	240	12 500	293	2900
	~255	12 000	298	3000
	314·5*	5600	298	3000
	~240	9200	290	2100
	~255	9000	298	2200
	315*	5100	*ca.* 262 very low minimum	
	~240	10 500	286	1650
	308†	6800	293	1800
			ca. 262 very low minimum	

* Each tetrachlorophenol can be approximately determined from absorbance at 315 nm taking ϵ as 5600 or 5100.
† Here Δ 308 nm (alkali–acid) = ϵ 5600.

Figure 6.25 Absorption spectra of 2,3- and 3,4-dichlorophenols. (A) 3,4-dichlorophenol in 0·1 N NaOH; (B) 3,4-dichlorophenol in 0·1 N HCl; (C) 2,3-dichlorophenol in 0·1 N NaOH; (D) 2,3-dichlorophenol in 0·1 N HCl (from Azouz *et al.*, 1955)

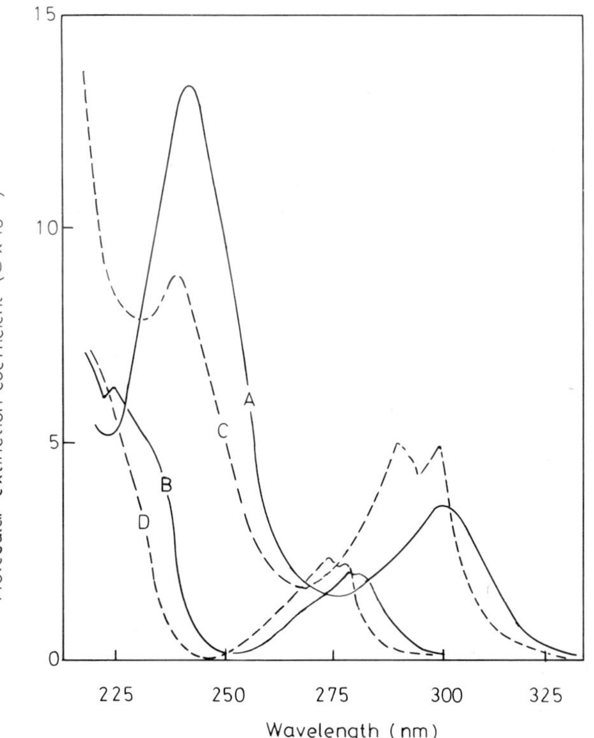

Figure 6.26 Absorption spectra of 2,5-dichlorophenol. (A) in 0·1 N NaOH; (B) in 0·1 N HCl (from Azouz *et al.*, 1955)

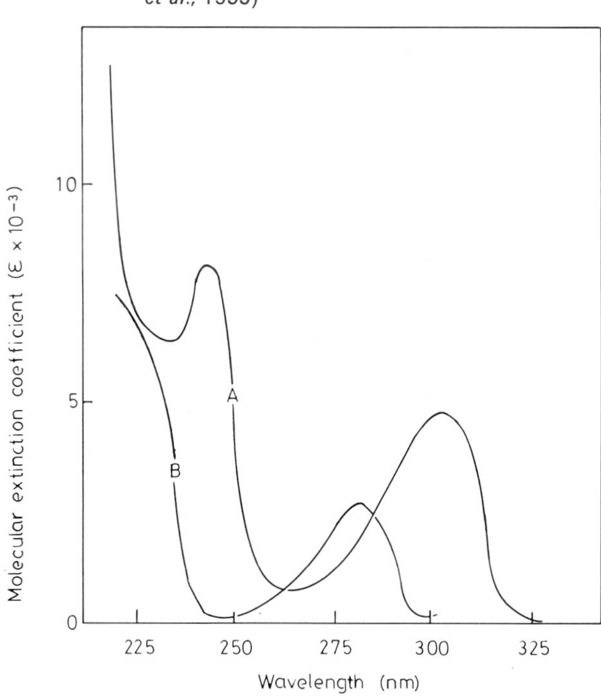

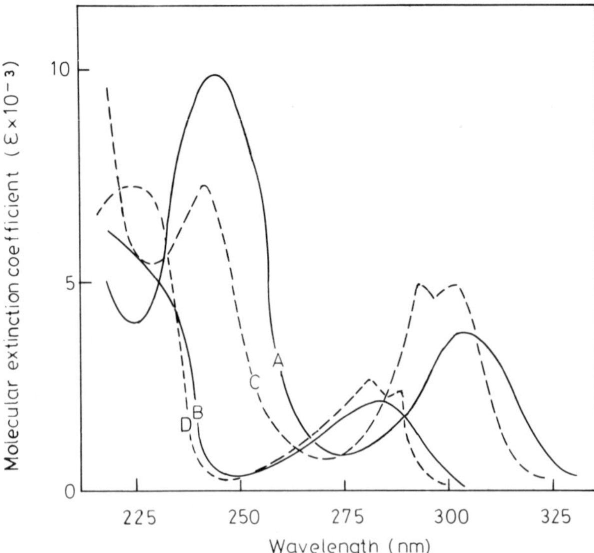

Figure 6.27 Absorption spectra of 2,4- and 3,5-dichlorophenols. (A) 2,4-dichlorophenol in 0·1 **N** NaOH; (B) in 0·1 **N** HCl; (C) 3,5-dichlorophenol in 0·1 **N** NaOH and (D) in 0·1 **N** HCl (from Parke & Williams, 1955)

Figure 6.28 Absorption spectra of 2,6-dichlorophenol. (A) in 0·1 **N** NaOH and (B) in 0·1 **N** HCl (from Parke & Williams, 1955)

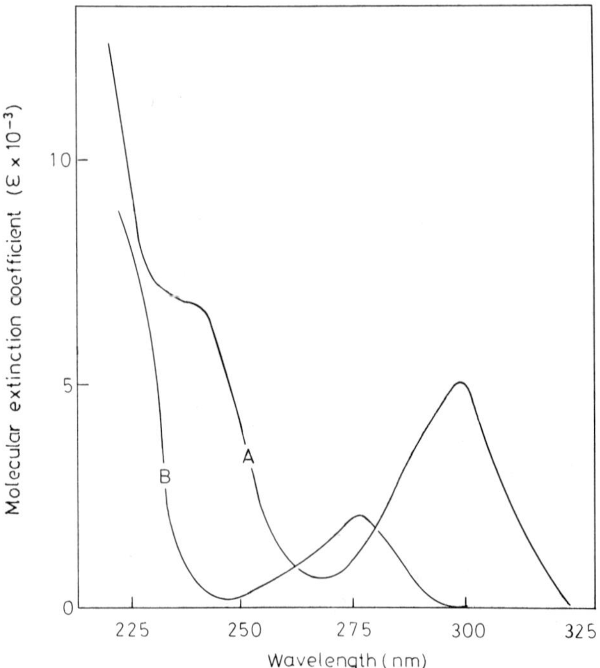

two dichlorophenol concentrations can be confirmed from the alkaline solution because at 304 nm both constituents have ϵ values near 3400; in addition at 285 nm in 0·1 **N** HCl both exhibit an ϵ value of 2000.

Azouz *et al.* (1955) found that *o*-dichlorobenzene fed to rabbits was in the main oxidized to 3,4-dichlorophenol and excreted as glucuronic and sulphuric acid conjugates (*ca.* 30% of the dose). About 9% of the dose was converted to 2,3-dichlorophenol and roughly 4% to catechols. A little 3,4-dichlorophenylmercapturic acid was also formed. The dye already referred to gave a red derivative with 2,3-dichlorophenol but not with its 3,4-isomer and so could be determined independently. The acid-hydrolysed urine (2 vols., plus conc. HCl, 1 vol.; refluxed for 3 hours) was steam-distilled and a portion used for the colour test. The absorption at 244 nm was measured in 0·1 **N** NaOH and in 0·1 **N** HCl ($\Delta\epsilon_{244}$ for the pure 3,4-isomer *ca.* 12 800; correction for the 2,3-isomer $\Delta\epsilon = 8250-200 = 8050$). As an independent check both isomers in alkali show at 304–5 nm, ϵ *ca.* 3100 and at 282 nm in acid both show ϵ 1950 so that the total is readily accessible (see Fig. 6.25).

p-Dichlorobenzene is largely excreted as 2,5-dichlorophenol which shows $\epsilon_{max.}$ 8500 at 241 nm in alkali, but in acid ϵ is 500 (see Fig. 6.26). The amount present in the steam distillate from acid hydrolysed urine is estimated from $\Delta\epsilon$ at 241 nm (alkali and acid).

Figure 6.29 Spectra of 2,3,4- and 3,4,5-trichlorophenols. (A) 2,3,4-trichlorophenol in 0·1 **N** NaOH and (B) in 0·1 **N** HCl; (C) 3,4,5-trichlorophenol in 0·1 **N** NaOH and (D) in 0·1 **N** HCl (from Jondorf *et al.*, 1955)

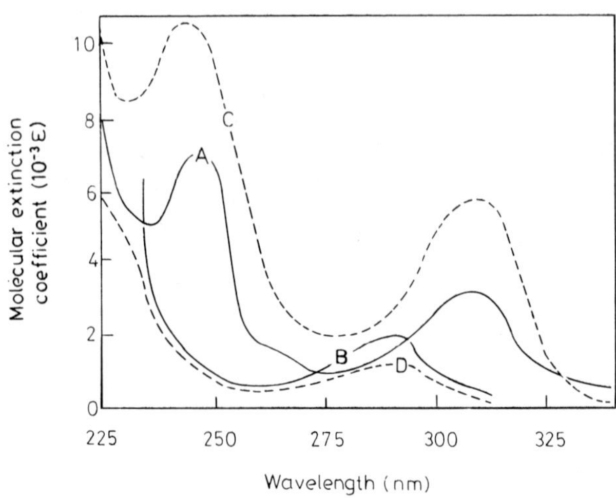

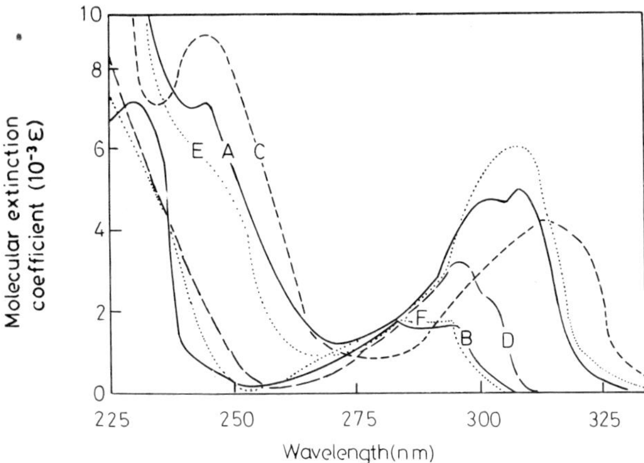

Figure 6.30 Spectra of 2,3,5-, 2,4,5- and 2,3,6-trichlorophenols. (A) 2,3,5-trichlorophenol in 0·1 N NaOH and (B) in 0·1 N HCl; (C) 2,4,5-trichlorophenol in 0·1 N NaOH and (D) in 0·1N HCl; (E) 2,3,6-trichlorophenol in 0·1 N NaOH and (F) in 0·1 N HCl (from Jondorf *et al.*, 1955)

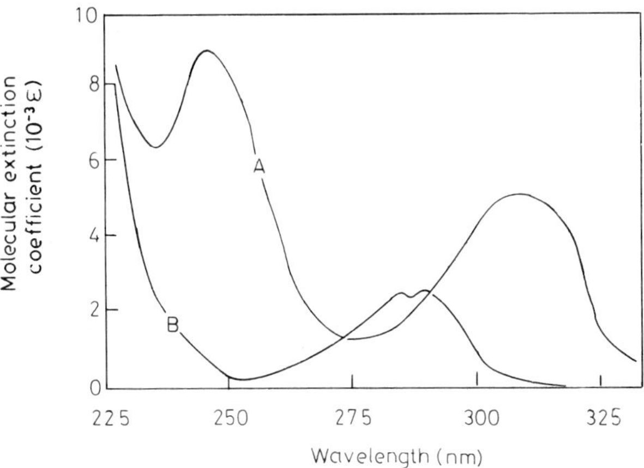

Figure 6.31 Spectra of 2,4,6-trichlorophenol. (A) in 0·1 N NaOH and (B) in 0·1 N HCl (from Jondorf *et al.*, 1955)

Jondorf *et al.* (1955) applied similar spectroscopic considerations to the analysis of phenols produced from 1,2,3-, 1,2,4- and 1,3,5-trichlorobenzenes (see Figs. 6.29, 6.30 and 6.31). In a later paper (Jondorf *et al.*, 1958) they described what occurred when polychlorobenzenes were administered.

Tetrachlorobenzenes given orally to rabbits may be eliminated (*a*) in the expired air, (*b*) in the faeces; they may undergo some dechlorination in the intestinal tract and absorbed material may be oxidized to phenols and excreted in the urine either free or conjugated.

In a much earlier study of the fate of benzene itself, Parke and Williams (1950) had placed rabbits in a respiration chamber and after administering benzene orally or by injection had passed the expired air into a nitrating mixture and had determined the benzene eliminated unchanged by measuring the *m*-dinitrobenzene formed. A sensitive procedure depends on the interaction of *m*-dinitrobenzene and the enolic form of a ketone in the presence of alkali. When methyl ethyl ketone was used and the experimental conditions were laid down in detail, the colour produced was linear from 0–40 μg (*m*-dinitrobenzene)/10 ml and nearly linear from 40 to 80 μg/10 ml. Over a period of 15–30 hours about 40% of the dose was eliminated unchanged at 0·25 and 0·5 g/kg body weight and 64% at 1 g/kg. The same authors (1953) carried out similar work with [$^{14}C_1$] benzene and showed that after a single oral dose 43% was exhaled unchanged within 2–3 days after dosing. Some 30% of the dose appeared in the urine as conjugated phenols.

After this digression on metabolites in expired air we may now revert to the tetrachlorobenzenes. These compounds are steam-volatile and tend to sublime, the 1,2,3,4- and 1,2,3,5-isomers more readily than the 1,2,4,5-. This suggested that the two first-mentioned isomers might be exhaled appreciably. The 1,2,3,4-isomer is more readily metabolized oxidatively than the 1,2,3,5-isomer. Unchanged tetrachlorobenzenes exhaled in the expired air were measured from the spectra shown by ethanolic solutions obtained at intervals of 6 or 12 hours over a period of 3 to 5 days after dosing. Although the positions of the peaks were occasionally slightly different from those of authentic chlorobenzenes a picture emerged that in broad terms was quite clear. Inspection of Table 6.10 shows that the 1,2,3,4-isomer could be recovered unchanged from faeces, tissues and breath. The expired air for all three tetrachlorobenzenes also contained dechlorinated derivatives. Figure 6.32 illustrates how the expired air after administering 1,2,4,5-tetrachlorobenzene gave definite indications of the presence of di- and trichlorobenzene as well as unchanged material. Figure 6.33 shows very clearly how the urinary metabolites of the same isomer changed with time.

Another aspect of the metabolism of benzene is the possibility that muconic acid should be formed by ring fission. Small amounts of *trans-trans*-muconic acid are in fact excreted in the urine. Porteous & Williams

Table 6.10 Fate of isomeric tetrachlorobenzenes

Isomer	Phenols: Tetrachloro phenols	Phenols: Other phenols	Unchanged tetrachlorobenzene in: Faeces	Unchanged tetrachlorobenzene in: Tissues	Unchanged tetrachlorobenzene in: Breath	Other chlorobenzenes in breath	Total
1,2,3,4-	43	< 1	5	10	8	2	68
1,2,3,5-	5	5	14	23	12	9	68
1,2,4,5-	2	5	16	48	2	10	83

(Dose 0·5g/kg orally. Excretions during 6 days after dosing).

Figure 6.32 Absorption spectrum in ethanol of the expired air (of a rabbit) collected for 12 hours after an oral dose of 1·5 g of 1,2,4,5-trichlorobenzene compared with the spectra of 1,3-di-, 1,2,4-tri- and 1,2,4,5-tetrachlorobenzene in ethanol. Ordinates for curve (A) are arbitrary. Expired air: (A) 255, 265, 270, 277–278, 285, 294 nm. Chlorobenzenes: absorption peaks

1,2,4,5-tetra-(B)					
—	—	—	276,	285,	294
1,2,4-tri-(C)					
—	—	270,	278,	287,	—
1,3-di-(D)					
256,	263,	270,	278,	—	—

(from Jondorf *et al.*, 1958)

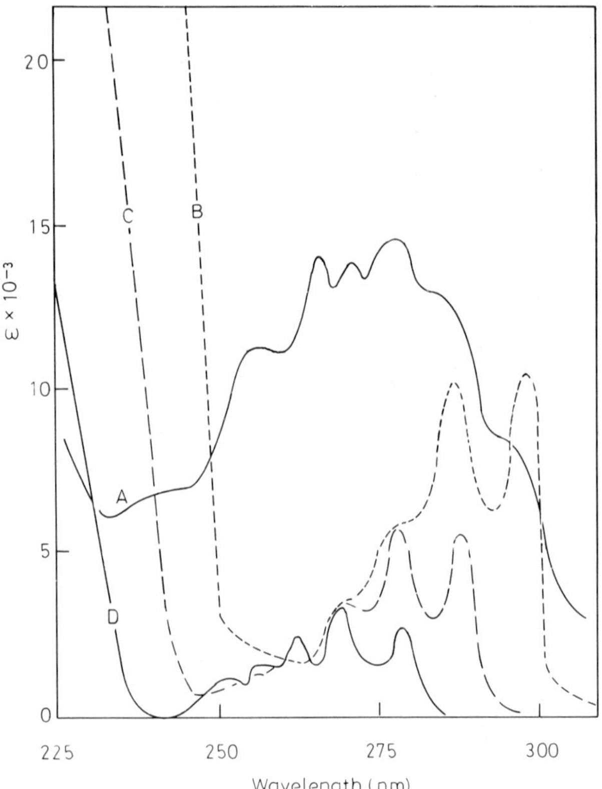

Figure 6.33 Ultraviolet spectra in 0·1 N NaOH of the steam-volatile products of the acid-hydrolysed urines from a rabbit which has been dosed with 1,2,4,5-tetrachlorobenzene (1·5 g). (A) is that from urine of the first day after dosing and the main peak is at 298 nm, which corresponds to 2,5-di- or 2,3,5-trichlorophenol. (B) is that of the second day after dosing and the main peaks are at about 298 and 308 nm, and correspond to a mixture of the components of curves (A) and (C). (C) is that of the sixth day after dosing and the main peak is at 308 nm and corresponds to the 2,3,5,6-tetrachlorophenol (from Jondorf *et al.*, 1958)

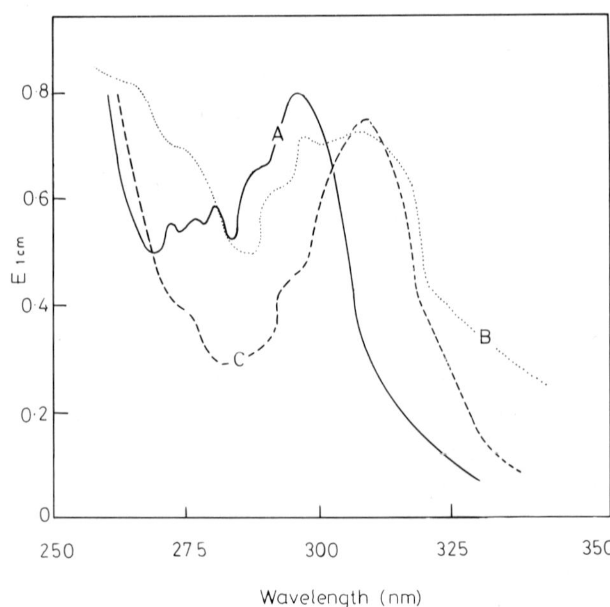

(1959) prepared *cis-cis* muconic acid and observed for it in ethanol a broad band extending from 230 to 290 nm with $\lambda_{max.}$ 258·5 nm and $\epsilon_{max.}$ 19 880. *Trans-trans*-muconic acid showed $\lambda_{max.}$ 259 nm, $\epsilon_{max.}$ 27 100 in ethanol and 261 nm, $\epsilon_{max.}$ 24 600 in 2% NaOH; this isomer can be absorbed on a short column of alumina, and after washing with water, quantitative elution occurs with 2% NaOH. Rabbit urine, however, contained an interfering substance absorbing at 260 nm.

6.4.3 *m*-Dinitrobenzene Metabolism

This compound is a poison which causes anaemia, liver damage and cerebral paralysis. Parke (1961) administered *m*-dinitro-[^{14}C]-benzene to the rabbit and found that much of it was rapidly excreted in the urine as *m*-nitroaniline, *m*-phenylinediamine, 2-amino-4-nitrophenol and 2,4-diaminophenol with smaller amounts of other metabolites. The fact that dinitrophenols were not excreted appreciably was established as follows: (1) the urine from rabbits given an oral dose of 0·5 g of *m*-dinitrobenzene was collected over two days, hydrolysed by refluxing with 0·5 vol. of conc. HCl. The hydrolysed urine was continuously extracted with ether; (2) the ether extract was then treated twice with 2 **N** NaOH and the alkaline extractthen acidified and re-extracted with ether. This procedure, more fully described by Parke (1961), should have concentrated any phenols. The absorption spectra in acid and alkali were determined and compared with the spectra of 2,4-, 2,6- and 3,5-dinitrophenol in 0·1 **N** NaOH and in O·1 **N** HCl. There was so little resemblance that the presence of these phenols in the original urine was quite improbable.

	0·1 N NaOH		0·1 N HCl	
	$\lambda_{max.}$ nm	$\epsilon_{max.}$	$\lambda_{max.}$ nm	$\epsilon_{max.}$
2,4-dinitrophenol (OH; NO_2; NO_2)	226	10 300	261	13 600
	258	7200	~300	9400
	358	14 800	—	—
	~400	10 000	—	—
2,6-dinitrophenol (NO_2; OH; NO_2)	~258	8000	248	10 800
	428	8200	345	5400
3,5-dinitrophenol (OH; NO_2; NO_2)	268	13 000	232	15 300
	400	2800	340	3100

References

6.1

American Petroleum Institute Project No. 44, Spectra 807–860 (No. 856).

Araki, G. & Murai, T. (1954). *J. Chem. Phys.*, **22**, 954.

Bailey, A. S., Bryant, K. C., Hancock, R. A., Morrell, S. H. & Smith J. C. (1947). *J. Inst. Petrol.*, **33**, 503.

Beale, R. N. & Roe, E. M. F. (1951). *J. Chem. Soc.*, 2884.

Booth, J. & Boyland, E. (1949). *Biochem. J.*, **44**, 361.

Boyland, E. & Levi, A. A. (1935). *Biochem. J.*, **29**, 2679.

Boyland, E. & Wolfe, G. (1950). *Biochem. J.*, **47**, 64.

Brunke, J. & Poppe, E. J. (1960). *Chem. Ber.*, **93**, 2572.

Capper, N. S. & Marsh, J. K. (1926). *J. Chem. Soc.*, 724.

Clar, E. (1929). *Ber. dtsch chem. Ges.*, **62**, 350, 1574. (*See also* **65**, 503, 846,; **69**, 607, 1671; **73**, 81, 104); (1950). *Spectrochimica Acta*, **4**, 116; (1952). In *Aromatische Kohlenwasserstoffe*, Springer-Verlag, Berlin, 2nd edition.

Coulson, C. A. & Rushbrooke, G. S. (1940). *Proc. Cambridge Phil. Soc.*, **36**, 193.

Coulson, C. A. & Longuet-Higgins, H. C. (1947). *Proc. Roy. Soc.* A, **191**, 39.

Falk, H. L. & Steiner, P. E. (1952). *Cancer Research*, **12**, 30.

Friedel, R. A. & Orchin, M. (1951). In *Ultraviolet Spectra of Aromatic Compounds*. John Wiley, New York and London. (*See also* Orchin, M. & Friedel, R. A. (1949). *J. Amer. Chem. Soc.*, **71**, 3002.)

Hellbronner, E., Frolicher, U. & Plattner, P. A. (1949). *Helv. Chim. Acta.*, **32**, 2479.

Henri, V. & Laszlo, H. de (1924). *Proc. Roy. Soc.* A, **105**, 662.

Jaffé, H. A. & Orchin, M. (1962). In *Theory and Application of Ultraviolet Spectroscopy*. John Wiley & Sons Inc., New York and London.

Jones, R. N. (1943). *Chem. Reviews*, **32**, 1; (1945). *J. Amer. Chem. Soc.*, **67**, 2021.

Klevens, H. B. & Platt, J. R. (1949). *J. Chem. Phys.*, **17**, 470.

Lijinsky, W. (1960). *Anal. Chem.*, **32**, 684. (1961). *Ibid.*, **33**, 810.

Lijinsky, W. & Shubik, P. (1964). *Science. N.Y.*, **145**, 53; (1965). *Toxicol. & Applied Pharmacol.*, **7**, 337, also *Food and Cosmetics Toxicology*, **3**, 145.

Mayneord, W. G. & Roe, E. M. F. (1935). *Proc. Roy. Soc.* A, **152**, 299; (1937). *Ibid.* A, **158**, 634.

Moffit, W. (1953). *J. Chem. Phys.*, **22**, 320.

Morton, R. A. & de Gouveia, A. J. A. (1934). *J. Chem. Soc.*, 911.

Mosby, W. L. (1953). *J. Amer. Chem. Soc.*, **75**, 3348.

Philips, J. P., Lyle, R. E. & Jones, P. R. (1969). Eds. *Organic Electronic Spectral Data*. Vol. V. Interscience Publishers, New York and London.

Platt, J. R. (1949). *J. Chem. Phys.*, **17**, 484; (1951). *Ibid.*, **19**, 263.

Plattner, P. A. & Hellbronner, E. (1947). *Helv. Chim. Acta.* **30**, 910; (1948). *Ibid.*, **31**, 804.

Robertson, W. W. & King, A. D. Jr. (1961). *J. Chem. Phys.*, **34**, 2190.

Rudenberg, K. & Parr, R. G. (1951). *J. Chem. Phys.*, **19**, 1268.

Sawicki, E. & Miller, R. R. (1958). *Anal. Chem.*, **30**, 109. (1964). *Microchem. J.*, **8**, 257.

Scholtz, M. & Treibs, W. (1961). *Z. Elektrochem.*, **60**, 120.

Van Duuren, B. L. (1958). *J. Natl. Cancer Inst.*, **21**, 1, 623. (1960). *Ibid.*, **25**, 53; *Anal. Chem.*, **32**, 1436. Also *Nat. Cancer Inst. Monograph*, No. 9.

Zimmermann, H. & Joop, N. (1960). *Z. Electrochem.*, **61**, 1219.

General Articles, Symposia, etc.

Gunther, F. A. & Buzzetti, F. (1965). *Residue Reviews*, **9**, 90.

Haenni, E. O. (1968). *Residue Reviews*, **24**, 41.

National Cancer Institute. Symposium on analysis of carcinogenic air pollutants. (1962). Eds. E. Sawicki and F. Cassell. U.S. Government Printing Office, Washington. (*See also* references under authors.)

Shubik, P. & Hartwell, J. L. (1957). *U.S. Public Service*, Publication 149.

Spectroscopy and Fluorescence Spectra

Haenni, E. O. & Hall, M. A. (1960). *J. Assoc. Official Agric. Chemists*, **43**, 92. (*See also* (1962), *ibid.*, **45**, 59 and 67.)

Hartwell, J. J. (1951, Supplement 1957). Survey of compounds which have been tested for carcinogenic activity. U. S. Department of Health.

Lijinsky, W. A., Chesaut, A. & Rama, C. R. (1962). *Chicago Med. School Quart.*, **21**, 49.

Mazee, W. M., Gersmann, H. R. & van der Wiel, A. (1966). *Food Cosmetic Toxicol.*, **4**, 17.

Moore, G. E., Monkman, J. L. & Katz, M. (1962). *Nat. Cancer Inst. Monograph*, No. **9**, 153.

Oser, B. L. & Oser, M. (1960). *Toxicol. & Applied Pharmacol.*, **2**, 441. (1962). *Ibid.*, **4**, 70.

Oser, B. L. (1962). *Residue Reviews*, **1**, 1.

Sawicki, E., Hauser, T. R. & Stanley, T. W. (1960 *c*). *Internat. J. Air Pollution*, **2**, 253.

Smokes and Exhaust Gases

Bonnett, J. & Neukomm, S. (1954). *Helv. Chim. Acta.*, **39**, 1724; (1957), *Oncologia* **10**, 124.

Commins, B. T. (1958 *a*). *Analyst*, **83**, 386; (1958 *b*). *J. Air Pollution*, **1**, 14; (1962). *Nat. Cancer Inst. Monograph*, No. **9**, 225.

Analytical

Falk, H. L. & Steiner, P. E. (1952). *Cancer Research*, **12**, 30.

Grimmer, G. (1965). *J. Chromatog.*, **20**, 89.

Hoffman, D. & Wynder, E. L. (1962). *Nat. Cancer Inst. Monograph.*, **9**, 91.

Howard, J. W. & Haenni, E. O. (1963). *J. Assoc. Official Agric. Chemists*, **46**, 933. (*See also* (1966), *ibid.*, **49**, 595.)

Kuratsune, M. (1956). *J. Nat. Cancer Inst.*, **16**, 1485.

Lijinsky, W. A., Donisky, J., Mason, G., Ramahi, H. Y. & Safair, T. (1963). *Anal. Chem.*, **36**, 952.

Lijinsky, W. A., Rama, C. R. & Keeling, J. (1961). *Anal. Chem.*, **33**, 810.

Lyons, M. J. & Johnson, H. (1957). *Brit. J. Cancer*, **11**, 554.

Nelson, L. B. & Stormont, D. J. (1960). *Chem. & Ind.*, 561.

Ramsey, L. L. (1967). *J. Assoc. Offic. Anal. Chemists*, **50**, 1913.

Sawicki, E. *et al.*, (1960 *a*). *Anal. Chem.*, **32**, 810; (1960 *b*). *J. Air Pollution*, **2**, 273. (1964). *Chemist-Analyst*, **53**, 6, 24, 28, 56, 88.

6.2

Braude, E. A. & Forbes, W. F. (1955). *J. Chem. Soc.*, 3776.
Bridges, J. W., Creavan, P. J. & Williams, R. T. (1965). *Biochem. J.*, **96**, 872.
Creaven, P. J., Parke, D. V. & Williams, R. T. (1965). *Biochem. J.*, **96**, 879.
Förster, T. (1950). *Z. Elektrochem.*, **54**, 531.
Gillam, A. E. & Hey, D. H. (1939). *J. Chem. Soc.*, 1170
Katritsky, A. R. & Simmons, P. (1960). *J. Chem. Soc.*, 4091.
Keiffer, F. & Rumpf, P. (1954). *Comptes rend. Acad. Sci. Paris*, **238**, 360, 700.
Klevens, H. B. & Platt, J. R. (1949). *J. Chem. Phys.*, **17**, 470.
Pickett, L. W., Hoeflich, N. J. & Liu, T. (1951). *J. Amer. Chem. Soc.*, **73**, 4862.
Pickett, L. W., Walter, G. F. & France, H. (1936). *J. Amer. Chem. Soc.*, **58**, 2296.
Wenzel, A. (1953). *J. Chem. Phys.*, **21**, 403.
Williams, R. T. & Bridges, J. W. (1964). *J. Clin. Path.*, **17**, 371.

6.3

Beale, R. N. & Roe, E. M. F. (1952). *J. Amer. Chem. Soc.*, **74**, 2302.
Beale, R. N. & Roe, E. M. F. (1953). *J. Chem. Soc.*, 2755.
Beaven, G. H. (1958). In *Steric Effects in Conjugated Systems.* Chapter 3. Ed. G. W. Gray. Butterworths, London.
Beaven, G. H. & Johnson, E. A. (1959). *Spectrochem. Acta.*, **14**, 67.
Braude, E. A. (1949). *J. Chem. Soc.*, 1902.
Braude, E. A. & Forbes, W. F. (1955). *J. Chem. Soc.*, 3776.
Braude, E. A. & Sondheimer, F. (1955). *J. Chem. Soc.*, 3754, 3773.
Braude, E. A. & Timmins, C. J. (1955). *J. Chem. Soc.* 3766.
Calvin, M. (1939). *J. Org. Chem.*, **4**, 256.
Calvin, M. & Alter, H. W. (1951). *J. Chem. Physics*, **19**, 765.
Dewar, M. J. S. (1952). *J. Amer. Chem. Soc.*, **74**, 3341.
Gray, G. W. (1958). Ed. *Steric Effects in Conjugated Systems.* (Chapters by G. H. Beaven, C. C. Barker, M. J S. Dewar, W. F. Forbes, E. S. Waight and R. L. Erskine.) Butterworths, London.
Newman, M. S. (1956). Ed. *Steric Effects in Organic Chemistry.* J. Wiley, New York.
O'Shaughnessy, M. T. & Rodebush, W. H. (1940). *J. Amer, Chem. Soc.*, **62**, 2986.
Pauling, L. (1939). *Fortschr. Chem. Org. Naturstoffe*, **3**, 203.
Pauling, L. (1949). *Helv. Chim. Acta.*, **32**, 2241.
Pinckard, J. H., Wille, B. & Zechmeister, L. (1948). *J. Amer. Chem. Soc.*, **70**, 1938.
Rao, C. N. R. (1967). In *Ultraviolet and Visible Spectroscopy.* 2nd Edition. Butterworths, London.
Smakula, A. & Wassermann, A. (1931). *Z. Physikal Chem.*, **155A**, 353.

6.4

Azouz, W. M., Parke, D. V. & Williams, R. T. (1952). *Biochem. J.*, **50**, 702; (1953), *Biochem. J.*, **55**, 146; (1955). *Biochem. J.*, **59**, 410.
Hocking, B. (1950). *Sci. Agric.*, **30**, 183.
Jondorf, W. R., Parke, D. V. & Williams, R. T. (1955). *Biochem. J.*, **61** 512; (1958). *Biochem. J.*, **69**, 181.
Ostrow, J. D., Jandl, J. H. & Schmid, R. (1962). *J. Clin. Invest.*, **41**, 1628.
Parke, D. V. (1961). *Biochem. J.*, **78**, 282.
Parke, D. V. & Williams, R. T. (1950). *Biochem. J.*, **46**, 236; (1953), *ibid.*, **55**, 337; (1955), *ibid.*, **59**, 415.
Porteous, J. W. & Williams, R. T. (1949). *Biochem. J.*, **44**, 56.
Simkover, H. G. & Shenfelt, R. D. (1952). *J. Econ. Ent.*, **45**, 11.
Smith, J. N., Spencer, B. & Williams, R. T. (1950). *Biochem. J.*, **47**, 284.
Spencer, B. & Williams, R. T. (1950). *Biochem. J.*, **47**, 279.
Williams, R. T. (1947). In *Detoxication Mechanisms. The Metabolism of drugs and allied organic compounds* (1st edn.). Chapman & Hall, London. (1959), In *Detoxication Mechanisms. The Metabolism of drugs, toxic substances and other organic compounds.* (2nd edn.). Chapman & Hall, London.

7. Carotenoids and Related Substances

7.1 Introduction

The lipochromes or fat-soluble pigments constitute a large class of unsaturated substances widely but irregularly distributed in micro-organisms, plants and animals. Much progress has been made in isolating them and establishing their structure; indeed very many have been synthesized by routes that confirm the structures arrived at by other methods. The principal lipochromes are the carotenoids containing forty carbon atoms. Many are hydrocarbons each having a long chain of conjugated double bonds to which the colour is due, β-carotene $C_{40}H_{56}$ being perhaps the best known. In a second large class of carotenoids each member contains one or more oxygen atoms in the form of hydroxyl, keto or epoxy groups and here perhaps zeaxanthin $C_{40}H_{56}O_2$ [$C_{40}H_{54}(OH)_2$] is as typical as any.

Systematic research has been carried out for nearly half a century and carotenoids have turned out to be numerous and varied. In some cases proof of structure was fraught with difficulty and some important members of the family were allocated structural formulae which, though consistent with the information until recently available, were shown to be erroneous when the techniques of today were applied. It is however likely that the procedures for establishing structures are now adequate and residual uncertainties have been or are being rapidly being eliminated.

Substantial progress has been made in elucidating biosynthetic pathways and metabolic routes and it is possible that quite a large number of quantitatively minor carotenoids present in tissues are 'left-overs' in biosynthesis so that the attempt to identify a function for each and every carotenoid may be to chase a will o' the wisp and to waste effort. Such a view cannot, however, be allowed to obscure the main issue, which is that some probable roles of carotenoids are only incompletely understood and hitherto unsuspected roles may well emerge.

The precursor function of those carotenoids that can be used by animals as provitamins A has proved to be the starting point for fundamental investigations including those on the chemistry of the visual processes. The role of carotenoids in the photosynthetic cycles characteristic of chloroplasts is now becoming clearer. These two approaches to the biological functions of carotenoids encourage further exploration.

The field of carotenoid studies has become very broad and in the present account scant justice can be done to many aspects; the emphasis here will be very largely on spectroscopic methods and even with this restriction much important work must be left out.

In the terpenoid field the 'building block' is the branched 5-carbon isoprenoid unit. Acetyl co-enzyme A is the starting point for the biosynthesis of mevalonic acid from which isopenentylpyrophosphate can be formed:

$$\underset{\displaystyle CH_2}{CH_3\overset{}{C}}\!\!\!\!\!\!\!\!\!\!\quad\;\; CH_2CH_2O.\text{—}P\text{—}O\text{—}P$$

$CH_3C(=CH_2)CH_2CH_2O.$—P—O—P

The steps leading to geranyl pyrophosphate and farnesyl pyrophosphate have been elucidated and the biosynthesis of squalene, of cholesterol and of the $C_{45}H_{73}$ side-chain of plastoquinone or the $C_{50}H_{71}$ side-chain of ubiquinone (Q-10) are no longer controversial.

The above biosynthetic pattern would suggest that a C-20 compound geranylgeranyl pyrophosphate could dimerize (tail to tail) to give lycopersene $C_{40}H_{66}$ analogous to squalene $C_{30}H_{50}$.

Lycopersene does not seem to accumulate naturally as does squalene and there is even uncertainty as to whether it is actually formed, but there is no doubt about the natural occurrence of phytoene which can be regarded as a 15,15′-dehydrolycopersene. The simplest true carotenoid is perhaps the tomato pigment lycopene $C_{40}H_{56}$.

Phytoene, by loss of two hydrogens, could yield phytofluene and the product would have five conjugated double bonds instead of three. Loss of two more hydrogens would result in ζ-carotene, which has seven conjugated double bonds. A further loss of two hydrogens could give neurosporene with nine conjugated double bonds and a final loss of two hydrogens could give lycopene with eleven conjugated double bonds. The absorption spectra shown by the series of compounds agree with this sequence of structural differences.

Mass spectrometry has shown that in the phytoene–lycopene series fragmentation can occur at the bis-allylic single bond adjacent to the chromophoric unit of conjugated double bonds. Thus in lycopene itself fission at C8 or 8′ suggests that fragments M-137 and M-69 are to be expected, and in fact are found. Phytoene (m/e 544, $C_{40}H_{64}$) gives a prominent fragment at m/e 329 (M-205) while phytofluene (m/e 542) has a fragment m/e 337, i.e. M-205, and another at m/e 405 (M-137). ζ-Carotene (m/e 540) has m/e 403, i.e. M-137 while neurosporene (m/e 538) shows m/e 401, i.e. M-137 and m/e 469, M-69. Recently, 7,8,11′,12′-tetrahydrolycopene has been detected so that there may in some organisms be a biosynthetic alternative to ζ-carotene.

β-Carotene and some closely related substances are synthesized in the green parts of plants and can be isolated therefrom. The most convenient source however is the carrot root in which the proportions of carotenoid congeners are favourably low, so that isolation of β-carotene is relatively easy. The caroten-

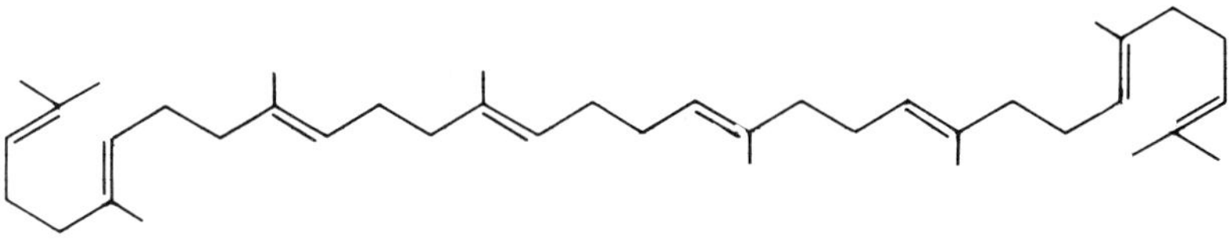

Lycopersene

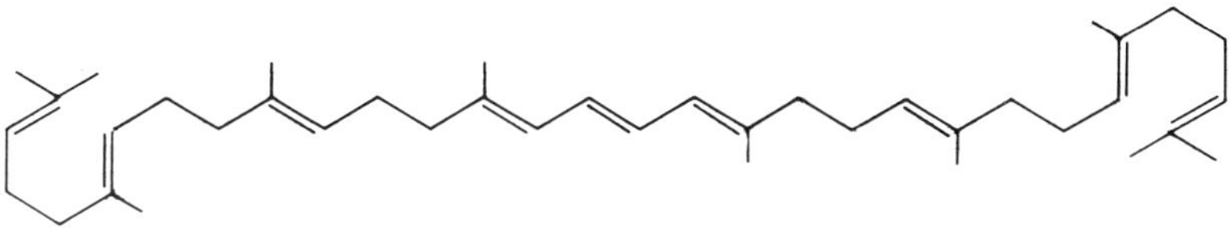

Phytoene

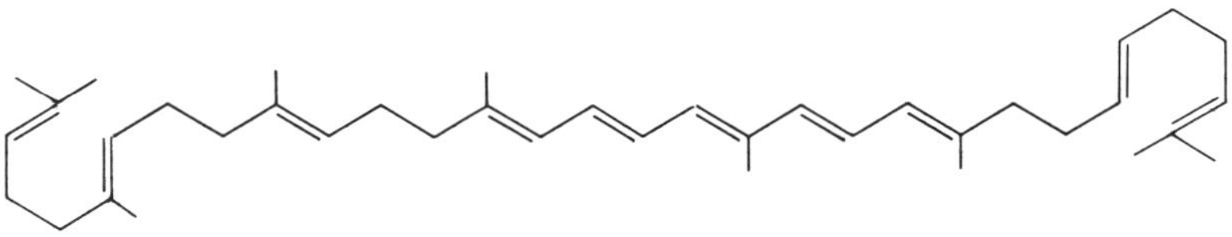

Phytofluene

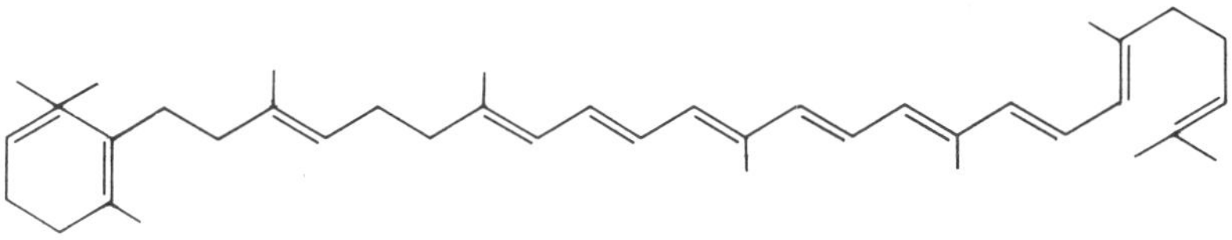

7,8,11,12-Tetrahydrolycopene

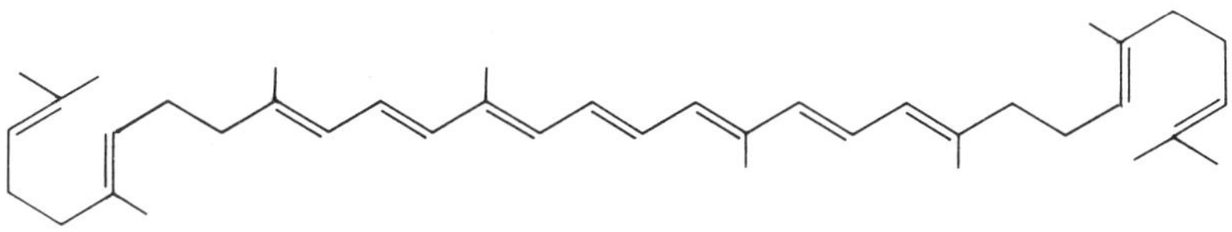

ζ-Carotene 7,8,7′,8′-Tetrahydrolycopene

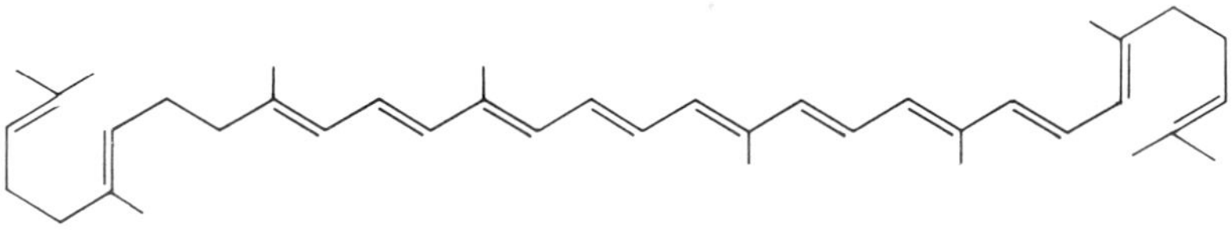

Neurosporene

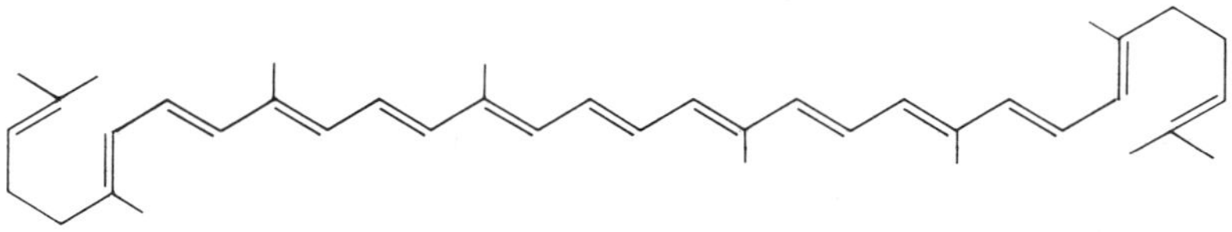

Lycopene

Table 7.1 Lycopene and its precursors.

	$\lambda_{max.}$ nm and ($\epsilon_{max.} \times 10^{-3}$)				Infrared cm^{-1}			n.m.r. τ			
Phytoene	276 (32·4)	286 (41·2)	197 (27·8)		1666	1632	966	8·41	8·35	8·25	
Phytofluene	317 (27)	332 (56)	348 (88·5)		1667	1631	961	8·41	8·35	8·22	
(all-*trans*)	368 (85)										
(a *cis* form, Neo B)	~249 ~318	331·5	347·5	367							
7:8, 11:12-Tetrahydro-lycopene	354 374	394·5	418·5		(Mass spectrum m/e 540·4699 required 540·4695 m/e 335 (M-205))						
ζ-Carotene (all *trans*) 7:8, 7′:8′-Tetrahydro-lycopene	361 (40·6)	379·5 (90)	400·5 (138)	424·5 (138)		966		8·42	8·35	8·20	8·09
(a *cis* form, Neo A)	285 296	357	375 396	419							
Neurosporene (all *trans*)	398 (57·6)	416 (100)	440 (155)	470 (155)	1631	961		8·40	8·32	8·19	8·05
	(55·1)*	(86·9)	(151)	(151)							
Lycopene	446 (120)	472 (185)	505 (169)		1447	1370	965	8·48	8·31	8·18	8·03

For comparison: Squalene $\tau = 8{\cdot}42$, 8·35.
Methyls on poly-ene chain (*a*) in-chain 8·01–8·10, (*b*) end-of-chain 8·19–8·25.
Methyls on isolated double bonds (*a*) *cis* 8·31–8·36, (*b*) *trans* 8·39–8·42.
* A second preparation.

oids are extracted by means of benzene or carbon disulphide and crystalline material can be obtained on addition of methanol or ethanol. Green plant tissues are digested with alkali and the unsaponifiable extract is partitioned between light petroleum and 90% aqueous methanol (Karrer & Jucker, 1948). The epiphasic carotene goes into the light petroleum and the hypophasic 'xanthophylls' into the methanol layer. Crude preparations are purified by chromatography on hydrated lime or alumina.

7.1.1 *Cis-trans* isomerism in Carotenoids

Gillam & El Ridi (1935, 1936) subjected β-carotene to repeated adsorption on alumina and observed the gradual formation of an artefact which they designated pseudo α-carotene. The process seemed to be reversible and a similar effect was observed for α-carotene:

$$\alpha\text{-carotene} \rightleftharpoons \text{neo } \alpha\text{-carotene}$$

The changes were not however due directly to the adsorbent, because Zechmeister & Tuzson (1939) found them to occur when solutions of carotenoids were kept at room temperature. In the case of lycopene there was a time-regulated drift of the absorption maxima towards shorter wavelengths. Carter & Gillam (1939) agreed that artefacts accumulated during prolonged handling of solutions. Rise of temperature accelerated the approach to equilibrium, refluxing being optimal. At room temperature addition of minute amounts of iodine to solutions of carotenoids greatly speeded up the attainment of equilibrium in daylight (Zechmeister & Tuzson, 1939). Meanwhile Strain (1938) had observed heterogeneity induced in xanthophylls by heat which was to some extent reversible. Gillam & El Ridi regarded the phenomena as examples either of *cis-trans* isomerism or of migration of a terminal olefinic linkage out of conjugation (hence the name pseudo α-carotene) and Zechmeister & Cholnoky (1937) added keto-enol tautomerism as a possibility. The explanation destined to survive further study was, however, geometrical isomerism. In several cases the number of chromatographically separable artefacts became too numerous for any other explanation.

The total number of theoretically possible *cis-trans*-isomers in any carotenoid is obviously large. Thus in an unsymmetrical conjugated system having n olefinic linkages the number N of isomers is 2^n. For symmetrical systems in which n is an odd number

$$N = 2^{(n-1)/2} \times (2^{(n-1)/2} + 1)$$

and when n is an even number

$$N = 2^{(n/2)-1} \times (2^{n/2} + 1)$$

On this basis when $n = 8$ the number of isomers for symmetrical and unsymmetrical molecules will be

136 and 256 respectively and when $n=9$, N will be 272 and 512. The number observed is however always much smaller than the 'theoretical'. Zechmeister (1962) enumerated restricting factors; thus the mixture of isomers would be more numerous for the non-symmetrical than the symmetrical molecules and in fact α- and γ-carotenes yielded more complex mixtures under comparable conditions than the symmetrical β-carotene. It was found also that natural all-*trans* carotenoids tended to isomerize to mono-*cis* and di-*cis* forms but not to poly-*cis* (Zechmeister & Petracek, 1952). Another restriction came from a distinction drawn by Pauling (1939) between 'hindered' and 'unhindered' double bonds in carotenoids. In β-carotene there are olefinic linkages which differ from the other double bonds all of which have a methyl substituent:

(i) at 15,15′

(ii) at 7,8 and 7′,8′

(iii) at 11,12 and 11′,12′

(iv) CH_3

In the grouping

CH
CH

the formation of an isomer by rotation is free from the steric hindrance that occurs with

CH_3
H H
C C C C C
H H

This does not mean that hindered *cis*-isomers can never be formed but an energy barrier always has to be overcome. In the case of β-carotene the ring double bonds are excluded and of the 9 remaining conjugated double bonds 5 are unhindered and 4 are hindered. The 5 would lead us to expect 20 possible unhindered isomers to be much more likely to occur than the hindered ones. In fact about 12 were found to occur in solutions.

Table 7.2 Calculated numbers of *cis-trans*-isomers of some representative carotenoids.

	Number n of conjugated olefinic linkages not in ring systems	Theoretical numbers of geometrical isomers	
		Total	Hindered
(*a*) Symmetrical molecules			
Lycopene	11	1056	72
β-Carotene	9	272	20
Zeaxanthin	9	272	20
(*b*) Unsymmetrical molecules			
Lycoxanthin	11	2048	128
γ-Carotene	10	1024	64
α-Carotene	9	312	32
Cryptoxanthin	9	512	32
Lutein	9	512	32
Astacene	9	272	20

The theory advanced by Pauling has stood the test of time for rearrangements starting from all-*trans* carotenoids but the 'ban' on hindered forms has been overcome by direct synthesis (Karrer's group: cf. Garbers *et al.*, 1952, 1953, 1954). Hindered forms of vitamin A (retinol) are referred to later (p. 391).

Cis forms of carotenoids are individual substances differing from all-*trans* forms in respect of adsorption, affinity, melting point, rotatory power and absorption spectra. The colour of carotenoids depends in the main on the number of conjugated double bonds. In lycopene there are 11 aliphatic conjugated double bonds whereas in β-carotene the two terminal double bonds are incorporated in β-ionone (cf. cyclohexene) rings. The spectrum of lycopene compared with that of β-carotene shows a marked bathochromic effect. γ-Carotene represents a half-way stage. The vibrational structure (i.e. the sharpness and persistence of narrow bands) is much more pronounced for lycopene than for β-carotene. The *gem*-dimethyl group at position 1 in β-carotene and similar compounds causes interference with the 7,8-double bond, diminishing the fine structure and reducing $\epsilon_{max.}$ to some extent. Introduction of a conjugated carbonyl group (as in astacene) or a conjugated double bond in the ring as in 3,4-dehydro-β-carotene also diminishes spectral resolution.

Zechmeister and his colleagues have shown for numerous carotenoids that formation of *cis*-isomers reduces the intensity of absorption in the visible and may affect the resolution adversely. The peaks are displaced a little in the direction of shorter wavelengths and the change is often written as '$\lambda_{max.}$ shift' expressed in nm. In the case of lycopene (Magoon & Zechmeister, 1957), about 40 clearly defined *cis*-isomers have been recorded. Zechmeister (1962) notes that of these some were obtained by rearrangement of all-*trans* lycopene, others by a partial *cis-trans* rearrangement of the poly-*cis* lycopene, others were isolated from plant products and a few were obtained by stereo specific synthesis. In every case the absorption peaks were on the short wave side of those characteristic of all-*trans* lycopene.

Another important aspect of the *trans-cis* rearrangement concerns selective absorption in the ultraviolet region. In the region 300–380 nm the all-*trans* forms of carotenoids tend to display comparatively weak absorption but the *cis*-isomers exhibit variable *cis*-peaks depending on the carotenoid and the particular *cis*-isomer. In non-polar solvents a solution of all-*trans* lycopene to which a minute amount of iodine had been added produced *cis* absorption at 345 and 360·5 nm on exposure to daylight but not in the dark. The molecular extinction coefficients in that region rose by up to 5400.

This *cis*-peak tends to occur some 141–143 nm on the short wave side of the longest wavelength maximum for the all-*trans* form. Thus for β-carotene the *cis* peak is at 338 nm, for α-carotene at 331 nm, for γ-carotene at 349 nm, for lycopene near 360–362 nm and for ζ-carotene at 296 nm. In zeaxanthin *cis*-isomers the ultraviolet peak occurs at 336 nm, in lutein at 331 nm (all in hexane or light petroleum). The intensity of absorption at the *cis*-peak depends on the particular isomer and does not run parallel with the $\lambda_{max.}$ shift.

Braude & Waight (1954) referred to the main high intensity absorption in the visible as due to the 'full-chromophore' and the *cis*-absorption as due to the 'half-chromophore'. So long as 'half' is not interpreted literally this designation is useful. Lewis & Calvin (1939) introduced the concept of 'partial oscillations' and Mulliken (1939) developed a theory which attached importance to the total length of the molecule; the more elongated the isomer in a conjugated polyene, the greater the intensity of absorption at the longest wavelength transition, such as $N \rightarrow V$, transitions being polarized along the major axis of the molecule. *Per contra* the more *cis*-like the molecule, the weaker the $N \rightarrow V$ spectra. Such ideas were developed by Zechmeister *et al.* (1943) and are discussed in Zechmeister (1962). Dale (1957) further analysed the free electron model for poly-enes.

The intensities for all mono-*cis*-isomers should lie between the values for the central mono-*cis*-forms and the all-*trans* forms. This is confirmed by a large body of information. To a first approximation the area under a *cis* peak varies as the square of the distance between the centre of the conjugated system and the mid-point of the straight line between its two ends. Dale (1954, 1957) proposed a rule to the effect that in a poly-ene with n conjugated double bonds the wavelength of a minor band (including a *cis* peak) is near that of a main band of a poly-ene with s double bonds, s depending on the nature of the isomer. Thus synthetic central mono-*cis*-β-carotene has a very strong *cis* peak ($\epsilon_{max.}$ 43 000) at 338 nm (cf. phytofluene $\lambda_{max.}$ 332, 348 and 367·5 nm in hexane, retinol 325 nm).

In order to establish beyond doubt the *cis-trans*-configuration of a given isomer it is necessary to compare the visible peaks with those of the all-*trans* form, to note $\epsilon_{max.}$ values, and the intensity of the *cis*-peak, especially in relation to that of the synthetic central mono-*cis*-peak. It is also desirable to examine the infrared spectrum in relation to the substantial literature (see Isler, p. 202).

This kind of study is much more than an academic exercise on laboratory artefacts because (*a*) *cis* and poly-*cis*-carotenoids can occur naturally and (*b*) *cis-trans* isomerism applies to many substances other than carotenoids, e.g. unsaturated acids (p. 61) retinol and retinal (p. 384). Most carotenoids however occur naturally as all-*trans* forms but there are some interesting exceptions.

The ripe 'tangerine' tomato is yellow instead of red and Le Rosen *et al.* (1942) obtained from the fresh tomatoes more than 20 mg/kg of a new pigment which, because it tended to replace lycopene, was designated prolycopene. It belongs to a class of poly-*cis* carotenoids of which several examples are now known. No poly-*cis* α- or β-carotenes have been found to occur naturally and indeed nearly all are hydrocarbons related to lycopene in the sense that at least one half the molecule has an open chain structure. Prolycopene crystallizes easily and is unusually heat-stable. The absorption spectrum in hexane shows

$\lambda_{max.}$ 438 nm with inflections at 418 and 465 nm. $\epsilon_{max.}$ at 438 nm is unusually low (*ca.* 103 000). The $\lambda_{max.}$ shift from all-*trans* lycopene 473 → 438 is large but no *cis* peak is displayed. If the fresh solution is left to stand in the presence of a trace of iodine the spectrum changes drastically. There is a large bathochromic shift and the resolution improves, $\epsilon_{max.}$ rises to nearly 150 000 and marked *cis* peaks appear (Fig. 7.1). Stepwise isomerization revealed considerable complexity but very little all-*trans* lycopene was obtained. The infrared spectrum of prolycopene showed clear bands at 7·25 μm and 13·15 μm. (1379 and 760 cm^{-1}), attributed to methylated and unmethylated *cis* double bonds. These bands gradually weakened as isomerization proceeded.

The prolycopene molecule is regarded by Zechmeister (1962) as having 'a straight overall shape and containing four or five *cis* double bonds'. The evidence would fit a penta-*cis* model in which the central 15,15′ double bond and four other sterically unhindered double bonds were *cis*, but the presence of one or two hindered *cis* double bonds cannot be excluded.

7.1.2 Fucoxanthin

This, which occurs in brown algae (Phaeophyceae) and in diatoms, is more plentiful than perhaps any other carotenoid. It was first obtained pure in 1914 (Willstätter and Page). From 1930 onwards the problem of its structure was attacked by Karrer, Kuhn, Heilbron and their colleagues but technical difficulties prevented them from arriving at the right answer, which was in any case a surprising one.

Quite recently new evidence has become available (Jensen, 1961, 1966) and a team led by Weedon (Bonnett *et al.*, 1969) has settled the matter. This work is of interest not only because fucoxanthin is a very important substance, but also because of the nature of the argument.

The infrared absorption spectrum of fucoxanthin displays bands at 957 cm^{-1} (C—H out-of-plane deformation of disubstituted *trans*-olefinic linkages) 3615 cm^{-1} (OH); 1660, 1740 cm^{-1} (one or more C=O groups, conjugated, and either unconjugated C=O or ester groups). There was also a band at 1927 cm^{-1} assigned to allene absorption (cf. Celmer & Solomons, 1953, reporting on mycomycin) and since observed in foliaxanthin (neoxanthin) by Cholnoky *et al.* (1966).

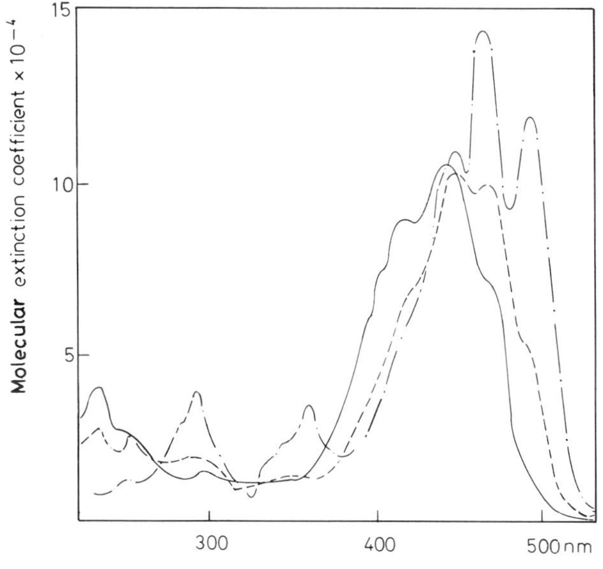

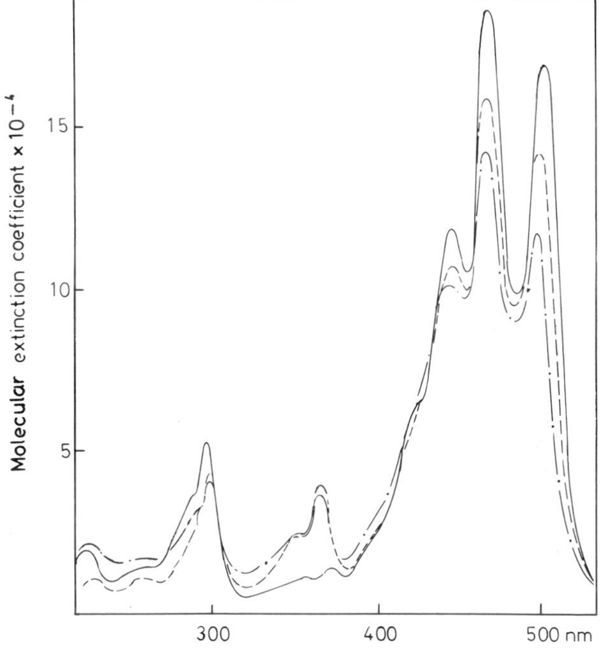

Figure 7.1 (a) (top) Absorption spectra in hexane of : ——— prolycopene; —·—·— prolycopene after iodine catalysis; — — — crystallizable stereoisomer. (b) Absorption spectra in hexane of: – all-*trans* lycopene; --- after refluxing in dark for 45 minutes; —·—· after iodine catalysis at room temperature (after Zechmeister *et al.*, 1943)

The absorption spectrum in the visible region shows a broad band ($\lambda_{max.}$ 450 nm in ethanol) with no fine structure. On reduction (lithium aluminium hydride) fucoxanthin forms a mixture of fucoxanthols with well-resolved peaks (380, 400, 423, 451 nm in ethanol) agreeing in position with those characteristic of conjugated octaenes. Fucoxanthin gives a mono-acetate. Perhydrofucoxanthin, however, gives infrared bands and n.m.r. shifts, indicating that fucoxanthin itself contained one acetate group. The acetoxy grouping had not previously been found in a naturally occurring carotenoid and it seemed that fucoxanthin might contain 42 carbon atoms instead of the usual 40 (Jensen, 1966; Bonnett *et al.*, 1964). Precision mass spectrometry applied for the first time in the carotenoid field gave a molecule ion (M) which corresponded uniquely with $C_{42}H_{58}O_6$. Ions were recorded at M-18, M-18-18, M-18-60 and M-18-18-60, a pattern consistent with the presence of two hydroxy groups and an acetoxy group in the parent fucoxanthin.

Partial oxidation with zinc permanganate gave a complex mixture which was separated by chromatography and the products were studied in respect of ultraviolet absorption before and after reduction with sodium borohydride. Two products, a tetraene and a pentaene dialdehyde, were identified.

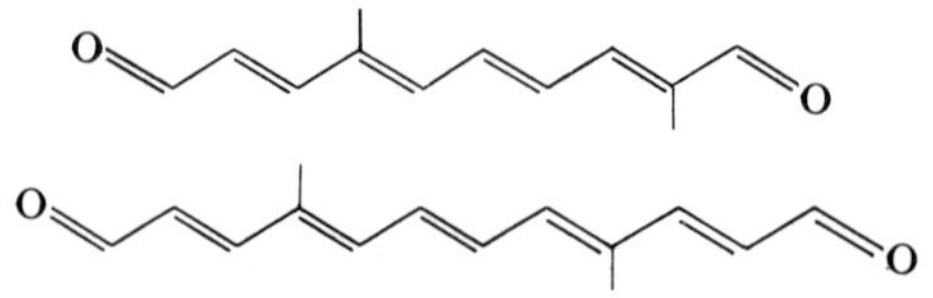

A C_{25} and a C_{27} aldehyde were also isolated and a pair of doublets in the n.m.r. spectra at 6·33 and 7·40 were shown to be due to a methylene group adjacent to the C-8 carbonyl group (Takahashi, 1964) and the 5–6 epoxide was identified largely by exclusion.

Some small amounts of allenes survived partial oxidation, among them the ketone

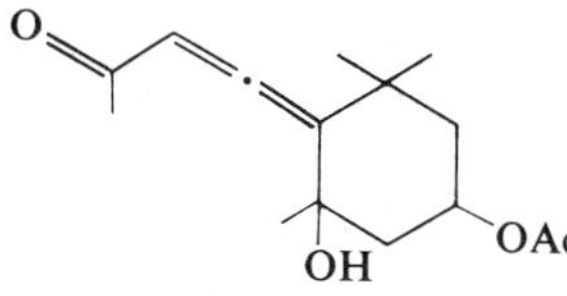

with $\lambda_{max.}$ 225 nm (cm^{-1} 3450; 1945, allene; 1730, ester; 1675, conjugated ketone). The intensity and location of the allene group indicated that it was conjugated with the ketonic group and the ultraviolet absorption was consistent with this view. The n.m.r. spectrum showed for τ 4·24 olefinic proton; 7·9 methyl ketone; 8·04 acetate and 8·59 ($\times 2$), 8·86 c-methyl groups. Mass spectrometry gave $C_{15}H_{22}O_4$ unambiguously and the fragmentation pattern fitted the formula. Hence the position in the molecule of the allene group was made virtually certain. This ketone is emitted when the grasshopper *R. microptera* is disturbed (Meinwald *et al.*, 1968).

In the fucoxanthols the epoxy group persisted but the ketonic group at C-8 was reduced to a secondary alcohol group. The molecular formula $C_{40}H_{58}O_5$ and a prominent ion at m/e 448 (M-170) supported the allotted positions for the keto and epoxy groups in fucoxanthin.

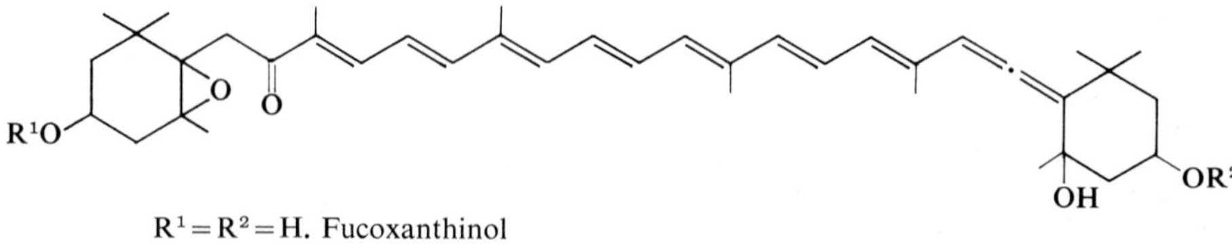

$R^1 = R^2 = H$. Fucoxanthinol
$R^1 = H$ $R^2 = Ac$. Fucoxanthin
$R^1 = Ac$. $R^2 = Ac$. Fucoxanthin acetate

	Fucoxanthinol		Fucoxanthin
$\lambda_{max.}$ nm	452 ϵ $8{\cdot}69 \times 10^4$		508, 478–450 ϵ c 10^{-4}
(ethanol)		(CS_2)	10·7, 13·4
(light petroleum)	476, 448, 425,	(hexane)	476, 450, 420,
max. cm^{-1}	3480, 3450, 1920, 1655, 1608, 978		3615, 1927, 1740, 1660, 957
n.m.r.	τ=9·03, 8·98, 8·92, 8·78, 8·65, 8·18, 8·05, 8·02 and 7·43, 6·33 (doublets)		τ=9·03, 8·87, 8·83, 8·63, 8·43, 8·37, 8·00, 7·98 7·92
m/e	616 ($C_{40}H_{56}O_5$) 598·400 (M-18); $C_{40}H_{54}O_4$ requires 598·402		m/e 658·4233 M $C_{42}H_{58}O_6$ requires 658·4233 640·4113 (M-18) 580·3908 (M-18-60)

There can be no doubt that neoxanthin, with its allene structure, is a very widespread carotenoid. How it arises and whether or not it has special or unique functions remains to be settled by further research.

Allenic carotenoids have also been obtained from sea urchins (*Paracentrotus lividus*) by Galasko *et al.* (1969). Earlier work (Lederer, 1935, 1938) had indicated the presence of carotenoids in the gonads and the coelomic epithelium. The 'xanthophylls' were recently separated by thin layer chromatography. The most strongly adsorbed pigment was identified as fucoxanthinol, previously known only as a product derived in the laboratory from fucoxanthin by reduction to fucoxanthols ($LiAlH_4$), followed by selective oxidation to fucoxanthinol. The isolated sea urchin product gave fucoxanthinol diacetate. Some fucoxanthin was also isolated together with larger amounts of a new allenic ketone designated paracentrone.

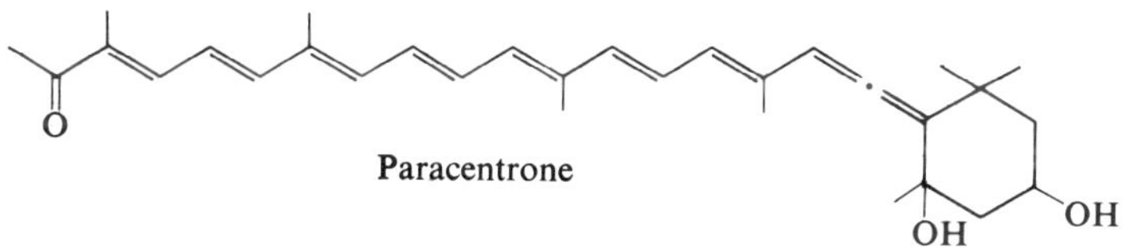

Paracentrone

	Paracentrone	Paracentrone monocetate
$\lambda_{max.}$ nm	480, 456, 437	
$\epsilon_{max.} \times 10^{-4}$	8·47	
$_{max.}$ cm^{-1}	3610, 3420	$_{max.}$ cm^{-1} 3602, 3420
(CCl_4)	1924 (—C=C=C—)	**1923, 1728** $_{max.}$ cm^{-1} (ester C=O)
	1660, 1607 (=C—C=O)	1654, 1600
	1046 (C—O)	1023
	963 (trans CH=CH—)	954
n.m.r.	τ=8·93(3H) 8·66(6H)	
	8·19(3H) 8·06(3H)	
	8·02(6H) 7·65(3H) (methyl ketone)	
m/e	462·313	m/e 504·323
M^+ $C_{31}H_{42}O_3$ requires 462·313		M^+ $C_{33}H_{44}O_4$ requires 504·374

Fragmentation patterns fit above formulae.

7.1.3 Neoxanthin

This is a universal pigment of green leaves known since 1938 (Strain, 1938; Strain *et al.*, 1944; Goodwin & Jamikorn, 1954). Analysis suggested $C_{40}H_{56}O_4$ and Curl & Bailey (1950–1957) favoured a structure containing an epoxy group and three hydroxyls. Goldsmith & Krinsky (1960) advanced the view that one hydroxyl occurred in the same ring as the epoxy group with two hydroxyls in the other ring. Neoxanthin undergoes a hypsochromic change (18 nm) on addition of acid in small amounts; this was interpreted as conversion of a 5,6-epoxide to a 5,8-furanoid oxide.

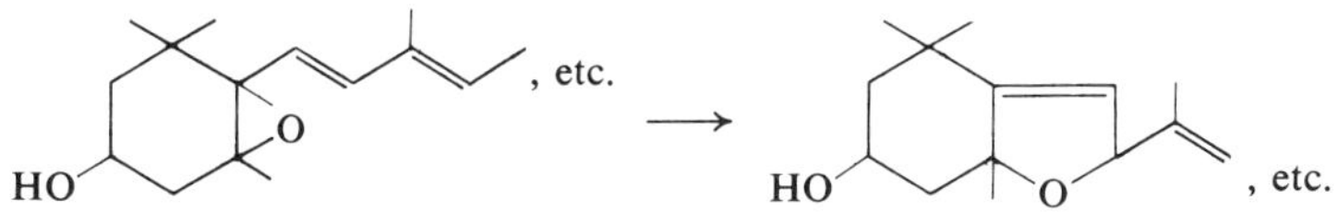

Cholnoky *et al.* (1956) isolated from paprika a pigment which they designated foliaxanthin which turned out later to be the same as neoxanthin (Cholnoky *et al.*, 1966; Mallams *et al.*, 1967; Jensen & Jensen, 1965). The name neoxanthin is retained and the related furanoid oxide is called neochrome (Cholnoky *et al.*, 1969). The visible absorption spectrum of neoxanthin shows well-defined structure consistent with a conjugated nona-ene chromophore. It readily forms a diacetate and on treatment with ethereal hydrogen chloride yields neochrome, which on reduction with lithium aluminium hydride gives zeaxanthin. Cholnoky *et al.* (1967) had shown that epoxides and furanoid oxides on reduction under appropriate conditions yield the parent carotenoid, so that in this case the formation of zeaxanthin established the carbon skeleton and the location of the two acetylatable hydroxyl groups in neochrome (also known as foliachrome).

It is clear from the new evidence that neoxanthin contains the allene group in positions 6′–8′.

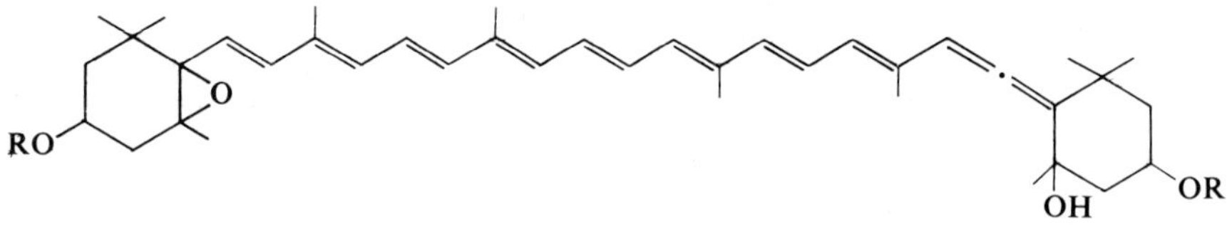

R = H neoxanthin
R = Ac. neoxanthin diacetate

	Neoxanthin	Neoxanthin diacetate
$\lambda_{max.}$ nm	467, 439, 416, (ethanol) 478, 448, 423, (benzene)	478, 448, 423, (benzene)
$\epsilon_{max.} \times 10^{-4}$	1·33, 1·2, 0·93, (ethanol)	
max. cm^{-1}	3597 1923 965	3590, 1923, 1720, 970
m/e	600·419	684
$C_{40}H_{56}O_4$ requires 600·418		$C_{44}H_{60}O_6$ requires 684
582 (M-18), 567 (M-18–15)		
564 (M-18-18), 546 (M-18-18-18)		
43		
(Cholnoky *et al.*, 1969)		

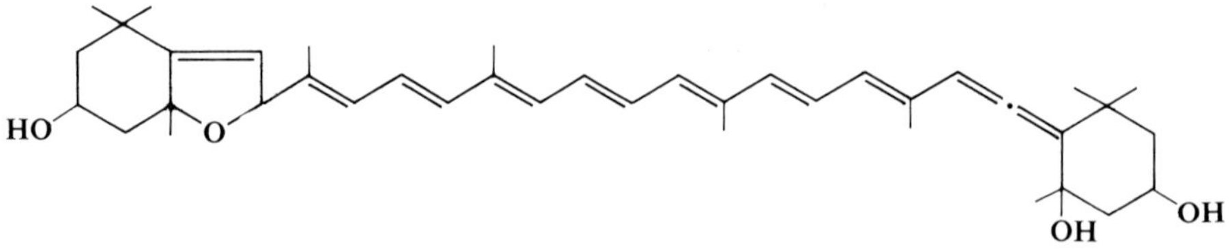

	Neochrome	Zeaxanthin
$\lambda_{max.}$ nm	451, 424, 401 (ethanol)	491, 463, 440 ($CHCl_3$)
	460, 431, 407 (benzene)	517, 482, 450 (CS_2)
max. cm^{-1}	3595, 1923, 970	3608, 965 cm^{-1}
m/e	600·415	($CDCl_3$) 8·94 8·29, 8·06
M	$C_{40}H_{56}O_4$ requires 600·418	m/e 568·426
	582 (M-18), 567 (M-18-15)	M $C_{40}H_{56}O_2$ requires 568·428
	564 (M-18–18), 520 (M-80)	
	508 (M-92), 484 (M-18-18-80)	
	221, 181, 105, 91 and 43	(cf. Loeber *et al.*, 1971)

The study of the compound is complicated by the occurrence of geometrical isomers. It seems however that Strain's neoxanthin and that now obtainable from maple leaves is the all-*trans* form. *Cis*-neoxanthins display a small shift of the absorption maxima on iodine-catalysed stereo mutation (Curl & Bailey, 1957). The nature of the *cis*-isomers has been discussed convincingly (Cholnoky *et al.*, 1969).

7.1.4 Alloxanthin

A new class of carotenoids has been obtained from flagellates of the algal class Cryptophyceae (Haxo & Fork, 1959; Chapman & Haxo, 1963; Allen *et al.*, 1964; Chapman, 1966). The principal member of the group is alloxanthin, $C_{40}H_{52}O_2$, which has been shown to be a diacetylenic analogue of zeaxanthin. The absorption spectrum in the visible is displaced a little in the direction of longer wavelengths compared with zeaxanthin. Bonnett *et al.* (1969) found in alloxanthin a weak band in the infrared at 2167 cm^{-1} due to the acetylenic linkages. The nuclear magnetic resonance spectrum showed all the expected CH_3 shifts but the protons at C-7 and C-8 which in vitamin A give $\tau = 3{\cdot}8$ were absent. The allylic methyl at C-4 gave the normal shift $\tau = 7{\cdot}72$. Two minor congeners were designated monodoxanthin and crocoxanthin respectively.

The giant scallop *Pecten maximus* yields pectenoxanthin and the tunicate *Halcynthia papillosa* gives cynthiaxanthin (Lederer, 1934, 1938; Nishibori, 1960; Tsuchiya & Suzuki, 1960). Campbell *et al.* (1967) showed that both pigments together with another from the mussel (*Mytilus edulis*) are identical with alloxanthin. Thus 'cynthiaxanthin' gave m/e 564·3945, $C_{40}H_{52}O_2$ requires 564·3967, diacetate 648·4204, $C_{44}H_{50}O_4$ requires 648·4178. A new acetylenic carotenoid pectenolone formed a diacetate $\lambda_{max.}$ 463 nm (ethanol). In the infrared region it gave 2175 (C≡C) 1730 (OAc) 1674 (C=O) 967 cm^{-1} (*trans* CH=CH); m/e was 664 and 604·3941, i.e. (M-60) corresponded with $C_{44}H_{56}O_5$ required 604·3916 for M-60.

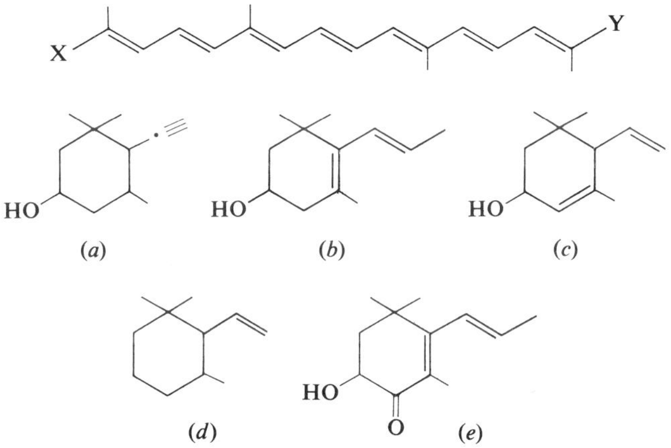

X=Y=(*a*) alloxanthin
X=(*a*), Y=(*c*) monadoxanthin
X=(*a*), Y=(*b*) diatoxanthin
also X=Y=(*a*) cynthiaxanthin, pectenoxanthin
X=(*a*), Y=(*e*) pectenolone
X=(*e*), Y=(*e*) astaxanthin
X=Y=(*b*) zeaxanthin
X=(*a*), Y=(*d*) crocoxanthin

Yamaguchi (1957, 1958, 1960) discovered carotenoids with aromatic end groups in the Japanese sea sponge *Reniera japonica*. A pigment isolated from Mycobacterium phlei (Grundman & Takeda, 1937) and designated leprotene proved to be identical with isorenieratine. Photosynthetic sulphur bacteria were found also to contain 'aromatic' carotenoids including β-isorenieratine, chlorobactene, hydroxy-chlorobactene and okenone (Cooper *et al.*, 1963; Liaaen-Jensen, 1965; Liaaen-Jensen *et al.*, 1964).

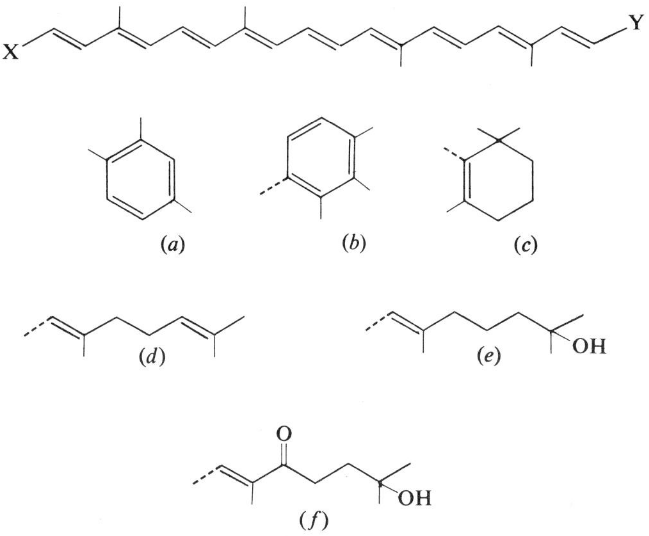

X=Y=(*a*) isorenieratine or leprotene
X=(*a*), Y=(*b*) renieratine
X=Y=(*b*) renierapurpurin
X=(*a*), Y=(*c*) β-isorenieratine
X=(*a*), Y=(*d*) chlorabactene
X=(*a*), Y=(*e*) hydroxychlorobactene
X=(*b*), Y=(*f*) okenone

The aromatic methyls can be detected by n.m.r. spectroscopy (Weedon, 1965), giving rise to τ 7·37, 7·73, 7·78 and the OMe to 6·84. The difference between 1,2,5- and 1,2,3-trimethyl substitution is reflected in the visible absorption spectra.

7.1.5 Loroxanthin

A new carotenoid loroxanthin has been obtained by Aizetmüller *et al.*, 1969. It was found in the green algae *Scenedesmus obliqus* and *Chlorella vulgaris* and is related to lutein, the C19 methyl group having been converted to CH_2OH (cf. Walton *et al.*, 1970). It forms a triacetate and with methanol and hydrogen chloride mono- and dimethyl ethers. The absorption spectrum in the visible is almost identical with that of lutein (solvent, ethanol). The mass spectrum indicates a molecular weight of 584·4233 ($C_{40}H_{56}O_3$). The infrared absorption (KBr disc) shows cm^{-1} 3380 (associated hydroxyl), 1040, 1023 (one non-allylic secondary alcohol group and two allylic hydroxyls), 1570 (conjugated C=C) 967 (*trans* CHR=CHR), 831 ($CHR=CR_2$). The n.m.r. CH_3 values include 8·15+8·17 (2 or 3) 8·25, 8·30, 8·45, 9·03, 9·20 (one each) and 8·97 (2) (of lutein). Fully deuterated loroxanthin $C_{40}D_{53}(OH)_3$ and $C_{40}H_{53}(OD)_3$ gave the expected molecular weights as did the triacetate and the methyl ethers. The location of the 'in chain' CH_2OH was decided partly by the fragmentation pattern of loroxanthin itself and also by the mass fragmentation of the triacetate.

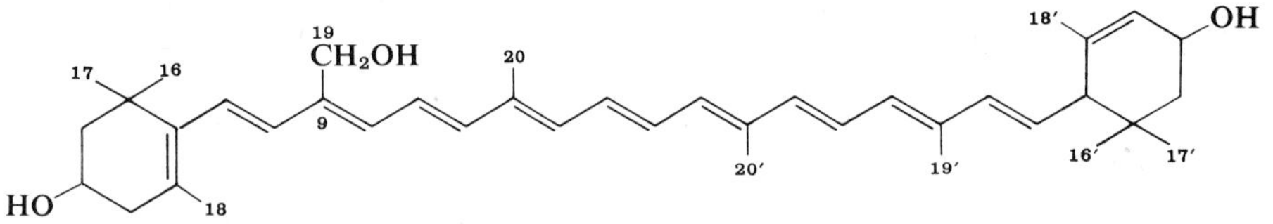

Loroxanthin (7-hydroxylutein)

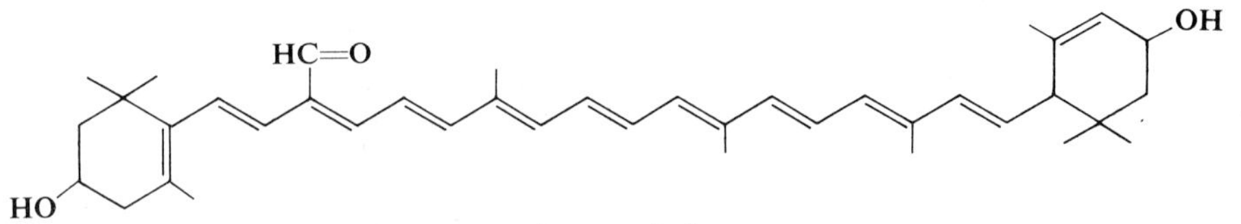

Loroxanthal

Loroxanthin is readily oxidized to loroxanthal by converting CH_2OH to CHO. It shows $\lambda_{max.}$ 475 nm (inflection 500 nm) in light petroleum. The mass spectrum, infrared absorption and τ values confirm the structure.

7.1.6 Oscillaxanthin

Oscillaxanthin is another carotenoid of blue-green algae. The structure has been established as related to that of 3,4,3′,4′-tetradehydrolycopene which would endow it with 15 conjugated double bonds, but the end groups are reduced to give a 1,2,1′,2′-tetrahydro-3,4,3′,4′-dehydrolycopene with 13 conjugated double bonds. Glycoside hydrolysis showed the presence of two L-rhamnosyl moieties. The hydroxyls of the sugar residues were acetylated to give a per-acetate, the molecular weight of which from the mass spectrum was 1144 ($C_{64}H_{88}O_{18}$). The tetrol (oscillol) is obtained in glycoside hydrolysis. Hertzberg & Liaaen-Jensen (1969) and Francis *et al.* (1970) described a new product similar to oscillaxanthin which is actually oscillol-2,2′-di(O-methyl-methylpentoside). The combination of modern techniques is here very effective, spectroscopy being able to identify the chromophore and the mass fragmentation patterns helping to identify the sugar residues.

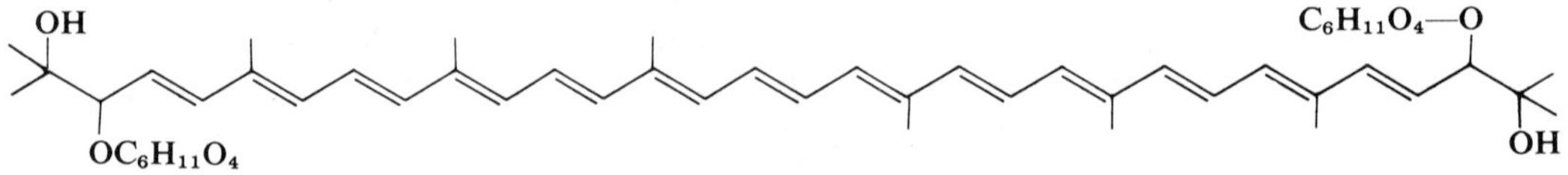

Oscillaxanthin

Spirilloxanthin (rhodoviolascin) was the earliest carotenoid with a methoxyl group to be characterized (Karrer & Solmssen, 1935, 1936; Karrer *et al.*, 1938) and the position of the methoxyls was made probable by interpreting absorption spectra and confirmed by synthesis (Surmates & Ofner, 1963) and the structure of a monomethyl analogue was ascertained on the basis of infrared absorption followed by total synthesis (Schneider & Weedon, 1967). Spirilloxanthin has 13 conjugated double bonds and the synthetic product (in $CHCl_3$) has peaks at 397, 480, 510 and 548 nm and 3,4-dehydrolycopene has peaks at 373, 390, 470, 499 and 533 nm while oscillaxanthin has 397, 480, 510 and 548 nm (Hertzberg & Liaaen-Jensen, 1969).

OCH_3 OCH_3

Spirilloxanthin

7.1.7 Rubixanthin

This substance was first isolated from *Rosa rubiginosa* by Kuhn & Grundman (1934). Its absorption spectrum in acetone shows a very intense main peak at 462 nm (ϵ 160 600) with subsidiary peaks at 439 and 492 nm (Liaaen-Jensen *et al.*, 1964). Gazaniaxanthin (from *Gazania rigens*) is very like rubixanthin and has been obtained from various sources (see Zechmeister & Schroeder, 1943). It also has $\lambda_{max.}$ 462 nm in acetone but $\epsilon_{max.}$ is lower (142 400) only to be raised by 5% when hexane is the solvent. Brown & Weedon (1968) concluded that both these carotenoids had the structure shown below, but while the rubixanthin is all-*trans*, gazaniaxanthin is the 5′,6′ *cis* isomer.

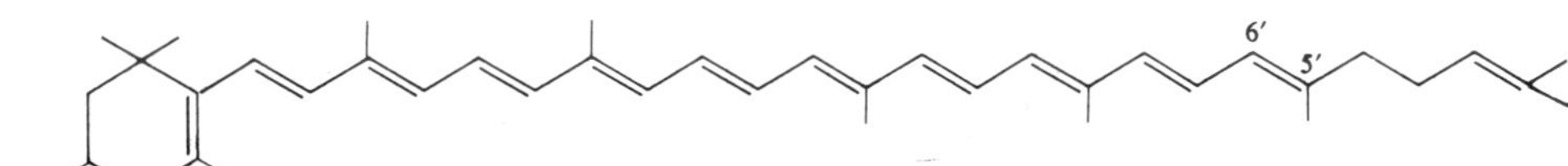

Rubixanthin all-*trans*
Gazaniaxanthin 5′6′ *cis*

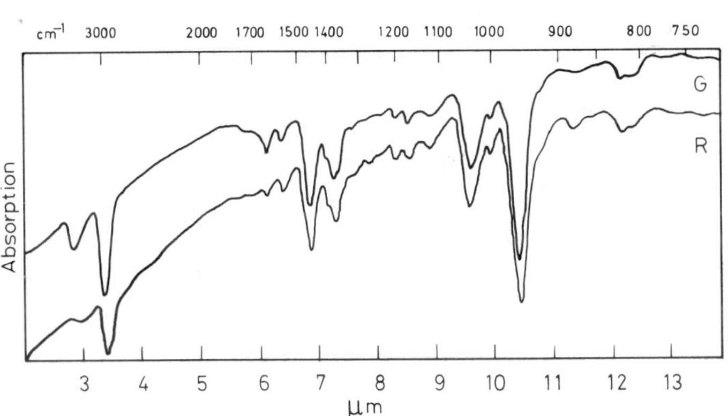

Figure 7.2 Infrared spectra of rubixanthin (R) and gazaniaxanthin (G) in KBr-pellet (from Arpin & Liaaen-Jensen, 1969)

Catalysis by iodine isomerizes the latter to the former and after chromatography rubixanthin can be obtained with the correct melting point and absorption spectrum. Arpin & Liaaen-Jensen (1969) confirmed the structures using mass spectra, infrared absorption and n.m.r. The latter (see Brown & Weedon) indicated small differences at τ 7·9 attributed to the allylic CH_2 group. The original literature should be consulted for interpretation of n.m.r. using other substances for comparisons. The infrared spectra are nearly but not quite identical (see Fig. 7.2).

The woody nightshade *Solanum dulcamara* contains lycoxanthin and lycophyll which were formerly believed

HOH_2C 16 15 15′ 6′ 5′ 16′

Lycoxanthin

to be the 3-hydroxy and 3,3′-dihydroxy derivatives of lycopene but are now known (Markham & Liaaen-Jensen, 1968; Cholnoky *et al.*, 1968) to be the 16-ol and the 16,16′-diol respectively. The spectroscopic evidence put the presence of a terminal CH_2OH group beyond doubt. The ripe berries of *Shepherdia canadensis* contain as the main carotenoids lycopene and methyl-apo-6′-lycopenoate. [Lycopene $\lambda_{max.}$ (in acetone) 446, 472, 504 nm, $E^{1\%}_{1cm}$ 472 nm, 2840, $\epsilon_{max.}$ 152 000; methyl-apo-6′-lycopenoate 448, 471, 503 nm, ϵ_{471} 122 000 or 107 000 in light petroleum.]

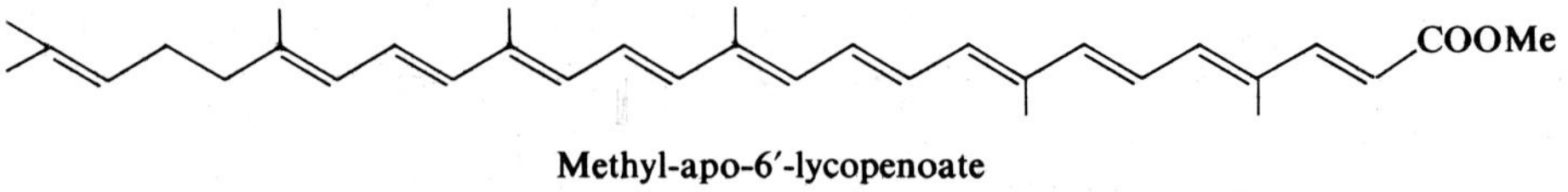

COOMe

Methyl-apo-6′-lycopenoate

7.1.8 Myxoxanthophyll

This was first isolated by Heilbron & Lythgoe (1936) and is characteristic of blue-green algae. Hertzberg & Liaaen-Jensen (1967, 1969) describe it as a mixed (1′,2′-dihydro-3′,4′-didehydro-3,1′-dihydroxy-γ-2′yl)-glycoside in which rhamnose is the predominant sugar moiety and a hexose is the minor component. The mass spectrum corresponded with m/e 730·4852 ($C_{46}H_{66}O_7$) for myxoxanthin; losses of 92 and 106 mass units were characteristic and a fragment m/e 566·4114 indicated loss of $C_6H_{12}O_5$. The parent substance showed $\lambda_{max.}$ 478 nm, $\epsilon_{max.}$ 15 800 with subsidiary peaks at 450 and 510 nm in acetone, and 462, 488 and 452 nm in benzene. The structure shown below is consistent with the ultraviolet and infrared absorption, the mass spectrum and the n.m.r. data. Phleixanthophyll (from *Mycobacterium phlei*) is a tertiary D-glucoside which on hydrolysis loses the sugar moiety with introduction of a double bond to yield 3,4-didehydrotorulene. A 4-ketophleixanthophyll is also present in *M. phlei.* Both phleixanthophyll and the 4 keto derivatives yield tetra-acetates, which are separable. As would be expected from the structures, the absorption spectra of myxoxanthophyll and phleixanthophyll in the visible region are almost identical.

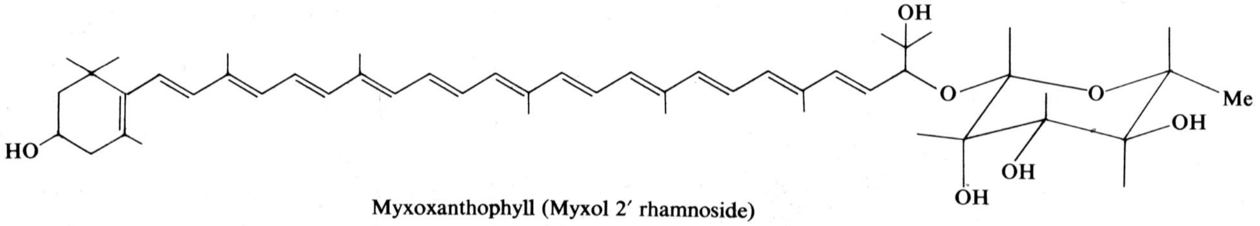

OH HO O O Me OH OH OH

Myxoxanthophyll (Myxol 2′ rhamnoside)

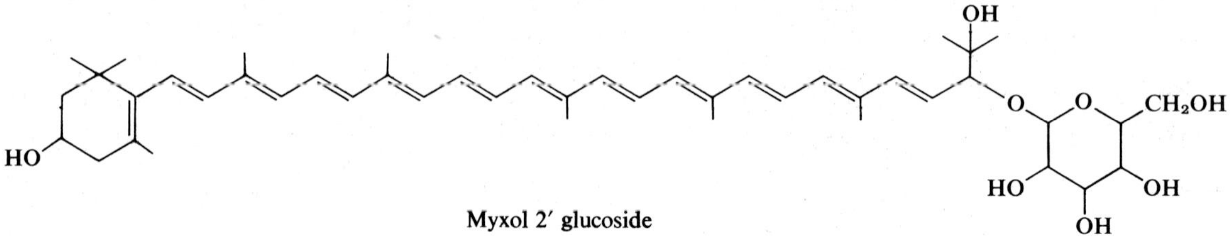

OH HO O O CH_2OH HO OH OH

Myxol 2′ glucoside

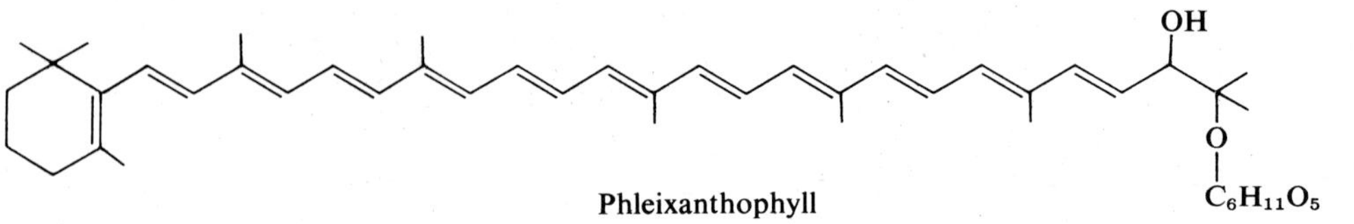

OH O $C_6H_{11}O_5$

Phleixanthophyll

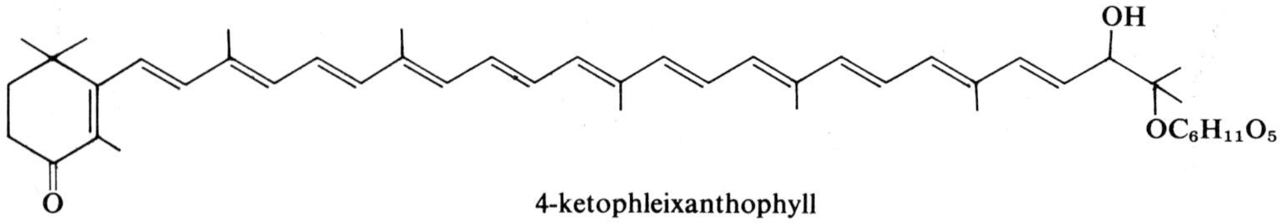

OH $OC_6H_{11}O_5$ O

4-ketophleixanthophyll

The carotenoid field continues to produce surprises. Thus the gram-positive *Flavobacterium dehydrogenans* grown under illumination has as its principal carotenoid a C_{50} compound made up of a familiar C_{40} unit with C_5 units at positions 1 and 1′ (Weeks, 1966; Jensen & Weeks, 1966; Weeks & Garner, 1967). The compound, 1,1′-di(3-hydroxy-methyl-but-2-en-yl)-ε-carotene or decaprenoxanthin is one of several carotenoids produced under different cultural conditions.

When the organism was grown on a synthetic medium in the light with continuous shaking, decaprenoxanthin accounted for 99·9% of the carotenoid but the total was not great. When the medium also contained a little asparagine (0·1%) and yeast extract (Difco, 0·01%) the total carotenoid increased 13·3-fold but decaprenoxanthin accounted for 99%, the remainder being 0·5% each of neurosporene and a complex P420. When the yeast extract was raised to 0·05% the total carotenoid reached about 24 times the original amount, but the major carotenoids were now phytoene (56·7%), decaprenoxanthin (36·4%), phytofluene (4·8%) with small amounts of others. Raising the amount of yeast extract to 0·1% reduced the total but left the proportions similar: phytoene (60·2%), decaprenoxanthin (32·6%), phytofluene (5%). Addition of diphenylamine to the medium reduced the yield substantially. The proportions were phytoene (46%), phytofluene (6%), P420 complex (12%), decaprenoxanthin (35%). When a nutrient broth was used, growth was only twice the original amount (synthetic medium) but the principal carotenoids were phytoene (39·6%), phytofluene (7·6%), ζ-carotene (2·8%). P420 complex (9·4%), decaprenoxanthin (38·9%).

The further investigation (Jensen, 1967; Jensen *et al.*, 1968; Weeks *et al.*, 1969) has been greatly assisted by mass spectroscopy and absorption spectra. The structure of decaprenoxanthin is shown below.

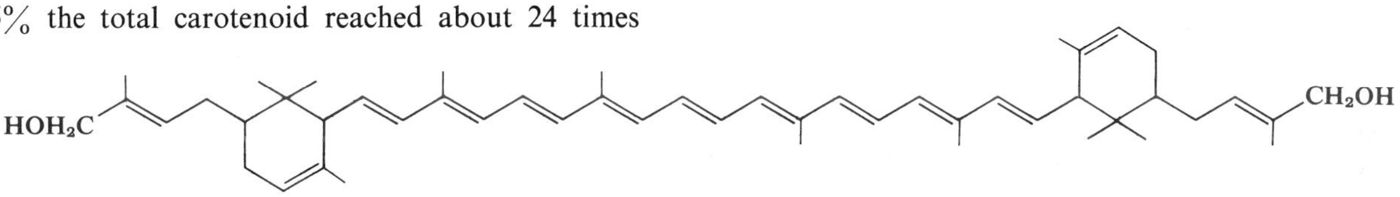

Decaprenoxanthin

The absorption spectrum is consistent with 9 conjugated double bonds and the molecular formula with 4 unconjugated double bonds.

High resolution mass spectra determined in Weedon's laboratory by Weeks *et al.* (1969) established the behaviour of authentic phytoene, phytofluene, ζ-carotene, neurosporene and lycopene. Comparisons with the compounds isolated from *F. dehydrogenans* established beyond doubt the identity of the latter. N.M.R. data on phytoene from this source gave signals at τ 8·43 8·36 and 8·27 consistent with (*a*) six methyls *cis* to chain methyls; (*b*) two methyl groups *trans* to chain methyls and two end-of-chain methyls. Allylic-CH_2-groups gave signals at τ 8·03 and 7·95 integrated to about 25 protons. These data corresponded exactly with those of authentic phytoene.

The chromatographic fraction P420 was characterized by mass spectra into P373, P422 and monohydroxy P439 (Weeks & Garner, 1967). P373 turned out to be III, a C_{45} derivative of δ-carotene, nonaprenaxanthin. One half of the molecule is identical with half the symmetrical decaprenaxanthin molecule and the other half is identical with a phytoene moiety.

[P373 (III) molecular ion m/e 624–528 ($C_{45}H_{68}O$); m/e 484 (624·528–140); m/e 419 (624·528–205) fission at 11′,12′ (phytoene phytofluene and ζ-carotene, 7,8,11,12-tetrahydrolycopene fragment at 11, 12 or 11′, 12′). The chromophore extends from 7,8 to 13′, 14′, 6 conjugated double bonds, ($\lambda\lambda_{max.}$ all *trans* ~337, 354, 373 and 394 nm in hexane)]. The compound can be written 2-(3-hydroxymethyl-but-2-enyl) 7,8′, 11′,12′-tetrahydro-ε-ψ-carotene, nonaprenaxanthin. It shows a bright yellow fluorescence under ultraviolet illumination.

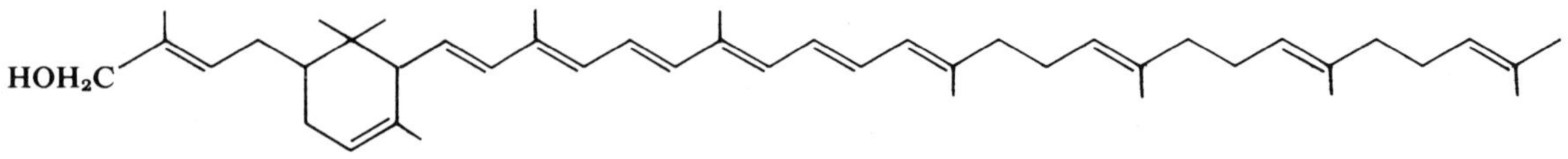

HOH₂C

Nonaprenoxanthin (III)

P422 shows an absorption spectrum corresponding with 8 conjugated double bonds. The molecular ion 622·511 corresponds to $C_{45}H_{66}O$ and the structure is made up of one half of decaprenoxanthin and half of ψ-carotene (or the distal terminal part of neurosporene. $\lambda_{max.}$ all *trans* ~377, 399, 422, 449 nm in hexane. The fragmentation supports this 622·5–140=482; M-137=485 fission at 7′,8′.

P439 (monohydroxy) is decaprenoxanthin in which one CH_2OH has been replaced by CH_3. Fission at 2,3,16 gives a fragment 140 and at 2′,3′, 5′,6′ a fragment 124. The molecular ion is m/e 688·565 ($C_{50}H_{72}O$) and m/e values at 548·5 and 564·5 are shown.

The absorption peaks at 265, ~394, 414, 438 and 447 nm in hexane are as expected the same as those of decaprenaxanthin.

It is perhaps worthwhile at this stage to recapitulate a little to put the physical properties of carotenoids in perspective. The absorption spectra in the visible and ultraviolet regions are determined by several factors: (i) the number of conjugated double bonds (including the carbonyl group), (2) the stereochemical configuration, (3) the additive effect of separated chromophoric groups and (4) the possibility of partial chromophores. Finally, (5) the nature of the solvent is a significant variable.

Numerous attempts have been made at generalizing the spectroscopic properties of poly-enes. Some have used the longest wavelength peak of all-*trans* poly-enes. Thus Lewis & Calvin (1939) found that $\lambda^2 = kn$ for poly-enes where $n < 6$ but Nayler & Whiting (1955) felt that even for dimethyl poly-enes no wholly satisfactory equation had been proposed. Braude (1950) using the oscillator strength

$$f = 4{\cdot}3 \times 10^{-9} \int \epsilon d\bar{\nu} \quad \text{(p. 137)}$$

found that $\epsilon_{max.}$ was roughly 20 000 n.

For the diphenylpoly-enes the Lewis–Calvin equation was modified to $(\lambda_{max.})^2 = A + Bn$ and a linear relationship was indeed observed for the $n=1$ to $n=7$ compounds. Hirayama (1955) modified the equation further to take the form:

$$(\lambda_{max.})^2 = A - BC^N$$

where N is related to n and to the groups present in the series of molecules. Jaffé & Orchin (1962) summarized the position concerning the spectra of poly-enes with full references to theoretical treatment on the basis of the free electron model.

Hirayama (1955) used the equation to cover the positions on the wavelength scale of the main peaks of all-*trans* poly-enes

$$(\lambda_{max.}\ \text{nm})^2 = (A - B \times 0{\cdot}92^N)10^4\ \text{nm}^2$$

where A and B are constants (depending on the type of poly-ene and the solvent used in measuring spectra) and N is determined by the number of conjugated double bonds and the mode of substitution, i.e. $N = n_1 + n_2 + n_3$ obtained from Table 7.3 in which one conjugated double bond is given a value of unity.

Vetter *et al.* (1971) pointed out that the formula when applied for example to a poly-ene of infinite length leads to a value of 608 nm. Though imperfect, the equation does make allowance for the fact that

Table 7.3 Hirayama's empirical constants.

Type of compound	A	B	Solvent
Poly-ene	36·62	38·72	Hexane
	36·98	39·10	Ethanol
	38·60	40·72	Chloroform
α-carbonyl poly-ene	39·78	39·33	Light petroleum
αα′-dicarbonyl-poly-ene	37·69	34·67	Light petroleum

Increments (n) for substituents and conjugated double bonds

	n
One conjugated double bond	+1·0
Alkyl or hydroxyalkyl	+0·1
Carboxyl group	+0·3
Furanoid oxide	0
αβ-Epoxy alkyl	+0·2
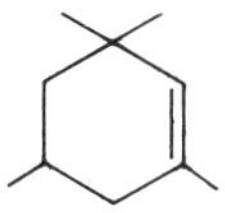	−0·8
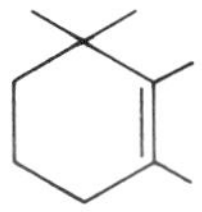	−1·0
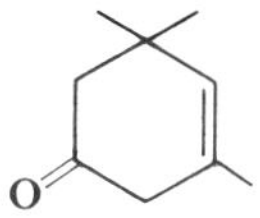 O	−0·9
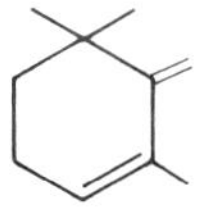	−0·6
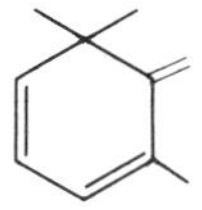	−0·2
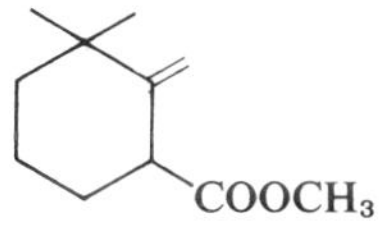 $COOCH_3$	−0·4

not all the conjugated double bonds are fully effective, e.g. the double bond in a β-end group. Thus both β-carotene and lycopene possess 11 conjugated double bonds, but in the former two are in the β-ionyl rings. The peaks (in light petroleum) are: for β-carotene ~425, 449, 475 nm and for lycopene 445, 470 and 505 nm; γ-carotene falls, as expected from its constitution, between the two. Similarly ϵ-carotene, δ-carotene and lycopene have 9, 10 and 11 conjugated double bonds and the principal peak moves from 438 to 456 to 470 nm with rising values for $\epsilon_{max.}$ (see Table 7.4).

In comparing β-carotene with longer chain homologues the following results are obtained:

	Conjugated double bonds	$\lambda_{max.}$ nm
β-carotene	9 + 2 (in end groups)	272, ~349, 425, 449, 470
Decapreno-β-carotene	13 + 2	325, ~400, 470, 497, 530
Dodecapreno-β-carotene	17 + 2	370, ~425, 438, 505, 533, 568

The complexity of the spectra raises problems. For instance β-carotene in the all-*trans* form has an absorption band between 250 and 300 nm, $\lambda_{max.}$ *ca.* 272 nm, $\epsilon_{max.}$ 2500 (in light petroleum). There is also an inflection consistent with a weak partially masked peak at about 345 nm. β-Carotene is obtainable as all-*trans* in a state of high purity which precludes attributing the bands to *cis*-impurities. In the case of dodecapreno-β-carotene the spectrum exhibits superimposed on the three banded absorption 505, 533, 568 nm, another group of three bands 420, 438 (and *ca.* 459 masked) nm, and a third group 355, 370 (and perhaps 385 masked) nm with feeble peaks near 324 and 265 nm.

Dale (1954, 1957) argued that the minor bands, corresponding to transitions to higher unoccupied orbitals, are overtones. In an aliphatic poly-ene with n-conjugated double bonds, the maximum of the λ_x band will be very near to the peak of the λ_1 band of a poly-ene with n/x double bonds and the λ_2 band would correspond with the λ_1 band of a poly-ene with $n/2$ conjugated double bonds. Schwieter *et al.* (1969) plotted n against λ for dodecapreno-β-carotene using $n=18$ for the long wave absorption (17+2 β-end groups counted as one). The points lay on a smooth curve in which for $n/2$, $\lambda=438$ nm; for $n/3$, $\lambda=370$ nm; for $n/4$, $\lambda=324$ nm and for $n/6$, $\lambda=281$ nm. The overtones of even order should according to Dale be forbidden in linear all-*trans* compounds but allowed in angular molecules such as *cis* isomers. It could be argued that at the great dilutions used in measuring the spectra *cis*-isomers could be formed to reach an equilibrium. If such a view is unacceptable it becomes necessary to note that $\epsilon_{max.}$ for the long wavelength bands of dodecaprene-β-carotene is *lower*

Table 7.4 Absorption maxima of selected carotenoids.
(Solvent is light petroleum unless otherwise stated)

	End groups	$\lambda_{max.}$ nm	$\epsilon_{max.} \times 10^{-5}$
α-Carotene	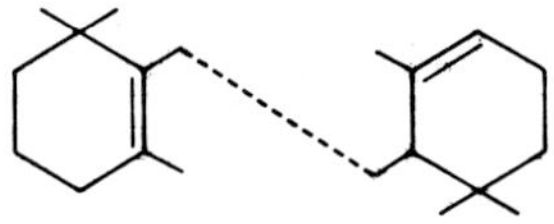	422 444 473	1·018 1·501 1·351
β-Carotene	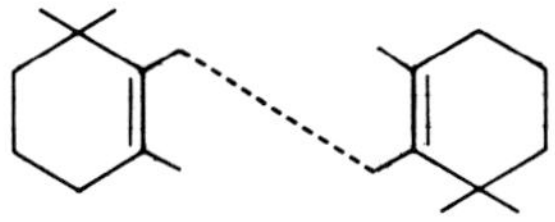	273 453 481	0·205 1·389 1·215
γ-Carotene	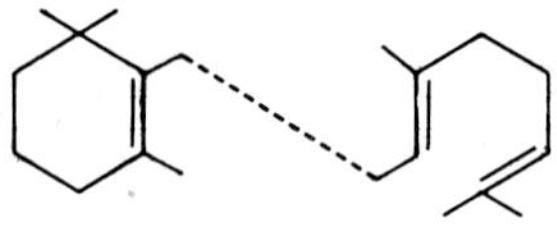	437 462 494	1·015 1·662 1·458
δ-Carotene	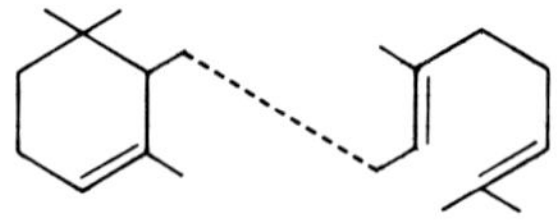	281 431 456 489	0·448 1·093 1·763 1·562
ε-Carotene	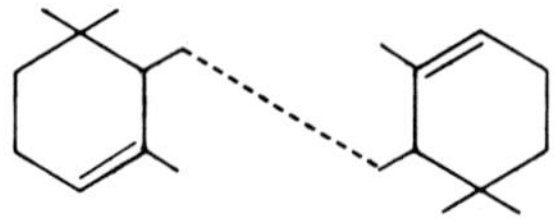	266 416 440 470	0·361 1·040 1·672 1·680
Lycopene	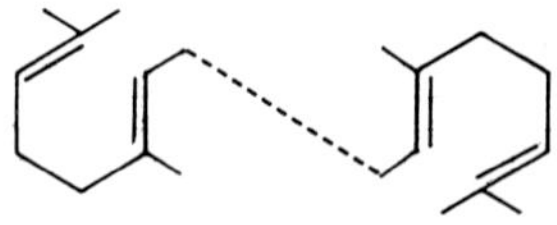	446 472 505	1·206 1·850 1·690
Bis-dehydro-β-carotene	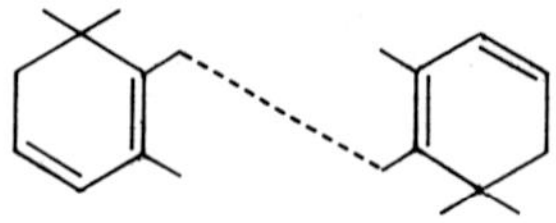	471	1·280
β-Zeacarotene (7′,8′-dihydro-γ-carotene)	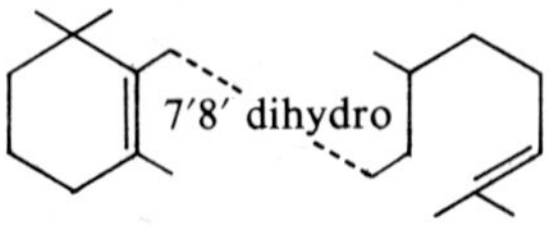 7′8′ dihydro	406 428 454	0·893 1·360 1·240
Lutein (xanthophyll)	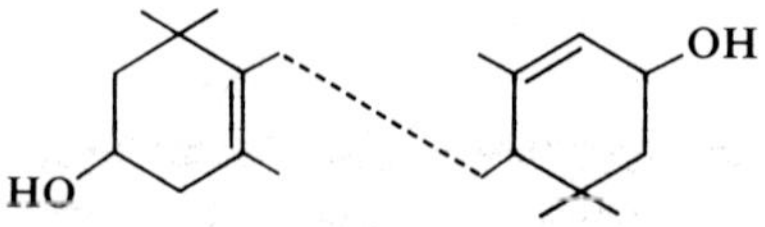 HO, OH	420 448 478	— — —
Zeaxanthin	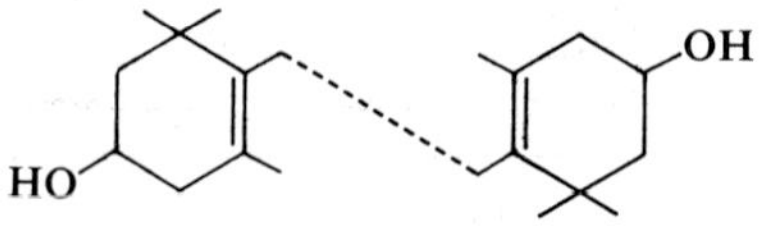 HO, OH	452 480	1·335 1·185
Cryptoxanthin	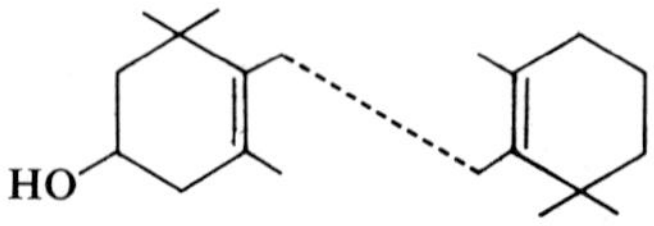 HO	452 480	1·308 1·096
β-Carotene-epoxide	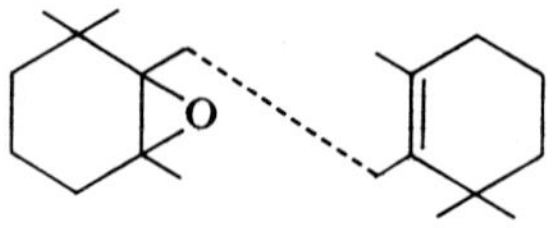 O	447 478	— —

Table 7.4 *continued*

	End groups	$\lambda_{max.}$ nm	$\epsilon_{max.} \times 10^{-5}$
β-Carotenediepoxide	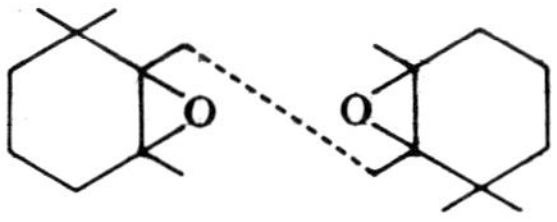 O O	443 471	— —
Citroxanthin	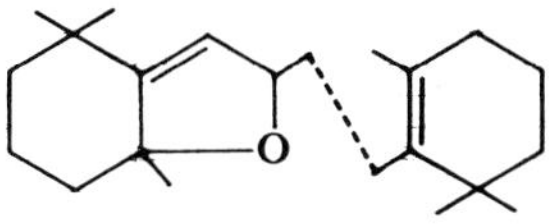 O	427 450	— —
Echinenone	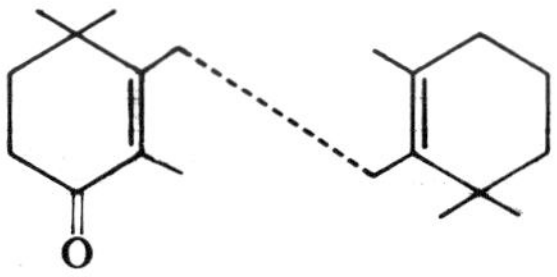 O	472	1·122
Canthaxanthin	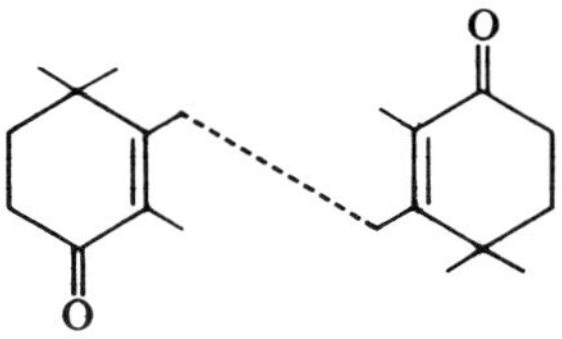 O O	466	Benzene 1·241
Astaxanthin	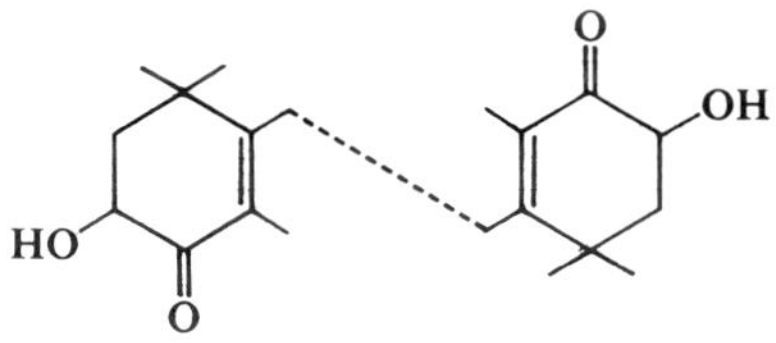 O OH HO O	476 493 513	In Pyridine — —
Torularhodin	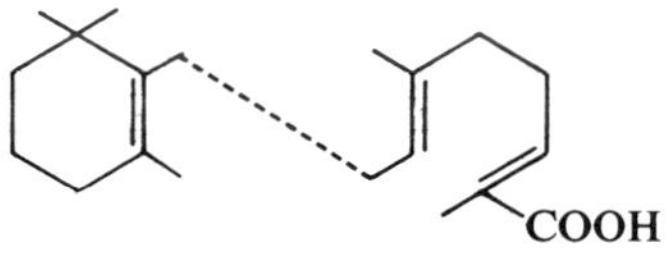 COOH	507	1·151
	Structures		
Crocetin (8,8′-diapocarotene-8,8′-dioic acid)	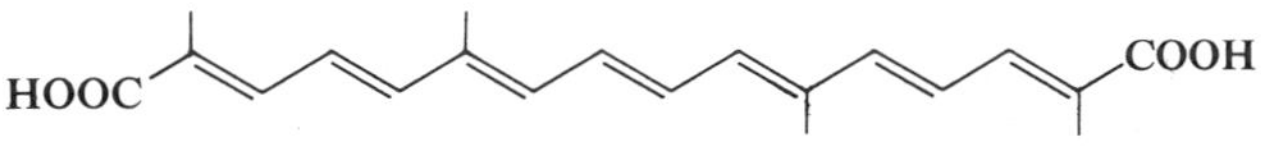 HOOC COOH	411 436 464	0·942 1·466 1·414
Bixin	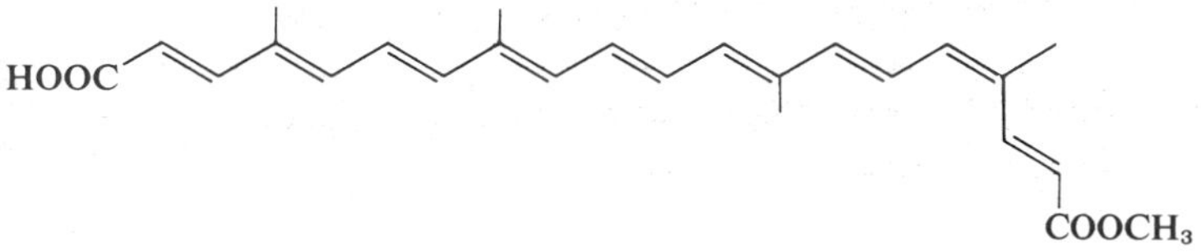 HOOC $COOCH_3$	427 452 484	— — —

than that for decapreno-β-carotene and that with the exception of a peak at 325 nm subsidiary bands in the latter are less developed (or more fully hidden). These results suggest that in the dodeca compound only some of the molecules have the conjugated double bond system fully effective ($\epsilon_{max.}$ is *ca.* 175 000 and could be expected to be significantly higher). Portions of the remaining molecules may exhibit the electronic transitions for a nona-ene, a hexa-ene and a tri-ene.

Zechmeister carried out many studies on *cis-trans* isomerism in the carotenoid family and summarized his findings in a valuable book (1962).

The absorption spectrum of a fresh solution of all-*trans* β-carotene after refluxing showed falls in intensity of absorption at 450 nm and at 275 nm and a rise in absorption at 340 nm. Catalysis by iodine enhanced the change. The peak at 340 nm was designated a *cis* peak. The absorption behaviour of lycopene is a good example in that the main band together with partially resolved absorption near 295 nm decreases in intensity as *cis* isomers appear, while resolved absorption at 350–360 nm becomes stronger. β-Carotene isomers make up a 'set' of pure substances, the 15-*cis* form among them. Its long wavelength absorption shows a decrease in intensity from 155 000 (for all-*trans*) to 92 300 and at 275 from 24 000 to 13 400, while the 340 nm absorption rises from < 10 000 to 52 500. This new absorption is approximately what would be expected in location and intensity for a penta-ene, accepting that if 40% of the molecules fail to display the full conjugation effect they may well display twice the effect of a penta-ene. The 9-*cis* isomeride does not show the absorption to be expected from the angular structure at position 10; there was a smaller fall in intensity for the visible absorption and a very small rise at 340 nm (Jaffé & Orchin, 1962).

Zechmeister (1962) discussed the complicated problem of the numerous naturally occurring (and synthetic) stereo-isomeric forms of lycopene. Prolycopene, which is obtainable in a pure state (Le Rosen & Zechmeister, 1942), shows a highly anomalous spectrum (see p. 48).

There remain aspects of carotenoid absorption spectra that are not understood. The 272 nm band of β-carotene belongs to the all-*trans* form just as much as the visible bands, while the 340 nm band with its varying intensity is characteristic not of one *cis* isomer only but of several. The precise effects of di-, tri- or poly-*cis* isomerism are not clear. It may well be that much of the absorption exerted by small moieties is masked and that the effects of dienes, trienes and tetraene chromophores continue to be speculative. Whether the idea of 'overtones' adds much to the concept of partial chromophores remains to be seen. The general position is however rather more tidy for *cis-trans* isomerism in the vitamin A field.

A main contribution of infrared spectroscopy to the advancement of carotenoid studies has been in the recognition of allene groups (e.g. the 1928 (or 1956) cm^{-1} peak in fucoxanthin) and of acetylenic groups. Alloxanthin, with disubstituted acetylenic groups at the 7,8 and 7′,8′ positions, shows a sharp peak at 2170 cm^{-1}. Zechmeister (1962) concluded that the *trans* grouping C—CH=CH—C gave rise to a sharp singlet at 965 cm^{-1} and that in the central mono-*cis* isomer of β-carotene this peak was weakened, but a doublet 966 and 955 cm^{-1} was present. The double peak also occurs in retro compounds. *Cis*-compounds possess a peak at 779 cm^{-1}. A tri-substituted double bond in the *cis* form often shows peaks at 830 and 1380 cm^{-1}. It is useful to distinguish methylated and unmethylated ethylenic linkages.

Mass spectrometry in the carotenoid field has given complicated results but the voluminous literature has been digested and assembled in a very useful way (Schwieter *et al.*, 1969; Vetter *et al.*, 1971). Appropriate methods of introducing the sample into a high resolution mass spectrometer allow in almost all cases the molecular weight to be ascertained accurately enough to fix the elemental composition. Unsubstituted carotenes show strong molecular ion peaks at 12 eV, and at 70 eV a very large number of fragments with m/e < 250 are seen. Carotenoids with substituted end groups have the disadvantage that the molecular ion peak is often missing or only feebly indicated. Complexities arise from thermal decomposition of the sample and from fragmentation via electron impact. The temperature variable is very significant for the carotenoids because if the vapour pressure is adequate the stability will be low. The best results are obtained when the ion-chamber is kept at 250°C and the heated sample reaches at least 200°C. The fragmentation at 70 eV frequently results in so many peaks that multistage processes must be postulated and structural interpretation then becomes difficult. In the case of β-carotene (M.W.536) there are notable peaks at 567, 444, 430 and 378. The peaks m/e 444 (M-92), m/e 430 (M-106) and m/e 378 (M-158) represent the loss of toluene, xylene and dimethylnaphthalene.

Kuhn & Winterstein (1933) had studied the pyrolysis at 200–300°C of carotenoids *in vacuo* and had found among the volatile products toluene, *m*-xylene (perhaps traces of *o*-xylene), 2,6-dimethylnaphthalene and *inter alia m*-toluic acid from carotenoids possessing carboxylic groups. Ionene was obtained from β-carotene (Mader, 1964). From such work it follows that thermal degradation must be a major factor in mass spectrometry. The aromatic compounds arise from the poly-ene chain, and Enzell *et al.* (1968 *a*, *b*) noted that they appeared whenever the carotenoid had 8 or more conjugated double bonds. The ratio of the abundances of the M-92 and M-106 ions varied with the length of the conjugated chain, M-106 gaining at the expense of M-92 as the chain length grew longer. Schwieter *et al.* (1969) observed that the results for the C_{50} and C_{60} β-carotene analogues supported the conclusion.

The mass numbers of peaks obtained from carotenoid end groups have been measured (Enzell *et al.*, 1968 *a*; Vetter *et al.*, 1971) and the large body of now well-organized information enables investigators to by-pass some of the difficulties of interpretation. Structural studies on new carotenoids have certainly become that much easier (cf. Enzell, 1969) than was once the case.

Proton magnetic resonance spectroscopy was first applied to carotenoids in 1960 by Weedon's group (Barber *et al.*, 1960). The literature has since become very copious but a number of good reviews are available (cf. Schwieter *et al.*, 1969; Vetter *et al.*, 1971). Work has been done with instruments operating at 60, 100 and 220 MHz and at least 70 poly-enes have been examined under varying conditions. The increase in chemical shift differences by a factor of 2·2 for the 220 MHz compared with the 100 MHz instrument allows resolution into singlets or multiplets of the olefinic portion of the spectrum in the region 5·5–7·5 p.p.m. The attributions for β-carotene lycopene and many more difficult compounds are now beyond dispute. The special value of the 220 MHz instrument has been amply vindicated.

N.m.r. signals of carotenoids in CCl_4

Gem CH_3	
β ring	8·85–8·95
CH_3 \ C=CH / CH_3	8·31–8·38
CH_3 in chain	8·03
CH_3 at C-5	8·26
CH_2 (nonallylic)	8·5–8·6
CH_2 (allylic)	7·65, 7·77, 7·89, 7·93
$(CH_3)_2C$=*CH*—	4·85–5·0
CH=*CH* conj.	3·2–4·2
*CH*OH	*ca.* 6·1–6·2
OH	6·52

P.m.r. values (τ) for β-carotene and lycopene

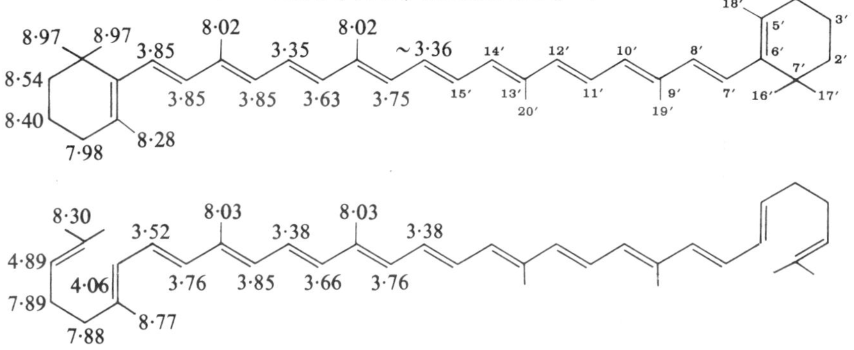

8·97 8·97 3·85 8·02 3·35 8·02 ~3·36 14′ 12′ 10′ 8′ 18′ 4′ 5′ 3′ 6′ 2′ 7′
8·54 3·85 3·85 3·63 3·75 15′ 13′ 20′ 11′ 9′ 19′ 7′ 16′ 17′
8·40 8·28 7·98

8·30 3·52 8·03 3·38 8·03 3·38
4·89 4·06 3·76 3·85 3·66 3·76
7·89 8·77 7·88

7.1.9 Carotenoproteins

Ovoverdin is a green protein present in the eggs of the common lobster (*Homarus vulgaris* or *H. gammarus* L). The pigment was studied in some detail by Kuhn & Sörensen (1938), who concluded that the green colour was due to astaxanthin (3,3′-dihydroxy-4,4′-diketo-β-carotene; see p. 159) bound to the protein. Earlier, Kuhn & Lederer (1933) had ground lobster eggs with sand and water and had obtained a dark green colloidal solution with $\lambda_{max.}$ 500 nm accompanied by end-absorption in the red region. The pigment was thrown out of solution by adding ammonium sulphate to saturation. The precipitate was redissolved in water and then adsorbed on aluminium hydroxide. Ovoverdin was eluted from the aluminium hydroxide by means of 0·4 saturated ammonium sulphate and reprecipitated by means of 0·65 saturated ammonium sulphate. The whole sequence of operations—adsorption, elution and reprecipitation—was repeated. This proved to be an effective method of fractionating this and other protein preparations. Ovoverdin contained carotenoid and protein in the ratio 1/242. From the molecular weight of astaxanthin (596) this corresponds with a minimum molecular weight of 144 000 for the protein. The pigment undergoes cleavage to a reddish carotenoid and protein by treatment with either alcohol or acetone or on gentle heating. Mineral acid or alkali precipitates protein and the colour changes from green to red, but acetic acid effects the colour change without throwing the protein out of solution. The red carotenoid can be obtained by direct extraction of the green eggs with acetone and can then be transferred to light petroleum after dilution of the acetone with water. When the light petroleum extract is shaken with 90% ethanol practically all the colour passes into the alcohol from which the carotenoid is precipitated by water as minute crystals readily recrystallized from pyridine. Kuhn & Lederer (1933) described it as an 'ovo ester' and found that it yielded astacene on treatment with alkali. It is, however, astaxanthin whereas astacene is 3,3′, 4′4-tetraketo-β-carotene; the action of alkali is not a saponification but the oxidation of a double α-ketol.

It is plausible to regard ovoverdin as the di-enolic form of astaxanthin, linked at 3,4 and 3′,4′ with two protein units of molecular weight *ca.* 72 000. Wyckoff however (1937) isolated ovoverdin from the eggs of *H. americanus* and obtained a molecular weight of 300 000 by a sedimentation method. Cheesman *et al.* (1966) isolated and characterized crustacyanin from the carapace of the common lobster. In earlier work Wald *et al.* (1948) had extracted a protein pigment ($\lambda_{max.}$ 625 nm) by prolonged treatment with an acid citrate buffer solution. Zagalsky & Cheesman (1963) obtained an electrophoretically homogeneous blue pigment with $\lambda_{max.}$ 632 nm (see also Jencks & Buten, 1964; Ceccaldi, 1965; Ceccaldi & Allemand, 1964, 1965; Ceccaldi *et al.*, 1966, 1967).

Crustacyanin occurs in the outer layer of the endocuticle and the blue parts are cut up and ground to a fine powder, avoiding excessive heating. The powder is suspended in 10% EDTA solution at pH 7·5 allowing 4 grams of solid for each gram of carapace. After stirring for 3 days in a cold room at 4°C, the suspension is filtered through a filter-aid (Hyflo Supercel) and the blue filtrate is treated with solid ammonium sulphate to 0·5 saturation. The pigment is filtered off and dissolved in 0·05 **M** neutral phosphate buffer. The absorption spectrum shows peaks at 630 and 278 nm with minor maxima at 410 and 370 nm and inflections at 605 and 320 nm. The ratio for absorption intensities at 278 and 630 nm varies between 0·8 and 3·0.

Cheesman *et al.* (1966) described the purification of crustacyanin and its separation from purple ($\lambda_{max.}$ 605 nm) and yellow ($\lambda_{max.}$ 410 nm) congeners. A small amount of a second blue pigment (278, 317, 367 and 616 nm) was seen. Crystacyanin crystallizes slowly from ammonium sulphate at 0·3 saturation in a cold room. Freeze-dried crustacyanin is very hygroscopic and the red solid quickly turns blue on exposure to moist air; it is a simple caroteno-protein free from hexose or hexosamine. The absorption spectrum in phosphate buffer, pH 7 (Fig. 7.3), shows:

$\lambda_{max.}$ nm	278	320	270	633 ± 2
$E^{0\cdot1\%}_{1cm}$	1·15			3·62
$\epsilon_{max.}$	41 000			132 800

and the molecular weight per mole of carotenoid is 35 700 ± 500. The colour is due to astaxanthin combined with the apoprotein.

When dialysed for 2 days against several changes of distilled water, the crustacyanin solution changes colour from blue to purple and the absorption spectrum shows displaced bands (633 → 595 nm, 370 → 362 nm, 320 → 315 nm) but addition of inorganic salts (e.g. NaCl to 0·075 **M**) restores the spectrum almost completely. The effect is mainly due to cations. However, when the dialysis is allowed to continue the purple solution ($\lambda_{max.}$ 278, 360, 585 nm) no longer regenerates the blue colour on addition of salts. Jencks & Buten (1964) called this purple material β-crustacyanin and the native blue pigment α-crustacyanin. (See Zagalsky, 1967; Zagalsky *et al.*, 1970.)

The molecular weight of α-crustacyanin measured in gel filtration experiments came out as 650 000, corresponding to about 18 astaxanthin molecules per molecule of protein. The initial dialysis effects a dissociation into sub-units of molecular weight *ca.* 35 000 each attached to one astaxanthin molecule. This (reversible) dissociation is also effected by 3 **M** urea, 1 **M** potassium thiocyanate or 10% (v/v) acetone. After the carotenoid has been removed from crustacyanin with acetone, the resulting apoprotein appears to have a molecular weight of about 20 000 but starch gel electrophoresis shows the protein so obtained to be quite heterogeneous. Nevertheless a blue crustacyanin can be regenerated by the inter-

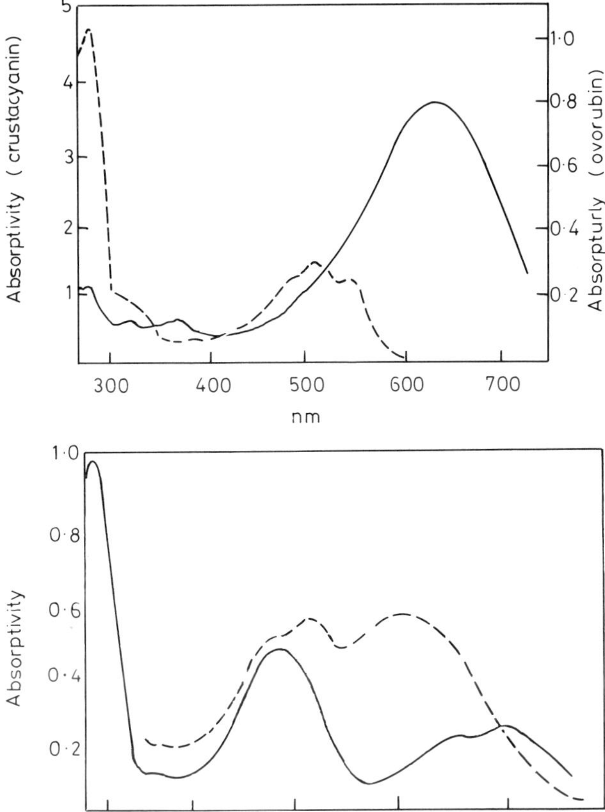

Figure 7.3 (a) (top) Crustacyanin (1 mg/ml) in 0·2 **M** phosphate buffer, pH 7 (solid curve), and ovorubin in water (1 mg/ml), dashed curve. The prosthetic group of each protein is astaxanthin. (Astaxanthin in light petroleum shows a single maximum in the visible region at 470–475 nm.) (after Cheesman, Zagalsky & Ceccaldi, 1966). (b) Solid curve: ovoverdin (1 mg/ml) in 0·05 **M** phosphate buffer, pH 7. Dashed curve: egg protein of *Eupagurus bernhardus* (concentration uncertain) in 0·1 **M** Na_2HPO_4 (after Zagalsky, 1964, and Cheesman & Prebble, 1966). The prosthetic group of ovoverdin is astaxanthin; that of *Eupagurus* protein is an ester of astaxanthin. (Both proteins are labile and the spectra are not very reproducible)

action of this protein and astaxanthin. This suggests that the quaternary structure of crustacyanin is induced by the interaction and that the caroteno-protein sub-units acquire a protein configuration favouring aggregation to a micellar structure with a hydrophobic carotenoid core.

The 633 nm band responsible for the blue colour is thus a property of a complex macromolecule which under appropriate conditions is very stable. The sub-units appear to consist of two different protein molecules held at the 3,4- and 3′,4′-positions of astaxanthin; a molecular weight of 20 000 is merely a mean for the two unequal proteins. The blue colour of α-crustacyanin is maintained only by a mass action (cationic) effect on the configuration of the sub-units. In environments of low ionic strength α-crustacyanin and its first degradation product α′-crustacyanin appear to be in equilibrium. The irreversible product is β-crustacyanin.

Cheesman (1958) had previously studied ovorubin, a chromoprotein present in the eggs of a gastropod mollusc *Pomacea canaliculata*. This is an amphibious fresh water prosobranch snail which lays vividly pigmented egg masses above the water level on plants or solid objects. To obtain the pigment the eggs are homogenized and then centrifuged at 10 000 g, giving a rather turbid red solution ($\lambda_{max.}$ 427, 510, 545 or 480 nm). The 427 nm band is due to a haemochromogen similar to helicorubin (cf. Keilin, 1956). Ovorubin is responsible for the bands at 480, 510 and 545 nm.

Purified ovorubin is a caroteno-glycoprotein of molecular weight *ca.* 335 000, calculated from the amount of carotenoid component. This is regarded as an ester (or possibly an ether) derived from astaxanthin. The carbohydrate polysaccharide component represents about one-fifth of the molecular weight. The carotenoid and the glycoprotein can be dis-associated but the pigment regenerates when the two are mixed. Astaxanthin is one of the most plentiful of all carotenoids and is particularly characteristic of crustaceans. Much remains to be done, however, before a full assessment of its biological roles can be attempted.

The visual pigments rhodopsin and iodopsin, etc., are not true caroteno-proteins but there are interesting similarities (see p. 545).

7.2 Sporopollenin*

Zetzsche (1932) coined the word sporopollenin from sporonin and pollenin. It refers to the highly resistant wall materials found in the spores of pteridophytes and the pollen grains of gymnosperms and angiosperms. The implication was that such compounds belonged to a chemical 'family'. Shaw & Yeadon (1964) advanced structural formulae for 'Recent sporopollenin' and it became likely that some carboniferous spore coats of *Lycopsida* (Potonié & Rehnelt, 1969) contained materials derived from sporopollenin by slow aromatization.

Zetzsche (1932) used fresh and fossil pollen and spore walls, and after drastic extraction processes he was left with the chemically insoluble sporopollenin in amounts varying with the species. The sporopollenins of fossil palynomorphs seem to persist for at least 10^8 years and perhaps longer. Their survival is regarded as the fundamental assumption of palynology but they do not persist completely unchanged. 'Sporin' is a name given to fossil-aromatized sporopollenin.

Southworth (1969) studied the ultraviolet absorption of the spore wall and obtained evidence that 'sexine' and 'nexine' were not identical, but the interpretation of the spectra was not easy.

A sporopollenin unit as postulated by Shaw & Yeadon (1964)

As postulated by Potonié & Rehnelt (1971)

A 'sporin' structural unit after aromatization

A C_{16} structural unit

Brooks & Shaw (1968) advanced the idea that sporopollenins were derived from carotenoids and this view has been widely accepted (cf. Potonié & Rehnelt, 1971). Zetzsche made calculations based on analyses indicating a C_{90} unit. Shaw (1971), summarizing the present state of knowledge, noted the presence of many OH groups, C—CH_3 groups and of substantial unsaturation. Vigorous degradation processes resulted in large amounts of simple C_3—C_6 dicarboxylic acids. Shaw and his colleagues found ozonolysis to be the most informative degradation process but

* Pollen, produced and discharged from the anthers of flowers, consists of microscopic grains within which two male germ cells are formed. The pollen grain has a wall composed of thick extine outer layers and a thin inner intine layer. The grain is sometimes regarded as the microspore of seed plants. The intine consists mainly of cellulose and hemicellulose.

Palynology (or pollen analysis) refers to the identification and characterization of pollen grains in media such as peat, lake mud or even honey. It includes identification of pollen in the atmosphere and in geological deposits and covers the identification of pteriodophyte spores (e.g. of mosses, ferns and other non-flowering plants). Palynology throws light on Quaternary geology and botanical development since very early times.

they decided to study the *formation* of sporopollenin as well as its degradation.

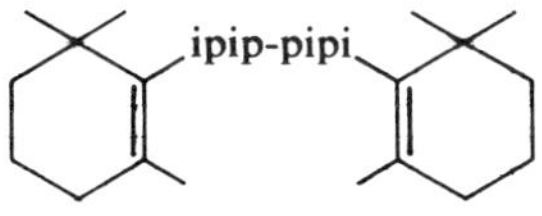

ipip-pipi

β-carotene

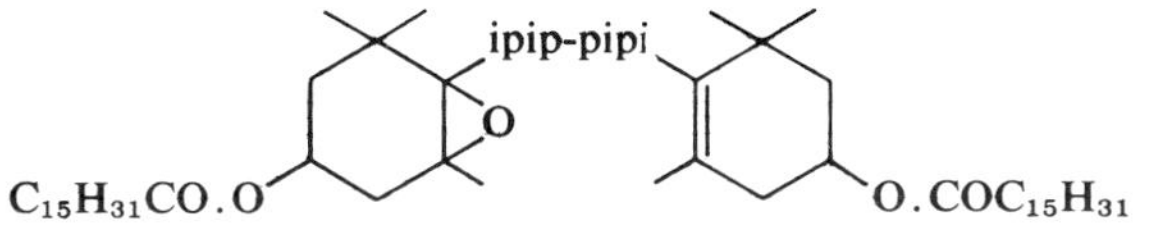

ipip-pipi

O

$C_{15}H_{31}CO.O$ $O.COC_{15}H_{31}$

Antheraxanthin dipalmitate

Oxidative copolymerization

↓

Sporopollenin, containing

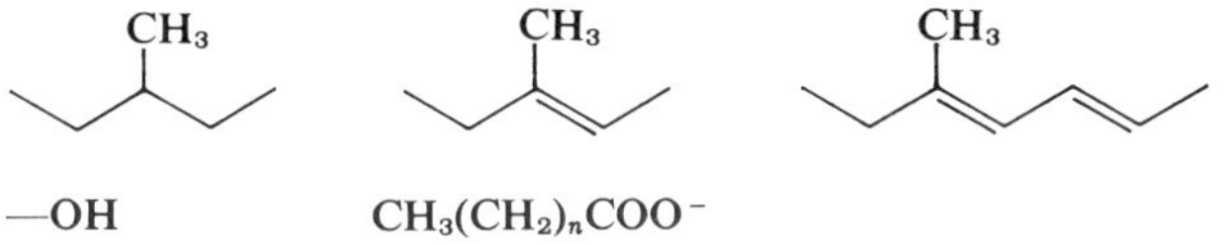

CH_3 CH_3 CH_3 and

—OH $CH_3(CH_2)_nCOO^-$

O C $CH_2(CH_2)_nCH_3$

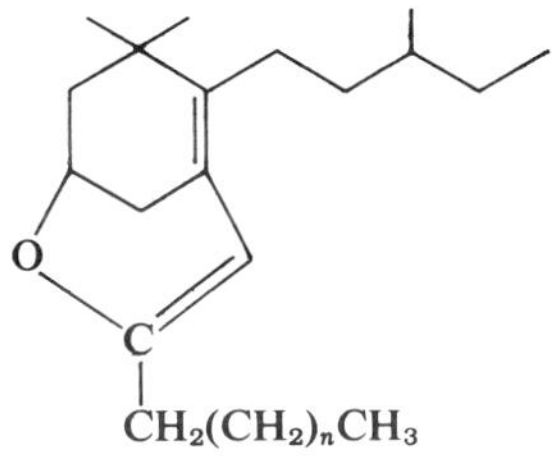

They chose to work on *Lilium Henryii*, partly because the anthers are large and also because much is known about the development of pollen in this plant. It was established that carotenoids were the only real candidates as precursors of sporopollenins. Brooks & Shaw (1968) noted in the formation of the anthers that a very sharp increase in carotenoid biosynthesis occurred when the exine was being laid down just before the formation of the actual pollen grain. They then studied the oxidative polymerization of various carotenoids with different catalysts and found that a methylene chloride solution of 'purified' mixed *Lilium* carotenoids in the presence of oxygen and an ionic catalyst (e.g. boron trifluoride) gave an insoluble oxygen-rich polymer in good yield. It closely resembled the natural sporopollenin from *L. henryii.* Of interest here is the comparison of infrared spectra.

The natural and 'synthetic' sporopollenins, after fusion with potash, yielded an identical mixture of phenolic acids (Brooks & Shaw, 1968). Polymers were also obtained from *β*-carotene and retinyl palmitate.

Brooks & Shaw (1968) also noted a marked similarity between sporopollenin and kerogen, the organic material present in Pre-Cambrian sediments. They also produced spectroscopic evidence to indicate that degradation products of Green River Shale kerogen and sporopollenin contained isoprenoid compounds consistent with a common carotenoid-precursor origin. The argument has however been contested (Cooper & Murchison, 1970). Brooks (1971) demonstrated that the various sporopollenins exhibit extremely similar infrared absorption spectra with no peaks between 2900 and 1700 cm^{-1}. When sporopollenin samples were heated (in the absence of air) at various temperatures between 200° and 450°C for 60 hours, the residues showed a slow reduction in hydroxyl absorption (3500 cm^{-1}) and a gradual increase in ethylene

$$\left(\rangle C = C \langle\right)$$

absorption around 1600 cm^{-1} with evidence of aromatization (600–700 cm^{-1}) at the higher temperatures.

Van Gijzel in a series of papers (summarized, 1971 *a*) has made very good use of ultraviolet fluorescence microphotometry in the study of fresh and fossil pollen and spores of varied origins. The technical aspects of fluorescence microspectroscopy are of great interest (van Gijzel, 1966, 1971 *b*, *c*). Excitation at 365 nm is followed by recording fluorescence at wavelengths between 400 and 700 nm, although exposure to 'microscopical' ultraviolet light can effect photochemical change which results in a fluorescence 'fading'. The fluorescence spectra of pollen grains show maxima at 470 and 620 nm. Much careful observational work on the fluorescence of pollen materials prepared in various ways has been done and although the interpretation of the findings is still rather tentative there is no doubt that the technique will prove to be useful.

The link between carotenoid biosynthesis and the formation of sporopollenin is a surprising development, important for botany and for palynology. Some chromophoric units are indispensable for the fluorescence and structural studies will in the future be of the greatest interest.

Pollen or spores were fixed in different ways and the pellet obtained by centrifugation was mixed with melted 2% agar, allowed to solidify and then cut into blocks which were dehydrated with acetone and propylene oxide and embedded in Epon. Sections were mounted in glycerol and examined using a Leitz u.v. microspectrograph (Ruch, 1960). A 4·5 second exposure of sections one micron thick gave a spectrum measurable by means of a recording microdensitometer.

Using *Lilium humboldtii*, the absorption spectra (formaldehyde fixation) of capita and foot layer showed $\lambda_{max.}$ 310 nm and $\lambda_{min.}$ 254 nm. No endexine is present. *Artemisia pycnocephala*, both tectum and foot layer, had $\lambda_{max.}$ 290 nm, $\lambda_{min.}$ 250 nm; here the endexine is very thin. In *Ambrosia trifida* the endexine had a broad plateau 280–290 nm, $\lambda_{min.}$ 250 nm. Sporopollenin forms with acetic anhydride acetyl sporopollenin. In *L. humboldtii* the absorption peak was shifted from 310 to 290 nm, but in other species the three exine layers had rather similar spectra.

In spores of *Lycopodium taxifolium* the walls had the absorption peak at 280 nm ($\lambda_{min.}$ 250–254 nm), and no distinction could be made between an inner granular layer and an outer lamellate layer.

Lignified xylem tracheids from three species showed $\lambda_{max.}$ 281 nm, $\lambda_{min.}$ 260 nm while onion root tips cortex showed the expected $\lambda_{max.}$ at 262–265 nm, $\lambda_{min.}$ 240–245 nm.

From Southworth's results sporopollenin covers a class of substances and not a specific molecule. All the ektexine layers of a given pollen grain type show qualitatively similar spectra although the absolute intensities and even the relative intensities at $\lambda_{max.}/\lambda_{min.}$ are variable. A distinction between sporopollenins and lignins is however evident. The absorption spectra of tracheid walls closely resemble those of pine lignin or oak lignin in dioxan ($\lambda_{max.}$ 280 nm, $\lambda_{min.}$ 260 nm; Schubert, 1965). Absorption in this region is probably associated with the presence of aromatic acids and indeed Shaw (1966) identified *m*- and *p*-hydroxybenzoic, protocatechuic, syringic and vanillic acids among the products obtained from *Lycopodium taxifolium* spore walls. It seemed very improbable that such acids were present in angiosperm sporopollenin, which must therefore be very different from lignin.

The experiments on onion root tips indicated that the fixation procedures did not result in spectroscopic artifacts. It is true that the chromophore of β-ionone

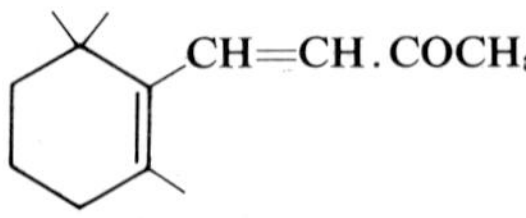

$$\text{(cyclohexene ring)}\text{—CH=CH.COCH}_3$$

could account for a peak at 296 nm of considerable intensity (ϵ 11 000) and that acids of the type

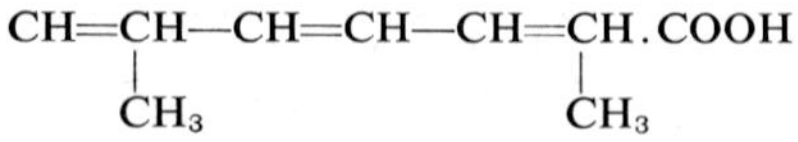

$$\text{CH=}\underset{\text{CH}_3}{\text{C}}\text{H—CH=CH—CH=}\underset{\text{CH}_3}{\text{C}}\text{H.COOH}$$

could have selective absorption near 310 nm, but this is as yet no more than a plausible explanation of Southworth's findings.

Two points can however be made; firstly, ultraviolet microspectrography is now a very serviceable tool, and secondly, progress into the nature of sporopollenin is being made, but more needs to be done.

Jonker (1971) summarized the state of knowledge on this topic. It has very broad implications.

References

7.1

Aitzetmüller, K., Strain, H. H., Svec, W. A., Grandolfo, M. & Katz, J. J. (1969). *Phytochem.*, **8**, 1761.

Allen, M. B., Fries, L., Goodwin, T. W. & Thomas, D. M. (1964). *J. Gen. Microbiol.*, **34**, 259.

Arpin, N. & Liaaen-Jensen, S. (1969). *Phytochem.*, **8**, 185–193.

Barber, M. S., Davis, J. B., Jackman, L. M. & Weedon, B. C. L. (1960). *J. Chem. Soc.*, 2870.

Bonnett, R., Spark, A. A., Tee, J. L. & Weedon, B. C. L. (1964). *Proc. Chem. Soc.*, 419.

Bonnett, R., Mallams, A. K., McCormick, A., Spark, A. A., Tee, J. L. & Weedon, B. C. L. (1969). *J. Chem. Soc.*, C, 429.

Braude, E. A. (1950). *J. Chem. Soc.*, 379.

Braude, E. A. & Waight, E. S. (1954). In *Progress in Stereochemistry*. Ed. W. Klyne. Butterworth, London.

Brown, B. O. & Weedon, B. C. L. (1968). *Chem. Commun.* 382.

Campbell, S. A. *et al.* (1967). *Chem. Commun.*, 941.

Carter, G. P. & Gillam, A. E. (1939). *Biochem. J.*, **32**, 1325.

Ceccaldi, H. J. & Allemand, B. (1964). *Rec. Trav. Stat. mar. Endoume*, **36**, 60.

Ceccaldi, H. J. & Allemand, B. (1965). *Rec. Trav. Stat. mar. Endoume*, **37**, 27, 53.

Ceccaldi, H. J., Cheesman, D. F. & Zagalsky, P. F. (1966). *C. r. Soc. Biol.*, **160**, 587.

Ceccaldi, H. J., Daumas, R. & Zagalsky, P. F. (1967). *C. r. Soc. Biol.*, **161**, 1111.

Ceccaldi, H. J. & Zagalsky, P. F. (1967). *Comp. Biochem. Physiol.*, **21**, 435.

Celmer, W. D. & Solomons, I. A. (1953). *J. Amer. Chem. Soc.*, **75**, 3430.

Chapman, D. J. (1966). *Phytochem.*, **5**, 1331.

Chapman, D. J. & Haxo, T. (1963). *Plant Cell Physiol.*, **4**, 57.

Cheesman, D. F. (1954). *Biochem. J.*, **58**, xxxviii.

Cheesman, D. F. (1956). *J. Physiol.*, **131**, 3P.

Cheesman, D. F. (1958). *Proc. Roy. Soc.* B, **19**, 571.

Cheesman, D. F. & Prebble, J. B. (1966). *Comp. Biochem. Biophys.*, **17**, 929.

Cheesman, D. F., Zagalsky, P. F. & Ceccaldi, H. J. (1966). *Proc. Roy. Soc.*, B, **164**, 130.

Cholnoky, L., Györgyfy, K., Nagy, E. & Panczel, M. (1956). *Nature, Lond.*, **178**, 410.

Cholnoky, L., Györgyfy, K., Szabolcz, J., Weedon, B. C. L. & Waight, E. S. (1966). *Chem. Comm.*, 404.

Cholnoky, L., Szabolcz, J. & Waight, E. (1968). *Tetrahedron Lett.*, 1931.

Cholnoky, L., Szabolcz, J. & Toth, G. (1967). *Ann. Chem.*, **708**, 218.

Cholnoky, L. *et al.* (1969). *J. Chem. Soc.*, 1256.

Cooper, R. D. G., Davis, J. B. & Weedon, B. C. L. (1963). *J. Chem. Soc.*, 5637.

Curl, A. L. (1956). *Food Res.*, **21**, 689; (1960) *J. Agric. Food Chem.*, **8**, 356.

Curl, A. L. & Bailey, G. F. (1950). *Agric. Food Chem.*, **4**, 156.

Curl, A. L. & Bailey, G. F. (1954). *Agric. Food. Chem.*, **2**, 685.

Curl, A. L. & Bailey, G. F. (1954). *Food Res.*, **22**, 13, 323.

Curl, A. L. & Bailey, G. F. (1957). *J. Agric. Food Chem.* **5**, 605.

Dale, J. (1954). *Acta. Chem. Scand.*, **8**, 1235; (1957). *Ibid.*, **11**, 265.

Enzell, C. R. (1969). *Pure & Appl. Chem.*, **20**, 497.

Enzell, C. R., Francis, G. W. & Liaaen-Jensen, S. (1968 *a*, *b*). *Acta. Chem. Scand.*, **22**, 727, 1054.

Francis, G. W., Hertzberg, S., Anderson, K. & Liaaen-Jensen, S. (1970). *Phytochem.*, **9**, 629.

Galasko, A. *et al.* (1969). *J. Chem. Soc.*, C, 1264.

Garbers, C. F., Eugster, C. H. & Karrer, P. (1952). *Helv. Chim. Acta.*, **35**, 1850; (1953). *Ibid.*, **36**, 562, 828; (1954). *Ibid.*, **37**, 382.

Garbers, C. F. & Karrer, P. (1953). *Helv. Chim. Acta.*, **36**, 1783.

Gillam, A. E. & El Ridi, M. S. (1935). *Nature, Lond.*, **136**, 914; (1936). *Biochem. J.*, **30**, 1735.

Goldsmith, T. H. & Krinsky, N. I. (1960). *Nature, Lond.*, **188**, 491.

Goodwin, T. W. & Jamikorn, M. (1954). *Biochem. J.*, **57**, 376.

Grundman, C. & Takeda, Y. (1937). *Naturwiss.*, **51**, 482.

Haxo, F. T. & Fork, D. C. (1959). *Nature, Lond.*, **184**, 1047.

Heilbron, I. M. & Lythgoe, B. (1936). *J. Chem. Soc.*, 1376.

Hirayama, K. (1955). *J. Amer. Chem. Soc.*, **77**, 373.

Hertzberg, S. & Liaaen-Jensen, S. (1967). *Acta. Chem. Scand.*, **21**, 15; (1969). *Phytochem.*, **8**, 1259, 1281.

Isler, O. (1971). *Carotenoids*, Birkhauser Verlag, Basel & Stuttgart. [A book with 12 Chapters by various authors—developed from and extending the book by Karrer & Jucker (1948)].

Jaffé, H. H. & Orchin, M. (1962). In *Theory and Applications of Ultraviolet Spectroscopy*. Chapter II. J. Wiley, New York, London.

Jencks, W. P. & Buten, B. (1964). *Arch. Biochem. Biophys.*, **107**, 511.

Jensen, A. (1961). *Acta. Chem. Scand.*, **15**, 1605; (1966). *Norw. Inst. Seaweed Res. Report*, No. 31, Tapir, Trondheim; (1966). *Acta. Chem. Scand.*, **20**, 1728.

Karrer, P. & Jucker, E. (1948). In *Carotenoide*. Birkhauser, Basle. (1950). English translation by E. A. Braude, Elsevier, Amsterdam.

Karrer, P. & Solmssen, U. (1935). *Helv. Chim. Acta.*, **18**, 1306; (1936). *Ibid.*, **19**, 3, 1019.

Karrer, P., Solmssen, U. & Koenig, H. (1938). *Helv. Chim. Acta.*, **21**, 454.

Keilin, J. (1956). *Biochem. J.*, **64**, 663.

Kuhn, R. & Grundmann, C. (1934). *Ber. dtsch. chem. Ges.*, **67**, 339, 1133.

Kuhn, R. & Lederer, E. (1933). *Ber. dtsch. chem. Ges.*, **66**, 488.

Kuhn, R. & Sörensen, N. A. (1938). *Ber. dtsch. chem. Ges.*, **71**, 1879, 1882; (1938). *Z. Angew Chem.*, **52**, 465.

Kuhn, R. & Winterstein, A. (1933). *Ber. dtsch. chem. Gres.*, **66**, 1733.

Lederer, E. (1934). In *Les Carotenoids des Plantes.* Hermann, Paris; (1935). *Comptes Rend. Acad. Sci.*, **201**, 300; (1938). *Bull. Soc. Chim. Biol.*, **20**, 554, 567.

Le Rosen, A. L. & Zechmeister, L. (1942). *J. Amer. Chem. Soc.*, **64**, 1075.

Lewis, G. N. & Calvin, M. (1939). *Chem. Revs.*, **25**, 273.

Liaaen-Jensen, S. (1964). *Acta. Chem. Scand.*, **18**, 1562; (1965). *Ibid.*, **19**, 1025, 1166; (1967). *Ibid.*, **21**, 1972.

Liaaen-Jensen, S. & Jensen, A. (1965). In *Progress in the Chemistry of Fats and other Lipids.* Vol. VIII, Part 2, p. 129. Ed. R. T. Holman. Pergamon Press, Oxford.

Liaaen-Jensen, S. Hertzberg, S., Weeks, O. B. & Schurpeter, U. (1968). *Acta. Chem. Scand.*, **22**, 1171.

Liaaen-Jensen, S. & Weeks, O. B. (1966). *Tidskr. Kjomi Bergv. Mettalurgi*, **26**, 130.

Liaaen-Jensen, S., Hegge, F. & Jackman, L. M. (1969). *Acta Chem. Scand.*, **18**, 1703.

Mader, I. (1964). *Science, N.Y.*, **144**, 533.

Magoon, E. F. & Zechmeister, L. (1957). *Arch. Biochem. Biophys.*, **68**, 263.

Markham, M. C. & Liaaen-Jensen, S. (1968). *Phytochem.*, **7**, 839.

Mallams, A. K. *et al.* (1967). *Chem. Commun.*, 301 and 484.

Meinwald, J. *et al.* (1968). *Tetrahedron Lett.*, **25**, 2959.

Mulliken, R. S. (1939). *J. Chem. Phys.*, **7**, 14.

Nayler, P. & Whiting, M. C. (1955). *J. Chem. Soc.*, 3042.

Nishibori, K. (1961). See *Chem. Abstr.*, **55**, 25073.

Pauling, L. (1939). *Fortscr. Chem. Org. Naturstoffe*, **3**, 203.

Schneider, D. F. & Weedon, B. C. L. (1967). *J. Chem. Soc.*, 1686.

Schwieter, U., Englert, G., Rigassi, N. & Vetter, W. (1969). In *Carotenoids other than Vitamin A.* Butterworths, London, and in *Pure & Appl. Chem.*, **20**, No. 4.

Strain, H. H. (1938). *Carnegie Institute Publication* No. 490, Washington; (1938). *J. Biol. Chem.*, **123**, 425.

Strain, H. H., Manning, W. M. & Harden, G. (1944). *Biol. Bull.*, **86**, 169.

Surmates, J. D. & Ofner, A. (1963). *J. Org. Chem.*, **28**, 2735.

Takahashi, T. (1964). *Tetrahedron Lett.*, 565.

Tsuchiya, Y. & Susuki, Y. (1960). See *Chem. Abstr.* **54**, 18801.

Vetter, W., Englert, G., Rigassi, N. & Schwieter, U. (1971). In *Carotenoids.* Ed. O. Isler. Birkhauser, Basel.

Wald, G., Nathanson, N., Jencks, W. P. & Tarr, E. (1948). *Biol. Bull. Woods Hole*, **95**, 249.

Walton, T. J., Britton, G., Goodwin, T. W., Diner, B. & Moshier, S. (1970). *Phytochem.*, **9**, 2545.

Weedon, B. C. L. (1965). In *Chemistry & Biochemistry of Plant Pigments.* Ed. T. W. Goodwin. Academic Press, London.

Weeks, O. B. (1966). *Tidskr. Kjemi Bergv. Metallurgi.*, **26**, 132.

Weeks, O. B., Andrews, A. G., Brown, B. O. & Weedon, B. C. L. (1969). *Nature, Lond.*, **224**, 879.

Weeks, O. B. & Garner, R. J. (1967). *Arch. Biochem. Biophys.*, **121**, 35.

Wiltstätter, R. & Page, H. J. (1914). *Annalen.*, 4041, 237.

Wyckoff, R. W. G. (1953). *Science, N.Y.*, **22**, 497.

Yamaguchi, M. (1957). *Bull. Soc. Chem. Japan*, **30**, 111, 979; (1958). *Ibid.*, **31**, 51, 739; (1960). *Ibid.*, **33**, 1560.

Zagalsky, P. F. & Cheesman, D. F. (1963). *Biochem. J.*, **89**, 21 P.

Zagalsky, P. F. (1967). *Biol. Rev. Cambridge Phil. Soc.*, **42**, 131.

Zagalsky, P. F., Ceccaldi, H. J. & Daumas, R. (1970). *Comp. Biochem. Physiol.*, **34**, 579.

Zechmeister, L. (1962). In *Cis-trans isomeric carotenoids, Vitamin A and arylpoly-enes.* Springer-Verlag, Vienna.

Zechmeister, L. & Cholnoky, L. (1937). *Liebigs Ann.*, **530**, 291.

Zechmeister, L., Le Rosen, A. L., Schroeder, W. A., Polgar, A. & Pauling, L. (1943). *J. Amer. Chem. Soc.*, **65**, 1940.

Zechmeister, L. & Schroeder, W. A. (1943). *J. Amer. Chem. Soc.*, **65**, 1535.

Zechmeister, L. & Tuzson, P. (1936). *Ber. dtsch chem. Ges.*, **69**, 1878; (1937). *Ibid.*, **70**, 1966.

7.2

Brooks, J. (1971). In *Sporopollenin.* Eds. J. Brooks, P. R. Grant, M. Muir, P. van Gijzel & G. Shaw. Academic Press, London, New York.

Brooks, J. & Shaw, G. (1968). *Grana Palynologica*, **8**, 227; (1968). *Nature, Lond.*, **219**, 532.

Cooper, B. S. & Murchison, D. G. (1970). *Nature, Lond.*, **227**, 195.

Jonker, F. P. in Brooks, J. (1971). p. 686.

Pontonié, R. & Rehnelt, K. (1969). *Bull. Soc. Roy. Sa. Liège*, **38**, 259; (1971). In *Sporopollenin.* Eds. J. Brooks, P. K. Grant, M. Muir, P. van Gijzel & G. Shaw. Academic Press, London, New York.

Ruch, F. (1960). *Z. Wiss. Mikrosk.*, **64**, 453.
Schubert, W. J. (1965). In *Lignin Biochemistry*. Academic Press, New York, London.
Shaw, G. & Yeadon, A. (1964). *Grana Palynologica*, **5** (2), 247.
Shaw, G. & Yeadon, A. (1966). *J. Chem. Soc.*, **16**,
Shaw, G. (1970). In *Phytochemical Phylogeny*, p. 31. Ed. J. Harborne. Academic Press, London, New York; (1971). In *Sporopollenin*. Eds. J. Brooks, P. K. Grant, M. Muir, P. van Gijzel & G. Shaw. Academic Press, London, New York.
Southworth, D. (1960). *Grana Palynologica*, **9**, 191; (1969). *Ibid.*, **9**, 5.
Van Gijzel, P. (1966). *Leitz. Mitt. Wiss. Tech.*, **3**/**7**, 206.
Van Gijzel, P. (1971 *a*). In *Sporopollenin*. Eds. J. Brooks, P. K. Grant, M. Muir, P. van Gijzel & G. Shaw. Academic Press, London, New York.
Van Gijzel, P. (1971 *b*). *Leitz. Mitt. Wiss. Tech.* V. (1971 *c*). *Zeitschr. Wiss Mikrosk. Microsk Techn.* Cited in *Sporopollenin*.
Zetzsche, F. (1932). Sporopollenins in *Handbuch der Pflanzer Analyse*, Vol. 3. Ed. D. Klein. Vienna.

8 Aminoacids, Proteins and Enzymes. Fluorescence

8.1 Aminoacids and Proteins

Clupeine is a polypeptide of molecular weight near 4000 and it contains no aromatic aminoacids. At pH values of 7·5, 10·5 and 12 it displays no absorption maxima in the region 225–290 nm. In the region 260–280 nm the end-absorption flattens out and then falls to a very low value at 300 nm. The molecular extinction coefficient $\epsilon_{260-280\ nm}$ is about 50 and as the polypeptide contains 40 aminoacid residues, ϵ per peptide link is of the order 1. Even this low value may be due in part to scattering of light and the evidence is clear that the peptide link *per se* contributes negligibly to the 280 nm absorption by proteins.

Most of the naturally occurring aminoacids exhibit negligible absorption in the region 210–300 nm, the exceptions being the aromatic aminoacids phenylalanine, tyrosine and tryptophan, together with cystine which has significant end-absorption beginning at *ca.* 300 nm and displays a shallow maximum at 250 nm. To a first approximation, the ultraviolet absorption spectra of proteins represent a summation of the contributions of the constituent aminoacids. As has been shown, the peptide linkage —CO.NH— gives rise to absorption so weak in the region 250–280 nm that this contribution is usually negligible.

Tryptophan shows peaks at 220 nm, ϵ 35 000 and 195 nm ϵ 20 000; tyrosine has maxima at 223 nm, ϵ 8500 and 194 nm, ϵ 47 000 while phenylalanine has peaks at 209 nm, ϵ 9000 and 188 nm, ϵ 58 000. Cystine has moderately strong end-absorption e.g. $\epsilon_{200\ nm}$ *ca.* 3000, $\epsilon_{185\ nm}$ *ca.* 10 000. Histidine has a peak at *ca.* 212 nm, ϵ 6000. Most of the other aminoacids have low ($\epsilon < 1000$) end-absorption to 195 nm rising steeply beyond that point. Measurements of absorption spectra at wavelengths between 185 and 200 nm are never easy; stray light can be a problem and the presence of impurities can have serious effects.

If the absorption spectra of proteins were truly additive some analytical determinations would be easier than they are. In fact, however, the small departures from additivity (the perturbations of aminoacid side-chains) are interesting in a constitutive sense. Within globular or helical proteins some of the absorbing aminoacids may be less exposed than others or they may participate in intramolecular bonding. The perturbations can sometimes be profitably displayed as difference spectra. Anomalous spectrophotometric titrations may contribute considerably to the elucidation of structure in individual proteins. Fluorescence and phosphorescence in aminoacids and proteins can provide new tools for structural investigations.

Taken in conjunction with other techniques,

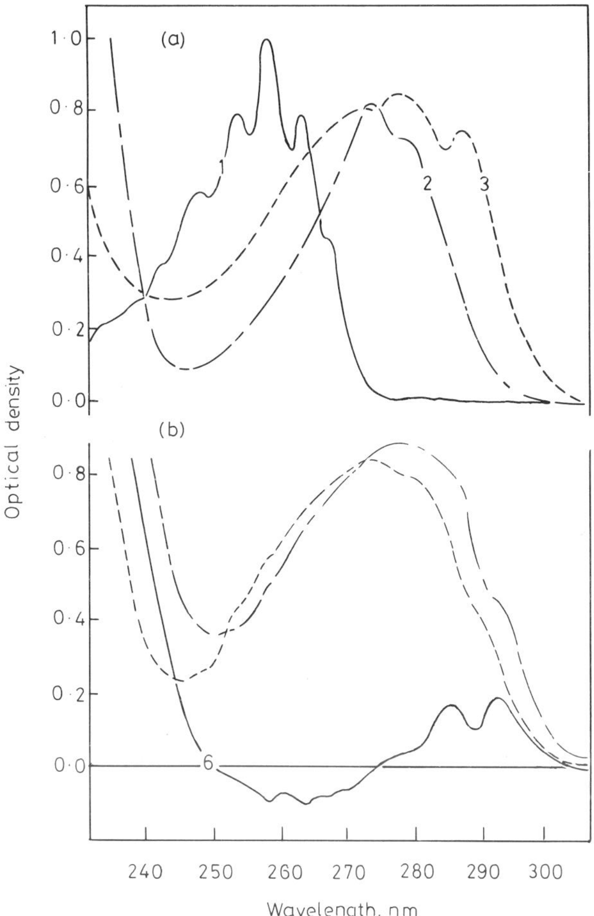

Figure 8.1 (*a*) Ultraviolet absorption spectra of the aromatic aminoacids. (1) 5×10^{-5} **M** phenylalanine: (2) $6{\cdot}4\times10^{-4}$ **M** tyrosine; (3) $1{\cdot}6\times10^{-4}$ **M** tryptophan. (*b*) Comparison of the spectrum of 2×10^{-5} **M** pepsin with that of a mixture of the aromatic aminoacids (dashed curve) in the proportions in which they occur in pepsin. (6) the difference spectrum obtained with the pepsin solution in the sample beam and the aminoacid mixture in the reference beam (from Yanari & Bovey, 1960)

spectrophotometry has a substantial part to play in protein studies. Although a great deal remains to be done at various levels of investigation and interpretation, it cannot be doubted that the foundations for future work have been well laid.

8.1.1 Absorption Spectra of Sulphur-containing Aminoacids

L-cystine in 0·01 **N** hydrochloric acid shows a weak absorption peak *ca.* 250 nm, $\epsilon_{max.}$ *ca.* 350 with a minimum near 236 nm, $\epsilon_{min.}$ 300. From 236 to 200 nm there is continuous end absorption rising to ϵ *ca.* 3000 at 200 nm. In L-cystinyl-D-cystine (Otey & Greenstein, 1954), D-cystinyl-L-cystine and L-cystinyl-L-cystine none of the three different absorption spectra corresponds with the sum of two cystine curves despite the fact that the disulphide group is the sole chromophore at 230–270 nm. The curve for L-cystinyl-diglycine is more intense than that of L-cystine but agrees better with a curve in which the ϵ values for the dipeptides with two cystine residues have been halved (see Fig. 8.2).

The absorption spectrum of cysteine is subject to complications. In the first place simple thiols are susceptible to oxidation in air in the presence of small amounts of some cations to give (via a mercaptide) disulphides. In proteins, however, the oxidation of sulphydryl groups to disulphide groups may be strongly inhibited. The technical difficulties were overcome by Benesch & Benesch (1955) in an im-

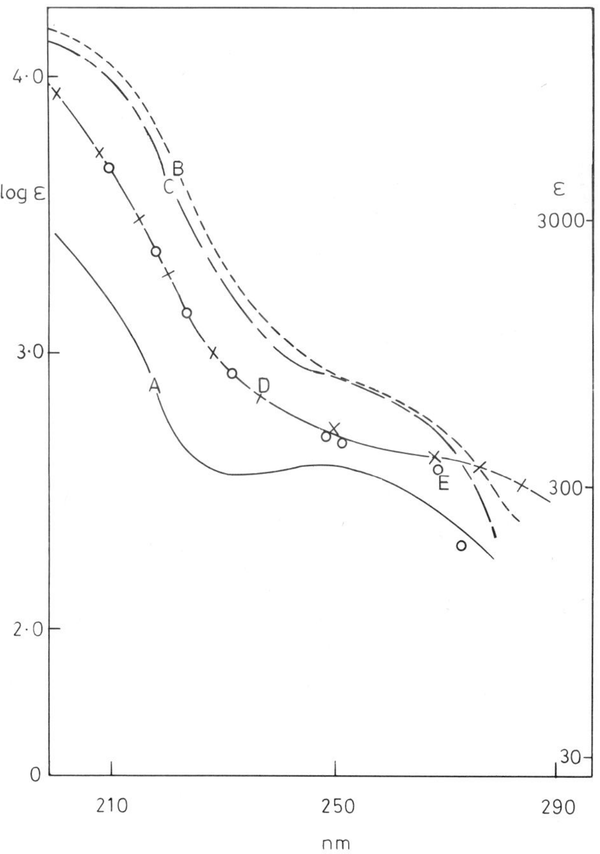

Figure 8.2 Absorption spectra of (A) L-cystine; (B) L-cystinyl-D-cystine and D-cystinyl-L-cystine; (C) Cyclo-L-cystinyl and L-cystinyl-L-cystine; (D) L-cystinyldiglycine; (E) half of L-cystinyl-diglycine (from Wetlaufer, 1962)

portant study. At different pH values between 7 and 12 the four main ionic species can be written:

$$\underset{\text{SH}\quad \text{NH}_3^+}{\text{CH}_2.\text{CH}.\text{COOH}} \qquad (\text{H}_3.\text{Cys.})^+$$

$$\updownarrows$$

$$(\text{H.Cys}^1) \quad \underset{\text{SH}\quad \text{NH}_3^+}{\text{CH}_2.\text{CHCOO}^-} \rightleftharpoons \underset{\text{SH}\quad \text{NH}_2}{\text{CH}_2.\text{CHCOO}^-} \quad (\text{H.Cys.})^-$$

$$(\text{H.Cys.}^1)^- \quad \underset{\text{S}^-\quad \text{NH}_3^+}{\text{CH}_2.\text{CHCOO}^-} \rightleftharpoons \underset{\text{S}^-\quad \text{NH}_2}{\text{CH}_2.\text{CHCOO}^-} \quad (\text{Cys.})^=$$

(Gorin, 1956)

The NH_4 and HS groups of cysteine have nearly the same tendency to transfer a proton to water and two tautomeric ions tend to be produced concomitantly.

When cysteine is titrated, new absorption in the region 230–250 nm appears (Fig. 8.3) and its intensity increases with increasing pH. The maximum (*ca.* 235 nm) is close to that of simple mercaptide ions. The small steady shift of $\lambda_{max.}$ from 230 to 236 nm is a secondary effect due to overlapping of NH_2 and HS ionizations. Accepting that H.Cys. absorption is zero at 230 nm (H.Cys.[1] absorption is reduced tenfold when water as solvent is replaced by 92% ethanol), it seems probable that $\epsilon_{max.}$ $(\text{H.Cys.}^1)^-$ $= \epsilon_{max.}$ $\text{Cys.}^= = 4400$ at 230 nm.

Benesch & Benesch (1955) supplemented the titration data to obtain four microscopic K values contributing to the overall situation. Some of their assumptions were verified by Gorin & Clary (1960), who found the ratio of $^-\text{S.Cys.NH}_4^-$ to HS.Cys.NH_2 was approximately constant and near 2.

$$\begin{array}{ccc} (a)\ \text{HSRNH}_3^+ & \underset{}{\overset{K_A}{\rightleftharpoons}} & ^-\text{SRNH}_3^+\ (b) \\ \downarrow\uparrow K_B & & \downarrow\uparrow K_C \\ (c)\ \text{HSRNH}_2 & \underset{K_D}{\rightleftharpoons} & ^-\text{SRNH}_2\ (d) \end{array}$$

In this connection gelatin presents an interesting protein because it is very low in aromatic aminoacids. The increment in ϵ at 238 nm on thiol ionization is about 4500 (for tryptophan at 238 nm ϵ is *ca.* 2500 and relatively independent of pH), but $d\epsilon/d\lambda$ is large in that spectral region. Tyrosine shows ϵ 238 nm varying from 600 at pH 7 to 10 000 at pH 12. Spectrophotometric titration of protein SH groups is promising only for proteins like gelatin with very small aromatic aminoacid contents.

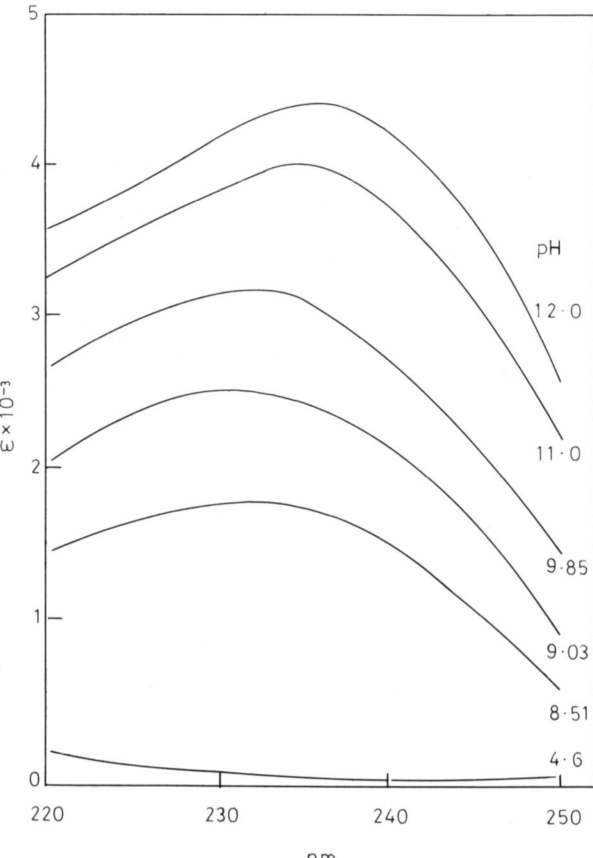

Figure 8.3 Absorption spectra of cysteine at different pH values $c = 1{\cdot}70 \times 10^{-1}$ M) (after Benesch & Benesch, 1954)

Thionein, a protein from equine renal cortex, can bind cadmium and zinc. Kagi & Vallee (1961) reported that it contains about 30% cysteine and that 95% of the sulphur in the protein is present as cysteine SH groups. There are no aromatic aminoacids in this very unusual substance. The absorption of thionein rises steadily from 250 to 200 nm.

Certain diatoms (e.g. *Navicula pelliculosa*) possess cell walls comprising a siliceous shell interlocked with an organic casing of unusual composition. In addition to the usual aminoacids there are sugars and polysaccharides, some sulphated, together with a number of unidentified ninhydrin-positive substances. One of the latter has been crystallized (m.p. 262°, $[\alpha]_D$—61·2°; $C_5H_9NO_4$). It gave colour tests similar to 3-substituted proline and its structure was shown to be that of a 3,4-dihydroxyproline (Nakajima & Volcani, 1969):

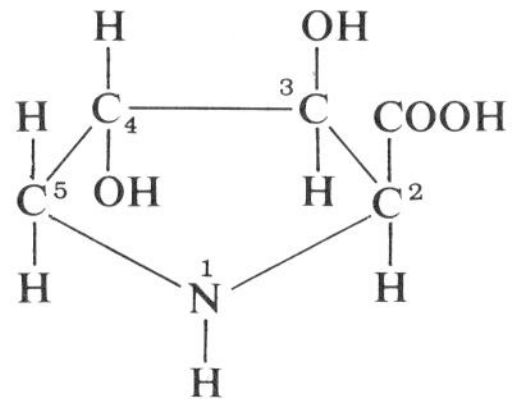

```
       H            OH
       |          3 |
   H   C4 ----------C   COOH
    \ /|            |  \ |
    C5 OH           H   C2
    |   \          /    |
    H    \    1   /     H
          \   N  /
              |
              H
```

2,3-*cis*,3,4-*trans* dihydroxyproline

Nuclear magnetic resonance spectra were obtained in D_2O using an Hz 220 Varian instrument with dimethylsilapentane sulphonate as internal reference.

There was no chemical shift due to a 3- or 4-methylene group. This indicated a 3,4-dihydroxyproline, confirmed by comparisons with 3- and 4-hydroxyprolines. Four isomerides of 3,4-dihydroxyproline are possible: two were synthesized (Hudson *et al.*, 1968; Mauger & Witkop, 1966) and excluded. The n.m.r. data indicated 2,3-*cis*, 3,4-*trans* dihydroxyproline.

```
        Hc   OD
DO····  |----|····HD
HB ◄  /        \ ◄CO2⁻
    ·    (+)     ·
 HA       N        HF
        /   \
       D     D
```

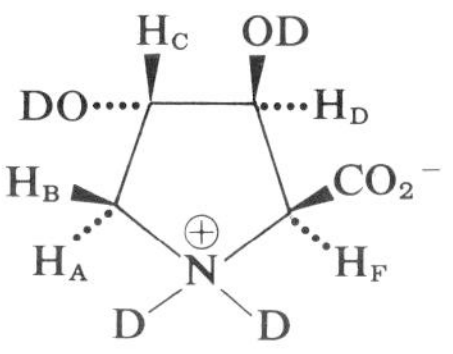

H_A = 3·36 p.p.m. (1 proton, doublet)
J_{AB} = 13 Hz
H_B = 3·72 p.p.m. (1 proton, double doublet)
J_{BC} = 3·5 Hz
H_C = 4·42 p.p.m. (1 proton, 1 doublet)
J_{CB} = 1 Hz
H_D = 4·48 p.p.m. (1 proton, doublet)
J_{AC} = 0·7 Hz
H_E = 4·39 p.p.m. (1 proton, doublet)
J_{DE} = 4 Hz

This is a comparatively simple case but it well illustrates the power of the new instruments.

Mass spectroscopy (Karle *et al.*, 1969) confirmed the structure. When 3,4-dehydro-DL-proline is oxidized by permanganate (Hudson *et al.*, 1968) it yields two *cis*-3,4-glycols. Comparison of the mass spectra of these with the *trans* glycol of L-proline from diatom cell walls showed a large M+1 peak at m/e=148. The exact mass for *trans* 3,4-dehydroxy proline was 148·060 (calculated 148·061). Fragmentation by loss of the carboxyl group led in both cases to a base peak at m/e 102. X-ray diffraction results confirmed the structure and settled the conformation. Atoms N, C(4), C(5) and C(2) are almost in a plane with C(3) above it. The carboxyl group is equatorial and the two hydroxyl groups axial to the ring. The molecule exists as a zwitterion with a negative charge on the carboxyl group and an extra protein on the nitrogen.

8.1.2 Ultraviolet Irradiation of Proteins

The absorption spectrum of the cystine (disulphide) chromophore shows a marked inflection at 245–250 nm, ϵ *ca.* 360 in neutral solution. In that region the phenylalanine absorption shows ϵ *ca.* 100, the tyrosine absorption is less than $\epsilon = 200$ and the tryptophan absorption, ϵ *ca.* 2000.

In the region around 250 nm the cystine absorption may be a significant contribution to the total, depending on the relative proportions of the three aromatic aminoacids and of cystine. It is, however, important that cystine residues are specially sensitive to irradiation. Thus when the Hg resonance line 253·7 nm is used to provide the light energy, the quantum yield for the process cystine → cysteine is 0·2 while the quantum yields for NH_3 and H_2S are 0·04 and 0·2 respectively. The overall quantum yield measured by the disappearance of cystine residues is 0·13 (see Table 8.1). On the other hand, the quantum yield for the production of ammonia from the aromatic aminoacids is only 0·002. Despite the more intense absorption shown by the aromatic aminoacids, the damage done at the cystine residues tends to overshadow that effected elsewhere (cf. Luse & McLaren, 1963).

The validity of the above argument, which is based on experiments with free aminoacids, can be verified by measuring the inactivation of enzymes by monochromatic light at 253·7 nm. Setlow (1955) plotted the quantum yield for inactivation against the percentage of half-cystine for 11 enzymes ranging from catalase and aldolase, which contain very little cystine, to RNAase and insulin which have about 6 and 12% of half-cystine respectively. The relationship is approximately linear whereas there is no correlation between quantum yields and the fraction of aminoacids made up by phenylalanine, tyrosine and tryptophan. Setlow (1960) also found that the quantum yield for inactivation of a number of enzymes varied with wavelength and was maximal at 250 nm. Cleavage of S—S bonds probably results in molecular rearrangement to give conformations unsuitable for effective enzyme-substrate interaction.

Granted that the cystine residues are the main

Table 8.1 Photochemical sensitivity of aminoacids in the 253·7 nm Hg radiation

	$\epsilon_{253\cdot7}$	ϕ	$\epsilon_{253\cdot7} \times \phi$	Conditions
Cystine	270	0·13	35·1	in 0·1 N HCl (N_2)
Tryptophan	2870	0·004	11·5	
Phenylalanine	140	0·013	1·8	
Tyrosine	320	0·002	0·6	
Histidine	0·24	0·003	0·0072	in water (N_2)
*[Peptide bond, e.g. acetyl alanine	0·2	0·05	0·01]	

*Based on McLaren & Shugar (1964), in proteins ϵ *ca.* 1·0 per CONH and $\phi_{(CONH)}=0{\cdot}005$.
(N_2) means stirring by a stream of pure N_2.

sites for ultraviolet damage to proteins, the question arises whether the less obvious damage is additive or whether the energy absorbed by, say, tryptophan can be transferred to do harm at some other locus in the protein molecule.

Luse & McLaren (1963) found for lysozyme, ribonuclease and trypsin quite good agreement by direct summation of effects calculated on the basis of quantum yields and percentage proportions of individual aminoacids (phenylalanine, tyrosine, tryptophan and cystine). In the case of chymotrypsin the calculated quantum yield for inactivation was double that observed. Setlow & Doyle (1957) had already pointed out that the action spectra for inactivation did not agree with the ultraviolet absorption spectra of proteins and indeed agreed rather well with the calculated contribution of the cystine moieties, especially in cystine-rich enzymes. There is thus little evidence of transfer of the energy absorbed by aromatic aminoacid residues to cystine residues. Titration of irradiated trypsin with *p*-chloromercuribenzoate showed a linear relation between enzyme inactivation and *p*-CMB-reacting groups. Trypsin contains 6 cystine moieties but they are not equally important for enzymic activity (Augenstein & Ghiron, 1961) and secondary disturbances in hydrogen bonding may affect protein configuration.

It is true that enzymes which, like aldolase, contain very little cystine are inactivated by ultraviolet light but the quantum efficiency shows less variation with wavelength and the process is imperfectly understood. It is interesting that the marked changes in ultraviolet absorption which accompany changes in pH are not in the case of lysozyme reflected in changes in quantum efficiency (253·7 nm, pH 3·6–12·0). It should be noted, however, that a good deal of work has been done at the one wavelength, chosen as a matter of convenience; irradiation at other wavelengths might well give more complex results.

There is no doubt that many proteins display altered absorption spectra as a result of irradiation. Sometimes there is aggregation resulting in light scattering; this can be detected by a correction procedure (p. 385) disclosing 'irrelevant' superimposed absorption linear between say 250 and 300 nm. Extrapolation should indicate whether the superimposed absorption follows the Rayleigh law. The changed absorption spectrum might involve absorption of oxygen but there is a need for more investigation.

The fact that energy transfer from absorbing aromatic aminoacids to cystine residues is not significant does not preclude transfer in other circumstances. Thus in the photodecomposition of haem–carbon monoxide complexes, the energy absorbed by the aromatic aminoacids is as effective as that absorbed by the iron-porphyrin in liberating carbon monoxide (Bucher & Kaspers, 1947). Similarly, radiation absorbed by protein can excite fluorescence in phycocyanin. A fluorescent dye (1-dimethylaminonaphthalene-5-sulphonyl-chloride) binds to free amino groups in proteins and light energy absorbed by the protein can be transferred to the dye and registered by the fluorescence induced. The most widely accepted view rests on the following considerations: firstly, the aromatic aminoacids are themselves capable of fluorescence (Shore & Pardee, 1956) and the emitted radiation lies within the absorption spectrum of the dyestuff (or the biological chromophore such as haem); secondly, in a 'conjugated' protein the distances between aromatic residues emitting fluorescence and the conjugate are small enough to allow energy transfer. This is consistent with the fact that the transfer depends quantitatively on the amount of dye attached to the protein.

The fluorescence spectra in neutral aqueous solution have been studied by Teale & Weber (1957) and are shown in Fig. 8.4. Tyrosine shows fluorescence

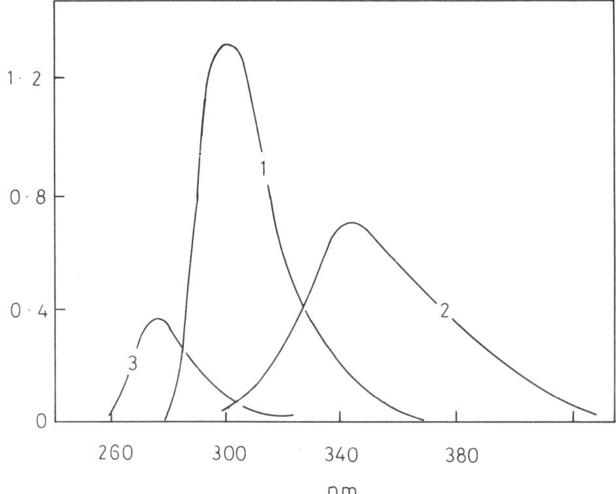

Figure 8.4 Fluorescence spectra in neutral aqueous solution. (1) Tyrosine; (2) Tryptophan; (3) Phenylalanine. Ordinate scale is relative number of quanta (from Teale & Weber, 1957)

from 280 to 360 nm with a good maximum at 303 nm and a quantum yield of 21%. Tryptophan fluorescence ($\lambda_{max.}$ 348 nm) extends from 300 to 400 nm with a 20% quantum yield, while phenylalanine fluorescence ($\lambda_{max.}$ 282 nm) covers the range 260–320 nm and a quantum yield of 4%. The excitation spectra agree with the absorption spectra, indicating that quantum yields do not depend on wavelength. The overlapping of the fluorescence spectra makes it probable that energy transfer can occur between aminoacid residues within the protein. The high quantum efficiency of photodissociation of carboxymyoglobin by ultraviolet light absorbed by aromatic aminoacids also indicates energy transfer.

Purified gelatin and the polypeptide polymyxin contain phenylalanine but not tyrosine or tryptophan. The fluorescence resembles that of phenylalanine but the quantum yield is smaller. Insulin, which contains tyrosine but not tryptophan, shows the tyrosine fluorescence in poor yield. When tyrosine and tryptophan are both present in a protein, the tryptophan can act as an inner light filter reducing the chance of absorption by tyrosine. This helps to explain why the fluorescence is characteristic only of tryptophan and in 8 **M** urea the peak is displaced to 350 nm.

Teale (1960) studied the ultraviolet-excited fluorescence of about 20 globular proteins and investigated possible quenching effects by measuring fluorescence yields of free tyrosine dissolved in water containing high concentrations of compounds like glycine, diglycine and tetraglycine. Contiguity of COOH groups (or even $COOCH_3$ groups) had a powerful quenching effect. Intraprotein interactions between aromatic aminoacid residues and adjacent aminoacid side-chains undoubtedly influence the ultraviolet absorption to a small but definite extent and such effects could have repercussions on energy transfer. Nevertheless Teale & Weber (1957) concluded that energy transfer from one excited aminoacid to another aminoacid acceptor had little effect on fluorescence intensity. Quenching was ascribed to hydrogen bonding between free hydroxyl groups from tyrosine to carboxylate groups, and to the idea that any breakdown of tertiary structure would tend to inactivate enzymes. The whole problem is discussed in some detail by Beaven (1961). Vladimirov (1959) held that a complete sequence of energy transfer

phenylalanine ⟶ tyrosine ⟶ tryptophan

was needed to explain his fluorescence studies.

8.1.3 Tyrosine and Tryptophan

Goodwin & Morton (1946) discussed the spectrophotometric determination of tyrosine and tryptophan in proteins, and decided that as an approximation many proteins could be treated for spectrophotometric analysis as tyrosine–tryptophan two-component systems (cf. Coulter *et al.*, 1936; Holiday, 1936; Holiday & Ogston, 1938). The advent of reliable photoelectric spectrophotometers proved of great assistance. The absorption spectra of tyrosine and tryptophan were determined in 0·1 **N** NaOH (tyrosine $\lambda_{max.}$ 293 nm, $\epsilon_{max.}$ 2300; tryptophan 280·5 nm, $\epsilon_{max.}$ 5250). The absorption curves were found to intersect at 294·4 nm (ϵ 2375) and at 257·15 nm (ϵ 2748). In a simple mixture the sum of the molar concentrations may be obtained from measurements at either point of intersection, but the 294·4 wavelength is the more suitable one because it lies near the tyrosine peak where $\Delta\epsilon/\Delta\lambda$ is minimal. In using these points of intersection rather than following the classical procedure of Vierordt, advantage is taken of the accuracy of modern photoelectric methods both on the wavelength and absorbance scales. An accurate reading at one other wavelength is necessary to determine the relative proportions of the two solutes. Thus if x is the sum of the concentrations in mol/l and y is the molar concentration of tyrosine, $x-y$ will be the molar concentration of tryptophan. At

Table 8.2 Ultraviolet absorption of aromatic aminoacids

	Phenylalanine			Tyrosine			Tryptophan		
	$\lambda_{max.}$	$\lambda_{min.}$ nm	ϵ	$\lambda_{max.}$	$\lambda_{min.}$ nm	ϵ	$\lambda_{max.}$	$\lambda_{min.}$ nm	ϵ
In acid	276		92	275		1 340	288		4 550
	263		152	223		8 200	278		5 550
	260		144	194		35 000	218		19 300
	258		195		245	170		242	2 000
	252		154						
	247		115	222 (pH 6)		7 800			
	242		61		209 (pH 6)	5 000			
		236	63						
In alkali	268		124	294		2 330	288		4 600
	264		160	240		11 050	280·5		5 430
	261		176						
	258		206		270	1 000	222		34 600
	252		172					244	1 900
	247		128		218 (pH 13)	5 000			
	242		84						
		236	62						

any wavelength other than a point of intersection, let ϵ (tyrosine) be A and ϵ (tryptophan) be B; the total absorption E would be

$$E = yA+(x-y)B$$

or

$$y = \frac{E-xB}{A-B}$$

Goodwin and Morton followed Holiday (1936) in selecting 280 nm as the suitable wavelength since this is $\lambda_{max.}$ for tryptophan ($\Delta\epsilon/\Delta\lambda$ is minimal) but 270 nm and 290 nm were equally suitable. Calibration with tyrosine–tryptophan mixtures showed very good additivity.

In protein solutions or hydrolysates in alkali there is irrelevant absorption, falling off with increasing wavelength, but still appreciable in the region 330–400 nm, where tyrosine and tryptophan exert no absorption of light. A simple procedure is to extrapolate linearly from 370 and 340 nm to 294·4 and 280 nm. Thus, as an illustration, if E 370 nm = 0·020 and E 340 nm is 0·028, ΔE is 0·008 and Δ nm is 30 and by extrapolation E 294 nm would be 0·040 and E 280 nm 0·044. The validity of such a linear extrapolation is checked by comparing tyrosine and tryptophan concentrations calculated from absorbance at the crossing point 294·4 nm and at different wavelengths, e.g. 270, 280 and 290 nm. The results will only agree if the extrapolation is near to the truth. The procedure was tried out on casein, zein and gelatin and has since been rather widely used.

Beaven & Holiday (1952) discussed the method in some detail. If a protein is dissolved in 0·1 **N** alkali the sample becomes completely alkali-denatured but the process is not instantaneous; in fact readings may need to be taken for up to 3 hours before becoming constant (Beaven, Holiday & Jope, 1950). Calibrations with synthetic peptides might be advantageous. They recommended the use of the Goodwin & Morton (1946) formula based on measuring the absorbance of a protein solution in 0·1 **N** alkali at 280 and 294·4 nm:

$$M_{(Tyr)} = (0{\cdot}592_{E294{\cdot}4} - 0{\cdot}263_{E280}) \times 10^{-3}$$

$$M_{(Try)} = (0{\cdot}263_{E280} - 0{\cdot}170_{E294{\cdot}4}) \times 10^{-3}$$

where the weight of the protein is known or estimated from N content, $E_{294{\cdot}4}$ and E_{280} are the extinction coefficients of the protein at 294·4 and 280 nm and M_{Tyr} and M_{Try} are the gram-molecules of tyrosine and tryptophan respectively in one gram of protein.

If the N or protein content of a solution is not known,

$$\frac{M_{Tyr}}{M_{Try}} = \frac{0{\cdot}592E_{294{\cdot}4} - 0{\cdot}263E_{280}}{0{\cdot}263E_{280} - 0{\cdot}170E_{294{\cdot}4}}$$

i.e. the molar ratio can be obtained.

Table 8.3 Tyrosine and tryptophan in proteins (grams aminoacid per 100g. protein)

Method*	Tyrosine C	Tyrosine S	Tryptophan C	Tryptophan S	Molar Ratio Tyr/try C	Molar Ratio Tyr/try S
Protein						
Insulin	12·3	12·7	0	0	—	—
Pepsin	8·5	8·3	2·4	2·4	3·8	3·9
Trypsin	7·8	7·6	3·0	3·7	2·4	2·3
Lysozyme	3·7	4·7	8·3	8·1	0·5	0·6
Ribonuclease	8·0	8·5	0	0	—	—
Actomyosin	5·5	8·0	1·4	1·3	4·4	6·9
Edestin	4·3	4·5	1·4	1·3	3·4	3·9
β-Lactoglobulin	3·6	4·6	2·1	2·2	2·0	2·3

* C = chemical analysis; S = spectrophotometric analysis (after Beaven & Holiday, 1952).

Bencze & Schmid (1957) proposed an alternative calculation procedure. They plotted the absorption spectrum of a solution of protein in 0·1 N alkali from about 275 to 298 nm. They ascertained the maxima (*ca.* 280 and 294 nm) and drew a line connecting the two peaks (tangentially). The absorbance at the higher of the two peaks was a measure of tyrosine plus tryptophan content and the slope of the line indicated the ratio of tyrosine to tryptophan. Purified tyrosine/tryptophan mixtures were used for calibration.

Wetlaufer (1962) examined pancreatic ribonuclease as a test case for additivity of absorption. This protein has a molecular weight of 13 580 and contains 6 tyrosyl and 4 cystinyl residues but no tryptophan. Its absorption maximum is at 275 nm and $\epsilon_{max.}$ 9800 is well verified. The tyrosine peak at 274·5 nm has $\epsilon_{max.}$ 1340 at pH 1 and the cystine absorption at 280 nm is $\epsilon = 150$. The calculation:

tyrosyl residues	$6 \times 1340 = 8040$
cystinyl residues	$4 \times 150 = 600$
	8640

The ratio ϵ observed/ϵ calculated is 1·13. Wetlaufer considers the experimental errors and concludes that $\epsilon_{max.}$ 1340 for tyrosine is reproducible in different laboratories (1370 ± 30) whereas $\epsilon_{max.}$ values for ribonuclease vary between 9000 and 10 600. There must at present be some doubt in the case of ribonuclease to what extent there is a departure from additivity.

Wetlaufer went further and compared the observed and calculated molecular extinction coefficients for 11 proteins using the findings of Beaven & Holiday (1952) for cystine, tyrosine and tryptophan.

It may be seen from Table 8.4 that the last protein, Δ^5-3-ketosteroid-isomerase, is clearly anomalous. It has neither tryptophan nor cystine and might have been expected to show very good agreement. For the other ten proteins the average ratio ϵ obs./ϵ calc. is 1·06 and it must be accepted that there is something to explain. Any allowance for absorption at 280 nm by the peptide link or by phenylalanine would tend to increase the discrepancy rather than reduce it.

Patchornik *et al.* (1958) found that N-bromosuccinimide could oxidize tryptophan moieties in proteins with simultaneous cleavage of the peptide link. Tyrosyl and disulphide groups were also oxidized (Ramachandran & Witcop, 1959; Peters, 1959) and this weakened the original suggestions that $E_{Try} = \Delta E_{280\,nm} \times 1{\cdot}31$ although it might well be valid for

Table 8.4 Observed and calculated $\epsilon_{max.}$ at $\lambda_{max.}$ 279 ± 2 nm for various proteins

	M.W. × 10^{-4}	Residues/mol. Try.	Residues/mol. Tyr.	Residues/mol. Cys.	ϵ obs./ϵ calc.
Carbonic anhydrase	2·8	6	8	0	1·13
Carboxypeptidase A	3·44	6	20	2	1·10
Chymotrypsinogen	2·51	7	4	5	1·12
α-Lactalbumin	15·55	5	5	4	0·93
β-Lactalbumin	3·77	5	8	4	0·96
Lysozyme	3·89	6	3	5	1·06
Papain	2·07	5	17	0	1·04
Ribonuclease	1·368	0	6	4	1·13
Bovine serum mercaptalbumin	6·6	2	20	18	1·07
Human serum mercaptalbumin	6·6	1	18	18	1·08
Δ^5-3-ketosteroid-isomerase	4·08	0	10	0	1·37

some peptides. In some proteins N-bromosuccinimide failed to effect peptide cleavage at about half the sites of attack (cf. Witkop, 1961). Excess of the reagent appears to work as follows:

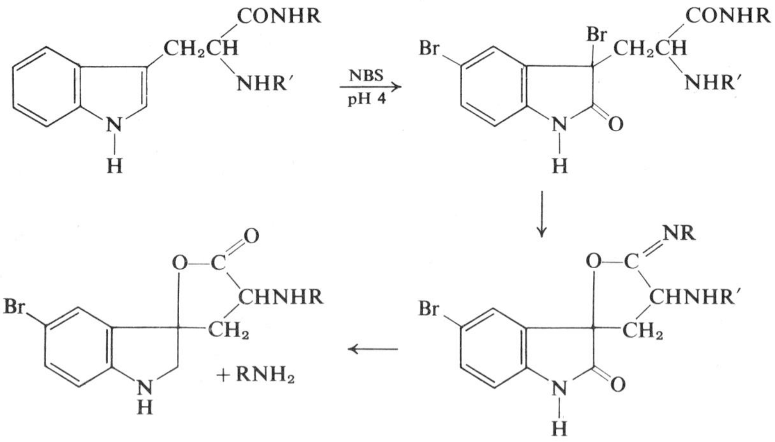

More recently N-bromosuccinimide has been used to serve a precise need. A spectrophotometric method for the determination of amino-terminal tyrosine residues has been worked out by Wilchek *et al.* (1968 *b*). The NH_2-terminal residues react with 4 equivalents of N-bromosuccinimide in acetic acid buffer or with excess of N-bromosuccinimide in formic acid, to form the chromophore of 5,7-dibromo-6-hydroxyindole-2-carboxamide.

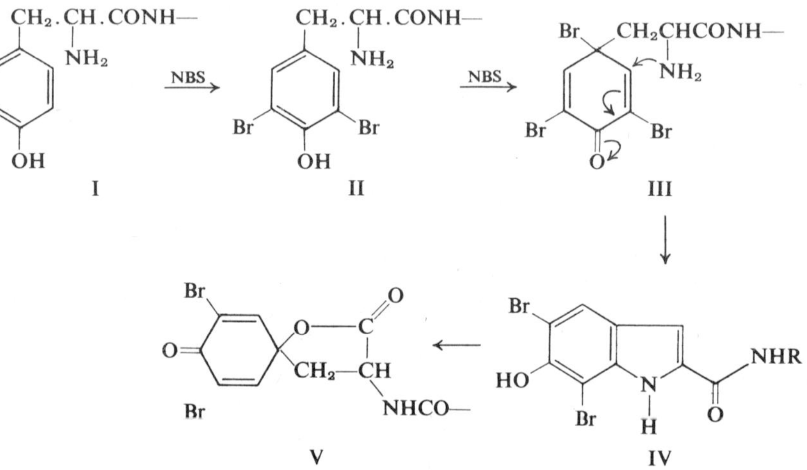

Various compounds containing the above chromophore were studied by Wilchek *et al.* (1968 *a*).

When R = H the following ultraviolet absorption is shown in 95% ethanol containing mineral acid:

$\lambda_{max.}$ nm	227	~245	313	~320
ϵ	21 380	15 140	16 600	16 200

and in 95% ethanol containing sodium hydroxide there is a displacement:

$\lambda_{max.}$ nm	273	355
ϵ	20 890	10 960

Using tyrosyl peptides the conversion to the 6-hydroxyindole derivative occurs reproducibly to about 25%. The value for $\epsilon_{max.}$ at 315 nm is taken as 17 000 for the chromophore and when trypsin-digested proteins are treated with N-bromosuccinimide in 0·1 N formic acid, $\epsilon_{315\ nm}$ amounts to about 4200 per tyrosine residue. The number of tyrosine residues is measured from the tyrosyl absorbance at 275 nm (ϵ 1370 in an acid medium). The ratio $E_{315\ nm}/E_{275\ nm}$ (4 equiv. NBS in acetic acid buffer) is maximal when $\epsilon_{315\ nm} = 4300–4700$. The use of formic acid buffer instead of acetic acid buffer permits excess of NBS, since formic acid does not interfere with the production of the 6-hydroxyindole but protects it from destruction. When the solution is made alkaline, the absorption peak is displaced to 340–350 nm and is about 20% more intense.

It is an advantage that formic acid allows addition of a considerable excess of NBS to proteins. Thus if undigested protein is placed in the control cell and the same amount of NBS is added to the protein digest and the control, any absorption at 315 nm produced by the action of NBS on interior residues is cancelled out. With trypsin-digested glucagon, one free α-amino group is found, and a 10 minute digest of globin also shows one terminal —NH_2. A 24-hour digest, however, shows $\epsilon_{max.}$ 8400 at 315 nm (two free α-amino groups) and this is confirmed by a doubling in alkali of $\epsilon_{max.}$ at 340 nm from 5300 to 10 700. For S-sulphochymotrypsinogen, the digest shows $\epsilon_{max.}$ at 315 nm at 12 300 consistent with 3 free α-amino groups belonging to tyrosyl residues.

If a tyrosine residue is within the peptide chain the N-bromosuccinimide gives rise to a bromodienone-lactone (V) $\lambda_{max.}$ 270 nm.

Edelhoch (1967) neatly circumvented two difficulties encountered in the spectrophotometric determination of tyrosine and tryptophan. The first obstacle is that acid hydrolysis of proteins destroys tryptophan while alkaline hydrolysis gives incomplete recovery. The second point is that the absorption spectra of the free aminoacids are not exactly reproduced in proteins. Hydrolysis is made unnecessary by the device of measuring absorbances in 6 **M** guanidine hydrochloride. The reference spectra are 'normalized' by using blocked tryptophanyl (N-acetyl-L-tryptophanamide) and blocked tyrosyl (glycyl-L-tyrosylglycine) as model substances (see Fig. 8.5.).

Figure 8.5 Absorption curves of (1) N-acetyl-DL-tryptophanamide; (2) glycyl-L-tyrosylglycine; (3) cystine in 6·0 **M** guanidine hydrochloride (pH 6·5)–0·02 phosphate buffer (from Edelhoch, 1967)

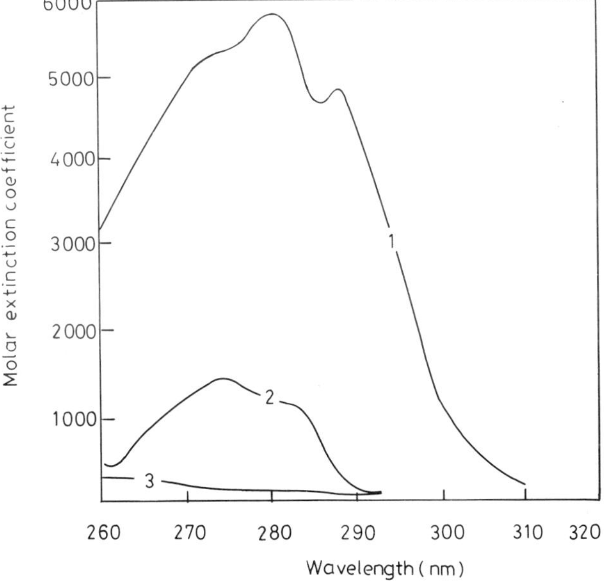

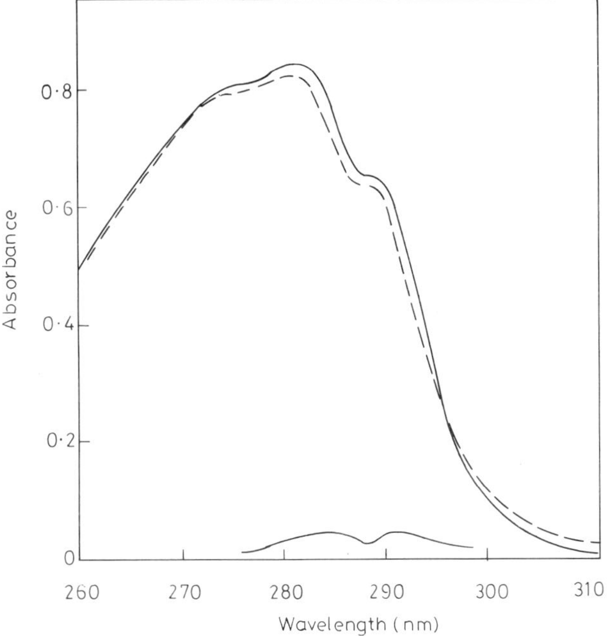

Figure 8.6 Comparison of absorption spectrum of chymotrypsinogen (solid line), $1{\cdot}61\times10^{-5}$ **M**, with solution of model compounds containing equiresidue concentrations of N-acetyl-DL-tryptophanamide, glycyl-L-tyrosylglycine, and cystine in 6·0 **M** guanidine hydrochloride (pH 6·5)–0·02 **M** phosphate buffer. Bottom curve is the absorption difference spectrum between protein and model solution (from Edelhoch, 1967)

Figure 8.7 Comparison of the absorption spectrum of trypsinogen ($2{\cdot}50\times10^{-5}$ **M**) (solid line) with solution of model compounds containing equiresidue concentrations of N-acetyl-DL-tryptophanamide, glycyl-L-tyrosylglycine, and cystine in 6·0 **M** guanidine hydrochloride (pH 6·5)–0·02 **M** phosphate buffer. Bottom curve is the difference spectrum between protein and model solution (from Edelhoch, 1967)

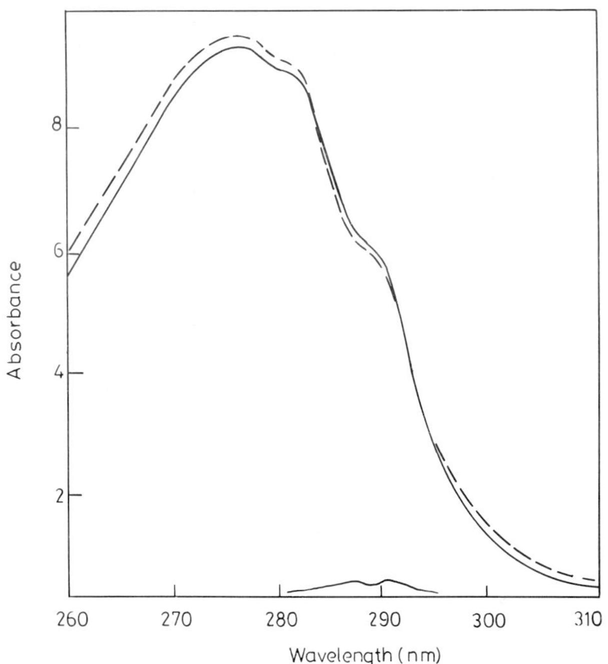

The following results show the significance of the blocking:

	$\lambda_{max.}$ nm	$\epsilon_{max.}$	$\lambda_{max.}$ nm	$\epsilon_{max.}$
L-tryptophan	280	5800	288·5	4800
N-acetyl-L-trypto-phanylamide	280·8	5690	289	4850
N-acetyl-L-trypto-phan	281	5500	289·2	4700
Acetyl-D-L-methionyl-tryptophan	281	5650	290·3	4810
L-tryptophanylglycine	279·8	5650	288·5	4700

(All in 6·0 **M** guanidine hydrochloride at pH 6·5, 0·01 **M** phosphate).

Under the same conditions the tyrosyl peptides showed:

	$\lambda_{max.}$ nm	$\epsilon_{max.}$	
Glycyltyrosylglycine	285·5	1470	
Acetyl-L-tyrosinamide	275·5	1490	
N-acetyl-L-tyrosylethyl ester	275·5	1500	
L-leucyl-tyrosine	275·5	1500	
L-tyrosine	275·5	1475	1515

The satisfactory reproducibility led to the following normalized quantities:

	ϵ values at		
Model substance	288 nm	280 nm	285·5 nm
N-acetyl-L-tryptophanyl-amide	4815	5690	—
Glycyl-L-tyrosylglycine	3815	1280	1500
Cystine	70–75	120	135–155

When these 'constants' were used to calculate the tryptophanyl and tyrosyl residues in proteins the results were:

	Molecular weight $\times 10^5$	Calculated		Found	
		try.	tyr.	try.	tyr.
Chymotrypsinogen	1·61	8	4	9	3·85
Trypsinogen	2·50	4	10	4·15	9·50
Lysozyme	2·44	6	3	5·95	3·55
Horse heart apomyoglobin	6·10	2	2	1·95	2·[illegible]
Ribonuclease	8·50	0	6	0	6·05

The tyrosyl content can be measured independently by the change in absorbance at 295 nm and 300 nm brought about by a change in pH from 6·5 to 12·5. For three tyrosine-containing peptides, ϵ_{295} is 2480 ± 40 (tyrosine 2590) and ϵ_{300} averages 2270 (2240, 2290, 2280) as against 2400 for tyrosine. Use of these normalized values leads to the following values for the number of tyrosyl residues per molecule of protein.

	Number/tyrosyl	Residues/molecule	
Protein	Expected	Found from	
		ϵ_{295}	ϵ_{300}
Chymotrypsinogen	4	4·05	4·25
Lysozyme	3	2·80	3·35
Horse heart apomyoglobin	2	2·05	2·10
Ribonuclease	6	6·00	6·20

The Edelhoch method has obvious convenience and advantage especially for determining tryptophyl residues.

8.1.4 Thyroxine

Gemmill (1955) used a spectroscopic method to determine the apparent ionization constants of the phenolic hydroxyl groups of thyroxine and related compounds (cf. Stenström & Goldsmith, 1926; Cranmer & Neuberger, 1943). The hydroxyl group is fully ionized in alkaline solution, unionized in acid solution and partially ionized at intermediate pH value. When ϵ is plotted against pH the curve shows a sharp change at the pK′. At a given wavelength the ϵ values for the acid and alkaline solution are averaged and the pH value where ϵ is at the mean is taken as pK′. The low solubility of thyroxine makes it necessary to use long cells in the spectrophotometer. Table 8.5 shows the absorption peaks

Table 8.5 Absorption maxima of thyroxine and related substances

Compound	0·04 N KOH solution		0·04 N HCl solution	
	$\lambda_{max.}$ nm	$\epsilon_{max.} \times 10^{-3}$	$\lambda_{max.}$ nm	$\epsilon_{max.} \times 10^{-3}$
L-Tyrosine	295	2·4	280	1·07
3,5-Di-iodo-L-tyrosine, $2H_2O$	310	5·92 (5·6)	285	2·73
DL-Thyronine	300	2·82	280	2·64
3,5-Di-iodo-DL-thyronine, HCl	305	3·33	285	3·21
3,3′,5-Tri-iodo-L-thyronine	320	4·66	295	4·09
Thyroxine (sodium)	325 (330)	6·21 (6·10)	295	4·16

Values obtained by Ginsel (1939) in brackets.

Table 8.6 Molecular extinction coefficients at wavelengths used for determination of pK′ values

Compound	$\lambda_{max.}$ nm	ϵ (in alkali) $\times 10^{-3}$	ϵ (in acid) $\times 10^{-3}$	mean $\times 10^{-3}$
Tyrosine	295	2·44	0	1·22
Di-iodotyrosine	310	5·92	0·320	3·12
Thyronine	305	2·69	0·27	1·48
Di-iodothyronine	310	3·17	0·41	1·79
Tri-iodothyronine	315	4·59	0·82	2·70
Thyroxine	325	6·21	0·79	3·50

and Table 8.6 the ϵ values used in obtaining pK′ values. The terminal hydroxyl group of thyroxine is over 50% ionized at physiological pH but the significance of this is not clear.

8.1.5 Insoluble Proteins

There are definite limits to the utility of ultraviolet absorption data in the study of insoluble proteins. A group led by Schauenstein and Kratky has published a good deal on silk (*Bombyx mori*) fibroin (Schauenstein, 1949, 1952; Schauenstein & Kratky, 1950; Schauenstein *et al.*, 1954). Fibroin can be dissolved in aqueous lithium bromide or in aqueous copper ethylene diamine and films can be cast from 'renatured' fibroin. The tyrosine residues of fibroin give rise to selective absorption at 270–310 nm; the absorption is intensified at high pH values and is displaced in the direction of longer wavelengths. Lucas *et al.* (1955, 1957) applied the procedure of Goodwin & Morton (1946) to the determination of tyrosine and tryptophan in various fibroins. The tyrosyl residues can undergo hydrogen bonding through the phenolic hydroxyl and the dissociation of these groups at high pH values, hindered by hydrogen bonding, is reflected in the absorption curves. Schauenstein's group observed that some tyrosyl residues had high pK values and that the proportion was greater in stretched than in unstretched films.

Fibroin also shows considerable absorption on the short wave side of 270 nm but its origin is perhaps debatable. Schauenstein attributes it to a —CH=N—OH chromophore obtained from peptide linkages by enolization

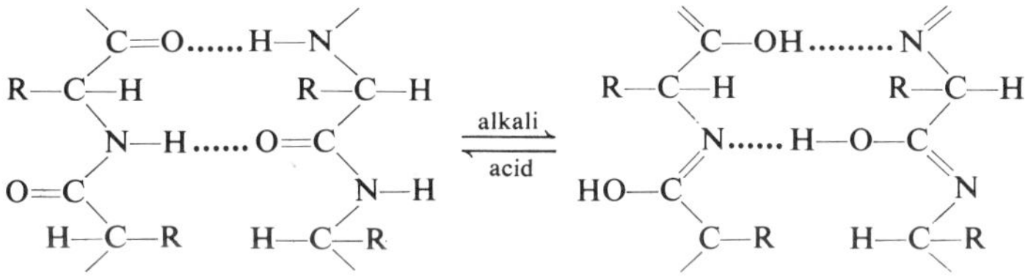

```
       \                /                        \\               //
        C=O......H—N                              C—OH.........N
       /            \                            /              \
  R—C—H            R—C—H                   R—C—H             R—C—H
       \            /          alkali           \              /
        N—H......O=C          ⇌                 N......H—O—C
       /            \          acid             //              \\
  O=C              N—H                      HO—C                N
       \            /                            \              /
  H—C—R        H—C—R                             C—R         H—C—R
       /            \                            /              \
```

to give 'peptenol' structures. This absorption is prominent in β-proteins with parallel peptide chains favouring interchain hydrogen bonding.

Glycyl tyrosine anhydride can take up a peptenol structure and the absorption spectrum is consistent with this:

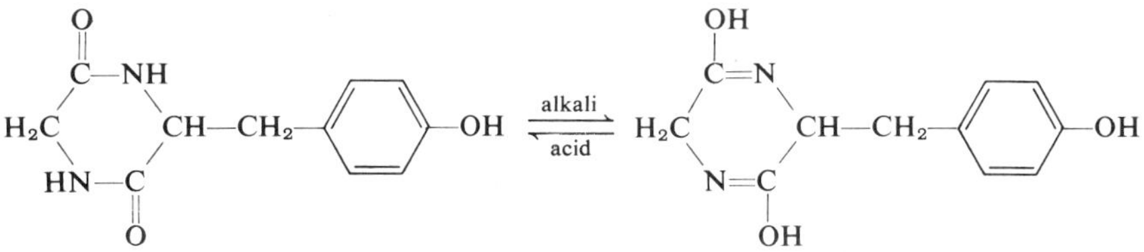

```
        O                                        OH
        ‖                                        |
        C—NH                                     C=N
       /    \                    alkali         /   \
  H2C        CH—CH2—C6H4—OH      ⇌        H2C        CH—CH2—C6H4—OH
       \    /                     acid          \   /
      HN—C                                       N=C
        ‖                                        |
        O                                        OH
```

The absorption in the region 250 nm shown by fibroin films is increased by a rise in pH and reversed by a decrease in pH. Stretched fibroin shows increased intensity of absorption, but fibroin dissolved in lithium bromide shows much less absorption at 250 nm at all pH values, in agreement with the accepted effect of the LiBr in reducing hydrogen bonding.

Both the familiar absorption in the region of 280 nm and the 250 nm absorption shown by oriented (stretched or rolled) fibroin exhibit dichroism. This has been interpreted as indicating that the tyrosyl residues lie at a minimum at 40° with the direction of the chains. Dichroism of the 280 nm absorption is only seen when there is dichroism of the 250 nm absorption.

It should be said that Beaven & Holiday (1952) disagreed with the 'peptenol' attribution of the 250 nm absorption. Doty & Guiduschek (1953) also rejected the 'peptenol' ascription largely because the role of light scattering had not been fully worked out. Unfortunately an acceptable alternative to the peptenol explanation seems to be lacking.

8.1.6 Ionization of Tyrosine

Since tyrosine looms large in the ultraviolet absorption of proteins it is necessary to consider its ionization in some detail. Many workers have contributed to this topic and the overall result is interesting and instructive.

Tyrosine in strongly acid solution contains three dissociable groups COOH (1), OH (2) and NH_3^+ (3). Martin *et al.* (1958) and Edsall *et al.* (1958) studied the ionization by spectrophotometric and electrometric titrations. Eight 'microscopic' forms of tyrosine are illustrated on p. 191 with twelve inter-relating ionization constants. Each of the three ionizing groups possesses four acidic 'micro'-constants since the readiness of each group to accept or donate a proton depends on state of the other two groups. In order to assist the study Martin *et al.* also examined tyrosine ethyl ester, O-methyltyrosine and O-methyltyrosine ethyl ester.

In a simple spectrophotometric titration molecular extinction coefficients are given the notation ϵ, the fractional degree of ionization is denoted by α, ϵ_0 is the molecular extinction coefficient when $\alpha=0$ and ϵ_1 when $\alpha=1{\cdot}0$ (assuming Beer's law to be valid at a fixed wavelength λ). Then,

$$\alpha=(\epsilon-\epsilon_0)/(\epsilon_1-\epsilon_0)$$

In a two-component system (displaying isosbestic points in the absorption spectra)

$$pK_1=pH-\log_{10}[\alpha/(1-\alpha)]$$

and the value of pK_1 does not significantly change with α.

In the case of tyrosine the experimental values for pK_1 rise slowly as α (calculated from the phenolate ion) increases. The ammonium ion can lose a proton in the same pH region. This is to be expected in an equilibrium involving H^+ and four species containing COO^-

	1 2 3			
(*a*)	− 0 0	COO^-	OH	NH_2
(*b*)	− − 0	COO^-	O^-	NH_2
(*c*)	− 0 +	COO^-	OH	NH_3^+
(*d*)	− − +	COO^-	O^-	NH_3^+

$$c \xrightarrow{k_{12}} d;\quad c \xrightarrow{k_{13}} b,\quad d \xrightarrow{k_{123}} a;\quad a \xrightarrow{k_{132}} b$$

The overall dissociation constants K_2 and K_3 follow:

$$K_2=k_{12}+k_{13}$$

$$K_2K_3=k_{12}k_{123}=k_{13}k_{132}$$

$$1/K_3=1/k_{123}+1/k_{132}$$

If α_{OH} represents the fraction of all the tyrosine molecules with a phenolic ion O^- it can be shown that

$$M_{OH}=\frac{[H^+].\alpha_{OH}}{1-\alpha_{OH}}$$

If M_{OH} is the ionization constant of OH if independent of NH_3^+ ionization

$$pM_{OH}=pH-\log_{10}\left[\frac{\alpha_{OH}}{1-\alpha_{OH}}\right]$$

when

$$\alpha_{OH}=\frac{k_{12}/[H^+]+k_{12}k_{123}/[H^+]^2}{1+(k_{12}+k_{13})/[H^+]+k_{12}k_{123}/[H^+]^2}$$

where pM_{OH} is an apparent pK for the phenolic ionization which varies with α_{OH}. By plotting pM_{OH} against α_{OH} a smooth curve is obtained between the values of pM_{OH} when α_{OH} approaches 0 (pk_{12} 9·63) and when α_{OH} approaches 1 (pk_{132} 10·4).

In pursuing this line of thought it was necessary to assume that (*d*) (− − +) and (*b*) (− − 0) had the same ϵ value at the wavelength λ.

$^-O-C_6H_4-CH_2CH(NH_2)COO^-$ − − 0   (*b*)

$^-O-C_6H_4-CH_2CH(NH_3^+)COO^-$ − − +   (*d*)

In fact in o-methyltyrosine the value of ϵ changes little with ionization of the NH_3^+ group. The value for (*b*) can be determined for tyrosine at pH values over 12·5 when (*b*) is overwhelmingly predominant, but ϵ for (*d*) (− − +) is much less accessible. For tyrosine at pH 13 $\lambda_{max.}=293$ nm, $\epsilon_{max.}$ 2390 and at pH 7 $\lambda_{max.}$ 275 nm has $\epsilon_{max.}$ 1400. N-trimethyltyrosine shows at pH 13 $\lambda_{max.}$ 292 nm, $\epsilon_{max.}$ 2350 and at pH 4·7 $\lambda_{max.}$ 274 nm, $\epsilon_{max.}$ 1340.

Tyrosine solutions were examined at 235–240 nm

at several pH values and for the cation (00+) $\epsilon_{235\,nm}$ was 1913 and for the dipolar ion (−0+) $\epsilon_{235\,nm}$ was 1448. The results led to $pK_1 = 2{\cdot}32 \pm 0{\cdot}05$. Accepting that chromophorically OCH_3 differed little from OH and $N(CH_3)^+$ from NH_3^+ in their effects on the rest of the molecule, progress could be made in interpreting the spectrophotometric results. In the case of tyrosinamide pK_1 (NH_3^+ group) was 7·48 and pK_2 (OH group) 9·89. Similarly tyrosine ethyl ester led to pK_1 7·33 and pK_2 9·80. O-methyl-tyrosine ethyl ester gave pK_1 7·31.

The following findings emerge for tyrosine: pK_1 2·34; pK_2 calc. 9·12, obs. 9·11; pK_3 calc. 10·20, obs. 10.13.

Polytyrosine and tyrosine copolymers were subjected to spectrophotometric titration (Katchalski & Sela, 1953; Sela & Katchalski, 1956), with the result that tyrosyl groups were reversibly ionizable without hindrance, indicating that tyrosyl-carboxylate bonding was negligible.

The value of pK_3 corrected to zero ionic strength is 10·15 for L-, D- and DL-tyrosine. For DL-tyrosine ethyl ester it is 10·09, glycyl-L-tyrosine 10·07, for L-glutamyl-L-tyrosine 10·33 and for leucyltyrosine 10·2. Proteins tend to give higher and more variable results. It is clear that in tyrosine itself the fact that the NH_2 and OH are close together in the same molecule affects the equilibria but the effects in proteins are not easily sorted out.

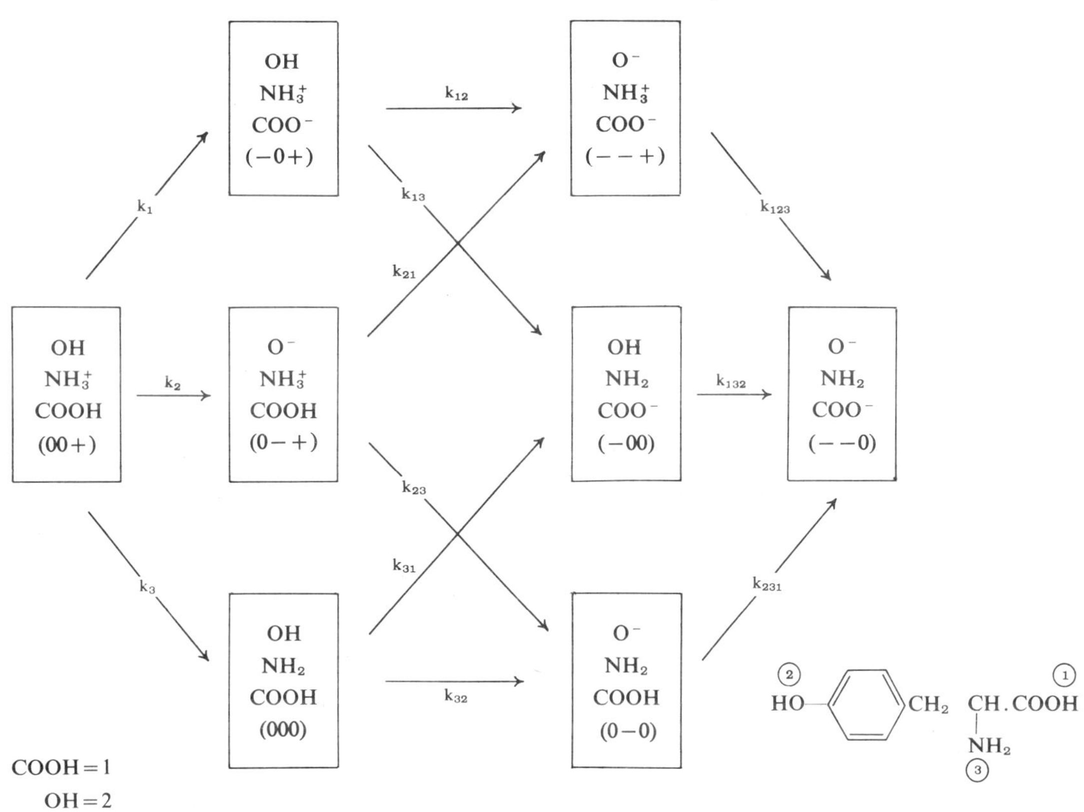

After Wetlaufer, D. B. (1962), based on Martin *et al.* (1958)

COOH		OH		NH_3^+	
pk_1	2·30	pk_2	9·42	pk_3	7·33
pk_{21}	2·51	pk_{12}	9·63	pk_{13}	9·28
pk_{31}	4·25	pk_{32}	9·80	pk_{23}	7·71
pk_{231}	4·49	pk_{132}	10·04	pk_{123}	9·69

O-methyl-L-tyrosine	pK_2	9·27
	pK_1	2·30
N-trimethyltyrosine	pK_2	9·75
	pK_1	1·5

8.1.7 Peptide Absorption in the Region 185–230 nm

The strong selective absorption due to the peptide link occurs in the far ultraviolet and is essentially the same as that of an N-substituted amide. The absorption curve shows a maximum at 190 nm and end-absorption extending towards longer wavelengths, *ca.* 240 nm. Masked by the end-absorption is a peak at about 220 nm. In an α-helical polypeptide (poly-γ-methyl-L-glutamate), Gratzer (1967) found three components to the far ultraviolet absorption: (*a*) $\lambda_{max.}$ 190 nm, perpendicularly polarized, $\pi^0 \rightarrow \pi^-$; (*b*) $\lambda_{max.}$ 208 nm (parallel polarization $\pi^0 \rightarrow \pi^-$); and (*c*) a peak at 222 nm, n—n^- band. The 208 nm band was characteristic of the α-helical form.

The peptide absorption in proteins is believed to be very sensitive to changes in conformation, but a number of side-chains also absorb in the region 185–230 nm and the reagents effecting denaturation also absorb in the same region. Circular dichroism and optical rotatory dispersion have proved very serviceable in studies of conformation, partly because the ORD curve of a polypeptide is determined predominantly by the absorption exerted by the peptide groups and their spatial relationships.

Although all peptides show selective absorption at 190–194 nm it has hitherto been difficult to profit from the fact because (*a*) spectrophotometers have not worked well enough at that region and (*b*) most solvent systems have displayed considerable irrelevant absorption.

Mayer & Miller (1970) have in principle overcome the technical difficulties. They used a Beckman DBG spectrophotometer with a deuterium lamp and a recorder, and using a new diffraction grating they connected the instrument to a source of continuously flowing high purity nitrogen dried over silica gel. The solvent in use was aqueous sodium fluoride (0·075 or 0·15 **M**). Absorption curves for purified bovine serum albumin and (L-alanyl)$_2$-L-alanine showed $\lambda_{max.}$ 193 nm and 192 nm respectively. In order to establish specific absorbance five proteins (trypsin, pepsin, haemoglobin, cytochrome c and bovine serum albumin) were examined (Table 8.7). The average value of $E^{1\%}_{1cm}$ was 686. Beer's Law was shown to be valid up to E=0·7 for 3 proteins and polyalanine. It will be seen that the specific absorbance at 191–194 nm is about 40–80 times that of the 280 nm band which is due to aromatic aminoacids and is of course variable from protein to protein.

In using absorption at 191–194 nm for monitoring protein solutions, interfering absorption is best eliminated by dialysing against aqueous sodium fluoride. Contamination by polysaccharides does not greatly influence absorption at 191–194 nm. (If an unknown protein is dried and weighed any carbohydrate will tend to reduce $E^{1\%}_{1cm}$ by simple 'dilution'.) Solutions should be as free as possible from light-scattering by particulate matter. The use of quite small glass (or plastic) vessels minimizes adsorption losses.

Although nitrogen purging is not indispensable, measurements at 190 nm are not far from the limit of performance in air. The replacement of a hydrogen

Table 8.7 Intensities of absorption of different proteins

Protein*	N% (micro-Kjeldahl)	$E^{1\%}_{1cm}$ 191–194 nm	$E^{1\%}_{1cm}$ 280 nm
Haemoglobin (human)	16·3	630	16·8
Cytochrome c (horse heart)	15·5	650	17·1
Plasma albumin (bovine)	—	650	8·20
Trypsin (bovine pancreas)	16·7	720	15·6
Pepsin (pig stomach)	14·3	780	14·1
Poly-DL-alanine	—	620	—
(L-alanyl)$_2$-L-alanine	—	660	—

$E^{1\%}_{1cm}$ absorbance of 1 cm layer of solution 10 mg/ml.

* Protein solutions were dialysed against distilled water and dried in the frozen state. Further drying was carried out over P_2O_5 in a vacuum desiccator to constant weight. Solutions (1000 μg/ml) were made in 0·15 M NaF. This solution was used for micro-Kjeldahl determinations. Threefold dilution (to 333 μg/ml exactly) permitted $E^{1\%}_{1cm}$ 280 nm to be measured. Dilution to 10 μg/ml allowed $E^{1\%}_{1cm}$ 191–194 nm to be measured (in 0·15 M NaF). The peptides were weighed out directly without drying, to constant weight.

lamp by a deuterium lamp provides additional light at 190 nm. It should be remembered that in a Beckman spectrophotometer the grating can deteriorate as time goes on and the mirrors must be cleaned (or preferably replaced) every few months.

The differences in specific absorbance shown by different proteins at 191–194 nm may be due to conformational differences (cf. Rosenheck & Doty, 1961).

8.1.8 Protein Meter

Bennett *et al.* (1970) described an effective protein meter based on the fact that nearly all proteins exhibit an absorption peak near 280 nm due to aromatic aminoacid moieties. A light source such as a hydrogen lamp or a mercury vapour lamp used in conjunction with a monochromator or an interference filter permits suitable radiation to be isolated for quantitative light absorption measurements (cf. Sullivan & Walsh, 1966). It is often desired to monitor chromatographic separations of proteins by measuring absorption at 280 nm on quite small volumes of liquid (e.g. 0·15 ml).

The element magnesium has a resonance line at 285·2 nm and this can be isolated from the light output by 'selective modulation' (Bowman *et al.*, 1966; Sullivan & Walsh, 1965, 1968). The radiation from a light source emitting the spectrum of magnesium is passed through a pulsating cloud of magnesium atoms mostly in the ground state. The 285·2 nm line is attenuated by passage through the cloud of ground state atoms; the effect is periodic because of the pulsation. The 285·2 nm line, and that line only, is modulated. The emergent beam falls on a photoelectric detector operating on a system responsive only to modulated signals. The details of the power supply and construction of a selective modulator are given by Bennett *et al.* (1970). The modulated radiation passes through a light filter (UG5 Schott u. Gen., Mainz) which has 80% transmission at 285·2 nm and is opaque to wavelengths between 450 and 650 nm. The system gives a pulse of 60 mA and *ca.* 2 msec duration at a repetition rate of 20 msec. The absorption of the resonance line varies from 50% to virtually zero. Small cells (1 cm × 0·3 cm diameter) can be used and the system is stable and almost noise-free. Absorbances are measured with a precision of ±0·25% and the instrument has a very small (~0·001 nm) spectral band pass.

This type of protein meter has considerable possibilities especially for easy monitoring of protein fractionations. The technique is moreover in principle applicable to the isolation of the resonance lines of other elements that are conveniently placed for other analytical problems.

8.1.9 Nitration of Tyrosyl Groups

Tyrosyl residues in proteins can be nitrated with tetranitromethane (Riordan *et al.*, 1965, 1966; Sokolovsky *et al.*, 1966). The characteristic nitrotyrosyl absorption peak ($\lambda_{max.}$ 428 nm) can be used for analytical purposes. Sokolovsky *et al.* (1967) reduced the nitro group to an amino group using sodium hydrosulphite ($Na_2S_2O_4$). The 428 nm peak rapidly disappeared, a new maximum appearing at 288 nm (for 3-aminotyrosine $\epsilon_{max.}$ 288 nm = 2800).

Lysozyme was nitrated and the nitrolysozyme contained 2·6 nitrotyrosyl residues and 0·4 tyrosyl residues per molecule. On reduction 2·5 moles of 3-aminotyrosyl residues per mole of lysozyme were obtained with no trace of nitrotyrosyl or any other changed residue. The absorption spectra of 3-aminotyrosyl residues show $\lambda_{max.}$ 288 nm at pH 7·4–9·0. In dilute hydrochloric acid $\lambda_{max.}$ is at 275 nm, $\epsilon_{max.}$ 1600. At pH 11·1 $\lambda_{max.}$ is displaced to 302 nm, $\epsilon_{max.}$ 4200. Between pH 4·7 and 11·1 there are isobestic points at 271 and 291 nm. The pK_1 values are 4·75 and 10·0 and resemble those of o-aminophenol, corresponding to the amino and hydroxyl groups respectively; pK_1 phenolic hydroxyl values found were: tyrosine 10·1, 3-aminotyrosine 10·0, 3-nitrotyrosine 7·2 and 3-acetylaminotyrosine 9·22. Introduction of the amino group into tyrosyl residues constitutes a new and promising approach to a number of problems.

2-Hydroxynitrobenzylbromide (HNBB) is another reagent which gives serviceable absorption spectra for tryptophanyl and cystinyl residues.

In the case of RNAase T_1 (from takadiastase) the molecule has one tryptophanyl residue and no methionine. This RNAase reacted with the reagent in 50-fold excess to give a new band $\lambda_{max.}$ 412 nm, $\epsilon_{max.}$ 18 900 corresponding with one tryptophanyl residue (position 59) being present.

DNAase (bovine pancreatic desoxyribonuclease, fraction A) has 16 tyrosine residues. Acetylation with N-acetylimidazole brought about a total loss of activity against calf thymus DNA. Only 7 or 8 of the tyrosines were acetylated and activity was restored by treatment with hydroxylamine. Hence acetylation did not effect an irreversible structural rearrangement. A spectrophotometric titration of native DNAase showed that 8–9 tyrosine residues were ionized in the pH range 7–11. If, however, DNAase is treated with tetranitromethane, a monomeric and a dimeric fraction can be separated. In the former a single tyrosine residue is converted to 3-nitrotyrosine. Peptides were isolated from tryptic and chymotryptic digests and the compositions and yields indicated that nitration had occurred at a single primary site in the DNAase molecule (Hugli & Stein, 1971).

DNAase can be estimated by ultraviolet absorption at 280 nm ($E^{1\%}_{1cm}$ 12·3). The 3-nitrotyrosine content can be determined at 428 nm (ϵ 4200; Riordan *et al.*, 1967). The number of O-acetyltyrosine groups formed was also obtained spectrophotometrically (Riordan *et al.*, 1965) by means of the difference value (Myers & Glazer, 1971) of $\epsilon_{278} = 1200$ per mole of tyrosine acetylated. The free sulphydryl groups were also determined spectrophotometrically (Price *et al.*, 1969).

A specific Ca^{2+} binding site on DNAase was destroyed by the nitration of the single tyrosine residue. Calcium ions no longer stabilized the nitrated enzyme against chymotryptic digestion or mercapto-ethanol reduction of an essential disulphide linkage. Optical rotatory dispersion on native and nitrated DNAase showed that the latter did not undergo the Ca^{2+}-induced conformation change shown by the former. Other experiments showed that a second calcium ion-binding site exists, but the specific Ca^{2+} binding site and the substrate site must be in close proximity in the enzyme molecule (cf. Price *et al.*, 1969).

As long ago as 1943 Cranmer and Neuberger concluded that tyrosine residues in ovalbumin are not free to ionize in the native state because the protein configuration is restricted by its tertiary structure.

The tyrosine residues in proteins are now recognized as either 'normal' or 'buried', 'normal' meaning that there is full exposure to solvent molecules and 'buried' meaning substantially hindered accessibility. The 'either/or' situation is however a considerable over-simplification (cf. Laskowski, 1966; Williams & Laskowski, 1965).

Kurihara *et al.* (1963) proposed cyanuric fluoride (CyF) as a tyrosine-modifying reagent. It can be regarded as a very reactive acid halide which acts readily on tyrosine residues. The ultraviolet absorption of tyrosine is greatly reduced and shifted to shorter wavelengths on reaction with CyF. Other aminoacids either are not affected or there is no change of absorption on the long wave side of 290 nm. The tyrosine-CyF product shows feeble absorption at pH 12, amounting at 290 nm to 3% of that of tyrosine itself and cyanuric acid is transparent on the long wave side of 285 nm. CyF distinguishes in insulin between two residues which react at pH 9, one which becomes accessible at pH 12 and one which is attacked only after 3 hours at pH 12·5.

Gorbunoff (1967) studied ribonuclease ($\epsilon_{max.}$ at 278 nm, 9700 at pH 7, 0·1 **M** phosphate buffer and 10% dioxan) and also α-lactalbumin and β-lactoglobulin. Spectrophotometric titrations were made (using the protein solutions at pH 7 for reference purposes) and equilibrium was reached in 1 hour. The difference spectra showed $\lambda_{max.}$ 295 nm for ribonuclease, 299 nm for α-lactalbumin and 297·5 nm for β-lactoglobulin. The number of ionized tyrosine residues in ribonuclease was calculated using ϵ 2630 for the phenoxide ion and for the other two proteins 2300.

In using cyanuric fluoride, excess of reagent is added to a solution of selected pH and the interaction occurs very quickly and is followed by slow hydrolysis of excess reagent to cyanuric acid. This must be complete or irrelevant absorption will occur. The pH is then adjusted to 13 and the ultraviolet absorption is measured between 290 and 340 nm. At pH 7 CyF does not change the absorption in this region. Times of reaction will vary and need to be checked by independent spectrophotometric titration of the untreated protein under study. The number of moles of tyrosine residues to react under given conditions of pH and temperature was calculated from the absorbance change at the peak (295, 299 or 297·5 nm) divided by the product of molar concentration of protein and the molar extinction coefficient for the phenoxide ion in the given protein. The spectrum was measured against a pH 7 standard and it gave the number of moles of unreacted tyrosine residues. Alternatively the spectra could be measured versus a pH 13 reference solution to give the number of tyrosine residues that had undergone reaction.

It was found that ribonuclease has six tyrosine residues, three of which are 'buried'. At pH $>1{\cdot}5$ denaturation results in their exposure. At pH 9·5 and at 25° the number of reactive tyrosine residues is 1·8, at pH 11·7 it is 2·8 and does not exceed 3 at pH 12, and at 3° the respective values are 1·0, 1·7 and 2·4. From this it is clear that the three 'buried' tyrosines

Table 8.8 States of tyrosine residues in purified proteins (Numbers of tyrosine residues)

Protein	pK	Titrated groups: Reversible	Titrated groups: Irreversible	Action of tyrosinase	Reactive to cyanuric fluoride
Ribonuclease	9·9	3	3	1	1+1+1
α-Lactalbumin	10·4	4	1	5	3+1
β-Lactoglobulin	10·9	3	1	0	2+1

are not identical. A similar study showed that α-lactalbumin contains one unreactive and four reactive residues. Of the four one was less reactive than the others. β-Lactoglobulin contains one unreactive and three reactive residues, one of which had lower reactivity.

8.1.10 Difference Spectra

Many workers have found it useful to study the small displacements of absorption spectra in proteins by means of difference spectra (Laskowski *et al.*, 1956; Scheraga, 1957; Donovan *et al.*, 1958). Suppose that an acidified solution of a protein at, say, pH 1 is placed in one cell and the reference cell contains the same protein at the same concentration but without added acid; the difference ΔE (or Δϵ when the molecular weight is known) can be measured against changing wavelength. Either solution can be taken as the reference solution but it is convenient to make the choice so that Δϵ is positive.

Beaven (1961) considers the 'true difference spectra' should refer to two solutions containing identical solutes but differing in concentration. When the difference is of the order 5–10% the performance of many spectrometers is good if one solution is placed in the sample beam and the other in the reference beam (Hiskey, 1949; Bastian, 1949). The concentration of one of two solutions should be accurately known. The advantage of this type of difference spectrum lies in the fact that provided stray light errors and fluorescence errors are avoided, it gives greater accuracy than subtraction of one independently measured spectrum from another. Qualitatively the difference spectrum will be the same as the reference spectrum.

If, however, the absorption spectrum of a solute such as tyrosine is measured at a definite pH, e.g. 1·0, it is possible to obtain a new curve in which ΔE (or Δϵ) is plotted against Δλ. Δϵ/Δλ will be zero at a maximum or minimum in the absorption curve, and will be maximal at wavelengths at which the absorption is rising or falling most steeply. First differential or derivative spectra can be recorded (Martin, 1957; Olson & Alway, 1960) and for tyrosine there are maxima at *ca.* 286 and 276 and minima at 278 and 263 nm. Tryptophan has maxima in its differential spectra at 289 and 282 nm with minima at 285·7 and 266 nm. Inflections in the original absorption curves are well displayed in the Δϵ/Δλ curve and indeed this type of curve helps to reveal fine structure (Giese & French, 1955). When the perturbation of a simple absorption curve involves a shift of the absorption envelope as a whole either to shorter wavelengths (blue shift) or longer wavelengths (red shift) with little change in intensity of absorption the difference curve will closely resemble the differential curve. Real situations are however more complicated, as changes occur on both the wavelength scale and the absorbance scale.

Yanari & Bovey (1960) compared solutions of glycyl-L-phenylalanine in which the pH was nearly neutral in the sample beam and 1 to 1·5 in the reference beam. The dfference spectrum showed narrow 'bands' in the region 250–275 nm. Similar difference spectra were obtained for glycyl-L-tyrosine and glycyl-L-tryptophan. A difference spectrum for ribonuclease showed a good resemblance with the tyrosine difference spectrum whereas chymotrypsin had peaks resembling both tryptophan and tyrosine and ovalbumin had indications of all three amino acids (Fig. 8.8).

Figure 8.8 (*a*) Difference spectra of the individual aromatic aminoacids and (*b*) of three proteins; solution of pH 6 to 7 in sample beam and solution of pH 1 to 1·5 in reference beam. (1) 10^{-2} M glycyl-L-phenylalanine; (2) $2{\cdot}5\times10^{-3}$ M glycyl-L-tyrosine; (3) 5×10^{-4} M glycyl-L-tryptophan (from Yanari & Bovey, 1960)

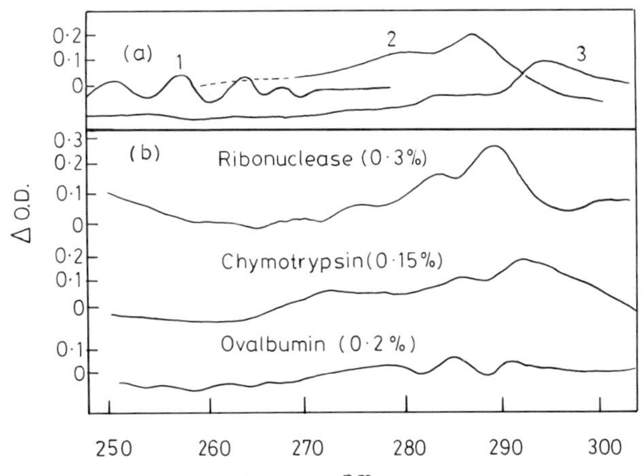

Donovan *et al.* (1961) studied the absorption of tryptophan at different pH values between 1·7 and 10·3 using a solution at pH 1·2 for reference. The difference curves show $\lambda_{max.}$ 290–294, 287, *ca.* 252 and 230 nm with minima at 264–267, 235 and 222–224 mn. The absorption of tyrosine at pH 6 and 13 and the difference curve are shown in Fig. 8.9.

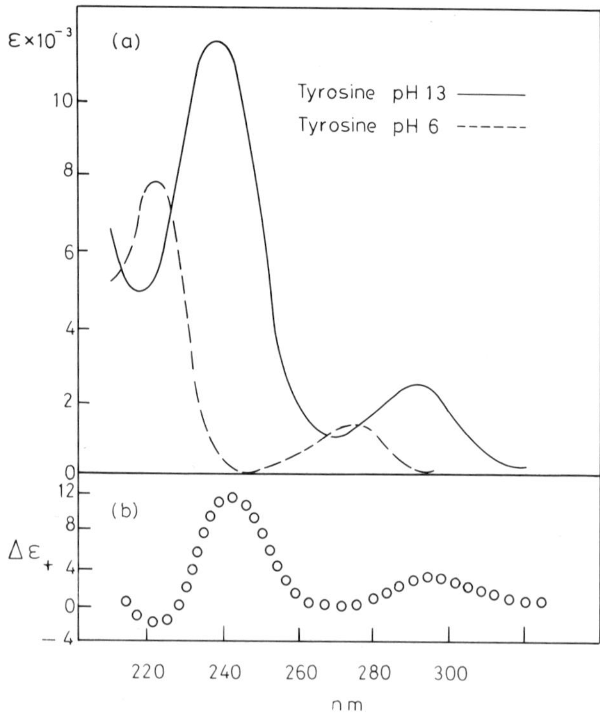

Figure 8.9 (*a*) Spectra of tyrosine at pH 13 and 6; (*b*) difference spectrum

O-methyltyrosine provides an interesting example. At pH 1·08 $\lambda_{max.}$ 273·9 nm, ϵ 1356 and as the pH rises to 5·7, 9·6 and 11·5 the ϵ values remain virtually unchanged on the short wave side of 270 nm but rise on the long wave side and there is a red shift. The difference spectrum pH 5·7 versus pH 1·08 shows two peaks corresponding with steeply rising absorption (287 nm and 283 nm) with a good minimum at 279 nm corresponding with the inflections in the directly measured spectra. The difference spectrum pH 11·5 versus pH 1·08 shows maxima at 289 and 284 nm and a minimum at 279 nm (Fig. 8.10). O-methyltyrosine is of course a simple example since the methoxyl group does not ionize (cf. Wetlaufer *et al.*, 1958; Chervenka, 1959); it is important, however, since it illustrates how ionization of the amino group can

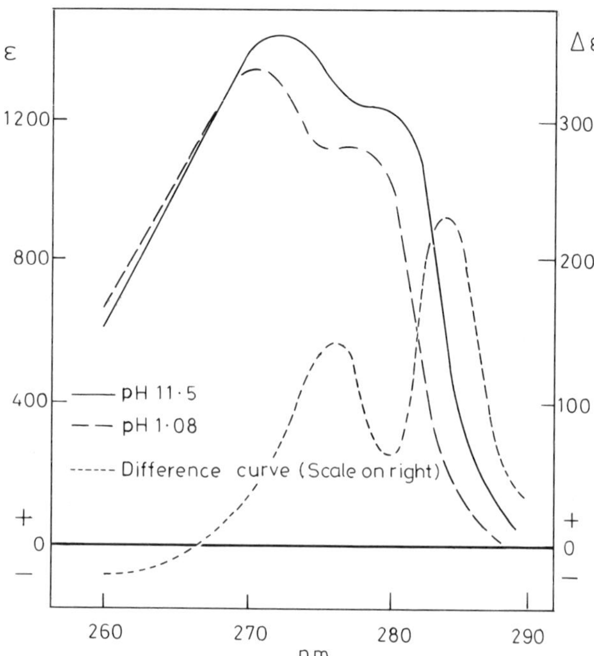

Figure 8.10 Spectra of O-methyltyrosine at pH 11·5 and 1·08 and the difference curve (right-hand scale) (after Wetlaufer *et al.*, 1958)

disturb the spectrum. In tyrosine the larger effect of phenolic ionization largely masks this effect. Wetlaufer *et al.* compared glycyl-O-methyltyrosine with O-methyltyrosine and found rather surprisingly that the perturbing effect of amino ionization shown by the latter did not occur with the former. The perturbation of an adjacent amino group decreases as the distance between NH_2 and COOH increases.

Proteins undergo modification in the presence of moderate concentrations of salts or of urea and the shifts that occur at first can at higher concentrations be reversed. The effects however can be somewhat complex (Bigelow & Geschwind, 1960). These workers studied the effect of various concentrations of urea on the absorption shown by ribonuclease at 287 nm (see Fig. 8.15). From 0 to 4 **M** urea, ϵ_{287} shows a linear increase with concentration of urea and if the line is extrapolated to 8 **M** it reaches $+\epsilon=400$, an amount corresponding with that expected for three tyrosyl groups. But between 4 **M** and 8 **M** urea the rise is not linear and $\Delta\epsilon$ decreases finally to −1750. This effect corresponds to the transfer of three anomalous tyrosyl groups from a state inaccessible to the solvent to full accessibility. The value of −1750 is made up as follows: (*a*) increase of ϵ for three normal tyrosyl groups (at 8 **M** urea) +400; (*b*) decrease on exposure of three anomalous tyrosyls and (*c*) increase of 400 for the three newly exposed groups in 8 **M** urea, so that the initial state for the exposed tyrosyls is −2550 (cf. Sarfare & Bigelow, 1967).

Oxidized ribonuclease (performic acid) has no buried or inaccessible tyrosyl groups and $\Delta\epsilon_{287}$ (8 **M** urea) is actually 880, not far from 2×400. (See also Nelson & Hummel, 1962.)

Bigelow (1961) went further and tried to connect spectroscopic changes with individual tyrosyl residues. Comparisons were made with native ribonuclease at pH 7 and at 25°. Denaturation with alkali normalized residues A, B and C, $\Delta\epsilon_{287}$ −2660 as did 8 **M** urea, and 10 **M** lithium bromide $\Delta\epsilon_{287}$ −2600 and −2700 respectively. In 5 **M** lithium bromide, residues A and B were normalized ($\Delta\epsilon_{287}$ −1700) and at lower neutral pH, residues A and B were normalized at temperatures over 43°C. (See also Scheraga, 1961.)

Brandts (1964) used difference spectra as a tool for exploring the thermodynamics of protein denaturation. Working with chymotrypsinogen, he considered reversible temperature denaturation:

N (native) ⇌ D (denatured)

$K=[\mathrm{D}]/[\mathrm{N}]$ K being the thermal equilibrium constant

Figure 8.11 Difference spectrum for alkali-denatured ribonuclease at neutral pH (ribonuclease concentration = 1·44 mg per mol). ○, uncorrected data; ●, corrected data; ----, correction for light scattering (from Bigelow, 1961)

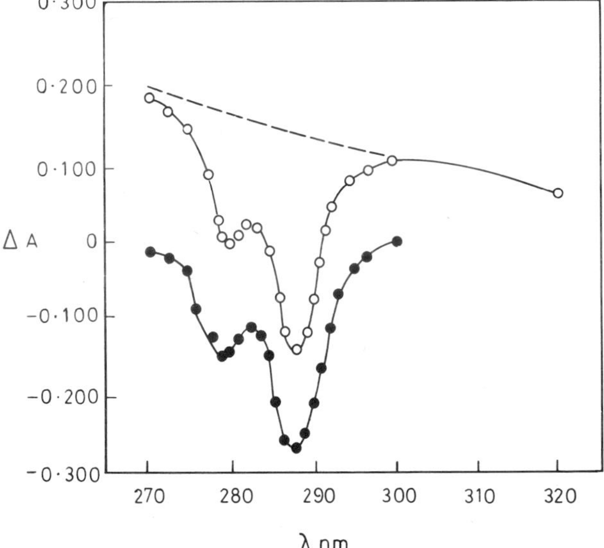

When ϵ_{N} stands for $\epsilon_{293\,\mathrm{nm}}$ for the native protein, ϵ_{D} for the denatured protein and ϵ for a solution in the transition region

$$K=\frac{\epsilon-\epsilon_{\mathrm{N}}}{\epsilon_{\mathrm{D}}-\epsilon}$$

or using difference spectra

$$K=\frac{\Delta\epsilon-\Delta\epsilon_{\mathrm{N}}}{\Delta\epsilon_{\mathrm{D}}-\Delta\epsilon}$$

where all the extinction coefficients were reduced by the same amount by subtracting the extinction coefficient for some arbitrary standard solution. The standard free energy of denaturation is

$$\Delta F^{0}=-RT\ln K$$

and the enthalpy change

$$\Delta H^{0}=-R\frac{\mathrm{d}(\ln K)}{\mathrm{d}(1/T)}$$

In chymotrypsinogen stabilization depends nearly equally on hydrogen bonding and hydrophobic bonds, whereas instability reflects a large conformational entropy of unfolding. Clearly 'anomalous' groups with unusually high (or low) dissociation constants in the native protein attain a normal dissociation constant in the denatured protein and strong localized interactions in the native protein tend to disappear. Changes in long-range electrostatic attractions must also occur in the N⇌D reaction.

Ribonuclease also undergoes reversible thermal denaturation involving drastic conformational change to give a very disordered product. Brandts (1965) and Brandts & Hunt (1967) measured $\Delta\epsilon_{287\,\mathrm{nm}}$ at five pH values and different temperatures. The results

Figure 8.12 Thermal transition of ribonuclease at pH 1·13 to 3·15, studied by difference spectroscopy. The difference ($-\Delta\epsilon$) is plotted relative to the native protein at the pH shown. $\Delta\epsilon$ does not reach zero because complete conversion to the native protein does not occur at 0° (from Brandts, 1965; Brandts & Hunt, 1967)

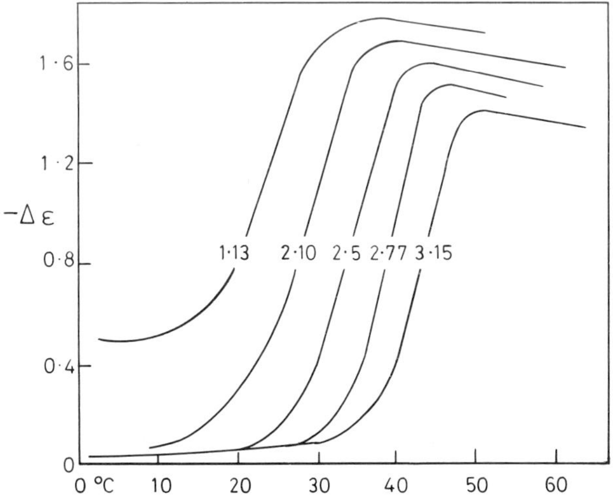

were plotted (Fig. 8.12), the differences always being related to the ϵ values for the native protein at the indicated pH. At 41°C thermal denaturation is about 90% complete, and it is useful to plot the fraction of the total change which occurs in going from the native to the denatured state.

8.1.11 Measurement of Accessibility of Protein Chromophores by Solvent Perturbation of their Ultraviolet Absorption Spectra

Solvent effects on the spectra of proteins are well known. Thus addition of a transparent substance to aqueous solutions of aromatic compounds generally results in a small change in extinction coefficient, usually a rise. As an example, the spectra of acetyl-L-tyrosine ethyl ester in water and in a 1·5% solution of β-Schardinger dextrin (cycloheptamylose) may be compared. The phenolic ring is thought to be embedded in a cavity of the dextrin molecule and in fact a stoichiometric (1:1) inclusion complex is formed. The so-called solvent effect is here a change in the absorption spectrum of the phenolic ring structure in a 'carbohydrate' environment. Addition to water of sucrose or glycerol results in qualitatively similar changes. Bigelow (1960) and Herskovitz & Laskowski (1960, 1962) used this kind of approach in studying globular proteins. As has been seen, bovine pancreatic ribonuclease has six tyrosyl residues, three of them readily and reversibly ionizable, and three ionized irreversibly and only at high pH. The three abnormal tyrosyls are 'buried' in the interior of the molecule. Addition of a neutral perturbant such as glycerol to native ribonuclease should perturb only the exposed tyrosyls. But ribonuclease denatured in 8 **M** urea by having its disulphide bonds cleaved behaves like a random coil and all six of its tyrosyls are readily and reversibly ionizable, with normal pK values (Harrington & Sela, 1959; Scheraga & Rupley, 1962). Addition of glycerol should then perturb all the tyrosyls.

Solvent perturbation difference spectra of the native protein should be half as large as that of the fully denatured derivative. A simple experimental arrangement of cells in tandem (Fig. 8.13) allows this to be tested. The spectrophotometer acts here as an analog computer; it subtracts the spectrum of the reference (standard) solution from that of the sample and displays the answer—the solvent perturbation difference spectrum—on the recorder.

Herskovitz & Laskowski (1962) have used solvent

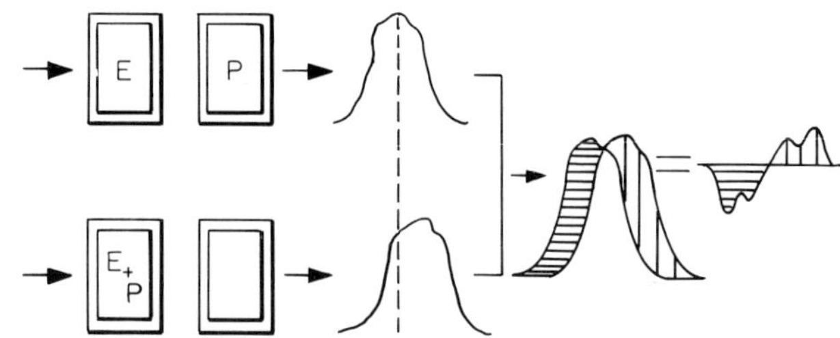

Figure 8.13 Schematic diagram of an experiment designed to measure the solvent perturbation difference spectrum of an enzyme E by a perturbant P. Note that in the tandem double cell in the reference compartment the perturbant and the enzyme are not allowed to come in contact, while in the tandem double cell in the sample compartment contact is allowed. This is the only difference encountered by the reference and sample beams

perturbation to locate the chromophoric residues within protein molecules. As a preliminary step they recorded the perturbation spectra for the N-acetyl-ethyl esters of phenylalanine, tyrosine, O-methyl-tyrosine and tryptophan, using polyethyleneglycol as the perturbant. The difference spectra were recorded 10–30 minutes after mixing. They proceeded to examine crystalline bovine serum albumin and human serum albumin. Both proteins display the normal ultraviolet absorption with $E^{1\%}_{1cm}$ 6·67 for BSA and 5·30 for HSA in aqueous solution. The probable primary structures in respect of aromatic aminoacids are (residues per mole):

	Tyrosine	Tryptophan	Phenyl-alanine
BSA	21	2	26
HSA	18	0·6	33

On this basis a 'model' for BSA would have the composition:

N-acetyl-L-tyrosine ethyl ester	$1{\cdot}8 \times 10^{-3}$ **N**
N-acetyl-L-tryptophan ethyl ester	$1{\cdot}7 \times 10^{-4}$ **N**
N-acetyl-L-phenylalanine ethyl ester	$2{\cdot}2 \times 10^{-3}$ **N**

The 'model' mixture should undergo maximal perturbation whereas the undenatured protein should undergo less if chromophore groups are buried.

Another approach would be to compare the perturbations brought about on the native protein with those effected on the protein denatured in different ways. A strong solution of urea (8 **M**) has the advantage that it helps to keep proteins in solution. A convenient measure of solvent perturbation is the quan-

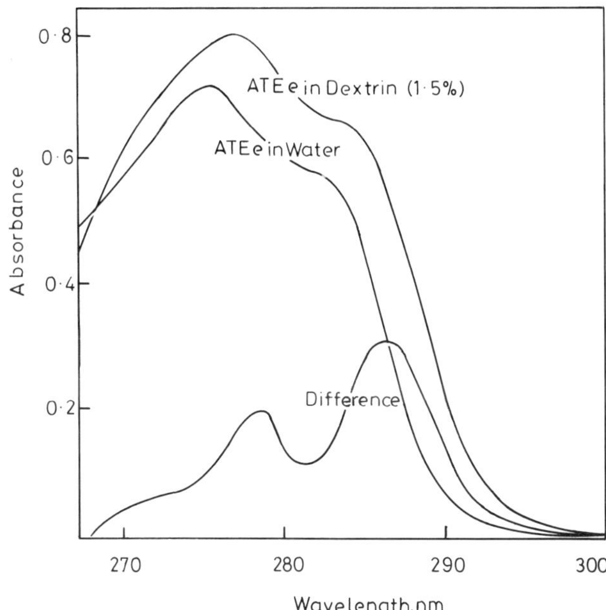

Figure 8.14 Ultraviolet spectra of acetyl-L-tyrosine ethyl ester in water and in 1·5% solution of β-Schardinger dextrin. The difference spectrum between these two solutions is also shown (from Laskowski, 1966)

tity $\Delta\epsilon/\epsilon$ where ϵ is the molar absorption coefficient of the native protein at 277–279 nm and $\Delta\epsilon$ is the molecular difference extinction coefficient.

Difference spectra for denatured serum albumin (BSA) and the model compound analogue

Perturbant	BSA: In 8 **M** urea $\Delta\epsilon/\epsilon$	BSA: Reduced (TGA)* in 8 **M** urea $\Delta\epsilon/\epsilon$	Model mixture
20% sucrose	0·035–0·036	0·035	0·03
20% glycerol	0·041	0·043	0·04
20% ethylene glycol	0·04–0·045	0·05	0·048
20% polyethylene glycol	0·059	0·074	0·082
20% dimethylsulphoxide	—	—	—
100% dimethylsulphoxide	0·077–0·085	0·112	0·15

* TGA: thioglycollic acid.

Careful denaturation of the protein is probably to be preferred to the use of the 'model' mixture although the two methods for ascertaining the perturbation when the aromatic residues are fully exposed are often in good agreement. It is quite important that the molecular extinction coefficient for tryptophan is about four times as great as that for tyrosine, so that large errors in $\Delta\epsilon/\epsilon$ for model mixtures could occur when the tryptophan content is not accurately known. In the case of BSA at pH 4–8, the native protein exhibits $\Delta\epsilon/\epsilon$ 0·1 to 0·11 as against 0·36–0·37 for the denatured protein (using in both cases 20% sucrose as the perturband). On this basis about 70% of the tyrosyl residues in the native protein are inaccessible to the perturbing actions of sucrose. If the pH is lowered to 3·8 about one-fifth of the 'buried' groups are exposed.

Good solvents for the method are aqueous sucrose, glycerol and ethylene glycol. They do not react strongly with the tyrosyl groups and are less prone than DMSO to disaggregate micelles.

It is now clear that a substantial proportion of the tyrosyl residues in bovine serum albumin and human serum albumin are buried in the folds of the native protein. At neutral and slightly acid pH values different perturbants behave quite similarly. In the case of BSA it has been shown (Tanford & Roberts, 1962) that at alkaline pH values *all* the tyrosyl residues are revers-

Figure 8.15 $\Delta\varepsilon$ versus urea concentration. ———, $\Delta\varepsilon$ observed for ribonuclease (3 tyrosyls exposed); ......, linear extrapolation; ----, oxidized ribonuclease (6 tyrosyls exposed). Reference solution, no urea

(Exposure to water of 3 buried tyrosyls	−2550
Water—8 **M** urea, 3 tyrosyls	+400
Water—8 **M** urea, second 3 tyrosyls	+400
	−1750

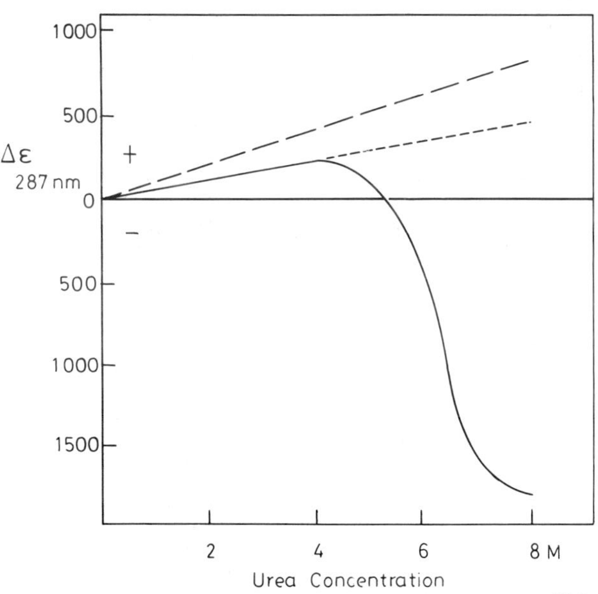

ibly ionized. It may well be that the protein undergoes a marked structural change at pH values greater than 7·5; an alternative explanation may be that the formation of tyrosylate ions precedes 'exposure'. On the other hand, there can be little doubt that over the pH range 4·3–3·8 serum albumins undergo considerable changes in conformation (as manifested by changed viscosity) and about half the tyrosyl residues are 'exposed'.

Foster (1960) proposed a model for BSA in which the peptide chain is folded into four flat sub-units stacked one on top of the other with three interfaces. Isomerizations are visualized as sub-unit separations. Herskovitz & Laskowski (1960) considered that three to five key tyrosyl groups occur on one interface between sub-units. 'Small' solvents like dimethylsulphoxide separate the sub-units and expose the three to five tyrosyl residues.

Leonard & Foster (1961) measured changes in optical rotation as a function of pH in BSA in the presence of strongly bound anions such as CNS^- and ClO_4^-. Two distinct stages were noted, the second corresponding with unmasking the three to five tyrosyl groups.

8.1.12 Ultraviolet Spectrophotometry of Serum Proteins

Strickland *et al.* (1960) used electrophoresis to separate 11 protein fractions from 12 normal blood sera and examined them all for tyrosine and tryptophan content by ultraviolet spectrophotometry. The variations in estimated tyrosine and tryptophan among the different protein fractions as well as the variations between corresponding fractions from different sera were quite wide. All the fractions were probably mixtures rather than pure proteins. The molar absorptivities obtained were:

Tyrosine	ϵ_{280nm}	1570	$\epsilon_{294 \cdot 4nm}$	2440
Tryptophan	ϵ_{280nm}	5210	$\epsilon_{294 \cdot 4nm}$	2440

so that

$$M_{tyr.} = 0 \cdot 588\epsilon_{294 \cdot 4nm} - 0 \cdot 275\epsilon_{280nm}$$
$$M_{try.} = 0 \cdot 275\epsilon_{280nm} - 0 \cdot 16\epsilon_{294 \cdot 4nm}$$

where ϵ refers to absorptivities at the indicated wavelengths and $M_{tyr.}$ and $M_{try.}$ are the millimolar concentrations of tyrosine and of tryptophan in the sample solution.

This investigation demonstrates the complexity of the problem. Although the major constituent of each electrophoresis fraction is probably known correctly, incomplete separations detract quite considerably from the value of the approach. Nevertheless this work seemed to show that the albumin fraction of serum from women was richer in both tyrosine and tryptophan than that from men.

Problems of buried tyrosyl residues have been tackled indirectly by studying the absorption spectra of tyrosine derivatives in non-polar solvents and by measuring circular dichroism under carefully selected conditions (Horwitz *et al.*, 1970; Strickland *et al.*, 1970; Edelhoch *et al.*, 1968; Strickland *et al.*, 1970; Strickland *et al.*, 1972).

As background the spectra of *p*-cresol and *p*-methylanisole were compared with those of *L*-tyrosine and O-methyl-*L*-tyrosine. The best resolution was obtained using perfluorinated hexane as an inert solvent. Figure 8.16 shows how well the vibrational structure

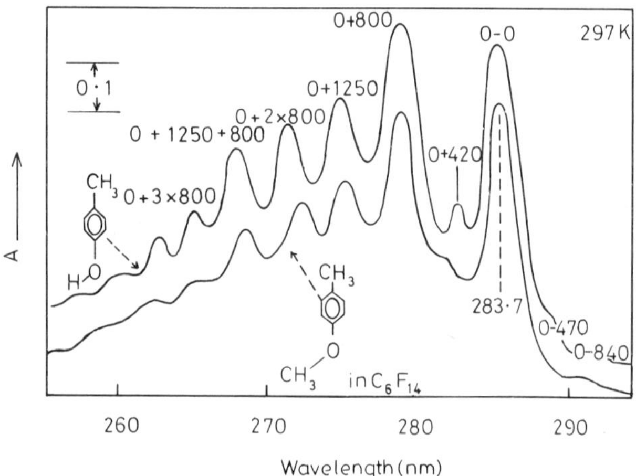

Figure 8.16 Absorption spectra of *p*-cresol and *p*-methylanisole dissolved in perfluorinated hexane. The spectrum of *p*-cresol has been offset to separate the records. One centimetre path length (from Strickland *et al.*, 1972)

is displayed even at 297 K. The 0—0 peak is at 283·7 nm and the structure shows frequency differences of $n \times 800$ cm^{-1}, 1250 and *ca.* 450 cm^{-1}.

To facilitate the work **N**-stearyl-*L*-tyrosine-*n*-hexyl ester and **N**-acetyl-**O**-methyl-*L*-tyrosine ethyl ester were prepared and the effects of hydrogen bonding

$HO-C_6H_4-CH_2CH(NHCOC_{17}H_{35})COOC_6H_{13}$

$CH_3O-C_6H_4-CH_2CH(NHCOCH_3)COOC_2H_5$

were ascertained. Methyl-cyclohexane was used as an inert solvent. **N**-stearyl-*L*-tyrosine-*n*-hexyl ester has the advantage of being soluble in both polar and non-polar solvents but is prone to aggregate. The corresponding O-methyl ester and **N**-acetyl-**O**-methyl-*L*-tyrosine ethyl ester do not aggregate so readily and are more suitable for investigation at low temperatures. Addition of low concentrations of dioxan, **N**-**N**-dimethylacetamide, 1-butanol or methanol to methylcyclohexane solutions caused a red shift (1–4 nm) and a 10–25% increase in the dipole strength. The interaction with the phenolic hydroxyl and proton acceptors like dioxan or dimethylacetamide reflects shifts caused by hydrogen bonding (cf. Nagakura & Baba, 1952; Takahashi *et al.*, 1967). The use of **N**-**N**-dimethylacetamide as a source of the carbonyl group which can form a hydrogen bond with phenolic hydroxyl provides a model for understanding a red shift of the tyrosine absorption in proteins.

Circular dichroism in the near ultraviolet provides a powerful tool (Strickland *et al.*, 1969). The CD intensity was calibrated by an aqueous solution of α-10-camphorsulphonic acid ($\Delta\epsilon = 2{\cdot}2\ M^{-1}\ cm^{-1}$ at 290 nm; Cassim & Yang, 1969) where $\Delta\epsilon$ is the molar extinction coefficient for left circularly polarized light minus that for right circularly polarized light. In calculating $\Delta\epsilon$ values at different temperatures the concentration increase on cooling was allowed for (Korver & Bosma, 1971; Passerini & Ross, 1953). Beer's law was then valid over at least a 5-fold concentration range.

The absorption spectrum of **N**-Ac-**O**-Me-*L*-Tyr. ethyl ester showed λ_{max} 274 nm, $\epsilon_{max.}$ 1400 in water, 277 nm, 1750 in EPA [ether, isopentane, ethanol 5:2:2 v/v] and 277 nm, 1850 in methylcyclohexane. The two **N**-stearyl compounds showed virtually the same spectroscopic properties. Figure 8.18 however shows that even at 297 K there was definite evidence of extended vibrational structure and at 225 K this was markedly enhanced. In EPA the solution of **N**-Ac-**O**-Me-*L*-Tyr. ester remained clear at 140 K and at 77 K, so that it was possible to measure absorption curves and circular dichroism. At room temperature a small negative dichroism occurs above 273 nm; this becomes stronger at 215 K, and at 140 K the increase in negative rotatory strength is more than 10-fold and the vibrational structure is fully developed.

The **N**-stearyl-*L*-tyrosine-*n*-hexyl ester dissolved in methylcyclohexane is strongly concentration-dependent both in absorption and CD (Fig. 8.19). The sub-

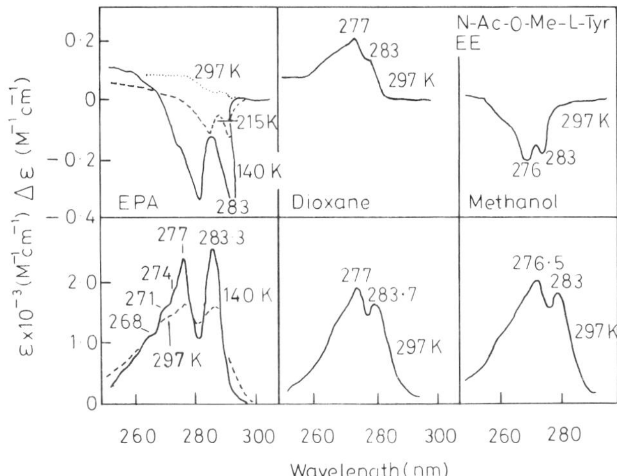

Figure 8.17 CD (top) and absorption (bottom) spectra of **N**-Ac-**O**-Me-*L*-Tyr. ethyl ester dissolved in EPA (left), dioxan (middle) and methanol (right). The rotatory strength of **N**-Ac-**O**-Me-*L*-Tyr. ethyl ester (27 mM) in EPA at 77 K is approximately the same as that observed at 140 K. The dipole strength of **N**-Ac-**O**-Me-*L*-Tyr. ethyl ester is $1{\cdot}3\times10^{-36}$ c.g.s. in all three solvents (from Strickland *et al.*, 1972)

Figure 8.18 CD (top) and absorption (bottom) spectra of **N**-Ac-**O**-Me-*L*-Tyr. ethyl ester dissolved in methylcyclohexane at 297 and 225 K. The concentration had to be kept less than 0·4 mM to prevent aggregation from occurring at 225 K. **N**-Ac-**O**-Me-*L*-Tyr. ethyl ester did not show any tendency to aggregate in this solvent at 297 K. The spectral half-intensity band widths were 1 nm for the CD measurements and 0·2 nm for the absorption measurements (from Strickland *et al.*, 1972)

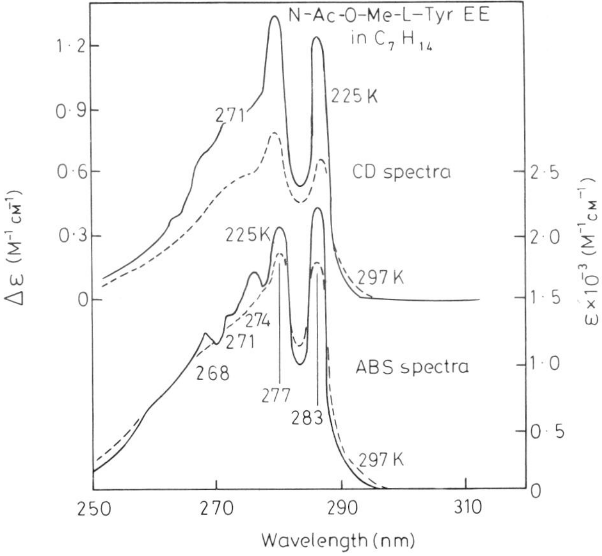

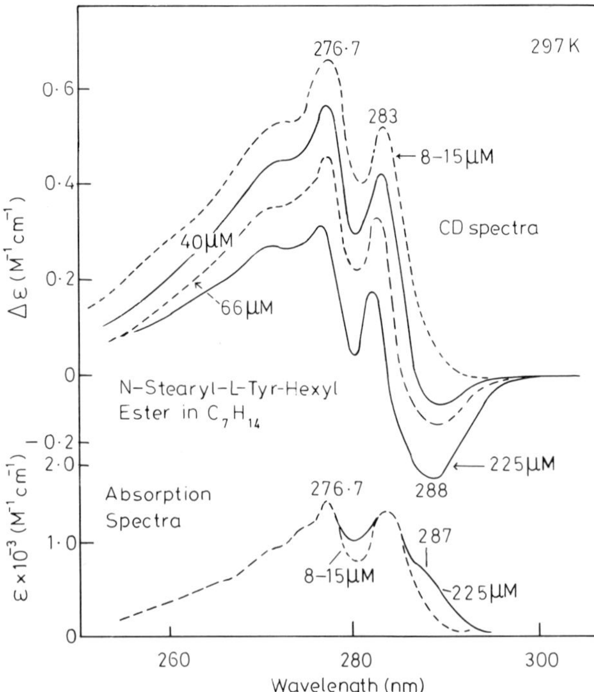

Figure 8.19 Effect of concentration upon the CD and absorption spectra of **N**-stearyl-*L*-tyrosine-*n*-hexyl ester dissolved in methylcyclohexane at 297 K. CD spectra are presented in terms of an apparent $\Delta\varepsilon$ calculated from $\Delta A = \Delta\varepsilon \times c \times l$, where c is the concentration of tyrosine derivative, l is the path length and A is the absorbance for left circularly polarized light minus that for right circularly polarized light (from Strickland *et al.*, 1972)

Figure 8.20 Effects of 1-butanol (left), *p*-dioxane (centre) and **N,N**-dimethylacetamide (right) upon the absorption spectra of **N**-stearyl-*p*-tyrosine-*n*-hexyl-ester (9 to 10 μ**M**) dissolved in methylcyclohexane. Solid lines were traced from instrument record to reveal the noise level. 0 **M** = zero concentration of additive. Path length was 5 cm. The reference cuvette had the same solvent composition as the sample cuvette. Solvent *versus* solvent base-lines were flat. The spectral half-intensity band width was 0·4 nm (from Strickland *et al.*, 1972).

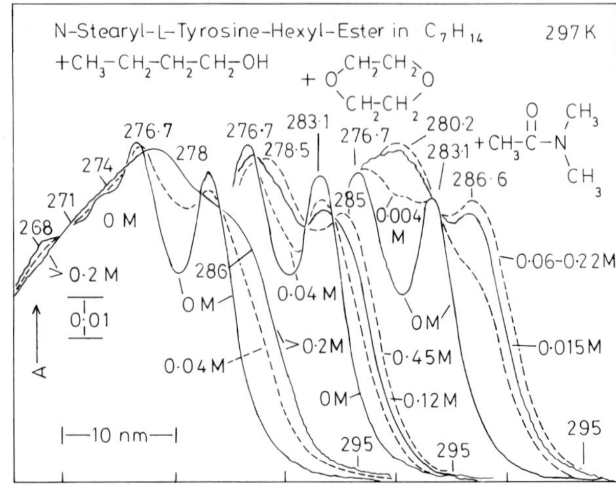

stance begins to associate at concentrations above 15 μM. Dimerization appears to be responsible for appearance of a new band at 287–288 nm (4 nm shift for the 0—0 band).

The effects of hydrogen bonding were studied as shown in Fig. 8.20. The excellent approach to isosbesticity supports the view that in each case a new entity is being formed, a view strengthened by the fact that the process reaches completion at a concentration of 30 mM **N**-**N**-dimethylacetamide when all the hydroxyl groups of the **N**-stearyl compound are hydrogen-bonded to carbonyl. The value of $\epsilon_{max.}$ at 280·2 nm was about 10% higher than that of the control at 276·7 nm. The dipole strength of the band was enhanced some 25% after hydrogen bonding to the dimethylacetamide. Strictly comparable changes occurred in the CD spectra. When butanol was used as the hydrogen bonding agent, the vibrational structure was blurred and the exact red shift became conjectural although clearly smaller than with the dimethylacetamide; perhaps more than one hydrogen-bonded species is formed. With **N**-stearyl-**O**-methyl-*L*-tyrosine-*n*-hexyl ester or **N**-stearyl-*L*-phenylalanine-*n*-hexyl ester the vibrational structure was not blurred, the dipole strength was unchanged and rather high concentrations of butanol were needed before the rotatory strength was greatly changed. An alcohol could thus form hydrogen bonds both with the hydroxyl group and the amide oxygen. Butanol perhaps interacts with **N**-stearyl-*L*-phenylalanine ethyl ester at quite numerous sites but there is no doubt about the overall importance of the phenolic hydroxyl group in hydrogen bonding. It greatly increases the tendency of tyrosine derivatives to associate in organic solvents, but this is not an interaction between two tyrosyl hydroxyls. The aggregation is stronger than for *p*-cresol and the shift is to the red whereas with *p*-cresol it is in the opposite sense. It is thus likely that the hydroxyl of one tyrosine is hydrogen-bonded to the carbonyl oxygen of another.

The large effect of cooling on the negative rotatory strength of **N**-Ac-**O**-Me-*L*-Tyr. ethyl ester is attributed to conformational motility (cf. Fretto & Strickland, 1971). The hydrogen bonding measured by equilibrium constants in the **N**-stearyl-*L*-tyrosine ester is strong and this seems to enhance dipole strength (Bailey *et al.*, 1968).

Lens Protein

The lens of the human eye tends to become yellow

with advancing age. Pirie (1971) has described a photochemical change occurring at the tryptophan moieties in protein and paralleled by photochemical change in free tryptophan. She used sunlight in her experiments, largely because of the possibility that exposure to the sun's rays has a part in causing cataract. Not all the protein of the lens is soluble in water, and the proportion of insoluble protein rises with increase of age. However, all the lens protein is soluble in 6 **M**-guanidinium chloride.

In the presence of air, protein exposed to sunlight undergoes oxidative cleavage of the indole ring of tryptophan yielding N′-formylkynurenine. Sunlight is not rich in the ultraviolet radiation absorbed by tryptophan but there is significant overlap between the emission spectrum of (terrestrial) sunlight and the absorption of tryptophan. The glass tubes used by Pirie transmitted 17% of light of 295 nm and 41% of light of 310 nm.

When solutions in 6 **M** guanidinium chloride of tryptophan or tryptophan-containing proteins are exposed to sunlight, they become yellow-brown and absorption at 280 nm diminishes while there are rises at 260 and 320 nm. This indicates the formation of N′-formylkynurenine with $\lambda_{max.}$ 321 nm and 260 nm. $\epsilon_{max.}$ 3750 and 10 980 (Knox & Mehler, 1950), and the absence of a rise at 360 nm points to the absence of kynurenine. [Ribonuclease, which is tryptophan-free, does not show these changes although it does turn yellow.] Formylkynurenine treated with 1 **M** HCl (20 min. at room temperature) yields kynurenine ($\lambda_{max.}$ 360 nm). On the other hand treatment of formylkynurenine with sodium borohydride reduces the oxo group with disappearance of the 260 and 321 nm peaks.

	X	Y
N′-formylkynurenine	CHO	O
Kynurenine	H	O
γ-(-2-aminophenyl)homoserine	H	HOH
γ-(2-formylaminophenyl)homoserine	CHO	HOH

8.1.13 G. Actin

On addition of salt, globular (G) actin changes to the fibrous (F) form, a linear aggregate of globular units arranged in a double helix. G-actin as usually prepared contains for each molecule of protein one Ca^{++} ion and one tightly bound ATP molecule which is converted to ADP during polymerization. Loss of bound nucleotide (e.g. by removal of the calcium by chelating agents) leads to a loss of capacity to form F-actin. Other nucleotides, including ADP, can replace ATP and in this case (ADP) polymerization occurs without liberation of inorganic phosphate.

West *et al.* (1967) added 1 mM EDTA to ATP-G-actin and observed a difference spectrum with a strong maximum at 256 nm, a double peak at 281 and 292 nm and a broad band with $\lambda_{max.}$ 350 nm (see Fig. 8.21).

The difference spectrum at 280–295 nm is due to changes in the environment of tyrosyl and tryptophyl residues. The 256 nm peak is probably due to the purine ring, so that a change in the nucleotide absorption is brought about by the EDTA. When EDTA was added to ITP-actin not only was the ITP band (249 nm) shown but the 256 nm band was also seen in the difference spectrum. This fact is difficult to explain convincingly.

Different portions of the difference spectra develop at different rates. It seems that the first step is the setting free of the nucleotide accompanied by changes

Figure 8.21 Ultraviolet difference spectrum of ATP-G-actin resulting from the addition of 1 mM EDTA. Final concentration of ATP-G-actin 1·3 mg/ml, in 1 mM Tris or TES, pH 7·4, 2°C (from West *et al.*, 1967)

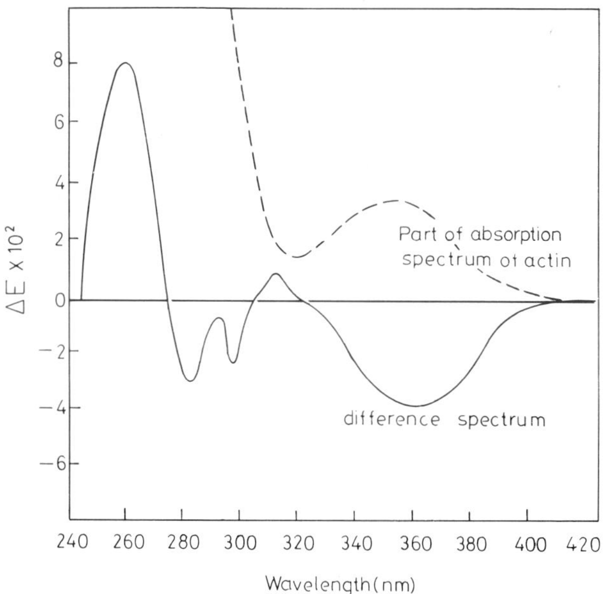

in absorption both for the nucleotide itself and for the tyrosyl and tryptophyl residues. The protein conformation is more slowly adjusted and further changes in selective absorption occur at 280–300 nm.

The peak in the absorption spectrum of fresh actin which is seen at 350 nm only to vanish on denaturation so far lacks explanation. The possibility that it is due to a charge-transfer effect in a tryptophan–nicotinamide complex is ruled out by the fact that actin does not contain nicotinamide or related compounds (Fisher & Cross, 1966).

8.1.14 Optical Rotatory Dispersion

The investigation of optical rotatory dispersion has thrown a good deal of light on protein conformation and it holds much promise for the future. Measurements are usually made in the visible region and in the near ultraviolet down to 300 nm. The specific rotation $[\alpha]_\lambda$ is given by

$$[\alpha]_\lambda = \frac{100}{dC}\, d_\lambda$$

where d is the light path thickness (cm) and C is concentration in g/100 ml and d_λ is positive when the plane of polarization is rotated anticlockwise. Drude's equation:

$$[\alpha]_\lambda = \frac{A}{\lambda^2 - \lambda_c^2}$$

fits the rotatory dispersion (near ultraviolet and visible region) for many unconjugated proteins, A being a constant and λ_c a dispersion constant corresponding to all the transitions producing the rotation and occurring in the ultraviolet. Yang & Doty (1957) plotted $[\alpha]_\lambda\lambda^2$ against $[\alpha]_\lambda$ to give a straight line of slope λ_c^2 and intercept A on the $[\alpha]_\lambda\lambda^2$ axis. The straight line indicates simple dispersion but many polypeptides in the α-helical conformation display non-linear curves. A quantity $[m^1]_\lambda$, the reduced mean residue rotation, is useful in comparing the rotatory power of polymers and is given by:

$$[m^1]_\lambda = \left(\frac{3}{n^2+2}\right)\frac{M}{100}\,[\alpha]_\lambda$$

where M is the mean residue weight and n the refractive index of the medium at wavelength λ.

Moffitt & Yang (1956) used the following equation to elucidate the significance of the experimental data:

$$[m^1]_\lambda = a_0\frac{\lambda_0^2}{\lambda^2-\lambda_0^2} + b_0\left(\frac{\lambda_0^2}{\lambda^2-\lambda_0^2}\right)^2$$

where $[m^1]_\lambda$ is the mean residue rotation, corrected for refractive index, at any wavelength λ, and λ_0 is given a constant value, 212 nm being the usual wavelength chosen.

The quantity b_0 is empirically useful for distinguishing between α-helical and random coil structures. Thus (Tanford, 1968), polypeptide chains known to be entirely right-handed α-helices have large negative values for b_0, of the order $-650°$ for 100% helix, whereas structures randomly coiled have b_0 near to zero. It can be concluded that when b_0 is far removed from zero, random coil structures are highly improbable.

Greenfield *et al.* (1967) studied a poly-L-lysine which could in solution exist in any of three different conformations, a right-handed α-helix, an inter- or intrastrand pleated-sheet hydrogen bonded β-structure and a random coil structure:

(*i*) α-helical form poly-L-lysine HCl 0·01–0·02 in water, pH 11–13·3, d = 1 mm, 22°; (*ii*) β-structure poly-L-lysine sample similar to α-helical form heated to 50° for 12–25 min, then cooled to 22°; (*iii*) random coil form poly-L-lysine HCl 0·02–0·06 in 0·2 **M** NaCl or H_2O, pH 4·7–5·0, d = 1 mm, 22°.

The optical rotatory dispersion curves were obtained in the region 190–250 nm (see Fig. 8.22); the α-helical form showed a pronounced peak at 198 nm, $[m^1]$-70 × 10^3 with a minimum at 233 nm, $[m^1]$-14 × 10^3. The β-form showed a weaker broad band $\lambda_{max.}$ 205 nm, $[m^1]$-29 × 10^3 with a flat minimum at 230 nm, $[m^1]$-6 × 10^3, while the random coil conformation showed a deep minimum at 205 nm, $[m^1]$-21 × 10^3, with a very small additional Cotton effect at 235 nm. Greenfield *et al.* (1967) used a computer to calculate theoretical combinations of the three structures

$$[m^1]_{\text{theoretical}} = [([m^1]_{\alpha-\text{helix}}a + [m^1]_{\beta\text{-form}})b]c + [m^1]_{\text{random}}^{d}$$

a, b, c and d were varied so that $a+b=1{\cdot}0$ and $c+d=100$ where $ac=\%$ helix, $bc=\%$ β and $d=\%$ random.

The generated ORD curves were, in respect of general shape, determined by d. As the α-helical form becomes random the peak in the ORD curve shows a blue-shift because the random form has a peak below 190 nm. The β-form has a peak at 205 nm and this results in a red-shift.

The ORD curve for lysozyme in 0·2 **M** NaCl was compared with the generated curves and it seemed that lysozyme had a secondary structure made up of approximately 20% α-helix, 32% β-structure and 48% random coil. Sperm whale myoglobin (Harrison &

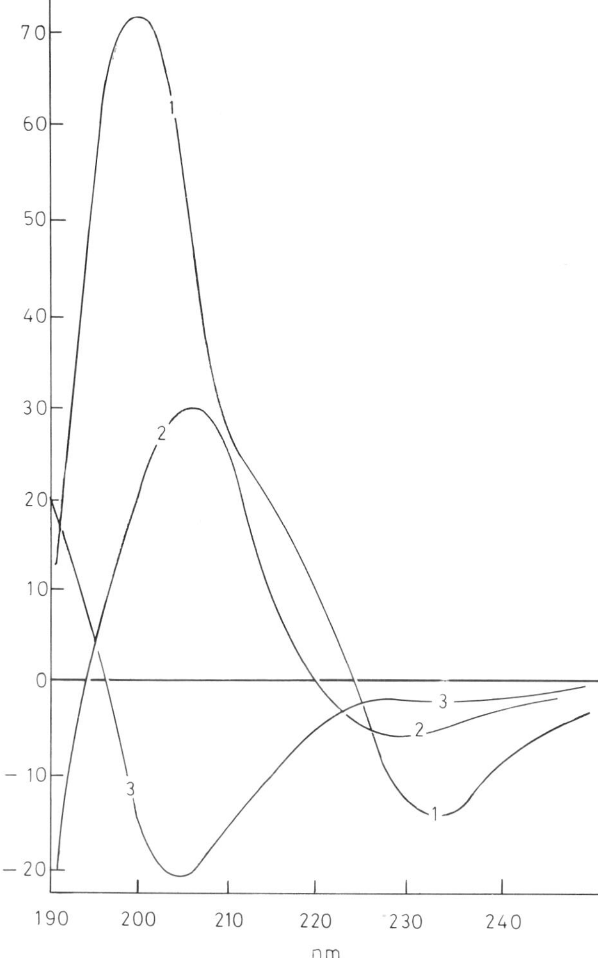

Figure 8.22 Optical rotatory dispersion curves for poly-L-lysine; (1) α-helical; (2) β-structure; (3) random chain. A very small Cotton effect near 235 nm is missing from curve (3) (from Greenfield *et al.*, 1967)

Blout, 1964) shows ORD fitting best the computed curve for 54% α-helix, 36% β-structure and 10% random coil. Compared with X-ray measurements, the β-structure was over-estimated and the α-helical and random forms under-estimated and the ORD analysis failed to take into account all the variables in proteins.

Excitation of an optically active electronic transition (ultraviolet absorption) results in anomalous rotatory dispersion or Cotton effect. Absorption peaks of isolated transitions correspond with the inflection where the rotation is zero. Helical polypeptides and proteins show two Cotton effects attributed to the peptide link. A negative Cotton effect at *ca.* 225 nm corresponds with a transition common to N-substituted amides ($n \rightarrow \pi^*$ transition) (Glazer & Rosenheck, 1962). The magnitude of the negative peak at 233 nm depends on the conformation; $[m]_{233} = -13^\circ \times 10^3$ in α-helical polypeptides and $-2^\circ \times 10^3$ in the random coil form (Simmons *et al.*, 1961). The peptide absorption maximum at 192 nm gives rise to a large positive Cotton effect $[m]_{198} = -80^\circ \times 10^3$ for the right-handed helix (Blout *et al.*, 1962).

The work on poly-L-lysine suggests that an absence of maxima or minima between 210 and 250 nm may be a useful (but not infallible) sign of random coiling. This generalization applies also to circular dichroism.

Tanford (1968) presents a good case in that the optical rotations of randomly coiled polypeptides are substantially accounted for by additive contributions from all the constituent aminoacids (remembering that the terminal amino acids will differ from internal aminoacid residues). The equation

$$[\phi']_\lambda = n[m^1]_\lambda = \sum n_i [m_i^1]_\lambda$$

indicates that the optical rotation should be a simple summation where $[m_i^1]$ is the intrinsic residue rotation of an amino acid moiety (internal) and n_i is the number of such moieties. The summation must include all the aminoacids present (see also Tanford *et al.*, 1967). Intrinsic residue rotations for internal aminoacid moieties belonging to a randomly coiled polypeptide chain (in dilute aqueous salt solution at 25°) have been tabulated (Table 8.9) by Tanford (1968) for the wavelengths 300, 400 and 589 nm. The figures for terminal aminoacid residues are less firmly based but this is not a serious handicap as the terminal units contribute only a very small fraction of the total rotation.

A simple use of the equation of Moffitt and Yang follows from the fact that b_0 can be accepted as zero in a randomly coiled polypeptide. If λ_0 is 212 nm

$$[m^1]_{300\ \mathrm{nm}} = a_0 \left(\frac{212^2}{300^2 - 212^2} \right) = a_0 \left[\frac{44\,944}{45\,056} \right]$$

hence

$$[m^1]_{300\ \mathrm{nm}} = a_0$$

This allows full use to be made of the equation and measurements on suitable polypeptides are likely in time to lead to more accurate evaluations of the contributions of the individual aminoacids.

A special situation can arise with ORD for proteins where changes in aromatic aminoacid absorption accompany changes in conformation. Cotton effects

Table 8.9 Intrinsic residue rotations for internal residues of a randomly coiled polypeptide chain in dilute aqueous salt solutions at 25° C

Amino acid	$[m_1^1]$ values at λ in nm 589	400	300
(*a*) Based on reliable experimental data			
Tyrosine	+140	+320	+920
Glycine	0	0	0
Arginine (H^+)	−85	−219	−530
Glutamate (COO^-)	−85	−219	−530
Lysine (H^+)	−103	−269	−677
Alanine	−130	−333	−830
(*b*) Based on less certain data			
Phenylalanine	0	+20	+125
Histidine (H^+)	−43	−112	−218
Serine	−42	−103	−300
Proline	−212	−556	−1418
(*c*) Approximations			
Leucine	−170	−445	−1130
(valine and isoleucine assumed equal to leucine)			
Cysteine (SH)			
Threonine	−60	−160	−420
Methionine			
Aspartate (COO^-)			

(After Tanford, 1968, modified).

can then occur in the region 265–300 nm. Carbonic anhydrase affords a good example (Rosenberg, 1966; Beychok *et al.*, 1966).

Advantage has been taken of the effect of high concentrations of guanidine hydrochloride (Tanford *et al.*, 1967). Under such conditions many proteins exhibit optical rotatory dispersions in which b_0 is practically zero (random coils). The quantity a_0 in the Moffitt and Yang equation is very variable (see Table 8.10).

The intrinsic residue rotations of aminoacid moieties in Table 8.9 apply quite well to concentrated guanidine hydrochloride solutions and the rotations calculated and observed for individual proteins agree very well at 300, 400 and 589 nm. The differences in optical rotation curves for proteins in which —S—S— groups are intact and reduced are small but real and the findings indicate that in concentrated guanidine hydrochloride proteins are randomly coiled whether the —S—S— links are intact or broken.

Similar ORD studies on proteins denatured by 8 **M** urea (Schellman & Schellman, 1958; Tanford, 1968) show that random coils are predominant as with guanidine hydrochloride, but that the concentration of urea can often be inadequate to complete the transition, particularly when the natural conformation tends to be made more stable by —S—S— links.

Using the experience gained, Tanford (1968) sum-

Table 8.10 Optical rotatory dispersion in concentrated aqueous Guanidine hydrochloride

Protein	Disulphide bonds reduced a_0	b_0	Disulphide bonds intact a_0	b_0
Lysozyme	−449	0	−494	−22
Insulin	−470	+4	−563	−4
Pepsinogen	−507	+6	−525	+1
Ribonuclease	−521	−4	−511	−1
Serum albumin	−587	+8	−628	+6

(From Tanford *et al.*, 1967).

marized the work on α_g- and β-casein which showed that the latter is randomly coiled in dilute salt solution. Among other applications is the explanation of change of solvent, e.g. various proportions of dioxan in the medium at pH 3 containing β-lactoglobulin.

Poly-L-lysine at pH 11·2 at 25° is in the α-helical conformation. Heating to temperatures below T_b causes the transition to the random form, a process reversed on cooling. Above T_b the concentration of random chains is sufficient to allow their association to give a β-structure, which is more stable and grows in size at the cost of unravelling the helix (Davidson & Fasman, 1967).

Martin & Bhatnagar (1967) used absorption spectra and ORD to explore the unfolding of proteins in urea, guanidinium chloride and 2-chloroethanol solutions.

Tinoco and Cantor (1970) have recently devoted a substantial review article to the application of optical rotatory dispersion and circular dichroism to the study of biopolymers with particular reference to proteins, nucleic acids, membranes, ribosomes and viruses.

8.1.15 Infrared Absorption of Proteins

(Cf. Bellamy, 1958; Rao, 1963). In proteins the prominent infrared bands are associated with the amide group and reference is often made to 'amide I' and 'amide II'. A full interpretation of the infrared spectra also clearly requires the effects of acidic side-chains to be identified.

In approaching the more complex spectra it will be convenient first to consider the information available about some polyaminoacids and then to consider the data on natural peptides and proteins.

A beginning may be made with polyglycine, which exists in two forms, I and II, differing in chain conformation. In polyglycine I no hydrogen bonding C—H...O=C is possible and there are two bands at 2850 and 2923 cm^{-1} respectively corresponding to the modes of unperturbed CH_2 groups. In the case of polyglycine II there are found bands in the 3 μm region: 2848, 2935 cm^{-1}, i.e. very near those shown in polyglycine I, and two displaced bands at 2800 and 2980 cm^{-1}. Some, but not all, of the CH_2 groups can form C—H...O=C bands. The lowering of one band (2850 to 2800 cm^{-1}) and the raising of the other (2923 to 2980 cm^{-1}) indicates that the force constant of the unassociated C—H bond is increased above its normal value. Cooling to −170° has little effect on the CH_2 stretching modes of polyglycine I but the comparable bands in polyglycine II are shifted towards the unperturbed frequencies, i.e. to 2815 and 2974 cm^{-1}. Simultaneously the peptide frequencies change in a manner consistent with a strengthening of the NH...O=C hydrogen bond. The evidence supports the idea that in polyglycine II the perturbed frequencies originate from attractive interaction of the type CH...O=C.

Poly-L-arginine shows a strong amide A band at 3290 cm^{-1} with evident background absorption in the 3 μm region. Progressive exchange of H in the ionized guanidinium group $—NH—C(NH_2)_2^+$ by deuterium atoms slowly reduces the intensity of absorption in this region (see Fig. 8.23). There is little or no deuteration of the peptide NH groups under the experimental conditions since the amide II band at 1547 cm^{-1} persists without change and there is only a small change in the amide II peak at 1450 cm^{-1}. The deuteration difference spectrum ($\Delta D = D_{undeuterated} - D_{deuterated}$) shows two strong peaks at 3350 and 3140^{-1} due in all probability to the asymmetric and symmetric NH_2 stretch vibrations of the original guanidinium group. The corresponding bands in the deuterated material are displaced to

Figure 8.23 Poly-L-arginine sulphate. Film cast from formic acid at room temperature. Deuteration from the vapour phase at 35°C and about 80% R.H. Specimen at 0% R.H. during tests. (*a*) Spectra at successive stages of exchange: (1) undeuterated; (2) 1 min.; (3) 10 min.; (4) 1 h. (*b*) Deuteration difference spectra ($\Delta D = D_{undeut} - D_{deut}$).

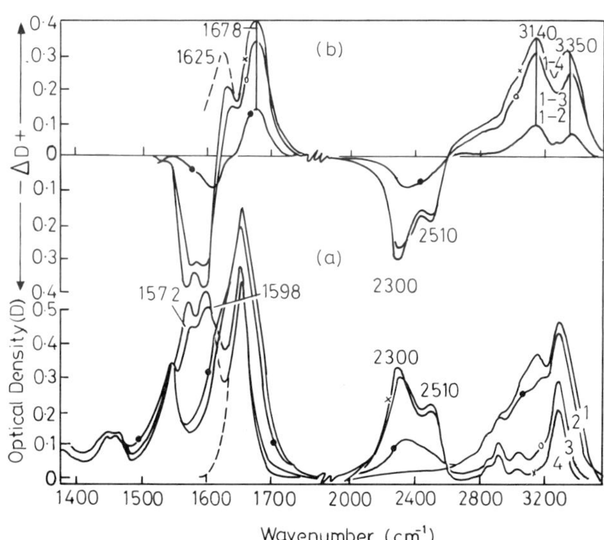

2510 and 2300 cm^{-1}. In the 6μm region the undeuterated poly-L-arginine shows the amide I peak at 1655 cm^{-1} with other absorption superimposed. Deuteration and difference spectra resolve the selective absorption into a doublet 1625, 1678 cm^{-1} and the fully deuterated guanidinium shows a displacement to 1752 and 1598 cm^{-1}. (Crystalline guanidine chloride shows bands at 1640 and 1670 cm^{-1} and the deuterated analogue 1587 cm^{-1}.)

Protamine is very unusual in having an arginine content of about 66%. Protoamine sulphate shows from deuteration difference spectra that NH_2 bands occur at 3350 and 3160^{-1}. H → D exchange results in gradual replacement of 1625 and 1670 cm^{-1} peaks by 1575 and 1600 cm^{-1}.

Lysozyme with 9% of arginine has too much absorption near 3 μm exerted by other aminoacids for the guanidinium effects to show up. In the 6 μm region the 1575, 1600 cm^{-1} doublet shows up in the deuterated product.

Poly-L-lysine as hydrobromide exhibits a strong region of absorption near 3000 cm^{-1} as well as the expected absorption near 3300 cm^{-1} (Fig. 8.24). On deuteration this absorption gradually decreases and is replaced by a new selective absorption near 2200 cm^{-1}. The difference spectra indicate that the absorption with peaks at 3000 and 2800 cm^{-1} and apparently several weak bands down to 2400 cm^{-1} are all associated with the NH_2^+ group of the lysyl side-chain. The ND_2^+ band *ca.* 2200 cm^{-1} appears also to show some structure. In the undeuterated polypeptide two rather weak bands, 1500 and 1600 cm^{-1}, are also associated

Figure 8.24 Poly-L-lysine hydrobromide. Film cast from water at room temperature. Deuterated from the vapour phase at 35°C and about 80% R.H. during tests. (*a*) Spectra at successive stages of deuteration: (1) undeuterated; (2) 2 min.; (3) 20 min. (b) Corresponding deuteration difference spectra ($\Delta D = D_{undeut} - D_{deut}$). (from Bendit, 1967)

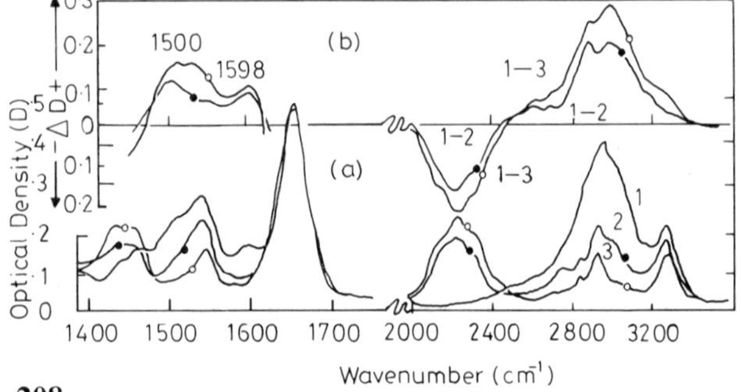

with NH_2^+ (asymmetric and symmetric NH_2 deformation modes, sometimes called aminoacid I and II bands). These bands are best revealed by deuteration difference spectra. This is specially true for histone, which may have 27% of lysine without displaying the characteristic lysine bands in the undeuterated material.

Poly-DL-phenylalanine shows ring stretching vibrations at 1604, 1585 and 1495 cm^{-1} with a peak at 1455 cm^{-1} shown very sharply. The strongest band (1495 cm^{-1}) is seen in insulin but most proteins contain too little phenylalanine for it to show up.

Poly-DL-tryptophan shows a peak at 3400 cm^{-1} due to the stretching vibration of the NH link in the indole group. This band disappears on deuteration. Pyrrole and indole show in dilute solution a corresponding band at 3490 cm^{-1}. The NH (ring) group of tryptophan may be concerned in hydrogen bonding.

Poly-DL-phenylalanine has ring stretching vibration, revealed by deuteration as a doublet (*ca.* 3200, 3400 cm^{-1}) but the explanation is uncertain.

Poly-L-tyrosine shows phenolic absorption strongly at 1515 cm^{-1} (weaker peaks at 1615, 1595, 1440 cm^{-1}). On deuteration of the phenolic hydroxyl the ring vibrations persist virtually unchanged. The deuteration difference spectrum around 3 μm shows two OH stretching peaks 3380, 3220 cm^{-1}. In proteins rich in tyrosine, e.g. insulin and some keratins, the 1515 cm^{-1} peak is seen in the undeuterated substance.

Care is necessary in ascribing bands to groups. Thus in glutamine and poly-L-glutamine there are asymmetric and symmetric NH_2 stretch vibrations at 3430 and 3190 cm^{-1}. The deuteration difference spectrum confirms this. Gliadin, which has a high amide content, shows the two bands at 3430 and 3180 cm^{-1} respectively on the basis of deuteration difference spectra.

The contributions of the COO^- group derived from glutamic or aspartic residues occur at 1565 and 1400 cm^{-1} while the COOH group shows in protein a peak at 1715–1710 cm^{-1}, agreeing closely with previous figures for simpler compounds.

The infrared spectra of natural polypeptides and of proteins reflect in the main the bands characteristic of the peptide link. The study of related simple compounds and particularly of substituted amides has resulted in the assignments shown in Table 8.11 (cf. Miyazawa *et al.*, 1958). The effect of hydrogen bonding is to diminish the main stretching frequencies but

Table 8.11 Infrared absorption bands of the peptide group

Designation	Mode	Frequency cm^{-1}	Intensity	Hydrogen-bonded cm^{-1}
Amide I	N—H stretching	3460	v.s.	3300
	C—O stretching 80%	1690	v.s.	1650
	C—H stretching	2940	s.	—
	N—H stretching (overtone)	6900	m.	6670
Amide II	N—H in-plane bending 60%	1550	v.s.	—
	C—N stretching 40%			
Amide III	C—N stretching 40%	1250	s.	—
	N—H in-plane bending 30%			
Amide IV	OCN bending out-of-plane	630	m.	—
Amide V	N—H bending out-of-plane	725	m.	—
Amide VI	C═O bending out-of-plane	600	m.	—
Combination bands				
	N—H stretching + amide I	4970	m.	[3300 + 1650 = 4950]
	N—H stretching + amide II	4800	m.	[3300 + 1550 = 4850]
	N—H stretching + amide III	4650	m.	[3300 + 1250 = 4550]

v.s. = very strong; s. = strong; m. = moderate intensity.
(From Miyazawa, Shimanouchi & Muzushima, 1958.)

to increase the bending frequencies. The —OH stretching frequency drops from *ca.* 3600 to 3400 cm^{-1} and the —SH frequency from 2250 to 2150 cm^{-1}. In polypeptides selective absorption due to amino-acid side-chains is often seen as weaker bands superimposed on the spectrum of the peptide link. Thus for NH_2, three bands (3225, 3370 and 5050 cm^{-1}) are associated with antisymmetric and symmetric stretching and a stretching-deformation respectively. The carboxylate ion —COO^- displays 1410 and 1570 cm^{-1} bands due to antisymmetric and symmetric stretching respectively. The carbonyl stretching occurs at 1700 in the carboxyl group and at 1740 cm^{-1} in the ester group. The —SH stretching is at 2150 cm^{-1} and CH_2 bending can contribute to the 1440 and 1370 cm^{-1} absorption.

Hydrogen atoms covalently bound to carbon do not exchange with hydrogen atoms of the medium whereas those linked to nitrogen, sulphur or oxygen may do so. If a fully deuterated protein is dissolved in water a slow exchange can be followed experimentally (Krause & Linderström-Lang, 1955). The exchanging hydrogens are principally those concerned in the hydrogen bonds of the α-helical structure.

Deuteration of the peptide link considerably changes the absorption frequencies of the peptide bands. Thus the H-bonded NH stretching frequency changes from 3200 cm^{-1} (NH) to 2440 cm^{-1} (ND) while the amide II band at 1535 cm^{-1} (NH) shifts to 9650 cm^{-1} (ND) and the amide V band from 700 cm^{-1} (NH) to 530 cm^{-1} (ND) (Beer *et al.*, 1959). $[\alpha[NH/N+H]^{\frac{1}{2}}$, i.e. deuteration, decreases the frequency by 1·37.] The extent of deuteration can be measured from the diminution of the band at 1535 cm^{-1}. The amide II absorption at 1650 cm^{-1} (C═O stretching) is unaffected by deuteration and can be used as a reference frequency in measuring difference spectra.

Ordinary water (H_2O) has strong absorption bands in the region 2800–3800 cm^{-1} (3·57 to 2·63 μm) and 1750–1500 cm^{-1} (5·71–6·6 μm) which mask any peptide or protein bands in the two regions. Heavy water (D_2O) has strong bands at 1250–1100 cm^{-1} (8–91 μm) and 2800–2200 cm^{-1} (3·57–4·55 μm) but the amide I and II bands are but little masked.

Blout *et al.* (1961) found that deuteration exchange was very slow (hours) in the α-helical form of poly-L-glutamate but quick (minutes) in the random coil form. Elliott & Hanby (1958) found poly-γ-benzyl-L-glutamate to behave similarly. Englander (1963) studied ribonuclease using tritium exchange.

Infrared absorption spectra of proteins have been studied by Bamford *et al.* (1953, 1954, 1956) and reviewed by Sutherland (1952). As would be expected. frequencies associated with NH and CO were readily identifiable but proved in the main unimportant because they were common to so many proteins. Minor changes have been correlated with the occurrence or absence of hydrogen bonding, inter or intramolecular. Thus in fibrous proteins like silk the insoluble β-structure is intermolecularly bonded while the α-structure, whether helical or randomly dispersed, is in the main intramolecularly hydrogen bonded. Polarized infrared radiation can be used to show up the dichroism of absorption bands measured by comparing the absorption when the electric vector of the incident radiation is perpendicular or parallel to the fibre axis (or direction of stretch). Fibroins in the β-state tend to show dichroism, the chains being oriented with the sample. In the α-state fibroin preparations are not dichroic and when attempts are made to orientate them they change to the β-state (cf. Ambrose & Elliott, 1951).

There are technical difficulties in studying fine filaments and some good work has been done on sections of silk suture drawn by hand from silk glands. It is more convenient to work with films of fibroin cast from solutions in aqueous lithium bromide, although the orientation may then be very different from that of the natural product.

The suture displays an oriented β-structure (Badger *et al.*, 1953). The dichroism at 3300 cm^{-1} and 1640 cm^{-1} is strong and perpendicular to the length of the suture. In *Bombyx mori* fibroin suture the β- and α-CO stretching frequencies were at 1636 and 1664 cm^{-1} while the β- and α-NH deformation frequencies were at 1528 and 1560 cm^{-1}. The 1664 and 1560 cm^{-1} bands did not polarize and were ascribed to the fully disoriented portion while the 1528 and 1636 cm^{-1} bands were highly dichroic and could be attributed to an oriented extended β-structure, probably crystalline. From these results Badger *et al.* (1953) estimated the relative amounts of the α- and β-forms. With certain reservations about two-thirds was β-form. Drucker *et al.* (1953) found a similar proportion for the part of fibroin precipitated with pancreatic enzymes and thought to correspond with the crystalline moiety.

Table 8.12 Infrared bands in fibrous proteins

Frequency	α-Configuration cm^{-1}	β-Configuration cm^{-1}	β-Configuration Dichroic sense
N–H deformation	1545	1525	$\parallel$
CO stretching	1660	1640	$\perp$
N–H stretching	3300	3300	$\perp$
CO combination	4600	4530	$\parallel$
N–H combination	4830	4825	$\parallel$

Other work (Lenormant, 1956; Elliott *et al.*, 1957) on films cast from solution of fibroin in lithium bromide or thiocyanate shows that the α-form (in such films) is devoid of helical structure and hydrogen bonding is largely intramolecular. The α-form is readily converted to the β-form but not completely.

Go *et al.* (1956) studied a synthetic polyglycyl-L-alanine of molecular weight 53 900 as a Nujol mull and compared it with the crystalline portion of *B. mori* silk fibroin. The resemblance over the whole region 625–5000 cm^{-1} was very close. Zahn & Schnabel (1957) synthesized a racemic peptide with the sequence Ser. Gly. Ala. Gly. Ala. Gly. already known (Lucas *et al.*, 1957) to be important in the make up of the crystalline fibroin. Again the infrared spectrum reproduced quite closely that of the natural product. The 1015 cm^{-1} band which occurs in polyglycine was not shown by the crystalline part of silk fibroin (Asai *et al.*, 1955).

The foregoing is a very bare outline of a large field of work [cf. Lucas *et al.*, (1958); Angell & Sheppard (1957); Bendit (1967); Fraser (1960); Krimm (1967); Krimm *et al.* (1967); Randle & Whiffen (1955)].

Hvidt & Kagi (1963) investigated the interaction of alcohol dehydrogenase and nicotineadeninedinucleotide (NAD or NADH) and found that some 14 peptide hydrogen atoms per mole of coenzyme ceased to exchange, whereas addition of substrate had no such effect.

Infrared difference spectra are undoubtedly useful in this area of investigation but interpretation requires caution and parallel work using other methods is often desirable (see Weber & Teale, 1965).

Spectroscopic studies using different parameters have together thrown much light on the conformations of histones, nucleohistones and protamines. The nuclei of cells of higher organisms contain roughly equal amounts of DNA and the basic proteins known as histones with much smaller amounts of other proteins. The biological role of histones is not fully known but one suggestion is that the union of DNA and histone assists the coiling of DNA into the compact form of the chromosome. An alternative view (Stedman & Stedman, 1950) is that histones suppress genetic activity in differentiated tissue. Many workers assembled evidence as to possible mechanisms and it is easy to oversimplify the problem. From the standpoint of the present discussion, histone studies afford good examples of physical methods of uncovering conformation changes.

Histone from calf thymus tissue has been separated into five major fractions (Philips, 1964; Murray, 1964, 1965). Optical rotations were measured at nine wavelengths between 296·7 and 578 nm (Bradbury *et al.*, 1965) and the rotations were analysed according to the Moffitt equation, λ_0 being placed at 212 nm and b_0 values being calculated (see p. 204). Infrared spectra of insoluble fractions and insoluble portions of fractions were recorded. Freeze drying was found to produce artefacts and its use was given up.

The solvent 2-chloroethanol favours formation of helices in proteins and some of the fractions when examined in this solvent gave b_0 values in the Moffitt ORD equation consistent with 46% while other fractions fitted 65–70% of helical content. In aqueous solutions some histone fractions were entirely in the random form and others largely so. Addition of salt brought about a small increase in helix content (10–20%). The two fractions having the lowest helical proportion in 2-chloroethanol or salt solution were the richest in proline.

As has been noted (p. 207) infrared spectroscopy has been used to indicate conformation in the polypeptide backbone (amide I 1630 cm^{-1} and amide II 1530 cm^{-1} for extended β-forms; α-helical and random coils, overlapping bands, amide I 1640–1660 cm^{-1} amide II 1540–1560 cm^{-1}, wider for the random coil than the α-helix). Moreover the random coil form and the helical forms can be distinguished by measuring the rate at which hydrogen and deuterium atoms exchange.

Ferretti (1967) examined poly-β-methyl-L-aspartate and poly-L-leucine in trifluoroacetic acid/deuteriochloroform solvent mixtures. The amide NH and α-CH n.m.r. peaks in solution of partially helical polymer were made up of two components—one component being assigned to protons in the helical form and the other to protons in the random coil form.

Bradbury, Crane-Robinson & Rattle (1967) in a similar study found the α-CH peaks to be a more straightforward indicator of conformation than the NH peak. Bradbury *et al.* (1967) concluded from all the evidence that lysine-rich histone fractions can form extended chain conformations. They investigated partial DNPs (native nucleoproteins) remaining after selective removal of histone fractions and found that all the labile proteins in the lysine-rich fractions were rapidly deuterated while for most of the other fractions deuteration was slow. These results again indicate that the lysine rich fractions are in extended chain conformation and the other fractions contain a good deal of the α-helical form. It was concluded that there is a stabilizing interaction between the tightly bound histones and DNA and this depends on (*a*) ionic linkages between the basic residues and the negatively charged phosphate groups and (*b*) hydrophobic interactions between the non-polar residues and the lower section of the deep grove of the DNA molecule known from other evidence to exist.

These are significant applications of spectroscopic methods but newcomers to the field need to study the voluminous original literature with due care.

8.1.16 Tobacco Mosaic Virus

This virus has an important place in science because it has been a very convenient tool in many types of research. There is now an immense literature and anything that can be written here must look inadequate to specialists.

The virus can be obtained as a particle, rod-like in shape and *ca.* 300 nm long. It consists of protein (94·8%) and nucleic acid (5·2%), the latter being deeply embedded in the former. The molecular weight of the particle is put at $39{\cdot}4\times10^6\pm2\%$ and that of the virus RNA at $2{\cdot}05\times10^6\pm2\%$. The protein yields sub-units of molecular weight 17 530 and there are 2130 of these in the TMV particle. The RNA residue has 6340 nucleotides with adenine, guanine, cytosine and uracil in the proportions 0·29, 0·26, 0·18 and 0·27 on a molar basis. The nucleic acid initiates infection but the protein potentiates the action strongly and is responsible for the serology. Different strains of TMV occur and several mutants are available. The amino-acid compositions of the proteins have been determined and numerous differences have been noted, although the size of the protein sub-unit varies little. From the point of view of spectra, it is interesting that two strains have Phe, Tyr and Try in the molar proportions 8:5:3 while in another the proportions are 8:6:2. The following review articles contain much to supplement the treatment possible here: Anderer (1963), Caspar (1963), Fraenkel-Conrat & Narita (1958), and the book by McLaren & Shugar (1964) is particularly valuable on photochemical aspects.

Many of the procedures discussed elsewhere in the present book find applications in the TMV field. The measured ultraviolet absorption shows a peak near 260 nm but there is also substantial non-selective absorption in the region 300–400 nm due to scattering. An approximate estimate of the contributions due to turbidity, to nucleotide absorption and to protein absorption is possible (McLaren & Takahashi, 1958). The nucleotide (RNA) contribution is of the same order as that of the protein despite the difference in percentage composition. The tryptophan 'notch' in the spectrum is masked, but can be revealed by studying the absorption when the plane of polarization of light is parallel to the virus axis (Seeds & Wilkins, 1950). With viruses in general (most having a greater proportion of nucleic acid) scattering can mask nearly all the selective absorption, but Oster (1955) found that by using virus particles suspended in a glycerol solution having a refractive index similar to that of the particles, the scattering could be very largely reduced and the true absorption spectrum recorded.

Intact TMV is degraded by detergents such as sodium dodecyl sulphate which separates RNA from protein but there is some denaturation of protein sub-units. Alkali (pH 10·5) suitably buffered, is more gentle in its effects but perhaps the least damaging procedure is to use cold 67% acetic acid. Fraenkel-Conrat & Narita (1958) measured the absorption spectrum of the solution (after removing precipitated nucleic acid by centrifugation). The preparation was apparently identical with that obtained by alkaline cleavage followed by ammonium sulphate precipitation.

In 67% acetic acid $\lambda_{max.}$ appears at 275 nm with $E^{1\%}_{1cm}$ 11·8 with the 290 tryptophan 'notch' at $E^{1\%}_{1cm}$ 9·0. In neutral solution $E^{1\%}_{1cm}$ 275 nm was 12·2 and $E^{1\%}_{1cm}$ 290 nm 8·8, with a minimum at $E^{1\%}_{1cm}$ 250 nm 5·3. At pH 2·2, $E^{1\%}_{1cm}$ 275nm 11·4 and $E^{1\%}_{1cm}$ 247nm 4·55. The acid solution compared with the neutral solution exhibits a negative difference spectrum with $\lambda_{max.}$ near 290 nm, resembling the difference curve obtained with ribonuclease. This implies that tyrosine-carboxylate-hydrogen bonding occurs. When the spectrum of the intact virus at pH 8 is compared with that of the virus sulphate in an acid medium in both cases degraded by sodium dodecyl sulphate, the difference spectrum is in the opposite sense, but the nucleic acid behaves similarly, leaving the situation difficult to interpret.

The protein sub-unit yields peptides, the aminoacid composition of which has been in part determined by the fluorodinitrobenzene method. The TMV protein aggregates are readily broken down by low concentrations (0·2–1·8 **M**) of urea (Smith & Lauffer, 1961) or by 0·02–0·05 **M** tetra-*n*-butylammonium ion.

TMV-protein (partially dissociated) was shown to exhibit α-helical folding by Simmons and Blout (1960) using optical rotatory dispersion. The sub-units in small aggregates appeared to have 25–35% helix content but the interpretation of the results on the full polymer is less conclusive. The results of Fraser (1952) and Beer (1958) on infrared dichroism shown by oriented intact TMV also support the view that in the sub-units α-helical moieties are roughly perpendicular to the particle axis.

The structure of TMV virus has emerged substantially from electron microscopy and X-ray diffraction analysis (Barrett *et al.*, 1972) but many problems remain to be solved. Anderer (1963) studied the structural features of the virus surface, an important matter for many reasons including the fact that serological reactions are surface phenomena. Carboxypeptidase releases threonine only from native TMV, but a mutant yields Thr, Ala and Leu so that here three residues of the C-terminal sequence are exposed. There is enzymatic and chemical evidence that no lysine and no tyrosine residue is in an exposed position in native TMV (cf. Scheele & Lauffer, 1969).

Perham (1973) made use of the reactivity of functional groups for investigating the topography of TMV and ingeniously applied it to mutants with additional lysine residues in the coat protein. Thus a mutant (E66) has lysine replacing asparagine at position 140 and another mutant (B13*a*) has two extra lysines replacing respectively glutamine at position 9 and asparagine at position 33 and a third mutant (B13*b*) as only the lysine substitute at position 9. It was already known (Fraenkel-Conrat & Colloms, 1967; King & Perham, 1971; Perham & Richards, 1968; Perham & Thomas, 1971) that lysine-68 in intact TMV reacts readily with imidoesters whereas lysine-53 is unreactive; neither lysine reacts with FNDB in the intact virus but both react in the isolated protein. Hence lysine-53 is 'buried' and lysine-68 is partially 'accessible' in TMV.

Perham used an imidoester methyl picolinimidate (Benisek & Richards, 1968) and the extent of reaction with protein amino groups was monitored spectrophotometrically at 262 nm. The results showed that the lysine at position 68 in the wild type TMV and in mutants B13*a* and B13*b* reacts with the reagent, whereas lysine-33 in the mutant B13*a* did not, although in the isolated coat protein the lysine residue was modified. The lysine residue of E66 at position 140 reacts readily with the imidoester, suggesting clearly that asparagine at position 140 in the wild type TMV must be near the 'surface'. In mutant 13*b* the new lysine at position 9 was accessible to the reagent. When intact TMV of the wild type was treated with *p*-iodobenzenesulphonyl chloride, no reaction with lysine residues at positions 33 or 68 in the sub-unit was detected, but complete modification of tyrosine-139 was achieved.

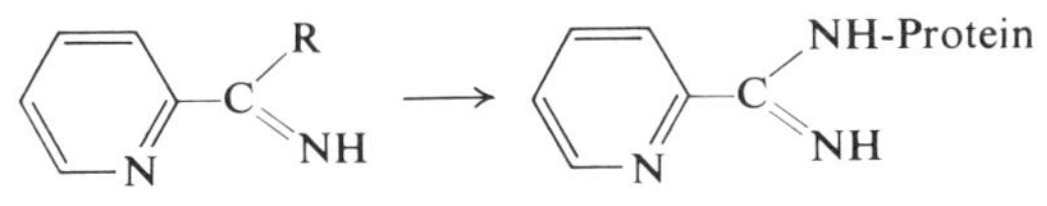

A further point is that differential reactivity of the lysine residues in TMV and its mutants permits the attachment of heavy metal labels at chemically defined positions for subsequent X-ray diffraction analysis.

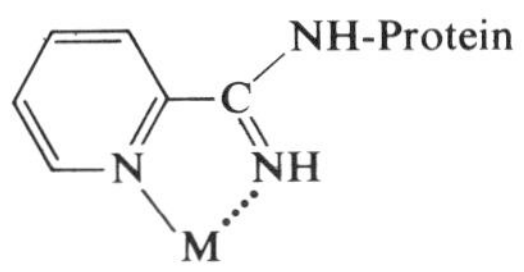

Riley & Perham (1973) synthesized methyl 5-iodo pyridine-2 carboximidate (methyl iodopicolinimidate) and studied its reaction with amino groups both in model compounds and performate-oxidized insulin. The extent of the interaction was readily measured by difference spectroscopy. Modification of lysine residues inhibits tryptic cleavage at the site of interaction. Tyrosine residues in protein can be nitrated (using tetranitromethane) and the 3-nitrotyrosine residues can be reduced (by dithionite) to 3-aminotyrosine. Methyl iodopicolinimidate reacts specifically at pH 5 with the aromatic amino group, yielding finally a 2-arylbenzoxazole residue.

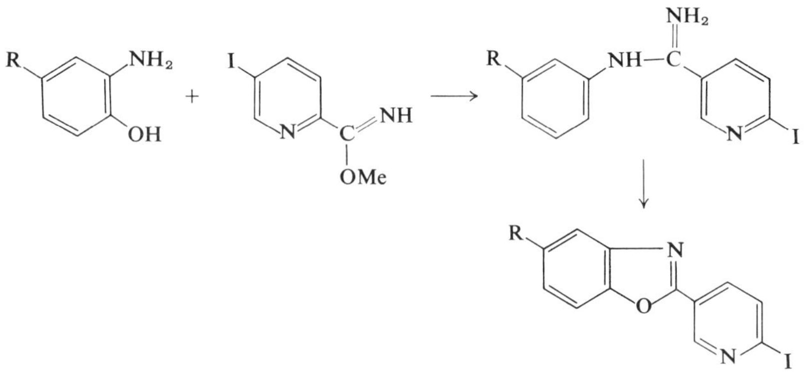

8.1.17 Carbonic Anhydrase of Erythrocytes

Lindskog (1960) found bovine red cell carbonic anhydrase to be chromatographically and electrophoretically heterogeneous. Nyman (1961) obtained three forms, A, B and C, and Rickli & Edsall (1962) provided confirmation (see also Laurent *et al.*, 1965). The three isoenzymes have molecular weight near 30 000 and one zinc atom per molecule; each has about 260 aminoacid residues in a single chain without —S—S— linkages. Human carbonic anhydrase shows A and B to be similar but different from C. Isoenzymes from various species show broad resemblances in respect of the aminoacid compositions of B and C types. All contain substantial amounts of aromatic aminoacids (residues out of 260: phenylalanine 10–13, tyrosine 7–9, tryptophan 5–7). With one exception they contain a single cysteine residue, none being detected in bovine isoenzyme B.

The zinc is rather firmly bound since Tupper *et al.* (1952) found no exchange with radioactive zinc in the medium over one month at pH 7, but it can be removed with chelating agents (citrate ion or 1,10 phenanthroline). The apoenzyme is inactive but regains its full efficacy when zinc is restored. Cobalt (II) also confers activity on the apoenzyme, the artificial holoenzyme showing $\lambda_{max.}$ 550 nm and at pH 8 additional maxima at 618 and 640 nm (probably due to tetracovalent cobalt).

The enzymes are compact molecules with buried reactive side-chains. This applies particularly to the histidine some 7 out of 9–12 residues being initially

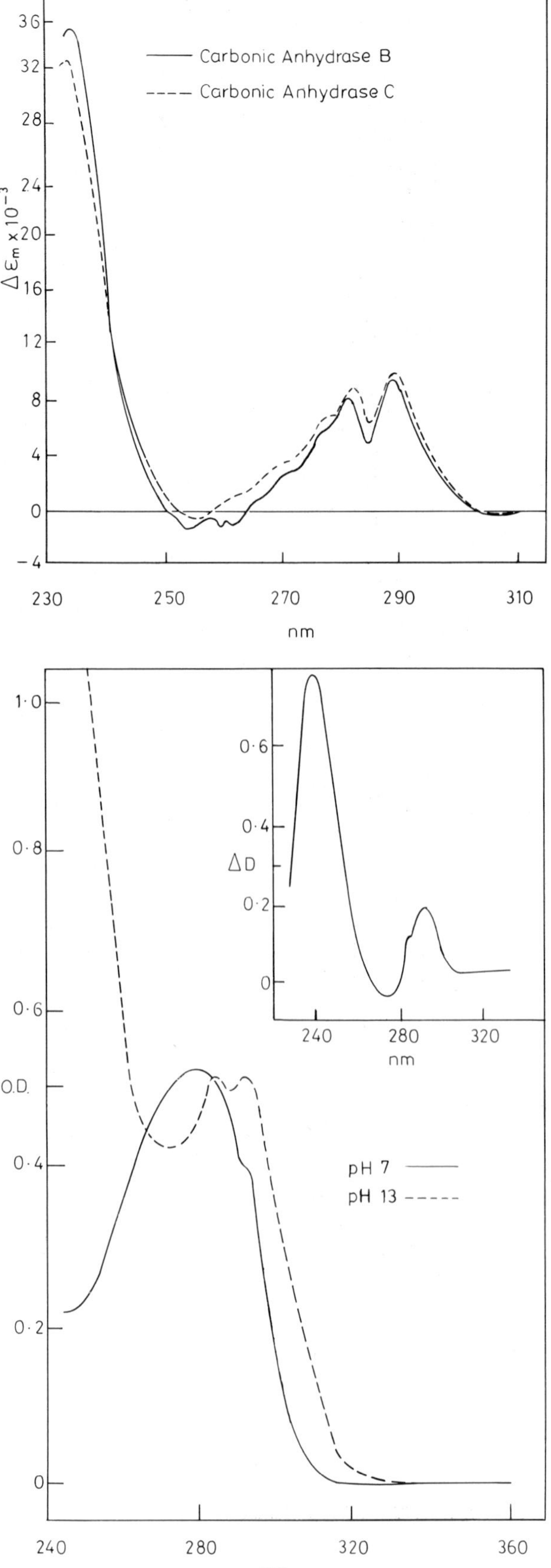

Figure 8.25 Acid difference spectra of carbonic anhydrases B and C at room temperature and an ionic strength of 0·05. Sample, pH 7·06 (phosphate buffer); reference, pH 1·49 (0·05 **N** HCl). Concentration of enzyme B, 0·605 mg/ml; concentration of enzyme C, 0·619 mg/ml. A was converted to $\Delta\varepsilon_m$, with $A_{280}^{1\%}$=16·3 for enzyme B and $A_{280}^{1\%}$=17·8 for enzyme C, and a molecular weight of 30 000 for both (from Riddiford, 1965)

inaccessible to solvent; but at pH 4 they become exposed and can bind protons. Riddiford also found that most of the tyrosine residues in B and C were buried, thus in the native form of enzyme B, only 2 or 3 out of 8 residues and in enzyme C, 3 out of 9 were reversibly titratable. The difference spectrum (Riddiford, 1965) resulting from acid denaturation shows $\lambda_{max.}$ 291·5 nm, ($-\Delta\epsilon$ 7000–8000) characteristic of tryptophan residues. Nearly all the tryptophan is buried inside the native enzyme and when unfolding occurs in acid solution exposure follows.

When carbonic anhydrases are treated with urea or guanidine hydrochloride the usual type of change appears. This is measured by observing the difference spectrum between the treated enzyme and the native form at pH 7. Three peaks appear as denaturation occurs, a very strong peak at 235 nm and weaker ones at 285 and 291 nm, all with negative values for $\Delta\epsilon$, the absorption of the native protein being stronger than that of the denatured protein throughout. Always, however, isoenzyme B resisted the denaturing reagent (e.g. 1 **M** guanidine hydrochloride) for several hours whereas isoenzyme C unfolded very quickly ($\Delta\epsilon$ 291 nm 7500; 285 nm 7000; 235 nm 43 000). The exposure of tyrosine and tryptophan appears to occur at the same time and the change at 235 nm, though less easily understood, probably reflects exposure of all three aromatic aminoacids as well as change in the polypeptide as a whole.

Edsall (1968) summarized the results of studies on the circular dichroism of carbonic anhydrase isoenzymes B and C (cf. Beychok, 1966; Beychok *et al.*,

Figure 8.26 Ultraviolet absorption spectra of human carbonic anhydrase B, at pH 7 and at pH 13. The protein concentration is 0·0326 g/100 ml. The inset is the alkaline difference spectrum, taking the protein solution at pH 7 as a reference, at a concentration of 0·0271 g/100 ml. The maximum in the difference spectrum in this region is at 245 nm, and the corresponding value of $\Delta\varepsilon_m$ is 87 000 M^{-1} cm^{-1} (from Riddiford, 1964)

1966). The results confirm and extend the findings based on ultraviolet absorption.

Riddiford (1964) obtained the spectra shown in Fig. 8.26. Spectrophotometric titrations at 1°, 10° and 25°C starting from iso-ionic undenatured enzyme (B) were consistent with three types of tyrosyl residues. Up to pH 11·5 four tyrosyl groups were substantially normal although the apparent pK values for some were higher than was expected. At pH values above 11·5 the remaining three or four tyrosyl groups ionized very slowly and there was a parallel loss of enzymatic activity. At pH 13 all the tyrosyl groups were at once ionized and a reverse spectrophotometric titration gave absorbances at 295 nm, much higher than those for the forward titration at pH $>10{\cdot}5$. At pH values $<10{\cdot}5$ the forward and reverse curves agreed.

Riddiford (1965) studied carbonic anhydrase C as well as B. $\Delta\epsilon_M$ per tryptophanyl group at 291·5 nm in the acid difference spectra was assigned to changes in the environment of the tryptophanyl groups and the values for $\Delta\epsilon$ were 1420 for enzyme B and 1330 for enzyme C with a marked tendency for the protein to unfold more completely as the ionic strength is lowered. The $\Delta\epsilon$ values are higher than those found by Yanari & Bovey (1960) for a range of proteins including lysozyme, which indicates that in native carbonic anhydrase the tryptophan moieties are 'buried' inside the molecule but become exposed with time and an appropriate pH.

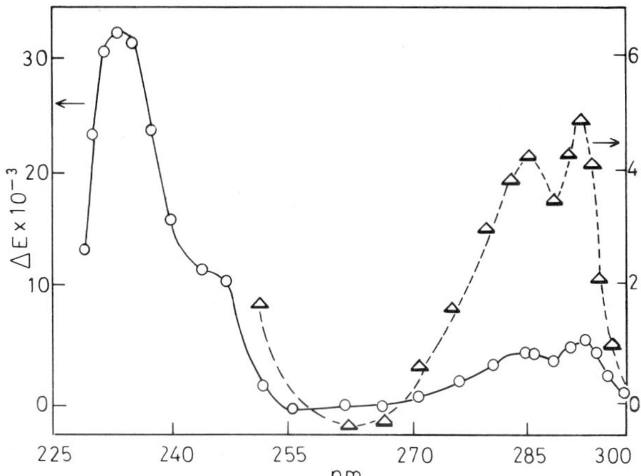

Figure 8.27 Acid difference spectra of carbonic anhydrase B. The dashed line (right-hand ordinate scale) is from measurements made on the Zeiss spectrophotometer. The sample solution is native enzyme in phosphate buffer, pH 7·5, $\Gamma/2=0{\cdot}15$. The solid line (left-hand ordinate scale) shows similar measurements made on another preparation over a wider range of wavelengths with Carey model 14 spectrophotometer (from Rickli *et al.*, 1964)

Figure 8.28 Ultraviolet absorption of carbonic anhydrase B between 185 and 230 nm for the native and for the acid-denatured enzyme. The scale marks 0 and 100 at 190, 197 and 205 nm indicate the calculated values of molar absorptivity (E_m) for a random coil and a completely helical polypeptide chain of the aminoacid composition of carbonic anhydrase B, calculated by the method of Rosenheck and Doty (from Rickli *et al.*, 1964)

8.1.18 Acetylcholine Esterase

Acetylcholine esterase is a most important enzyme in the control of electrical activity of excitable membranes of nerve and muscle cells (Nachmansohn, 1964, 1966). The enzyme has been purified (Leuzinger & Baker, 1967) and its absorption spectrum has been measured (Leuzinger *et al.*, 1968). The solubility in distilled water is very low, making it difficult to measure extinction coefficients directly. Accordingly the enzyme was dissolved in either 0·02 **M** ammonium carbonate or acetate and dialysed for two days against the same salt solutions. The dialysed enzyme was centrifuged (5 min., 30 000 rev/min) and the optical density measured at 280 nm. An aliquot was freeze dried, heated (130°, 20 min.) and cooled in a desiccator over P_2O_5 to constant weight. The outcome was $E^{1\%}_{280\ nm}=16{\cdot}1$. The absorption curves are shown in Fig. 8.29. It will be noticed that near 280

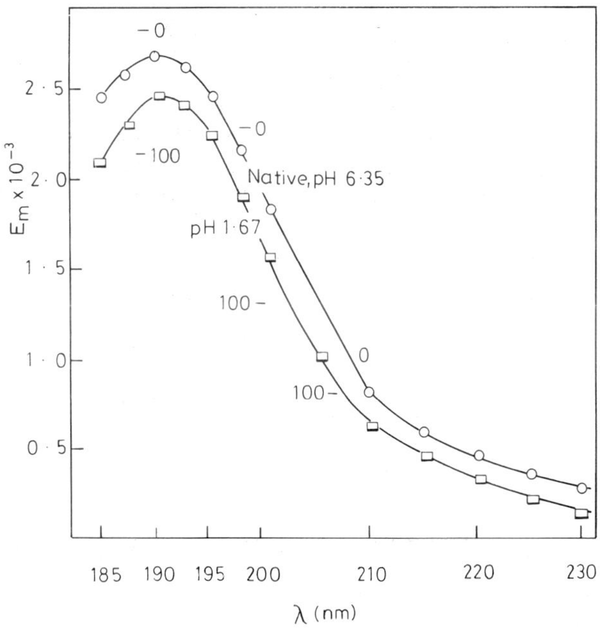

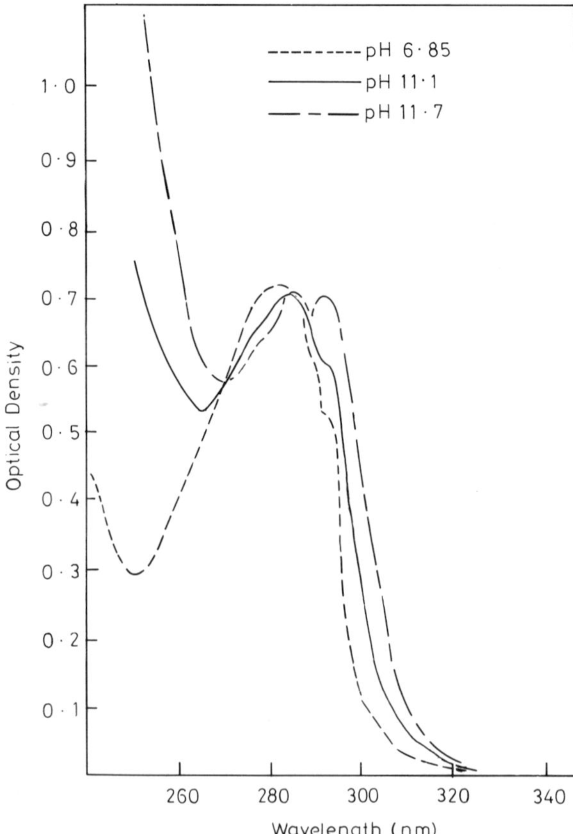

Figure 8.29 Ultraviolet absorption spectra of acetylcholine esterase in 0·02 **M** potassium phosphate at pH 6·85, 11·1 and 11·7 (from Leuzinger *et al.*, 1968)

Figure 8.30 Difference spectrum of $1{\cdot}98\times10^{-4}$ **M** indolylethylnicotinamide versus $1{\cdot}98\times10^{-4}$ **M** tryptamine hydrochloride and $1{\cdot}98\times10^{-4}$ **M** nicotinamide methochloride (in two cells in tandem). Solvent, methanol

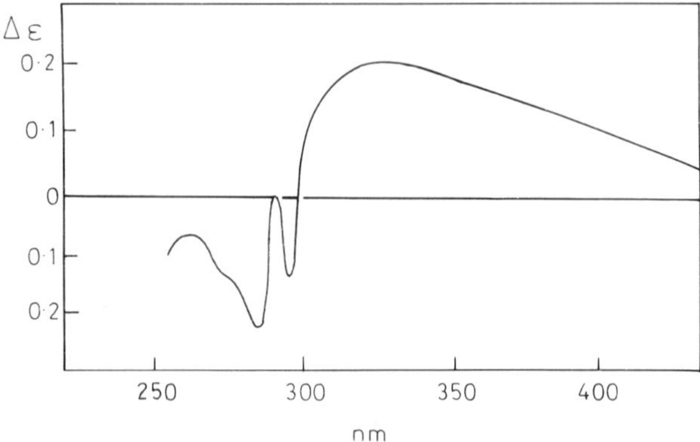

nm the absorbance is independent of pH; there are in fact isosbestic points at 282 and 268 nm.

8.1.19 Glyceraldehyde-3-phosphate Dehydrogenase

This has glutathione as a prosthetic group and thiol esters are used by the enzyme, which is inactivated by iodoacetic acid. Racker & Krimsky (1952) examined the absorption spectrum of the enzyme plus nicotine adenine dinucleotide and found a flat region of absorption extending from 320 to 380 nm. When the spectrum of the enzyme plus NAD plus iodoacetic acid was measured it was found to be appreciably weaker from 300 to 400 nm and a difference curve showed an absorption maximum near 360 nm. The fully reduced dehydrogenase (having maximal activity without added glutathione) when treated with 2 equivalents of NAD^+ showed an effect that could not be increased by further addition of NAD. Iodoacetate (or *p*-chloromercuribenzoate) acted as inhibitor and absorption at 340 nm was markedly diminished.

Shifrin (1964) used indolylethylnicotinamide as a model for interactions between the tryptophan moiety of dehydrogenases and the nicotinamide portion of NAD^+. The absorption spectrum of indolylethylnicotinamide shows a very broad unresolved band between 310 and 430 nm attributable to a charge-transfer transition. This extended absorption (Fig. 8.30) is very similar to that shown in the spectrum of glyceraldehyde-3-phosphate dehydrogenase. Charge-transfer complexes in biochemical processes have been discussed by Mulliken (1952) and Kosower (1956). The term implies a process whereby an electron is partially or wholly transferred from one component of a complex to another. Indole has an electron donor property and the pyridinium ring is an acceptor, so that the broad absorption noted by Racker & Krimsky (1952) for the mixture of NAD^+ and glyceraldehyde-3-phosphate dehydrogenase reflects a charge-transfer transition (Kosower, 1956). It was observed by Cilento & Giusti (1959) that on adding indoles to 1-benzyl-3-carboxamide pyridinium chloride, a yellow colour was formed as in the enzyme–co-enzyme complex and they showed that it was due to a charge-transfer process ($\lambda_{max.}$ 370–380 nm, $\epsilon_{max.}$ 500–1410 depending on the indole derivative). Shifrin (1964) prepared the model compound indolylethylnicotinamide and examined its absorption spectrum and that of its dihydroderivative (Fig. 8.31).

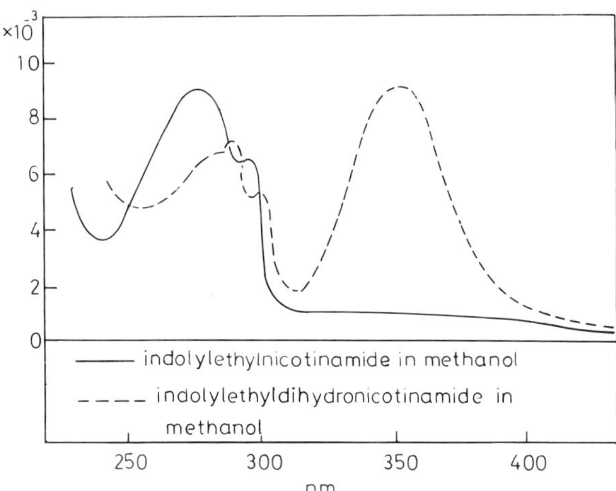

Figure 8.31 Absorption spectra of indolylethylnicotinamide and indolylethyldihydronicotinamide in methanol

If there had been no interaction between the indole and pyridinium moieties, the absorption would have been the sum of those due to tryptamine hydrochloride and nicotinamide methochloride. A difference spectrum was therefore measured as follows: Compensating beam: two cells in tandem one containing $1{\cdot}98 \times 10^{-4}$ **M** tryptamine hydrochloride and the other $1{\cdot}98 \times 10^{-4}$ **M** nicotinamide methochloride. Sample beam: two cells, one containing $1{\cdot}98 \times 10^{-4}$ **M** indolylethylnicotinamide and the other methanol.

The difference spectrum exhibits a positive portion indicating a new transition and the negative values indicate where the compensating system absorbs more intensely than the complex. The new absorption from 300 to 450 nm arises from an electron transfer from the indole to the pyridinium moiety. If $\lambda_{max.}$ is taken as *ca.* 325 nm, $\epsilon_{max.}$ is about 1000. The negative portion reveals a hypochromic effect with peaks at 277·5 and 290·5 nm and a zero or null point at 287 nm. The nearness of the two chromophores effects electron transfer states as well as 'locally' excited states.

When fairly concentrated solutions of tryptamine hydrochloride and nicotinamide benzylochloride were mixed (0·75 **M** aqueous solutions 1:1 v/v) a yellow solution was obtained and aliquots were diluted to give final concentrations 0·01 to 0·75 **M**. The absorbance A divided by the concentration C was plotted against A for various wavelengths (325, 350, 370, 400 nm) and a linear relation emerged in each case. Such deviations from Beer's law are to be expected since the ϵ values vary as the square of the concentration of the donor–acceptor pair. Addition of a saturated solution of potassium cyanide in methanol destroyed the yellow colour (i.e. absorption 310–450 nm) and a new band appeared at 340 nm ϵ *ca.* 7000 and changes in the short wave region indicated that the pyridinium ring absorption had vanished but the indole absorption was unaffected. The charge-transfer transition disappeared in neutralizing the positive charge of the pyridinium ion.

The absorption shown by the reduced compound $\lambda_{max.}$ 356 nm for the complex is almost exactly additive (Fig. 8.32). The fluorescence spectrum shows that

Figure 8.32 A/C=Absorbance (A) divided by the concentration (C) of the mixture of nicotinamide benzylchloride and tryptamine hydrochloride. 0·75 **M** solutions were prepared and mixed 1:1. The solution was diluted to give concentrations between 0·01 and 0·75 **M** and absorbances were measured at particular wavelengths (after Shifrin, 1965)

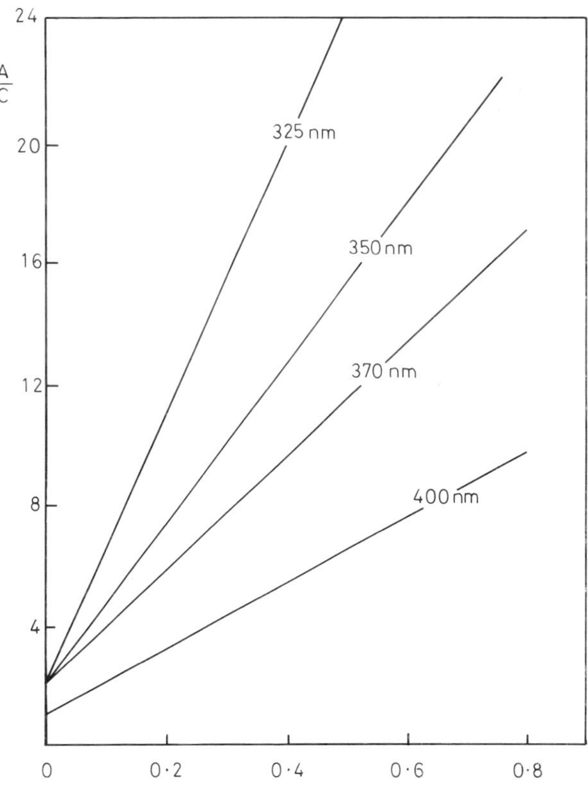

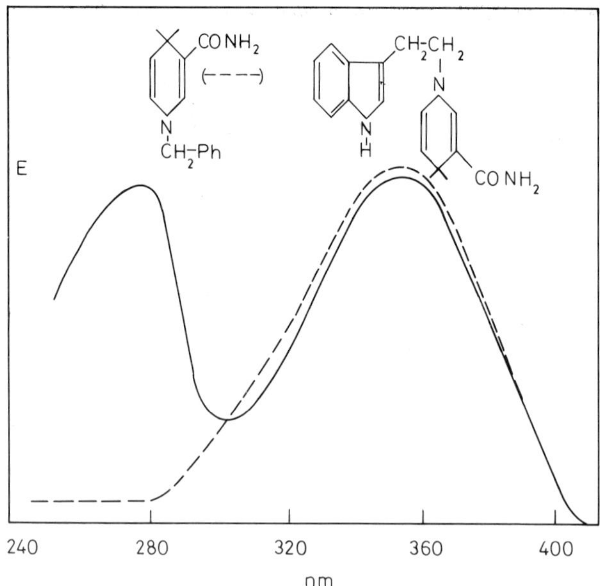

Figure 8.33 Fluorescence excitation spectrum of indolyldihydronicotinamide (solid curve) and 1-benzyl-1,4-dihydronicotinamide (dashed curve) in methanol. Emission wavelength constant at 460 nm (after Shifrin, 1965)

electronic excitation is transferred here by 'inductive resonance' or 'sensitized fluorescence'. A methanol solution of the complex can be excited at 290 nm and at this wavelength, 90% of the absorption can be assigned to the indole moiety but the fluorescence is all due to the nicotinamide portion.

The question arises as to how the two compounds of the charge-transfer complex are orientated. The hypochromic effect (shown by the negative part of the difference spectrum) agrees nicely with the idea that the 'spike' at 287 nm in indole arises from a transition with a moment at right angles to the main 280 nm transition (Weber, 1960). Of these transitions, only the 280 nm should be responsible for the hyperchromy since this will be parallel to that the pyridinium ring, while the other should give null interaction since the oscillation would be at right angles (see Fig. 8.33).

Although absorption spectra fail to reveal interaction between the indole and dihydronicotinamide moieties, there is no doubt about the transfer of energy of excitation of the indole moiety to the fluorescing molecular species. The rate of transfer varies as the sixth power of the distance between donor and acceptor (Förster, 1948). When E=excitation, R_0 is the distance corresponding to 50% energy transfer and r is the distance between donor and acceptor while j is the exponent of the distance dependence. $E=(R_0/r)^j/[(R_0/r)^j+1]$ and by plotting $\log(E^{-1}-1)$ against $\log r$, the slope of the line is j; R_0 is given by the value of r at $E=0{\cdot}5$. Latt *et al.* (1965) contrived an energy transfer system with a relatively fixed donor–acceptor system.

Stryer & Haughland (1967) accepted Förster's (1948) proposal of an energy transfer by a dipole–dipole resonance interaction between the energy donor and the acceptor chromophore, the conditioning factors being an overlap between the emission spectrum of the energy donor and the absorption spectrum of the acceptor. Latt *et al.* had shown that transfer could occur over distances of about 30 Å. Stryer & Haughland ingeniously varied the distance and devised what they called a solid phase spectroscopic ruler. They synthesized by modern methods a series of oligomers of L-proline (n=1 to 12) with an α-naphthyl group at the carbonyl end of the polypeptide and a dansyl group at the imino end.

The synthesis of a diprolyl (n=2) member of the series is illustrated opposite. The energy donor can be regarded as the naphthyl semicarbazide with $\lambda_{max.}$ 295 nm $\epsilon_{max.}$ *ca.* 6600 and an emission peak at 350 nm, while the acceptor-dansyl-(L-propyl)-hydrazide absorbs maximally at 255 nm and 340 nm ($\epsilon_{max.}$ 1500 and 4200 respectively). It will be noted that the following requirements for a satisfactory experiment are fulfilled: (*a*) the absorption spectra allow selective excitation of either chromophore, (*b*) the emission spectra are clearly separated, (*c*) the emission of the naphthyl group overlaps the absorption spectrum of the dansyl chromophore, (*d*) the energy donor has an appreciable quantum yield of 0·6 for the model compound 1-acetyl-4(1-naphthyl)semicarbazone, (*e*) the energy acceptor fluoresces (510–540 nm) to make the energy transfer readily detectable (see Fig. 8.34).

Stryer & Haughland recorded the efficiency of electronic excitation energy transfer for derivatives of oligomers of poly-L-proline which separated the donor and acceptor moieties by 12 to 46 Å, the efficiency decreasing from 100% to 16% and at 34·6 Å reaching 50%. The value of j (5·9±0·3) agreed well with the requirements of Förster's theory. The phenomenon thus serves as an excellent spectroscopic 'ruler' and it affords a potential tool for studying 'space' phenomena in enzymes.

NMe$_2$

H—N—C—CO—N—C—COO—CH$_2$— RESIN + →

SO$_2$Cl

NMe$_2$

SO$_2$—N—C—CO—N—C—COOCH$_2$— RESIN + $H_2N.NH_2$ →
(20% in EtOH)

NCO

→ etc. —CO—NH—NH$_2$ + →

NMe$_2$

SO$_2$—N—C—CO—N—C—CO—NH—NH—CO—N—
H

Synthesis of dansyl-(L-prolyl)$_2$-α-naphthyl

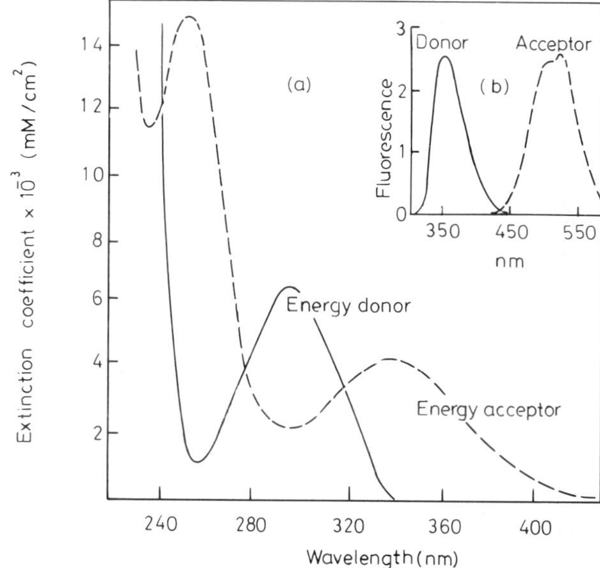

Figure 8.34 (*a*) Absorption spectra of the energy donor, 1-acetyl-4(1-naphthyl)semicarbazone and the energy acceptor, dansyl-(L-prolyl)$_2$-hydrazine methane. (*b*) Emission spectra of the energy donor and the energy acceptor (from Stryer & Haughland, 1967).

8.2 Fluorescence

An important step was taken by Stokes (1852) who realized that when fluorescence occurs, absorption of light at one wavelength is followed by emission at a longer wavelength. This distinguished fluorescence from simple light scattering. The emission of light normally persists after excitation has ceased for a very short period (of the order of nanoseconds, 10^{-9} sec.) but it can exceptionally last for 10^{-4} sec. Phosphorescence on the other hand may persist for 10^{-4}–10 sec. The term photoluminescence includes Rayleigh scattering, Raman scattering and both phosphorescence and fluorescence (Förster, 1951; Udenfriend, 1962; Ehrenberg & Theorell, 1962).

When photons of ultraviolet or visible light are absorbed the molecules of the absorbing entity are excited to higher electronic states, e.g. absorption within the band of longest wavelength will usually excite from the singlet ground state to the lowest excited state at one of its vibrational sub-levels and in time-span of 10^{-12} or 10^{-13} sec. The lowest vibrational state of the excited molecule may be reached. From then the return to the ground state may either be 'radiationless', i.e. the energy will be dissipated as heat (long wavelength emission), or it may coincide with the emission of fluorescence and the molecule may reach one of the vibrational levels of its initial electronic state. Stokes' Law states that the emission will be at wavelengths longer than those absorbed and indeed the fluorescence spectrum is often described as akin to a mirror image of the absorption spectrum (cf. Reid, 1964).

The Franck–Condon principle, first applied to very simple molecules, indicates that the configuration of nuclei cannot change significantly during an electronic transition since the period for the former is of the order 10^{-12} seconds as against 10^{-15} seconds for the electronic transition. The most probable transitions involve no change in vibrational kinetic energy. Other transitions described as 'forbidden' have smaller probabilities and weak absorption or emission. The relation between absorption and emission is given by

$$\int \epsilon \, d\bar{\nu} = \frac{N}{2{\cdot}303 \times 8\pi c} \cdot \frac{1}{\bar{\nu}^2} \cdot \frac{g_u}{g} \cdot \frac{1}{\tau_N}$$

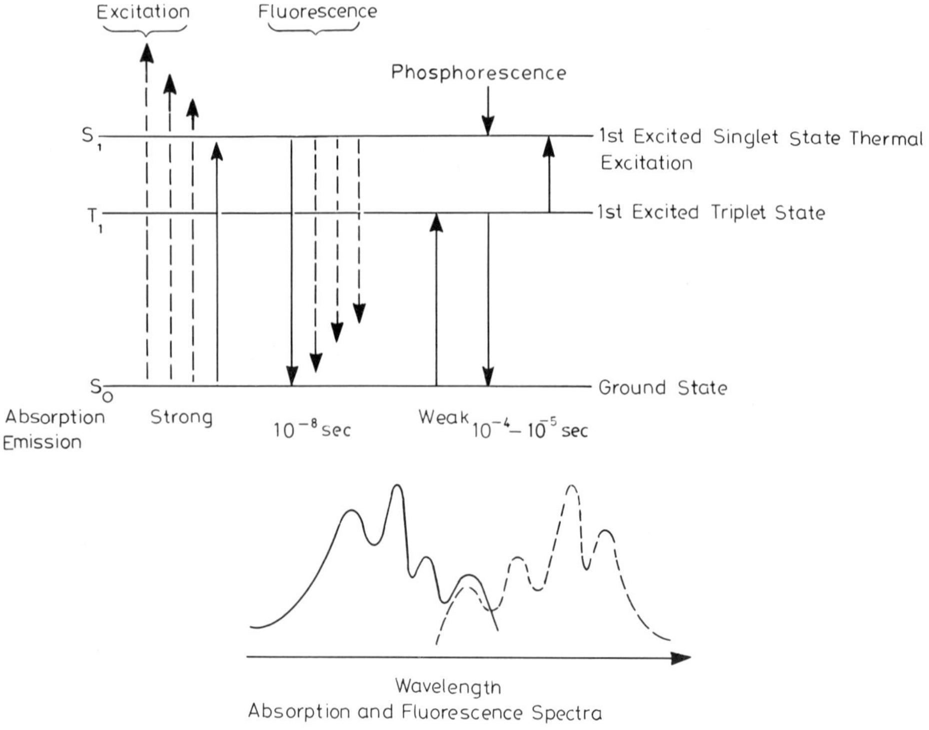

Figure 8.35 Energy states and light absorption and emission

where ϵ is the molecular extinction coefficient and $\bar{\nu}$ is the absorption peak (cm^{-1}). N is the Avogadro number and τ_N is the natural life-time of the excited state while g_u and g_l are the multiplicities of the excited and unexcited states respectively (1 for singlets, 3 for triplets). The equation applies to the particular situation in which the quantum yield of fluorescence is unity and the duration of the excited state depends on the transition probability. In practice, however, excitation of fluorescence often involves the use of larger photons (shorter wavelengths) and the absorbing entity may be raised to higher levels of electronic excitation. Internal rearrangements may, however, shed some or even much of the energy as heat and then result in the normal fluorescence at the expected or even reduced intensity. Hence the larger photons may sometimes result in lower quantum yields.

In phosphorescence the life of the excited state is longer and the excited molecule may become a bi-radical in which electron spins are not all paired. If a pair of electrons acquires parallel spin the molecule can then have a 'triplet' electronic level. The molecule is metastable, the triplet state is lower than the excited singlet state and the return to the ground state is delayed by the low probability of the transition. The energy change is smaller than for the fluorescence emission and the wavelengths for phosphorescence emission are on the long wave side of the fluorescence.

Changes whereby excited molecules return to the ground state without fluorescing are termed quenching processes; they can be intra or intermolecular. The absence of fluorescence depends on molecular structure in the sense that 'radiationless' energy conversions between electronic states are facilitated and the life-time of the excited state is diminished to the point that fluorescence is weakened or abolished. Theories of intramolecular quenching aim at rationalizing empirical relationships between chemical structure and fluorescence (cf. Pringsheim, 1949;

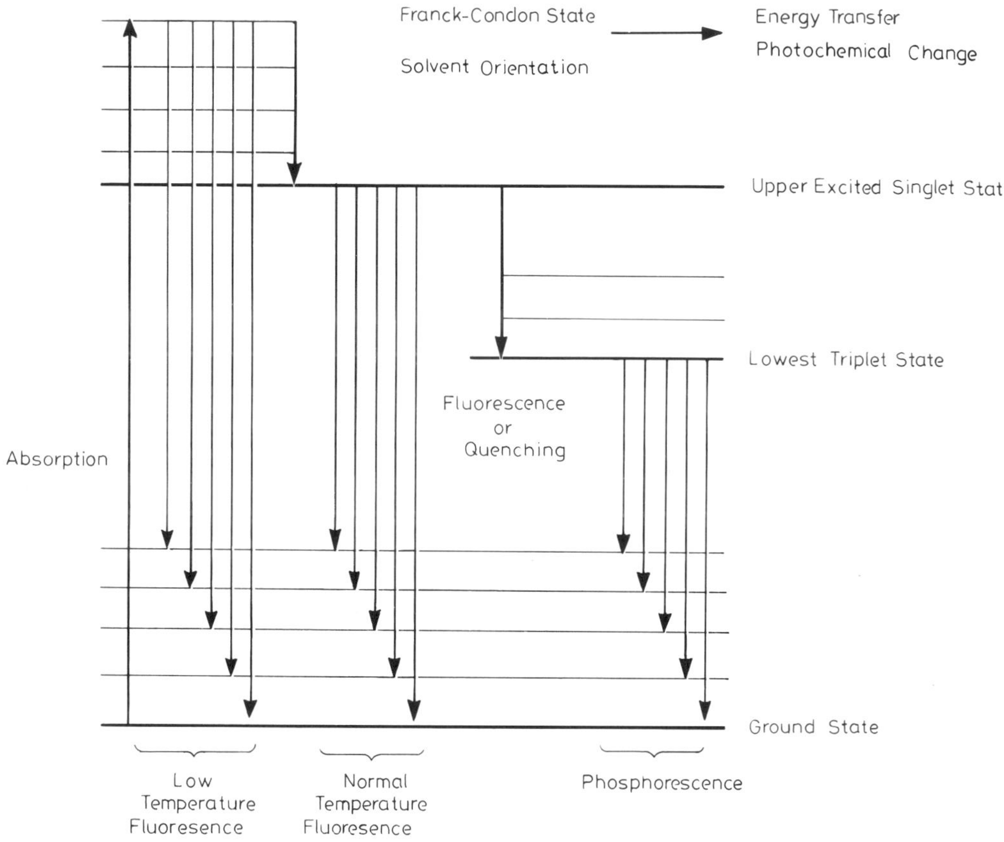

Figure 8.36

West, 1956) but with complicated organic substances prediction is not easy. Intermolecular quenching can result from collisions between an excited molecule and the perturbing molecule, often a solvent molecule. This is called dynamic quenching whereas static quenching depends on the formation of a complex of relatively long life with its own properties. Such a complex might be non-fluorescent.

Fluorescence spectra record the photon distribution of the fluorescence excited by monochromatic light absorbed by single substances. Fluorescence excitation spectra record the intensity of light emitted at one wavelength when the wavelength of excitation is varied. True excitation spectra would follow if the number of quanta falling on the cell were kept constant as the wavelength of excitation changed.

In measuring an emission spectrum the total number of quanta may be put at q and the recording instrument will register $dq/d\lambda$ and the curve will be given by plotting $K\,dq/d\lambda$ (relative quantum output) against λ. The area under the emission curve is a measure of q.

Since 'true' spectra are not obtained directly because most spectrofluorimeters have light sources, and detector systems with their own wavelength dependencies, the result is that each instrument produces apparent excitation and fluorescence spectra, which though possessing undoubted if limited usefulness really need to be 'corrected' to give 'absolute' or reproducible spectra (Chapman *et al.*, 1963).

The absorption spectra of the simpler organic substances that fluoresce agree very well with their true excitation spectra (Parker & Rees, 1962) and the quantum yield does not vary with the wavelength of excitation. This does not, however, apply to protein–prosthetic group complexes when the absorption spectrum of the complex may differ from the excitation spectrum of the 'fluorophore' (cf. Velick, 1961). Some authors prefer to plot on a wave-number scale where

$$dq/d\bar{\nu} = \lambda^2(dq/d\lambda)$$

Modern spectrofluorimeters allow automatic recording of corrected spectra (Perkin Elmer Model 236, Turner Model 210 and Aminco Bowman). Chen *et al.* (1967) discuss the problems of correction in an authoritative manner. Udenfriend (1962) deals with many experimental problems in a standard work.

The *quantum yield* ϕ is the measure of fluorescence efficiency. It is given by the ratio of the number of photons emitted to the number of photons absorbed. It may be measured directly or indirectly. The indirect method demands a reference standard for fluorescence quantum yield, which must itself be determined by an absolute method. Melhuish (1961, 1964) gave the following values for ϕ: quinine (in 0·1 or 1 N H_2SO_4) 0·55 (366 nm excitation); anthracene (in ethanol) 0·27; acridone (in ethanol) 0·83; proflavin (in water pH_4) 0·34. Chen (1967) found: 0·50 for quinine in sulphuric acid (348 nm excitation). Fluorescein (in 0·1 N NaOH) shows 0·85 (verified by Parker & Rees, 1960). The fluorescence of quinine excited by the mercury line 366 nm is rather widely used as a standard despite a disconcerting variation with wavelength.

Teale & Weber (1957) found $\phi = 0{\cdot}20$ and 0·21 for tryptophan and tyrosine in water. This provides a convenient standard of comparison for proteins but temperature sensitivity (*ca.* 2% per degree around 15–20°) is a disadvantage. Chen *et al.* (1967) put the probable accuracy of this quinine standard as 10% and regard Weber & Teale's values as erring on the high side.

The classification of compounds as fluorescent or non-fluorescent raises interesting issues. As Bowen & Wokes (1953) stressed, benzene and anthracene are highly fluorescent whereas the related substances pyridine and phenazine are completely non-fluorescent. Lewis & Kasha (1944) observed that many substances that are weakly fluorescent or non-fluorescent at room temperature became highly luminescent at very low temperature (liquid nitrogen) provided that the solvent sets to a 'glass'. This fluorescence resembles phosphorescence in that it has a long life-time and its position on the wavelength scale is more distant from that of the exciting radiation than is the case for normal fluorescence; in addition it displays more fully developed fine structure. The absorbing molecule is excited to a singlet level and then passes to a triplet level, losing excess vibrational energy by collision and then emitting long wave radiation as it returns to the ground state. If, as for example in the case of nitrobenzene, a long-lived triplet state of excitation soon follows excitation to the singlet level, frequent collisions (e.g. with solvent molecules) may cause it to shed energy before it can emit fluorescence. Nitrobenzene seems to be non-fluorescent for this reason.

Another aspect of inhibition of fluorescence is the phenomenon of self-quenching shown by solutions of fluorescent substances at concentrations greater than *ca.* 1 mg/ml. A relatively simple example is afforded by solutions of anthracene in benzene where molecules of solute in the first singlet state of excitation colliding with unexcited anthracene molecules produce a diradical which can either dissociate or form dianthracene. Many substances (e.g. dyes) exhibit concentration-dependent polymerization in solution which is detectable by changes in absorption spectra. Dimeric forms may be less fluorescent than the monomers or even non-fluorescent, moreover they may compete with the monomers for the radiations capable of exciting fluorescence in the monomers. They can thus act as 'built-in' light filters reducing the apparent quantum efficiency of fluorescence. In aqueous solutions the more complex aspects of concentration quenching are more evident than in less polar solvents and there is much to indicate that hydrogen bonding promotes the dissipation of electronic energy. Thus *o*-hydroxybenzophenone has a protective role in plastics exposed to light. It is, however, necessary to distinguish between self-quenching and simple reabsorption as the fluorescent light passes through a moderately concentrated solution. For instance, such a solution of fluorescein will display strong fluorescence on frontal exposure only in the first fraction of a millimetre of the light path. Clearly dilute solutions should be used and the excitation should be effected by carefully chosen monochromatic light. The design of modern spectrofluorimeters takes these difficulties into account.

Special problems arise when fluorescence must be measured for types of molecules not readily freed from all contaminants. Quenching by 'foreign' substances can occur by electron transfer, a classical example (Bowen *et al.*, 1949) being the quenching of the fluorescence shown by anthracene crystals contaminated by 0·1% of phenazine. When the excitation energy reaches a phenazine molecule it can be degraded through the triplet level and the fluorescence of anthracene is totally quenched. In many problems involving fluorescence as an analytical tool electron transfer difficulties certainly do arise but can often be evaded by working with very dilute solutions. Many biochemical issues have however required sophisticated research on absorption and fluorescence in complex systems before the potential value of fluorescence methods has been made actual.

The analysis of quenching processes affords a good basis for the discussion of the relationship between chemical structure and fluorescence (cf. Williams, 1959). It is a fact that whole classes of organic substances display strong electronic absorption but do not show serviceable fluorescence. The principal reason for this is 'internal conversion' of the energy of excitation. Ideally the reversal of electron spin from the excited singlet state to a triplet state leads to phosphorescence but the process is commonly very inefficient and there is no such emission. In some cases the excited electronic state permits a bond to break, thus butadiene absorbing at *ca.* 210 nm (140–160 kcal./mole) has its weakest link (C=C) with a bond strength of 125 kcal/mole and displays the phenomenon of predissociation. On the other hand decapentaene, with its conjugated system and non-localized electrons, absorbs at *ca.* 300 nm. The quantum (*ca.* 95 kcal/mole) is smaller than the bond strength (C=C) and fluorescence occurs, whereas with butadiene it does not. Predissociation provides an alternative energy-shedding route. Substituents which effect bathochromic shifts frequently permit fluorescence to appear as predissociation becomes less plausible. Many writers have discussed the relation between chemical structures and fluorescence (Förster, 1951; Reid, 1957; Udenfriend, 1962; Williams, 1959). The survey by Williams assembled in a very useful way the work of his own group in the context of a wide literature.

8.2.1 Fluorescence and Chemical Constitution

Saturated aliphatic compounds do not fluoresce but if ultraviolet emission is included with emission in the visible region, fluorescence occurs in a great range of compounds. Substances with ultraviolet chromophoric groupings due to the π electrons of conjugated double bonds or the mobile electrons of the benzene ring tend to be fluorescent. Substitution in the benzene ring has however marked consequences. Thus benzene is itself only weakly fluorescent but introduction of an amino group increases the emission while a nitro group abolishes it. The excitation and emission maxima on aqueous solutions of benzene (269 and 291 nm respectively) and of aniline (290 and 345 nm) are all in the ultraviolet, but the aniline emission is about 50 times as intense as that of benzene. Nitrobenzene does not measurably fluoresce in the ultraviolet or visible.

The amino group 'activates' and increases the freedom of π electrons whereas the nitro group tends to withdraw electrons from the ring to increase their localization. The amino group is familiar as *ortho-para*-directing and the nitro group as *meta*-directing. As a first approximation, *meta*-directing groups tend to reduce or abolish fluorescence. Bridges & Williams (1962, 1963) point out, however, how this needs to be qualified. With aniline and phenol, fluorescence is detectable at concentrations of 0·01 μg/ml, whereas 500 μg/ml is needed with chlorobenzene. Bromo- and iodobenzene are non-fluorescent as is the phenoxide ion. The only known *meta*-directing group compatible with the occurrence of ultraviolet fluorescence is the cyanide group. This is, however, a special case because the electrons of the —C≡N bond can interact with the π-electron cloud of the phenyl group to produce conditions favourable to fluorescence.

In disubstituted benzene derivatives the results are not yet fully predictable. Some examples are consistent with competitive effects; thus in *p*-nitroaniline, which is non-fluorescent, the nitro group has overcome the effect of the amino group. Benzenesulphonamide is non-fluorescent but *p*-aminobenzenesulphonamide (sulphanilamide) is five times as intensely fluorescent as aniline. Benzoic acid is non-fluorescent but a role of the carboxyl group as a suppressor of fluorescence seems to be reversed in salicylic and anthranilic acids which are respectively more strongly fluorescent than phenol or aniline. Changes in pH can greatly influence fluorescence. While phenol at pH 7 displays a fluorescence peak at 310 nm the phenate ion $C_6H_5O^-$ at pH 12 is non-fluorescent, although anisole fluoresces at both pH values. Salicylic acid over the pH range 4 to 12 is 100 times more intensely fluorescent than at pH 1 (Rosen & Williams, 1961).

Aniline affords an interesting example with implications for biochemical problems. Below pH 2 there is no fluorescence and the anilinium cation $C_6H_5NH_3^+$ is predominant. As the pH is allowed to increase, the intensity of fluorescence ($\lambda_{max.}$ 350 nm) parallels the appearance of the molecular aniline (pK_a 4·5) and becomes constant at pH 7·5 to 8·0. At pH values beyond 12 the fluorescence rapidly decreases as the aniline anion $C_6H_5NH^-$ is formed. Monomethylaniline with its replaceable hydrogen behaves similarly but dimethylaniline continues to fluoresce ($\lambda_{max.}$ 363 nm) at pH values over 12. In many types of compound (e.g. indoles: Udenfriend *et al.*, 1955) fluorescence spectra are more responsive to change of pH than are their ultraviolet absorption spectra—as a result of excited state protonation (see below).

White (1959) contrasted aniline and acetanilide, indole and quinoline, 3-hydroxyquinoline and quinoline, and 1,4-dihydroxynaphthalene with 1,4-naphthoquinone. In each pair only the first named compound was fluorescent. Other examples given included tyrosine and di-iodotyrosine, aniline and the anilinium ion, phenol and its simple esters, the second of each pair being non-fluorescent.

Phenylalanine ($\lambda_{max.}$ 260 nm) like benzene shows weak absorption and weak fluorescence, while tyrosine ($\lambda_{max.}$ 275 nm) is strongly fluorescent. Conjugated polyenes like carotenoids, retinol and dehydroretinol have very intense absorption spectra with $\epsilon_{max.} > 5 \times 10^4$ and absorption peaks in the near ultraviolet and visible (see p. 384). The fluorescence characteristics of these classes of compounds are discussed elsewhere (p. 48, 229).

Phenol excited at the wavelength of its absorption peak (270 nm) exhibits a fluorescence maximum at 420 nm with intensity high at pH 1 and negligible at pH 13. The phenolate ion ($\lambda_{max.}$ 287 nm) is thus non-fluorescent. On the other hand anisole, which cannot ionize, fluoresces strongly over the whole pH range 1–14.

Dihydric phenols such as quinol, catechol and resorcinol fluoresce in the unionized state. *p*-Hydroxybenzoic acid must ionize at both the carboxyl and hydroxyl groups if it is to be fluorescent, whereas *p*-hydroxyphenylacetic acid shows the opposite behaviour:

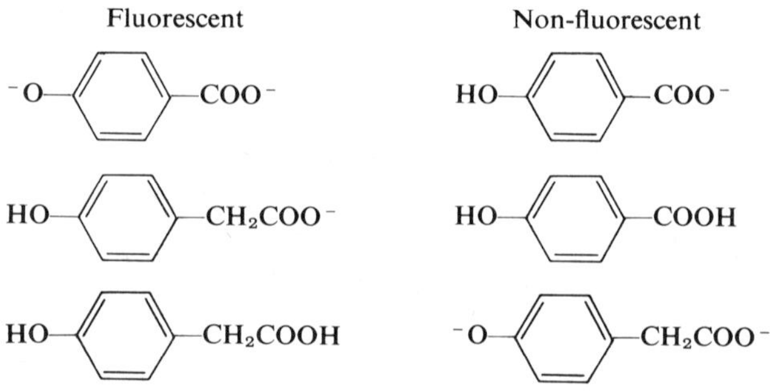

Fluorescent	Non-fluorescent
$^-O-C_6H_4-COO^-$	$HO-C_6H_4-COO^-$
$HO-C_6H_4-CH_2COO^-$	$HO-C_6H_4-COOH$
$HO-C_6H_4-CH_2COOH$	$^-O-C_6H_4-CH_2COO^-$

The naphthols (both α- and β-) are fluorescent in the ionized state and the neutral molecules are non-fluorescent, but as Williams pointed out, the possibility of resonance may help to explain the difference in behaviour between phenol and naphthols:

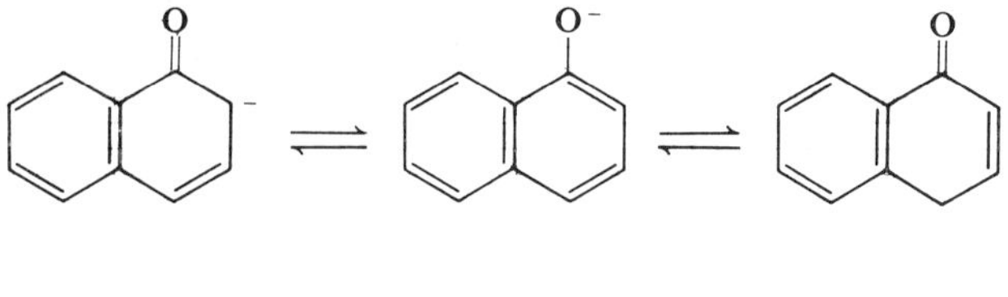

Bridges *et al.* (1963) made an instructive study of the fluorescence of 3-hydroxypyridine. The ultraviolet absorption spectra (Metzler & Snell, 1955) show considerable changes with variations in pH:

3-Hydroxypyridine

	$\lambda_{max.}$ nm	$\epsilon_{max.} \times 10^{-3}$
Alkaline solutions	234	10·2
	298	4·5
Neutral solutions	246	4·7
	313	3·0
Acidic solutions	222	3·3
	283	5·9

Bridges (1964) observed excited state ionization with this compound. A pK_a of 4·8 corresponds with the change from the cation to the zwitterion and a second pK_a of 7·8 indicates conversion of the zwitterion to the anion. At pH values between 5 and 7·8 the excitation peak is at *ca.* 315 nm and the fluorescence peak is at 390 nm. At pH values of *ca.* 8·5 to 14 the excitation peak is lower (*ca.* 310 nm) and the fluorescence peak is at 360 nm, the operative entity being the anion. When fluorescence intensity is plotted against pH, a new pK_a is shown at about 0·95. The new species shows an excitation peak at 297 nm and an emission peak at 317 nm and is ascribed to excited state ionization:

	Absorption	Fluorescence
Normal fluorescence	$R.OH \xrightarrow{h\nu_1} R^*OH$	$\longrightarrow R.OH + h\nu_2$
	$RO^- \xrightarrow{h\nu_3} R^*O^-$	$\longrightarrow RO^- + h\nu_4$

Excited state ionization

$$R.OH \xrightarrow{h\nu_a} R^*OH \xrightarrow{\text{ionization}} R^*O^-$$

$$R^*O^- \longrightarrow RO^- + h\nu_b \xrightarrow{H^+} R.OH$$

Bridges *et al.* (1966) gave the new pK_a as H_0 −3·7 to −3·3 (using Hammett's (1940) scale). The ionization occurs during the period of the excited state and reversion to the ground state involves fluorescence and reprotonation. As Förster had appreciated, the ionic form does not exist in the ground state, i.e. the fluorescence is due to the phenomenon of excited state ionization. Here the phenol has become a much stronger acid.

Weller (1952) derived an equation

$$pK_a - pK_a^* = \frac{\Delta E - \Delta E^1}{2{\cdot}303RT}$$

where ΔE and ΔE^1 are transition energies from the ground state to the first excited singlet state of the acid and anion (or conjugate base). Bridges *et al.* (1966) replaced R by the Boltzmann constant k and E by hc/λ where λ is wavelength in cm and rewrote the equation with λ in nm:

$$pK_a - pK_a^* = \frac{10^7 . hc}{2{\cdot}30\, kT} \cdot \left(\frac{\lambda^1 - \lambda}{\lambda\lambda^1}\right)$$

where λ is the absorption or fluorescence peak for the acid and λ^1 is the absorption or fluorescence peak for the anion or conjugated base. Putting $T = 293$ K and inserting numerical values of constants,

$$pK_a - pK_a^* = 21\ 187\left(\frac{\lambda^1 - \lambda}{\lambda\lambda^1}\right)$$

In the case of 3-hydroxypyridine,

$$pK_a - pK_a^* = 21\ 187\left(\frac{313 - 283}{313 \times 283}\right)$$

$$4{\cdot}8 - pK_a^* = 7{\cdot}02$$

$$pK_a^* = -2{\cdot}22$$

Bridges *et al.* (1966) obtained a value for pK_a^* of −3·0 by direct fluorometric titration.

Leeman *et al.* (1963) expressed the difference between excitation maxima and fluorescence maxima in wave numbers (cm^{-1}) and Bridges *et al.* (1966) compared these 'Stokes' differences for 3-hydroxypyridine and pyridoxine.

3-Hydroxycoumarin fluoresces weakly from pH −1 to 9 in the unionized form while the phenolic ion (pH 9–12) shows a new emission peak at 465 nm. In strongly alkaline solutions the fluorescence is appreciably stronger. On the other hand 7-hydroxycoumarin (umbelliferone) fluoresces very strongly. The fluorescence shown over the pH range 9–14 is due to the anionic form since it occurs at pH values greater than the pK_a. Below pH 9 the excitation maximum is displaced corresponding with the change in $\lambda_{max.}$ for the unionized form but the fluorescence peak remains at 455 nm. Although the absorbing species is unionized, the fluorescence is that of the ionized molecule so that excited state ionization must occur. Creaven *et al.*

(1965) think that in very acid solution a hydrogen bonded dimer may be formed.

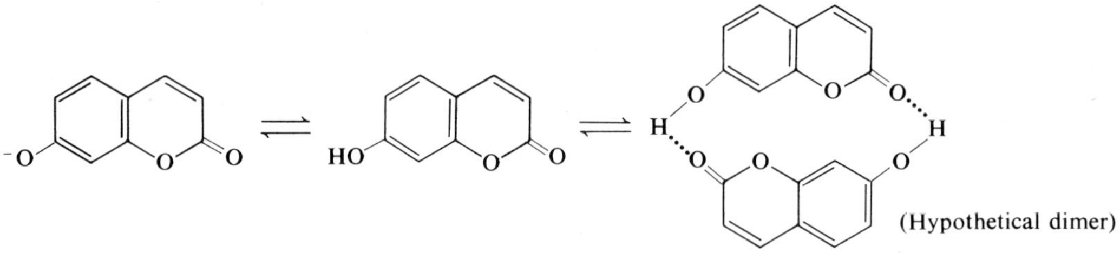

(Hypothetical dimer)

$\lambda_{exc.}$ nm	375	340	355
$\lambda_{fl.}$ nm	455	385	475
pH	9–14	(in ethanol)	0–1

Solvent effects are only moderately predictable. In fluorescence work carefully purified solvents should be used (see Udenfriend, 1962) and special care should be taken to remove traces of detergents or cleaning agents from glassware. Cells are better cleaned with nitric acid rather than with chromic acid. Aqueous solutions have been used more frequently than organic solvents, perhaps on grounds of convenience rather than intrinsic advantages.

Van Duuren (1963) found the excitation maxima for indole to be at 285 nm in several solvents but the fluorescence peaks (in nm) were: water 350; ethanol 330; dioxan 310; benzene 305; cyclohexane 297. The emission maximum is moved to longer wavelengths as the dielectric constant increases and exerts its influence on the electrons. The fluorescence of indole-3-acetic acid in water (360 nm) or ethanol (340 nm) is seen at 5 μg/ml but cyclohexane or benzene solutions show no detectable fluorescence at that concentration.

Bridges (1963) measured the intensity of fluorescence of sulphanilamide at 2 μg/ml in various solvents with excitation by radiation in the region 270–305 nm. The fluorescence ($\lambda_{max.}$ 350 nm) was best shown in ethanol or water. Intensities up to 70% of that shown by aqueous solution occurred with propanol, isopropanol, methanol, *n*-butanol, ethyl ether, formamide and 1,2-dichloroethane. Hexane, *n*-pentanol and dimethylformamide were 42–48% as effective while acetone, chloroform, carbon tetrachloride, nitromethane and *p*-xylene abolished the fluorescence.

Fluorescence shown by substances containing conjugated double bonds has been referred to before (p. 128). A good example is vitamin A which with $\lambda_{max.}$ 325 nm in absorption shows an excitation peak at 327 nm and a strong fluorescence with a peak in the green at 510 nm (Hagins & Jennings, 1950). Polycyclic hydrocarbons like biphenylnaphthalene, anthracene, benzpyrene, etc. have longer sequences of 'double bonds' and more electrons. Many substituted derivatives are much more strongly fluorescent than the corresponding benzene derivatives.

The fluorescent properties of 3-pyridol already mentioned are enhanced in the vitamin B_6 group.

Stokes displacements (cm^{-1}) *for hydroxypyridines*

	3-Hydroxypyridine		Pyridoxine
Normal cation	(8 N HCl)	5750	5210 cm^{-1}
Dipolar ion	(pH 6)	6240	4920 cm^{-1}
Excited state cation	(3 N HCl)	9830	9420 cm^{-1}

The larger displacements for the excited state fluorescence mean that smaller quanta are emitted and that energy is used in the actual excitation process. The behaviour of 3-hydroxypyridine is reproduced to a large extent in pyridoxine, pyridoxal and pyridoxamine.

The strongest fluorescence in pyridoxol and pyridoxamine are shown by the dipolar ions whilst for pyridoxal and pyridoxal phosphate the anions are the most effective forms. Williams & Bridges (1964) attribute maximum fluorescence to the cyclic hemiacetal of pyridoxal and to the tri-anion of the aldehyde hydrate of pyridoxal phosphate. A weaker fluorescence ($\lambda_{max.}$ 525 nm) is thought to be due to the ionized free aldehyde form of pyridoxal phosphate (see p. 436).

Heterocyclic compounds containing, oxygen and sulphur show fluorescence phenomena, suggesting an order NH— —O— —S— in enhancing electron effects. When there are two hetero atoms as in thiazoles and oxazoles, the joint effects are roughly additive but the nature of the substituents is decisive. Diphenyloxazole (PPO: Ott *et al.*, 1957) fluoresces but the nitro derivative does not. The well-known substance POPOP [2,2′-*p*-phenylene-bis-(5-phenyloxazole)], which is widely used in scintillator counting, has sustained conjugation of double bonds and a very high fluorescence efficiency.

Fluorescence intensity is the resultant of numerous factors. Leaving aside those due to instrumentation some of the more important ones are concentration, pH, solvents and buffers, and temperature. Excitable species may be light-sensitive and conditions of observation can be relevant to the outcome.

Solvents which, like benzene, toluene, xylene, phenol and acetone, absorb ultraviolet light are often effective as quenchers. Williams & Bridges (1964) cite the case of β-naphthol in aqueous solution where the fluorescence is quenched by acetone in a manner linearly related to the acetone concentration (Hymie *et al.*, 1960). Self-quenching or concentration quenching is of course the most familiar inner filter effect.

The presence of oxygen in solutions frequently gives rise to marked quenching which has had alternative explanations. A practical point is that it is often profitable to bubble nitrogen through a fluorescent solution when quantitative results are being sought. Parker & Barnes (1957) actually determined oxygen by its concentration-dependent quenching of the fluorescence of a borate–benzoin complex.

Creaven *et al.* (1965) made a spectrofluorometric study of the 7-hydroxylation of coumarin by liver microsomes but first established the properties of 3- and 7-hydroxycoumarin, and of 7-hydroxy- and 7-methoxy-4-methylcoumarin.

Liver microsomes acting on biphenyl produce 2- and 4-hydroxybiphenyl. The excitation and fluorescence behaviour are shown in Figs. 8.37 and 8.38. With 2-hydroxybiphenyl the fluorescence changes at pH 1 whereas excitation does not change until pH 9·5. The pK_a of the excited molecule is 1·5 so that it is then 10^8 times stronger as an acid than the unexcited form. At pH 6 the 4-hydroxy compound fluoresces at 340 nm and the 2-hydroxy isomer at 415 nm so that they can be determined independently. Williams & Bridges (1964) observed excited state ionization for 3-pyridol (p. 237), pyridoxol, pyridoxal, pyridoxamine and pyridoxal phosphate, as well as *p*-anisidine, *p*-toluidine and 2,2′-dihydroxybiphenyl. The phenomena had

Table 8.13 Fluorescence of 3-and 7-hydroxycoumarins

pH ranges	Excitation $\lambda_{max.}$ nm	Fluorescence $\lambda_{max.}$ nm	Relative fluorescence intensities	At pH
(*a*) 3-Hydroxycoumarin (5 µg/ml)				
−1*–9	320	380	0·005	6
0–12	350	465	0·002	10·4
13–14	340	460	0·09	13
14		None		
(*b*) 7-Hydroxycoumarin (0·1 µg/ml)				
−1*		None		
0–1	335	475	17·5	1
2–9	335	455	44·0	7
0–14	375	455	100‡	10
15†		None		
Ethanol	340	385		
(*c*) 7-Hydroxy-4-methylcoumarin (0·1 µg/ml)				
0	330	480	7·5§	0
(*d*) 7-Methoxy-4-methylcoumarin (0·1 µg/ml)				
0	330	385	5·2§	0

* 10 N HCl.
† 10 N KOH.
‡ Intensity constant for pH 9·4–10·5 and taken as 100.
§ Not maximum intensities.

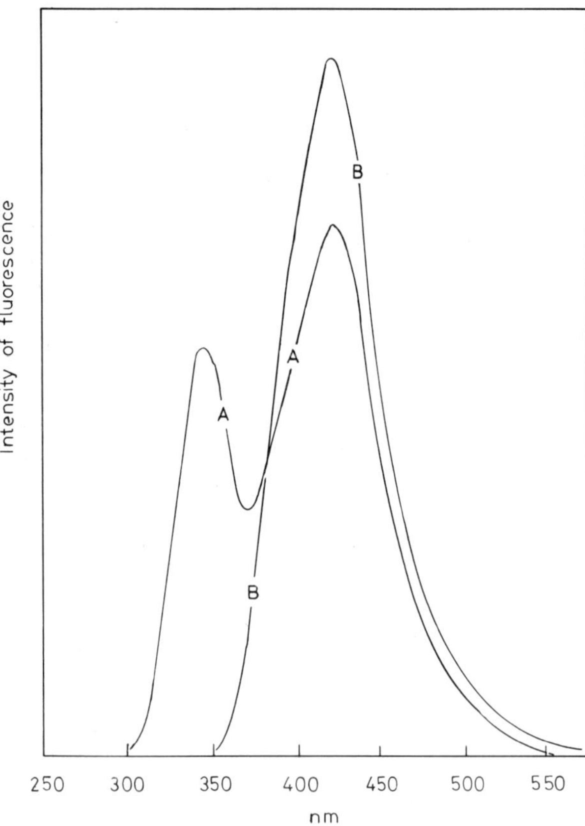

Figure 8.37 Tracing of the fluorescence of a mixture of 2- and 4-hydroxybiphenyl(2-isomer, 3 μg/ml ; 4-isomer, 5 μg/ml in water) A, at pH 6 ; B, at pH 13 (after Creaven, Parke & Williams, 1965)

Figure 8.38 Fluorescence intensity ($\lambda_{exc.}$ 295 nm; λ_{fl} 415 nm) of 2-hydroxybiphenyl at pH 5·5 in the presence of different amounts of 4-hydroxy-biphenyl : ○, none ; △, 0·18 μg/ml ; ●, 0·36 μg/ml ; ▲, 0·48 μg·ml (after Creaven, Parke & Williams, 1965)

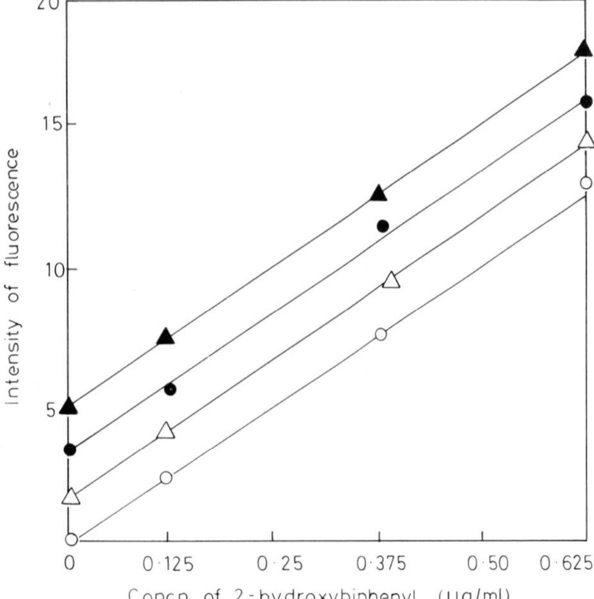

previously been seen in 1- and 2-naphthol and 1- and 2-naphthylamine (Foster, 1950; Hercules & Rogers, 1959).

Some substances are subject to photochemical change when exposed to ultraviolet light. Quinine sulphate decomposes at 0·01 μg/ml but not significantly at 1 μg/ml unless the light source is a very intense one. Although 3-pyridol is stable, pyridoxol at pH 6·8 (5 μg/ml. λ (excitation) 295 nm, λ fluorescence 400 nm) loses 90% of its fluorescence intensity in 90 seconds. The other members of the vitamin B_6 family also undergo photodecomposition with loss of fluorescence. Other cases are known in which photochemical changes result in fluorescent decomposition products (Feigl *et al.*, 1955; Brodie *et al.*, 1947).

An important development in spectrofluorometry depends upon the use of light pulses of very short duration down to 4 nanoseconds with rise and fall times of about 1·6 nanoseconds and relatively high radiated power (up to 140 watts) in different regions depending on the light source. Thus a deuterium lamp (TRW Instruments, California) generates a continuous spectrum (200–350 nm), a nitrogen lamp acts as a line source in the visible and near ultraviolet regions (297·8 to 399·8 nm) while a krypton lamp generates a line spectrum in the near infrared region (750 to 900 nm). Interference filters or absorption filters can be used as necessary. The availability of facilities for controlled short-duration pulses paved the way to improved measurement of fluorescence and phosphorescence decay times between 1·7 nanoseconds and 10 milliseconds. With suitable sample holders decay times can also be measured at very low temperatures.

Chen *et al.* (1967) measured τ (the mean duration of the excited state after excitation had ceased, or fluorescence decay time) for some 48 compounds dissolved in water. The list included proteins (2–5 nanoseconds) and co-enzymes of the pyridine nucleotide, flavin and pyridoxal groups (4–5 nanoseconds). The flash lamps had repetition rates of 1–5 kHz. Decay times for proteins vary but as a rule exceed that of free tryptophan. The lengthening of times could be due in part to tyrosine–tryptophan energy transfer or to intramolecular quenching. Some compounds that form protein complexes (e.g. porphyrins, anthranilic acid and naphthalene sulphonates) have rather long lasting fluorescence and the complexes they form also have high τ values. When fluorescence polarization is used to study protein structure it is an advantage to

Table 8.14 Fluorescence characteristics of selected organic compounds

Substance	Concentration and solvent **M**	Excitation wavelength nm	Emission maximum nm	Decay time nanoseconds	Quantum yield %
[9-Aminoacridine	water or ethanol; can be used as a standard substance			—	98–99]
Quinine sulphate	10^{-5} in	250, 350	450	19·5	—
	0·1 N H_2SO_4	365	466	19·2	—
			(318, 348)*		
Naphthalene	10^{-4} in	310	325	98	10, 12†
	hexane	265	325, 336	96	23
			(266, 276, 286)		
Anthracene	10^{-5} in	360	425	4·9	29
	hexane	254	380, 400, 428	4·9	36
			(325, 340, 360, 376)		
Indole[1]	$8{\cdot}5 \times 10^{-3}$	265	309	3·7	—
	in water‡	280	330	2·7	45
			(270, 276, 287)		
Skatole[2]	$7{\cdot}6 \times 10^{-3}$ in water‡	290	370	6·4	42
Tryptophan	$4{\cdot}9 \times 10^{-3}$ in water‡	280	350	2·6	—
Serotonin	10^{-4} in water‡	295	340	2·7	—
Resorcinol	10^{-5} in water‡	265	315	4·7	—
Salicylic acid	10^{-5} in water‡	310	>490	3·9	—
Anthranilic acid	10^{-5} in water‡	300	405 (335)	8·4	—
3-Hydroxyanthranilic acid	10^{-5} in water‡	320	415	10·9	—
Pyridoxal	10^{-4} in water‡	325	390	4·2	—
Tyrosine	10^{-3} in water‡	275	310	2·6	—
Biacetyl§	5×10^{-2}	390	475	10	0·3
Riboflavin	10^{-5} in water‡	370	520	4·2	26

* Absorption peaks in brackets.
† In ethanol.
‡ 0·01 **M** tris Cl^-, pH 7 in water.
§ Oxygen removed by bubbling with N_2.
[1] 1 mg/ml. [2] 1 mg/ml is a saturated solution at 24°C.

have a fluorescence decay-time agreeing approximately with protein rotational relaxation-time.

Non-fluorescent compounds can often be combined with other substances to form fluorescent products that can be used for analytical purposes. The combination of simple carbonyl compounds and anthrone is a case in point.

Anthrone exists in a non-fluorescent ketonic form in equilibrium with its enol anthranol. It reacts with simple aldehydes like formaldehyde and acetaldehyde and ketones like acetone in the presence of concentrated sulphuric acid to produce fluorescent compounds. It also reacts with glycol or glycerol as shown. The absorption spectra of the new compounds reflect solvent-solute interactions.

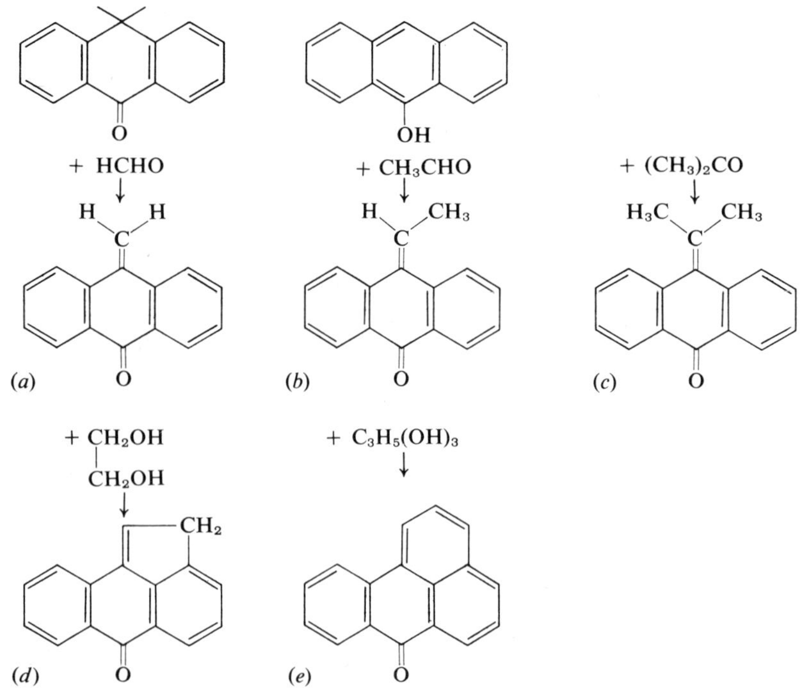

+ HCHO → (a)

+ CH_3CHO → (b)

+ $(CH_3)_2CO$ → (c)

+ CH_2OH–CH_2OH → (d) CH_2

+ $C_3H_5(OH)_3$ → (e)

The fluorescence properties in concentrated sulphuric acid are as follows:

Compound	Absorption $\lambda_{max.}$ nm	Fluorescence $\lambda_{max.}$ nm	Stokes shift (cm^{-1})
(*a*)	490	568 strong	2800
(*b*)	485	576 weak	3260
(*c*)	475	537 strong	2430
(*d*)	480	525–570 strong	1790–3290
(*e*)	495	580 strong	2960

Symmetrical fluorescent compounds emit more intensely than unsymmetrical, e.g. the condensation product with acetone fluoresces stronger than that with methyl ethyl ketone and the Stokes shift is larger for the unsymmetrical compounds.

Methyleneanthrone (*a*) and dimethylmethyleneanthrone (*b*) afford interesting compounds (Fujimori, 1955):

	Absorption spectra			
	In ethanol		In benzene	
Compound	$\lambda_{max.}$ nm	$\epsilon_{max.}$	$\lambda_{max.}$ nm	$\epsilon_{max.}$
(*a*)	390–400	*ca.* 1000	385	*ca.* 600
(*b*)	395	1450	395 420	900
(*c*)	360–420	1400	360–390	2100
(*d*)	390	6200	385	14 200

	Fluorescence spectra			
	In ethanol		In benzene	
Compound	$\lambda_{max.}$ nm	intensity	$\lambda_{max.}$ nm	intensity
(*a*)	490, 510, 600	strong	430	weak
(*b*)	490, 540, 550	strong	550 600	weak
(*c*)	480, 520	strong	600	weak
(*d*)	470	strong		

From these results it is clear that the absorbing entities in the different solvents are not the same and that solvent–solute complexes must be postulated. The fluorescence spectra are different in location and intensity. With suitable preliminary work fluorescence can here be used in qualitative and quantitative analysis. Thus the fluorescence of (*a*) is violet in benzene, greenish yellow in ethanol and orange in sulphuric acid.

Dopa (3,4-dihydroxyphenylalanine) is obtained from tyrosine by oxidation and is biochemically important as a precursor of melanin, adrenaline and noradrenaline as well as being valuable in the treatment of Parkinson's disease. It exhibits non-specific fluorescence ($\lambda_{max.}$ (exc.) 285 nm, $\lambda_{max.}$ (fl.) 325 nm) but it can be separated from catecholamines by chromatographic methods and determined fluorometrically after oxidation. Dopamine condenses with ethylene diamine giving a highly fluorescent derivative or it can be oxidized to a dihydroxyindole

(HO)(HO)$C_6H_3CH_2CH_2NH_2$ → (O)(O)-indoline (CH_2–CH_2–N–H) $\xrightarrow{NaHSO_3}$ (HO)(HO)-indole (CH=CH–N–H)

$\lambda_{exc.}$ nm 345
$\lambda_{fl.}$ 410

(Carlsson & Waldeck, 1958; Drujan *et al.*, 1959). Dopa gives similar fluorescence and a chromatographic separation of dopamine and dopa may be necessary.

The oxidation of adrenaline and noradrenaline can be effected in various ways (ferricyanide pH 6, MnO_2, pH 6·5 and iodine). At pH 3·5 adrenaline only is oxidized with iodine and at pH 6·5 both adrenaline and noradrenaline are oxidized (Crout, 1957), while after ingenious separation of the two substances, Shore & Olin (1958) were able to measure noradrenaline only by iodine oxidation at pH 5. The trihydroxyindoles are prepared from the iodoaminochromes with the aid of alkaline ascorbic acid:

$\lambda_{max.}$ exc.	$\lambda_{max.}$ fl. nm		
395	505	(nor-adrenaline)	pH 6·5
410	520	(adrenaline)	pH 3·5

The condensations with ethylene diamine give different products for adrenaline and noradrenaline (Wallerstein *et al.*, 1947; Natelson *et al.*, 1949) and Weil-Malherbe & Bone (1952) devised an analytical procedure which has been much used. The product derived from adrenaline is shown below (Harley-Mason & Laird, 1958) but the explanation of the noradrenaline reaction is more debatable. The fluorescence maxima (noradrenaline 480–495 nm, adrenaline 525–530 nm) are different but separations are desirable (see Fig. 8.39). Nevertheless with excitation at 420 nm fluorescence can be measured at both 510 and 580 nm. At 510 nm the noradrenaline derivative fluoresces at twice the intensity shown by the adrenaline derivative, while at 580 nm the intensities are exactly reversed. Simultaneous equations set up by calibration permit both substances to be determined.

Chen (1968) used 4-methylumbelliferone as a fluorescent pH indicator. The spectra are as follows:

	Absorption		Emission, quantum yield	
	$\lambda_{max.}$ nm	$\epsilon_{max} \times 10^{-4}$	$\lambda_{max.}$ nm	%
0·1 M NaOH anionic form	360	3·66	457	69
0·01 M HCl neutral form	320	2·87	474	70
(Isosbestic point	332	2·17)	—	—

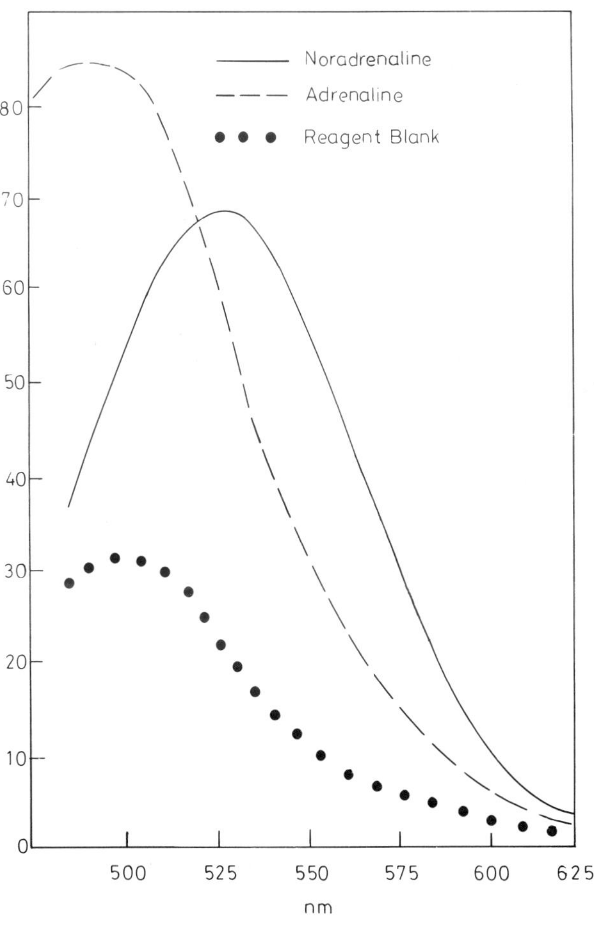

Figure 8.39 Fluorescence spectra of ethylene diamine-condensation products with noradrenaline (solid curve), adrenaline (dashed curve) and reagent blank (dotted curve). $\lambda_{exc.}$ 436 nm bases 0·2 μg/ml (from Mangen & Mason, 1957)

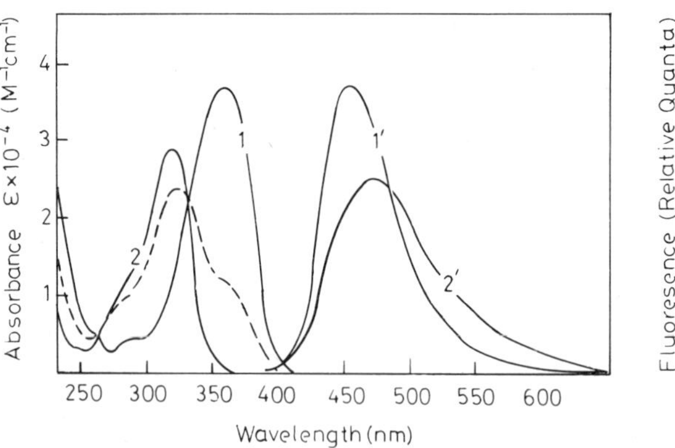

Figure 8.40 Spectra of 4-methylumbelliferone. 1 and 1′ are the absorption and corrected emission spectra in 0·01 **M** NaOH ; 2 and 2′ are the corresponding spectra in 0·01 **M** HCl. The dotted line is the absorption spectrum in potassium phosphate buffer, pH 7 (from Chen, 1968)

The fluorescence is blue in alkali and pale blue in acid.

It will be noticed that at 360 nm absorption by the cationic form is very low and by the anionic form very strong (see Fig. 8·40). For titrations with alkali using 4-methyl umbelliferone as indicator, excitation by light of 360 nm would be effective while titrations with acid are followed by excitation in the region 300–320 nm (Fig. 8.41). The mid-point of the transition is at pH 7·6 whether it is ascertained from fluorescence or ultraviolet absorption. The decay times of the anionic and neutral forms were similar (6·8 and 6·9 nanoseconds) and the quantum yields were in good agreement.

Figure 8.41 Titration curves of 4-methylumbelliferone, (A) Fluorimetric titrations with excitation at 360 nm (—●—) and 300 m (—○—) observed at 460 nm. (B) Absorptiometric titration (from Chen, 1968)

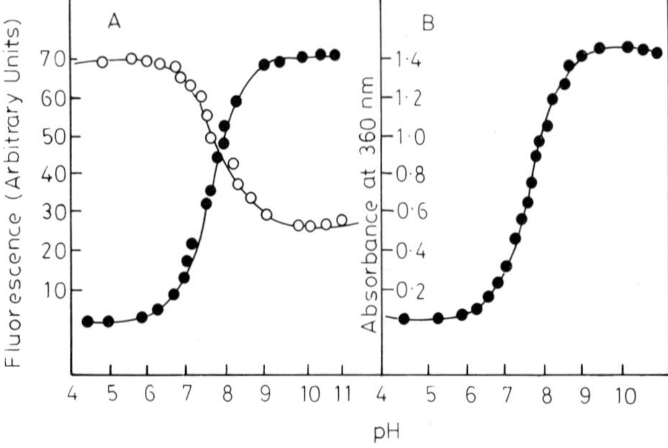

Chen found this indicator valuable in following the pH change induced in the hydration of carbon dioxide catalysed by carbonic anhydrase.

Chen & Kernohan (1967) found that ox erythrocyte carbonic anhydrase combines with DNSA (5-dimethylaminonaphthalene-1-sulphonamide) to form a highly fluorescent complex in which 1 molecule of DNSA is bound per molecule of protein (dissociation constant $2{\cdot}5\times10^{-1}$ **M** at pH 7·4).

The fluorescence of free DNSA ($\lambda_{max.}$ 580 nm) has a low quantum yield (5·5%) whereas the complex shows an emission peak at 468 nm with a high quantum yield (84%). The blue shift is explained on the basis that the $—SO_2NH_2$ group of the ligand loses a proton on binding and that the site on the protein molecule is very hydrophobic. It seems that carbonic anhydrase possesses a site with a special affinity for sulphonamide

The ultraviolet absorption of carbonic anhydrase is preponderantly due to 7 tryptophan moieties in the protein molecule and 85% of the absorbed quanta is transferred to the DNSA molecule—a very high transfer efficiency. The 'average distance' in the complex between the DNSA and tryptophan residues is about 16 nm. The probability of energy transfer between two molecules is expressed in terms of critical transfer distance R_0 for which resonance transfer is 50% complete:

$$R_0=\left(\frac{1{\cdot}66\times10^{-33}\times\tau J_\nu}{n^2\bar{\nu}_0^2}\right)^{1/6}$$

where τ is the decay time of the donor fluorescence $\bar{\nu}_0$ is the mean of the donor emission and lowest energy absorption peaks (in cm^{-1}) J_ν is the overlap integral and n the refractive index. For carbonic anhydrase (tryptophans) n is put at 1·6, $\tau=2{\cdot}6$ nanoseconds (by direct measurement) and $J_\nu=6{\cdot}01\times10^{10}$ cm^3 m $\mathbf{M}^{-2}$, $\bar{\nu}_0=32{\cdot}8\times10^3$ cm^{-1}. R_0 came out at 2·13 nm. The tryptophan fluorescence in the carbonic anhydrase was quenched by the single DNSA molecule to the extent of 73%. This figure, being less than the energy transfer efficiency of 85%, suggests that the seven tryptophan residues have different fluorescence efficiencies. The DNSA is probably so bound that energy transfer occurs most readily from the tryptophan residues that are relatively less fluorescent. Moreover the carbonic anhydrase, which is roughly spherical ($d=ca.$ 5·1 nm), must have most of the tryptophans and the $—SONH_2—$ binding site very much 'inside' the molecule. Roth (1969) has published an important review of the fluorometric assay of enzymes with full documentation (423 references).

8.2.2 Aromatic Amino acids

Phenylalanine exhibits very weak fluorescence but tyrosine and tryptophan and many of their derivatives have characteristic and rather strong fluorescence. Only under special conditions is it worth while to attempt the fluorometric determination of phenylalanine because the value of ϵ at 260 nm (the fluorescence excitation optimum wavelength) is only 2000 and the fluorescence efficiency is low (4%). Both factors are much more favourable in the cases of tyrosine and tryptophan (Teale & Weber, 1957).

Dopa (3,4-dihydroxyphenylalanine) is excited (285 nm) to fluoresce maximally at 325 nm, but this effect is non-specific and analytically without value. Decarboxylation to dopamine followed by condensation with ethylene diamine yields a highly fluorescent product. Alternatively, dopamine can be oxidized with an o-quinone to 5,6-dihydroxyindole which is strongly fluorescent ($\lambda_{exc.}$ 345, $\lambda_{fl.}$ 410 nm).

Tyrosine fluoresces maximally in the pH range 4–9 (White, 1959) and with excitation in the ultraviolet, (200–340 nm, $\lambda_{max.}$ 275 nm) fluorescence is strong between 260 and 410 nm, $\lambda_{max.}$ 303 nm (Duggan & Udenfriend, 1956). Tryptophan shows an excitation maximum at 287 nm, and a fluorescence peak at 348 nm.

If proteins are subjected to acid hydrolysis the tryptophan is almost entirely destroyed, giving non-fluorescent products, whereas the tyrosine survives the treatment. Hydrolysis, [protein 20–25 mg, heated in 7 N H_2SO_4 in an open tube in an autoclave for 20 hours at 2 atmospheres; hydrolysate neutralized by NH_4OH, and made up to 25 ml with water] followed by suitable dilution (e.g. 10-fold with phosphate buffer pH 8·0) allows the fluorescence to be determined reproducibly. Known amounts of tyrosine added to mixtures of aminoacids can be accurately measured in fluorescence calibrations. Duggan & Udenfriend obtained good agreement with published values for the tyrosine in insulin, bovine serum albumin and β-lactoglobulin. Velick obtained confirmatory results for the tyrosine in triosephosphate dehydrogenase. Although the method is satisfactory when applied to isolated proteins, there is often interference when it is extended to tissues.

When proteins are subjected to alkaline hydrolysis the losses of tryptophan are small (5–10%). Although tryptophan fluoresces over a wide pH range the intensity rises to a quite sharp maximum at pH 11 and under the best conditions for exciting and measuring its fluorescence, interference by tyrosine is negligible even when there is much more tyrosine than tryptophan. Provided that the whole procedure has been calibrated using known amounts of added tryptophan, the losses occurring in hydrolysis can be allowed for. 'Recoveries' of tryptophan added to aminoacid mixtures are quantitative (excitation 287 nm, fluorescence peak 348 nm). Udenfriend regards 'the fluorometric assay of tryptophan' as 'now the simplest and most sensitive of all available procedures. Because of its simplicity it is also precise'. Insulin, which contains no tryptophan, registered zero, and the results with bovine serum albumin and β-lactoglobulin confirmed published values within experimental error.

The possibility of interference has prompted investigation of highly fluorescent coloured products made from tyrosine and 1-nitroso-2-naphthol (Waalkes & Udenfriend, 1957). The reaction product was in the first instance used in a direct colorimetric assay but the fluorescence method (excitation peak 460 nm, fluorescence peak 570 nm) allows measurement of tyrosine down to 1 μg. The procedure is particularly suitable for the determination of tyrosine in tissues but it does not distinguish between tyrosine and tyramine; this is not however usually a serious source of error since the latter is rarely more than a very minor constituent. Tyramine can in any case be separated from tyrosine by chromatography or by partitioning between solvents. Fluorescence methods can be adapted to the task of determining free and conjugated tyramine in urine (Oates *et al.*, 1960) which is sometimes clinically informative (see Carlsson *et al.*, 1958).

Adrenaline undergoes conversion to a trihydroxyindole, adrenolutine, which is highly fluorescent. An early fluorometric method was described by Gaddum & Schild (1934) but many difficulties were later encountered (see Udenfriend, 1962). The technical problems have been overcome and elucidated (Crout, 1959; Crout & Sjoerdsma, 1960). Many other authors have contributed to the work and it is now possible to determine adrenaline (epinephrine) and noradrenaline (norepinephrine) in urine, plasma and tissues and to distinguish between the free catecholamines and the

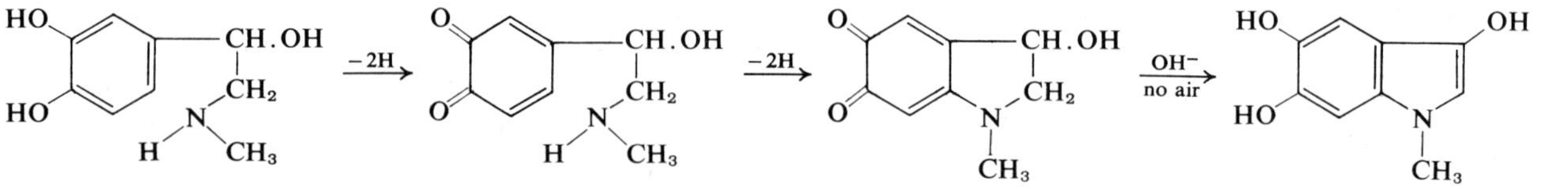

Adrenaline → Adrenalino-quinone → Adrenochrome → Adrenolutine (3,5,6-trihydroxy-1-methylindole)

acid-labile conjugated forms. Essentially the procedure consists of forming adrenolutin and its methylated analogue. At pH 6·5 (excitation peak 395 nm, fluorescence peak 505 nm) noradrenaline is measured, at pH 3·5 (excitation peak 410 nm, fluorescence peak 520 nm) adrenaline is measured. Careful calibrations are necessary and the latest literature should be consulted on entering the field.

Another procedure is to condense the catecholamines with ethylenediamine (Weil-Malherbe & Bone, 1952; Weil-Malherbe, 1959), the probable mechanisms being shown below. The fluorescence maxima for adrenaline and noradrenaline are at 525 and 485 nm respectively.

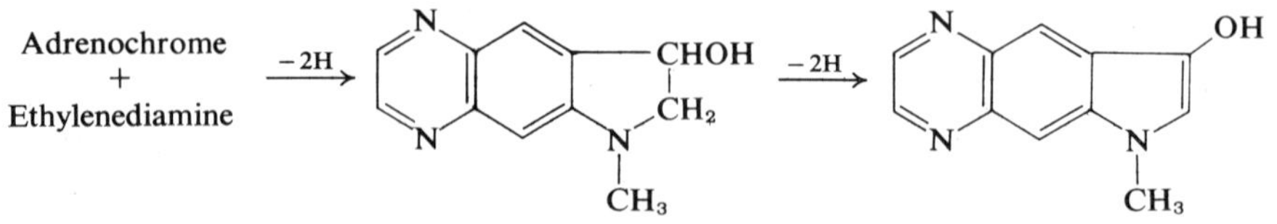

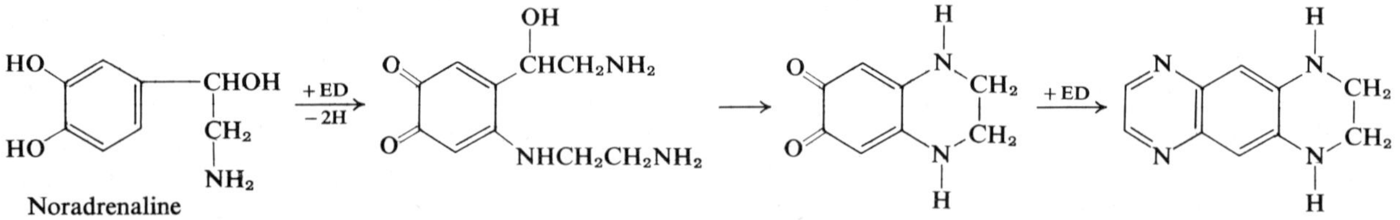

Adrenochrome + Ethylenediamine

Noradrenaline

Hess & Udenfriend (1959) have described an effective method for determining tryptamine in urine or in tissues. The amine is extracted by an organic solvent and condensed with formaldehyde to give tetrahydronorharman which is oxidized via dihydronorharman to norharman (excitation 365 nm, fluorescence peak 440 nm). There is little interference and the calibrations are satisfactory (see also Sjoerdsma *et al.*, 1960).

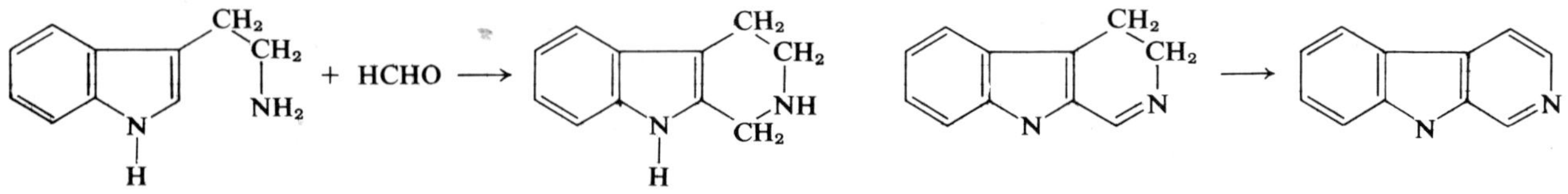

Tetrahydronorharman → 3,4-Dihydronorharman → Norharman

Table 8 15 Fluorescence intensities of 5-hydroxy- or 5-methoxy-indoles after using extraction methods appropriate for each compound

Compound	Single compound*	Extracted mixed indoles†	Mixed indoles pineals‡	Pineal contents	
				ng/pineal	ng/mg
Serotonin	100	130	295	80–300	100–350
N-Acetylserotonin	100	99	104	< 15	< 21
5-Hydroxytryptophan	114	100	114	< 10	< 14
5-Hydroxyindole-3-acetic acid	100	109	115	5–15	7–21
5-Methoxyindole-3-acetic acid	100	95	91	< 5	< 7
Melatonin	100	103	113	< 5–20	< 7–28

* 500 ng/tube; † 500 ng/tube; ‡ Two adult rat pineal glands.

Satoh & Price (1958) have discussed the relevance of fluorescence methods to the metabolism of tryptophan via the kynurenine pathway with particular reference to the fluorometric determination of kynurenic and xanthurenic acids.

5-Hydroxyindoles are of considerable biochemical importance. Thus tryptophan may be hydroxylated at position 5 and then by decarboxylation yield serotonin (5-hydroxytryptamine). In neutral solution $\lambda_{max.}$ in absorption is near 295 nm and there is strong ultraviolet fluorescence $\lambda_{max.}$ 330 nm. In the presence of increasing amounts of hydrochloric acid, the 330 nm fluorescence declines and a new emission with $\lambda_{max.}$ 550 nm reaches a peak intensity at 3 N HCl. This change is reversible and does not occur with unsubstituted indoles (Udenfriend *et al.*, 1955). The large Stokes shift 295–550 nm is consistent with excited state ionization. Udenfriend *et al.* (1958) describe fluorometric methods of determining 5-hydroxyindoles in urine, blood, etc. (see Udenfriend, 1962, 1969). An alkaline homogenate of tissue is saturated with salt and extracted with *n*-butanol. The extract is mixed with dilute aqueous acid and heptane partitioned so that serotonin passes into the aqueous acid and can be assayed. In most tissues the serotonin content is not more than about 2 μg/g, but rat pineal gland has about 65 μg/g. In human carcinoid tumour it rises to 0·3 to 1·5 mg/g. Banana pulp is a good source (20–30 μg/g) and the peel (50–150 μg/g) even richer.

Quay (1963) described methods for the selective extraction and measurement of six 5-hydroxy- and 5-methoxyindoles that occur in animal tissues. The problem centres on the fact that the 540–550 nm fluorescence is shown by all the 5-hydroxyindoles so that separation methods are essential. EDTA (ethylenediamine tetra-acetic acid disodium salt) and ascorbic acid are used to protect the indoles and promote 'recovery'. In each case after separation fluorescence is measured for solutions in 3 N HCl. Melatonin (N-acetyl-5-methoxy tryptophan), a hormone from the pineal gland, condenses with *o*-phthalaldehyde to give a product fluorescing several times more intensely than the corresponding product from serotonin.

Quay (1963) obtained results shown in Table 8.15.

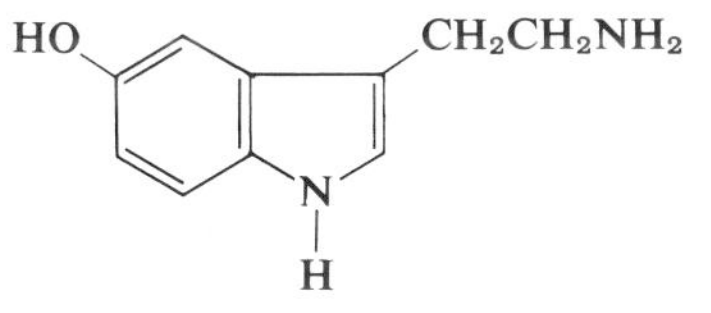

HO … $CH_2CH_2NH_2$ … N–H

Serotonin

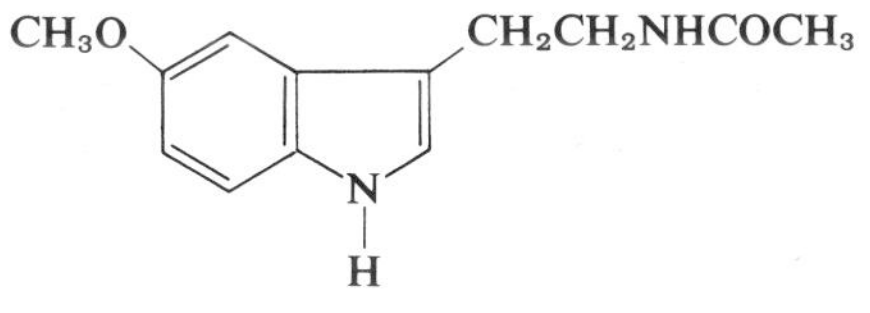

CH_3O … $CH_2CH_2NHCOCH_3$ … N–H

Melatonin

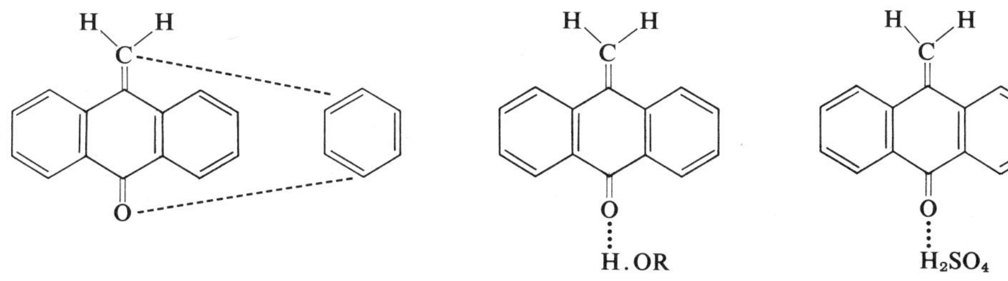

H H C … O

H H C … O … H.OR

H H C … O … H_2SO_4

It is thus clear that the rat pineal gland is particularly rich in serotonin. There are difficulties in determining melatonin and the literature should be consulted.

Mann *et al.* (1961) examined the semen of man, bull, ram, boar and dog by chromatographic and spectrofluorometric methods. 5-Hydroxytryptamine if present at all was in minute quantities. The uterus-stimulating effect of human seminal plasma is not due to this substance. Tryptophan, though present in semen, is not converted to 5-hydroxytryptamine on incubation. When 5-hydroxytryptophan is added to semen it is oxidatively deaminated but not decarboxylated. This investigation affords a good example of fluorescence as a tool.

The ultraviolet fluorescence of proteins arises from absorption by tyrosine and tryptophan, but when both of these acids are present in a protein there is a strong tendency for the fluorescence to arise from tryptophan only (Teale, 1960; Konev, 1967). Haemocyanins, for instance, contain about twice as many tyrosine moieties as tryptophans (Ghiretti-Magaldi *et al.*, 1966) and the fluorescence of haemocyanin from the mollusc *Levantina hierosalina* is due to tryptophan. Deoxygenation can enhance this fluorescence fourfold (Shaklai & Daniel, 1970) so that binding of oxygen at the copper atoms has a quenching action on tryptophan fluorescence. Bannister & Wood (1971), noting that thiocyanate reversibly liberates oxygen from oxyhaemocyanin (Rombauts & Lontie, 1960), used this to follow the fluorescence of haemocyanin from the whelk *Murex trunculus*.

The haemocyanin and apohaemocyanin showed $E^{0\cdot1\%}_{1\text{cm}}$ 280 nm = 1·39 at pH 9·2. The copper–oxygen band in the near ultraviolet showed $\lambda_{max.}$ 348 nm and the absorbance of a haemocyanin solution equilibrated with air was taken to indicate saturation. The intensity of this band diminished as potassium thiocyanate was added to displace oxygen. The quantum yield for a neutral solution of L-tryptophan was taken as 0·2 (Teale & Weber, 1957). Fluorescence excitation was maximal at 292 nm and extended from about 225 to 325 nm (half-width at 266 and 302 nm for oxyhaemocyanin). The emission for oxyhaemocyanin, haemocyanin in 185 m**M** KCNS and for apohaemocyanin was maximal at 338 nm and the half-width about 315–318 and 365–375 nm, extending from 290 to 450 nm.

Although haemocyanin has a molecular weight about 9 million this is an aggregate of smaller units. In denaturing solutions, a sub-unit of molecular weight *ca.* 220 000 is obtained and there is electrophoretic evidence of a 50 000 unit containing two copper atoms and binding one oxygen molecule. The spectrophotometric method of analysis indicated that there were 17·3 tyrosine and 7·7 tryptophan moieties or 8·4 tryptophan residues by the Edelhoch (1967) method. Ghiretti-Magaldi (1966) reported 7 tryptophan and 17 tyrosine per 50 800 unit.

The haemocyanin was excited by monochromatic radiations 270, 280 and 290 nm and fluorescence emission was measured at 405, 325 and 340 nm for each exciting wavelength. The results were analysed following the methods of Weber (1961). Addition of thiocyanate to oxyhaemocyanin enhanced the fluorescence measured at 340 nm in a manner linearly proportional to the degree of deoxygenation, the maximum enhancement being $360 \pm 9\%$. There is, however, overlap in the ultraviolet region between the copper–oxygen absorption band and the tryptophan fluorescence band. Some quenching can thus occur in oxyhaemocyanin. When a quantitative correction for this 'trivial' reabsorption was made, the maximum enhancement of fluorescence by the action of KCNS became $300 \pm 3\%$. The enhancement obtained when apohaemocyanin was examined reached 513%.

Granted that enhancement is here due to exposure of the tryptophan moieties to a more hydrophilic environment (Konev, 1967), subtraction of the ultraviolet absorption of apohaemocyanin from that of oxyhaemocyanin shows a main difference peak at 348 nm with two positive peaks at 287 and 296 nm compatible with increased exposure of tryptophan to the solvent.

Binding of copper to the protein and binding of thiocyanate and of oxygen to the copper decrease tryptophan fluorescence in haemocyanin by about 30, 40 and 80% respectively. The site at which one oxygen molecule is bound in haemocyanin contains two copper atoms (Manwell, 1964). The copper oxygen absorption (320–380 nm, $\lambda_{max.}$ 348 nm) and the tryptophan fluorescence (318–375 nm) display almost complete overlapping and efficient transfer of excitation energy from tryptophan to the copper–oxygen site would appear to explain the quenching of tryptophan fluorescence in haemocyanin at pH 7 (Shaklai & Daniel, 1970). Tryptophan must be at or near the binding site. Radiation-less transfer is plausible but other processes have been discussed.

8.2.3 Laser Spectroscopy

Diffusion coefficients of biological macromolecules in solution lead to molecular weights by applying Svedberg's equation which uses sedimentation coefficients and partial specific volumes. The measurement of diffusion coefficients (Gostling, 1956; Strassburger & Reinert, 1969) is slow and difficult and not very accurate. Foord *et al.* (1970) devised a new method, which is accurate and uses small volumes of dilute solutions, and depends on digital autocorrelation of scattered photons from a laser beam. The procedure has been applied to determining diffusion constants at various concentrations of haemocyanin from the whelk *Murex trunculus*.

A solution of macromolecules exhibiting Brownian motion scatters light and the spectrum of the scattered light is related to the diffusion coefficients (cf. Dubin *et al.*, 1967). Statistical analysis of photon-counting techniques and digital techniques enabled Jakeman *et al.* (1968) to obtain the spectrum of the scattered light. Helium–cadmium (441·6 nm) and helium–neon (632·8 nm) lasers were used. The autocorrelation method was for calibration purposes applied to the measurement of diffusion coefficients of the relatively small proteins bovine serum albumin (BSA) and lysozyme with satisfactory results. Wood *et al.* (1971) measured diffusion coefficients ($D_{20\,w} \times 10^7$ cm^2 sec^{-1}) for 5 different gastropod haemocyanins with consistent results ($D = 1{\cdot}0\text{–}1{\cdot}05 \times 10^7$).

Sedimentation coefficients were also determined ($S^0_{20\,w} = 102{\cdot}7$), the corresponding molecular weights being close to $8{\cdot}9 \times 10^6$. The partial specific volume (calculated from aminoacid composition) was 0·726 for *M. trunculus* and 0·729 by pyknometry.

The Svedberg equation:

$$M = \frac{RTs}{D(1 - \bar{v}_{p})}$$

where $\bar{v}$ is the partial specific volume, leads to M $= 8{\cdot}8 \times 10^6$.

This application of laser spectroscopy in its more technical aspects emanates from the Royal Radar Establishment, Malvern and from the standpoint of gastropod haemocyanins, from Professor W. H. Bannister's group at the Royal University of Malta.

8.2.4 Vitamin B_6,

Pyridoxic acid and its lactone are urinary metabolites of members of the vitamin B_6 family. The lactone

COOH; HO; CH_2OH; H_3C; N

is obtained from pyridoxic acid by heating with acid and at pH 9 is it very strongly fluorescent. Since the acid and its lactone have the same excitation and emission spectra, the greater intensity shown by the lactone seems to be due to a higher quantum efficiency of fluorescence (Huff & Perlzweig, 1944). Chromatographic methods of obtaining pyridoxic acid from urine have been described (Reddy *et al.*, 1958). The excitation is best done at 350 nm and fluorescence is maximal *ca.* 450 nm. The normal human urinary excretion is 3–5 μmoles/day (0·56–0·92 mg).

Pyridoxine extracted from tissues and concentrated by chromatography may be oxidized by potassium permanganate to pyridoxic acid and converted to the lactone by heating with acid. Pyridoxamine may be converted to pyridoxine with nitrous acid and the analysis is continued as for pyridoxine (Fujita *et al.*, 1955, *a*, *b*, *c*). Pyridoxal in animal or plant tissue is concentrated by chromatography on Amberlite and selectively eluted with 1 N NaOH. The important step is an oxidation with ammoniacal silver nitrate to convert pyridoxal to pyridoxic acid which is then converted to the lactone. The analytical operations as described by Fujita *et al.* are somewhat complicated but essentially straightforward. Calibrations with authentic materials are clearly desirable. Each component is eventually determined via the intense fluorescence of the lactone. Fujita & Fujino (1955) showed how blood and urine could be analysed for pyridoxine, pyridoxamine, pyridoxal and pyridoxic acid (see also Udenfriend, 1962, p. 253).

Coursin & Brown (1958) worked on the native fluorescences so as to estimate three B_6 compounds in blood.

Table 8.16 Spectroscopic properties in the vitamin B_6 group

		Pyridoxine	Pyridoxamine	Pyridoxal
pH				
6·75	Absorption maxima nm	330	315	330
6·75	*Excitation peaks nm	340	335	330
	Fluorescence peaks nm	400	400	385
	Relative intensities	44	56	35

* Phosphate buffer; concentrations 10^{-6} M.
Addition of a very small amount of 30% hydrogen peroxide destroys pyridoxal but not pyridoxine or pyridoxamine. Exposure to ultraviolet light at the high dilution used for fluorescence measurements speedily destroys all three substances (e.g. quartz mercury vapour lamp, 3 cm from cell, 5 minutes).

Coursin & Brown obtained fluorescence at 385–400 nm from blood fractions and argued that since pyridoxic acid shows a fluorescence peak at 430 nm, none was present. Fujita & Fujino on the other hand found pyridoxic acid in blood, no pyridoxine and pyridoxamine greatly predominant over pyridoxal.

Bonavita (1960) found that both pyridoxal and its 5-phosphate reacted with cyanide in alkaline solution to form cyanhydrins:

Cyanhydrins	
Pyridoxal	$\lambda_{max.}$ exc. 358 nm, $\lambda_{max.}$ fl. *ca.* 430 nm (pH 9–10)
Pyridoxal-5-phosphate	$\lambda_{max.}$ exc. 315–318 nm, $\lambda_{max.}$ fl. 420 nm

With appropriate choice of pH and excitation wavelength each could be determined without interference from the other.

8.2.5 Fluorescence of Steroids

Oestrogens, because of their phenolic nature, exhibit 'native' ultraviolet fluorescence, but show remarkable differences in intensity (Duggan *et al.*, 1957). Thus excitation in the case of three substances, oestrone, oestriol and oestradiol, is maximal at 285 nm and the fluorescence peak is at 325–330 nm. The intensity of the native fluorescence of oestradiol is about 8 times that of oestrone and use has been made of this (Roberts & Siino, 1963). Equilin shows rather weak native fluorescence ($\lambda_{exc.}$ 290, $\lambda_{fl.}$ 345 nm) while equilenin fluoresces very strongly at $\lambda_{max.}$ 370 nm. On heating oestrogens with concentrated sulphuric acid a yellow-green fluorescence is seen. The fluorophors can be extracted (Ittrich, 1960) into a solvent consisting of acetylenetetrabromide containing nitrophenol (2%) and ethanol (1%). Excitation at 540 nm gives strong fluorescence measured at 560 nm and the sensitivity is of the order 0·01 μg of oestrone, oestradiol or oestriol/2 ml of solvent. This method has been strongly recommended (Hobkirk & Metcalfe-Gibson, 1963). In order to differentiate between the three compounds, preliminary chromatographic separations are necessary. Mahesh (1964) used chloroform as solvent ($\lambda_{exc.}$ 530 nm, $\lambda_{fl.}$ 670 nm).

The application of fluorescence to most problems of steroid analysis depends essentially upon production of derived fluorophors by chemical change. Many years ago empirical procedures had led to serviceable colour tests, e.g. for cholesterol; thus concentrated sulphuric acid produces coloured products with cholesterol (Salkowski test) and the Liebermann–Burchard reaction in which cholesterol reacts with a sulphuric acid–chloroform–acetic anhydride reagent had given reasonably reproducible results. Festenstein (1955), working in the writer's laboratory, examined the absorption spectra of numerous steroids treated with sulphuric acid. The results were very complex and the proportions of mixed products varied with concentrations, time and temperature. Unfortunately the work was not published, largely because the process could not be interpreted and reproducibility was less than satisfactory. Kalant (1958) had similar experience and Braunsberg & James (1960) agreed that the experimental conditions had to be established empirically and needed to be strictly defined and rigidly adhered to. This applied equally to methods based on selective absorption or fluorescence (see p. 379).

The Liebermann–Burchard reaction was adapted to histochemistry by Albers & Lowry (1955) and to the determination of cholesterol in very small amounts (0·02 ml) of serum (McDougal & Farmer, 1957; Carpenter *et al.*, 1957). Fluorescence here has the advantage of possessing high sensitivity.

Fluorometric methods for determining free and combined oestrogens in urine have been described (Jailer, 1948; Bauld *et al.*, 1960) but according to Udenfriend (1962) the best procedure is due to Nakao & Aizawa (1956). The urine (50 ml) is heated with hydrochloric acid to hydrolyse combined oestrogens and the solution is extracted with ether; the ether is removed and replaced by benzene and the solution is subjected to adsorption chromatography on alumina. Elution is carried out with portions of 1, 5 and 30% methanol in benzene, oestrone coming through first and oestriol last. The solvent-free fractions are taken up in ethanol (0·1 ml), 90% H_2SO_4 is added and the solution heated at 80° for 10 minutes. The solution is diluted with 65% sulphuric acid and the fluorescence is measured under standardized conditions (see p. 372).

Fluorescence methods for determining progesterone, pregnenolone and testosterone have been discussed by Udenfriend (1969) but a good deal depends on the efficacy of the preliminary isolation procedures.

It is often convenient to determine cortisol (17α-hydroxycorticosterone or hydrocortisone) and corticosterone jointly by a reaction applicable to both and to assay aldosterone sepa.ately. Steroids (except aldosterone) which have a hydroxyl group at position 11 produce fluorescent products after interacting with sulphuric acid. On a scale such that the fluorescence intensity of cortisol is 100, that of corticosterone is 252 and oestradiol 23 (Mattingly, 1962). Fluorescence is excited at 470–475 nm and measured at 520–530 nm. In plasma the corticosterone concentration is less than 10% of the cortisol concentration. The determination of aldosterone is difficult and recent original literature should be consulted (see also Udenfriend, 1969).

Berlman (1971) has published a second edition of a valuable handbook of fluorescence spectra of aromatic molecules. It comprises about 100 pages devoted to a general discussion of theory, models, chromophores and compounds with 255 references. The remainder of the work consists of graphs showing absorption spectra and fluorescence spectra for about 200 compounds. The collection includes a large number of benzene derivatives, polyphenyls, naphthalene, anthracene, phenanthrene, fluorene derivatives, perylene, chrysene and other polycyclic hydrocarbons. There are also a large number of indole derivatives and a few dyes. The book has a valuable appendix with a large selective bibliography covering many specialized aspects of fluorescence studies.

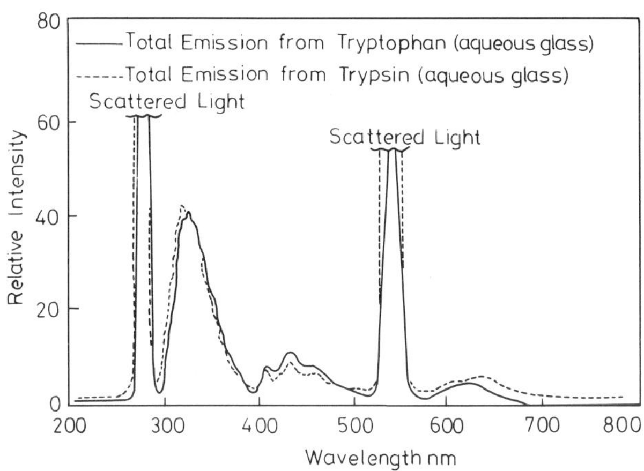

Figure 8.42 The total emission from tryptophan and trypsin in aqueous glass excited with light of 278 nm. The intensity of the two curves should not be compared, since the instrumental factors for the two spectra were not the same. Temperature 77 K (from Nag-Chaudhuri & Augenstein, 1964)

8.2.6 Phosphorescence and Further Aspects of Fluorescence in Relation to Proteins

An excellent review has appeared (Roth, 1969). Nag-Chaudhuri & Augenstein (1964) made a study of the effects of environment on excited states of amino-acids and proteins. They prepared EPA (diethyl ether–isopentane–ethanol 5:5:2) saturated with dry HCl gas (0·32 **M**) and used as a solvent this material diluted with EPA to give a 0·1 **M** HCl solution which set to a clear glass at 77 K. Another solvent used was 0·55% glucose in water. Rigorously dried powders were also examined using an Amino phosphorimeter.

For each of the test substances two kinds of spectra were obtained, one for the phosphorescence alone and the other for the total emission at 77 K. Thus the phosphorescence of tyrosine from an aqueous glass showed $\lambda_{max.}$ 395 nm and a fluorescence peak at 300 nm, while tryptophan under similar conditions showed a phosphorescence peak at 435 nm and a fluorescence peak at 330 nm. The fluorescence peak for tryptophan occurred at 350 nm at room temperature. The emission peaks for trypsin coincided with those for tryptophan. On the other hand, the fluorescence peaks for RNAase and tyrosine in powder form both fell at 305 nm and agreed fairly well with those observed in aqueous glucose. The mean decay times of phos-

Table 8.17

Compound	Wavelength of fluorescence maximum, nm			Wavelength of phosphorescence maximum, nm		
	Powder	EPA +HCl	0·5% glucose	Powder	EPA +HCl	0·5% glucose
Phenylalanine	295 (258)[a]	285 (258)	290 (258)	415 (258)	380 (258)	385 (258)
Tyrosine	305 (280)	305 (260)	300 (260)	450 (260)	395 (278)	395 (260)
Tryptophan	330 (278)	325 (278)	330 (278)	490 (278)	435 (278)	435 (278)
Trypsin	330 (280)	—	330 (280)	445 (280)	—	435 (280)
RNase	305 (260)	—	290 (260)	425 (260)	—	415 (260)

[a] The values given in parentheses indicate the exciting wavelength used.

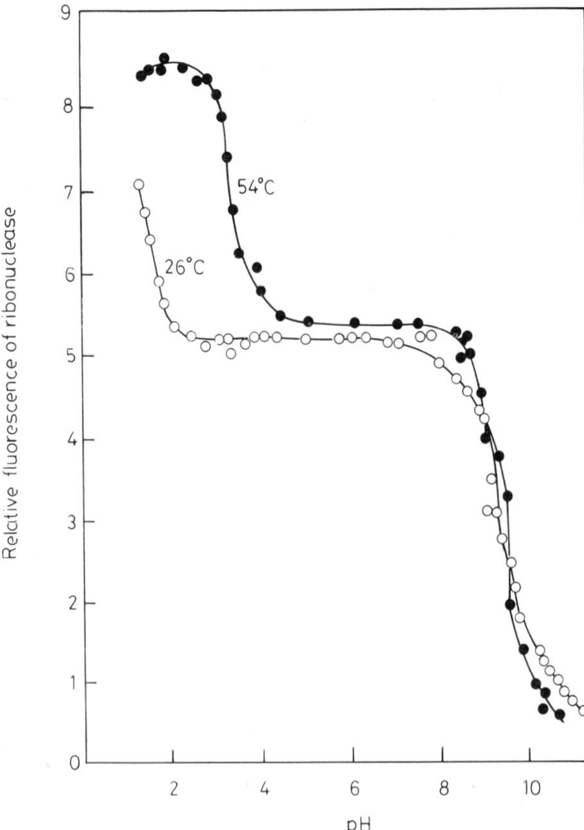

Figure 8.43 The relative intensity of fluorescence at 303 nm of an aqueous solution of ribonuclease (0·2 mg/ml) as a function of pH, measured at two different temperatures. The wavelength of the exciting light was 268 nm (from Gally & Edelman, 1964)

phorescence in EPA and aqueous glucose were: phenylalanine *ca.* 5·5 sec., tyrosine 2·5 sec., tryptophan 6 sec. and for trypsin in aqueous glucose *ca.* 6·1 sec. and RNAase *ca.* 1 sec.

Gally & Edelman (1962, 1964) found that when tyrosine in water (20 μg/ml) was excited by light of wavelength 268 nm the emission extended from 280 to 370 nm with the fluorescence peak at 303 nm. The intensity was progressively reduced by rising temperature, the value at 85° being about half that at 25°. The fluorescence of aqueous tryptophan (4 μg/ml) excited by light of wavelength 280 nm extended from 300 to 360 nm ($\lambda_{max.}$ 350 nm), the intensity at 85° being one-sixth of that at 25°. In the case of tyrosine the decrease was linear with temperature (when the intensity at 25° was put at 100, that at t° was given by 118·65–0·746 t). The decrease was due to a fall in the quantum yield of fluorescence.

Gally and Edelman also studied ribonuclease (which contains tyrosine but no tryptophan) and five other proteins (pepsin, ovalbumin, lysozyme, human γ-globulin and bovine plasma albumin). They observed in each case a nearly linear decrease in fluorescence intensity as the temperature was raised, but no two proteins behaved exactly the same (see Fig. 8.43). In the case of bovine serum albumin a sharp decrease in fluorescence intensity over the range 30–40°C indicated a change in conformation. RNAase also gave evidence of a temperature-dependent conformational change. Steiner *et al.* (1964) concluded that wide variations in quantum yields of tryptophan-containing proteins indicate that the fluorescence efficiency of the tryptophan moiety is very susceptible to differences in the intramolecular environment. Polarization spectra of different proteins add to the evidence. Conformational changes may be monitored by changes in fluorescence intensity although detailed interpretation may be difficult.

The emission of tryptophan was also investigated by Steele & Szent Gyorgyi (1957, 1958) and by Isenberg & Szent Gyorgyi (1968). There were two types of emission from tryptophan in frozen water systems: (*a*) excitation within the 280 nm absorption band gave rise to a phosphorescence (relatively long-lived) due to decay from a metastable triplet state and (*b*) excitation at 340 nm resulted in emission having a much shorter life and attributed to an association (or at any rate interaction) of tryptophan molecules. Further experiments were carried out using tryptophan solutions (5×10^{-3} **M** to 10^{-2} **M**) in a supersaturated solution of glucose (up to 21 g glucose/10 ml water). On cooling to −78°C a clear 'glass' was obtained. A broad region of emission ($\lambda_{max.}$ 470 nm) with rapid decay followed excitation (action spectrum peak 360 nm). The emission was viscosity-dependent and also varied with tryptophan concentration.

It was noted that an ox lens exposed to near ultraviolet radiation emitted a brilliant blue light at room temperature as well as at low temperature (liquid nitrogen) but there was no afterglow. If the lens was exposed to shorter wavelength ultraviolet radiation a long-lasting afterglow was seen and the kinetics of decay were consistent with emission from a metastable triplet state. Examination of ox lens, bovine serum albumin, trypsin and myosin showed in each case a fluorescence peak at 333–345 nm both at 298 K and 77 K. The phosphorescence showed its peak at 440–470 nm.

Table 8.18 Fluorescence of 1·0 μmole/ml solutions of aminoacids

Activation Fluorescence		305 445		275 305		320 nm 405 nm		Fluorescence maxima nm	
Aminoacid	pH	1·75	11·5	1·75	11·5	1·75	11·5	1·75	11·5
Cystine		—	—	—	—	1·30	0·86	360	340 410
Methionine		—	1·33	—	—	—	1·46	360	362 412
Phenylalanine		—	—	1·2	1·4	—	—	288	288
Tyrosine		1·4	5·7	211	4·9	2·2	33	305	408
Tryptophan*		10	170	9	6	2·9	56	352	362

* The fluorescence at optimal wavelengths in acid at 290–352 nm is 479, and in alkali at 297–362 nm is 1919.
Aminoacids not listed showed negligible or no fluorescence.
Absence of figures indicates no fluorescence.

Insulin (which has phenylalanine and tyrosine residues but no tryptophan) was studied at 77 K and the exciting radiation was varied from 260 to 450 nm. With excitation of absorbed light (*ca.* 280 nm) there was some emission, possibly due to fluorescence (290–340 nm) caused by the aromatic aminoacids, but under no circumstances was phosphorescence observed nor was any short-lived long wavelength emission excited by 340 nm radiation. This work underlines the significance of tryptophan residues in the fluorescence and phosphorescence of proteins.

Fluorescence studies can sometimes be applied to protein hydrolysates but care is needed lest artefacts should prove misleading.

Ledvina & LaBella (1970) examined the fluorescence properties of twenty-two authentic aminoacids dissolved in dilute hydrochloric acid (pH 1·75) at a concentration of 1 μmole/ml. Readings were made at three wavelengths (see Table 8.18) and corrected for blanks. Sodium hydroxide was then added to make the pH 11·5 and fluorescence readings were repeated at the same wavelengths. In addition to the emission by aromatic aminoacids, significant fluorescence was shown by solutions of methionine and cystine and also arginine. Each aminoacid was further subjected to conditions simulating those obtaining in the acid hydrolysis of proteins [300-fold excess of 6 N HCl fluorimetric grade; N_2 atmosphere; 110°; 54 hours; HCl then removed under suction at 42°]. The 'hydrolysed' samples were made up to 1 μmole/ml and the pH adjusted to 2·7. Fluorescence was again measured at pH 1·75 and 11·5.

It is important that even after prolonged treatment with 6 N HCl fluorescence remains quite low for most

Table 8.19 Fluorescence of 'acid hydrolysed' solutions (0·1 μmole/ml) of aminoacids

Activation Fluorescence		305 445		275 305		320 nm 405 nm		Fluorescence maxima nm	
Aminoacid	pH	1·75	11·5	1·75	11·5	1·75	11·5	1·75	11·5
Isoleucine		—	—	1·2	—	1·36	—	360 390	360 400
Serine		1·18	—	—	—	0·91	1·05	360 412	360 400
Cysteine		1·02	—	—	—	1·82	—	390 357	— —
Cysteic acid		1·01	—	—	—	1·56	—	357 390	—
Methionine		1·34	0·92	—	—	2·50	1·91	360 390	358 396
Histidine		9·7	1·26	0·80	—	2·68	1·38	445	357 400
Arginine		—	—	—	—	1·77	1·01	355	356 400
Phenylalanine		—	—	3	2·9	—	—	288	288 330
Tyrosine		7·2	17	332	—	5·4	100	305	410
Tryptophan		394	395	5·6	—	1235	1276	387	388

aminoacids despite some rise. Histidine produces an artefact with a fluorescence peak at 445 nm. With tyrosine the fluorescence at pH 1·75 is enhanced at 305 nm and in alkali the 410 nm fluorescence peak becomes significantly intensified. Treatment of tryptophan with 6 **N** HCl for 54 hours produces numerous new fluorescing entities, including some responding to activation at 320 nm.

A more detailed study of the complete fluorescence spectra of tyrosine and tryptophan shows that the spectrum of the latter (but not the former) is considerably changed as a result of heating with 6 **N** HCl. Untreated tryptophan (0·25 μmole/ml, pH 2·66) has strong activation peak at 228 nm and a weak one near 355 nm while the fluorescence spectrum displays its main peak at 355 nm and a weak one at 660–700 nm. After acid treatment a double peak (292 and 320 nm) activation curve gives rise to a strong fluorescence with a single peak near 383 nm. The 'hydrolysis' of histidine brings about appreciable fluorescence but in absolute terms it is always feeble.

Gel filtration and paper chromatography reveal (*a*) that tyrosine is largely unchanged but yields one artefact, presumably an oxidation product, and (*b*) that tryptophan solutions after acid treatment contain at least four fluorescent substances one of which is apparently kynurenine. There is however no evidence of unchanged tryptophan. The strongly fluorescent tryptophan artefacts might perhaps permit determination of tryptophan in hydrolysates without recourse to preliminary alkaline hydrolysis.

Certain complications are noted by Ledvina & LaBella (1970). In hydrolysates of peptides and proteins fluorescence quenching is always a possibility (cf. Cowgill, 1963, 1968) and an intermolecular transfer of the excited state from the phenyl group to the indole group has been noted (Weber, 1960). Moreover fluorescent aldehydes and biphenyl, derived from structural proteins like collagen and elastin, can be fluorescent and may indicate moieties that are integral parts of proteins (Keeley *et al.*, 1969; LaBella *et al.*, 1967; Anderson, 1966). This work has interesting possibilities for the future.

McCaman & Robins (1962) found that ninhydrin and peptides interacted to give fluorescent materials. Samejima *et al.* (1971 *a*, *b*) showed that phenylacetaldehyde was formed by the action of ninhydrin on phenylalanine and that the aldehyde could combine with additional ninhydrin or peptides or any primary amine to yield extremely fluorescent products. Weigele *et al.* (1972 *a*, *b*) investigated the structures and found that three substances were formed: (2) in 70% yield, (3) in 22% yield and (4) in 8% yield. The enamine made from 1,3-diphenyl-1,2-propane dione reacted with ethylamine to give the fluorophor.

$PhCH_2CHO$ + (1) $\xrightarrow{RNH_2}$ (2) 70%; (3) 22%; (4) 8%

(3) $\xrightarrow{HCl,\ acetone}$ HOOC–C$_6$H$_4$ pyrrolinone (N, O, OH, Ph)

[(3) $\xrightarrow[KOH]{MeOH,\ O_2}$ (2)]

R = Et

(2): $\lambda_{max.}$ nm 275 386; $\epsilon \times 10^{-3}$ 18·9 6·0; Fl. exc. 288, 395 nm; emission 485 nm

(3): $\lambda_{max.}$ nm 240 291 372; $\epsilon \times 10^{-3}$ 29·8 19·8 5·8; Fl. exc. 302, 377 nm; emission 480 nm

(4): $\lambda_{max.}$ nm 267 386; $\epsilon \times 10^{-3}$ 20·1 5·9; non-fluorescent

Experiments with model substances indicated a route to the fluorophor.

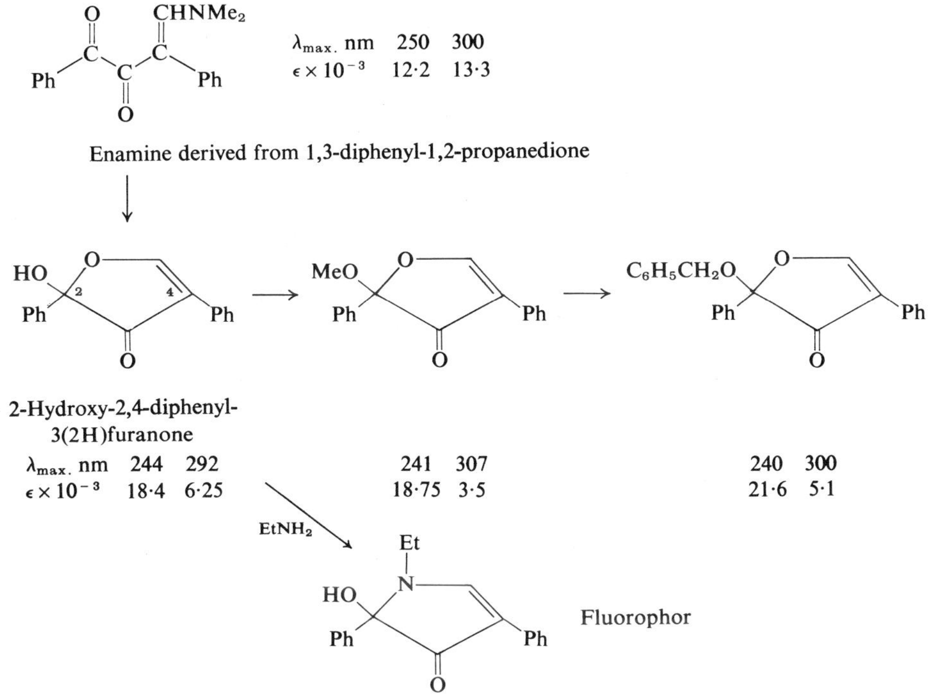

$\lambda_{max.}$ nm 250 300
$\epsilon \times 10^{-3}$ 12·2 13·3

Enamine derived from 1,3-diphenyl-1,2-propanedione

2-Hydroxy-2,4-diphenyl-3(2H)furanone

	2-Hydroxy-2,4-diphenyl-3(2H)furanone	MeO derivative	$C_6H_5CH_2O$ derivative
$\lambda_{max.}$ nm	244 292	241 307	240 300
$\epsilon \times 10^{-3}$	18·4 6·25	18·75 3·5	21·6 5·1

$EtNH_2$

Fluorophor

They went on to synthesize a new reagent, 4-phenylspiro[furan-2(3H), 1′-phthalan]-3,3′-dione to which they gave the trivial name *fluorescamine*. The compound reacts directly with primary amines to form the same fluorescent substances as were formed in the ninhydrin-phenylacetaldehyde reaction:

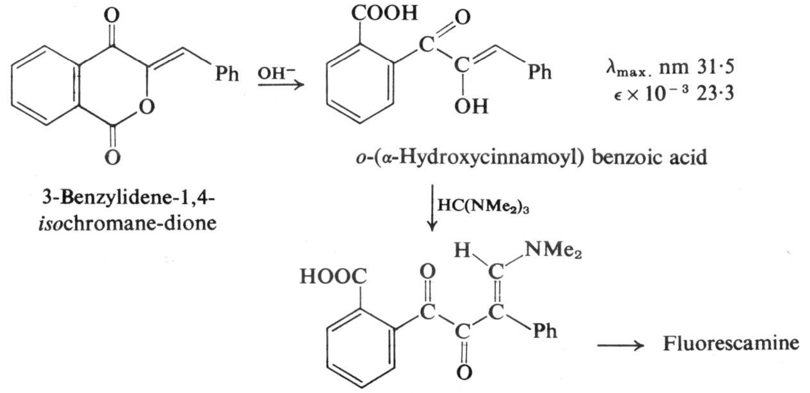

OH^-

$\lambda_{max.}$ nm 31·5
$\epsilon \times 10^{-3}$ 23·3

3-Benzylidene-1,4-*iso*chromane-dione

o-(α-Hydroxycinnamoyl) benzoic acid

$HC(NMe_2)_3$

→ Fluorescamine

The reaction between primary amines and fluorescamine at pH 9 occurs at room temperature and is virtually instantaneous; moreover excess reagent is destroyed in a few seconds. Neither the fluorescamine itself nor its hydrolysis products exhibits fluorescence.

For analytical purposes the fluorescamine, dissolved in a non-hydroxylic, water-soluble solvent such as acetone, is added to a primary amine in a solution buffered to pH > 7. The fluorophor is formed in a fraction of a second and excess of fluorescamine disappears within one minute. The fluorescent products are however stable for several hours.

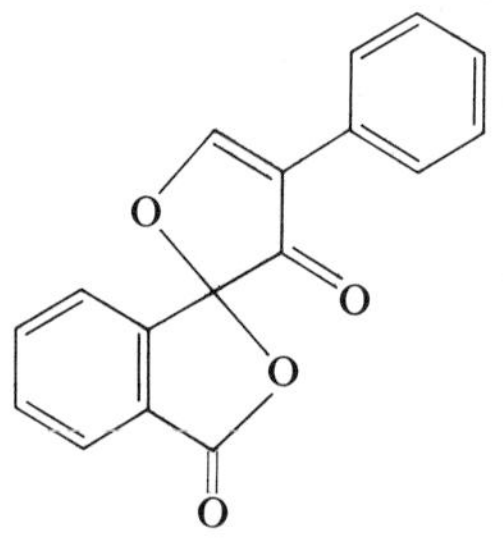

$+RNH_2$

Fluorescamine

$\lambda_{max.}$ nm	235	276	284	306
$\epsilon \times 10^{-3}$	25·9	3·95	4·1	3·8 (in ether)
cm^{-1}	1810	1745	1722	

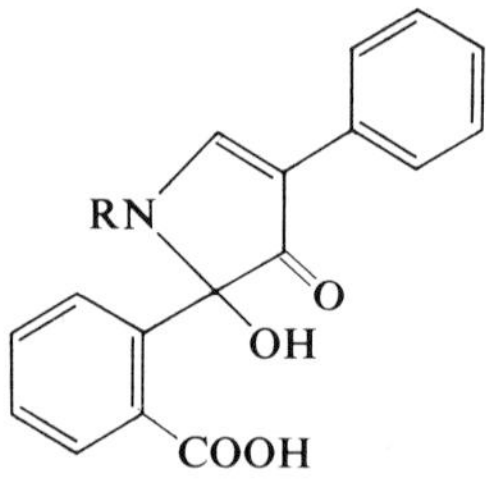

Fluorophor

$\lambda_{max.}$ nm 269 380
$\epsilon \times 10^{-3}$ 15·0 5·45, e.g. when
$R=CH_2(CONHCH_2)_4COOH$
(fluorescamine + pentaglycine + triethylamine in aqueous acetonitrile)
excitation maxima *ca.* 275 and 380 nm
emission 400–600 nm, peak at 475 nm

Fluorescamine does not react with proline or hydroxyproline but these two aminoacids can be converted into primary amines and reaction then occurs. The procedure is extremely sensitive and lends itself to automation; as little as 50 picomole of each aminoacid can be determined. The aminoacids of hydrolysed ribonuclease were determined by Udenfriend *et al.* (1972) on 1 μg of material. Ammonia produces relatively little fluorescence with fluorescamine.

Peptides tend to give with fluorescamine a higher fluorescence than the sum of the effects of the constituent aminoacids. They do this at pH 7 whereas the aminoacids require pH 9 to engender their full effect. By choosing appropriately buffered solutions the analyst can take advantage of this difference in behaviour. In fact the relative fluorescent intensities of different fluorescamine products are constant over the pH range 4 to 10; the variations depend on reactivity at different pH values and on differences in quantum yield.

Among the peptides that yield strongly fluorescing derivatives with the reagent are oxytocin, vasopressin, bradykinin, angiotensin, glucagon and insulin. Manual assays can be used to detect 0·5 μg of protein and automated procedures can be made much more sensitive.

The fluorophors are excited maximally at 390 nm and the emission peak is at 475 nm.

8.2.7 Scintillators

(See Kobayashi & Maudsley, 1969; Kerr *et al.*, 1957; Ott *et al.*, 1959.) Scintillators are fluorescent substances which emit radiation, usually in the ultraviolet and visible regions, on exposure to the ionizing particles emitted by radioactive isotopes. The scintillations are detected by a photomultiplier or better by a pair of photomultipliers operating in coincidence.

A number of organic crystals can be used for the spectrometry of β-radiation. Among the substances used are naphthalene, anthracene, *trans*-stilbene and terphenyl with their emission maxima at 348, 448, 384 and 391 nm, respectively. Anthracene has a high light output and *trans*-stilbene has a very short decay time while terphenyl has negligible self-absorption. These scintillators are suitable for α-counting and each advantage can be exploited.

Liquid scintillation counting is much used in biochemical work. A fluorescent material Y is dissolved in an aromatic solvent X. The energy W of a β-particle, dissipated in the scintillator is converted into a scintillation pulse of amplitude L at the anode of the photomultiplier. The β-particle produces excited molecules and ions from the solvent; most of them dissipate their energy thermally, but a small proportion retain the acquired energy, which is rearranged producing SX_1, its lowest excited electronic singlet state. The number A so formed varies as W, i.e. $A=sW$ where s is a solvent conversion factor (s is of the order one S_1X excited molecule per 100 eV of emitted β-particle energy). Putting rx as the excitation life-time, a fraction f of A transfers excitation energy to solute molecules forming S_1y, the lowest excited electron state of the solute Y. The number of molecules so formed is B where

$$B=fA=sfW$$

The quantity f is the solvent–solute energy transfer quantum efficiency; it varies with solvent and with the solute concentration but is independent of the nature of the solute (over a wide range of suitable substances). The solute fluorescence has a finite life-time t during which a fraction q of the B excited mole will limit its characteristic radiative:

$$qB = stqW = P$$

where P is the number of fluorescence photons making up a scintillation. The quantity q is known as the solute fluorescence quantum efficiency and depends on the nature of the solute. The photons display a distribution of energies corresponding to the fluorescence spectrum of the solute.

A fraction G of the scintillation photons strikes the photomultiplier cathode, G being the light collection factor so that $G = P' = stqGW$ and a fraction mK registers at the dynode of the photomultiplier so that

$$N = mKP' = stqmGKW$$

mK is the mean photoelectric quantum efficiency (photo-electron per photon) averaged over the solute fluorescence spectrum. It is made up of two factors: K, the maximum photoelectric quantum efficiency at the peak of the photocathode spectral response, and m, a spectral matching factor connecting the photocathode spectral response and the solute fluorescence spectrum.

Of the variables, s, f and q are properties of the scintillator. The solvent conversion factor s is arbitrarily taken as 100 for toluene. It varies between 82 and 107 for the following: mesitylene, anisole, benzene, ethylbenzene, phenylcyclohexane and xylene. The solvent–solute energy transfer efficiency f depends on the molar concentration of the solute, thus for toluene it ranges from 0·11 at 10^{-4} **M** to 0·55 at 10^{-3} **M**, 0·92 at 10^{-2} **M**, and 0·99 at 10^{-1} **M**. For a variety of scintillation substances q is between 0·7 and 0·9 and t is of the order 1–2 ns.

PPO Fluorescence
$\lambda_{max.}$ 365 and 381 in toluene

Non-fluorescent (NO_2)

POPOP Fluorescence 420 and 440 nm in toluene
Quantum efficiency nearly 1

This subject is very clearly discussed in the catalogue of the Koch–Light Laboratories Ltd. (Scintillator Division, Colnbrook, Bucks.) and both absorption and fluorescence spectra for a considerable range of scintillators are given.

References

8.1

Ambrose, E. J. & Elliott, A. (1951). *Proc. Roy. Soc.* A, **206**, 206.

Anderer, H. A. (1963). *Adv. Protein Chem.*, **18**, 1.

Angell, C. L., Sheppard, N., Yamaguchi, Y., Shimanouchi, T., Miyazawa, T. & Mizushima, S. (1957). *Trans. Faraday Soc.*, **53**, 589.

Asai, M., Tsuboi, M., Shimanouchi, T. & Mizushima, S. (1955). *J. Phys. Chem.*, **59**, 322.

Augenstein, L. G. & Ghiron, C. A. (1961). *Proc. Nat. Acad. Sci., U.S.A.* **47**, 1530.

Badger, R. M., Pullin, A. D. E. & Rubalcava, H. (1953). *Research Report No.* 8, Gates and Crellin Laboratory University of California.

Bailey, J. E., Beaven, G. H., Chignall, D. A. & Gratzer, W. B. (1968). *Eur. J. Biochem.*, **7**, 5.

Bamford, C. H., Elliott, A. & Hanby, W. E. (1956). In *Synthetic Polypeptides.* Academic Press, New York and London.

Bamford, C. H., Brown, L., Elliott, A., Hanby, W. E. & Trotter, I. F. (1953). *Nature, Lond.*, **171**, 1149; (1954). *Nature, Lond.*, **173**, 27.

Barrett, A. N. *et al.* (1972). *Cold Spring Harbour Symp. Quant. Biol.*, **36**, 433.

Bastian, R. (1949). *Anal. Chem.*, **21**, 972.

Beaven, G. H. (1961). *Adv. Spectroscopy*, **2**, 331.

Beaven, G. H. & Holiday, E. R. (1952). *Adv. Protein Chem.*, **7**, 319.

Beaven, G. H., Holiday, E. R. & Jope, E. M. (1950). *Faraday Soc. Discussions*, **9**, 406, 492.

Beaven, G. H. & Johnson, E. A. (1961). In *Molecular Spectroscopy, Methods and Applications in Chemistry.* Ed. G. H. Beaven, E. A. Johnson, H. A. Wills and R. G. J. Miller. Macmillan, London.

Beer, M. (1958). *Biochem. Biophys. Acta*, **29**, 423.

Beer, M., Sutherland, G. B. B. M., Tanner, K. N. & Wood, D. L. (1959). *Proc. Roy Soc.*, A, **249**, 147.

Bellamy, L. J. (1958). *The Infrared Spectra of Complex Molecules*. 2nd Edition, p. 238. Methuen & Co. London, Wiley & Sons, Ind., New York.

Bencze, W. L. & Schmid, K. (1957). *Anal. Chem.*, **29**, 1193.

Bendit, E. G. (1966). *Biopolymers*, **4**, 539; (1967). *Ibid.*, **5**, 525; (1967). In *Symposium on Fibrous Proteins, Australia*, p. 386. Ed. W. G. Crewther. Butterworths, London.

Benesch, R. E. & Benesch, R. (1955). *J, Amer. Chem. Soc.* **77**, 587.

Benisek, W. F. & Richards, F. M. (1968). *J. Biol. Chem.*, **243**, 4267.

Bennett, P. A., Sullivan, J. V. & Walsh, A. (1970). *Anal. Biochem.*, **36**, 123.

Beychok, S. (1966). *Science, N.Y.*, **154**, 1288.

Beychok, S., Armstrong, J. M., Lindblow, C. & Edsall, J. T. (1966). *J. Biol. Chem.*, **241**, 5150.

Bigelow, C. C. (1960). *Compt. rend. trav. lab. Carlsberg sér. chim.*, **31**, 305; (1961). *J. Biol. Chem.*, **236**, 1706.

Bigelow, C. C. & Geschwind, I. I. (1960). *Compt. rend. trav. lab. Carlsberg sér. chim.*, **31**, 183.

Blout, E. R., de Loze, C. & Asadourian, A. (1961). *J. Amer. Chem. Soc.*, **83**. 1805.

Bowman, J. A., Sullivan, J. V. & Walsh, A. (1966). *Spectrochim. Acta.*, **22**, 205.

Bradbury, E. M., Crane-Robinson, C., Goldman, H., Rattle, H. W. E. & Stephens, R. M. (1967). *J. Mol. Biol.*, **29**, 507.

Bradbury, E. M., Crane-Robinson, C. & Rattle, H. W. E. (1967). *Nature, Lond.*, **216**, 862.

Bradbury, E. M., Crane-Robinson, C., Phillips, D. M. P., Jones, E. W. & Murray, K. (1965). *Nature, Lond.*, **205**, 1315.

Brandts, J. F. (1964). *J. Amer. Chem. Soc.*, **86**, 4291, 4302; (1965). *J. Amer. Chem. Soc.*, **87**, 2759.

Brandts, J. F. & Hunt, L. (1967). *J. Amer. Chem. Soc.*, **89**, 4826.

Bucher, T. & Kaspers, J. (1947). *Biochem. Biophys. Acta*, **1**, 21.

Caspar, D. L. D. (1963). *Adv. Protein Chem.*, **18**, 37.

Carlsson, A., Lindquist, M., Magnusson, T. & Waldeck, B. (1958). *Science, N.Y.*, **127**, 471.

Cassim, J. Y. & Yang, J. T. (1969). *Biochemistry*, **8**, 1947.

Chervenka, C. H. (1959). *Biochim. et Biophys. Acta.*, **31**, 85.

Cilento, C. & Giusti, P. (1959). *J. Amer. Chem. Soc.*, **81**, 3801.

Coulter, C. B., Stone, F. M. & Kabat, E. A. (1936). *J. Gen. Physiol.*, **19**, 739.

Cranmer, J. L. & Neuberger, A. (1943). *Biochem. J.*, **37**, 302.

Crout, R. J. (1957). In *Standard Methods in Clinical Chemistry*, Vol. 3, p. 62. Ed. D. Seligson. Academic Press, London and New York; (1959). *Pharmacol. Revs.*, **11**, 296.

Crout, R. J. & Sjoerdsma, A. (1960). *Circulation*, **22**, 516.

Davidson, B. & Fasman, G. D. (1967). *Biochemistry*, **6**, 1616.

Donovan, J. W., Laskowski, M. Jr. & Scherago, H. A. (1961). *J. Amer. Chem. Soc.*, **83**, 2686; (1958). *Biochim. Biophys. Acta.*, **29**, 455.

Doty, P. & Guiduschek, E. P. (1953). In *The Proteins*, Vol. 1A, p. 393. Ed. H. Neurath and K. Bailey. Academic Press, London and New York.

Drucker, B., Hainsworth, R. & Smith, S. G. (1953). *J. Textile Inst.*, **44**, T. 420.

Duggan, D. E. & Udenfriend, S. (1956). *J. Biol. Chem.*, **223**, 313.

Edelhoch, H. (1967). *Biochemistry*, **6**, 1948.

Edelhoch, H., Lippoldt, R. E. & Wilchek, M. (1968). *J. Biol. Chem.*, **243**, 4790.

Edsall, J. T. (1968). *The Harvey Lectures Series* 62 (1966–7), p. 191.

Edsall, J. T., Martin, R. B. & Hollingsworth, B. R. (1958). *Proc. Nat. Acad. Sci. U.S.A.*, **44**, 505.

Elliott, A. & Hanby, W. E. (1958). *Nature, Lond.*, **182**, 654.

Elliott, A., Hanby, W. E. & Malcolm, B. R. (1957). *Nature, Lond.*, **180**, 1341.

Englander, S. W. (1963). *Biochemistry*, **2**, 798.

Ferretti, J. A. (1967). *Chem. Commun.*, 1030.

Fisher, H. F. & Cross, D. J. (1966). *Science, N.Y.*, **153**, 414.

Foster, J. F. (1960). In The *Plasma Proteins*. Ed. F. W. Putman.

Fraenkel-Conrat, H. & Colloms, M. (1967). *Biochemistry*, **6**, 2740.

Fraenkel-Conrat, H. & Narita, K. (1958). In *Symposium on Protein Structure*, Ed. Neuberger, A. Methuen, London & Wiley, N.Y.

Fraser, R. D. B. (1952). *Nature, Lond.*, **170**, 491.

Fraser, R. D. B. (1960). In *A Laboratory Manual of Analytical Methods of Protein Chemistry*, Vol. 2, Chapter 9. Ed. P. Alexander and R. J. Block. Pergamon Press, Oxford.

Fretto, L. & Strickland, E. H. (1971). *Biochem. Biophys. Acta.*, **235**, 473.

Gaddum, J. H. & Schild, H. (1934). *J. Physiol. London*, **80**, 9P.

Gemmill, C. L. (1955). *Arch. Biochim. Biophys.*, **54**, 359; (1956 *a*). *Ibid.*, 63, 177; (1956 *b*). *Ibid.*, **63**, 192.

Ginsel, L. A. (1939). *Biochem. J.*, **33**, 428.

Giese, A. T. & French, C. S. (1955). *Appl. Spectroscopy*, **9**, 78.

Glazer, A. N. & Rosenheck, K. (1962). *J. Biol. Chem.*, **237**, 3674.

Go, Y., Noguchi, J., Asai, M. & Hayakana, T. (1956). *J. Polymer Sci.*, **21**, 147.

Goodwin, T. W. & Morton, R. A. (1946). *Biochem. J.*, **40**, 628.

Gorbonoff, M. J. (1967). *Biochemistry*, **6**, 1606.

Gorin, G. (1956). *J. Amer. Chem. Soc.*, **78**, 767.

Gorin, G. & Clary, C. W. (1960). *Arch. Biochem. Biophys.* **90**, 4045.

Gratzer, W. B. (1967). *Proc. Roy. Soc.* A, **297**, 163.

Greenfield, N., Davidson, B. & Fasman, G. D. (1967). *Biochemistry*, **6**, 1630.
Harrington, W. F. & Sela, M. (1959). *Biochim. Biophys. Acta*, **31**, 427.
Harrison, S. C. & Blout, E. R. (1964). *J. Biol. Chem.*, **240**, 299.
Herskovitz, T. T. & Laskowski, M. Jr. (1960). *J. Biol. Chem.*, **235**, PC. 57; (1962). *Ibid.*, **237**, 2481.
Hess, S. & Udenfriend, S. (1959). *J. Pharmacol. Exptl. Therap.*, **127**, 175.
Hiskey, C. F. (1949). *Anal. Chem.*, **21**, 1440.
Holiday, E. R. (1936). *Biochem. J.*, **30**, 1795.
Holiday, E. R. & Ogston, A. G. (1938). *Biochem. J.*, **32**, 1166.
Horwitz, J., Strickland, E. H. & Billups, C. (1970). *J. Amer. Chem. Soc.*, **92**, 2119.
Hudson, C. B., Robertson, A. V. & Simpson, W. R. J. (1968). *Austr. J. Chem.*, **21**, 769.
Hugli, T. E. & Stein, W. H. (1971). *J. Biol. Chem.*, **246**, 7191.
Hvidt, A. & Kagi, J. H. R. (1963). *Compt. rend. trav. lab. Carlsberg sér. chim.*, **33**, 497.
Kagi, J. H. R. & Vallée, B. L. (1961). *J. Biol. Chem.*, **236**, 2435.
Karle, I. L., Daly, J. W. & Witkop, B. (1969). *Science, N.Y.*, **164**, 1401.
Katchalski, E. & Sela, M. (1953). *J. Amer. Chem. Soc.*, **75**, 5284.
King, L. & Perham, R. N. (1971). *Biochemistry*, **10**, 981.
Korver, O. & Bosma, J. (1971). *Anal. Chem.*, **43**, 1119.
Kosower, E. M. (1956). *J. Amer. Chem. Soc.*, **78**, 3497.
Knox, W. E. & Mehler, A. H. (1950). *J. Biol. Chem.*, 419.
Krause, I. M. & Lindestrøm-Lang, K. (1955). *Compt. rend. trav. lab. Carlsberg sér. chim.*, **29**, 367.
Krimm, S. (1968). In *Symposium on Fibrous Proteins, Australia* 1967, p. 85. Ed. W. G. Crewther. Butterworths, London.
Krimm, S., Kuroiwa, K. & Rehane, T. (1967). In *Conformation of Biopolymers*, p. 439. Ed. G. N. Ramachandran. Academic Press, New York.
Kurihara, K., Horinishi, H. & Shibata, K. (1963). *Biochim. Biophys. Acta.*, **39**, 379.
Laskowski, M. Jr. (1966). *Fed. Proc.*, **25**, 20.
Laskowski, M. Jr., Widom, J. W., McFadden, K. L. & Scheraga, H. A. (1956). *Biochim. Biophys. Acta.*, **19**, 581.
Latt, S. A., Cheung, H. T. & Blout, E. R. (1965). *J. Amer. Chem. Soc.*, **87**, 975.
Laurent, G., Castay, M., Marriqu, C., Garcon, D., Charrel, M. & Derrien, Y. (1964). *Bull. Soc. Chem. Biol.*, **46**, 603.
Lenormant, H. (1956). *Trans. Faraday Soc.*, **52**, 549.
Leonard, W. J. & Foster, J. F. (1961). *J. Biol. Chem.*, **236**, 2662.
Leuzinger, W. & Baker, A. L. (1967). *Proc. Nat. Acad. Sci. U.S.A.*, **57**, 446.
Leuzinger, W., Baker, A. L. & Cauvin, E. (1968). *Proc. Nat. Acad. Sci. U.S.A.*, **59**, 620.
Lindskog, S. (1960). *Biochem. Biophys. Acta.*, **39**, 218.
Lucas, F., Shaw, J. T. B. & Smith, S. G. (1955). *J. Textile Inst.*, **46**, T. 440; (1957). *Biochem. J.*, **66**, 468; (1958). *Advances in Protein Chemistry*, **13**, 107.
Luse, R. A. & McLaren, A. D. (1963). *Photochem. Photobiol.*, **2**, 343.
McLaren, A. D. & Shugar, D. (1964). *Photochemistry of Proteins and Nucleic Acids*. Pergamon Press, Oxford, London.
McLaren A. D. & Takahashi, W. N. (1959). *Biochim. Biophys. Acta*, **32**, 355.
Martin, A. E. (1957). *Nature, Lond.*, **180**, 231.
Martin, C. J. & Bhatnagar, G. M. (1967). *Biochemistry*, **6**, 1638.
Martin, R. B., Edsall, J. T., Wetlaufer, D. B. & Hollingsworth, B. R. (1958). *J. Biol. Chem.*, **233**, 1429.
Mauger, A. B. & Witkop, B. (1966). *Chem. Rev.*, **66**, 47.
Mayer, M. M. & Miller, J. A. (1970). *Analyt. Biochem.*, **36**, 1.
Miyazawa, T., Shimanouchi, T. & Mizushima, S. (1958). *J. Chem. Phys.*, **29**, 611.
Moffit, W. & Yang, J. T. (1956). *Proc. Nat. Acad. Sci. U.S.A.*, **42**, 596.
Mulliken, R. S. (1952). *J. Amer. Chem. Soc.*, **74**, 812.
Murray, K. (1964). In *The Nucleohistones*, p. 21. Ed. J. Bonner and P. O. Holden Day, San Francisco; (1965). *Ann. Rev. Biochem.*, **34**, 209.
Myers, B. & Glazer, A. N. (1971). *J. Biol. Chem.*, **246**, 412.
Nachmansohn, D. In *New Perspectives in Biology* (1964). Ed. M. Sela. B.B.A. Library, Vol. 4, p. 176. Elsevier Publishers, Amsterdam.
Nachmansohn, D. (1966). *Ann. N.Y. Acad. Sci.*, **137**, 877.
Nagakura, S. & Baba, H. (1952). *J. Amer. Chem. Soc.*, **74**, 5693.
Nakajima, T. & Volcani, B. E. (1969). *Science, N.Y.*, **164**, 1400.
Nelson, C. A. & Hummel, J. P. (1962). *J. Biol. Chem.*, **237**, 1567.
Nyman, P. O. (1961). *Biochim. Biophys. Acta.*, **52**, 1.
Oates, J. A., Jr., Gillespie, L., Udenfriend, S. & Sjoerdsma, A. (1960). *Science, N.Y.*, **131**, 1890.
Olson, E. C. & Alway, C. D. (1960). *Anal. Chem.*, **32**, 370.
Oster, G. (1955). In *Physical Techniques in Biological Research*. Ed. G. Oster & A. W. Pollister, Vol. 1. Academic Press, N.Y.
Otey, M. C. & Greenstein, J. P. (1954). *Arch. Biochem. Biophys.*, **53**, 501.
Passerini, R. & Ross, I. G. (1953). *J. Sci. Instrum.*, **30**, 274.
Patchornik, A., Lawson, W. B. & Witcop, B. (1958). *J. Amer. Chem. Soc.*, **80**, 4747.
Perham, R. N. (1973). *Biochem. J.* (*Mol. aspects*), **131**, 119.
Perham, R. N. & Richards, F. M. (1968). *J. Mol. Biol.*, **33**, 795.
Perham, R. N. & Thomas, J. O. (1971). *J. Mol. Biol.*, **62**, 415.

Peters, T. (1959). *Compt. rend. trav. lab. Carlsberg sér. chim.*, **31**, 227.

Phillips, D. M. P. (1964). In *The Nucleohistones*, p. 46. Ed. Bonner J. Holden-Day. San Francisco.

Pirie, A. (1971). *Biochem. J.*, **125**, 203.

Price, P. A., Stein, W. H. & Moore, S. (1969). *J. Biol. Chem.*, **244**, 924, 929.

Ramachandran, L. K. & Witcop, B. (1959). *J. Amer. Chem. Soc.*, **81**, 4028.

Racker, E. & Krimsky, I. (1952). *J. Biol. Chem.*, **198**, 731.

Randle, R. R. & Whiffen, D. W. (1955). In *Molecular Spectroscopy*, p. 111. Ed. G. Hill. Institute of Petroleum, London.

Rickli, E. E. & Edsall, J. T. (1962). *J. Biol. Chem.*, **237**, PC 258–260.

Riddiford, L. M. (1964). *J. Biol. Chem.*, **239**, 1079; (1965). *Ibid.*, 240, 168.

Riley, M. & Perham, R. N. (1973). *Biochem. J.*, **131**, 625.

Riordan, J. F., Sokolovsky, M. & Vallee, B. L. (1966). *J. Amer. Chem. Soc.*, **88**, 4104; (1967). *Biochemistry*, **6**, 3609.

Riordan, J. F., Wacker, W. E. C. & Vallee, B. L. (1965). *Biochemistry*, **4**, 1758.

Rosenheck, K. & Doty, P. (1961). *Proc. Nat. Acad. Sci. U.S.A.*, **47**, 177.

Rosenberg, A. (1966). *J. Biol. Chem.*, **241**, 5119, 5126.

Sarfare, P. S. & Bigelow, C. C. (1967). *Canad. J. Biochem.*, **45**, 651.

Satoh, K. & Price, J. M. (1958). *J. Biol. Chem.*, **230**, 781.

Schauenstein, E. (1949). *Monatsh.*, **58**, 820, 843; (1952). *Österr Chemiker-Ztg.*, **53**, 187.

Schauenstein, E. & Kratky, O. (1950). *Z. Naturforsch.*, **56**, 281.

Schauenstein, E., Trieber, E., Berndt, W., Gelbinger, W. & Zima, H. (1954). *Monatsh.*, **85**, 120.

Scheele, R. B. & Lauffer, M. A. (1969). *Biochemistry*, **8**, 3597.

Schellman, C. J. & Schellman, J. A. (1958). *Comp. tend. trav. lab. Carlsberg sér. chim.*, **30**, 463.

Scheraga, H. A. (1957). *Biochim. Biophys. Acta.*, **23**, 196; (1961). In *Protein Structure*, Academic Press, New York

Scheraga, H. A. & Rupley, J. A. (1962). *Adv. Enzym.*, **24**, 161.

Seeds, W. E. & Wilkins, M. F. (1950). *Disc. Faraday Soc.*, **13**, 251.

Sela, M. & Katchalski, E. (1956). *J. Amer. Chem. Soc.*, **78**, 3986.

Setlow, R. B. (1955). *Biochem. Biophys. Acta.*, **16**, 44; (1955). *Ann. N.Y. Acad. Sci.*, **59**, 469.

Setlow, R. B. & Doyle, B. (1957). *Biochim. Biophys. Acta.*, **24**, 27.

Shifrin, S. (1964). *Biochem. Biophys. Acta.*, **81**, 205.

Shore, V. G. & Pardee, A. B. (1956). *Arch. Biochem. Biophys.*, **60**, 100.

Simmons, N. S. & Blout, E. R. (1960). *Biophys. J.*, **1**, 55.

Simmons, N. S., Cohen, C., Szent-Gyorgi, A., Wetlaufer, D. B. & Blout, E. R. (1961). *J. Amer. Chem. Soc.*, **83**, 4705.

Sjoerdsma, A., Oates, J. A., Zaltman, P. & Udenfriend, S. (1960). *New Engl. J. Med.*, **263**, 585.

Smith, C. E. & Lauffer, M. A. (1961). *Biophys. Soc. meeting Abstr. SB2.* See also Lauffer in *Molecular Bases of Neoplasia* (1962). Univ. Texas Press, 180.

Sokolovsky, M., Riordan, J. F. & Vallee, B. L. (1966). *Biochemistry*, **5**, 3582; (1967). *Biochem. Biophys. Res. Commun.*, **27**, 20.

Stedman, E. & Stedman, E. (1950). *Nature, Lond.*, **166**, 780.

Stenström, W. & Goldsmith, N. (1926). *J. Phys. Chem.*, **20**, 1683.

Stokes, G. (1852). *Phil. Trans.*, **143**, 463.

Strickland, E. H., Horwitz, J. & Billups, C. (1969). *Biochemistry*, **8**, 3205.

Strickland, R. D., Mack, P. A., Podleski, T. R. & Childs, W. A. (1960). *Anal. Chem.*, **32**, 199.

Strickland, E. H., Wilchek, M., Horwitz, J. & Billups, C. (1970). *J. Biol. Chem.*, **245**, 4168.

Strickland, E. H., Wilchek, M., Horwitz, J. & Billups, C. (1972). *J. Biol. Chem.*, **247**, 579.

Stryer, L. & Haughland, R. P. (1967). *Proc. Nat. Acad. Sci., U.S.A.*, **58**, 719.

Sullivan, J. V. & Walsh, A. (1965). *Spectrochim. Acta.*, **21**, 719; (1966). *Ibid.*, **22**, 1843; (1968). *Appl. Optics.*, **7**, 1271.

Sutherland, G. B. B. M. (1952). *Advances in Protein Chem.*, **7**, 291.

Takahasi, F., Karoly, W. J., Greenshields, J. B. & Li, N. V. (1967). *Can. J. Chem.*, **45**, 2033.

Tanford, C. (1968). *Advances in Protein Chem.*, **23**, 155.

Tanford, C., Kauchara, K., Lapanje, S., Hooker, T. M. Jr., Zarlengo, M. H., Salahuddin, A., Anne, A. K. C. & Takagi, T. (1967). *J. Amer. Chem. Soc.*, **89**, 5023.

Tanford, C. & Roberts, J. M. (1962). *J. Amer. Chem. Soc.*, **74**, 1509.

Teale, F. W. J. (1960). *Biochem. J.*, **76**, 381.

Teale, F. W. J. & Weber, G. (1957). *Biochem. J.*, **65**, 476.

Tinoco, I. & Cantor, C. R. (1970). *Methods of Biochemical Analysis.* Vol. 18, pp. 81–203. Ed. D. Glick, Interscience, N.Y. & London.

Tupper, R., Watts, R. W. E. & Wormall, A. (1952). *Biochem. J.*, **50**, 429–432.

Udenfriend, S. (1962). *Fluorescence Assay in Biology and Medicine.* Academic Press, New York, London.

Udenfriend, S., Weissbach, H. & Brodie, B. B. (1958). In *Methods of Biochemical Analysis.* Ed. D. Glick, Vol. 6, p. 95.

Velick, S. F. (1958). *J. Biol. Chem.*, **233**, 1455.

Vladimorov, Y. A. (1959). *Bull. Acad. Sci. URSS. Ser. Phys.*, **23**, 86.

Waalkes, T. P. & Udenfriend, S. (1957). *J. Lab. Clin. Med.*, **50**, 733.

Weber, G. & Teale, F. W. J. (1965). In *The Proteins.* Ed. H. Neurath & K. Bailey. Vol. III, p. 445. Academic Press, London & New York.

West, J. J., Nagy, B. & Gergely, J. (1967). *Symposium on Fibrous Proteins, Australia*, p. 164. Ed. Crewther. W. G. Butterworths, London, 1968.

Wetlaufer, D. B. (1962). *Advances in Protein Chem.*, **17**, 303.

Wetlaufer, D. B., Edsall, J. T. & Hollingsworth, B. R. (1958). *J. Biol. Chem.*, **233**, 1421.

White, A. (1959). *Biochem. J.*, **71**, 217.

Williams, E. J. & Laskowski, M. (1965). *J. Biol. Chem.*, **240**, 3580.

Wilchek, M., Spande, T., Milne, G. & Witcop, B. (1968 *a*). *Biochemistry*, **7**, 1777.

Wilchek, M., Spande, T. & Witcop, B. (1968 *b*). *Biochemistry*, **7**, 1787.

Witcop, B. (1961). *Advances in Protein Chem.*, **16**, 221.

Yang, J. T. & Doty, P. (1957). *J. Amer. Chem. Soc.*, **84**, 4235.

Yanari, S. & Bovey, F. A. (1960). *J. Biol. Chem.*, **235**, 2818.

Zahn, H. & Schnabel, E. (1957). *Liebig's Annalen*, **604**, 62.

8.2

Albers, R. W. & Lowry, O. H. (1955). *Anal. Chem.*, **27**, 1829.

Anderson, S. O. (1966). *Acta Physiol. Scand.*, **66**, Suppl. 263.

Bannister, W. H. & Wood, E. J. (1971). *Comp. Biochem. Physiol.*, **40B**, 7.

Bauld, W. S., Givner, M. L., Engel, L. L. & Goldzieher, J. W. (1960). *Can. J. Biochem. & Physiol.*, **38**, 213.

Berlman, I. B. (1971). *Fluorescence spectra of aromatic molecules*. Academic Press, New York, London.

Bonavita, V. (1960). *Arch. Biochem. Biophys.*, **88**, 366.

Bowen, E. J., Mikiewicz, E. & Smith, F. W. (1949). *Proc. Phys. Soc. London*, A, **62**, 26.

Bowen, E. J. & Wokes, F. (1953). In *Fluorescence of Solutions*. Longmans Green, New York.

Braunsberg, H. & James, V. H. T. (1960). *Anal. Chem.*, **27**, 1829.

Bridges, J. W. (1964). In *Methods of Polyphenol Chemistry*. Pergamon Press, Oxford, p. 59.

Bridges, J. W., Creaven, P. J., Davies, D. S. & Williams, R. T. (1963). *Biochem. J.*, 65P.

Bridges, J. W., Davies, D. S. & Williams, R. T. (1966). *Biochem. J.*, **98**, 451.

Bridges, J. W. & Williams, R. T. (1962). *Nature*, **196**, 59.

Brodie, R. B., Udenfriend, S., Dill, W. & Champkin, T. (1947). *J. Biol. Chem.*, **168**, 319.

Carlsson, A., Lindquist, M., Magnusson, T. & Waldeck, B. (1958). *Science*, **127**, 471.

Carlsson, A. & Waldeck, B. (1958). *Acta Physiol. Scand.* **44**, 293.

Carpenter, K. J., Gotsis, A. & Hegsted, D. M. (1957). *Clin. Chem.*, **3**, 233.

Chapman, D., Lloyd, D. R. & Prince, R. A. (1963). *J. Chem. Soc.*, 3645.

Chen, R. F. (1968). *Analytical Letters*, **1** (7), 423.

Chen, R. F. & Kernohan, J. C. (1967). *J. biol. Chem.*, **242**, 5813.

Chen, R. F., Vurek, G. G. & Alexander, N. (1967). *Science N.Y.*, **156**, 949.

Coursin, D. B. & Brown, V. C. (1958). *Proc. Soc. Exp. Biol. Med.*, **98**, 315.

Cowgill, R. W. (1963). *Biochem. Biophys. Acta.*, **100**, 36.

Cowgill, R. W. (1968). *Biochim. Biophys. Acta.*, **168**, 417, 431.

Creaven, P. J., Parke, D. V. & Williams, R. T. (1965). *Biochem. J.*, **96**, 879; also 390.

Crout, J. R. (1959). *Pharmacol. Revs.*, **11**, 296.

Crout, J. R. (1959). In *Standard Methods of Clinical Chemistry*. Vol. 3. Ed. D. Seligson. Academic Press, New York.

Crout, J. R. & Sjoerdsma, A. (1960). *Circulation*, **22**, 516.

Drujan, B. D., Sourkes, T. L., Layne, D. S. & Murphy, G. F. (1959). *Can. J. Biochem. Physiol.*, **37**, 1153.

Dubin, S. B., Lunacek, J. H. & Benedek, G. B. (1967). *Proc. Nat. Acad. Sci. U.S.A*, **57**, 1164.

Duggan, D. E., Boroman, R. L., Brodie, B. B. & Udenfriend, S. (1957). *Arch. Biochem. Biophys.*, **68**, 1.

Duggan, D. E. & Udenfriend, S. (1956). *J. biol. Chem.*, **223**, 313.

Ehrenberg, A. & Theorell, H. (1962). In *Comprehensive Chemistry*, Vol. 3. Chapter VI. Ed. M. Florkin & E. H. Stotz, Elsevier, Amsterdam, New York.

Edelhoch, H. (1967). *Biochemistry*, **6**, 1948.

Feigl, F., Feigl, H. E. & Goldstein, D. (1955). *J. Amer. Chem. Soc.*, **77**, 4162.

Festenstein, G. N. (1955). Liverpool, Ph.D. Thesis.

Foord, R., Jakeman, E., Oliver, C. J., Pike, E. R., Blagrove, R. J., Wood, E. & Peacocke, A. R. (1970). *Nature, Lond.* **227**, 242.

Förster, T. (1950). *Z. Elektrochem.*, **54**, 531.

Förster, T. (1951). In *Fluorescenz Organischer Verbindungen*. Vandenhock and Rupprecht, Gottingen.

Fujimori, E. (1955). *Rept. Inst. Ind. Sci. Univ. Tokyo*, **4**, No. 3.

Fujita, A. & Fujino, K. (1955). *J. Vitaminol (Osaka)*, **1**, 290.

Fujita, A., Fujita, D. & Fujino, K. (1955 *b*, *c*). *J. Vitaminol. (Osaka)*, **1**, 275, 279.

Fujita, A., Matsuura, K. & Fujino, K. (1955 *a*). *J. Vitaminol (Osaka)*, **1**, 267.

Gaddum, J. H. & Schild, H. (1934). *J. Physiol. London*, **80**, 9P.

Gally, J. & Edelman, G. M. (1962). *Biochem. Biophys. Acta*, **60**, 499; (1964). *Biopolymer Symposium No. 1. Quantum aspects of polypeptides and polynucleotides.* Ed. M. Weissbluth. Interscience Publishers, New York (see also Edelman, G. M. & Gally, J. (1962). *J. Exp. Med.*, **116**, 207).

Ghiretti-Magaldi, A., Nuzzolo, C. & Ghiretti, F. (1966). *Biochemistry*, **5**, 1943.

Gostling, L. T. (1956). *Adv. Protein Chem.*, **11**, 429.

Griffith, J. S. (1956). *Proc. Roy. Soc.*, **A 235**, 23.

Hagins, W. A. & Jennings, W. H. (1959). *Disc. Faraday Soc.*, **27**, 180.

Hammett, H. (1940). In *Physical Organic Chemistry*. McGraw-Hill, New York.

Harley-Mason, J. & Laird, A. H. (1958). *Biochem. J.*, **69**, 59P.

Hercules, D. M. & Rogers, L. B. (1959). *Spectrochem. Acta*, **14**, 393.

Hess, S. & Udenfriend, S. (1959). *J. Pharmacol. Exptl. Therap.* **127**, 175.

Hobkirk, R. & Metcalfe-Gibson, A. (1963). In *Standard Methods of Clinical Chemistry*, Ed. Seligson, D. Vol. 4, p. 65. Academic Press, New York & London.

Huff, J. W. & Perlzweig, W. A. (1944). *J. biol. Chem.*, **155**, 345.

Hymie, I., Vecerek, B. & Wagner, J. (1960). *Cas. Lek. Ces.*, **99**, 88.

Isenberg, I. & Szent-Gyorgyi, A. (1958). *Proc. Nat. Acad. Sci. U.S.A.*, **44**, 519.

Ittrich, G. Z. (1960). *Acta Endocrinol.*, **35**, 34.

Jailer, J. W. (1948). *J. Clin. Endocrin.*, **8**, 564.

Jakeman, E., Oliver, C. J. & Pike, E. R. (1968). *J. Physiol.*, **A1**, 406.

Kalant, H. (1958). *Biochem. J.*, **69**, 79.

Keeley, F., LaBella, F. S. & Queen, G. (1969). *Biochem. biophys. Res. Commun.*, **34**, 1569.

Kerr, V. N., Hayes, F. J. & Ott, D. G. (1957). *Internat. J. Appl. Radiotherapy Radiation Isotopes*, **1**, 184.

Kobayashi, Y. & Maudsley, D. V. (1969). In *Methods of Biochemical Analysis*. Ed. D. Glick. Interscience Publishers, New York.

Konev, S. V. (1967). In *Fluorescence and Phosphorescence of Proteins and Nucleic Acids*. English Trans. Plenum Press, New York.

LaBella, F. S., Keeley, F., Vivian, S. & Thornhill, D. (1967). *Biochem. biophys. Res. Commun.*, **26**, 748.

Ledvina, M. & LaBella, F. S. (1970). *Anal. Biochemistry*, **36**, 174.

Leeman, H. G., Stich, K. & Thomas, M. (1963). In *Progress in Drug Research*. Ed. E. Jucker. Vol. 6, Birkhauser, Basel, p. 152.

Leirulov, M. & Meister, A. (1954). *J. biol. Chem.*, **209**, 327.

Lewis, G. N. & Kasha, M. (1944). *J. Amer. Chem. Soc.*, **66**, 210, see also Kasha, M. (1947). *Chem. Rev.* **41**, 401.

McCaman, M. V. & Robins, E. (1962). *J. Lab. Clin. Med.*, **59**, 885.

McDougal, D. B. & Farmer, H. S. (1957). *J. Lab. Clin. Med.*, **50**, 485.

Mahesh, V. B. (1964). *Steroids*, **3**, 647.

Mandeles, A. *et al.* (1954). *J. biol. Chem.*, **209**, 327.

Mangen, G. F., Jr. & Mason, J. W. (1957). *Science*, **126**, 562.

Mann, T., Seamark, R. F. & Sharman, D. F. (1961). *Brit. J. Pharmacol.*, **17**, 208.

Manwell, C. (1964). In *Oxygen and the Animal Organism*. Ed. F. Dickens & E. Neil. Pergamon Press, Oxford, p. 49.

Mattingly, D. (1962). *J. Clin. Pathol.*, **15**, 374.

Mehhuish, H. (1961). *J. Phys. Chem.*, **65**, 229. (1964). *J. Opt. Soc. Amer.*, **54**, 183.

Metzler, D. E. & Snell, E. E. (1955). *J. Amer. Chem. Soc.*, **77**, 2431.

Nag-Chaudhuri, J. & Augenstein, L. (1964). *Biopolymers Symposium No. 1. Quantum aspects of polypeptides and polynucleotides*. Ed. M. Weissbluth. Interscience Publishers, New York.

Nakao, T. & Aizawa, Y. (1956). *Endocrinol. Japan*, **3**, 92.

Natelson, S., Lugovoy, J. K. & Pincus, J. B. (1949). *Arch. Biochem.*, **23**, 157.

Oates, J. A., Gillespie, L., Udenfriend, S. & Sjoerdsma, A. (1960). *Science, N.Y.* **131**, 1890.

Ott, D. G. *et al.* (1959). *Nucleonics*, **17**, 106.

Parker, C. A. & Barnes, W. J. (1957). *Analyst*, **82**, 606.

Parker, C. A. & Rees, W. T. (1962). *Analyst*, **87**, 83.

Pringsheim, P. (1949). In *Fluorescence and Phosphorescence*. Interscience Publishers, New York.

Quay, W. B. (1963). *Anal. Biochem.*, **5**, 51.

Reddy, S. K., Reynolds, M. S. & Price, J. M. (1958). *J. biol. Chem.*, **233**, 691.

Reid, C. (1957). In *Excited States in Chemistry and Biology*. Academic Press, New York, Butterworths, London.

Reid, C. (1964). Radiation Res. Proc. Intern. Conference, Natich, Mass., 1963, 133–142. (See *Chem. Abst.*, 1964, **61**, 4669.)

Roberts, H. R. & Siino, M. R. (1963). *J. Plasm. Sci.*, **52**, 370.

Rombauts, W. & Lontie, R. (1960). *Arch. Int. Physiol. Biochim.*, **68**, 695.

Rosen, A. & Williams, R. T. (1961). *Bull. Photoelectric Spectrophot.*, **13**, 339.

Roth, M. (1969). In *Methods of Biochemical Analysis*, Ed. D. Glick. Vol. 17. Interscience Publishers, New York, p. 189.

Samejima, K., Daviman, W. & Udenfriend, S. (1971 *a*). *Anal. Biochem.*, **42**, 222.

Samejima, K., Daviman, W., Stone, J. & Udenfriend, S. (1971 *b*). *Anal. Biochem.*, **42**, 237.

Satoh, K. & Price, J. M. (1958). *J. biol. Chem.*, **230**, 781.

Shaklai, N. & Daniel, E. (1970). *Biochemistry*, **9**, 564.

Shore, P. A. & Olin, J. S. (1958). *J. Pharmacol. Exptl. Therap.*, **122**, 295.

Sjoerdsma, A., Oates, J. A., Zaltman, P. & Udenfriend, S. (1960). *New Engl. J. Med.*, **263**, 585.

Steele, R. H. & Szent-Gyorgyi, A. (1957). *Proc. Nat. Acad. Sci. U.S.A.*, **43**, 477; (1958). *Ibid.*, **44**, 540.

Steiner, R. F., Lippoldt, R. E., Edelhoch, H. & Frattali, W. (1964). *Biopolymers Symposium No. 1.* Quantum aspects of *polypeptides and polynucleotides.* Ed. M. Weissbluth. Interscience Publishers, New York.

Stokes, G. C. (1852). *Phil. Trans. Roy. Soc. London*, **A142**, 463.

Strassburger, J. & Reinert, K. E. (1969). *Jena Rev.*, **3**, 188.

Teale, F. W. J. (1960). *Biochem. J.*, **76**, 381.

Teale, F. W. J. & Weber, G. (1957). *Biochem. J.*, **65**, 476.

Udenfriend, S. (1962). In *Fluorescence Assay in Biology and Medicine*. Vol. I. Academic Press, New York and London.

Udenfriend, S. (1969). Vol. II, *ibid.*

Udenfriend, S., Bogdanski, D. F. & Weissbach, H. (1955). *Science, N.Y.*, **122**, 972.
Udenfriend, S., Stein, S., Bohlen, P., Daviman, W., Leimgruber, W. & Weigele, M. (1972). *Science, N.Y.*, **178**, 871.
Udenfriend, S., Weissbach, H. & Brodie, B. B. (1958). *Methods of Biochem. Anal.*, **6**, 95.
Van Duuren, B. L. (1963). *Chem. Rev.*, **63**, 325.
Velick, S. F. (1958). *J. Biol. Chem.* **233**, 1455.
Waalkes, T. P. & Udenfriend, S. (1957). *J. Lab. Clin. Med.*, **50**, 733.
Wallerstein, J. S., Alba, R. T. & Hale, M. G. (1947). *Biochem. Biophys. Acta.*, **1**, 175.
Weber, G. (1950). *Biochem. J.*, **47**, 114.
Weber, G. (1960). *Biochem. J.*, **75**, 335.
Weber, G. (1961). *Nature, Lond.*, **190**, 27.
Weigele, M., Blount, J. F., Tengi, J. P., Czaijkowski, R. C. & Leimgruber, W. (1972 *a*). *J. Amer. Chem. Soc.*, **94**, 4052.
Weigele, M., De Bernado, S. L., Tengi, J. P. & Leimgruber, W. (1972 *b*). *J. Amer. Chem. Soc.*, **94**, 5927.
Weil-Malherbe, H. (1959). *Pharmacol. Revs.*, **11**, 241.
Weil-Malherbe, H. (1960). *Biochim. Biophys. Acta.*, **40**, 351.
Weil-Malherbe, H. & Bone, A. D. (1952). *Biochem. J.*, **51**, 311.
Weller, A. (1952). *Z. Elektrochem.*, **54**, 42.
West, W. (1956). In *Techniques of Organic Chemistry*, Vol. IX. Ed. A. Weissberger, Interscience Publishers, New York.
White, A. (1959). *Biochem. J.*, **71**, 217.
Williams, R. T. (1959). *J. Roy. Inst. Chemistry*, **83**, 611.
Williams, R. T. & Bridges, J. W. (1964). *J. Clin. Path.*, **17**, 371.
Wood, E. J., Bannister, W. H., Oliver, C. J., Lontie, R. & Witters, R. (1971). *Comp. Biochem. Physiol.*, **40B**, 19.

9 Heterocyclic Compounds Including Nucleotides and Nucleic Acids

9.1 Heterocyclic Compounds

Saturated cyclic compounds containing an oxygen or a nitrogen atom absorb in the vapour state light of wavelength <200 nm. Thus ethylene oxide absorbs in the region accessible to vacuum grating spectroscopy using cells with fluorite windows, and shows two peaks at 143·5 and 157·2 nm. There are also two Rydberg series converging to the ionization potential at 10·8 eV Tetrahydrofuran also shows two bands, at 162–180 and 180–200 nm, respectively. Dioxan absorbs below 200 nm but when pure is highly transparent above 200 nm. Piperidine is said to absorb over the range 176 to 250 nm but the purified substance is very transparent and can even be used as a spectroscopic solvent. Pyrrolidine exhibits two bands, one at 162–187 nm and the other reaching 250 nm; there is a weak third band ($\lambda_{max.}$ 233 nm, ϵ *ca.* 300). Piperazine is transparent above 215 nm (Pickett *et al.*, 1951, 1953; Liu & Duncan, 1949).

Some at least of the large number of known unsaturated heterocyclic ring systems possess biochemical significance. The five-membered ring systems may be considered first and it is convenient to compare the parent substance with cyclopentadiene:

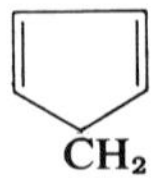

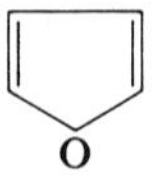

CH_2 O

$\lambda_{max.}$ nm	200	238·5	207	~220
$\epsilon_{max.} \times 10^{-3}$	10	3·4 (or 2·5)	9·1	2·1

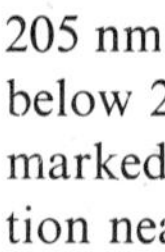

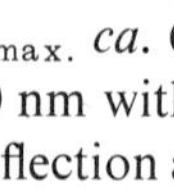

N–H S

$\lambda_{max.}$ nm	210	250	204	235
$\epsilon_{max.} \times 10^{-3}$	7·7 (or 15)	0·3		4·5 (in hexane) (or 7·1 in $C_6 12_{12}$)

There are discrepancies between published ϵ values (see Gillam & Stern, 1970; Rao, 1967).

Furan in the vapour state shows a band with $\lambda_{max.}$ 205 nm, $\epsilon_{max.}$ *ca.* 6300 and there is further absorption below 200 nm with definite resolution. There is also a marked inflection at 220 nm and perhaps some absorption near 250 nm with a very low intensity (ϵ *ca.* 1) in hexane or cyclohexane solutions. Furan is, however, much more transparent than cyclopentadiene in the region >200 nm. Pyrrole shows a peak at 210 nm, extreme estimates for the extinction coefficient being 31 600 for the vapour and 5100 for hexane solutions;

Table 9.1

		$\lambda_{max.}$ nm	$\epsilon_{max.} \times 10^{-3}$
Furan	α-CHO	227	3·0
		272	13·2
		278	15·8
		(in ethanol)	
	α-$COCH_3$	225	2·3
		265–270	12·9
	α-COOH	214	3·8
		242·5	10·7
	β-COOH	238	20
		265	4·0

(MacKinnen & Temmer, 1948; Beroza, 1952; Birch & Richards, 1956).

there is also weak absorption around 250 nm (ϵ *ca.* 300 in hexane). Thiophene in hexane has $\lambda_{max.}$ 235 nm, $\epsilon_{max.}$ 4500 but in isooctane $\lambda_{max.}$ is at 231 nm and ϵ is given as 7080 (Leandri *et al.*, 1955; Boig *et al.*, 1953). These differences in $\epsilon_{max.}$ are unexpected. Introduction of alkyl groups into the furan ring has little effect but carbonyl groups exert considerable bathochromic changes (see Table 9.1).

In the pyrrole series the introduction of alkyl groups causes the relatively weak absorption peak at 240 nm to vanish and the band near 210 nm undergoes a small blue shift. In acid, cryptopyrrole forms an unstable salt with $\lambda_{max.}$ 260 nm, $\epsilon_{max.}$ 5000.

Introduction of carbonyl or carbonyl groups results in new and enhanced absorption.

		$\lambda_{max.}$ nm	$\epsilon_{max.} \times 10^{-3}$
Pyrrole	α-$COCH_3$	251	4·1 (methanol)
		290	16·4
	α-CHO	252	5·0
		289·5	16·6
	α-COOH	228	4·5 (ethanol)
		260	12·6
	β-COOH	245	4·8 (ethanol)
	2,3-di-COOH	259	7·3 (ethanol)
	1,4-di-COOH	272	20·1
	1,2-di-COOH	282	4·2
N-methylpyrrole	α-CHO	255	5·3
		288·5	12·6
	α-$COCH_3$	245	4·9
		270	10·5
	α-COOH	236	6·3
		260·5	12·3

N-methylpyrrole closely reproduces the spectrum of the parent substance. Carbonyl groups in the α-position produce two new bands, the aldehydes and the ketones having a greater red shift than the acids. The long wavelength band near 290 nm is intense. The effect of substitution in the β position is illustrated above (see Marshall & Walker, 1951; Cookson, 1953, for further data).

The spectrum of thiophene in cyclohexane shows the 235 nm band with evidence of four vibrational levels and extremely weak inflections at 267, 274 and 300 nm but these may have arisen from impurity. The thiophene absorption displays well the bathochromic effects caused by certain substituents. With halogens in the α-position the absorption of the parent substance ($\lambda_{max.}$ 231 nm, ϵ 7100) is moved relatively little (235–236 nm for α-Cl and α-Br 243 nm for α-I) but the effects are larger for carbonyl derivatives.

		$\lambda_{max.}$ nm	$\epsilon_{max.} \times 10^{-3}$	[$\lambda_{max.}$ nm	$\epsilon_{max.} \times 10^{-3}$
Thiophene	α-CHO	265	10·5	*[260	11·4 (in alcohol)
		278·5	6·5	[284	1·8 (in alcohol)
	α-$COCH_3$	252	10·5	[259	10·3 (in alcohol)
		273	7·2	[294	75 (in alcohol)
	α-COOH	249	11·5	*[247	10·2 (in alcohol)
		268·5	8·2	[261	7·6
	α-OH	220	8·0	—	—
		263	2·8	—	—

(Shejnker *et al*, 1957; Pappalardo, 1959; Huebner et al., 1953).
* New data by Timmons & Blackburn in Gillam & Stern (1970). (See also Huebner *et al.*, 1953.)

Five-membered rings with two or more hetero-atoms are not unknown in natural products. The imidazoles, pyrazoles and isoxazoles all show absorption peaks in the region 207–211 nm of moderate

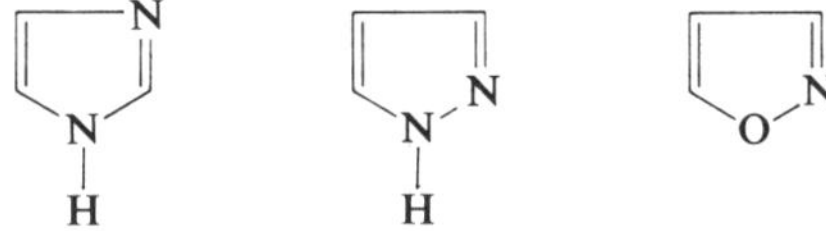

intensity ($\epsilon_{max.}$ *ca.* 5000). Introduction of methyl groups has only a very small bathochromic effect and relatively little effect on $\epsilon_{max.}$ (Leandri *et al.*, 1955). Thiazole shows peaks at 209 and 232 nm, $\epsilon_{max.}$ 2750 and 3550 in heptane and the 4-methyl derivative has its maximum at 243 nm and isothiazole (in water) has

rather similar absorption: $\lambda_{max.}$ 205, 244 nm, $\epsilon_{max.}$ 2950 and 6200. Acetylimidazole has its peak at 255 nm, ϵ 2500.

Heterocyclic compounds like pyridine and piperidine show in respect of ultraviolet absorption resemblances to benzene and cyclohexane. The molecules of pyridine and benzene with their double bonds display electrons (one from each atom forming the double bond) possessing much greater mobility than the others. The non-localized electrons (π electrons) have orbitals in layers above and below the cyclic system and overlapping little with the orbitals of the σ electrons shared in establishing the single bonds. The nitrogen heterocyclic compounds like pyridine have aromatic character in the same way as does benzene. The π electrons contribute to a delocalization energy in virtue of their range of movement and this tends to reduce chemical reactivity. Changes brought about by substituents are rapidly transmitted by the π electrons so that a new balance is struck.

Benzene as a symmetrical molecule has a uniform distribution of one π electron per carbon atom but the substituent in aniline results in partial localization. The lone pair of the amino group and the π electrons of the ring set up a new distribution shown below:

−0·02
0 0
−0·025 −0·025
−0·02
NH_2
+0·09

In benzene the net charge is one π electron for each carbon with no excess at any point. The amino group has a deficit of electrons equal to the sum of all the excess charges. This could also be depicted

1·02
1·00 1·00
1·025 1·025
1·02
NH_2
1·9

Dipole moment measurements lead to approximate values for excess charges and the molecular orbital method has been very important. The calculation of excess charges in heterocyclic molecules has been made by Longuet-Higgins & Coulson (1949), e.g. for pyridine

+0·18
+0·05 +0·05
+0·15 +0·15
N
−0·58

the nitrogen atom being 'electron-rich', i.e. attracting electrons while the 4 position has the greatest deficiency. The calculations for polycyclic molecules such as quinoline and isoquinoline are rather uncertain (cf. Albert, 1959).

The π-deficient nitrogen–heterocyclic compounds include pyridine, quinoline, isoquinoline, acridine, phenanthridine and their derivatives as well as the pyridazine, pyrimidine, pyrazine, sym. triazine and tetrazine families and the quinazoline, pteridine, phenazine and naphthapyridine groups. Of this great range a few groups are specially important for biochemistry. A general effect is an attraction by the ring nitrogen for π electrons which polarizes the molecule and results in low electron densities on the ring carbons. Thus pyridine has a dipole moment and its reactivity differs from that of benzene.

Entry of an amino group into pyridine has effects varying greatly with the position of the substituent. Thus 3-aminopyridine (β-amino) does not show a great change in pK and the ring nitrogen is still the main centre of basicity, but in 4-aminopyridine (γ-amino) the basicity increases by 10^4 (pK increases by 4 units) and 2-aminopyridine shows a similar increase. In the formation of cations these and many other heterocyclic compounds attach the first proton to the ring nitrogen (Hearn, Morton & Simpson, 1951).

Tautomerism can occur in many substituted nitrogen heterocyclic compounds and physical properties have greatly helped to elucidate structures. Different methods do not always lead to the same answer but much depends on the molecular environments. An early example was the demonstration by an X-ray crystal study (Knaggs & Lonsdale, 1940) that solid melamine has three primary amino groups:

NH_2
N N
H_2N N NH_2

Similarly aminopyrimidine molecules had primary amine groups forming H bonds with the ring N atoms of other molecules (White & Clews, 1956). Infrared investigations (Brown, Hoerger & Mason, 1955 *b*) confirmed this for 2- and 4-aminopyrimidine. Ultraviolet absorption studies in which methyl derivatives of fixed structure are compared with the parent substances are of great value for dilute solutions. This applies both to amino- and hydroxypyridines and pyrimidines (Anderson & Seeger, 1949).

9.1.1 Pyridine and its Derivatives

In the vapour state pyridine shows well-resolved absorption over a considerable wavelength range while in hexane or cyclohexane the main peak occurs near 256 nm ($\epsilon_{max.}$ variously recorded 1860–2090) with an inflection near 280 nm. The 256 nm band is assigned to a $\pi \rightarrow \pi^*$ transition. In ethanol $\lambda_{max.}$ at 257 nm is little changed but the vibrational fine structure disappears and $\epsilon_{max.}$ rises to about 2750. (Different workers give $\lambda_{max.}$ for neutral solutions at 253–257 nm with $\epsilon_{max.}$ between 2880 and 3600.) In acid solution protonation raises $\lambda_{max.}$ to *ca.* 5250 but in alkali it remains near 2750 (Halverson & Hirt, 1956; Brown & Mihm, 1955; Brown & McDaniel, 1955; Houghton & Cain, 1972).

Alkyl substitution influences both the location and intensity of the peak absorption (Ewing & Steck, 1946). In general, vibrational fine structure tends to be smoothed out. The position of the substituent is significant; the 2- and 3-methyl derivatives show in alkaline solution $\lambda_{max.}$ 262–263 nm, $\epsilon_{max.}$ 3100–3500 and in acid solution $\epsilon_{max.}$ rises to 5500–6600. These changes, though not very large, are characteristic; values are lower for 4-methylpyridine, $\lambda_{max.}$ 255 nm, $\epsilon_{max.}$ 2100 in alkali and $\lambda_{max.}$ 252·5 nm, $\epsilon_{max.}$ 4500 in acid (cf. Bliznynkov & Reznikov, 1955).

Introduction of a thiol group or a hydroxyl group in the 2 position results in a significant red shift (Table 9.2). Tautomerism complicates the interpretation (cf. Metzler & Snell, 1955). Thus, the classical method of comparison with *o*-methyl and *N*-methyl derivatives shows unmistakedly that the pyridone structure predominates

2-Methoxypyridine (N, OMe):

	$\lambda_{max.}$ nm	$\epsilon_{max.}$
pH 7	< 205	> 5300
	269	3230
pH 1	210	3550
	279	6200

2-Pyridone (N–H, =O):

	$\lambda_{max.}$ nm	$\epsilon_{max.}$
pH 6	224	7230
	293	5900

N-Methyl-2-pyridone (N–Me, =O):

	$\lambda_{max.}$ nm	$\epsilon_{max.}$
pH 5	226	6100
	297	5700

Similarly the amide (pyridone) form predominates in neutral solution for substitution at position 4.

4-Methoxypyridine (OMe):

	$\lambda_{max.}$ nm	$\epsilon_{max.}$
pH 2	235	9 500
pH 9	222	9 300

4-Pyridone (O, N–H):

	$\lambda_{max.}$ nm	$\epsilon_{max.}$
pH 1	239	11 800
pH 7	260	18 900

N-Methyl-4-pyridone (O, N–Me):

	$\lambda_{max.}$ nm	$\epsilon_{max.}$
pH 0	234	9 800
pH 7	253	14 800
pH 13	239	14 900
	260	2 200

(Mason, 1959).

When the hydroxyl group is at position 3 the following structures are possible

neutral molecule (OH, N):

$\lambda_{max.}$ nm	$\epsilon_{max.}$
246	4700(5720)
278	2320
313	3000(3060)

zwitterion (O^-, ^+N–H)

cation (OH, ^+N–H):

$\lambda_{max.}$ nm	$\epsilon_{max.}$
222	3300(3730)
283	5900(5840)

anion (O^-, N):

$\lambda_{max.}$ nm	$\epsilon_{max.}$
235	10 200
	(11 000
298	4 500
	(4 960

3-Methoxypyridine (OMe, N):

	$\lambda_{max.}$ nm	$\epsilon_{max.}$
pH 7	216	8320
	276	3960

3-Methoxypyridinium (OMe, ^+N–H):

	$\lambda_{max.}$ nm	$\epsilon_{max.}$
pH 2	224	4290
	284	6240

(For Ramon and infrared spectra, see Albert & Spinner, 1960; Spinner, 1960; Brown & Mason, 1956; Kracmar & Kraacmarova, 1968; also p. 209, p. 265.)

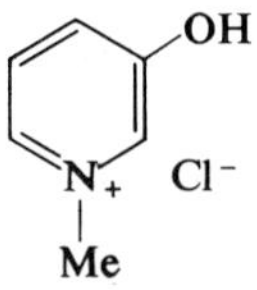

		$\lambda_{max.}$ nm	$\epsilon_{max.}$
1-Methyl-3-hydroxypyridinium chloride		249	8120
	pH 7	320	5810
		224	3790
	pH 2	288	5900

and the spectra indicate that the zwitterion and the 3-hydroxy compound are in equilibrium. The 1-methyl-3-pyridoxide ion shows $\lambda_{max.}$ 320 nm in water and there is a large displacement to 356 nm in 100% dioxan (Mason, 1959). Pyridine-2,3-diol shows $\lambda_{max.}$ 297 nm ($\epsilon_{max.}$ 8125) while the pyridine-3,4-diol has $\lambda_{max.}$ 273 nm, $\epsilon_{max.}$ 10 130.

Houghton & Cain (1972) determined 3-hydroxypyridine, pyridine-2,3-diol and pyridine-3,4-diol in mixtures by measuring intensities of absorption at three carefully chosen wavelengths, namely 273, 297 and 313 nm. At each wavelength one of the three substances showed much stronger absorption than the other two.

Let C_x, C_y and C_z stand for the concentrations of 3-hydroxypyridine, pyridine-2,3-diol and pyridine-3,4-diol respectively. The molecular extinction coefficients at pH 7·5 given below permit the solution of three simultaneous equations

$$E_{313} = C_x X_5 + C_y Y_3 + C_z Z_3$$
$$E_{297} = C_x X_4 + C_y Y_2 + C_z Z_2$$
$$E_{276} = C_x X_3 + C_y Y_1 + C_z Z_1$$

Mixtures of 4-hydroxypyridine and pyridine-3,4-diol were analysed as follows:

$$E_{254} = C_s S_1 + C_z Z_4$$
$$E_{273} = C_s S_2 + C_z Z_5$$

where C_s and C_z are concentrations of 4-hydroxypyridine and pyridine-3,4-diol respectively at pH 8·5. Similarly, mixtures of 3-hydroxypyridine and pyridine-2,5-diol were determined from

$$E_{276} = C_r R_1 + C_x X_1$$
$$E_{320} = C_r R_2 + C_x X_2$$

where C_x and C_r are the concentrations of 3-hydroxypyridine and of the 2,5-diol respectively.

Pyridine-4-carboxylic acid (isonicotinic acid) and the related methyl betaine have been studied by Black (1955). The acid at pH values >7 shows a single peak at 267 nm, $\epsilon_{max.}$ *ca.* 2700 (see Fig. 9.2). On the acid

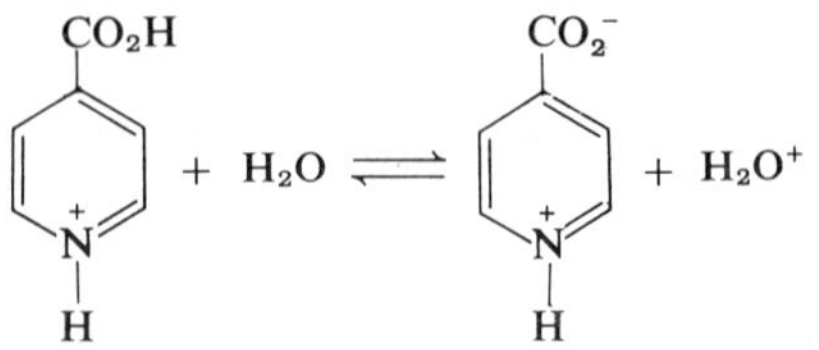

$$\text{4-}CO_2H\text{-pyridinium (N}^+\text{-H)} + H_2O \rightleftharpoons \text{4-}CO_2^-\text{-pyridinium (N}^+\text{-H)} + H_3O^+$$

Table 9.2 Hydroxypyridines

	pH	λnm	ε × 10⁻³	Symbol
3-Hydroxy-	6·0	276	2·3	X_1
		320	2·42	X_2
	7·5	273	1·895	X_3
		297	1·62	X_4
		313	3·02	X_5
4-Hydroxy-	8·5	254	16·5	S_1
		273	2·64	S_2
2,3-Diol	7·5	273	3·0	Y_1
		297	8·15	Y_2
		313	4·5	Y_3
2,5-Diol	6·0	276	1·3	R_1
		320	5·62	R_2
3,4-Diol	7·5	273	9·72	Z_1
		297	2·1	Z_2
		313	0·36	Z_3
	8·5	254	3·78	Z_4
		273	9·0	Z_5

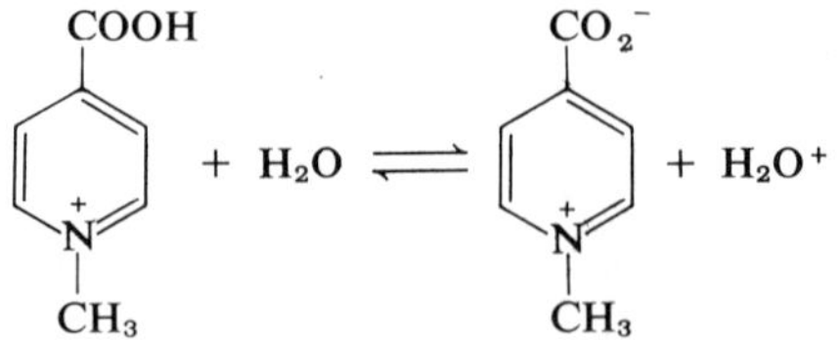

$$\text{4-COOH-1-}CH_3\text{-pyridinium} + H_2O \rightleftharpoons \text{4-}CO_2^-\text{-1-}CH_3\text{-pyridinium} + H_3O^+$$

side isosbesticity indicates a two-component system with $\lambda_{max.}$ 267 nm, $\epsilon_{max.}$ 4300 at 0·01 N HCl and $\lambda_{max.}$ 281 nm, $\epsilon_{max.}$ 4650 at 5·0 N HCl. The methyl betaine displays very good isosbestic points (Fig. 9.1) at pH 7 $\lambda_{max.}$ occurs at 264 nm and 219 nm ($\epsilon_{max.}$ 4540 and 8400) and in 5·0 N HCl $\lambda_{max.}$ is at 277 and 244 nm ($\epsilon_{max.}$ 9, 4500 and 9500 respectively). The equilibria are represented below and pK values can readily be calculated.

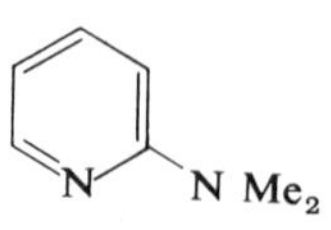

	$\lambda_{max.}$ nm	$\epsilon_{max.}$
2-(N Me_2)pyridine	305	2700

2-NH_2-pyridine		$\lambda_{max.}$ nm	$\epsilon_{max.}$
	pH 1	229	8900
		300	5700
	pH 9	229	9400
		287	3800
	In ether	293	4570

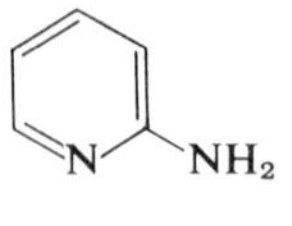

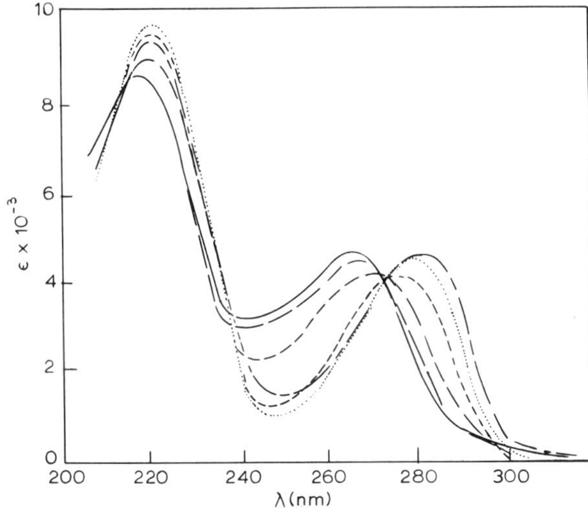

Figure 9.1 Ultraviolet absorption spectra of isonicotinic acid methyl betaine; ———, pH 7 buffer; — —, 0·1 N NaOH; - - - - -, 0·01 N HCl; — — — — —, 0·1 N HCl; — - — - —, 1·0 N HCl; . . . ., 5·0 N HCl (from Black, 1955)

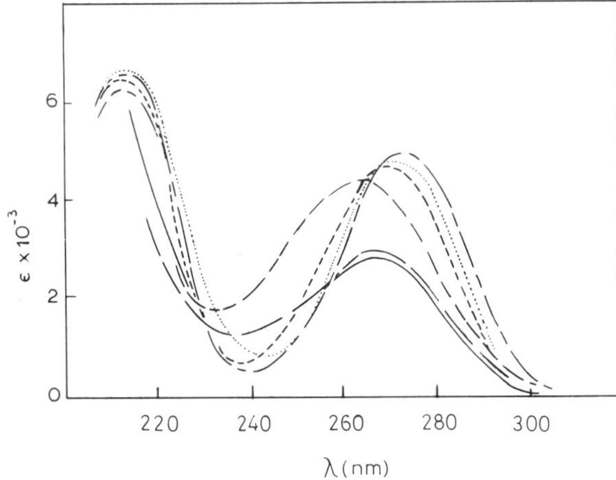

Figure 9.2 Ultraviolet absorption spectra of isonicotinic acid; ———, pH 7 buffer; — —, 0·1 N NaOH; — — — — —, 0·01 N HCl; - - - - - -, 0·1 N HCl; . . . ., 1·0 N HCl; — - — - —, 5 N HCl (from Black, 1955)

(1-Methyl-2-iminopyridine: N—Me, =NH)

	$\lambda_{max.}$ nm	$\epsilon_{max.}$
pH 14	251	11 700
	371	4 100
In cyclohexane	258	12 600
Fine structure	348 } 370 }	2 500
In ether	352	3160

(NH_2, pyridine N)

	$\lambda_{max.}$ nm	$\epsilon_{max.}$
pH 1	263	16 500
pH 12	241	14 000
	~265	2 400

(NH, N—Me)

	$\lambda_{max.}$ nm	$\epsilon_{max.}$
pH 13	268	16 500
In cyclohexane	275	15 850

Pyridine-*N*-oxide shows $\lambda_{max.}$ 203 and 254 nm ($\epsilon_{max.}$ 17 000 and 11 900) and in strong acid this band becomes much weaker ($\epsilon_{max.}$ 2900). Both the 3- and 4-methylpyridine-1-oxides shows similar strong absorption near 255 nm for the free base greatly weakened in strong acid. In 2-hydroxypyridine-*N*-oxides the pyridine structure is preponderant.

In the case of 2-aminopyridine the structure is mainly amino whereas the 4-amino derivative is largely imino (Mason 1960). The spectra of nicotine and related compounds are noted below and need little explanation.

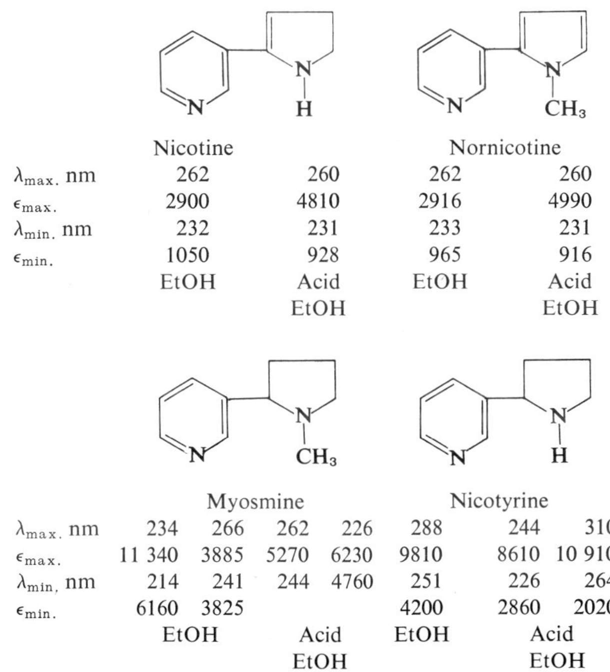

	Nicotine		Nornicotine	
$\lambda_{max.}$ nm	262	260	262	260
$\epsilon_{max.}$	2900	4810	2916	4990
$\lambda_{min.}$ nm	232	231	233	231
$\epsilon_{min.}$	1050	928	965	916
	EtOH	Acid EtOH	EtOH	Acid EtOH

	Myosmine				Nicotyrine		
$\lambda_{max.}$ nm	234	266	262	226	288	244	310
$\epsilon_{max.}$	11 340	3885	5270	6230	9810	8610	10 910
$\lambda_{min.}$ nm	214	241	244	4760	251	226	264
$\epsilon_{min.}$	6160	3825			4200	2860	2020
	EtOH		Acid EtOH		EtOH	Acid EtOH	

(M. L. Swain *et al.*, 1949; B. Witcop, 1954).

Mimosine [β-(1,4-dihydro-3-hydroxy-4-oxo-1-pyridyl)-alanine] can be extracted from the seeds of

Table 9.3 Absorption peaks of mimosine

Solvent	Water		0·01 M HCl		0·07 M NaOH	
	$\lambda_{max.}$ nm	$\epsilon_{max.} \times 10^{-3}$	$\lambda_{max.}$ nm	$\epsilon_{max.} \times 10^{-3}$	$\lambda_{max.}$ nm	$\epsilon_{max.} \times 10^{-3}$
Synthetic DL	283	12·6	277	7·4	309	11·0
Natural DL	283	12·9	277	7·2	309	11·0

Leucaena glauca or from freshly cut petioles of *Mimosa pudica*. It is an unusual aminoacid with pK_1 2·1, pK_2 7·2 [α-NH_3^+] and pK_3 9·2 (OH) (Spencer, 1962). Its absorption spectra are shown in Table 9.3.

The infrared absorption peaks (KBr disc) include (cm^{-1}) 3400 (broad, OH), 2900–2850 (broad, NH), 1640 (s. CO), 1588 (s. COO^-), 1530 s., 1490 (s. NH_3^+).

3,4-Dihydroxypyridine can exist in keto and enol forms and is mainly keto.

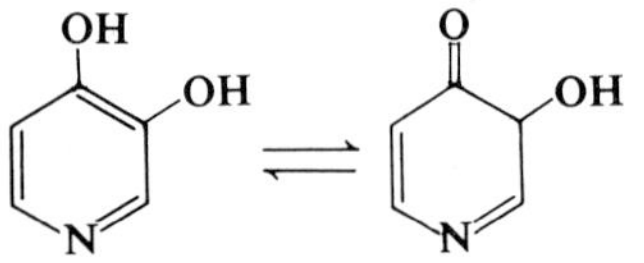

Its absorption curves (Fig. 9.3) are responsive to pH changes:

Water

$\lambda_{max.}$ nm	$\epsilon_{max.} \times 10^{-3}$
214	14·4
274	12·2
$\lambda_{min.}$ nm	
234	1·8

0·1 N HCl		0·01 N NaOH	
$\lambda_{max.}$ nm	$\epsilon_{max.} \times 10^{-3}$	$\lambda_{max.}$ nm	$\epsilon_{max.} \times 10^{-3}$
210	14·2	230	15·55
243	3·9	300	8·77
270	3·6		
$\lambda_{min.}$ nm		$\lambda_{min.}$ nm	
221	1·1	233	3·3
254	3·3		

Figure 9.3 Absorption spectra of 3,4-Dihydroxypyridine; solid curve, in water; dotted curve, in 0·1 N HCl; dashed curve, 0·1 N NaOH

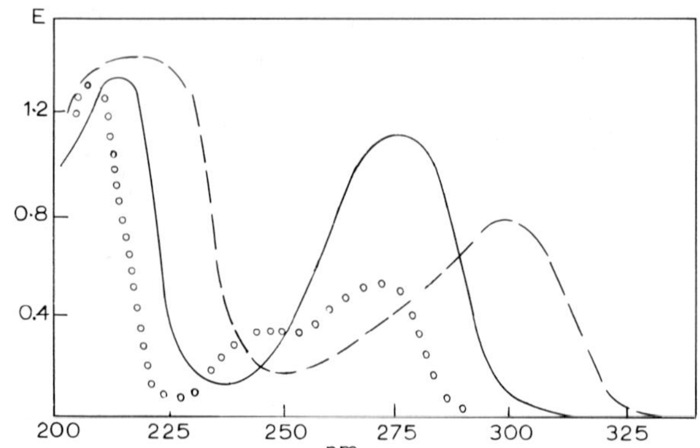

The principal infrared peaks are as follows: cm^{-1} 3509, 3425 (OH), 290, 2632–2500 (NH), 1626 (C═O), 1587, 1527, 1504, 1439 (ring stretching vibrations), 1361, 1319, 1242, 1220, 1160, 1117 (CH in plane deformations), 980, 939 (ring skeletal), 833, 806 (CH out of plane vibrations).

I am indebted to Professor L. Fowden for the ultraviolet absorption curve and to Dr. J. F. Pennock for an infrared curve.

Sigel & Brintzinger (1963) studied the metal ion complexes of 2-aminopyridine-1-oxide. The background of the work included the formation of complexes between adenosine and nucleotides, e.g. ATP and 3*d* divalent metal ions. According to Perrin (1960) adenosine-1-oxide permitted coordination complexes:

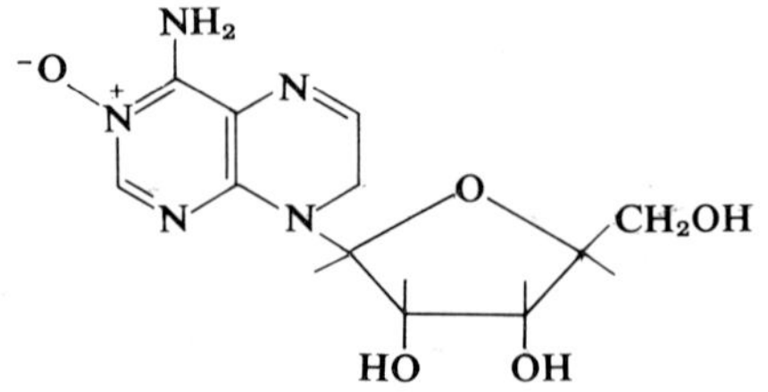

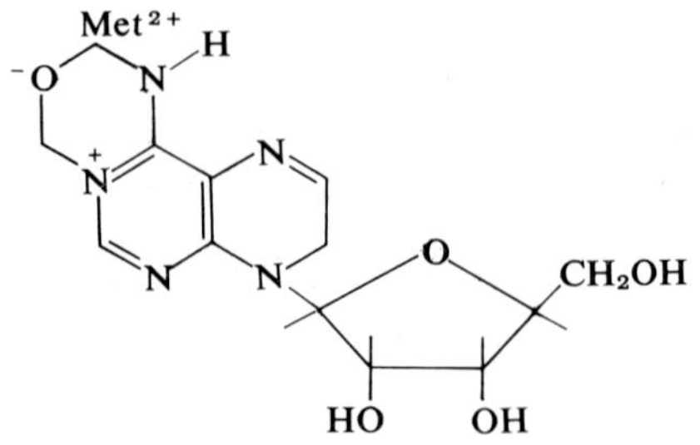

Complexes with 2-aminopyridine-1-oxide provided a simpler model:

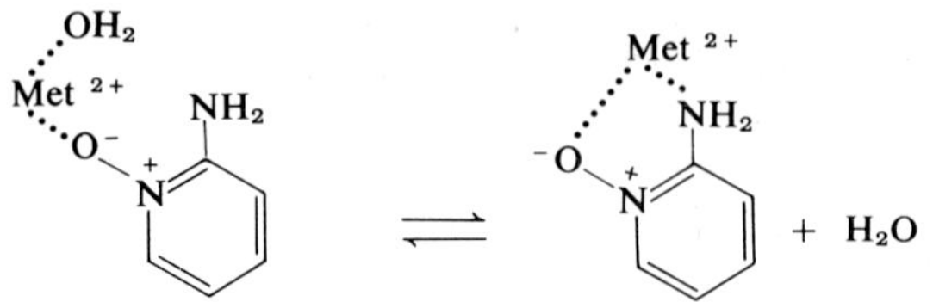

Met = metal

Two types of complex are in fact formed, one in which the ligand is in its neutral form and the other in which it is in its anionic form.

Sigel *et al.* (1963) found that the absorption spectrum of 2-aminopyridine-1-oxide (10^{-4} **M**) exhibited with concentrations of sodium or potassium hydroxide between 6 **M** and 20 **M** a family of curves exhibiting a near approach to isosbesticity. At the higher concentration of OH^- the isosbestic points were slightly displaced. In water the maxima fell at 308·5 and 237·5 nm, ϵ 4460 and 6150. The anion absorption with peaks at 266 and 356 nm had calculated $\epsilon_{max.}$ values of 14 810 and 4490 (NaOH) or 16 400 and 6250 (KOH). There was a linear relationship between intensity of absorption at 266 nm and NaOH activities. The activity constant K^H_{LH}

(L) Neutral molecule ⇌ (L⁻) Anion + H^+

was calculated and $pK^H_{LH(0\cdot1)} = 16\cdot74 \pm 0\cdot11$. The anion formed by loss of a proton of the 2-amino group can co-ordinate to Cu^{2+} and possibly to Fe^{3+} but other metal ions do not co-ordinate because hydrolysis occurs more easily than chelation. Infrared absorption of solid products confirmed the formation of a copper complex. Potentiometric determinations of the complex equilibrium constants showed that Cu^{2+} ions raise the apparent acidity of the ligand very greatly.

Sigel & Brintzinger (1963) had shown that the neutral form forms complexes with a range of metal ions (Ba^{2+}, Ca^{2+}, Mg^{2+}, Mn^{2+}, Fe^{2+}, Co^{2+}, Co^{2+}, Ni^{2+}, Cu^{2+} and Zn^{2+}). The absorption curves obtained for 2-aminopyridine-1-oxide (10^{-2} **M**) in $Cu(ClO_4)_2$ ($7\cdot5 \times 10^{-3}$ **M** to 0·15 **M**) all show at pH 3·8 a marked inflection at 390 nm and the relation

$$(1/[Cu^{2+}]_{total})/(1/E_{390(corr.)})$$

was linear and ϵ for the complex was 306 ± 10.

The 390 nm absorption shown by the complex formed between Cu^{2+} and 2-aminopyridine-1-oxide affords a good example of a charge-transfer band. It occurs at a wavelength where the absorption of the free ligand and the hydrated Cu^{2+} ion are negligibly small. There is much more in this study than can be discussed here.

In addition to the main $\pi \rightarrow \pi$ absorption bands, all monocyclic azines show weak selective absorption, having no counterpart in the spectrum of benzene. This absorptive process shows a shift to lower wavelengths with change from non-polar to polar solvents. It arises as a result of promotion of an electron from a lone-pair nitrogen orbital to a π orbital of the ring ($n \rightarrow \pi$ transition). Such transitions are best identified in the vapour state absorption spectra of pyridine, pyrimidine, pyrazine and *sym.*-tetrazine. They

Pyridazine

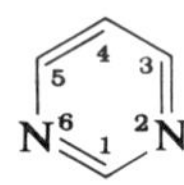

Pyrimidine

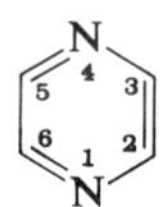

Pyrazine

exhibit a small red-shift on change from vapour to *cyclo*hexane solutions and when aqueous solutions are compared with *cyclo*hexane solutions there are blue-shifts in going from a non-polar to a polar solvent. In acid solutions the formation of a cation causes the $n \rightarrow \pi$ band to move further towards shorter wavelengths or to disappear; here the lone-pair electrons either bind a proton or are otherwise held more strongly in the orbital by a charged centre in the molecule. The magnitude of the blue-shifts of the $n \rightarrow \pi$ bands (*cyclo*hexane to aqueous solutions) affords an indication of electron-donating capacity in azines.

Table 9.4 shows the band origins (ν_{0-0}) in wavelength units and $\lambda_{max.}$ for *cyclo*hexane solutions with the peaks for aqueous solutions.

The $n \rightarrow \pi$ bands move towards the blue when electron-donating substituents are introduced but towards the red with electron-accepting substituents. On the other hand, both types of substituent effect bathochromic shifts in the stronger $\pi \rightarrow \pi$ bands. The longest wavelength absorptions of the azines and their derivatives show low values for $\epsilon_{max.}$ (*ca.* 200–1000). The low intensity is explained by spatial separation of the lone-pair and π-orbitals and by the fact that only the *s*-component of the *s*-*p*-hybrid lone-pair orbital contributes to the transition moment of $n \rightarrow \pi$ excitation. The blue-shift for aqueous solutions is explained on the basis of hydrogen-bonding in the water having a stabilizing effect on the ground state. The $n \rightarrow \pi$ bands move towards the blue with *o*- and *p*-directing substituents and to the red with *m*-directing groups. Many of the details of the spectra in Table 9.5 are satisfactorily accounted for in terms of orbital theory (Mason, 1959 *a*, *b*).

Table 9.4 Absorption spectra of monocyclic azines
($n \rightarrow \pi^*$ transitions)

Compound	Vapour state: Band origin	Vapour state: $\lambda_{max.}$ nm	Solvent: Cyclohexane: Origin	Solvent: Cyclohexane: $\lambda_{max.}$ nm	Solvent: Water: $\lambda_{max.}$ nm	Oscillator strength f
Pyridine	288	—	—	270	—	0·003
Pyrimidine	322	296	324·6	298·5	271·4	0·0069
Pyrazine	324	324	327·7	327·7	304·9	0·0104
Pyridazine	365·1	325	373	340	330	0·0058
sym. Triazine	317·3	270·3	323·6	272	260·1	0·021
sym. Tetrazine						
(*a*)	551·5	551·5	562·5	538·8	510·2	0·0042
(*b*)	—	—	—	320	305·1	0·001

Table 9.5 Absorption spectra of monocyclic azines and their derivatives in polar and non-polar solvents

Compound	Solvent	bands: $\lambda_{max.}$ nm	bands: $\epsilon_{max.}$	bands: $\lambda_{max.}$ nm	bands: $\epsilon_{max.}$	pK_a
Pyridine	C_6H_{12}	251	2 000	~270	~450	5·23
	pH 9	257	2 650	not seen		
2-Cyano-	C_6H_{12}	265	2 730	~278	340	0·26
	10 N H_2SO_4	267	7 520	not seen		
3-Cyano-	C_6H_{12}	265	2 200	~279	430	1·36
	4 N H_2SO_4	265	5 130	not seen		
4-Cyano-	C_6H_{12}	271	2 840	~290	~500	1·90
	2 N H_2SO_4	276	5 710	not seen		
Pyrazine	C_6H_{12}	260	5 600	328	1040	0·6
	EtOH	261	6 000	310	860	
	5 N H_2SO_4	266	7 300	not seen		
2-Methyl-	C_6H_{12}	266	5 700	320	830	1·47
	2 N H_2SO_4	276	6 680	not seen		
Pyridazine	C_6H_{12}	246	1 300	340	315	2·33
	EtOH	246	1 160	313	303	
	pH 0	238	1 610	not seen		
3-Methyl-	EtOH	251	1 300	310	400	
4-Methyl-	C_6H_{12}	252	1 296	331	375	2·92
	EtOH	250	1 370	303	350	
	pH 0·7	221	6 020	not seen		
3-Methoxy-	C_6H_{12}	272	1 990	327	326	2·52
	pH 0·05	269	2 010	not seen		
4-Methoxy-	C_6H_{12}	259	1 560	307	258	3·70
	pH 1	247	10 320	not seen		
	pH 7	254	2 570	~285	~390	
Pyrimidine	C_6H_{12}	243	2 030	298	326	1·30
	EtOH	243	2 920	280	373	
	4 N H_2SO_4	242	5 540	not seen		
2-Methoxy-	C_6H_{12}	264	4 180	~295	~400	<1
	pH 7	267	4 530	not seen		
4-Methoxy-	C_6H_{12}	248	3 100	~270	~274	2·5
	pH 7	248	3 370	not seen		
5-Hydroxy-	EtOH	276	5 330	not seen		
Sym. Tetrazine	C_6H_{12}	252	2 150	542	829	<0
				~320	26	
	pH 7	255	2 840	510	362	
				~305	~157	

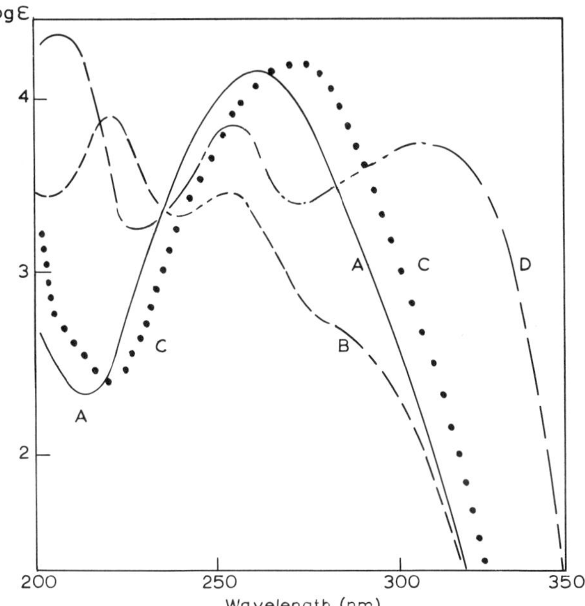

Figure 9.4 Absorption spectra of (A) 4-hydroxypyridazine (neutral molecule), pH 4·8; (B) 4-methoxypyridazine (neutral molecule), pH 7·0; (C) 1-methyl-4-pyridazine (neutral molecule); (D) 4-hydroxypyridazine-2-methochloride (zwitterion). pH 7·0 (from Mason, 1957)

The $\pi \longrightarrow \pi^*$ transitions of monocyclic hydroxyazines reflect the presence of various charged forms and tautomeric modifications. In general these hydroxy derivatives can exist as neutral enol, cation, anion and zwitterion or oxo-forms. Provided that bands of corresponding intensity are compared, the peaks occur in order of wavelength:

zwitterion > anion > cation > enol.

In the 4-hydroxy derivatives of pyridine and pyridazine low intensity inflections appear on the long wave side of the major more intense bands. The charged species absorb at shorter wavelengths and show higher ϵ values than the corresponding species of the isomeric derivatives but sometimes as in 4-hydroxypyrimidine there seems to be a complex mixture of absorbing entities.

Figures 9.4 to 9.11 illustrate the situation clearly.

Pyrazine in cyclohexane shows a peak near 258 nm of moderate intensity and some fine structure in the region 310–330 nm ($\epsilon < 1000$). In acid (pH 1) $\lambda_{max.}$ is displaced to 268 nm with but little change in intensity. Strong acid displaces both bands in the direction of shorter wavelengths but the peak at 246 nm is considerably more intense. Possibly the ion

$$H{-}\overset{+}{N}\langle\;\rangle\overset{+}{N}{-}H$$

is formed.

The 2-hydroxypyrazine over the pH range 5–12 shows a practically constant spectrum in which the 315–320 nm absorption of pyrazine is increased in intensity sevenfold. The *N*-methyl-2-oxo pyrazine has almost exactly the same absorption spectrum at pH 5·2, so that the keto form must predominate. The spectrum of 2-methoxypyrazine at pH 5 ($\lambda_{max.}$ 292 nm) is significantly different.

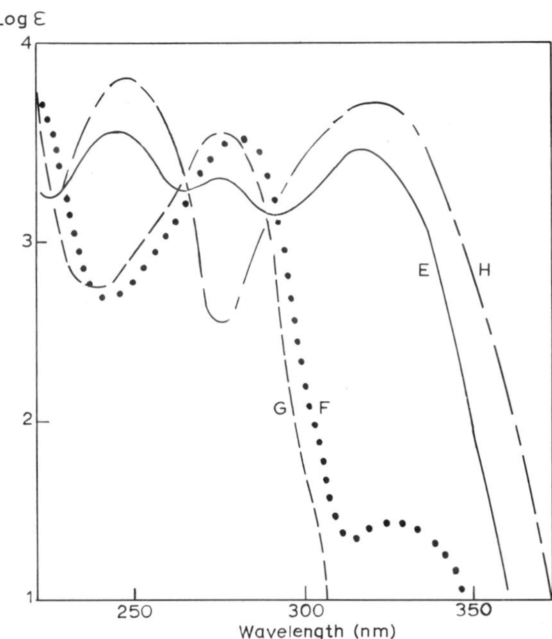

Figure 9.5 Absorption spectra of (E) 3-hydroxypyridine in aqueous buffer, pH 6·9; (F) 3-hydroxypyridine in alcohol; (G) 3-methoxypyridine (neutral molecule); (H) 3-hydroxypyridine methochloride (zwitterion), pH 7·0 (from Mason, 1957)

Figure 9.6 Ultraviolet spectra in cyclohexane of pyridazine (solid curve), 3-methoxypyridazine (dotted curve) and 4-methoxypyridazine (dashed curve)

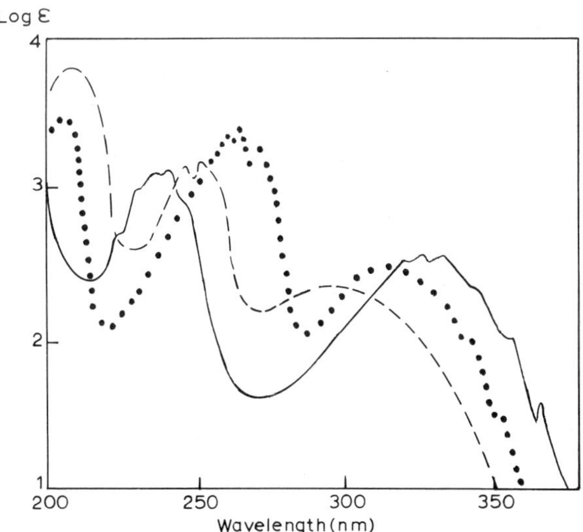

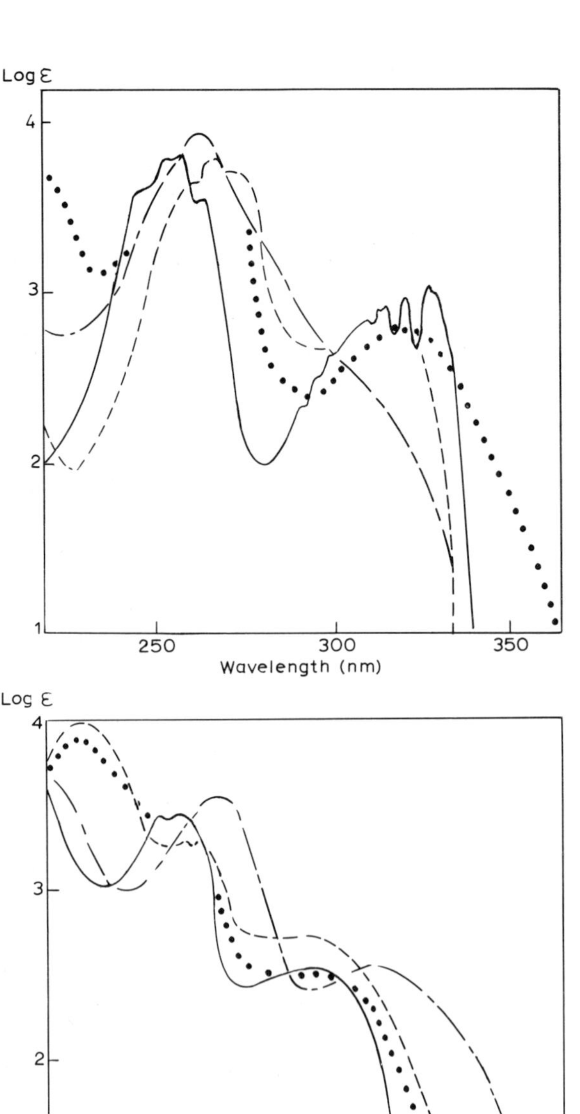

Figure 9.7 Ultraviolet absorption spectra of: ———, pyrazine in cyclohexane; —·—, pyrazine in 5N sulphuric acid; — — — —, 2,5-dimethylpyrazine; ...., dimethylpyrazine-2,4-dicarboxylate

Compound	Solvent	$\lambda_{max.}$ nm	$\epsilon_{max.}$
	Cyclohexane	258	5620
	or isopentane	313*	800
	* Fine structure 315, 320, 326.		
	2 N $HClO_4$	246	12 300
		283	1 180
	pH 1	268	6 310
	1 N NaOH	244	12 880
OH ⇌ O (N–H)	pH 5	221	8 800
		317	5 520
	pH 12	223	10 230
		342	6 310
	pH −3·1	223	10 230
		342	6 310
OCH_3	Cyclohexane	210	9120
		304	6920
	pH 5·1	292	5730
	pH −2·25	219	9120
		305	6920
O (N–CH_3)	pH 5·2	223	8500
		319	5620
	pH −3·1	226	9770
		344	6920
CH_2OH, CH_3	Water	274	4680
CH_3, CH_2OH	Water	274	6760
CH_3, CH_2OH, H_2N	Water	220	4170
		265	7240
		~300	1100

Figure 9.8 Ultraviolet absorption spectra in ethanol of: ———, pyrimidine; ......, pyrimidine-2-carboxylic acid; —·—·—, pyrimidine-4-carboxylic acid; ----, pyrimidine-5-carboxylic acid

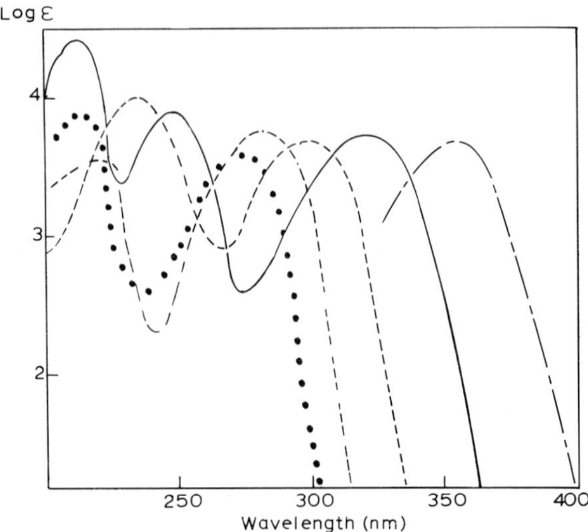

Figure 9.9 Ultraviolet absorption spectra of: ......, 3-methoxypyridine at pH 7; —·—·—, cation of 3-hydroxypyridine at pH 2; -------, anion of 3-hydroxypyridine at pH 13; ——— 3-hydroxypyridine 1-methohydroxide in water, pH 7; — — — —, 3-hydroxypyridine 1-methohydroxide in dioxan

Figure 9.10 Ultraviolet absorption spectra of: — — — — — —, 4-methoxypyridine at pH 9; . . . . . ., the cation, N HCl; ------, the anion, pH 13; ———, the zwitterion or amide form of 4-hydroxypyridine at pH 7

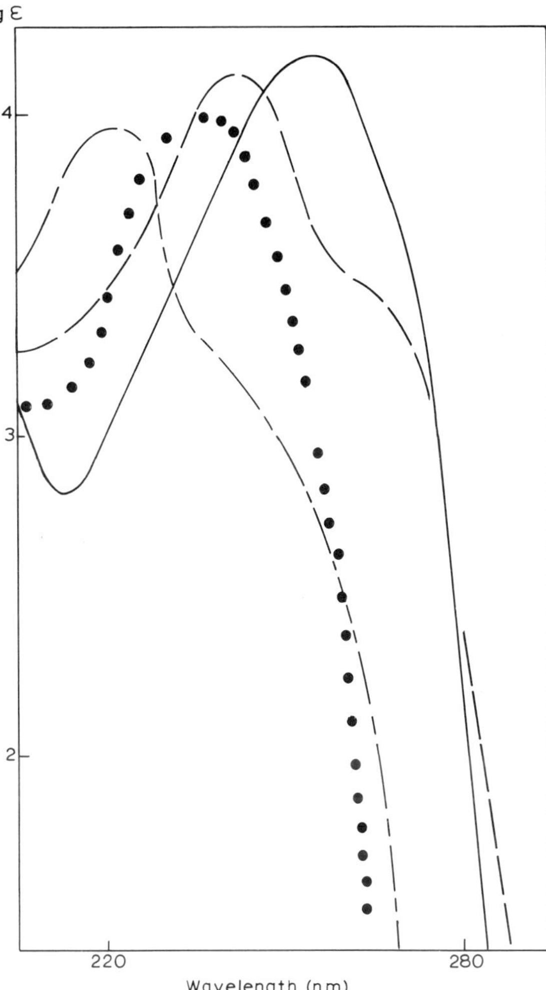

	Solvent	$\lambda_{max.}$ nm	$\epsilon_{max.}$
2-methylthiopyrazine (SCH_3)	pH 3·8	350	8510
		322	6170
pyrazine-2,3-dicarboxylic acid (COOH, COOH)	pH 7	281	2760
		~315	800
	pH 2·2	274	6850
		~310	800
		269	6170
	EtOH	314	700

The 2-aminopyrazine at pH 6·3 shows very similar absorption with peaks at 280 and 318 nm. The methyl amino derivative with $\lambda_{max.}$ at 242 and 332 nm shows a shift towards longer wavelengths which is carried further in the dimethylamino derivative. It seems virtually certain that 2-aminopyrazine does not have the imino structure. Alkyl substituents displace the main band to about 274 nm but the 2-amino-3,5-dialkyl derivative seems anomalous. Introduction of carboxyl groups displaces the main band but leaves the rather weak 315 nm absorption substantially unchanged. The 2,3-dihydroxy-5,6-dicarboxy derivative shows one intense band near 297 nm.

	Solvent	$\lambda_{max.}$ nm	$\epsilon_{max.}$
2-aminopyrazine (NH_2)	pH 0·1	230	11 200
		328	5 890
	pH 1	227	10 720
		326	5 750
	pH 6·3	230	10 000
		285	2 140
		318	4 900
2-methylaminopyrazine ($NHCH_2$)	pH 0·4	237	10 960
		331	5 130
	pH 7·2	242	10 960
		~285	800
		332	4 370

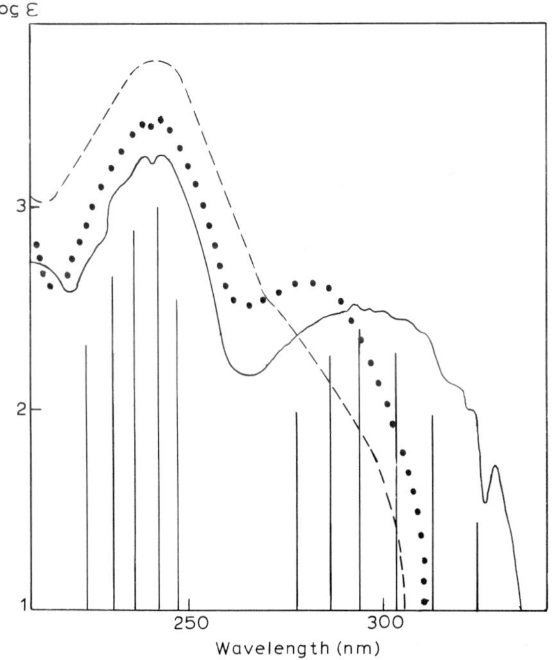

Figure 9.11 Electronic spectrum of pyrimidine; ———, in cyclohexane; . . . . . ., in neutral aqueous solution; ------, in 4N sulphuric acid (spectrum of cation). The vertical lines give one of the main progressions in the spectrum of pyrimidine vapour, the heights of the lines representing relative intensities only

Solvent	$\lambda_{max.}$ nm	$\epsilon_{max.}$
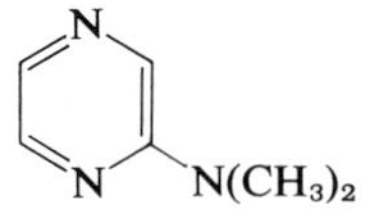 pH 0·3	244	10 000
	352	4 470
pH 7·2	252	13 500
	287	140
	346	4 370
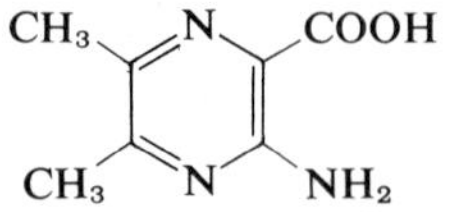 pH 7	248	10 000
	343	7 400
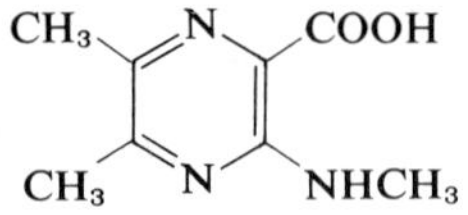	259	13 500
	365	6 760
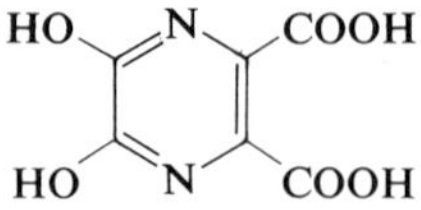 Water	297	16 600

In the 2,3-dimethyl-4-carboxy-2-amino derivative the dominant element is the 2-amino substituent, with the 2-aminomethyl derivative showing a further displacement of both bands and intensification of the 259 nm peak.

The N-heteroaromatic hydroxy-compounds exhibit tautomerism (Mason 1957 *a*, *b*). In the solid state, 2- and 4-hydroxypyridine (Gibson *et al.*, 1955), -quinazoline (Culbertson *et al.*, 1952), -pyrimidine (Brown & Short, 1953; Brown *et al.*, 1955 *a*, *b*) and 2-hydroxyquinoline (Gibson *et al.*, 1955) all exhibit infrared spectra indicating an amide structure. In chloroform solution compounds with a hydroxyl group α or γ to a ring nitrogen display the amide CO frequency (1630–1780 cm^{-1}) and the NH stretching vibration (3350–3500 cm^{-1}). When compounds contain a hydroxyl group that is neither α nor γ to a ring nitrogen, there is a sharp band at 3600 cm^{-1} due to O—H stretching vibration. If the OH is *peri* to the ring nitrogen the hydroxyl absorption (3395–3470 cm^{-1}) points to hydrogen bonding.

Compounds having a quasi *o*-quinonoid structure, e.g. the keto form of 2-hydroxypyridine, show the N—H stretching frequency at 3360–3420 cm^{-1} while the *p*-isomer (quasi-*p*-quinoid) has N—H vibrations 3415–3445 cm^{-1} and their analogues with 5-membered rings, e.g.

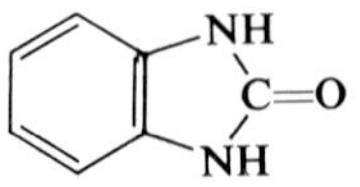

show a peak in the region 3440–3485 cm^{-1}. In general the C=O frequency for the quasi-*o*-quinonoid amides is greater than that for the quasi-*p*-quinonoid isomerides. In certain compounds, e.g. 4-hydroxypyrimidine, the hydroxyl is both α and γ to a ring nitrogen but the structure (*a*) is more likely than structure (*b*),

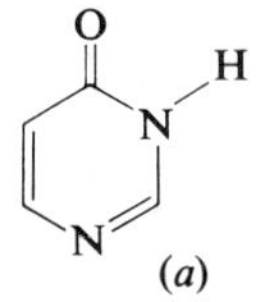

(*a*) 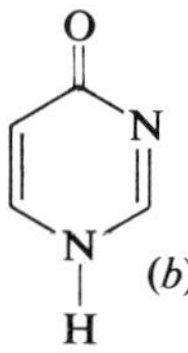 (*b*)

a conclusion supported by the ultraviolet absorption spectra for this compound, 4-hydroxyquinazoline and 4-hydroxypteridine.

Mason (1957 *a*, *b*) further shows that 2,4-dihydroxyquinoline and -isoquinoline in the solid state have a diamide (diketo) structure.

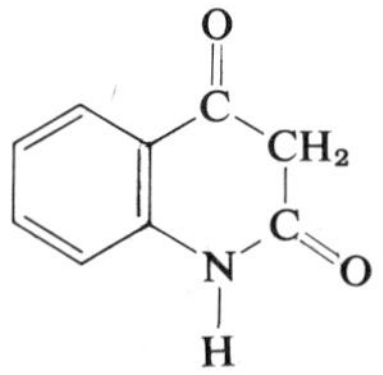

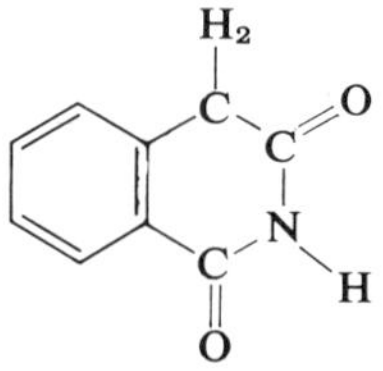

2-Hydroxyquinoline (carbostyril) is a classic example (Hartley & Dobbie investigated it in 1899) of the use of absorption spectra and in determining constitution (cf. Morton & Rogers, 1925; Ault *et al.*, 1935; Mason, 1957, 1960). Comparison of the spectra of the two methyl derivatives (Fig. 9.12) fixed the structure as the quinoline.

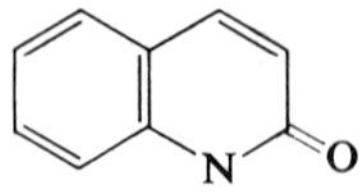

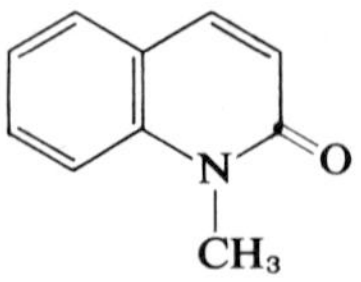

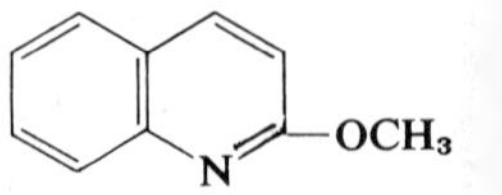

$\lambda_{max.}$ nm	269	327	270·5	328	308·5	322
$\epsilon_{max.}$	7000	6760	6610	6030	3700	4500

Quinolone spectra are also shown when the hydroxy substituent is in position 4

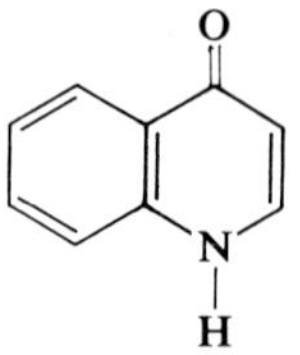

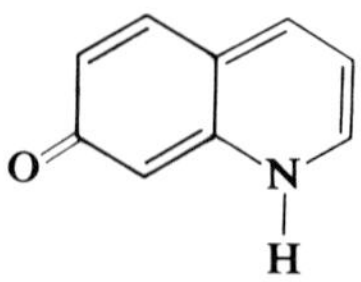

Table 9.6 Infrared maxima of α- and γ-N-heteroaromatic compounds in the N—H, O—H and C=O stretching vibration regions (solvent chloroform)
(s = strong, m = medium, w = weak)

Compound	N—H stretching cm⁻¹ In solution	N—H stretching cm⁻¹ In solid state		C=O stretching cm⁻¹ In solution		C=O stretching cm⁻¹ In solid state	
2-Hydroxypyridine	3398	3198 m	3165 s	1654		1650 s	
2-Hydroxypyrazine	3393	3257 m	3166 s	1730		1710 s	1662 s
3-Hydroxypyridazine	3387	3237 m	3194 s	1681		1678 s	1652 s
2-Hydroxypyrimidine	3394	3195 m		1765		1733 m	1647 s
2-Hydroxyquinoline	3386	3280 w	3252 m	1656		1648 s	
1-Hydroxyquinoline	3411	3287 w	3150 s	1658		1653 s	
2-Hydroxybenziminazole	3460	—	—	1722		1728 s	
8-Hydroxy-7-methylpurine	3444	—	—	1739		1742	
8-Hydroxy-9-methylpurine	3450	—	—	1744		1745	
4-Hydroxypyridine	3442	3200 m	3104 s	1638		1638 s	
4-Hydroxypyridazine	3430	3288 m	3078 m	1662		1660 m	1640 s
4-Hydroxyquinoline	3438	3226 m	3140 m	1645		1638 s	
4-Hydroxycinnoline	3422	3222 m	3190 s	1638		1610 s	
4-Hydroxypyrimidine	3390	3257	3200 s	1721		1716 m	1684 s
2,4-Dihydroxy-1-methylpyrimidine	3395	—	—	1712	1689	—	
2,4-Dihydroxy-3-methylpyrimidine	3422	—	—	1717	1661	—	
6-Hydroxy-7-methylpurine	3390	—	—	1702		1697	
6-Hydroxy-9-methylpurine	3388	—	—	1711		1679	

Table 9.7 Enolic N-heterocyclic hydroxy-compounds (solvent carbon tetrachloride)
(s = strong, m = medium, w = weak, b = broad)

Compound	O—H stretching vibrations (cm⁻¹) In solution	O—H stretching vibrations (cm⁻¹) In solid state	Double bond stretching vibrations (cm⁻¹) In solid state	
3-Hydroxypyridine	3595	2925–2500 m + b	1573 s	1476 s
3-Hydroxyquinoline	3591	2942–2525 m + b	1598 s	1469 s
5-Hydroxyquinoline	3599	2945–2530 m + b	1616 m	1580 s
6-Hydroxyquinoline	3601	2960–2500 m + b	1633 w	1577 s
7-Hydroxyquinoline	3597	2960–2525 m + b	1615 s	1533 s
8-Hydroxyquinoline	3412 b	3180 s + vb	1577 s	1507 s
4-Hydroxyisoquinoline	3603	2900–2500 m + b	1626 m	1582 s
8-Hydroxycinnoline	3440 b	3100 s + vb	1624 m	1582 m
6-Hydroxyquinazoline	3603	3125 s 2933–2600 m + b	1640	1582
8-Hydroxyquinazoline	3440 b	3145 s + vb	1624 m	1592

Table 9.8 Infrared maxima (in solid state) of some N-heteroaromatic hydroxy-compounds insoluble in chloroform in the N—H, O—H and C=C stretching vibration regions
(s = strong, m = medium, w = weak, b = broad)

Compound	N—H or OH stretching vibrations (cm⁻¹)	Double bond stretching (cm⁻¹)	
2,4-dihydroxypyridine	3230 w 2915–2500 m + b	1661 s	1635
5-Hydroxypyrimidine	2922–2532 m + b	1608 s	1568 s
2,4-Dihydroxyquinoline	3250 w 2908–2525 m + b	1662 s	1638 s
4,8-Dihydroxyquinoline	2921–2535 m + b	1621 m	1590 s
6-Hydroxycinnoline	3145 w 2875–2600 m + b	1640 m	1500 s
7-Hydroxycinnoline	2915–2620 m + b	1622 s	1565 s
2-Hydroxyquinazoline	3313 w 3194 s	1680 s	1608 s

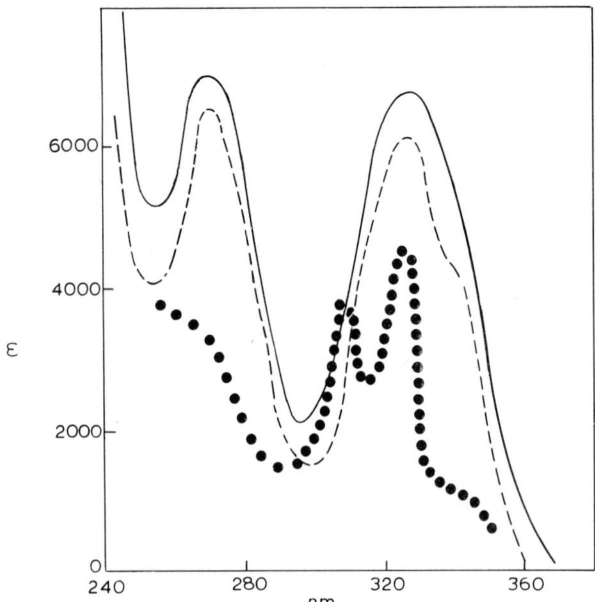

Figure 9.12 Absorption spectra of: ———, carbostyril; ------, N-ether; . . . . . ., O-ether (from Morton & Rogers, 1925)

The selective absorption of 1-methyl-4-quinolone (a compound of fixed structure) agrees very closely with that of the unsubstituted material, whereas the spectrum of 4-methoxyquinoline (Fig. 9.13) is quite different (Hearn *et al.*, 1951). On the other hand in 5-, 6- and 8-hydroxyquinolines the spectra of the neutral solutions undergo displacement in 0·1 N acid or alkali, as in simple phenols, while the absorption of 8-methoxyquinoline is very similar to that of the parent substance except for the absence of a bathochromic shift in alkali. This indicates that there is no quinolone contribution in this case.

The isoquinoline compound 1-isoquinolone (1-hydroxyisoquinoline) is also ketonic. Ewing & Steck (1946) argued that quinoline and isoquinoline compounds with hydroxyl introduced in the α or γ positions in the pyridine ring (relative to nitrogen) are ketonic in structure. The work of Hearn *et al.* supports this view with further evidence.

The spectra of aminoquinolines show that the 5- and 6-amino compounds with $\lambda_{max.}$ 355 nm, $\epsilon_{max.}$ *ca.*

Table 9.9 Quinoline derivatives

Compound	$\lambda_{max.}$ nm	$\epsilon_{max.}$	0·1 N acid $\lambda_{max.}$ nm	0·1 N acid $\epsilon_{max.}$	0·01 N alkali $\lambda_{max.}$ nm	0·01 N alkali $\epsilon_{max.}$
3-Hydroxy	234	25 000	240	25 000	235	19 000
	336	5 000	345	5 000	350	5 000
4-Hydroxy	243	19 500	252	19 500	254	15 900
	328	1 590	370	1 260	365	2 000
6-Hydroxy	233	15 850	247	15 000	245	15 900
	276	1 585	315	2 500	360	1 950
	333	1 950	344	2 000	—	—
7-Hydroxy	262	25 000	236	25 000	244	31 600
	325	1 590	274	1 590	360	3 160
	342	1 590	360	1 590	—	—
8-Hydroxy	242	25 100	252	25 100	254	15 900
	310	1 590	319	1 260	354	2 500
	—	—	360	1 260	—	—
Carbostyril	229	15 900	226	15 900	232	25 120
(2-quinolone)	269	3 980	271	3 980	—	—
	328	3 160	374	3 160	330	2 500
4-quinolone	233	12 600	227	19 500	234	12 600
	317	10 000	314	3 980	316	6 310
	331	12 600	—	—	—	—
2-Methoxy	210	31 600	—	—	—	—
	308	4 000	—	—	—	—
	322	5 000	—	—	—	—
4-Methoxy	225	63 000	301	4 000	287	4 000
	283	7 950	—	—	—	—

There are discrepancies concerning some of these compounds in the literature

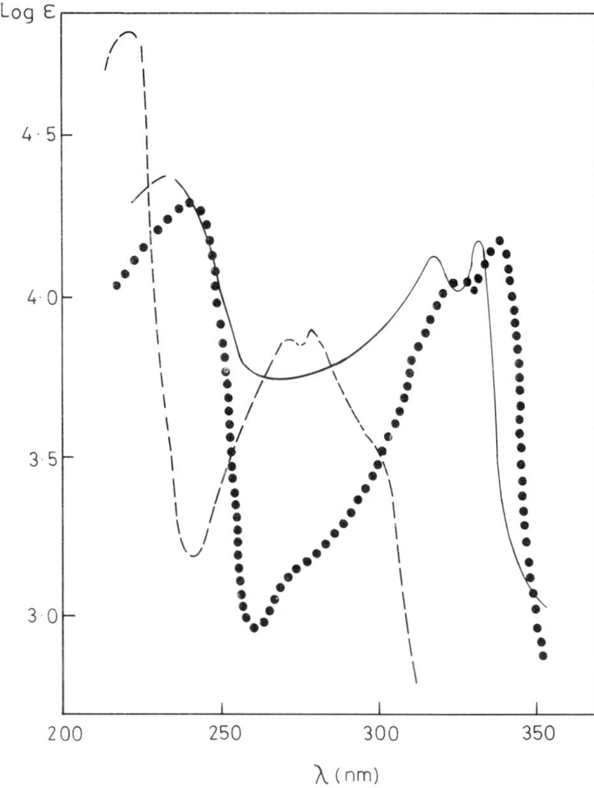

Figure 9.13 Absorption spectra of: ———, 4-hydroxyquinoline; ------, 4-methoxyquinoline; ......, 1-methyl-4-quinolone (from Hearn *et al.*, 1951)

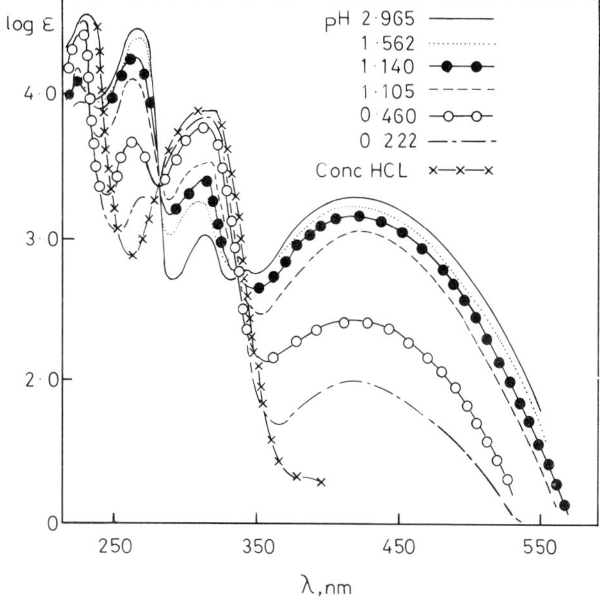

Figure 9.14 (*Above*) Absorption spectra of 5-aminoquinoline in aqueous hydrochloric acid. (*Below*) Absorption spectra of 6-aminoquinoline in aqueous hydrochloric acid

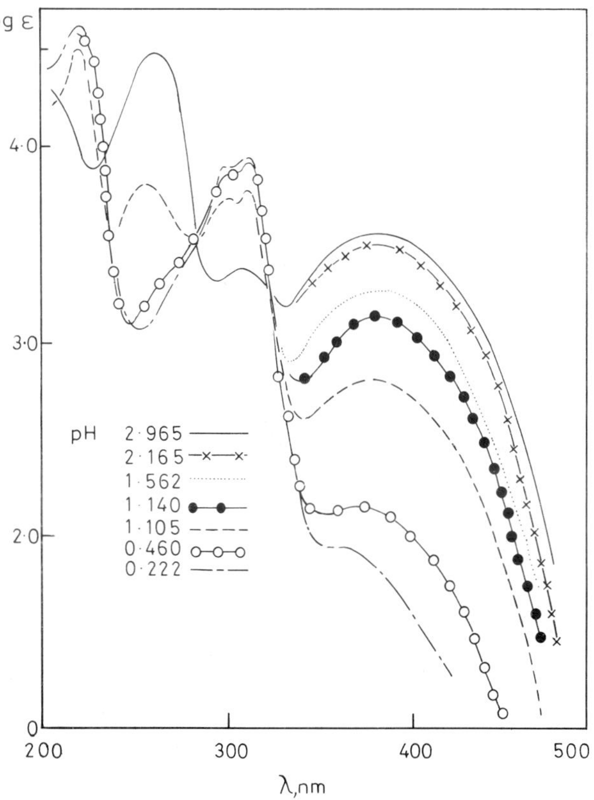

3160 differ considerably from those of 4-aminoquinolines (Fig. 9.14). In fact 4-aminoquinoline resembles 4-hydroxyquinoline, which is mainly quinolone and it seems highly probable that the former is amino in structure

NH_2 ⇌ NH

Hearn *et al.* also studied several cinnoline and quinazoline compounds, some of the spectra being shown in Figs. 9.15–9.18.

Diazanaphthalenes show relatively small effects due to replacing CH by N.

Cinnoline — Phthalazine — Quinazoline

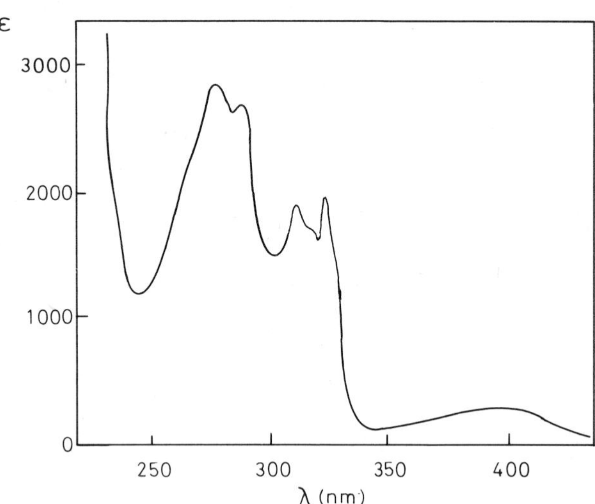

Figure 9.15 Absorption spectrum of cinnoline in cyclohexane (from Hearn *et al.*, 1951)

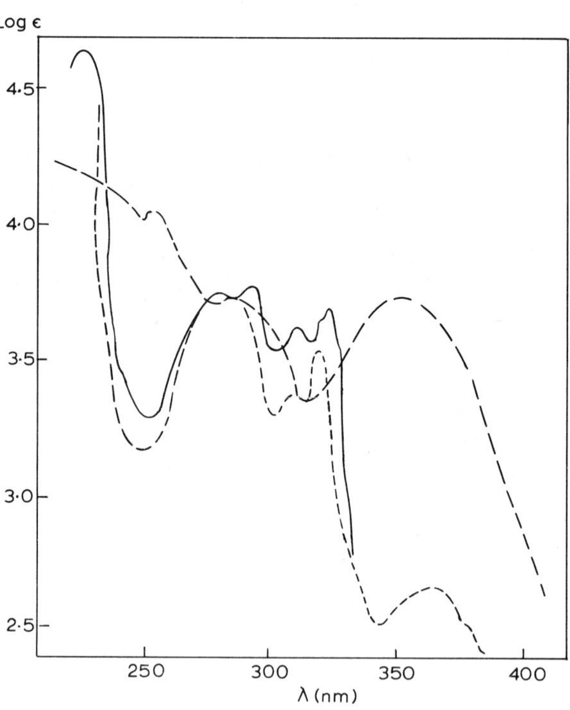

Figure 9.16 Absorption spectra of: ———, 4-phenoxycinnoline; ------, 4-ethoxycinnoline; — — —. 6-nitro-4-phenoxycinnoline (from Hearn *et al.*, 1951)

Table 9.9a Cinnoline derivatives

Compound	$\lambda_{max.}$ nm	$\epsilon_{max.}$	
4-methoxy-	224	37 150	
(in ethanol)	291	5 500	
	313	4 470	
4-Ethoxy-	280	6 100	
(in cyclohexane)	290·5	6 100	
	310	2 700	
	318	2 630	
	321·5	3 713	
	362·5	470	(also very weak bands 377–399 nm)
4-Phenoxy-	225	12 000	
(in cyclohexane)	282·5	6 300	
	292·5	6 700	
	310·5	4 370	
	323	4 800	
1-Methyl-4-cinnolone	251·5	8 910	
	352	12 900	
	369	12 900	
4-Hydroxycinnoline	237	11 200	
	255	8 910	
	262	6 760	
	284·5	2 750	
	296	2 700	
	338	12 800	
	343	12 600	
	352	12 300	
4-Hydroxy-3-carboxylic acid	263	7 400	(deep minimum 285 nm)
	339	12 600	
	~340	11 000	
	~354	9 300	
4-Amino-	240	12 600	(deep minimum near 275 nm)
	345	11 500	

Cinnoline itself shows a spectrum with moderately intense bands at 275·5 and 286 nm, $\epsilon_{max.}$ *ca.* 2800 and narrow bands at 308·5, 317 and 322·5 nm, with $\epsilon_{max.}$ *ca.* 2000, are considerably more intense than the corresponding bands in naphthalene. There is also a weak band at 390 nm (Fig. 9.15). The weak absorption at 390 nm in cinnoline is probably due to a contribution from the —N=N— group as a chromophore within the larger unit, 4-ethoxycinnoline in cyclohexane.

Phthalazine (Amstutz, 1952) has three bands grouped together: 252, 259 and 267 nm, $\epsilon_{max.}$ 4270, 4900 and 3900 respectively, with weak bands at 290 and 296·5 nm, $\epsilon_{max.}$ 13 and 11·5 respectively.

4-Hydroxyquinazoline might have any of the three structures (1), (2) and (3). The three methyl derivatives of fixed structure are shown as (1*a*), (2*a*) and (3*a*). In 1*a* the long wave absorption (beyond 300 nm) is significantly more intense than that of the parent substance so that (1) cannot be an important contributor. The resemblance between the absorption spectrum of (3*a*) and that of the parent substance is much closer and (3) must be the predominant form. By solving simultaneous equations, Hearn *et al.* (1951) argue that 4-hydroxyquinazoline is made up of 70± 2·5% of (3), 10±% of (2) and 20 ±% of (1).

In the case of 4-hydroxy-6-nitroquinazoline the evidence indicates the structure (4) but 4-aminoquinazoline has a quinazoline rather than a quinizolone structure.

(1) (2) (3)

(1*a*) (2*a*) (3*a*)

(4)

(4*a*)

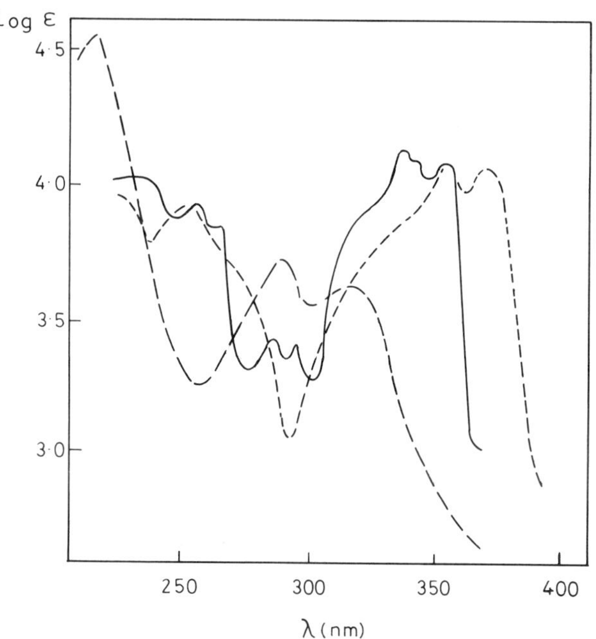

Figure 9.17 Absorption spectra of: ———, 4-hydroxycinnoline; — — —, 4-methoxycinnoline; ------, 1-methyl-4-cinnolone (from Hearn *et al.*, 1951)

Figure 9.18 Absorption spectra of: ———, 4-hydroxyquinazoline; —·—·—, 1-methyl-4-quinazolone; ------, 3-methyl-4-quinazolone; — —, 4-methoxyquinazoline (from Hearn *et al.*, 1951)

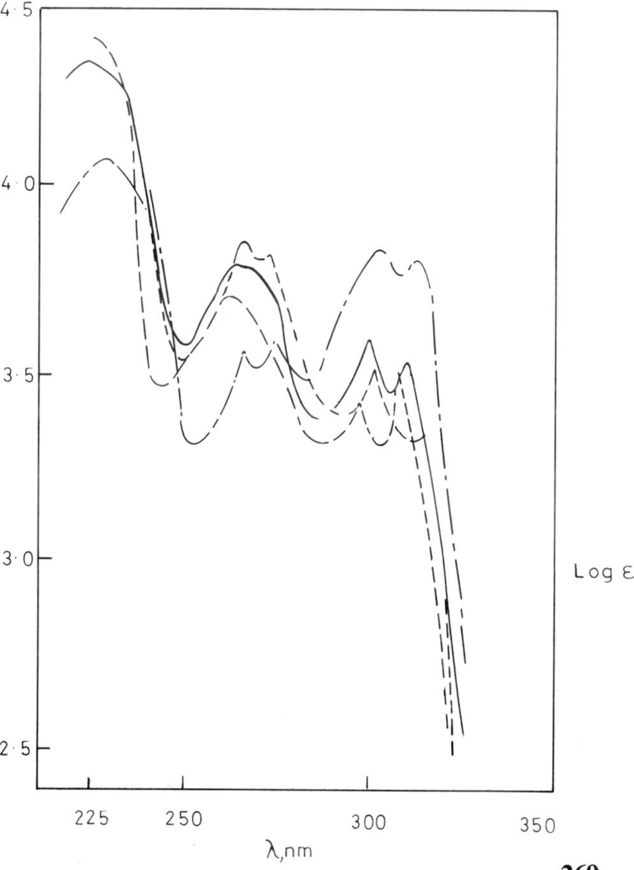

In the cinnoline series the 4-methoxy, ethoxy and phenoxy derivatives (7) all show characteristic selective absorption on either side of 300 nm with resolutions (in cyclohexane) and in the case of the 4-ethoxy compounds the very weak azo (—N=N—) absorption is seen between 377 and 399 nm. 1-Methyl-4-cinnolone (6) has two peaks (352 and 369 nm inflection *ca.* 330 nm) and a deep minimum (ϵ *ca.* 1000) near 293 nm. It has, moreover, a maximum near 250 nm where the alkoxy and phenoxy cinnolines have a marked minimum. The compound 4-hydroxycinnoline clearly resembles the cinnolone but some of the cinnoline is also present. There is evidently a tautomeric equilibrium with the cinnoline as the minor component. The carboxylic acid (9) is unequivocally a cinnolone and the 4-amino derivative has the imino cinnolone structure (8). The 6-nitro-4-hydroxy compound is also a cinnolone (10).

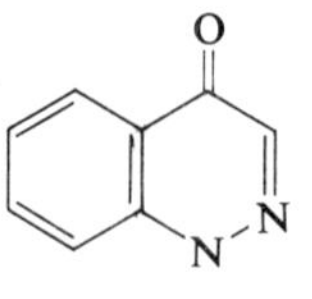

(5) 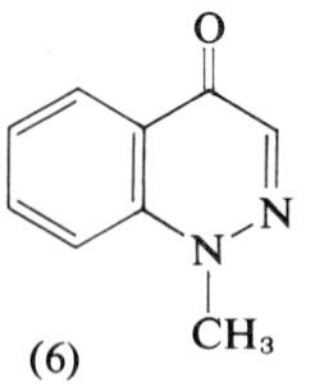 (6) 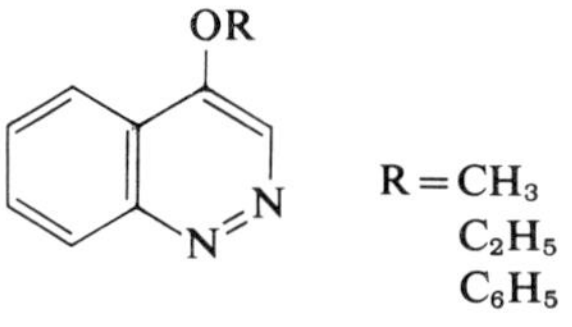 (7) R = CH_3, C_2H_5, C_6H_5

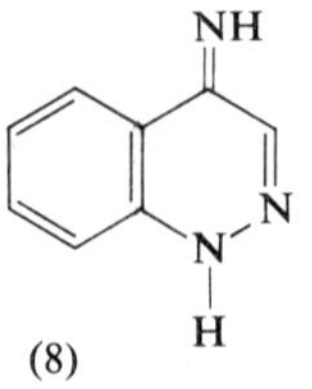

(8) 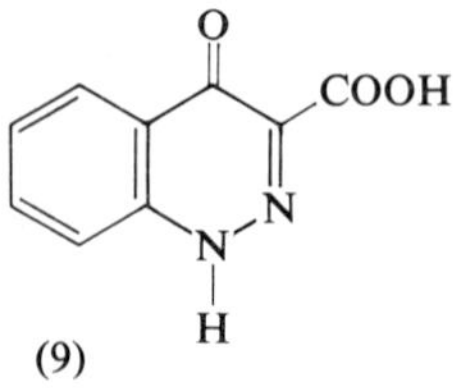 (9) 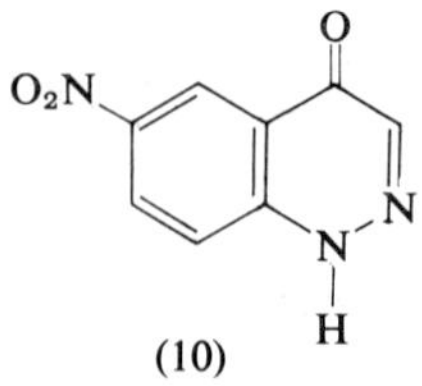 (10)

Quinazoline as a neutral molecule shows an ultraviolet absorption spectrum resembling that of the other diazanaphthalenes, but acidification effects an unusual blue-shift of the long wave band. Albert (1955) attributed this to a cation derived from a covalently hydrated form of quinazoline. Albert *et al.* (1961 *a*) noted that quinazoline can form either a normal or an abnormal mono-cation, the latter in an aqueous medium. In fact two different hydrochlorides could be prepared as solids; one, $C_8H_7ClN_2$, was hygroscopic and formed a hydrate stable at room temperature. The infrared spectrum of the hydrate showed extra peaks at 1474 and 1240 cm^{-1} consistent with covalently bound water. In an acid medium of low water content, such as dichloroacetic acid, quinazoline displayed a normal mono-cation absorption spectrum (with $\lambda_{max.}$ 330 nm) very like that of 4-methylquinazoline. This peak was absent from aqueous dilute acid solutions of quinazoline. In different mixtures of sulphuric acid and water evidence was obtained for three quinazoline molecular species. Albert *et al.* (1961 *b*) dissolved quinazoline in dilute aqueous acid and added excess of alkali, following the changes by rapid-flow absorption spectroscopy. Covalently hydrated quinazoline was formed as a transient species with a half-life of 9 seconds at pH 10 and subsequently yielded quinazoline itself. When alkali was added to 3-methyl quinazolinium iodide the hydrated form was stable:

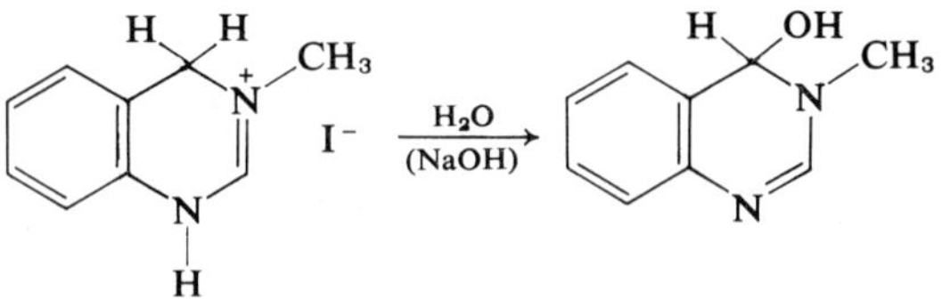

I^- $\xrightarrow[(NaOH)]{H_2O}$

4-Substituted quinazolines tend to form normal mono-cations even in dilute aqueous solution; neither 4-methyl nor 2,4-dimethylquinazolines shows any real departure from the pH 7 spectrum on acidification except that a dication is formed in sulphuric acid.

Dihydroquinazoline has the 3,4-dihydro structure but the effects of changes in pH are not fully understood. At pH 7 there is selective absorption near 210–215 nm, ϵ *ca.* 20 000 and at 276 nm, ϵ *ca.* 5500, while at pH 12·3 the latter band moves to 290 nm, ϵ 6460 (Armarego, 1961).

Chatterjee *et al.* (1954), Chakravarti *et al.* (1962) and Bhattacharyya *et al.* (1963) studied the alkaloids of *Glycosmis arborea* Correa, an Indian roadside shrub

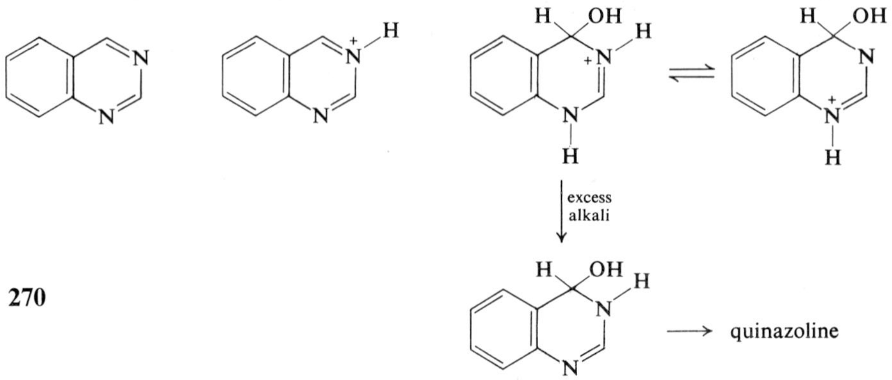

excess alkali

→ quinazoline

half life 9 sec.
pH 10

used in the Ayurvedic system of medicine as a febrifuge and anthelmintic. The species, also known as *Glycosmia pentaphylla* is referred to in India as the tooth brush plant. Chatterjee & Mazumdar (1954) isolated *inter alia* a product which they called glycosine with the structure 2-benzylidene-1-methyl-4-quinazoline (with $\lambda_{max.}$ nm 231, 268, 277, 306, $\epsilon \times 10^{-3}$ 21, 4·5, 5·0, 8·5). Chakravarti *et al.* (1962) recognized that their product arborine and glycosine were identical. On the basis of a detailed study of ultraviolet, infrared and n.m.r. spectra, they concluded that the tautomeric 2-benzyl-1-methyl-4-quinazolone structure was the stable form in the solid state or in solution over a wide pH range. They obtained glycosmicine and glycorine, 2- and 4-(1-methyl) quinazolones, with glycosminine and arborine. Metha *et al.* (1963) isolated vasicine (a broncho-dilator) and vasicinone from *Adhatoda vasica* Nees. The findings most relevant here are summarized below.

As a result of war-time studies concerned with antimalarials from plant sources, a Chinese remedy

Glycosmicine

EtOH		0·01 N NaOH	
$\lambda_{max.}$ nm	$\epsilon_{max.} \times 10^{-3}$	$\lambda_{max.}$ nm	$\epsilon_{max.} \times 10^{-3}$
219	50·1	223	50·1
244	9·33	—	—
311	1·29	313	5·6

cm^{-1} 1701, 1661, 1605, 1484, 1315

Glycorine

EtOH		0·01 N HCl	
$\lambda_{max.}$ nm	$\epsilon_{max.} \times 10^{-3}$	$\lambda_{max.}$ nm	$\epsilon_{max.} \times 10^{-3}$
269	38·9	—	—
278	4·47	282	4·79
306	8·13	295	5·37
317	6·92	304	4·9

cm^{-1} (KBr) hydrochloride 3320, 3113, 3067, 3022, 1392 (NH_4 salt) 1704, 1652, 1602, 1538

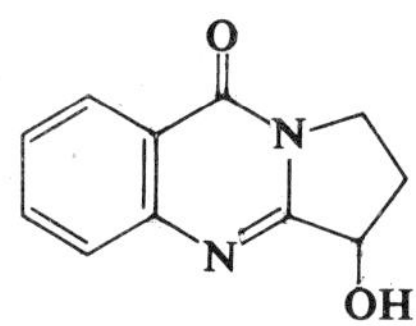

Vasicinone

H_2O
$\lambda_{max.}$ nm
227
272
302
315

cm^{-1} 3115, 2934, 1658, 1455, 1385, 774, 749, 720

Glycosminine

EtOH		
$\lambda_{max.}$ nm	$\epsilon_{max.} \times 10^{-3}$	cm^{-1} (KBr)
225	27·5	3356 (OH or NH)
265	8·91	1676, 1613 C=O
303	4·57	770 O-disubstituted benzene
312	3·71	748, 713 monosubstituted benzene ring

Vasicine

made from a saxifrage *Dichora febrifuga* was studied by Koepli *et al.* (1949). A substance $C_{16}H_{19}O_3N_3$ was isolated in two forms designated febrifugine and isofebrifugine. Both contained a 3 substituted 4-4-quinazolone moiety. The absorption spectra of 2-methyl-4-quinazolone, 3-allyl-4-quinazolone and of febrifugine were compared. There was a great similarity, especially between febrifugine and the 3-substituted 4-quinazolone. Indeed, the intensities of absorption measured in

Febrifugine $C_{16}H_{19}O_3$

In 0·1 N NaOH

$\lambda_{max.}$ nm	265, 275, 302, 313	
$\epsilon_{max.}$	8240	4640

In 0·1 N HCl

$\lambda_{max.}$ nm 272·5, ~295

terms of $E^{1\%}_{1cm}$ values were approximately consistent with a non-absorbing $C_8H_{14}NO_2$ moiety. The investigations of Chatterjee & Majumdar (1954), Chakravarti *et al.* (1962), Pakrashi *et al.* (1963) and Metha *et al.* (1963) afford very good examples of ultraviolet and infrared spectroscopy together with nuclear magnetic resonance spectroscopy applied to the characterization of natural products.

A group led by Williams (Ablondi *et al.*, 1952) obtained from hydrangea leaves an alkaloid ($C_{16}H_{19}N_3O_3$) which turned out to be identical with febrifugine and had the structure shown in the diagram. From the results of Grammaticakis (1961) 2-methyl-3-hydroxy (or amino) 4-quinazolinones show peaks at 269, 302 and 313 nm with ϵ values of 7940, 3980 and 3980 which agree quite well with those of Ablondi *et al.* (1952) for febrifugine.

The alkaloid is very active in relieving avian malaria

Table 9.10 Absorption maxima of quinazoline and some related substances

Compound	$\lambda_{max.}$ nm	$\epsilon_{max.} \times 10^{-3}$	$\lambda_{max.}$ nm	$\epsilon_{max.} \times 10^{-3}$	$\lambda_{max.}$ nm	$\epsilon_{max.} \times 10^{-3}$
Quinazoline	224	37·1	208	15·85	297	2·75
	271	2·5	260	8·13	309	2·29
	305	2·4	—	—	333	1·26
	(pH 7–9·8)		(pH 1)		(pH 0·9)	
2-Methylquinazoline	223	39·8	207	19·05	—	—
	268	2·57	258	8·91	—	—
	310	2·51	—	—	—	—
	320	2·04	—	—	—	—
	(pH 7)		(pH 1)			
4-Methylquinazoline	223	41·7	234	33·1	298	2·57
	270	2·82	270	2·95	328	2·85
	305	2·82	279	2·82	(in dichloracetic acid)	
	314	2·57	323	2·19	253	31·6
	(pH 7)		(pH 0·3)		305	3·55
					353	1·74
					(di-cation in strong acid)	
3,4-Dihydroquinazoline	212	18·2	224	12·3		
	217	17·8	290	5·75		
	280	4·9	—	—		
	(pH 7)		(pH 12)			
2-Methyl-4-quinazolinone	264	7·94				
	303	5·0				
	314	5·0				
	(EtOH)					
3-Methyl-4-quinazolone	230	22·4				
	267	6·76				
	276	6·61				
	301·5	3·63				
	313·5	2·88				
1-Methyl-4-quinazolone	230	11·5				
	260	3·98				
	277·5	4·27				
	306·5	6·92				
	317	6·17				
4-Methoxyquinazoline	225	26·3				
	261	5·12				
	297·5	3·02				
	300	3·47				
4-Phenoxyquinazoline	263·5	6·17				
	298	3·72				
	310	4·17				

CH_3 / NH_2

234	8·51
285	2·09
(in *n*-hexane)	

CH_2OH / NH_2

240	6·31
290	2·29

but has the disadvantage of being also a powerful emetic. A series of 16 papers by Williams and his colleagues has provided a great deal of information in the quinazoline field and this work illustrates how a molecular moiety showing no ultraviolet absorption may be quite difficult to characterize unequivocally.

9.2 Nucleic Acids, Nucleotides, Nucleosides and Pyrimidine and Purine Bases

The importance to biochemistry of nucleoproteins, nucleic acids and their derivatives can now scarcely be exaggerated. In extreme examples the proportion of protein in a nucleoprotein may be quite low, as in the heads of spermatozoa which contains some 30% of protamine and 60% of nucleic acid. A little free nucleic acid sometimes occurs, as in yeast. Many nucleoproteins, e.g. those from pancreas or thymus, contain more protein than nucleic acid. Very gentle treatment ruptures the protein–nucleic acid linkages.

Alkaline hydrolysis in the cold breaks nucleic acid down to nucleotides; further hydrolysis (e.g. with dilute ammonia) splits off phosphoric acid, yielding nucleosides. Further hydrolysis cleaves the nucleosides to give pentose sugars and purine or pyrimidine bases.

When yeast is extracted with alkali and the extract is acidified, nucleic acid is obtained. The following diagram illustrates the degradation:

The pyrimidines are joined to the carbon 1 of the pentose by a glycosidic link at position 3, while the purine–pentose linkage involves carbon 1 of the sugar and position 9 of adenine or guanine. Adenosine is thus adenine-9-D-riboside.

The nucleic acid from thymus gland yields on hydrolysis thymine instead of uracil together with adenine, guanine and cytosine and small amounts of 5-methylcytosine. The sugar is 2-deoxy-ribose. The nucleotide units from ribonucleic acid may be 2′-, 3′- or 5′- phosphates. Enzymatic hydrolysis of calf liver nucleic acid gives 5′-phosphates. The two main types of nucleic acid are written RNA (ribose nucleic acid) and DNA (deoxy-ribonucleic acid).

Neither the simple sugar residues nor the phosphate moieties exhibit selective absorption in the ultraviolet. An early example of the utility of spectrophotometry was provided by Gulland *et al.* (1934), who determined

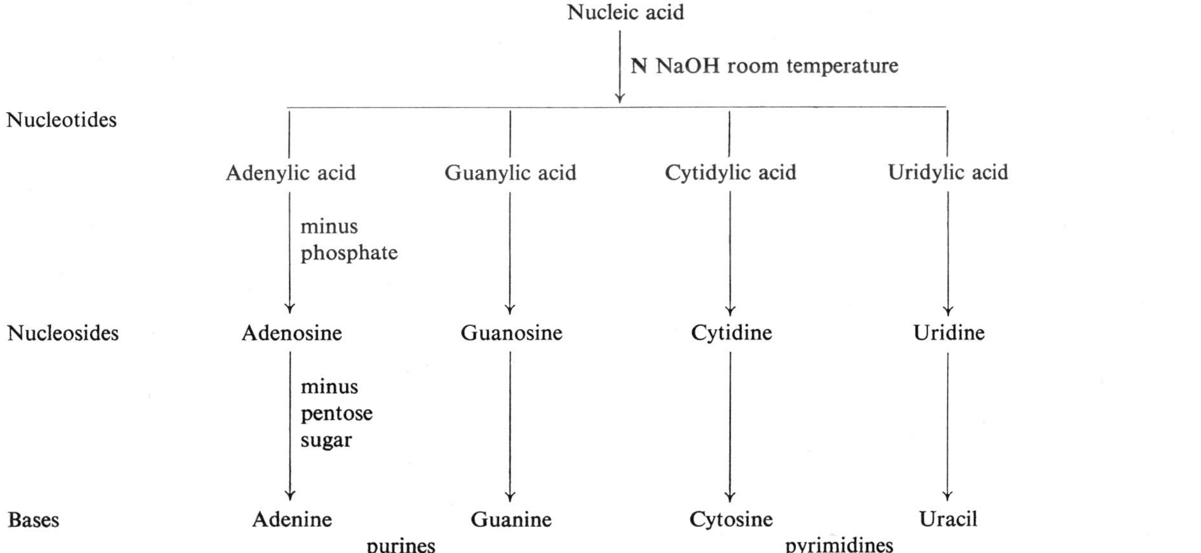

Table 9.11 Spectra of xanthosine and methylxanthosine

	pH 5 (HCl)		pH 10 (NaOH)	
	$\lambda_{max.}$ nm	$\epsilon_{max.} \times 10^{-4}$	$\lambda_{max.}$ nm	$\epsilon_{max.} \times 10^{-4}$
Xanthosine	238	0·78	247	0·86
	264	0·84	278	0·74
1-Methylxanthosine	266	1·02	241	0·85
	—	—	276	0·85
3-Methylxanthosine	271	1·0	273	1·2
7-Methylxanthosine	269	0·96	290	0·97
9-Methylxanthosine	235	0·73	247	0·93
	264	0·93	278	0·93

the spectra of 4-methylxanthines and compared them with the spectrum of xanthosine. The results very clearly favoured the base-sugar linkage at position 9.

The effects of increasing molecular complexity are illustrated in Table 9.12. (see also Ulbricht, 1962).

Table 9.12 clearly supports the view that the sugar and phosphate moieties act primarily as spectroscopic diluents since molecular extinction coefficients are reasonably constant. The anomalous behaviour of uracil in alkali will be referred to later.

The nucleoproteins exhibit the absorption due to the aromatic acids in the protein plus that of the nucleic acids. A fact for the analyst to note is that the intensity of absorption shown by nucleic acids is only about 65% of that to be expected from a summation of the absorptions due to the nucleotides present. This hypochromicity is in part due to intramolecular effects such as hydrogen bonding in a double helix.

The phenomenon in oligo- and polynucleotides, though not fully understood, has been put to good use. When the intensity of ultraviolet absorption of an oligonucleotide is lower than the sum of that of its constituent mononucleotides it displays hypochromicity; thus, if the absorption is 75% of that of the nucleotides, the hypochromicity is 25%. Alternatively the term hyperchromicity is used for the rise in intensity when oligonucleotides are compared with the mononucleotides. Here the rise is 25 in 75 or 33%.

The topic has a considerable literature (Hotchkiss, 1957; Michelson, 1958; Warner, 1957; Davis & Rich, 1958; Felsenfeld & Rich, 1957; Shugar, 1960). Michelson showed that compared with the mononucleotide, di-, tri-, tetra-, penta- and poly(13·6)adenylic acids showed hypochromicities rising from 15% to 36·7% in 0·2 N NaOH. In 0·1 N HCl the rise was slower but reached 22·9% for the polyadenylic acid. Polyguanylic acid displayed hypochromicity only in an acid medium. Hydrogen bonding could not be the sole cause. Warner showed that poly-I (polyinosinic acid) exhibited the effect very clearly, the polymer having absorption only

Table 9.12 Spectra of adenine, uracil and derivatives

	Acid solutions		Alkaline solution	
	$\lambda_{max.}$ nm	$\epsilon_{max.} \times 10^{-3}$	$\lambda_{max.}$ nm	$\epsilon_{max.} \times 10^{-3}$
Adenine	263	13·1	269	12·3
Adenosine	257	14·6	260	14·9
Adeninedeoxyriboside	258	14·1	—	—
Adenylic acid (2′)	257	14·4	259	15·4
Adenylic acid (3′)				
Adenylic acid (5′)	257	15·1	259	15·4
Adenosinetriphosphate	257	14·7	259	15·4
Uracil	259	8·2	284	6·15
Uridine	262	10·1	262	8·5
Uracildeoxyriboside	262	10·2	267	7·63
Uridylic acid (2′)				
Uridylic acid (3′)	262	9·9	261	7·3
Uridylic acid (5′)	261	9·7	—	—

half that of the mononucleotide. Rich and his colleagues showed that poly-A and poly-U each exhibited hypochromicity separately and that when they were mixed, there was interaction accompanied by a further substantial decrease in absorption.

Hotchkiss (1957) found that the absorption shown by DNA rose under the action of DNA-ase and that of RNA under the action of RNA-ase. This led to a method of determining DNA and RNA in mixtures. The absorption was measured first in neutral solution. RNA-ase was added first and after the reaction had been completed the ultraviolet absorption was remeasured (using appropriate blanks). Experience had shown that the increment was about one-third of the original absorption attributable to native RNA. Hence the increment multiplied by 3 gave the initial RNA absorption. DNA-ase (Mg^{++} needed) was then added to the solution and to the blank and the further increment measured after an appropriate time (the reaction is at first quite slow). The increment × 3 gives the initial absorption attributable to DNA. The sum of the RNA and DNA values is subtracted from the initial reading at 260 nm and the difference may arise from degradation products or irrelevant absorption. An optical density of 1·0 at 260 nm with a 1 cm light path corresponds to about 45 μg RNA or DNA per ml ($E^{1\%}_{1cm}$ 260 nm = 222).

The importance of RNA and DNA has immensely accelerated research in the broad nucleic acid field and the discovery of transfer RNA in 1958 has been followed by recognition of many and varied minor components (Goldwasser & Heinrikson, 1966; R. H. Hall, 1971).

Uridin-5-Oxyacetic acid

	$\lambda_{max.}$ nm	A*	$\lambda_{min.}$ nm	A*
pH 7	277·5	100	243	37·3
pH 2	273·4	103·4	241	34·5
pH 12	274·2	71	252	45·7

* Approximately relative absorbances.

Nakanishi *et al.* (1970) and Thiebe *et al.* (1971) isolated and characterized fully a base Y found in the phenyalanine specific to the RNA of yeast or wheat germ.

Base Y **Base Yt**

	$\lambda_{max.}$ nm	A*	$\lambda_{min.}$ nm	A*
pH 1–7	241	100	260	52
	281	74	—	—
pH 11	272	74	247	50
	295	59	—	—

* Approximately relative absorbances.

Now that primary sequences of tRNAs are ascertainable it is clear that the minor components occur at particular places in a clover-leaf structure (cf. Zachau, 1969). In fact the minor components fall into three categories depending on their locations: (i) those in the first position of the anticodon, (ii) those next to the 3′-OH end of the anticodon and (iii) minor components occurring elsewhere in the tRNA molecule. Nishimura (1972) describes the methods of detection using hydrolysis (RNAase T_2) followed by thin layer chromatography of the nucleotides. The spots are identified by ultraviolet illumination (Hg 254 or 365 nm) and cut out. The nucleotide is extracted with water and its ultraviolet absorption spectrum is measured. The various nucleotides are converted to the related nucleosides and the absorption spectra are redetermined. A large number of minor components have been isolated from tRNA; they include methylated purine nucleosides, 2′ O-methyl derivatives and deaminated derivatives; in addition there are isopentenyladenosine and thioadenosines and pyrimidine methylated nucleosides.

Hall *et al.* (1966) studied N-(γγ-dimethylallyl)adenosine obtained from soluble RNA of baker's yeast. The compound displays cytokinin activity (p. 299). Quantitatively it is a minor constituent since it accounts for only 0·1 mole % of the nucleotides. At pH 1 it has $\lambda_{max.}$ 265 nm, $\epsilon_{max.}$ 20 300 and at pH 7–12 $\lambda_{max.}$ is

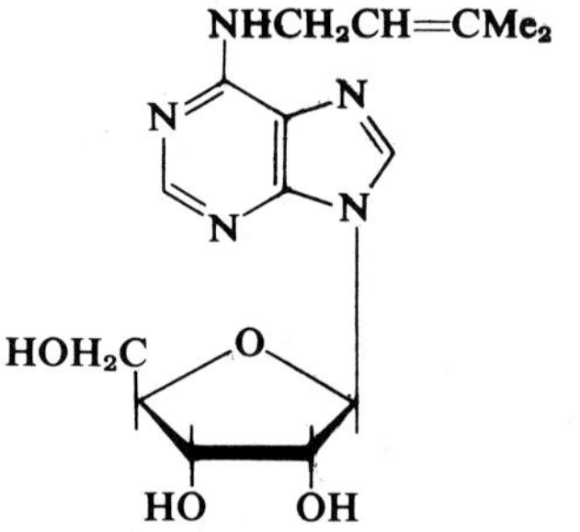

NHCH$_2$CH=CMe$_2$ [structure: N^6-isopentenyladenosine; HOH$_2$C, HO, OH]

at 269 nm, $\epsilon_{max.}$ 19 900. The n.m.r. spectrum shows τ 8·25, (6 protons, 2 vinyl methyl groups; τ 4·6, one vinyl proton; and τ 5·8, a multiplet 3 protons, CH$_2$ to N^6 and C4 proton of the ring. The mass spectrum indicates m/e 335 consistent with the structure and there is a prominent fragment corresponding with the free base (pH 1·5, $\lambda_{max.}$ 264 nm, $\epsilon_{max.}$ 14 750; pH 7, $\lambda_{max.}$ 270 nm, $\epsilon_{max.}$ 13 120; pH 11·8, $\lambda_{max.}$ 270 nm, $\epsilon_{max.}$ 11 700). This base can add water to the isopentenyl side-chain yielding a substance $C_{10}H_{15}N_5O$ with a spectrum consistent with an N^6-alkyl derivative of adenine (pH 7, $\lambda_{max.}$ 268 nm, $\epsilon_{max.}$ 17 700; pH 1, $\lambda_{max.}$ 272 nm, $\epsilon_{max.}$ 16 800; pH 11·5, $\lambda_{max.}$ 274 nm, $\epsilon_{max.}$ 17 300).

Hall (1964) measured the ultraviolet absorption spectra of N^6 (aminoacyl) adenosines from yeast RNA. The substances may have a single aminoacid or a peptide attached at N^6. At pH 6·4 there is a peak at 265 nm with an inflection near 278 nm; at pH 10 there is a single peak $\lambda_{max.}$ 277 nm while at pH 11·3 bands occur near 269, 278 and 295 nm with a shoulder at 305 nm. 0·1 N NaOH causes the main band to be at 295–303 nm. The spectra and the pH effects are characteristic and quite different from, for example, the behaviour of benzoyl adenosine.

Schweizer *et al.* (1970) isolated from yeast tRNA N[9-(β-D-ribofuranosyl)-purin-6-yl-carbamoyl]glycine with absorption spectra uniquely associated with the ureido-adenosine chromophore.

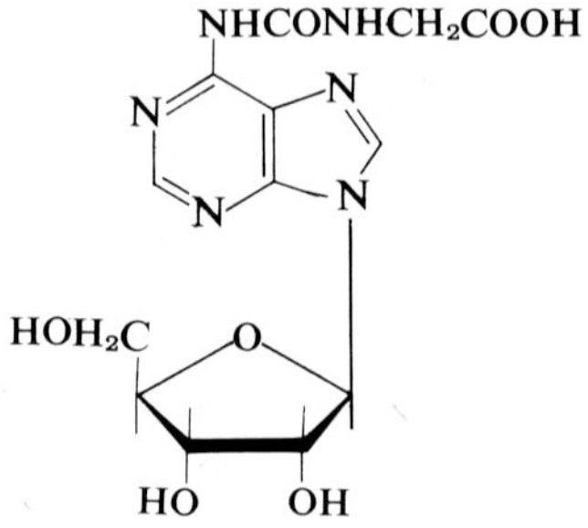

NHCONHCH$_2$COOH [structure: N-(purin-6-ylcarbamoyl)glycine riboside; HOH$_2$C, HO, OH]

	pH 1·2	pH 6·7	pH 10–11
$\lambda_{max.}$	276∼268	286∼276	269, 277, 297

High resolution mass spectroscopy indicated for the largest m/e 293·0775 (calculated for $C_{11}H_{11}N_5O_5$, 293·0760). Loss of glycine resulted in m/e 218·0562 (calculated for $C_8H_6N_6O_2$, 218·0552).

Chheda *et al.* (1969) isolated from yeast, *E. coli* and calf liver N(purin-6-yl carbanyl)threonine and the corresponding nucleoside. Both showed the expected spectroscopic properties.

Hall (1965) described a general procedure for isolating minor nucleosides from RNA hydrolysates and listed the known examples. Absorption spectra at pH 1·5, 7·0 and 11·5 are valuable in identifying the nucleosides (see, e.g. Fig. 9.19–9.22).

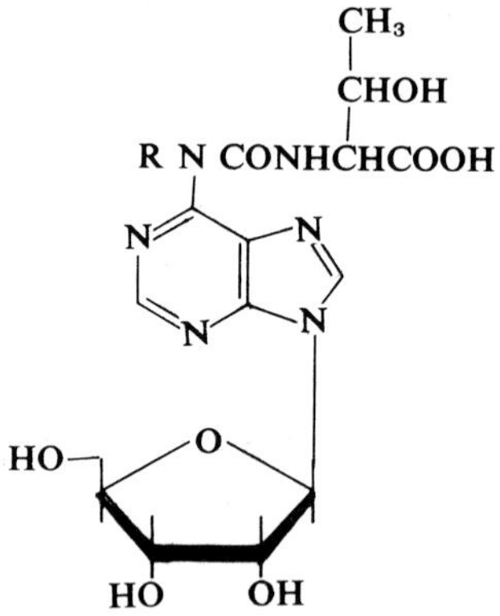

CH$_3$–CHOH–CHCOOH; R N CONH– [structure: purinyl carbamoyl threonine, R = H or ribose; HO, HO, OH]

R=H

	$\lambda_{max.}$ nm	$\lambda_{min.}$
pH 7	269	*ca.* 230
	∼275	—
0·1 M HCl	276	240
0·1 N NaOH	272	247
	282	290
	302	—

R=CH$_3$

	$\lambda_{max.}$ nm
pH 7	280*
	∼290
0·1 M HCl	290
0·1 N NaOH	280
	∼290

* Not very responsive to pH change.

9.2.1 Some Thio Compounds

Burrows *et al.* isolated from *E. coli* SRNA a compound displaying cytokinin activity. Mass spectroscopy indicated a molecular weight of 381·149 (calculated for $C_{16}H_{23}N_5O_4S$, 381·147). The structure

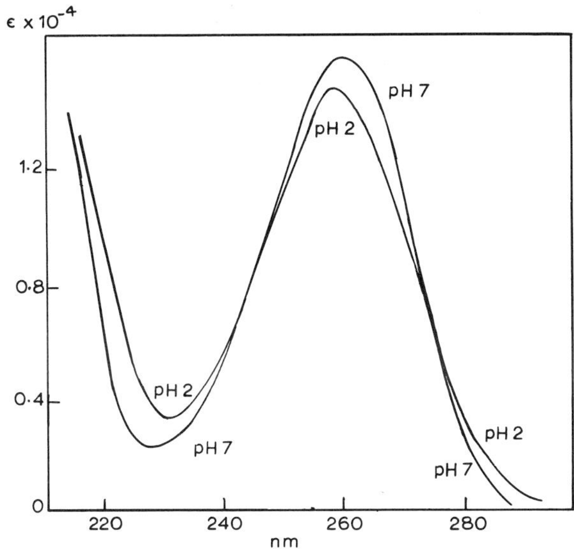

Figure 9.19 Absorption spectra of adenosine-5′-triphosphate (ATP) at pH 2 and pH 7

λ_{max} nm	ε_{max}	λ_{min} nm	ε_{min}	pH
257	14 700	230	3500	2
259	15 400	227	2500	7

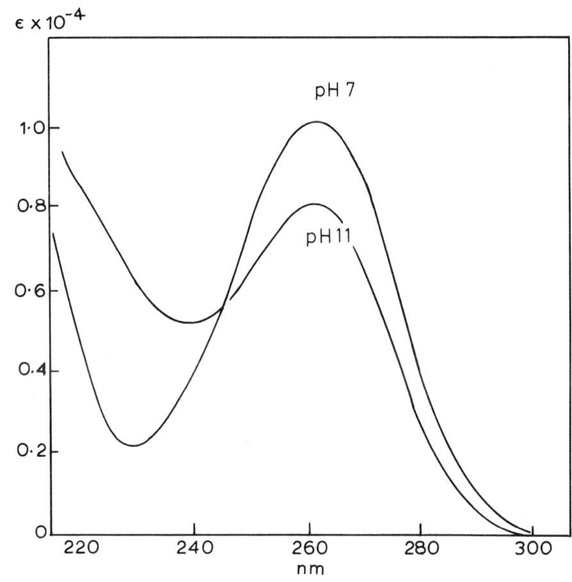

Figure 9.20 Absorption spectra of uridine-5′-triphosphate (UTP) at pH 11 and pH 7

λ_{max} nm	ε_{max}	λ_{min} nm	ε_{min}	pH
262	10 000	230	2100	7
261	8 100	239	5400	11

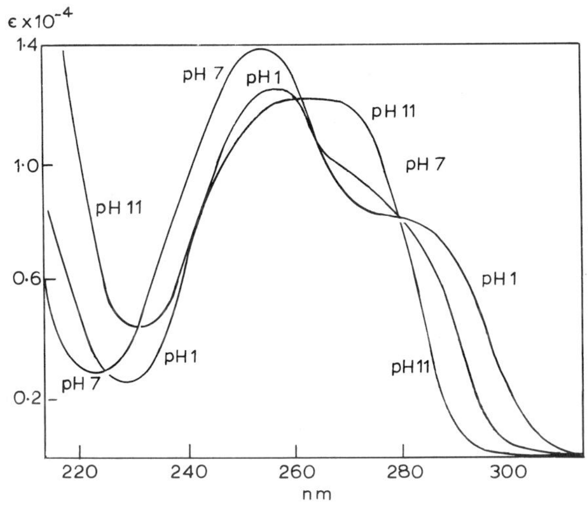

Figure 9.21 Absorption spectra of guanosine-5′-triphosphate (GTP) at pH 1, pH 11 and pH 7

λ_{max} nm	ε_{max}	λ_{min} nm	ε_{min}	pH
256	12 400	228	2800	1
252	13 700	223	3000	7
258	11 900	230	4300	11

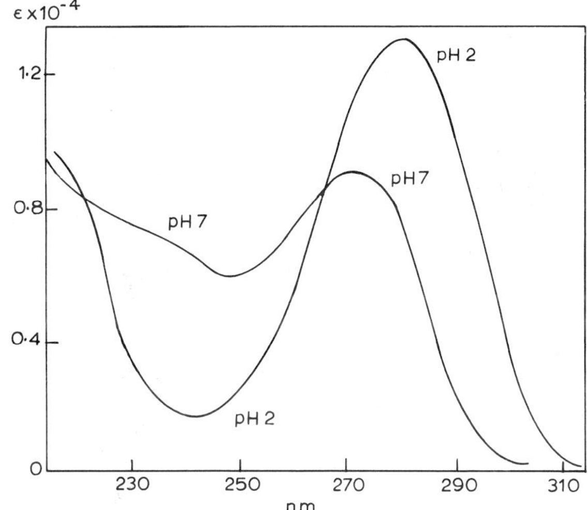

Figure 9.22 Absorption spectra of cytidine-5′-triphosphate (CTP) at pH 2 and pH 7

λ_{max} nm	ε_{max}	λ_{min} nm	ε_{min}	pH
280	13 000	241	1600	2
271	9 100	249	6300	7

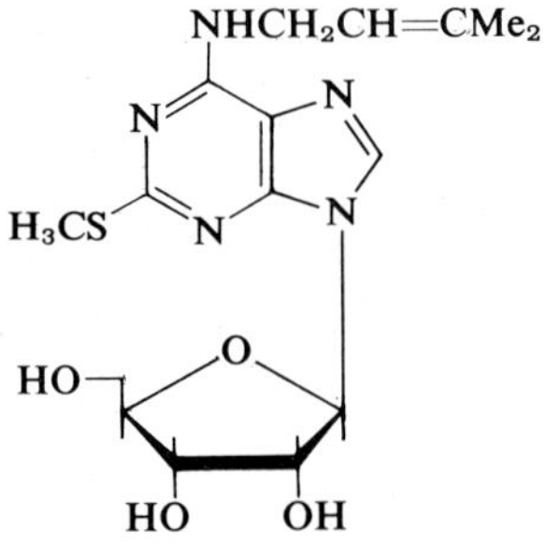

$NHCH_2CH{=}CMe_2$

N N

N H_3CS N N

HO O

HO OH

	EtOH		acid		alkali	
$\lambda_{max.}$ nm	244	283	246	286	243	283
$\epsilon_{max.} \times 10^{-3}$	25·3	18·0	18·6	16·1	24·9	18·0

(6-(3-methyl-2-butenylamino)-2-methylthio-9-β-D ribofuranosyl purine) was confirmed by synthesis.

In the infrared region the S—H stretching vibration occurs at 2600–2650 cm^{-1} and is weak. Alkyl mercaptan bands are at the top of the range (Bellamy, 1961; Murthy *et al.*, 1962) and thiophenol shows only a little dimerization. Thiosalicyclic acid shows evidence of chelation (2515 cm^{-}). The C—S stretch gives a moderately strong or weak vibration at 720–570 cm^{-1} (Trotter & Thompson, 1946).

The thioketones and aldehydes show absorption around 1025–1225 cm^{-1}, e.g. thiofenchone 1180, pyr-4-thiones 1100, 1075, thiopyr-4-thione 1070 cm^{-1}. Broadly the C═S stretching vibration is at 1125 ± 100 cm^{-1}. Substances containing the groups —N—C═S or

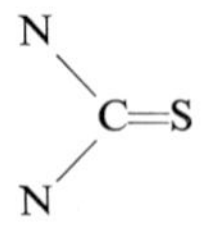

N
C═S
N

have presented difficult problems of assignment caused in the main by vibrational couplings. Absorption peaks occur at 1570–1395, 1420–1260, 1140–940 and 850–680 cm^{-1} but the interpretation is complex and inconclusive (Rao & Venkataraghavan, 1962). The absorption at 1225–1025 cm^{-1} seems to be due unequivocally to the —C═S stretching in the simpler compounds where coupling is least likely, but in —N—C═S the three bands noted above are sufficiently characteristic to be used in qualitative analysis although an easy solution of structural problems in the thiouracils can hardly be expected.

Replacement of —C═O by C═S to give thioaldehydes or thioketones displaces the weak carbonyl band (*ca.* 270 nm) to longer wavelengths. The lone pair of electrons in S is in a higher orbital than that of O in the C═O group because shielding of the nuclear charge is much more effective. Many thioketones are, however, unstable but in steroid thioketones the 'forbidden' weak carbonyl band has been displaced to 492 nm ($\epsilon_{max.}$ 12·9) with a second peak at 239 nm ($\epsilon_{max.}$ 8770), solvent dioxan (Djerassi, 1960). Thiocamphor and thiofenchone have similar spectra and ethyl dithioacetate is not very different (Rao *et al.*, 1961; Fabian & Mayer, 1964). The weak ($n \rightarrow \pi^*$ band) is again near 490 nm and the 240 nm band is strong.

	$\lambda_{max.}$ nm	$\epsilon_{max.}$
Thiocamphor	493*	12
	244	11 500
	214	4 170
Thiofenchone	488*	11
	240	10 000
	215	3 980
Ethyldithioacetate	460†	17·8
	305	12 000
Ethylenetrithiocarbonate	467†	—
	315	—
	451‡	83
	318	15 800

* In cyclohexane; † In heptane; ‡ In ethanol.

Cyclohexanethione shows a very weak carbonyl peak at 504 nm ($\epsilon_{max.}$ 10) and in thiobenzophenone there is a peak at 595 nm ($\epsilon_{max.}$ 177) and a strong peak at 315 nm ($\epsilon_{max.}$ 15 100) (see Brocklehurst & Burawoy, 1960; Jannssen, 1960; Oster, Citarel & Goodman, 1962, who give for thiobenzophenone $\lambda_{max.}$ 605 nm, $\epsilon_{max.}$ 66; 315 nm, ϵ, 17 000).

In thiophosgene the weak band is displaced to 572 nm. A different situation arises with compounds in which the —C═S chromophore is attached to nitrogen. A simple example is thioacetamide, where the shift from the carbonyl absorption is less great:

	$\lambda_{max.}$ nm	$\epsilon_{max.}$
Thiourea	~291	71
	254	12 600
Thioacetamide		
in ethanol	325	38
	265	13 840
in ether	358	17·8
	268	11 200
Thiazolidine thionene		
in ethanol	326	132
	276	15 140

(see Fehnel & Carmack, 1949; Evans & Gillam, 1943).

In uracil the feeble ($n \rightarrow \pi^*$) carbonyl absorption is masked. The absorption spectra of thiouracils are of interest here.

It must be noted the acetamide, urea, fenchone and camphor lack chromophores other than the carbonyl group. Thiocamphor and thiofenchone, in addition to the low intensity absorption ($n \rightarrow \pi^*$) at 490 nm, show strong absorption at about 240 nm, while ethyl dithioacetate has a new band at 305 nm, ϵ 12 000. Similarly the thioacetamide and thiourea peaks at 265 nm respectively are at high intensities. According to Rao these bands 'are probably due to the $n \rightarrow \sigma^*$ and $\pi \rightarrow \pi^*$ transitions of the thiocarbonyl group and on this basis the major C=S band readily undergoes bathochromic effects'.

In 2-thio-4-methylpyrimidine the weak absorption at 366 nm in acid or 388 nm in neutral solution testifies to the thione structure but the diminution in absorption intensity at 270–290 nm with increasing pH might suggest that a thione–thiol change cannot be excluded but infrared evidence tends to favour the thione. The explanation is more likely to be in terms of ionic forms. In the case of 4-thio-6-methylpyrimidine the thione structure is again very probably, the major band occurring at *ca.* 325–335 nm. (Possibly no search was made for a very weak C=S band in the visible region.) By inspecting Table 9.13 the reader will be able to draw his own provisional conclusions. To achieve a satisfactory account of the structures of all the thiopyrimidines further work is necessary but the facts so far gathered show for the mono- and di-sulphur analogues of uracil and thymine a very considerable element of additivity despite some discrepancies.

Table 9.13 Thiouracil and related compounds

	Solution	$\lambda_{max.}$ nm	$\epsilon_{max.} \times 10^{-3}$
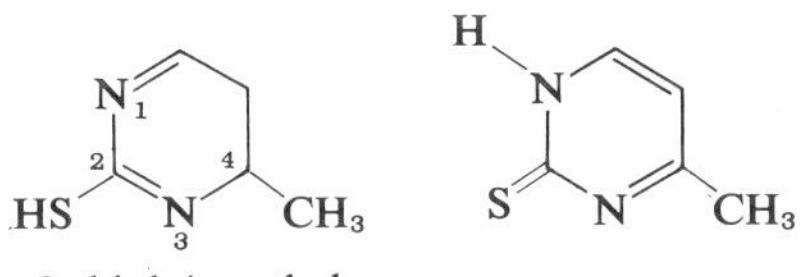 2-thiol-4-methyl	Acid	221	7·94
		285	25·12
		366	1·41
	Neutral	215	10·0
		277	9·05
		338	3·24
	Alkaline	269	17·0
		~300	2·51
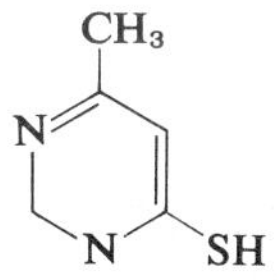 4-thiol-6-methyl	Acid	312	18·2
	Neutral	288	10·0
		322	10·98
	Alkaline	292	18·85
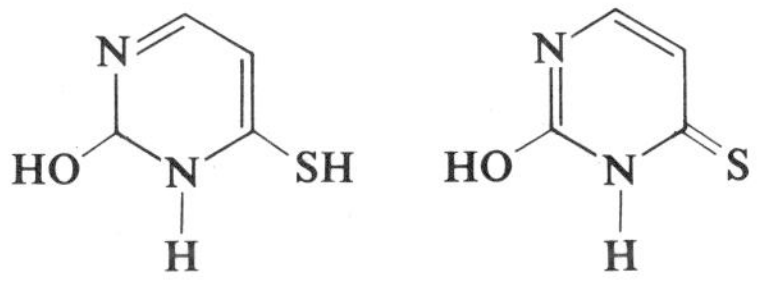 4-thiouracil	pH 1	242	3·47
		326	19·05
	pH 11	334	17·8
	pH 7	325	16·5
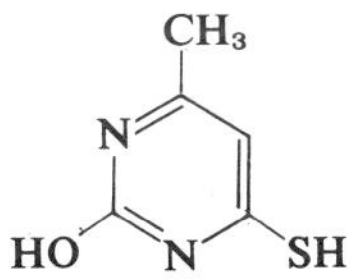 4-thiothymine	pH 1	331	13·3
	pH 7	332	11·0
	pH 11	341	11·0
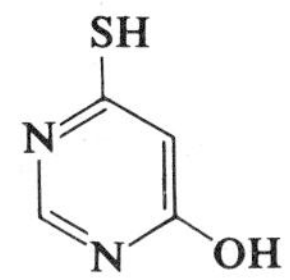 4-ol-6-thiol	pH 1	243	8·9
		304	17·4
	pH 11	239	17·8
		288	13·2

Table 9.13—*continued*

Compound	Solution	$\lambda_{max.}$ nm	$\epsilon_{max.} \times 10^{-3}$
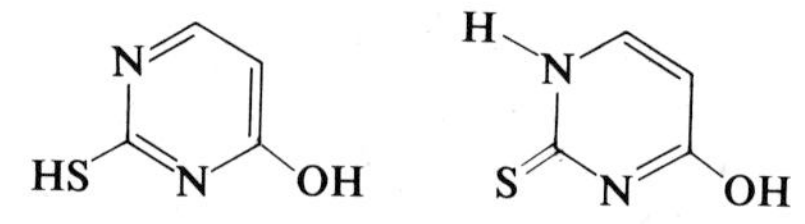 2-thiol-4-ol (2-thiouracil)	pH 1	273	13·8
	pH 7	272	10·6
	pH 11	233	10·96
		259	10·23
		310	6·9
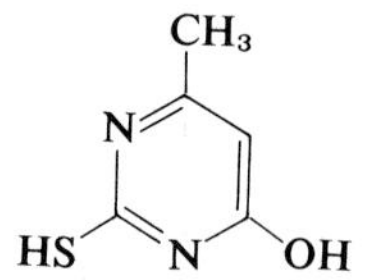 2-thiothymine	pH 1	280	19·0
	pH 7	280	17·0
	pH 11	260	16·6
		310	7·4
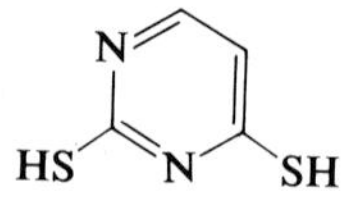 2,4-dithio (uracil)	pH 1	281	21·8
		341	11·0
	pH 7	275	17·2
		260	10·5
	pH 11	278	16·0
		360	8·5
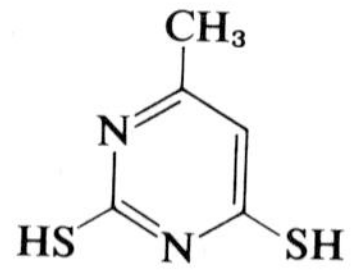 2,4-dithiothymine	pH 1	284	23·0
		342	12·9
	pH 7	283	18·6
		365	7·5
	pH 11	276	19·05
		354	10·96
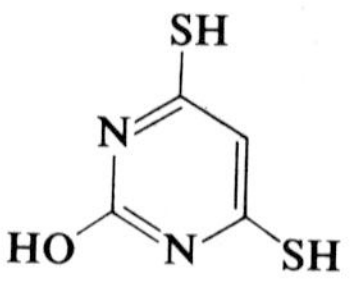 (2-hydroxy-4,6-dimercaptopyrimidine)	pH 1	244	5·6
		277	7·6
	pH 11	244	7·76
		278	8·32
		340	26·3
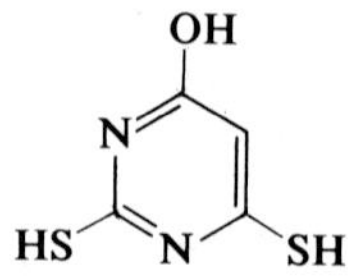 2,4-dithiobarbituric acid	pH 1	274	20·4
		308	13·5
	pH 11	238	15·1
		304	12·6
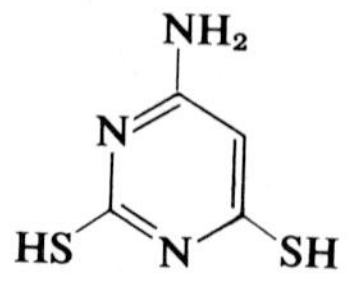 (4-amino-2,6-dimercaptopyrimidine)	pH 1	281	29·5
		326	26·3
	pH 11	247	17·0
		306	28·2
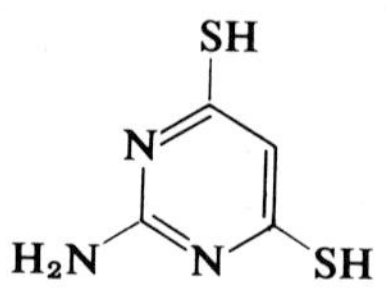 (2-amino-4,6-dimercaptopyrimidine)	pH 1	247	10·7
		278	11·5
		370	63·1
	pH 11	253	10·5
		325	25·7

Table 9.13—*continued*

	Solution	$\lambda_{max.}$ nm	$\epsilon_{max.} \times 10^{-3}$
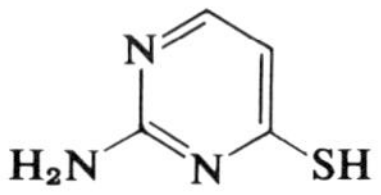 N, H_2N N SH	pH 1	256	4·79
		325	14·8
	pH 11	262	6·17
		311	13·2
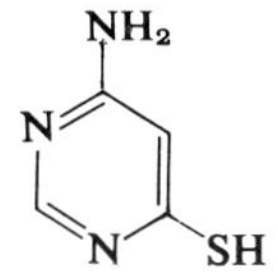 NH_2, N, N SH	pH 1	237	6·7
		303	28·8
	pH 11	244	7·08
		289	16·2
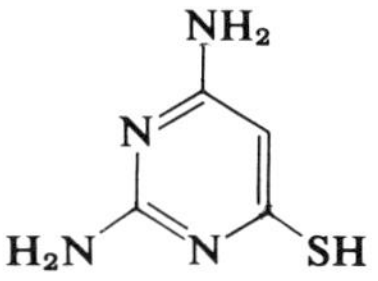 NH_2, N, H_2N N SH	pH 1	243	10·7
		323	33·1
	pH 11	299	19·0
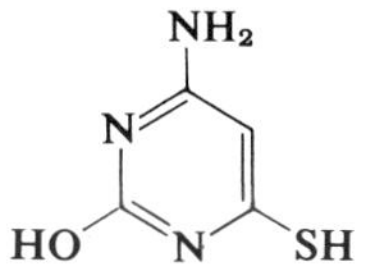 NH_2, N, HO N SH	pH 1	242	7·59
		317	40·7
	pH 11	303	43·7
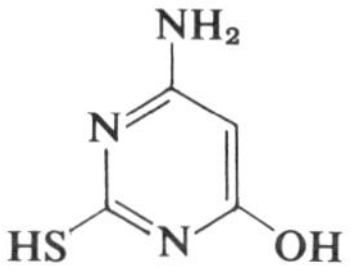 NH_2, N, HS N OH	pH 1	246	8·13
		276	20·0
	pH 11	241	17·8
		289	13·2
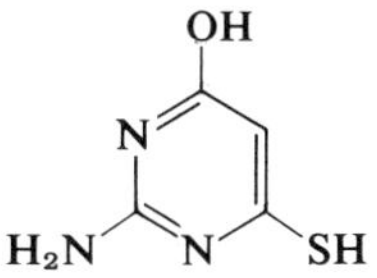 OH, N, H_2N N SH	pH 1	237	12·3
		309	26·3
	pH 11	296	19·1
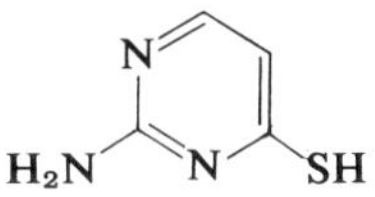 N, H_2N N SH	pH 1	256	4·79
		325	14·8
	pH 11	262	6·17
		311	16·2
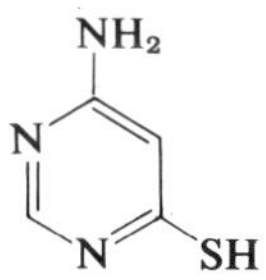 NH_2, N, N SH	pH 1	237	6·6
		303	28·8
	pH 11	244	7·08
		289	16·2
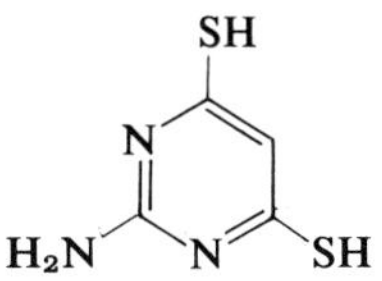 SH, N, H_2N N SH	pH 1	207	10·7
		278	11·5
		370	63·1
	pH 11	253	10·5
		325	25·7

9.2.2 Pyrimidines

The parent substance pyrimidine dissolves in cyclohexane and the solution shows a strong peak at 189 nm (ϵ 10 000), moderately strong ($\pi \rightarrow \pi^*$) absorption at 244 nm (ϵ *ca.* 2050) and weak absorption near 300 nm (ϵ 325). This weak ($n \rightarrow \pi^*$) absorption, which extends from 260 to 330 nm, has marked fine structure and in the 230–270 nm region there are peaks at 240 and 244 nm, with additional inflections. In ethanol peaks occur at 279 nm and 244 nm (ϵ 330 and 2400 respectively) and the incipient fine structure disappears. In water there are peaks at 267 and 240 nm (ϵ 400 and 3200) and in aqueous acid the main peak is at 243 nm, ϵ 5500 (see Table 9.14 for pH effects).

Introduction of methyl groups at positions 2 or 4 and of two methyls at positions 4 and 6 results in peaks at 244–248 nm with the ϵ values 2800, 3390 and 4300 respectively. Even for 2,4,6-trimethylpyrimidine the main band is at 251 nm (ϵ 4200) with the weaker band enhanced (292 nm, ϵ 1400). These compounds all exist in the pyrimidine form and do not exhibit tautomerism.

The infrared spectra of the aminopyrimidines display no sign of the presence of iminotautomers (Bellamy, 1958) and the ultraviolet absorption curves are confirmatory. 5-Hydroxypyrimidine also retains the pyrimidine structure, but this is not the case for the 2- and 4-'hydroxy' pyrimidines. Ketonization occurs and the 4-substituted compound is a pyrimidone and its sulphur analogue is a thione. Similarly the 2-'hydroxy' compounds are cyclic amides and thioamides respectively. Comparison of the ultraviolet curves for the 2-and 3-'OH' compounds with those of the related methoxy compounds reveals differences explicable only in terms of oxo groups in the parent substances. Infrared spectra (Thompson *et al.*, 1950) also indicate that keto forms are greatly predominant. The structures of the ionic forms appear to have been established by Spinner (1960) on the basis of infrared absorption and Raman spectra for alkaline and acidic solutions. When oxygen is replaced by sulphur, the stretching vibration C═S at 1140 cm^{-1} confirms a thione structure. Moreover the 2-SCH_3 derivative differs sharply from the thio amide in respect of ultraviolet absorption.

The 5,6-double bond is readily reducible, e.g. uracil yields dihydrouracil which shows only end absorption in the region below 250 nm (Batt *et al.*, 1954). This is consistent with an amide structure. Alkali opens the ring at the 3,4 bond but before this occurs a new peak appears at 230 nm. In 1,3-dimethyldihydrouracil a peak occurs at 222 nm with $\epsilon_{max.}$ *ca.* 5000.

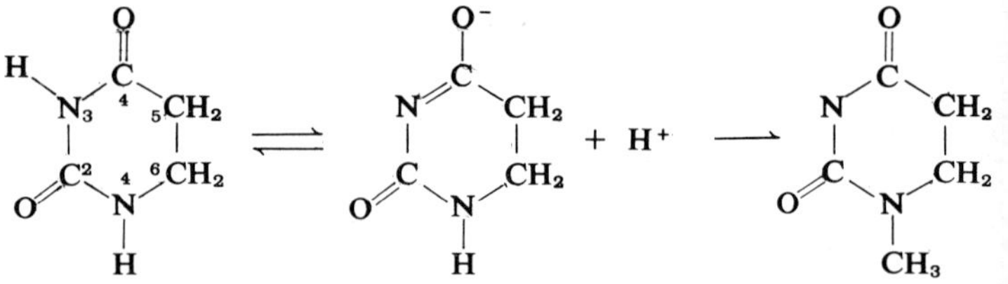

Table 9.14 summarizes the spectra of pyrimidine derivatives. Interpretation is sometimes equivocal.

Uracil in neutral or acidic solution shows $\lambda_{max.}$ near 260 nm with a keto group clearly at position 2 (cf. 3-methyluracil). Thymine like uracil in alkali shows a new band at 283–290 nm, ϵ 5000–6000 and in 3-methyluracil, $\epsilon_{max.}$ is even higher. It seems possible that enolization can occur, especially since 1-methyluracil in alkali (pH 13) shows practically no response to pH change ($\lambda_{max.}$ 265–268 nm). Orotic acid occupies an equivocal position (cf. Fox *et al.*, 1957).

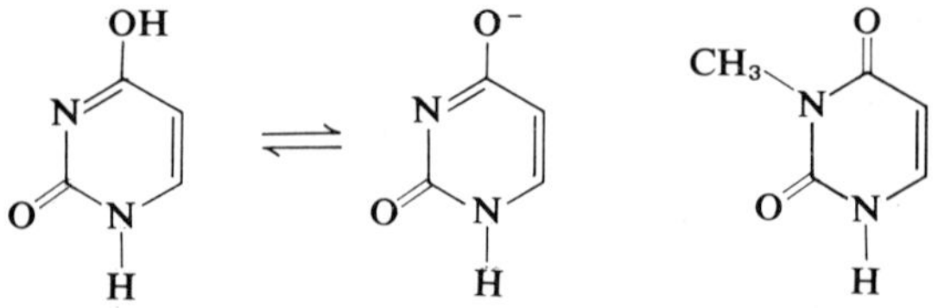

In the ultraviolet there are strong absorption peaks for 2-aminopyrimidine at 221 and 292–300 nm. Change of pH markedly influences intensities but not the location of $\lambda_{max.}$. The introduction of an amino group at position 4 has rather little effect on the spectrum of pyrimidine, but an amino group at position 5 causes a shift to longer wavelengths although it is not necessary to postulate tautomeric change.

The ultraviolet absorption spectra of cytosine at different pH values fall into two divisions: (*a*) over the pH range 1·0 to 7·2, a two-component system exists with isosbestic points near 257·5 and 220 nm and a sharp peak 276 nm at pH 1·0; (*b*) in the range pH 12–14 there are isosbestic points near 272 and 233·5 nm and a good peak is seen at 283 nm (Shugar & Fox, 1952).

The results have been interpreted in terms of the following equilibria:

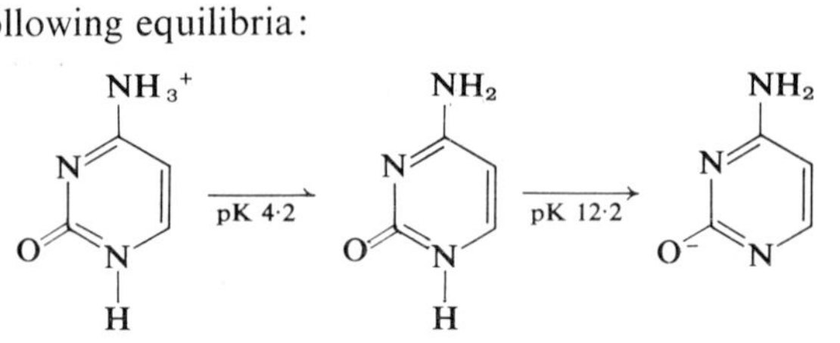

pK 4·2   pK 12·2

Table 9.14 Absorption peaks of pyrimidines

Substance		$\lambda_{max.}$ nm	$\epsilon_{max.} \times 10^{-3}$	Substance		$\lambda_{max.}$ nm	$\epsilon_{max.} \times 10^{-3}$
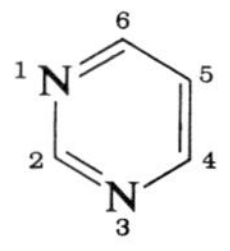	pH 3·6	~238	3·72	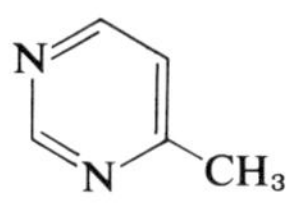		244	3·4
		242	4·37				
		248	3·31				
		~260–296	3·00–2·50				
	pH 8·9	~242	5·37				
Pyrimidine		246	5·89				
		~252	4·17				
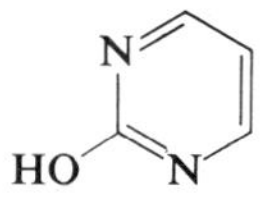	pH 7·2	212	10·8	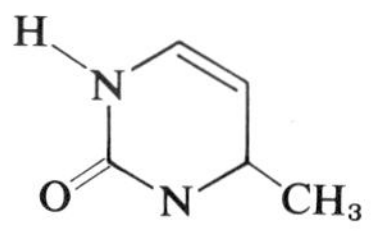	Acid	305	7·1
		298	4·75		Neutral	213	11·0
	2-Methoxy-					296	6·0
	pH 7·0	264	4·66		Alkaline	220	11·5
2-Hydroxy					Enol	290	5·8
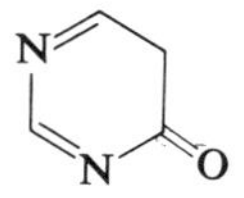	pH 6·2	233	7·32–7·5	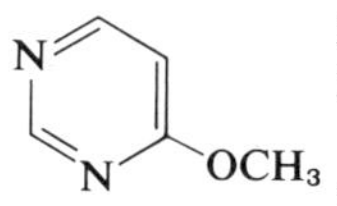	pH 6·95	247	3·35
		260	3·74		pH 0	227	7·7
	pH 13	227	11·1			~288	6·8
		263	3·28–3·5		Alkaline	288	7·7
	5 N H_2SO_4	234	9·84				
4-Hydroxy		251	2·92				
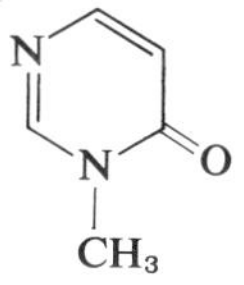	pH 5	221	6·8	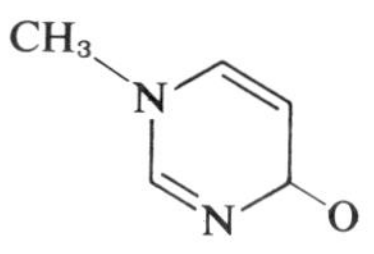	pH 6	240	14·6
		269	3·9		pH 0	229	10·2
	2·5 N HCl	226	9·08			~250	2·65
		258	2·94				
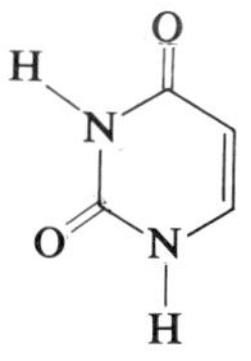	pH 1·0	259	8·14	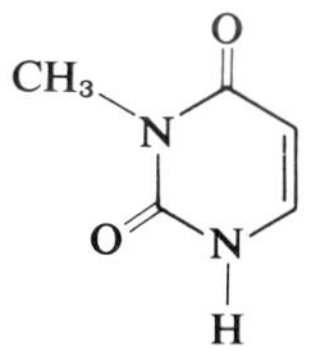	pH 1	265	9·33
	pH 7·2	{ 202	8·8		pH 7	268	9·33
		{ 259·5	8·2		pH 13	265	9·55
	pH 12	285	6·17				
	pH 13	283	6·03				
	Neutral	261	6·3				
Uracil	EtOH	260	6·3	1-Methyluracil			
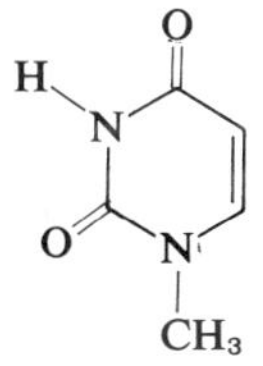	pH 1–7	259	7·24	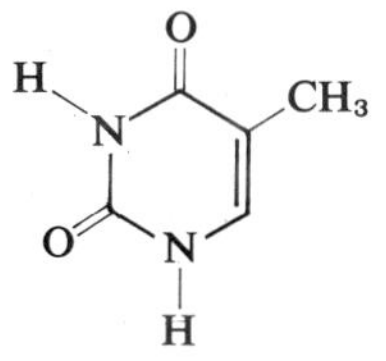	0·1 N HCl	262	7·8
	pH 9·7	253	5·75		n.s.g.*	266	7·24
	pH 13	283	10·72		pH 2	207	9·5
						264	7·89
					pH 12	291	5·44
					0·1 N NaOH	290	5·00
3-Methyluracil				Thymine			
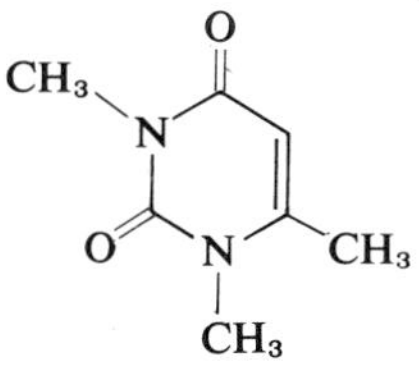	pH 7·2	267	?				

* n.s.g. = no solvent given. Bresnick *et al.*, 1960; Stimson & Reuter, 1943; Toth & Billes, 1968.

Table 9.14—*continued*

Substance		$\lambda_{max.}$ nm	$\epsilon_{max.}$	Substance		$\lambda_{max.}$ nm	$\epsilon_{max.}$
Orotic acid	pH 2	205	10·9		pH 1·6	253	8·0
		280	7·52		pH 13	222	7·0
	pH 12	286	5·98			263	7·5
2-Amino	pH 1	221	14·75	4-Amino	pH 0	~246	18·6
		302	4·0			268	5·0
	pH 7	224	12·6		pH 13	233	18·20
		292	3·15			268	5·0
	pH 11	221	14·70				
		302	4·57				
	EtOH	227	16·2		EtOH	236	20·0
		297	3·3			272	5·13
5-Amino	pH 1	253	14·45		pH 9·3	250	16·0
		332	3·71			286	3·6
	pH 7	236	11·5				
		292–298	3·09				
	EtOH	246	12·6				
		315	3·16				
Cytosine	pH 1	276	10·0		pH 1	280	9·6
	pH 2	210	9·7		pH 13	284	7·6
		274	10·2				
	pH 7·2	267	6·13				
	pH 12	270	0·03				
	pH 14	282	7·86				
	EtOH	267	5·50				
5-Methylcytosine	pH 1	210	12·0	4-Hydroxy-2-amino	pH 1	215	9·23
		282·5	9·79			257	6·92
	pH 7·2	210	14·2		pH 13	225	7·76
		274	6·23			275	6·31
	pH 14	289·5	8·05		EtOH	220	8·71
						286	8·31
					EtOH + HCl	255	6·61
					EtOH + alk.	274	6·76
	pH 7	268	10·0	4-Hydroxy-6-amino	EtOH	215	30·9
						256	6·46
					EtOH + HCl	208	20·9
						252	8·91
					EtOH + alk.	254	3·98
	Water	256·5	25·0				

This explanation has however been criticized (Dekker, 1960; Katritsky & Waring, 1962; Fikus *et al.*, 1962) and protonation is now seen as occurring on the nitrogen at position 1. Infrared absorption spectra support this view (cf. Brown & Teitei, 1964, 1965).

At pH 7 cytidine 2′3′-phosphate shows peaks at 234 and 270 nm (ϵ *ca.* 9000), 240 nm (ϵ *ca.* 8900), while the dihydro-derivative has a single peak $\lambda_{max.}$ 240 nm (ϵ *ca.* 8900) (cf. Fox *et al.*, 1953).

The infrared absorption spectra of nucleosides and

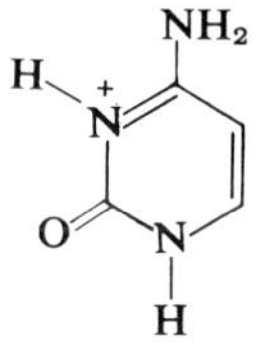

nucleotides show specially interesting bands in the region 5·55 to 6·7 μm (1800–1500 cm^{-1}) that arise from (in the main) in-plane vibrations of the base residues. Neither the sugar unit nor the phosphate group absorbs in this region.

In the pyrimidines themselves the ring-stretching vibrations fall for the parent substance at 1610, 1569, 1461 and 1400 cm^{-1} and for substituted compounds at 1555–1590, 1520–1565, 1400–1480 and 1375–1410. The C—H in-plane deformation vibrations occur at 1220, 1165, 1140 and 1021 cm^{-1} with ring skeletal vibrations near 991 cm^{-1} for the parent substance and for 2- or 4-substituted compounds. The out-of-plane vibrations are at 721 and 680 cm^{-1}. C—H stretching vibrations occur in derivatives at 3100–3000 cm^{-1} and non-bonded N—H stretching vibrations occur at 3450, 3400 cm^{-1} (Short & Thompson, 1952; Lord *et al.*, 1957; Wiley & Slaymaker, 1961). The 2- and 4-hydroxypyrimidines exist in keto forms (Thompson *et al.*, 1950), whereas hydroxyls in positions 5 and 6 retain enolic character (Tanner, 1956). Amino substituents do not form tautomers. The (ketonic) hydroxypyrimidines form cations and anions in aqueous solutions as the pH is varied, and Spinner (1960) favours the structures:

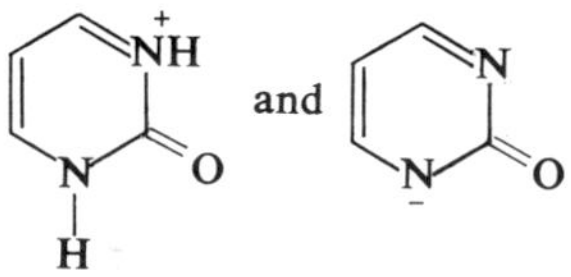

Cytidine has been studied in KBr discs, in nujol mulls and in D_2O. The solid, obtained from neutral solution and examined as a potassium bromide disc, shows four well-defined peaks at 1670, 1647, 1529 and 1500 cm^{-1} with inflections at 1670 and 1585 cm^{-1}. In D_2O at pD 7·0 there are peaks at 1652, 1615, 1527 and 1504 cm^{-1}. In the heavy water only NH_2 (or NH_3^+) groups of the cytosine moiety will be deuterated and all the peaks referred to will arise from ring vibrations involving C=C, C=N or C=O stretching. If the cytidine had been obtained from an alkaline solution, the KBr disc showed peaks at 1645, 1605, 1528 and 1500 with a shoulder at 1665 cm^{-1}, while the D_2O solution at pD 11·0 shows maximal 1652, 1615, 1582 and 1504 with an inflection at 1527 nm. If the solid was obtained from an acid solution, the KBr disc showed peaks at 1725, 1680 and 1535 cm^{-1} and a D_2O solution at pD 2 showed very similar peaks (1709, 1658, 1591 and 1515 cm^{-1}) at somewhat different relative intensities. The 1709 cm^{-1} band is a normal C=O stretching vibration and the 1658 cm^{-1} seems to be due to a C=N ring stretching. Ordinary cytidine (undeuterated, neutral) shows a strong 1603 cm^{-1} peak and a 1760 cm^{-1} inflection, whereas deuterated cytidine at pD 7 has a moderately strong band at 1615 cm^{-1}. It looks as if the NH_2 bending vibration is at 1670 and 1615 cm^{-1} corresponds with the 1603 cm^{-1} peak. The protonated form may be either

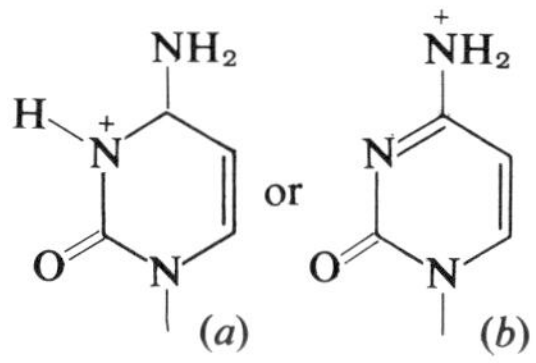

(*a*) or (*b*)

In pyridine ($\lambda_{max.}$ 1597 and 1581 cm^{-1}) protonation displaces the peaks to 1635 and 1610 cm^{-1}, whereas in aniline the stretching bands at 1601 and 1500 cm^{-1} are not displaced on protonation (see Shimanouchi *et al.*, 1964). By analogy structure (*a*) is preferred to (*b*). Aniline has a strong absorption at 1276 cm^{-1} and cytidine at 1285 cm^{-1}; both are probably due to the C—N stretching vibration (cf. Fox *et al.*, 1953).

Uridine is a diketo compound:

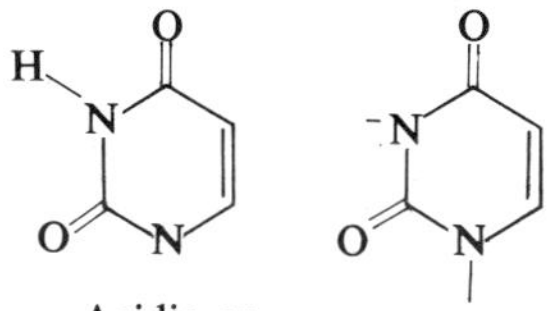

Acidic or neutral

Basic form

When the solid is obtained from neutral solution and examined as a KBr disc there is a good peak at 1681 cm^{-1} with inflections at 1700, 1668 and 1623 cm^{-1}. The nujol mull gives quite similar results. The curve for material obtained from acid solution is very similar but from alkaline solution there are peaks at 1635 and 1510 cm^{-1} with shoulders at 1605 and 1535 cm^{-1} and the deuterated form shows peaks at 1640 and 1508 cm^{-1} at pD 11·0. The removal of a proton at position 1 tends to delocalize electrons throughout the system of conjugated double bonds and the resultant weakening of double bonds accounts for the wave-number displacements. Polycitidylic acid (Miles, 1959, 1961) examined in neutral solution (D_2O) showed peaks at 1651 and 1617 cm^{-1} (cytidine 1652, 1615 cm^{-1}).

It could be argued that 2-thiouracil and 2-thiothymine contain the

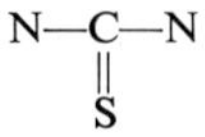

$$N{-}\underset{\underset{S}{\|}}{C}{-}N$$

chromophore of thiourea ($\lambda_{max.}$ 254 nm) displaced to 272 or 280 nm and that the feeble 291 nm peak is displaced and masked but enhanced in alkali to 310–312 nm. Now 4-thiouracil contains a different chromophore:

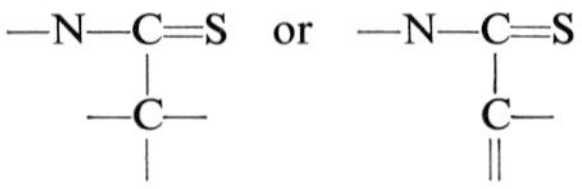

—N—C=S (with —C— below) or —N—C=S (with C— and = below)

and the absorption at 325–340 nm is relatively little affected by pH changes in either 4-thiouracil or 4-thiothymine (see also Schneider & Halverstadt, 1948).

The wavelength displacements in the 2,4-dithiouracil and 2,4-dithiothymine agree with expectations but the intensities are surprising. In the case of 2-mercaptobenzthiazole (Morton & Stubbs, 1939) the N-methyl derivative (stabilizing the thione structure)

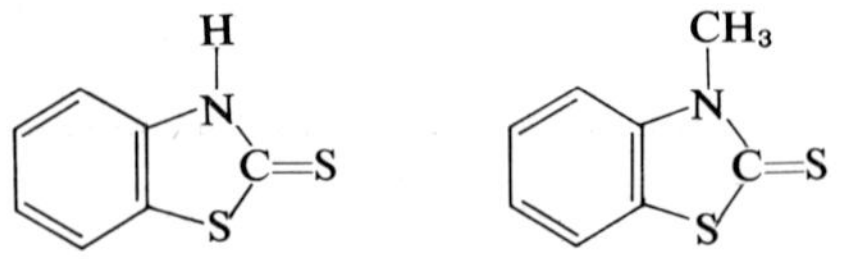

[Structures: benzothiazole-2-thione with N–H; benzothiazole-2-thione with N–CH_3]

shows $\lambda_{max.}$ 3245 nm as does the parent substance, both having $\epsilon_{max.}$ *ca.* 26 000 (see p. 88). Although the evidence is not conclusive it seems likely that 2-thiouracil in alkali can form a thiol whereas in 4-thiouracil the very small effect of change of pH points to a fixed structure.

Compound		$\lambda_{max.}$ nm	$\epsilon_{max.} \times 10^{-3}$
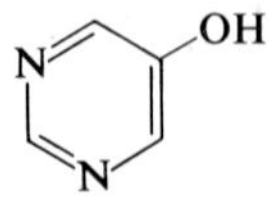 [pyrimidine, 5-OH]	pH 7	276	5·37
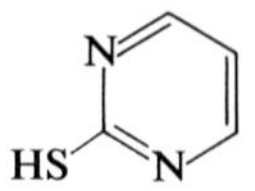 [pyrimidine, 2-SH]	pH 1	283	17·4
	pH 11	235	3·72
		269	13·20
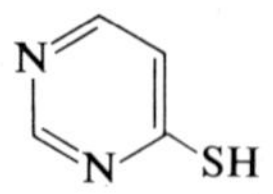 [pyrimidine, 4-SH]	pH 1	293	10·7
		310	10·7
	pH 11	293	13·2
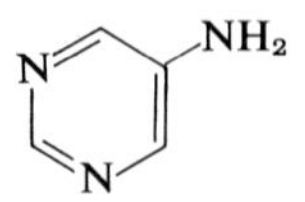 [pyrimidine, 5-NH_2]	pH 1	253	14·5
		332	3·72
	pH 7	236	10·96
		298	3·09

Compound		$\lambda_{max.}$ nm	$\epsilon_{max.} \times 10^{-3}$
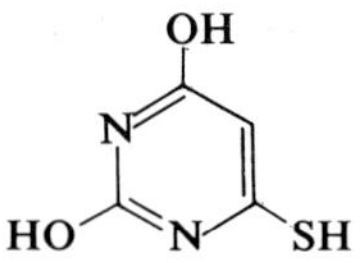 [pyrimidine, 6-OH, 2-HO, 4-SH]	pH 1	306	19·5
	pH 11	302	25·1
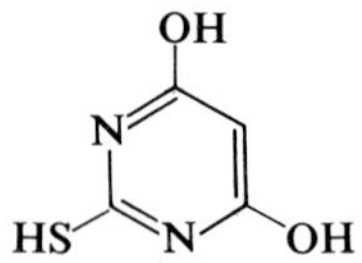 [pyrimidine, 6-OH, 2-HS, 4-OH]	pH 1	234	8·51
		282	21·4
	pH 11	235	13·8
		280	11·2
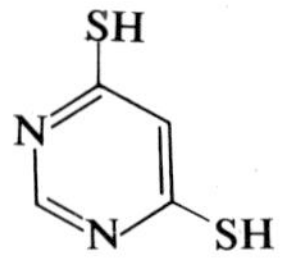 [pyrimidine, 4-SH, 6-SH]	pH 1	280	14·5
		370	57·5
	pH 11	268	24·55
		316	19·95
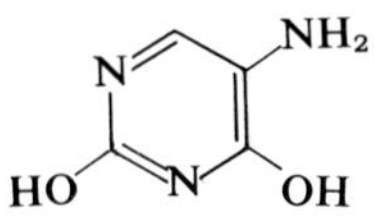 [pyrimidine, 5-NH_2, 2-HO, 4-OH]	pH > 10	216	19·5
		289	8·13

The ultraviolet absorption of barbiturates has been studied by Stuckey (1940, 1941, 1942), Fox & Shugar (1952), Goldschmidt *et al.* (1953) and Patterson *et al.* (1949).

Barbituric acid is potentially tautomeric but intermediate stages between the extremes shown below must be considered.

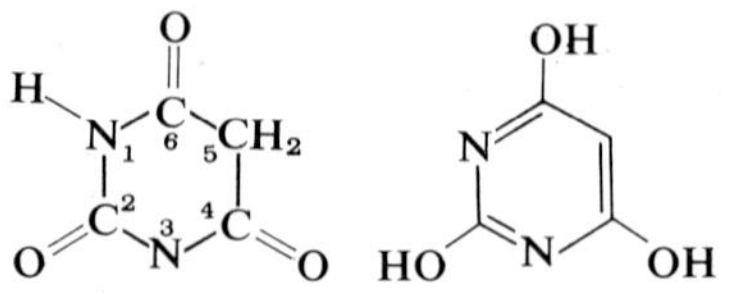

[Structures: barbituric acid keto form (H–N1, C6=O, 5 CH_2, C2=O, N3, C4=O); trihydroxypyrimidine form (OH, HO, OH)]

Stuckey (1940) noted an effect of dilution:

	Barbituric acid	
	$\lambda_{max.}$ nm	$\epsilon_{max.} \times 10^{-3}$
M/200 in water	257·5	2·88
M/4000 in water	258	8·91
M/40 000 in water	258	19·95
In N/20 HCl	255·5	0·603
In N/20 NaOH	258	19·95

	1-Methylbarbituric acid	
	$\lambda_{max.}$ nm	$\epsilon_{max.} \times 10^{-3}$
M/200 in water	259	1·90
M/4000 in water	259	7·24
M/40 000 in water	259	20·4
In N/20 HCl	258	0·295
In N/20 NaOH	259	20·4

1,3-Dimethylbarbituric acid		
	$\lambda_{max.}$ nm	$\epsilon_{max.} \times 10^{-3}$
M/200 in water	259·5	1·90
M/4000 in water	259·5	6·03
M/40 000 in water	259	20·4
In **N**/20 HCl	262	0·2
In **N**/20 NaOH	259	20·4

Similarity in behaviour of the three substances is unmistakeable. The dimethylbarbituric acid can only enolize in one position, namely by using a hydrogen from the active methylene group.

Assuming that the ϵ values shown above represent 100% of two different forms the percentage of keto form in water can be calculated:

	M/40 000	**M**/4000	**M**/200
Barbituric acid	0	57·1	88·2
1-Methylbarbituric acid	0	64·7	91·8
1,3-Dimethylbarbituric acid	0	70	91

All three compounds yield a similar enolic form and there is no evidence of a dienol. Although the agreement at **M**/4000 is imperfect it may be concluded that the triketo form absorbs with negligible intensity at 255–260 nm. Stuckey (1942 *a*) re-examined barbituric

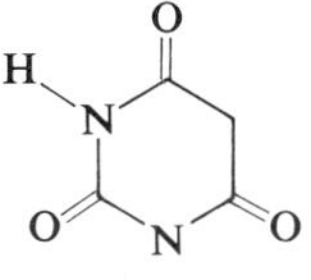

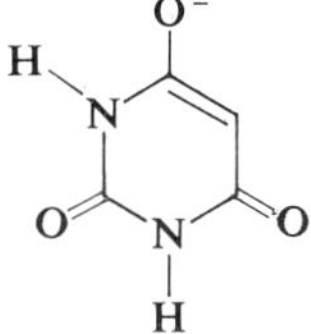

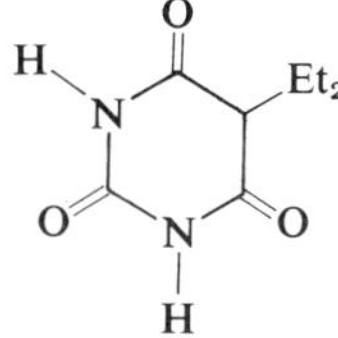

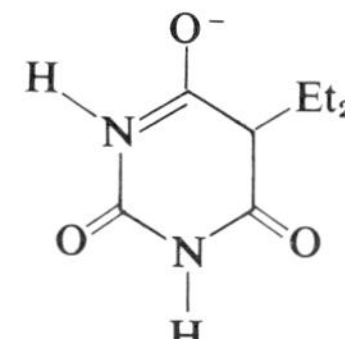

Barbituric acid

	Neutral	Anion pH 12	Cation pH 1	
$\lambda_{max.}$ nm	257	258	205	258
$\epsilon \times 10^{-3}$	25·0	21·5	10·4	0·6

Diethyl barbituric acid

Aqueous acid	Water	Alkali
212	239	255
9·15	10·6	7·95

acid and found $\lambda_{max.}$ 255 nm, $\epsilon_{max.}$ 550 for **N** HCl as well as for 0·1 **N** HCl while the peak for alkaline solutions at 258 nm showed $\epsilon_{max.}$ values 17 000, 20 000 and 25 000 for 1, 0·1 and 0·0001 **N** NaOH respectively. Moreover for an aqueous solution at extreme dilution (10^{-6} **M**) $\epsilon_{max.}$ was at 25 000.

The ionization constant of barbituric acid has been determined by electrical conductivity methods and Fredholm (1937) obtained $k = 1{\cdot}10 \times 10^{-4}$. Stuckey calculated the degree of ionization from

$$\alpha = \frac{\epsilon_{max.} - \epsilon_0}{\epsilon_\infty - \epsilon_0} \qquad (\epsilon_0 = 550,\ \epsilon_\infty = 25\,000)$$

and over the range **M**/200 (0·133) to **M**/40 000 (0·82) obtained excellent agreement with the α values calculated from Fredholm's constant. This established that the enolic form was ionized.

Both the 1-methyl and 1,3-dimethylbarbituric acids at $\mathbf{M} \times 10^{-6}$ dilution also showed $\epsilon_{max.}$ 25 000. The decrease to $\epsilon_{max.}$ 20 000 in stronger alkali is a significant observation which needs to be explained.

The ionic form of barbituric acid will be predominant but it may be that the excess of alkali favours formation of an ionized sodium enolate.

5-Methyl- and 1,5-dimethylbarbituric acid were also examined. The former shows in acid solution $\lambda_{max.}$ 260 nm, $\epsilon_{max.}$ 850 with $\lambda_{max.}$ 262 nm ϵ, 195 000 in **N**/10 sodium hydroxide. It is easily oxidized to give 5,5′-hydroxymethylbarbituric acid (5-methyldialuric acid) which displays no absorption maximum in alkaline solution. 1,5-Dimethylbarbituric acid very closely resembles the 5-methyl derivative ($\lambda_{max.}$ 264 nm, ϵ, 550 in acid solution and $\lambda_{max.}$ 268 nm ϵ, 19 200 in alkali).

Stuckey extended his work to barbitone and phenobarbitone where amido–imidol tautomerism is possible at two positions:

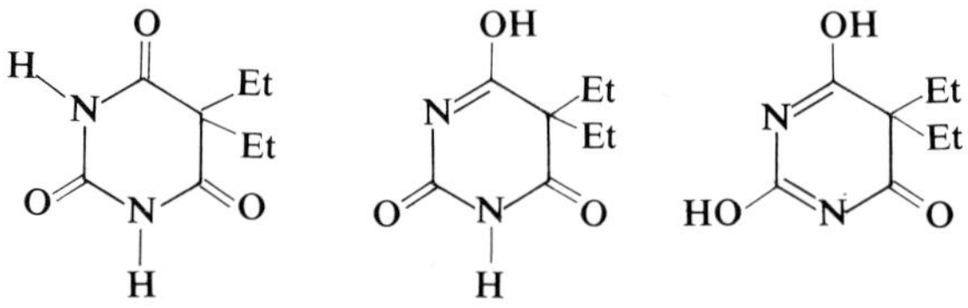

The absorption spectra for both barbitone and 1-methylbarbitone show in alkali $\lambda_{max.}$ 245 nm, $\epsilon_{max.}$ *ca.* 9000. In water or acid dilute solutions of barbitone have end absorption, the 1-methyl derivative has $\lambda_{max.}$ 220 nm, $\epsilon_{max.}$ *ca.* 8000 and 1,3-dimethylbarbitone $\lambda_{max.}$ 230 nm, $\epsilon_{max.}$ 7600. The latter in alkali shows only weak end absorption. Phenobarbitone

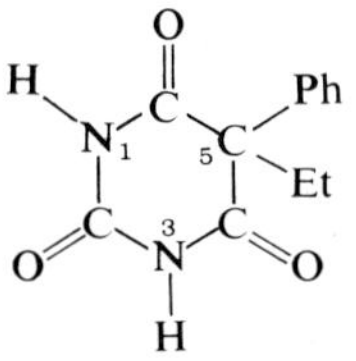

Phenobarbitone

and its substitution products behave as follows:

	Acid		Water M/4000		Alkali N/10	
	$\lambda_{max.}$ nm	$\epsilon \times 10^{-3}$	$\lambda_{max.}$ nm	$\epsilon \times 10^{-3}$	$\lambda_{max.}$ nm	$\epsilon \times 10^{-3}$
Phenobarbitone	end*		end*		256	8·8
1-Methylphenobarbitone	end		end		246	9·0
1,3-Dimethylphenobarbitone	230	7·6	288	7·6	end absorption	

* End absorption 240 nm.

The sharp distinction between the disubstituted and the monosubstituted phenobarbitones makes it clear that a mono-enol is formed in phenobarbitone and that it exists in dilute solution and in alkali as an ion. A large number of 5,5′-disubstituted barbituric acids all showed in N/10 alkali $\lambda_{max.}$ 246–256 nm, $\epsilon_{max.}$ 6500–9000 with the exception of the 5,5′-dimethyl compound which is anomalous in that it alone forms a stable disodium salt. Alloxan in acid exists as a cation with very weak absorption and as an anion in water, and dialuric acid shows a strongly absorbing anion in water:

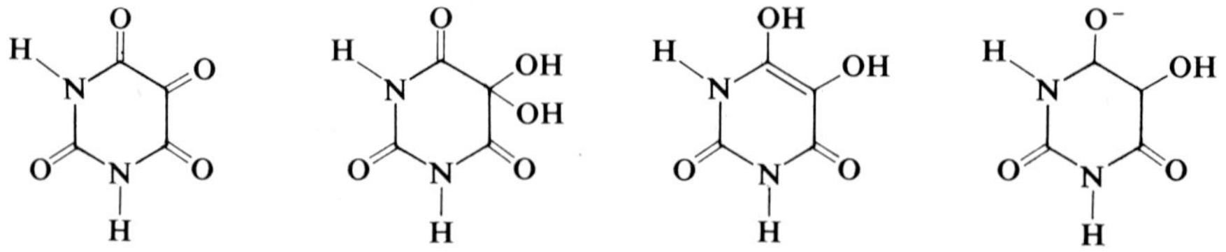

Alloxan

Hydrated (dihydroxy)

Dialuric acid

	Acid	Water	Acid	Water
$\lambda_{max.}$ nm	270	275	265	245
$\epsilon \times 10^{-3}$	3·2	16·5	0·094	4·8
			(cation)	(anion)

The significance of spectroscopic studies is perhaps best illustrated by considering the photochemistry of pyrimidine derivatives. The main actions of ultraviolet light on DNA depend on the conditions and are physical or chemical. Perhaps the most important effect is the formation of pyrimidine dimers which can inhibit DNA synthesis in bacterial cells and inactivate transforming DNA. These are the only defined actions of ultraviolet light with proved biological results. They do not, however, include photochemically induced mutations, the mechanisms of which remain imperfectly understood. The damage done to DNA in bacteria (and bacterial viruses) can be reversed by a light-sensitive enzymatic monomerization. Repair can also be effected by a cellular process of excising the dimerized moiety.

Heyroth & Loufbourow (1931) in pioneer studies irradiated uracil using a quartz-mercury arc and found that the selective absorption near 260 nm gradually fell to a very low value. Sinsheimer & Hastings (1949) and Sinsheimer (1954) found that the photochemical change induced in uracil, uridine and uridylic acid could be reversed on acidification (pH 1), inasmuch as the original absorption spectrum was restored. (To avoid side reactions it was necessary to exclude Hg lines of wavelengths <230 nm.) With uracil dissolved in water the quantum yield, measured by the rate of absorption at 260 nm, was low (*ca.* 0·05) and about half the changed uracil could be regenerated on acidification or on heating the neutral solution.

1-Methyluracil and 1,3-dimethyluracil behaved rather similarly but the latter showed particularly simple behaviour (Moore & Thompson, 1955, 1957). The main absorption band ($\lambda_{max.}$ 265 nm) of dimethyl uracil disappeared on irradiation and the product was practically transparent from 250 to 300 nm. There was an isosbestic point near 236 nm and the photoproduct showed end absorption only. On heating at pH 1 and 100°C for 5 minutes the original absorption was quantitatively restored. The photoproduct could be isolated and the uracil was then found to have taken up a molecule of water.

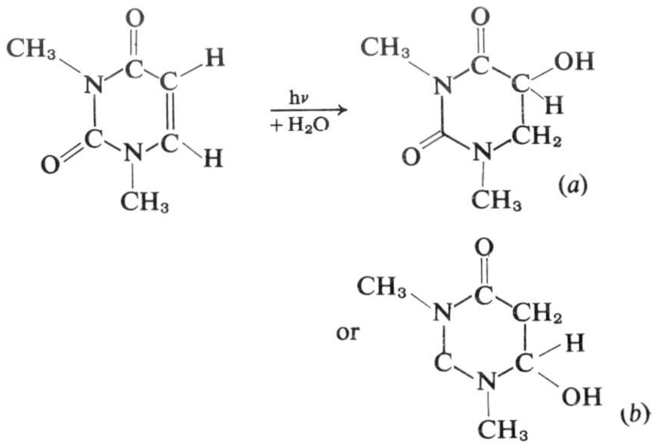

$$\xrightarrow[+H_2O]{h\nu}$$

(*a*)

or

(*b*)

It showed an infrared peak ν cm^{-1} 3355 (OH) absent from uracil, and the disappearance of the ultraviolet absorption was accounted for by entry of water at positions 5 and 6. Synthesis of 1,3-dimethyl-6-hydroxy-5-hydrouracil confirmed structure (*a*). The uracil reaction was thus shown (Moore, 1958) to be:

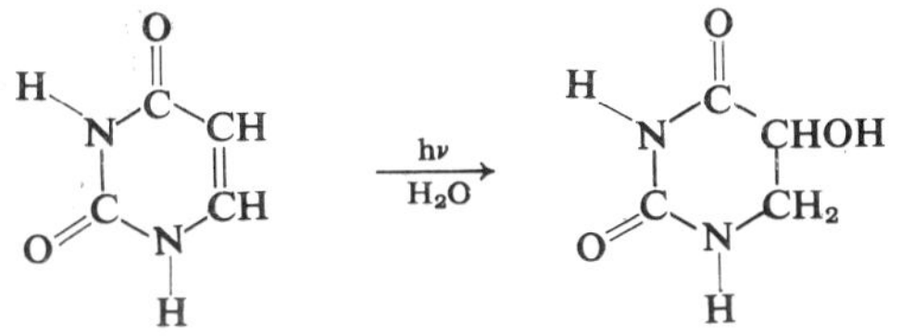

$$\xrightarrow[H_2O]{h\nu}$$

Uridine phosphates and uridylic acids behaved in very much the same way (Sinsheimer, 1954; Shugar & Wierzchowski, 1958 *b*) but derivatives of uracil which already had substituents in the 5 or 6 position often failed to show reversibility.

Beukers *et al.* (1959) made an interesting discovery when they observed that although dilute aqueous solutions of thymine do not undergo much change on irradiation at room temperature, the frozen solution, exposed to the mercury resonance line radiation (254 nm), underwent rapid change. The intensity of absorption at 260 nm decreased rapidly. The frozen solution contains small solid aggregates in which thymine molecules are sufficiently contiguous for an absorbed photon to form a dimer (Wang, 1961). The quantum yield is between 1 and 2 (Füchtbauer & Mazur, 1966). When the solution was thawed the same radiation effected a rapid increase in absorption intensity until it almost reached the original value. The quantum efficiency was between 0·5 and 1·0. The photoproduct, which itself showed only end absorption, was thermostable and resistant to pH changes. Unlike thymine it was insoluble in ethanol and after removing unchanged material it could be crystallized from water. Its molecular weight indicated a dimer in which the double bonds in the 5,6 positions were found to be

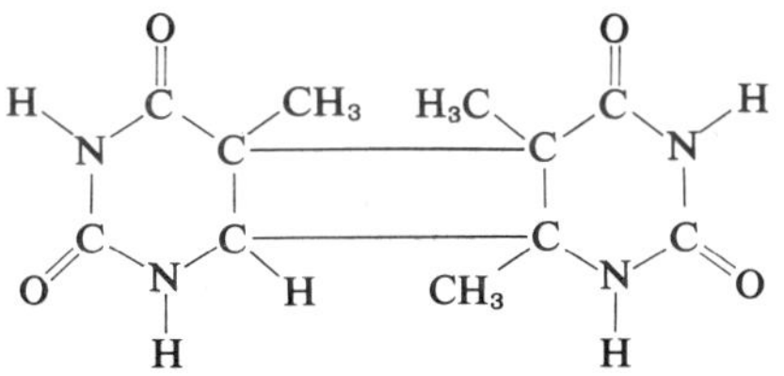

saturated to form a cyclobutane ring (Beukers & Berends, 1960, 1961; Wulff & Fraenkel, 1961). (It may be that the initial effect in weak aqueous solutions of uracil is also dimerization, followed by hydration.) Frozen solutions of uracil yielded a photodimer which

Smietanowska & Shugar (1961) showed to be a *cis* isomer.

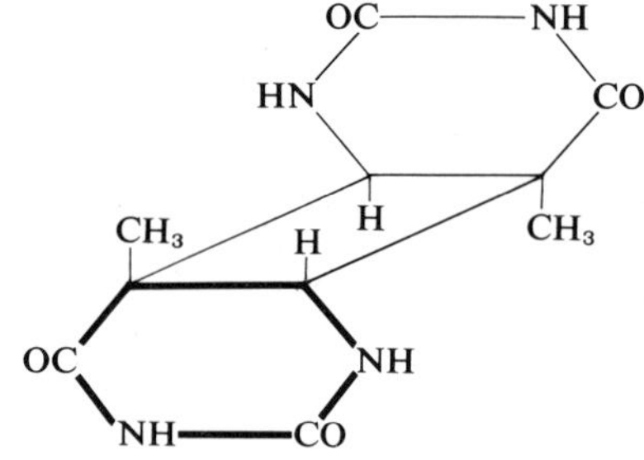

Possible structure of dimer from frozen thymidine

The failure to form thymine dimers by irradiating very dilute solutions at room temperature can be overcome by introducing a sensitizer or by the use of very concentrated solutions. It is interesting that aqueous solutions of polynucleotides form dimers on irradiation much more effectively at room temperature than in the frozen state, which indicates a greater capacity to attain the correct juxtaposition of molecules in the liquid state (cf. Deering & Setlow, 1963).

The formation of cyclobutane-type pyrimidine dimers in general

$$-\mathrm{Py}+-\mathrm{Py} \underset{h\nu''}{\overset{h\nu'}{\rightleftharpoons}} -\overline{\mathrm{Py}-\mathrm{Py}}$$

is reversible depending on the conditions and the wavelength of the light used.

The absorption spectrum of the thymine dimer varies with pH; in neutral solution there is end absorption only and with increasing alkalinity a new peak near 240 nm becomes gradually stronger, reaching $\epsilon_{max.}$ 6600 in N NaOH. Irradiation (254 nm) at pH 12 results in the appearance of the spectrum of thymine ($\lambda_{max.}$ 291 nm at pH 12) (see Fig. 9.23).

Irradiation of DNA has been shown (Blackburn & Davies, 1966; Varghese & Wange, 1967) to result in a thymine–thymine dimer having the structure indicated below. The absorption spectra help to explain the effects on the irradiation process of wavelength change. Irradiation of a thymine solution with any wavelength below 300 nm will produce some dimer and a steady state can be attained. At longer wavelengths (e.g 275 nm) dimerization will be favoured and at shorter wavelengths (e.g. 235 nm) 'free' thymine will be favoured.

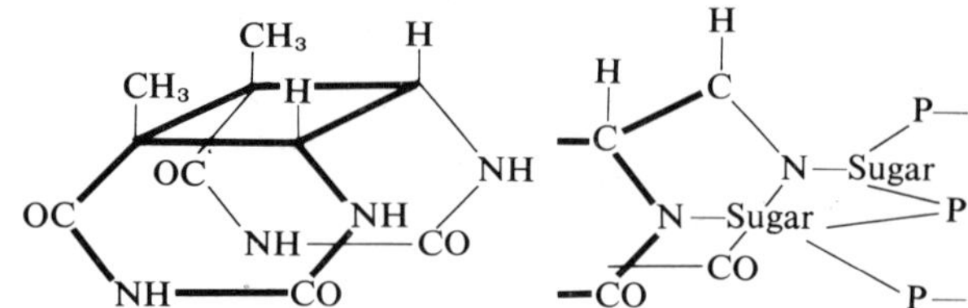

Proposed structure of dimer from frozen thymine thymidine and from DNA

Moreover the quantum yield for the monomerization process is much greater than that for dimerization. The situation was well illustrated by Wulff (1962) in experiments on the effects of quanta of 275 nm and 235 nm on DNA from *Escherichia coli.*

The dye proflavine binds to DNA and changes the polymer so as to stop the formation of dimers but not the fission of any already present.

$$-\mathrm{Py}+\mathrm{Py} \underset{\text{proflavine} + h\nu}{\overset{h\nu}{\rightleftharpoons}} -\overline{\mathrm{Py}-\mathrm{Py}}$$

Because of their characteristic absorption spectra dimers containing cytosine are more readily split than

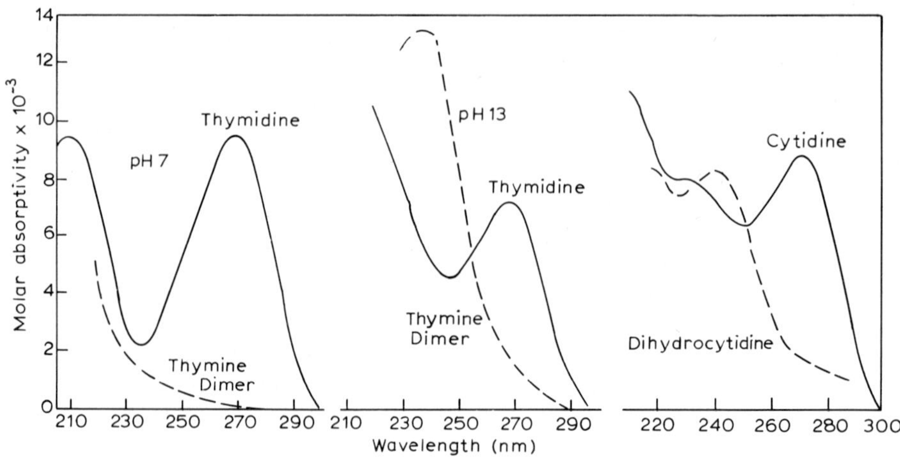

Figure 9.23 Absorption spectra of thymidine, thymine dimer, cytidine and dihydrocytidine. The deoxycytidine derivatives presumably have spectra similar to those of cytidine, and a cytosine dimer in a polynucleotide probably has a molecular absorptivity twice that of dihydrocytidine (from Setlow, 1966)

Figure 9.24 The relative absorbances of polynucleotides irradiated with a combination of long (265 or 280 nm), and short (239 or 240 nm) radiation. The absorbance increases produced by short wavelengths after long wavelengths irradiation are characteristic of dimer monomerization (from Setlow, 1966)

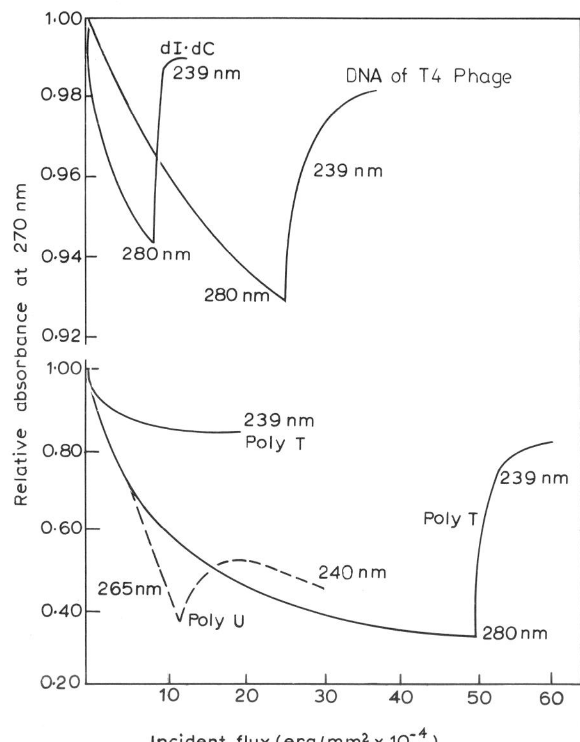

the thymine–thymine dimers. This phenomenon can be used to effect selective monomerization and to show that cytosine dimers also possess biological significance (Setlow & Setlow, 1967). Cytosine derivatives also form photoproducts which revert in the dark under the action of heat at pH 7 or at room temperature in acid or alkali. The photoproducts are unstable and actual isolation has not been achieved. Cytidine-3′-phosphate has a peak at 273 nm, $\epsilon_{max.}$ 8600 with a marked inflection near 225–230 nm, ϵ *ca.* 8000 and a peak near 203 nm, ϵ 14 000. The photoproduct has $\lambda_{max.}$ 240 nm $\epsilon_{max.}$ 13 000 and 205 nm, $\epsilon_{max.}$ 13 500. Irradiation greatly favours the photoproduct but regeneration on heating restores the original state (Wierzchowski & Shugar, 1961 *a*). The absorption curves for cytidine-2′,3′-phosphate and its dihydroderivative were measured by Janion & Shugar (1960):

	$\lambda_{max.}$ nm	$\epsilon_{max.} \times 10^{-3}$	$\lambda_{max.}$ nm	$\epsilon_{min.} \times 10^{-3}$
Cytidine-2′,3′-diphosphate	270	9·1	252	7·85
	232	8·95	225	8·7
Dihydrocytidine diphosphate	240	8·85	227	7·8

Dihydrocytosine also shows the 240 nm absorption,

$$\text{cytosine (NH}_2\text{, C, N, CH, CH, C, O, N–H)} \xrightarrow{2H} \text{dihydrocytosine (NH}_2\text{, C, N, CH}_2\text{, CH}_2\text{, C, O, N–H)}$$

but the 232 nm band is not seen in cytosine or 1-methylcytosine. The spectroscopic change brought about by irradiation could here be due to hydration or to dimer formation. The evidence obtained by the use of D_2O strongly favours hydration. A valuable and much more detailed account of the behaviour of cytosine, methylcytosine and their derivatives is given by McLaren & Shugar (1964).

Cytosine ($\lambda_{max.}$ 267 nm) on irradiation at pH 7 gives a photoproduct with $\lambda_{max.}$ 240 nm in water and 300 nm in ethanol. Reversibility is 50%. 1-Methylcytosine ($\lambda_{max.}$ 274 nm) gives a photoproduct with $\lambda_{max.}$ 240 nm in water and 310 nm in ethanol. Reversibility is again 50%. On the other hand 5-methylcytosine ($\lambda_{max.}$ 273·5 nm) gives a photoproduct with a peak at 285 nm (310 nm in ethanol) and the reaction is irreversible.

9.2.3 Purines

With four double bonds to be placed, purines can isomerize in numerous ways. The parent substance has a moderately strong absorption band in the middle ultraviolet region:

pH	$\lambda_{max.}$ nm	$\epsilon_{max.} \times 10^{-3}$
1	258 (260)	6·31 (6·0)
7	261 (263)	7·94 (8·0)
12	270 (271)	7·94 (7·6)

with a strong band on the short wave side of 220 nm. Published values for absorption peaks differ somewhat as is shown by the figures in brackets. Introduction of methyl groups at various positions effects only minor changes in the absorption. Mason (1954) examined a considerable range of mono- and di-substituted purines and advanced theoretical interpretations (see also Clark & Tinoco, 1965; Tanaka & Nagakura, 1966; Kwiatowski, 1968).

An important survey by Beaven *et al.* (1955) includes a large number of absorption curves, including many previously unpublished. A folder placed in a pocket in that volume contains metric scale drawings of the spectra of 8 bases at various pH values and of the corresponding nucleosides. The same curves appear in the text. Some of the curves were obtained by Shugar & Fox (1952). pK_a values were obtained from:

$$pK_a = pH - \log \frac{\epsilon_{HA} - \epsilon}{\epsilon - \epsilon_A}$$

where ϵ_{HA} and ϵ_A are the absorptivities of two ionic species and ϵ is the observed absorptivity at the stated pH.

Volkin & Cohn (1954) found it useful to measure absorbances at 250, 260, 280 and 290 nm, and the ratios A_{250}/A_{260}, A_{280}/A_{260}, A_{250}/A_{290} together with $\epsilon_{260\,nm}$ have been much used. The values are tabulated by Beaven *et al.* A large sheet made freely available by the California Corporation for Biochemical Research (3625 Medford Street, Los Angeles, 63, California) tabulates the main maxima and minima for the main bases, ribosides and nucleotides with values for ϵ_{260} and the ratios A_{250}/A_{260} and A_{280}/A_{260}.

A hydroxy substituent at position 2 exerts a large bathochromic effect and a hydroxyl at position 8 causes a small red shift while a hydroxyl at position 6 effects a blue shift. The amino substituent also has little effect at position 6 compared with positions 2 and 8.

Kalchar & Shafran (1947), finding ultraviolet absorption rather unspecific within the purines, made use of specific enzymes to attack individual naturally occurring purines. The resulting drop in selective absorption was used to make a quantitative determination. The residue was then submitted to attack by another specific enzyme and the fall in absorption at an appropriate wavelength was measured.

Infrared spectra of purines and related substances were studied by Blout & Fields (1948, 1950) and the results were summarized by Rao (1963).

The absorption of uric acid is characteristic and can be used in analysis. A mono-enol is probably the dominant form. The changes on spectra with pH have been discussed (Bergman & Dikstein, 1955). Caffeine can be determined spectroscopically but it is necessary to remove or correct for irrelevant absorption.

It is convenient to illustrate the relevance of spectroscopy to purines by discussing guanosine in some detail.

Guanine may be depicted in various tautomeric forms since three 'mobile' protons can be arranged in 20 different ways at the standard locations shown below.

$\lambda_{max.}$ nm	245–246	274–276
$\epsilon_{max.} \times 10^{-3}$	10·2–11·0	7·8–8·3

(at pH *ca.* 6)

Guanine cation (at pH 1)

$\lambda_{max.}$ nm	248–249	270–276
$\epsilon_{max.} \times 10^{-3}$	10·7–11·5	7·1–7·4

7-Methylguanine

$\lambda_{max.}$ nm	248	283
$\epsilon_{max.} \times 10^{-3}$	5·7–6·2	7·8

9-Methylguanine

$\lambda_{max.}$ nm	252	~270
$\epsilon_{max.} \times 10^{-3}$	10·2	9·3

Table 9.15 Absorption peaks of some purines

Substituent Solution	Neutral		Acidic (cation)		Alkaline (anion)	
	$\lambda_{max.}$ nm	$\epsilon_{max.} \times 10^{-3}$	$\lambda_{max.}$ nm	$\epsilon_{max.} \times 10^{-3}$	$\lambda_{max.}$ nm	$\epsilon_{max.} \times 10^{-3}$
2-Hydroxy	238	2·9	264	2·3	265	4·0
	315	4·9	322	6·5	312	6·75
2-Methoxy	246	2·6	—	—	—	—
	283	8·1	284	6·75	283	7·6
6-Hydroxy (Hypoxanthine)	249	10·5	248	10·5 (9·12)	262 (258)	11·0
8-Hydroxy	235	3·25	280	10·5	284	12·6
	277	11·0	—	—	—	—
6-Methoxy	252	9·8	254	10·0	261	9·8
6,8-Dihydroxy	257	12·0	—	—	271	13·8
	~280	5·7	—	—	—	—
2,6-Dihydroxy	269	11·20 (8·0)	265	7·94	241	8·32
(Xanthine)	—	(9·78)	—	—	278 (283)	8·91 (8·7)
6-Amino	260	13·5 (9·2)	262 (260)	13·0 (13·5)	267	12·0
2-Amino	~236	5·0	~235	6·5	~276	4·1
	305	6·0	325*	4·2	303	5·75
	—	—	314†	4·0	—	—
8-Amino	241	3·25	288	16·0	~230	9·5
	283	14·50	—	—	290	12·0
2-Amino-6-oxo	249	10·3	—	—	—	—
(Guanine)	273	7·5	—	—	—	—
		(See Table 9.16)				
2,6,8-Trihydroxy	238	9·8	231	7·94	—	—
(Uric acid)	293	12·2	285	12·6	296	12·6
1,3,7-Trimethylxanthine	272	10·47	—	—	—	—
(Caffeine)	278	10·9	—	—	—	—
3,7-Dimethylxanthine	274	10·23	—	—	234	7·08
(Theobromine)	—	—	—	—	273	10·2
2-Dimethylamino	223	26·0	228	33·0	232	25
	332	5·0	340	3·0	327	4·7
6-Dimethylamino	275	17·8	276	15·0	221	16·2
	—	—	—	—	281	17·8
1,3,9-Trimethylxanthine	237	9·77	—	—	—	—
	268	10·0	—	—	—	—

* Dication. † Monocation. (See also Whitehead & Travers, 1960)

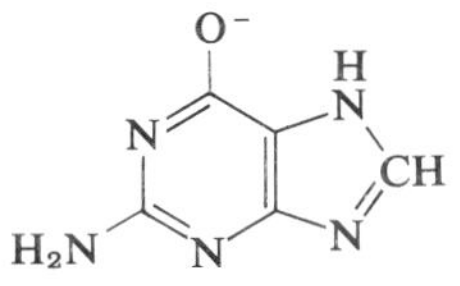

anion (at pH 10–11)

$\lambda_{max.}$ nm	243–246	273–277
$\epsilon_{max.} \times 10^{-3}$	6·0–8·5	8·0–10

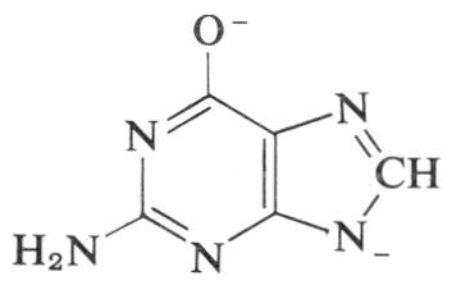

dianion (at pH 13)

$\lambda_{max.}$ nm	221	274
$\epsilon_{max.} \times 10^{-3}$	13·2	8·7–9·9

In the nucleotides with their attachment at position 9, 10 tautomers are possible. There is, however, strong evidence in favour of a single structure (Mason, 1963; Katritzky & Lagowski, 1963) and infrared studies on guanosine provide confirmation (Angell, 1961; Tsuboi *et al.*, 1962; Miles *et al.*, 1963). There is, however, another possibility of isomerism for the parent substance in that 7-H and 9-H forms can be envisaged. Indeed comparison of the absorption spectra of 7-methylguanine and 9-methylguanine indicates that both forms occur in guanine (Pfleiderer, 1961). The residual problems concern the structures of the anion and the cation. Guanine exhibits three pK_a values: pK_{a1} 3·0 (transition from cation to neutral species); pK_{a2} 9·3 (transition from neutral species to anion); pK_{a3} 12·3 (transition from anion to dianion).

In acid the protonation probably occurs on the imidazole ring.

Shapiro (1968) has assembled and interpreted the data on the absorption spectra of guanine derivatives using the work of Pfleiderer (1961), Beaven *et al.* (1955) and Mason (1954). The compound 3-methylguanine was not available to the earlier workers but its spectrum (Townsend & Robins, 1962; Elion, 1962) in acid is distinctive and helps to exclude the protonation at position 3 of guanine.

Ultraviolet absorption has been useful in comparing 'unknowns' with model guanine derivatives. Shapiro (1968) points out that there are risks in relying upon measurements at pH values around 12 because several

Table 9.16 Properties of guanine and its derivatives
(a=anion; n=neutral; c=cation; d=dianion)

	pK values	$\lambda_{max.}$ nm		$\epsilon_{max.} \times 10^{-3}$		
Guanine	—	221	274	13·2	8·7–9·9	*d*
	3·0	243–246	275	6·0–8·5	8·0–10·0	*a*
	9·3	245	275	10·2–11·0	7·8–8·3	*n*
	12·3	248·9	270·6	10·7–11·5	7·1–7·4	*c*
9-Methylguanine	2·9	~258	268	~10·2	11·2	*a*
	9·8	252	270	10·2	~9·3	*n*
	—	251	276	12·0	7·6	*c*
Guanosine	2·1–2·2	258–266	—	11·3	—	*a*
	9·2–9·5	253	~270	13·7	—	*n*
	—	257	~277	—	—	*c*
1-Methyl	3·1	249	273	10·2	8·1	*a*
	10·5	250	~270	10·7	~7·1	*c*
1-Methylguanosine	—	254	270	10·4	—	*n*
	—	~255	277	—	—	*c*
7,9-Dimethylguanine	7·2	252	282	5·9	6·6	*n*
	—	253	279	11·7	7·6	*c*
3-Methylguanine	—	273–274	—	13·0–13·7	—	*a*
	—	244·5	264·5	8·2	11	*c*
O^6-Methylguanine	—	246	284	4·5	7·9	*a*
	—	240	280	7·9	7·9	*n*
	—	286	—	11·2	—	*c*
O^6-Methylguanosine	—	248	278	11·1	10·5	*n*
	—	247	287	8·5	10·5	*c*
N^2-Methylguanine	3·3	245	279	8·2	7·3	*a*
	8·8–9·8	248	278	—	—	*n*
	12·8	251	277	12·3	6·9	*c*
N^2-Methylguanosine	—	258	—	—	—	*a*
	—	258	—	—	—	*c*
N^2-Dimethylguanine	—	245	277	9·2	8·0	*a*
	—	257	289	14·5	6·3	*c*
N^2-Dimethylguanosine	—	262	—	—	—	*a*
	—	265	—	15·1	—	*c*
7-Methylguanine	3·5	~240	280	~6·5	7·3–7·8	*a*
	10·0	248	283	5·7–6·2	7·8	*n*
	—	250	~270	10·6–11·0	0·9–7·1	*c*
7-Methylguanosine	—	275	—	9·7	—	*n*
	—	257	278	10·3–12·3	6·9–9·7	
9-Methylguanine	2·9	~258	268	~10·2	11·2	*a*
	9·8	252	~270	10·2	39·3	*n*
	—	251	276	12·0	7·6	*c*
1,7-Dimethylguanine	3·4	250	283	5·6	7·4	*n*
	—	252	~273	10·2	7·6	*c*
1,7-Dimethylguanosine	—	259	~280	11·3	—	*c*

examples of pK values between 12·5 and 13 are known. Nearly all guanine derivatives occur largely in cation form at pH 1·0. Table 9.16 is a selection of the data.

With the exception of 7,9-dimethylguanine (which protonates on N-1) all the derivatives accept a proton to form a cation (pK_a values 2–3·5). Dekker (1960) interpreted the spectra as indicating protonation on N-7 and further work has provided ample confirmation, while the idea that protonation occurs on the amino group is clearly wrong for 9-substituted guanines.

For guanines without a substituent at position 9 protonation probably occurs at N-9 and is certainly in the imidazole ring. In 'neutral' guanine derivatives with H at N-1 the dissociation displays a pK_a of 8·8–10. The anions of guanine and 7-methylguanine have similar ultraviolet absorption, and infrared spectra support the view that as there is no indication of carbonyl absorption, the structure resembles that of O-methylguanosine allowing the anion O^- to be placed at position 6. The anion for 1-methylguanine involves loss of a proton at N-7 or N-9 while 1,7- and 1,9-dimethylguanines do not dissociate so that the protons on the amino group lack acidic properties. When guanine moieties are present in oligonucleotides pK_a values are increased.

Hydrogen bonding is particularly important in guanine derivatives and a specific relationship towards cytosine is required for DNA with hydrogen bonding at two sites and perhaps a third (guanine NH_2 to O-2 of cytosine).

Kyoguku *et al.* (1966) used infrared spectroscopy to elucidate hydrogen bonding in derivatives of purines and pyrimidines. In order to obtain solubility in chloroform they prepared 2′,3′-benzylidene-5′-trityl derivatives and examined the infrared absorption in deuterochloroform. Guanosine and cytidine both separately showed clear evidence of association but together they formed much stronger complexes. A similar association occurred between adenine and uracil (Hamlin *et al.*, 1965) but tests for all the possible interactions of purine and pyrimidine pairs established that the only associating mixed dimer pairs were guanosine–cytidine and adenine–uracil, reinforcing the specificity expected for two stranded nucleic acids.

The non-polar solvent permitted the emergence of stretching vibrations for N—H groups involved directly in hydrogen bonding and elminated hydrogen bonding by the solvent. Figures 9.25 and 9.26 illustrate the association occurring in the cytidine and

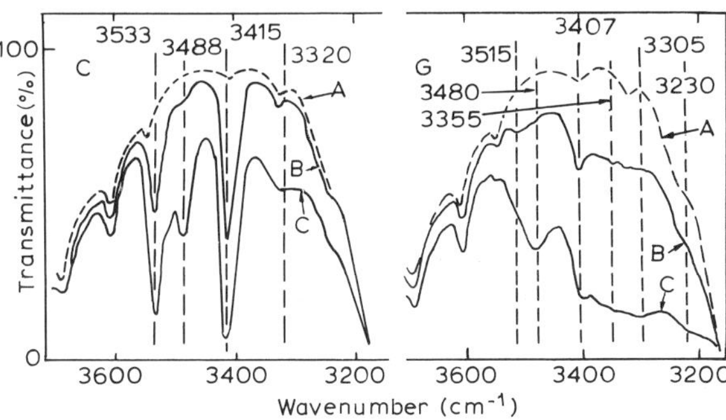

Figure 9.25 Infrared absorption spectra of cytidine (left) and guanosine (right) 2′,3′-benzylidine-5′-trityl derivatives in $CDCl_3$, 1 cm path length. (A) absorption spectrum of solvent; (B) spectrum of 0·0016 **M** solution; (C) spectrum of 0·008 **M** solution. Bands at 3448,3415 cm^{-1} for C and 3515,3407 cm^{-1} for G represent N–H stretch of the free amino group. 3488,3320 cm^{-1} (C) and 3480,3355,3305 and 3230 cm^{-1} (G) are association bands due to H-bonding of the NH_2 group (from Kyogoku *et al.*, 1966)

Figure 9.26 Absorption spectra of mixtures of cytidine (C) and guanosine (G) derivatives (path length 2·5 cm). Total concentration, 0·0016 **M**, molar concentrations as shown. The solvent absorption is taken as the base line, and the optical density plotted against wave number. Strong association is shown by the 50:50 mixture (from Kyogoku *et al.*, 1966)

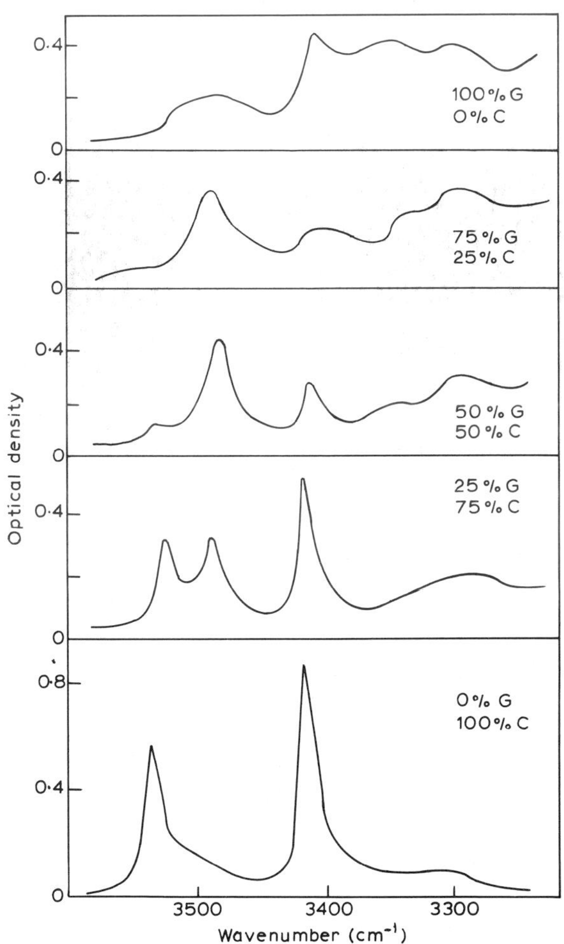

guanosine derivatives and the very strong association occurring with mixtures. Similarly the association constants of adenine with adenine or uracil with uracil are only 1/20 of the constant of adenine and uracil. N.M.R. spectra of mixtures in dimethylsulphoxide (Shoup *et al.*, 1966; Katz & Penman, 1966) indicate that simply substituted purines and pyrimidines interact to form base-paired complexes and that the specificity of pairing seems similar to that characteristic of polymer-monomer interactions in aqueous solution.

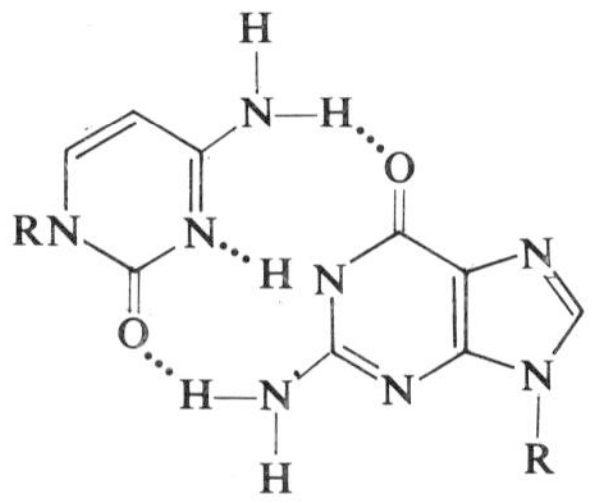

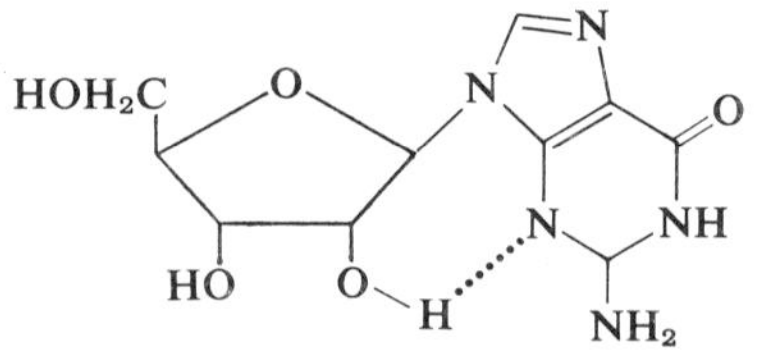

Purines do not give rise to fluorescence in the visible region but the advent of quartz fluorescence spectrometers allowed ultraviolet fluorescence to be measured (Duggan *et al.*, 1957). The excitation and fluorescence spectra of adenine and guanine were obtained by Agroskin *et al.* (1961) and by Udenfriend & Zaltman (1962). Adenine, adenosine, AMP, ADP and ATP all show excitation peaks at 272 nm and a fluorescence maximum at 390 nm in 5 N H_2SO_4. Adenine fluorescence is most intense, however, at pH 1. Guanine fluorescence $\lambda_{exc.}$ 275 nm, $\lambda_{fl.}$ 350–360 nm is much stronger than that of adenine and is specially strong at pH 11 when adenine is non-fluorescent.

Adenosine, AMP, ADP and ATP fluoresce in a strongly acid medium much more effectively than adenine. Quantum efficiencies are quite low ($\phi = 0{\cdot}003$ to 0·045 for purine at pH 10·6) and fluorescence maxima are all in the region 350–400 nm (Börresen, 1963).

Apart from thymine at pH > 10 the naturally occurring pyrimidines are non-fluorescent when exposed in aqueous solution to ultraviolet light at room temperature. Longworth *et al.* (1966) however, found that pyrimidines are fluorescent at 77 K when excited by radiation of wavelength 270 nm.

Compound	$\lambda_{max.}$ exc. nm	$\lambda_{max.}$ fl. nm pH 1	pH 7	pH 11
Cytosine	270	315	312	340–365
Thymine	270	—	316	320
Uracil	270	—	315	350
Cytidine	270	318	315	—
Thymidine	—	—	318	322
Uridine	—	—	318	350

A more serviceable application of fluorescence consists of labelling RNA with a fluorescent dye (Zamecnik *et al.*, 1960). The ribose of a terminal nucleoside yields a dialdehyde on periodate oxidation and this can condense with the amino group of a suitable dye so that a 'label' such as acriflavine can be attached (Churchich, 1963).

Ethidium bromide (see §9.2.5) can be used for fluorescence assay of DNA and RNA (cf. Udenfriend, 1969).

An ingenious indirect procedure for determining thymine was devised by Roberts & Friedkin (1958). Thymine (or 5-methylcytosine) is oxidized with

Table 9.17 Relative fluorescence intensities at different pH values

Compound	$\lambda_{max.}$ exc. nm	$\lambda_{max.}$ fl. nm	pH 13	pH 11	pH 7	pH 1	5 N H_2SO_4
Adenine	265	380	0	0	negligible	57	7
Adenosine	272	390	0	0	0	21	58
ATP	272	390	0	0	0	29	39
Guanine	275	350–360	150	680	10	200	50
Guanosine	285	390	6	16	0	200	140
Guanylic acid	285	390	5	7	0	165	125
Thymine	290	380	35	35	—	0	0
Purine	285	370	1050	940	40	negligible	

(After Udenfriend & Zaltman, 1962).

bromine water to liberate acetol ($CH_3CO.CH_2OH$), which when condensed with *o*-aminobenzaldehyde yields 3-hydroxyquinaldine. The procedure also applies to thymidine, its 5-phosphate and even to DNA, and the quinaldine moiety acts as a reproducible and comparable fluorophore.

9.2.4 Cytokinins

This topic is reviewed by Skoog & Leonard, 1968. Plant hormones of the auxin type (indoleacetic acid) promote root regeneration. Some, but not all, purines promote shoot regeneration at doses of 10 p.p.m. (Skoog & Miller, 1957). A synthetic purine derivative 6-(aminofurfuryl)purine was found to promote growth in isolated leaves and discs at much lower doses (10^{-3} to 1·0 p.p.m.). The substance was designated kinetin (Miller *et al.*, 1956; Kuraishi & Okumura, 1956; Miller, 1956; Scott & Liverman, 1956) and it was noted that growth occurred by cell enlargement. Kinetin and indoleacetic acid are now seen as mutually antagonistic; the former inhibits root regeneration in excised leaves and the latter inhibits shoot growth. The role of kinetin in terminating dormancy in buds and promoting growth of leaf and shoot is perhaps related to the maintenance of protein balance and the retention of soluble nutrient in the cell.

Activity in kinetin analogues depends on the substituent R. In kinetin itself R=C_4H_4O.

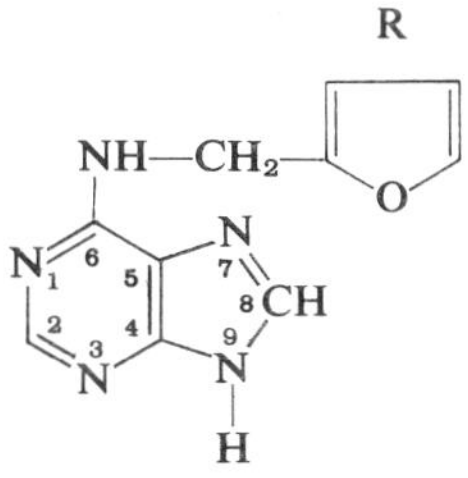

R

NH—CH_2—(furyl)

Kinetin

and marked activity persists when R=C_6H_5

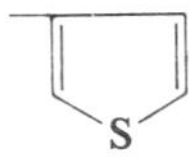

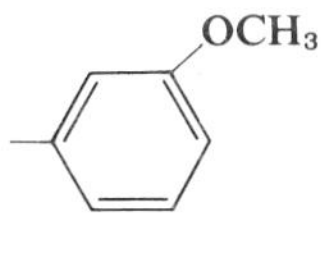

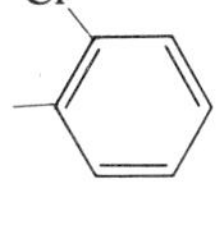

(2-thienyl)

(OCH_3-phenyl)

(Cl-phenyl)

but not when R=

(p-NH_2-phenyl)

(OCH_3, OCH_3-phenyl)

(p-Cl-phenyl)

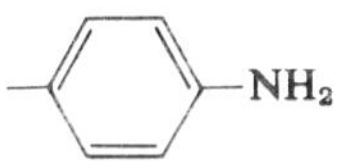

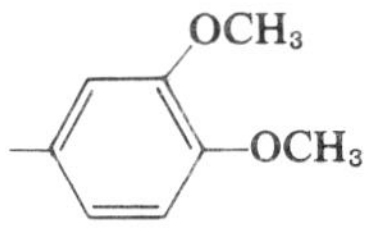

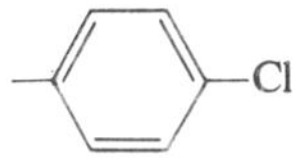

(Mothes, 1963).

Another *cytokinin* (a term due to Skoog *et al.*, 1965) was isolated from sweet corn kernels (1 mg/70 kg fresh tissue) and designated zeatin (Letham *et al.*, 1964, 1965; Shaw & Wilson, 1964). It was considerably more active than kinetin by the assay technique used. It resembles kinetin in structure R being

$$—CH_2{=}C(CH_3)—CH_2OH,$$

so that zeatin is 6-(*trans*-3-methyl-4-hydroxy-2-butenyl amino)purine.

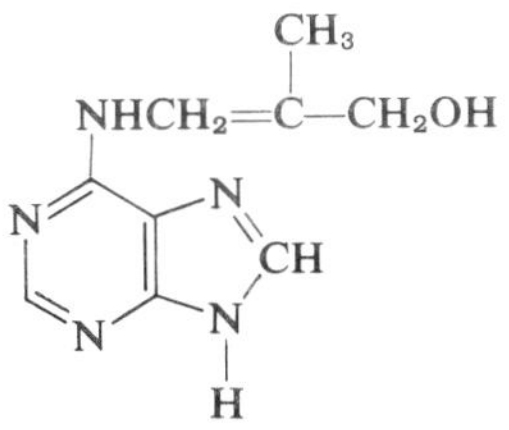

$NHCH_2{=}C(CH_3)—CH_2OH$ (purine)

Zeatin

Another widespread naturally occurring cytokinin is 6-($\gamma\gamma$-dimethylallylamino)purine [also designated N^6-(Δ^2 isopentenyl) adenine] and abbreviated 6-DMAAP. This is ten times more potent than kinetin.

The spectra of some relevant purine derivatives differ sufficiently to be useful diagnostically. 1-Benzylpurine shows a single absorption band with the following peak values: pH 1, 268 nm, pH 7, 276 nm, pH 11, 276 nm. In this respect it resembles 1-methyl-, 1-ethyl and 1-($\gamma\gamma$-dimethylallyl)purine (pH 1, 270 nm, $\epsilon_{max.}$ 7500; pH 7, 276 nm, $\epsilon_{max.}$ 7450; pH 11, 276 nm, $\epsilon_{max.}$ 7450). 6-Methylaminopurine shows $\lambda_{max.}$ 267 nm, $\epsilon_{max.}$ 15 800 at pH 2 to 7 and $\lambda_{max.}$ 273 nm, $\epsilon_{max.}$ 15 800 at pH 12. 3-Methyladenine shows $\lambda_{max.}$ 272–274 nm, $\epsilon_{max.}$ *ca.* 13 000 for the base and 15 850 in acid, while 1-methyladenine shows $\lambda_{max.}$ 268 nm,

$\epsilon_{max.}$ 10 000 at pH 7·2 and at alkaline pH values 270 nm, $\epsilon_{max.}$ 12–13 000 and at pH 4, 259 nm, $\epsilon_{max.}$ 11 800; 7-methyl- and 9-methyladenine have spectra very similar to that of the 3-methyl isomer. 6-(Farnesylamino)- and 6-(geranylamino) purines exhibit similar spectra in 95% ethanol: $\lambda_{max.}$ 269–270 nm, $\epsilon_{max.}$ *ca.* 20 000, $\lambda_{min.}$ 230 nm, $\epsilon_{min.}$ *ca.* 2000. In 0·1**N** HCl/ethanol $\lambda_{max.}$ is near 280 nm, with little change in ϵ and in 0·1 **N** alkaline ethanol there is a peak at 276 nm, *ca.* 20 000 with a shoulder at 284–285 nm and a minimum near 243 nm, $\epsilon_{min.}$ *ca.* 5350. In the case of 6($\alpha\alpha$-dimethylallylamino)purine the peaks occur in aqueous media at 269 (pH 7), 275 (pH 1) and 275, 283 nm, (pH 12) with $\epsilon_{max.}$ near 16 400 and $\epsilon_{min.}$ *ca.* 232; in neutral and acid solutions ϵ *ca.* 2250 and 240 nm, ϵ 3300 at pH 12.

In the case of 6-($\lambda\lambda$-dimethylallylthio)purine

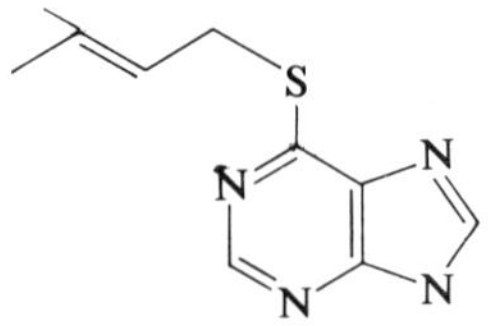

the absorption maxima fall at 215 and 292 nm ($\epsilon_{max.}$ *ca.* 14 000 and 19 500 respectively) in 95% ethanol (neutral and 0·1 **N** with respect to HCl). Alkali (0·1 **N** NaOH) leaves the main peak at 292 nm with little change in $\epsilon_{max.}$ but the minimum moves from 245 nm, $\epsilon_{min.}$ *ca.* 4000 to 252 nm, $\epsilon_{min.}$ 5400, a probably significant change (cf. Hecht *et al.*, 1970).

Tobacco pith tissue grown in culture produces N^6-(Δ^2 isopentenyl)adenosine, IPA (Hall *et al.*, 1966, 1967; Chen & Hall, 1969). Some cytokinins are shown below:

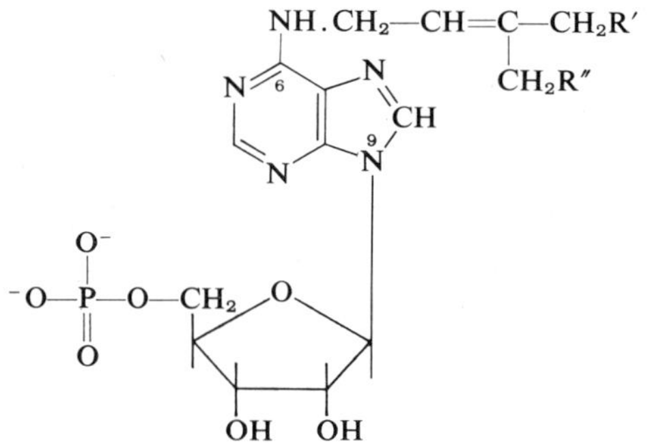

Ipa-5′-phosphate

R′ and R″	
H′ and H′	isopentenyladenosine IPA
H and OH	occurs in plants
OH and H	ribosylzeatin

Further experience shows that as many as eight cytokinins occur in sweet corn. Other sources include yeasts, fungi and bacteria.

Thus the fungus *Rhizopagon roseolus* contains zeatin and the 9-β-D-ribofuranoside (Miller, 1968). The identification of zeatin was verified by mass spectroscopy, the main m/e peaks being 219, 202, 188, 160, 148, 136, 135. The following ultraviolet absorption peaks were noted:

Solvent	Free base $\lambda_{max.}$ nm	Nucleoside $\lambda_{max.}$ nm
95% ethanol	260	268
0·1 **N** HCl	273–274	264
0·1 **N** NaOH	218, 273–274	217, 268

Table 9.18 Cytokinins from kernels of sweet corn

	Identity	*$\lambda_{max.}$ nm		Activity index
C_1	Zeatin	273·5 (pH1)	275 (pH 13)	70 (2 μM)
C_2	9-β-D-Ribofuranosylzeatin	265 (pH 1)	268·5 (pH 13)	53 (2 μM)
C_3	9-β-D-Ribofuranosylzeatin-5′ phosphate	265 (pH 2)	268·5 (pH 7)	51 (2 μM)
C_4†	6-(Monosubstituted amino) 9-substituted purine	265 (pH 2)	271 (EtOH)	50 (5 mg/l)
C_5†	6-(Monosubstituted amino) 9-substituted purine	266 (pH 2)	270 (EtOH)	18 (5 mg/l)
C_7†	6-(Monosubstituted amino) purine	273 (pH 2)	269 (EtOH)	30 (0·1 mg/l)

* Solvent water except where otherwise stated.
† Structure not certain.
Concentrations tested are given parenthetically.
(After Letham, 1966).

The nucleoside, zeatin 9-β-D-ribofuranoside, gave as the main mass spectral peaks m/e 351, 334, 320, 262, 248, 219, 202, 188, 160, 136, 119, 108, 57, 44 and 32.

Klämbt *et al.* (1966) isolated three 'cytokinins' from *Corynebacterium fascians*, a pathogen which brings about abnormality in plants. Helgeson & Leonard (1966) identified the fractions as 6-($\gamma\gamma$-dimethylallylamino)purine (6DMAAP) nicotinamide and 6-methylaminopurine. Pure nicotinamide was inactive and synthetic 6-methylaminopurine had a low specific activity. The major source of activity was the 6-DMAAP but other active substances were probably

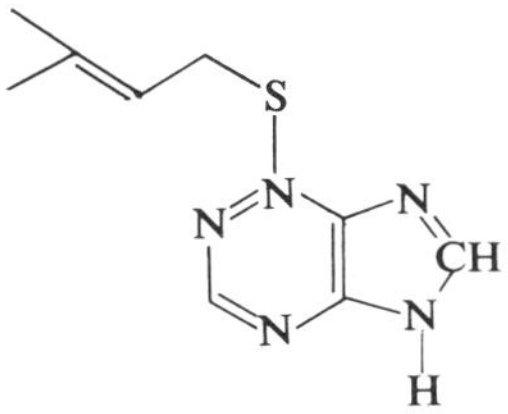

6-($\gamma\gamma$-Dimethylallylthio)purine

present. The isolation procedure was elaborate and towards the final stages the chromatography was controlled by measuring ultraviolet absorption. Concentrated fractions were treated with saturated picric acid solution and the picrates were recrystallized, the ultraviolet absorption determined and cytokinin activity determined by biological assay. The picrates showed $\lambda_{max.}$ 265–269 nm and $\lambda_{min.}$ 242–244 nm in water. The evidence that 6-DMAAP occurs naturally is here strong, indeed it is probably present as the nucleoside, 6-($\gamma\gamma$-dimethylallylamino)-9-β-ribofuranoside or its 5′-phosphate. The poor recovery of the initial activity may mean that other cytokinins were lost in manipulation. From the present standpoint a paper by Leonard *et al.* (1965) on the characterization of N_x, N_y-disubstituted adenines by ultraviolet absorption spectra is of considerable importance. A representative selection of N_x, N_y-dibenzyladenines was synthesized and the spectra were determined in neutral, acid and basic solution. The information so obtained makes it feasible to distinguish between possible N,N-disubstituted adenines (see also Jones & Robins, 1962, 1963; Pal, 1962; Leonard & Fujii, 1964). Townsend *et al.* (1964) surveyed the spectra of monosubstituted adenines.

Leonard *et al.* (1965) compared the spectra of eight N,N-dibenzyladenines in 95% ethanol, in 0·1 N NaOH in 95% ethanol and 0·1 N HCl in 95% ethanol. The CH_2 of the benzyl group insulates the phenyl and purine chromophores from each other and the contribution of the phenyl group is substantially masked and can for most purposes be neglected. The authors regard the general shape of the absorption curves as more serviceable than the brief statement of $\lambda_{max.}$ and $\epsilon_{max.}$ at different pH values.

The cytokinins have been shown to occur in transfer RNA. This is a most important development but whether or not it helps to explain the growth-promoting action of the free base is not yet clear.

In protein biosynthesis tRNA links the genetic information coded in messenger RNA as follows: it accepts the aminoacid specific to the individual tRNA and 'reads' correctly the codon in the genetic message and then transfers the aminoacid to the growing 'protein'.

It has been evident for some time that soluble RNA contains 'odd' or unusual bases in small amounts. Isopentyladenosine (IPA) occurs in yeast RNA (Hall *et al.*, 1966) and in particular it occurs in two serine-specific tRNAs (from brewer's yeast) having known primary sequences (Zachau *et al.*, 1966; Biemann *et al.*, 1966). Baker's yeast soluble-RNA was subjected to countercurrent fractionation (Kovoor & Klämbt, 1968) and various transfer RNA fractions were measured by intensity of absorption at 260 nm and assayed by a tobacco pith callus procedure. The evidence indicated at least two cytokinin-containing

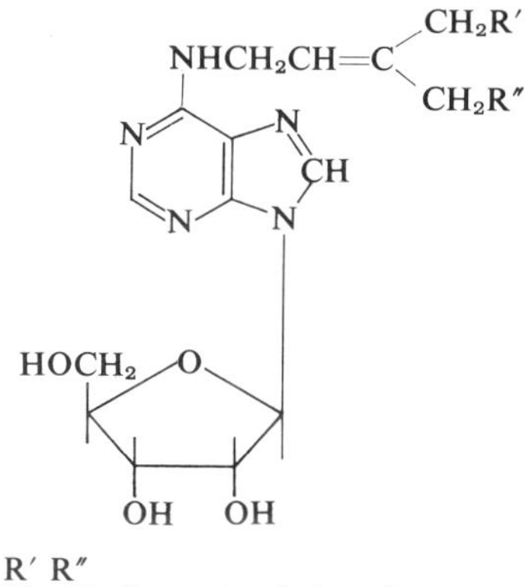

R′ R″
H H = Isopentenyladenosine
OH H = Ribosylzeatin
H OH = Isomer (in plants)

tRNAs. Hall (1968) identified IPA in plant and animal tissue as well as in yeast. Hall *et al.* (1967) has also isolated the *cis* isomer of ribosylzeatin [6-N(*cis*-4-hydroxy-3-, ethylbut-2-enylamino)-9-β-D-ribofuranozyl)purine] from the tRNA fractions of garden peas, spinach and young maize kernels. In seryl-tRNA of

Table 9.19 Ultraviolet maxima of 6-(Δ^2-isopentenylamino) purines

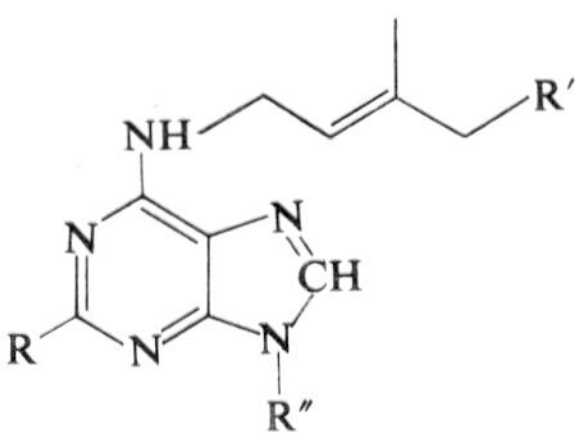

R	R′	R″	
Cl	H	H	= Ia
SCH_3	OH	H	= Ih
SCH_3	H	ribose (CH_2OH)	= IIc
Cl	OH	H	= Ie
NH_2	OH	H	= If
OH	OH	H	= Ig

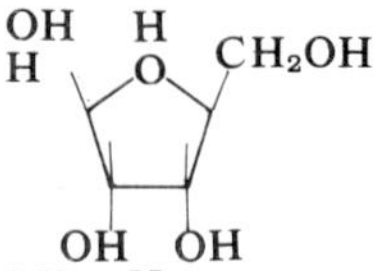

Compound	$\lambda_{max.}$ nm	$\epsilon \times 10^{-3}$	$\lambda_{max.}$ nm	$\epsilon \times 10^{-3}$	$\lambda_{max.}$ nm	$\epsilon \times 10^{-3}$
Ia	273	(15·1)	272	(18·6)	278	(17·2)
Ie	273	(17·1)	272	(19·2)	278	(17·8)
If	282	(12·9)	282	(12·3)	289	(11·8)
	248	(13·0)	248	(10·7)		
Ig	289	(16·1)	284	(12·0)	288	(17·5)
			243	(11·5)		
Ih	292	(17·4)	279	(17·5)	286	(16·5)
	252	(24·2)	241	(28·4)	231	(34·0)
IIc	282	(10·5)	272	(13·0)	272	(13·5)
			279	(sh)	279	(sh)

(Hecht *et al.*, 1970).

yeast IPA occurs adjacent to the anticodon and modification of IPA interferes with the mRNA——tRNA interaction (Fittler & Hall, 1966).

The Δ^2-isopentenyl side-chain of IPA in the tRNA of micro-organisms is derived from mevalonic acid but this, though probable, has not actually been demonstrated for plant and animal tissues. Nevertheless important possibilities of feed-back are possible since acetate and mevalonate participate in very many biosynthetic sequences.

It seems clear that the natural cytokinins are isoprenoid nucleosides (or isoprenoid purine bases) capable of stimulating cell proliferation and differentiation in plants. It may well be that IPA-like substances are necessary for such processes in animals. Certainly the link with tRNA molecules seems definite. Chemical or enzymatic changes may be induced in the purine moiety, synthetic nucleosides or bases may be provided and spectrophotometry may well find a place in following the results.

Hecht *et al.* (1970) studied the biological activity of 2-substituted compounds in the IPA and zeatin series. Burrows *et al.* (1968, 1969) had isolated and identified two nucleosides responsible for cytokinin activity in *E. coli* tRNA's. One had a methylthio-substituent at position 2 (ms2IPA) while the other was IPA [N^6-(Δ^2-isopentenyl)adenosine].

The possible importance of substitution in position 2 prompted the synthesis of various compounds followed by bioassays by the tobacco method. Introduction of OH at position 2 lowered activity in both the IPA and zeatin series, while insertion of $—NH_2$ or $—SCH_3$ groups had a smaller effect and Cl a negligible effect.

Mass spectrometry for the 2-substituted N^6-isopentenyladenines provided the first indication of the presence of the methylthio substituent. The major fragments resulted from stepwise splitting off of the sugar and side-chain moieties. The 2-amino-IPA lost fragments as follows: from the sugar portion to ions of m/e 261 $(M\text{-}89)^+$, 247 $(B+30)^+$ and 218 $(B+1)^+$, from the side-chain to give peaks at m/e 203, 175, 163, 150 and 134. The substituent at 2 was retained until rather low m/e values were reached. The ribosides and the free bases behaved similarly and mass spectroscopy is clearly particularly useful if 2-substituents are likely to prove important.

The modification of the base adjacent to the anticodon in $tRNA^{Tyr}$ seems to be an example (Gefter & Russell, 1969).

9.2.5 Ethidium Bromide (EB)

This was first used as a trypanocide (Dickenson *et al.*, 1953). Its interest as a reagent depends on an ability to bind to nucleic acids and various ways of using it have been reviewed (Balis, 1968). Much work has been done on the physical chemistry of nucleic acid–EB interaction (Le Pecq, 1971).

$R = C_2H_5$ Br^-

Ethidium bromide
(3,8-Diamino-6-phenyl-5-ethylphenanthridinium)

Ethidium bromide attaches to DNA or RNA at two different types of site. The first is called the 'fluorescent' site, while the second site, at which attachment is less strong, is mainly electrostatic; the fluorescent quantum efficiency is negligibly low for EB bound at this site.

EB is itself fluorescent but when it is bound at the first site there is a large enhancement. Thus, on binding to double-stranded DNA the quantum efficiency of fluorescence increases 21-fold and on binding to double-stranded RNA it increases 25-fold. The absorption spectrum undergoes a red shift on binding to nucleic acids at either type of binding site.

In water, EB shows $\lambda_{max.}$ 285 nm, ϵ 51 000 and $\lambda_{max.}$ 480 nm, ϵ 5300. In the presence of excess DNA the peaks occur at $\lambda_{max.}$ 300 nm, $\epsilon_{max.}$ 21 250 and $\lambda_{max.}$ 521 nm, ϵ 3900. With intermediate concentrations of DNA a series of curves is obtained and isosbestic points appear at 300, 392 and 512 nm (see Fig. 9.27). Commercial EB contains one molecule of methanol and $\epsilon_{480\,nm}$ corrected = 5450. Le Pecq (1971) points out that wavelengths can be found where free EB absorbs very much less strongly than bound EB. Apparent fluorescence intensity as measured in a spectrofluorometer for a given wavelength of excitation depends both on the quantum efficiency of fluorescence and the absorbance at that wavelength. By choosing favourable conditions, the fluorescent intensity for EB bound to DNA or RNA can be about 100 times greater than the fluorescent intensity of free EB under the same conditions. This is illustrated in Fig. 9.28, where the fluorescence excitation spectrum intensity of free EB is very low compared with that of DNA-bound EB. The fluorescence spectrum shows a very sharp peak near 590 nm and the quantum efficiency of fluorescence for bound EB is 0·14.

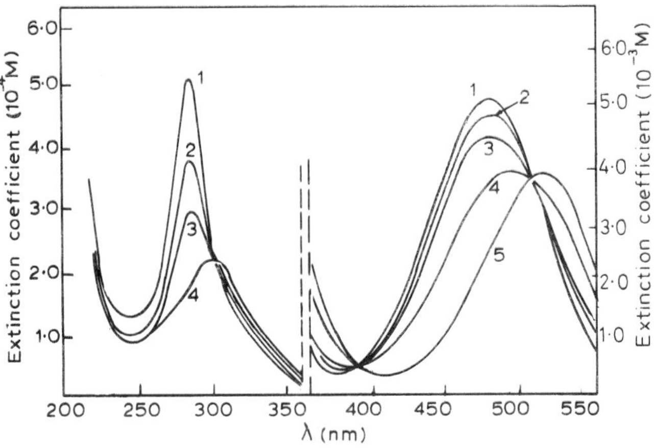

Figure 9.27 Absorption spectra of ethidium bromide. Curves 1 and 1′ show the EB absorption spectrum in aqueous solution curves 4′ and 5 show the absorption of EB in the presence of an excess of DNA, i.e. the absorption spectrum of DNA-bound EB. Curves 2′, 3′, 2, 3, 4 show the EB absorption in the presence of increasing concentrations of DNA; three isosbestic points appear (from Le Pecq, 1971)

Energy transfer experiments have indicated that the fluorescence depends on binding at an intercalation site, specific for double-stranded polynucleotides. In a mixture of poly A and poly U that contains single-

Figure 9.28 Fluorescence excitation and emission spectra of free and bound ethidium bromide. Curves 1 and 2, non-corrected fluorescence excitation spectra of free and DNA-bound EB. Curve 3, fluorescence emission spectrum of DNA-bound EB (from Le Pecq, 1971)

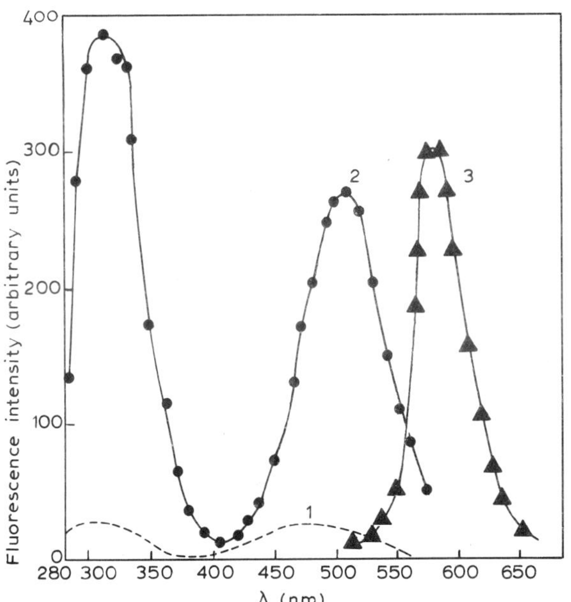

stranded and triple-stranded polynucleotides, the increase in EB fluorescence is specific to the double-stranded structure. On DNA there is one binding site per five nucleotides. Single-stranded RNA or DNA when folded in a 'hair-pin' structure can attach EB at the bend. The second binding site modifies the absorption spectrum of EB in exactly the same way as the first site and the same three isosbestic points are observed. The distinction between the two sites is made by spectrofluorometry. The affinity constant of EB at the second site falls markedly as the ionic strength increases with the result that in 0·1 **M** NaCl as the medium, binding is negligible.

The properties of ethidium bromide make possible a fluorometric method for determining DNA or RNA (Le Pecq & Paoletti, 1966). To an EB solution of concentration c_{EB} (at defined temperature, salt concentration and pH) nucleic acid is added stepwise. When the nucleic acid concentration is c and I_0 is the fluorescence intensity without nucleic acid and I that with added nucleic acid

$$I - I_0 = Kc$$

if c is small compared with c_{EB} and all the fluorescent sites are fully occupied. Fluorescence intensity and nucleic acid concentration are linearly related. Calibration with a standard nucleic acid solution allows K to be determined. The concentration of the standard nucleic acid solution is determined separately by spectrophotometry (Webb & Levy, 1958). Technical details are given by Le Pecq (1971). They include: salt concentration 0·1 **M** NaCl; 10^{-3} **M** EDTA and Tris-HCl; pH 7·5; EB concentration 10 μg/ml for nucleic acid concentrations 1–10 μg/ml. Any native DNA can be used for calibration (except for closed circular DNA). Single-stranded DNA obtained by denaturation can under 'renaturing' conditions form hair-pin like structures containing some double-stranded regions. These can be determined. RNA is in the main similar to the denatured DNA structure and undergoes similar 'renaturation'. Calibration with an identical RNA is necessary in exploiting the method. The whole procedure is simple and very sensitive.

Covalently closed-circular double-stranded DNAs occur widely in animal viruses and bacteriophages and many other sources and there is no doubt about their importance. They vary in size and some are very small. EB is intercalated between two base pairs and this is in a way equivalent to adding an additional base pair to the double helix. The helix must unwind sufficiently to leave room for the EB molecule. If there is a nick on the DNA to act as a swivel, one strand can rotate round the other. But for covalently closed-circular DNA there is no nick and swivelling cannot occur. Unwinding of DNA must then modify the tertiary turn number, unwinding of the helix by one turn resulting in one tertiary turn, but 30 EB molecules per circle are required to unwind by one turn (Crawford & Waring, 1967; Bauer & Vinograd, 1968).

Now EB decreases the buoyant density of DNA in a caesium chloride gradient and the density change varies directly as the amount of EB bound per DNA molecule (Le Pecq *et al.*, 1964). This is the basis for a quantitative procedure for separating DNAs. Since linear DNA and closed-circular DNA bind EB to different extents, they can be separated in a suitable gradient. This applies also to separating single-stranded and double-stranded RNA.

EB binding is measured by ultraviolet spectrophotometry with subtraction of DNA absorption. Advantage is taken of the high $\epsilon_{max.}$ at *ca.* 280 nm. The absorption at 280 nm of the EB–DNA solution is measured with a DNA solution of the same concentration in the compensating cell. If c_b and c_f are the concentrations of free and bound EB and e_b and e_f are the respective extinction coefficients

$$A_{280} = c_b e_b + c_f e_f$$

where A is observed absorbance and

$$c_b = \frac{c_t - A}{e_f - e_b}$$

where $c_t = c_b + c_f$, i.e. total EB concentration. As an alternative a fluorescence method can be used.

Let I_o be the fluorescence intensity for a pure EB solution of concentration c_o, I the intensity of fluorescence of the EB–DNA solution and F the factor for the fluorescence ratio: bound EB/free EB. Le Pecq & Paoletti (1967) showed that

$$c_b = \frac{I - I_o}{(F-1)k}$$

subject to certain assumptions.

Le Pecq (1971) describes ingenious extensions to the fluorometric assay of DNAase using linear DNA or closed circular DNA and for determining DNA ligase as well as polymerase activities.

References

Ablondi, F., Gordon, S., Morton, J. I. & Williams, J. H. (1952). *J. Org. Chem.*, **17**, with 14 other papers 1–156 (*Williams et al.*).

Agroskin, L. C., Korolev, N. V., Kulaev, I. S. & Pomoshchnikova, N. A. (1961). *Dokl. Acad. Nauk, SSSR*, **136**, 226.

Albert, A. (1955). *J. Chem. Soc.*, 2690.

Albert, A. (1959). *Heterocyclic Chemistry*, p. 20. Butterworths, London.

Albert, A., Armarego, W. L. F. & Spinner, E. (1961 *a*). *J. Chem. Soc.*, 2689; (1961 *b*). *Ibid.*, 5267.

Albert, A. & Spinner, E. (1960). *J. Chem. Soc.*, 1221, 1226, 1232; (1964). *Ibid.*, 1523.

Amstutz, E. D. (1952). *J. Org. Chem.*, **17**, 1508.

Anderson, L. C. & Seeger, N. V. (1949). *J. Amer. Chem. Soc.*, **71**, 340 and 343.

Angell, C. L. (1961). *J. Chem. Soc.*, 504.

Angier, R. B. & Curran, W. V. (1961). *J. Org. Chem.*, **26**, 1891.

Armarego, W. L. F. (1961). *J. Chem. Soc.*, 2697.

Ault, R. G., Hirst, E. L. & Morton, R. A. (1935). *J. Chem. Soc.*, 1653.

Austin, J. E. (1934). *J. Amer. Chem. Soc.*, **56**, 2141.

Badger, G. M. & Walker, I. S. (1956). *J. Chem. Soc.*, 122.

Balis, E. (1968). *Antagonists and Nucleic Acids. Frontiers of Biology.* Vol. **10**. North-Holland Publishing Co., Amsterdam.

Bates, R. W. (1954). In *Progress in Hormone Research*, **9**, 95.

Batt, R. D., Martin, J. K., Ploesser, J. M. & Murray, J. (1954). *J. Amer. Chem. Soc.*, **76**, 3663.

Bauer, W. & Vinograd, J. (1968). *J. Molec. Biol.*, **33**, 141.

Bauer, L., Wright, G. E., Mikrut, B. A. & Bell, C. L. (1965). *J. Heterocyclic Chem.*, **2**, 447.

Beaven, G. H., Holiday, E. R. & Johnson, E. A. (1955). In *The Nucleic Acids*. Vol. I, Chapter 14, p. 493. Ed. E. Chargaff and J. N. Davidson. Academic Press, New York.

Bellamy, L. J. (1958). *Spectrochim. Acta*, **13**, 60; (1961). In *Organic Sulfur Compounds*. Ed. N. Kharasch. Pergamon Press, Oxford.

Benitoz, A., Ross, L. O., Goodman, L. & Baker, R. R. (1960). *J. Amer. Chem. Soc.*, **82**, 4585.

Bergman, K. & Dikstein, S. (1955). *J. Amer. Chem. Soc.*, **77**, 691; (see also (1956)). In *Methods of Biochemical Analysis*, Vol. **6**, p. 79. Ed. D. Glick. J. Wiley Interscience, New York.

Beroza, M. (1952). *J. Amer. Chem. Soc.*, **74**, 1585

Beukers, R., Ylstra, J. & Berends, W. (1959). *Rev. Trav. Chim.*, **78**, 883.

Beukers, R. & Berends, W. (1960). *Biochem. Biophys. Acta.*, **41**, 550; (1961). *Ibid.*, **49**, 181.

Bhattacharyya, J. (1963). *Tetrahedron*, **19**, 1011.

Biemann, K., Tsunakawa, J., Sonnenbichler, J., Feldmann, H., Dütting, D. & Zachau, H. G. (1966). *Angew. Chem. Int. Edn.*, **5**, 590.

Birch, A. J. & Richards, R. W. (1956). *Aust. J. Chem.*, **9**, 241.

Black, M. L. (1955). *J. Physical Chem.*, **59**, 670.

Blackburn, G. M. & Davies, R. J. H. (1966). *Biochem. Biophys. Res. Commun.*, **22**, 704.

Blinzynkov, V. I. & Reznikov, V. M. (1955). *Zhur. Obschei Khim.*, **25**, 401.

Blout, E. R. & Fields, M. (1948). *J. Biol. Chem.*, **178**, 335; (1950). *J. Amer. Chem. Soc.*, **72**, 479.

Boarland, M. P. V. & McOmie, J. F. W. (1951). *J. Chem. Soc.*, 128. (1952). *Ibid.*, 5716.

Boig, F. S., Costa, G. W. & Osvar, I. (1953). *J. Org. Chem* **18**, 775.

Börresen, H. C. (1963). *Acta. Chem. Scand.*, **17**, 921.

Bresnick, E., Singer, S. & Hitchings, G. H. (1960). *Biochim. Biophys. Acta.*, **37**, 251.

Brocklehurst, P. & Burawoy, A. (1960). *Tetrahedron*, **10**, 118.

Brown, D. J. & Mason, S. F. (1956). *J. Chem. Soc.*, 3443; (1957). *Ibid.*, 682.

Brown, D. J., Hoerger, E. & Mason, S. F. (1955 *a*). *J. Chem. Soc.*, 4035; (1955 *b*). *Ibid.*, 211.

Brown, D. J. & Short, L. N. (1953). *J. Chem. Soc.*, 331.

Brown, D. J. & Teitei, T. (1964*a*). *Australian J. Chem.*, **17**, 567. (1964*b*). *J. Chem. Soc.*, 3204. (1965). *Ibid.*, 832.

Brown, H. C. & McDaniel, D. H. (1955). *J. Amer. Chem. Soc.*, **77**, 3752.

Brown, H. C. & Mihm, X. R. (1955). *J. Amer. Chem. Soc.*, **77**, 1723.

Burrows, W. J., Armstrong, D. J., Skoog, F., Hecht, S. M., Boyle, J. T. A., Leonard, N. J. & Occolowitz, J. (1968). *Science, N.Y.*, **161**, 691. (1969). *Biochemistry*, **8**, 3071.

Chakravarti, D., Chakravarti, R. N., Cohen, L. A., Dasgupta, B., Datta, S. & Miller, H. K. (1962). *Tetrahedron*, **16**, 224.

Chatterjee, A. & Majumdar, S. G. (1954). *J. Amer. Chem. Soc.*, **76**, 2459.

Chheda, G. B., Hall, R. H., Magrath, D. I., Mozejko, J. & Schweizer, M. P. (1960). *Biochemistry*, **8**, 3278.

Chen, C. M. & Hall, R. H. (1969). *Phytochem.*, **8**, 1687.

Churchich, J. (1963). *Biochem. Biophys. Acta*, **75**, 274; and *Biochemistry*, **2**, 781.

Clark, L. B. & Tinioco, I. (1965). *J. Amer. Chem. Soc.*, **87**, 11.

Cookson, G. H. (1953). *J. Chem. Soc.*, 2789.

Craig, D. P. & Short, L. N. (1945). *J. Chem. Soc.*, 419.

Crawford, L. V. & Waring, M. J. (1967). *J. Molec. Biol.*, **25**, 23.

Culbertson, H., Decuis, J. C. & Christensen, B. E. (1954). *J. Amer. Chem. Soc.*, **74**, 4834.

Davis, D. R. & Rich, A. (1958). *J. Amer. Chem. Soc.*, **80**, 1003.

Dekker, C. A. (1960). *Ann. Rev. Biochem.*, **29**, 453.

Deering, R. A. & Setlow, R. B. (1963). *Biochem. Biophys. Acta.*, **68**, 526.

Dickinson, L., Chantrill, D. H., Inkley, C. W. & Thompson, M. J. (1953). *Brit. J. Pharmacol.*, **8**, 139.

Djerassi, C. (1960). In *Optical Rotatory Dispersion*. McGraw-Hill, New York.

Duggan, D. E., Bowman, R. L., Brodie, B. B. & Udenfriend, S. (1957). *Arch. Biochem. Biophys.*, **68**, 1.

Elison, G. B. (1962). *J. Org. Chem.*, **27**, 2478.

Elison, G. B., Ide, W. S. & Hitchings, G. H. (1946). *J. Amer. Chem. Soc.*, **68**, 2137.

Evans, K. L. & Gillam, A. E. (1943). *J. Chem. Soc.*, 565.

Ewing, G. W. & Steck, E. A. (1946). *J. Amer. Chem. Soc.*, **68**, 2181.

Fabian, J. & Mayer, R. (1964). *Spectrochim. Acta.*, **20**, 299.

Fehnel, E. A. & Carmack, M. (1949). *J. Amer. Chem. Soc.*, **71**, 84.

Feldmann, H., Dütting, D. & Zachau, H. G. (1966). *Z. f. Physiol. Chem.*, **347**, 236.

Felsenfeld, G. & Rich, A. (1957). *Biochem. Biophys. Acta*. **26**, 457.

Fikus, M., Wierzchowski, K. L. & Shugar, D. (1962). *Photochem. & Photobiol.*, **1**, 325.

Fittler, F. & Hall, R. H. (1966). *Biochem. Biophys. Res. Commun.*, **25**, 441.

Fox, J. J., Cavalieri, L. F. & Chang, N. (1953). *J. Amer. Chem. Soc.*, **74**, 4315.

Fox, J. J. & Shugar, D. (1952). *Bull. Soc. Chim. Belges.*, **61**, 44.

Fox, J. J., Yung, N. & Wempen, I. (1957). *Biochem. Biophys. Acta.*, **23**, 295.

Fredholm, H. (1937). Thesis, Uppsala (see *Chem. Abstracts*, 1939).

Füchtbauer, W. & Mazur, P. (1966). *Photochem. Photobiol.*, **5**, 323.

Gefter, M. L. & Russell, R. L. (1969). *J. Mol. Biol.*, **39**, 145.

Gibson, J. A., Kynaston, W. & Lindsey, A. S. (1955). *J. Chem. Soc.*, 4340.

Gillam, A. E. & Stern, E. S. (1962). In *Introduction to Electronic Spectroscopy*. 2nd Ed. (1970). 3rd Ed., Ed. E. S. Stern and C. J. Timmons. Edward Arnold, London.

Goldwasser, E. & Heinrikson, R. L. (1966). *Prog. in Nucleic Acid Research*, **5**, 399. Ed. J. N. Davidson and W. E. Cohn. Academic Press, New York.

Goldschmidt, S., Lamprecht, W. & Helmreich, E. (1953). *Z. Physiol. Chem.*, **292**, 125.

Grammaticakis, P. (1961). *C. R. Acid. Sci., Paris*, **252**, 4011.

Gronowitz, S. & Hoffman, R. A. (1960). *Arkiv. fur Kemi.*, **16**, 459.

Gronowitz, S., Norman, B., Gestblom, B. O., Mathiasson B. & Hoffmann, R. A. (1964). *Arkiv. fur Kemi.*, **22**, 65, 82.

Gulland, J. M. (1938). *J. Chem. Soc.*, 1722.

Gulland, J. M. & Holiday, E. R. (1936). *J. Chem. Soc.*, 765.

Gulland, J. M., Holiday, E. R. & Macrae, T. F. (1934). *J. Chem. Soc.*, 1641.

Hall, R. H. (1965). *Biochemistry*, **4**, 66.

Hall, R. H. (1971). *The Modified Nucleosides in Nucleic Acid*. Columbia University Press, New York.

Hall, R. H., Robins, M. J., Stasinh J. & Thedford, R. (1966). *J. Amer. Chem. Soc.*, **88**, 2614.

Hall, R. H., Csxonka, L. C., David, H. & McLennan, B. (1967). *Science, N.Y.*, **156**, 69.

Hall, R. H. (1968). In *Biochemistry and Physiology of Plant Growth Substances*. Ed. F. Wightman and G. Setterfield. Runge Press, Ottawa, Canada.

Halverson, F. & Hirt, R. C. I. (1952). *J. Chem. Phys.*, **19**, 711.

Hamlin, R. M., Lord, R. C. & Rich, A. (1965). *Science, N.Y.*, **148**, 1734.

Hearn, J. M. Morton, R. A. & Simpson, J. C. E. (1951). *J. Chem. Soc.*, 3318.

Hecht, S. M., Leonard, N. J., Schmitz, R. Y. & Skoog, F. (1970). *Phytochem.*, **9**, 1173.

Helgeson, J. P. & Leonard, N. J. (1966). *Proc. Nat. Acad. Sci. Wash.*, **56**, 60.

Heyroth, F. F. & Loufbourow, J. R. (1931). *J. Amer. Chem. Soc.*, **53**, 3441.

Hotchkiss, R. D. (1957). In *Methods in Enzymology*, Vol. III, p. 708. Ed. S. P. Colowick & N. O. Kaplan. Academic Press, New York.

Houghton, C. & Cain, R. B. (1972). *Biochem. J.*, **130**, 879.

Huebner, C. F., Daissi, P. A. & Scholz, C. R. (1953). *J. Org. Chem.*, **18**, 21.

Inoue, Y., Furutacki, N. & Nakanishi, K. (1966). *J. Org. Chem.*, **31**, 175–8.

Janion, C. & Shugar, D. (1960). *Acta. Biochem. Polon.*, **7**, 309.

Janssen, M. J. (1960). *Rec. Trav. Chim.*, **19**, 454, 464; (1961). *Spectrochim. Acta.*, **17**, 475.

Jones, J. W. & Robins, R. K. (1962). *J. Amer. Chem. Soc.*, **84**, 1914; (1963). *Ibid.*, **85**, 193.

Kalckar, H. M. & Shafran, M. (1947). *J. Biol. Chem.*, **167**, 429.

Katritzky, A. R. & Lagowski, J. M. (1960). In *Heterocyclic Chemistry*. Academic Press, London.

Katritzky, A. R. & Waring, A. J. (1962). *Chem. & Ind.*, 695.

Katz, I. & Penman, S. (1966). *J. Mol. Biol.*, **15**, 220.

Kavoor, A. & Klämbt, D. (1968). In *Biochemistry and Physiology of Plant Growth Substances*. Ed. F. Wightman and G. Setterfield. Runge Press, Ottawa, Canada.

Klämbt, D., Thies, G. & Skoog, F. (1966). *Proc. Nat. Acad. Sci. Wash.*, **56**, 52.

Knaggs, I. E. & Lonsdale, K. (1940). *Proc. Roy. Soc.* A **177**, 140.

Koepli, J. B., Mead, J. F. & Brockman, J. A. (1949). *J. Amer. Chem. Soc.*, **71**, 1048.

Kracmar, J., Kraacmarova, J., Zyka, J. (1968). *Cesk. Farm.*, **17**, 68.

Kuraishi, S. & Okumura, F. S. (1956). *Botan. Mag. (Tokyo)*, **69**, 300.

Kwiatowski, J. S. (1968). *Theor. Chem. Acta.* **11**, 167. (See also *Bull. Akad. Pol. Sci. Ser. Chim.*, **16**, 51.)

Kyogoku, Y., Lord, R. C. & Rich, A. (1966). *Science, N.Y.*, **154**, 518.

Leandri, G., Mangini, A., Montanari, F. & Passerini, R. (1955). *Gazz. Chim.*, **85**, 769.

Leonard, N. J., Carraway, K. L. & Helgeson, J. P. (1965). *J. Heterocyclic Chem.*, **2**, 291.

Leonard, N. J. & Fujii, T. (1964). *Proc. Nat. Acad. Sci. Wash.*, **51**, 73.

Le Pecq, J. B. (1971). In *Methods of Biochemical Analysis*. Ed. D. Glick. Vol. 20. Interscience Publishers, New York, London.

Le Pecq, J. B. & Paoletti, C. (1966). *Analyt. Biochem.*, **17**, 100; (1967). *J. Molec. Biol.*, **27**, 87.

Le Pecq., J. B., Yot, P. & Paoletti, C. (1964). *C.R. Acad. Sci., Paris* **259**, 1786.

Letham, D. S. (1968). In *Biochemistry and Physiology of Plant Growth Substances*. Ed. F. Wightman and G. Setterfield. Runge Press, Ottawa, Canada.

Letham, D. S. (1966). *Life Sci.*, **5**, 551.

Letham, D. S. & Miller, C. O. (1965). *Plant Cell. Physiol.*, **6**, 355.

Letham, D. S., Shannon, J. S. & McDonald, I. R. (1964). *Proc. Chem. Soc.*, 230.

Liu, T. C. & Duncan, A. B. F. (1949). *J. Chem. Phys.*, **17**, 241.

Longuet-Higgins, M. S. & Coulson, C. A. (1949). *J. Chem. Soc.*, 971.

Longworth, J. W. (1962). *Biochem. J.*, **84**, 10P.

Longworth, J. W. *et al.* (1966). *J. Chem. Phys.*, **45**, 2930.

Lord, R. C., Martom, A. L. & Mill, F. A. (1957). *Spectrochim. Acta*, **9**, 113.

MacKinnen, G. & Temmer, O. (1948). *J. Amer. Chem. Soc.*, **70**, 3586.

McLaren, A. D. & Shugar, D. (1964). In *Photochemistry of Proteins and Nucleic Acids*. Pergamon Press, Oxford, London.

Marshall, J. R. & Walker, J. (1951). *J. Chem. Soc.*, 1004.

Mason, S. F. (1954). *J. Chem. Soc.*, 2071; (1957). *Ibid.*, 4874, 5010; (1959). *Ibid.*, 1253; (1960). *Ibid.*, 219; (1963). In *Physical Methods in Heterocyclic Chemistry*, Vol. 2, Chapter 7. Ed. A. R. Katritzky.

Metha, D. R., Naravane, J. S. & Desai, R. M. (1963). *J. Org. Chem.*, **28**, 445.

Metzler, E. E. & Snell, E. E. (1955). *J. Amer. Chem. Soc.*, **77**, 2431.

Michelson, A. M. (1951). In *Organic Chemistry of the Nucleic Acids* (see also (1951). *J. Cellular Comp. Physiol.*, **38**, Suppl. 1); (1958). *Nature, Lond.*, **182**, 1502.

Miles, H. T. (1959). *Nature, Lond.*, **181**, 1814; (1958). *Biochem. Biophys. Acta.*, **30**, 324; (1959). *Ibid.*, **45**, 196; (1960). *Ibid.*, **47**, 791.

Miller, C. O. (1956). *Plant Physiol.*, **31**, 318; (1961). *Ann. Rev. Plant Physiol.*, **12**, 395; (1968). In *Biochemistry and Physiology of Plant Growth Substances*. Ed. F. Wightman and G. Setterfield. Runge Press, Ottawa, Canada; (1961). *Proc. Nat. Acad. Sci. Wash.*, **47**, 170; (1965). *Ibid.*, **54**, 1052.

Miller, C. O., Skoog, F., Okumura, F. S., Von Saltzer, M. H. & Strong, F. M. (1956). *J. Amer. Chem. Soc.*, **78**, 1375.

Miles, H. T. & Frazier (1964). *Biochem. Biophys. Acta.*, **79**, 216.

Moore, A. M. & Thompson, C. H. (1955). *Science, N.Y.*, **122**, 594; (1957). *Canad. J. Chem.*, **35**, 163; (1968). *Ibid.*, **36**, 281.

Morton, R. A. & Rogers, E. (1925). *J. Chem. Soc.*, **127**, 2699.

Morton, R. A. & Stubbs, A. L. (1939). *J. Chem. Soc.*, 1321; (1940). *Ibid.*, 1347.

Mothes, K. (1960). *Naturwiss.*, **47**, 337; (1961). *Ber. deut. botan. Ges.*, **74**, 24; (1963). In *Comprehensive Biochemistry*, Vol. II, p. 159. Ed. M. Florkin and E. Stotz. Elsevier, Amsterdam.

Murthy, A. S. N., Rao, C. N. R., Rao, B. D. N. & Venkateswardu, P. (1962). *Trans. Faraday Soc.*, **58**, 855.

Nakanishi, K., Furatashi, N., Funamizu, M., Grunberger, D. & Weinstein, B. (1970). *J. Amer. Chem. Soc.*, **92**, 7617.

Nishimura, S. (1972). In *Progress in Nucleic Acid Research*, **12**, 49. Ed. J. N. Davidson & W. E. Cohn. Academic Press, New York.

Oster, C., Citarel, L. & Goodman, M. (1962). *J. Amer. Chem. Soc.*, **84**, 703.

Pakrashi, S., Bhattacharyya, J., Johnson, L. F. & Budzikiewicz, H. (1963). *Tetrahedron*, **19**, 1011.

Pal, B. C. (1962). *Biochemistry*, **1**, 558.

Pappalardo, G. (1959). *Gazz. Chim.*, **89**, 450.

Patterson, J. W., Lasarow, A., Lemm, F. J. & Levey, S. (1949). *J. Biol. Chem.*, **177**, 187.

Perrin, D. O. (1960). *J. Amer. Chem. Soc.*, **82**, 5642.

Pfleiderer, W. & Nubel, G. (1961). *Annalen*, **647**, 167; (1963). *Nature, Lond.*, **198**, 1193.

Philips, J. P. (1960). In *Organic Electron Spectral Data*. M. J. Kamlet Ed. Vol. 1; H. E. Ungnade Ed. Vol. 2. Interscience Publishers, New York.

Pickett, L. W., Corning, M. E., Wieder, G. M., Semenow, D. A. & Buckley, J. M. (1953). *J. Amer. Chem. Soc.*, **75**, 1618.

Pickett, L. W., Hoeflich, N. J. & Liu, T. C. (1951). *J. Amer. Chem. Soc.*, **73**, 4865.

Pitha, J., Jones, R. N. & Pithova, P. (1966). *Canad. J. Chem.*, **44**, 1045.

Rao, C. N. R. (1963). *Chemical Applications of Infrared Spectroscopy*. Academic Press, New York and London.

Rao, C. N. R. & Venkataraghavan, R. (1962). *Spectrochim. Acta.*, **17**, 541.

Rao, C. V., Balasubramanian, A. & Ramachandran, A. (1961). *J. Sci. Ind. Res.* (*India*), 20B, 382.

Roberts, B. W., Lambert, J. B. & Roberts, J. D. (1965). *J. Amer. Chem. Soc.*, **87**, 5439.

Roberts, D. & Friedkin, M. (1958). *J. Biol. Chem.*, **233**, 483.

Schneider, W. C. & Halverstadt, L. F. (1948). *J. Amer. Chem. Soc.*, **70**, 2626.

Schweizer, M. P., McGrath, K. & Baczynsko, L. (1970). *Biochem. Biophys. Res. Commun.*, **40**, 1046.

Scott, R. A. & Liverman, J. L. (1956). *Plant Physiol.*, **31**, 321.

Setlow, R. B. (1966). *Science*, *N.Y.*, **153**, 379; (1967). In *Comprehensive Biochemistry*. Vol. 27, p. 157. Ed. M. Florkin and E. H. Stotz. Elsevier, Amsterdam; (1968). In *Progress in Nucleic Acid Research and Molecular Biology*. Vol. 8, p. 257. Ed. J. N. Davidson and W. E. Cohn.

Setlow, R. B. & Setlow, J. K. (1967). *Nature*, *Lond.*, **213**, 907.

Shapiro, R. (1968). In *Progress in Nucleic Acid Research and Molecular Biology*. Vol. 8, p. 73. Ed. J. N. Davidson and W. E. Cohn.

Shaw, G. & Wilson, D. V. (1964). *Proc. Chem. Soc.*, 231.

Shejnker, U. N., Kushkin, V. V. & Postovski, I. Y. (1957) *Zhur. Fiz. Khim.*, **31**, 214.

Shimanouchi, T. & Harada (1969). *J. Chem. Phys.*, **41**, 2651.

Shimanouchi, T., Tsuboi, M. & Kyogoku, Y. (1965). In *Advances in Chemical Physics*. Vol. 7, Ed. J. Duchesne. Interscience Publishers, New York.

Short, L. N. & Thompson, H. W. (1952). *J. Chem. Soc.*, 168.

Shoup, R. H., Miles, H. T. & Becker, E. D. (1966). *Biochem. Biophys. Res. Commun.*, **23**, 194.

Shugar, D. (1960). In *The Nucleic Acids*. Vol. III, p. 39. Ed. E. Chargoff & J. N. Davidson. Academic Press, New York.

Shugar, D. & Fox, J. J. (1952). *Biochem. Biophys. Acta.*, **9**, 199 and 369.

Shugar, D. & Wierzchowski, K. L. (1958). *J. Polymer Sci.*, **31**, 269; *Bull. Acad. Polon. Soc. Cl. II*, **6**, 283.

Sigel, H. & Brintzinger, H. (1963). *Helv. Chim. Acta.*, **46**, 701.

Sigel, H. Brintzinger, H. & Erlenmeyer, H. (1963). *Helv. Chim. Acta.*, **46** 712.

Sinsheimer, R. L. (1954). *Rad. Res.*, **1**, 505.

Sinsheimer, R. L. & Hastings, R. (1949). *Science*, *N.Y.*, **110**, 525.

Skaric, V., Gaspert, B. & Jekunica, I. (1966). *Croat. Chem. Acta*, **38**, 1.

Skoog, F. & Leonard, N. J. (1968). In *Biochemistry and Physiology of Plant Growth Substances*. Ed. F. Wightman and G. Setterfield. Runge Press, Ottawa, Canada.

Skoog, F. & Miller, C. O. (1957). *Symposium Soc. Exp. Biol.*, **11**, 118.

Skoog, F., Strong, F. M. & Miller, C. O. (1965). *Science*, *N.Y.*, **148**, 532.

Smietanowska, A. & Shugar, D. (1961). *Bull. Acad. Polon. Soc. Cl. II*, **9**, 375.

Spinner, E. (1960). *J. Chem. Soc.*, 1226, 1232, 1237.

Stimson, H. M. & Reuter, M. A. (1943). *J. Amer. Chem. Soc.*, **65**, 153.

Stuckey, R. E. (1940). *Quart. J. Pharm. & Pharmacol.*, **13**, 312; (1941). *Ibid.*, **14**, 217; (1942). *Ibid.*, **15**, 370; (1942 *a*). **15**, 377.

Swain, M. L. *et al.* (1949). *J. Amer. Chem. Soc.*, **71**, 1341.

Tanaka, J. & Nagakura, S. (1966). *Theor. Chem. Acta.*, **6**, 320.

Tanner, E. M. (1956). *Spectrochim. Acta.*, **8**, 9.

Thiebe, R., Zachau, H. G., Baczynsko, L., Biemann, K. & Sonnenbichler, J. (1971). *Biochem. Biophys. Acta*, **240**, 163.

Thompson, H. W., Nicholson, D. & Short, L. N. (1950). *Discussions Faraday Soc.*, No. 9, 222.

Toth, A. & Billes, F. (1968). *Acta. Chimica Sci. Hung.*, **56**, 229.

Townsend, L. B. & Robins, R. K. (1962). *J. Amer. Chem. Soc.*, **84**, 3008; (1963). *Ibid.*, **85**, 242.

Townsend, L. B., Robins, R. K., Loepsky, & Leonard, N. J. (1964). *J. Amer. Chem. Soc.*, **86**, 5320.

Trotter, L. F. & Thompson, H. W. (1946). *J. Chem. Soc.*, 481.

Tsuboi, M., Kyogoku, Y. & Shimanouchi, T. (1961). *Biochem. Biophys. Acta.*, **55**, 1.

Tsuboi, M. T., Kyogoku, K. & Shimanouchi, T. (1962). *Biochem. Biophys. Acta.*, **55**, 1.

Udenfriend, S. (1962). In *Fluorescence Assay in Biology and Medicine*. Academic Press, New York, London.

Udenfriend, S. (1969). In *Fluorescence Assay in Biology and Medicine*. Vol. II. Academic Press, New York, London.

Udenfriend, S. & Zaltman, P. (1962). *Anal. Biochem.*, **3**, 49.

Ulbricht, T. L. V. (1963). In *Comprehensive Biochemistry*, **8**, 158. Ed. M. Florkin & E. H. Stotz. Elsevier, Amsterdam.

Varghese, A. J. & Wang, S. Y. (1967). *Nature*, *Lond.*, **213**, 909.

Volkin, E. & Cohn, W. E. (1954). In *Methods of Biochemical Analysis*. Ed. D. Glick. Vol. I, p. 287. Interscience Publishers, New York.

Wang, S. Y. (1961). *Nature*, *Lond.*, **190**, 690.

Warner, R. C. (1957). *J. Biol. Chem.*, **229**, 711.

Webb, J. N. & Levy, H. B. (1958). In *Methods of Biochemical Analysis*, Vol. 6. Ed. D. Glick. Interscience Publishers, New York, London.

White, N. E. & Clews, C. B. J. (1956). *Acta. Cryst.*, **9**, 586.
Whitehead, C. W. & Traverso, J. J. (1960). *J. Amer. Chem. Soc.*, **82**, 3971.
Wiley, R. H. & Slaymaker, S. C. (1957). *J. Amer. Chem. Soc.*, **79**, 2253.
Wierzchowski, K. L. & Shugar, D. (1961). *Acta. Biochim. Polon.*, **8**.
Witcop, B. (1954). *Experimenta*, **10**, 420.
Wulff, D. L. (1962). *Biophys. J.*, **3**, 355.
Wulff, D. L. & Fraenkel, G. (1961). *Biochem. Biophys. Acta.*, **51**, 332.
Zachau, H. G. (1969). *Angew Chem. Internat. Ed.*, **8**, 711.
Zachau, H. G., Dütting, D. & Feldmann, H. (1966). *Z. für Physiol. Chem.*, **347**, 212.
Zamecnik, P., Stephenson, M. & Scott, J. (1960). *Proc. Nat. Acad. Sci. U.S.A.*, **46**, 811.

10 Porphyrins, Bile Pigments and Cytochromes

10.1 Porphyrins

The porphyrins have a dominant place in biochemistry because (*a*) iron-porphyrin complexes constitute the prosthetic groups of haemoproteins such as haemoglobins, peroxidases and cytochromes and (*b*) other metalloporphyrins, e.g. the chlorophylls, are essential for photosynthesis in the green plant.

The chemistry of the porphyrins was already firmly established in 1940 due in large measure to the work of Hans Fischer and his colleagues (Fischer & Orth, 1937; Fischer & Stern, 1940). The biochemical aspects were assembled by Lemberg & Legge (1949) in an important book and others including Falk (1963) contributed later summaries, while Phillips (1963) reviewed the physiochemical properties of porphyrins.

The varied biological roles of metalloporphyrins depend to a great extent on physico-chemical properties influenced by the extensive electron delocalization characteristic of the conjugated system in the porphyrin ring system. The iron and magnesium atoms in the haemoproteins and the chlorophylls exhibit complex-formation by co-ordination. Spectroscopy has played an important part in the advance of knowledge in this field and is essential for further progress.

The parent substance is porphin ($C_{20}H_{14}N_4$) which has four pyrrole-like rings joined by four CH (methene) groups. The outer carbon atoms are numbered 1–8 as shown below. In the porphyrins all eight of these carbon atoms have substituents other than hydrogen, including methyl (M), propionic acid (P), vinyl (V), ethyl (E), but like groups do not occur as substituents of the same pyrrole nucleus.

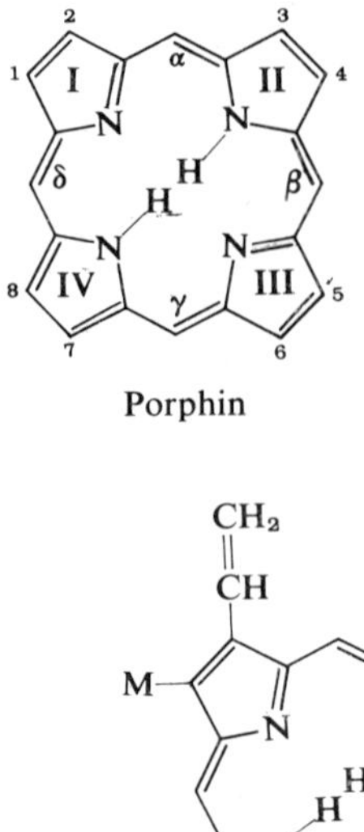

Porphin

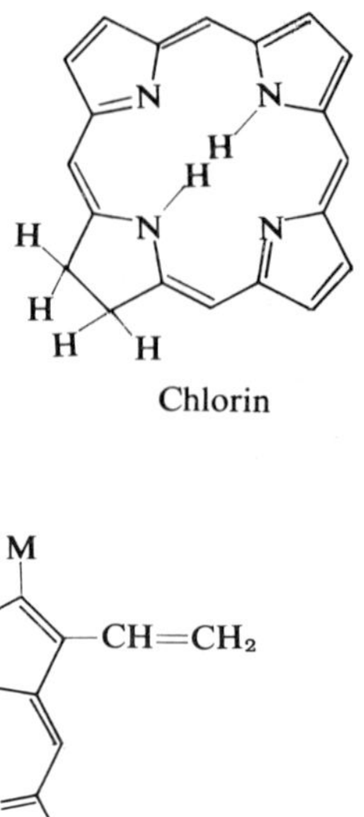

Chlorin

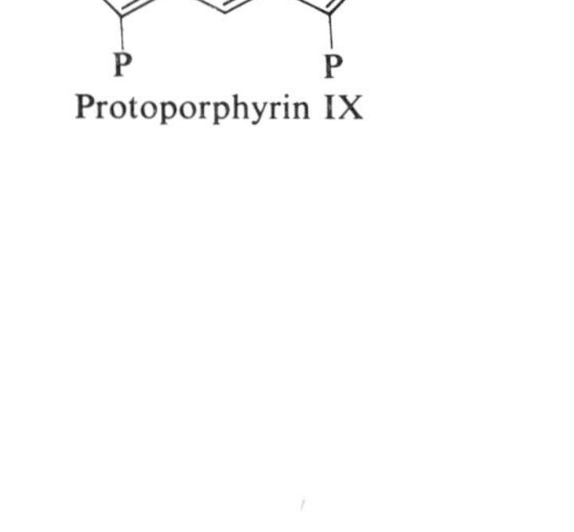

Protoporphyrin IX

Aetioporphyrin has four methyl and four ethyl substituents and (accepting the limitation that no pyrrole residue has two identical substituents) four position isomers are possible, and in fact Fischer was able to synthesize them all.

When the substituents are methyl (M) and propionic acid (P) four coproporphyrins can exist. They were given designations I, II, III and IV by Fischer. In the uroporphyrins I–IV, acetic acid (A) groups

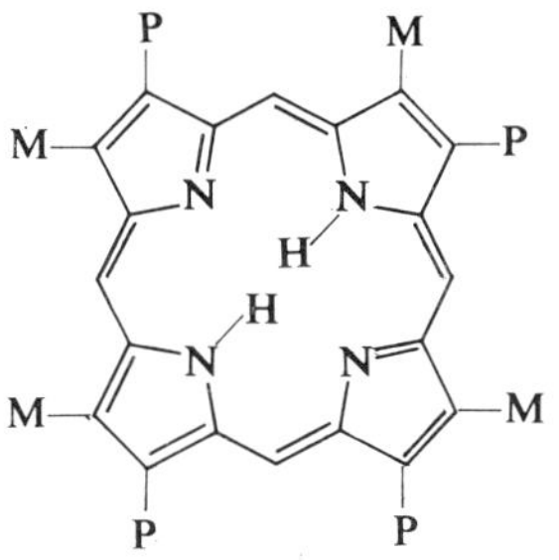

Coproporhyrin III

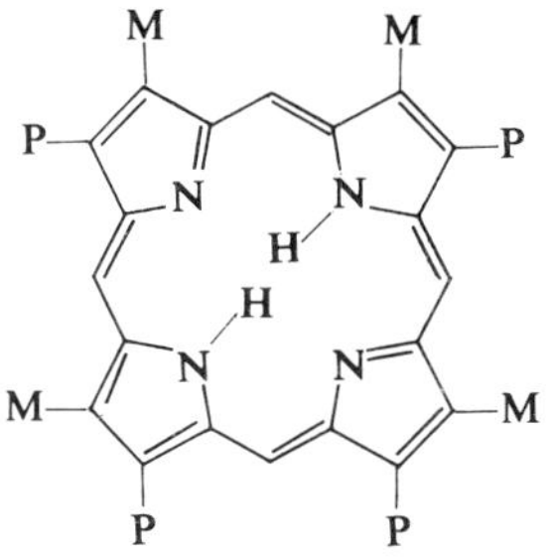

Coproporphyrin IV

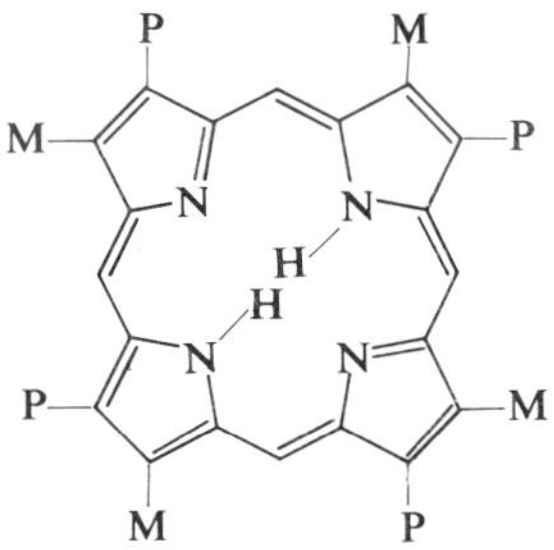

Coproporhyrin I

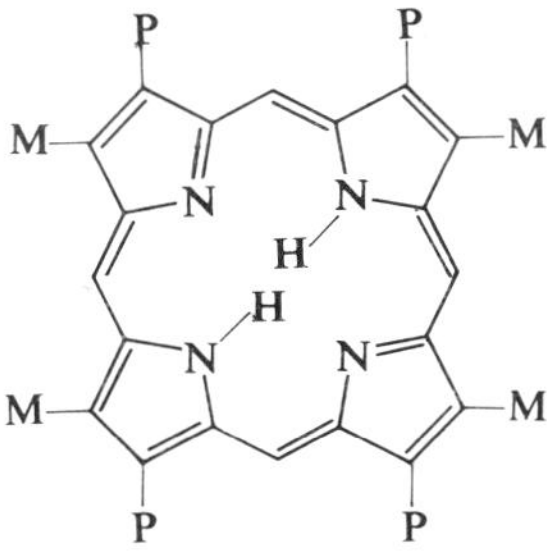

Coproporphyrin II

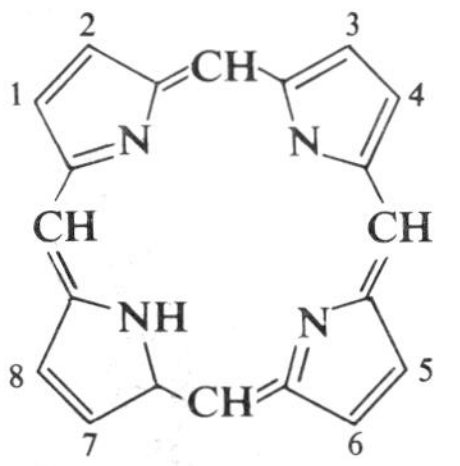

M = methyl. A = CH_2COOH. V = CH=CH_2. P = $CH_2.CH_2.COOH$

Uroporphyrin I 1, 3, 4, 7 V 2, 4, 6, 8 P.
Uroporphyrin III, 1, 3, 4, 8 V 2, 4, 6, 7 P.
Protoporphyrin IX 1, 3, 5, 8 M 2, 4 V 6, 7 P.

Table 10.1 Constitution of the porphyrins and spectroscopic properties

Compound	Formula	Substituents* M	E	V	P	A	HE	C	C.CH₃	Un-substituted	$\lambda_{max.}$ nm in 0·1 N KOH	$\lambda_{max.}$ nm in 25% HCl	$\lambda_{max.}$ nm in ether–acetic acid	Fluorescence peaks (nm) λ_{max} nm in dioxane	$\lambda_{max.}$ in 2 N HCl	
Porphin	$C_{20}H_{14}N_4$	No substituents										542	487, 517·5 560·5, 613 (in dioxane)			
Deutero-porphyrin	$C_{30}H_{30}N_4O_4$	4			2					2	535·5, 559·5 611·5	404, 548 591	494, 526 566, 621·5	626 ($CHCl_3$)		
Aetioporphyrin	$C_{32}H_{38}N_4$	4	4										496, 528, 566 595, 621 (in dioxane)			
Pyrroaetio-porphyrin	$C_{30}H_{34}N_4$	4	3							1		404, 549 591·5 (Pyrro-XV-monomethyl ester)	620			
Phylloaetio-porphyrin	$C_{29}H_{32}N_4$	4	2						1	1						
Phylloporphyrin	$C_{32}H_{37}N_4O_3$	4	2		1				1	1			502, 533·5 (γ-monomethyl ester 573·5, 627 in dioxan)			
Rhodoporphyrin	$C_{32}H_{35}N_4O_3$	4	2		1			1					635·5			
Haematopor-phyrin	$C_{34}H_{38}N_4O_6$	4			2		2					410·5, 551·7 (575·5), 595·8		625, 654† 673, 691	597, 619 653	634, 663† 683, 701
Protoporphyrin	$C_{34}H_{34}N_4O_4$	4		2	2						540, 591, 645	411, 557·2 (582), 602·5	502, 537 576, 632·5	605·5, *632·5* 669·5, 704	*604*, 626 655	638 in $CHCl_3$
Mesoporphyrin	$C_{34}H_{38}N_4O_4$	4	2		2						502, 538·5, 567·5, 618	404, *548·2* (572·5), 593	494·5, 528·5 567·5, 623·5			
Coproporphyrin	$C_{36}H_{38}N_4O_8$	4			4						503, 538·5 565·5, 617·5	405, *550·5* (574), 593·5	495, 528·5 568, 623·5	597, *623** 654, 671, 691	*595*, 618 652	
Uroporphyrin	$C_{40}H_{38}N_4O_{16}$			4	4						504, 539, 560·5 612	410·5, *553·5* 597		600, *626** 658, 676, 695	596, 619 653	

M = methyl. E = ethyl. V = vinyl CH=CH_2. P = propionic acid—CH_2CH_2COOH. A = acetic acid—CH_2COOH.

HE = α-hydroxyethyl CHOH.CH_3. C = carboxyl COOH. C.CH_3 = methyl on methine group.

* Ester. † In pyridine.

replace the methyl (M) groups of the coproporphyrins. For protoporphyrins with three types of substituent, fifteen isomerides are theoretically possible but one isomeric type (IX) is predominant in nature, e.g. protoporphyrin IX, which is 1,3,5,8-tetramethyl-2,4-divinylporphin-6,7-dipropionic acid. Porphyrins with carboxylic acid side-chains form esters with alcohols like methanol [5% (v/v) of concentrated sulphuric acid is added and the mixture is left to stand in the dark for 24 hours at room temperature. The esters are stable and soluble in organic solvents and the free porphyrins can be regenerated on saponification].

The tetrapyrrole pigments often display an ultra-violet peak around 265–280 nm (aetioporphyrin I 264 nm, ϵ 7800; rhodoporphyrin XV methyl ester 265 and 282 nm, ϵ 14 200 and 11 600). More important is a very intense band around 400 nm designated the Soret band. The porphins and related compounds exhibit four main absorption peaks in the visible region which are numbered I, II, III and IV starting from the long-wave end. These bands are of moderate intensity, about one order of magnitude weaker than the Soret bands, and are attributed to two distinct electronic transitions. One has its moment along the axis through the two hydrogen atoms on diagonally opposite nitrogen atoms (bands III and IV). The other has its moment perpendicular to this axis (bands I and II). Bands I and II become more intense when substitution makes the molecular less symmetrical.

Stern & Wenderlein (1934, 1935, 1936 *a*, *b*) helped to reduce to order the spectra in the visible region. Introduction of substituents has a bathochromic effect even when the groups (like CH_3 and C_2H_5) are not chromophores. A definite stage is reached with octasubstituted porphins, e.g. octaethyl- and octamethylporphin, aetioporphyrin I (1,3,5,7-tetramethyl-2,4,6,8-tetraethylporphin) and aetioporphyrin II (1,4,5,8-tetramethyl-2,5,6,7-tetraethylporphin) which all show similar spectra:

		Soret band	IV	III	II	I
Porphin	$\lambda_{max.}$ nm	397	490	520	564 569	617
	$\epsilon_{max.} \times 10^{-3}$	2·64	15·8	2·64	4·98 4·17	0·86 (in C_6H_6)
Octamethyl-	$\lambda_{max.}$ nm	402	499	533	570	624
	$\epsilon_{max.} \times 10^{-3}$	14·6	13·1	9·3	6·4	5·1
Octaethyl-	$\lambda_{max.}$ nm	—	496	528	566 656	621
	$\epsilon_{max.} \times 10^{-3}$	—	14·0	10·0	5·95 1·33	5·78
Aetioporphyrin I	$\lambda_{max.}$ nm	400	499	535	566	621 649
	$\epsilon_{max.} \times 10^{-3}$	15·3	12·4	9·1	6·00	4·4 1·1

The distribution of $\epsilon_{max.}$ values between the different maxima differs from that shown by porphin itself, but is characteristic of the octasubstituted products named above which show $\epsilon_{max.}$ rising in the order bands I–IV. This has been designated Aetio-type absorption.

When any of the positions 1 to 8 are occupied by substituents containing a carbonyl group (conjugated) as in

$$—COCH_3 \quad \text{or} \quad —C(=O)OCH_3$$

a notable change occurs in the broad shape of the absorption curve. There is a considerable increase in the relative intensity of Band III; the type of curve is designated Rhodo and rhodoporphyrin (1,3,5,8-

tetramethyl-2,4-diethyl-6-carbomethoxy-7-propionic acid) shows the effect:

		Soret band	IV	III	II	I
Rhodoporphyrin XI	$\lambda_{max.}$ nm	408	510	549	573	633
diethyl ester	$\epsilon_{max.} \times 10^{-3}$	2·10	10·6	17·2	9·4	1·5

One carbonyl effects the change but if a second group enters an 'opposite' pyrrole ring the 'Rhodo' characteristic is enhanced (see Fig. 10.1). On the other hand, if (as in diacetyldeuteroacetoporphyrin III) two $COCH_3$ groups occur in adjacent rings (I and II), the absorption spectrum reverts to the aetio type. When substitution occurs in the α, β, γ or δ methine positions a new distribution of intensities is seen and the curves are described as of the 'Phyllo' type in which band III is weaker than band II:

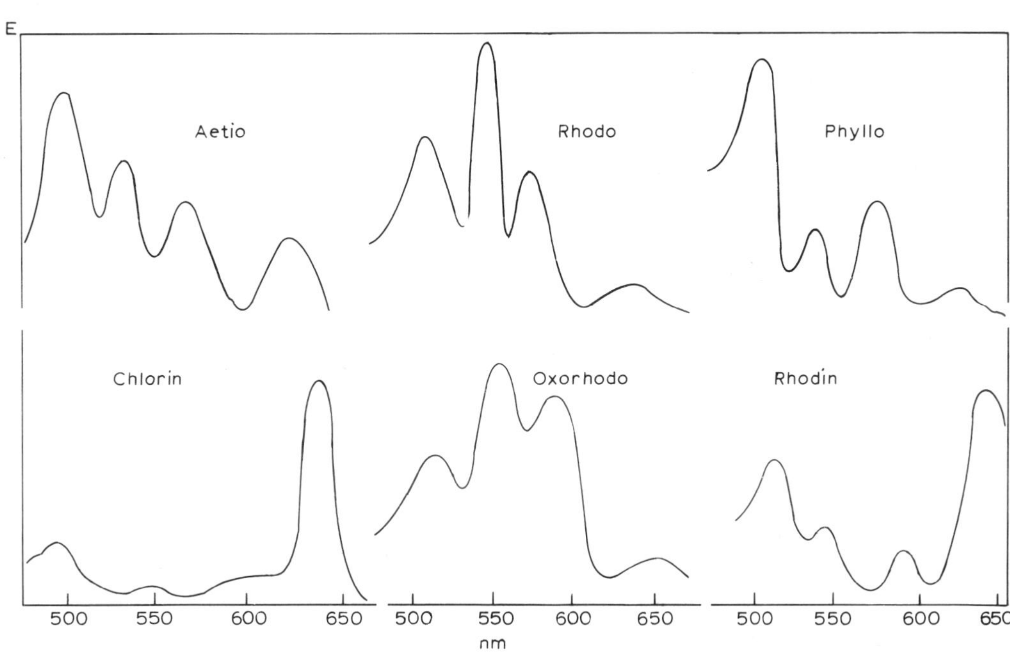

Figure 10.1 Absorption spectra of aetio-, rhodo-, phyllo-, chlorin-, oxorhodo- and rhodin-types

		Soret band	IV	III	II	I	
γ-Phylloporphyrin XV	$\lambda_{max.}$ nm	408	505	538	574	627	648
methyl ester	$\epsilon_{max.} \times 10^{-3}$	2·02	15·2	5·4	6·1	1·6	0·425

(IV > II > III > I)

Substitution in two of the α, β, γ or δ positions enhances the effect but if one of the positions 1 to 8 is already occupied by $COCH_3$ or $—COCH_3$, substitution of H at a bridge CH brings about a return to the aetio type.

Electron-attracting substituents ($CH{=}CH_2$, CHO, COOH, CH=CH.COOH and $COCH_3$) extend the conjugation of the ring system and have been classed as 'Rhodofying' (Lemberg & Falk, 1951). Non-rhodofying side-chains (CH_3, C_2H_5, etc.) give aetio-type spectra, one rhodofying substituent gives rise to rhodo-type curves and two such substituents on diagonally disposed pyrrole units give rise to oxorhodo-type curves in which peaks II and III are the strongest (order of intensities III > II > IV > I).

The chlorins are dihydroporphins containing one reduced pyrrole ring. In addition there are phorbins, in which an isocyclic ring joins the γ carbon atom, and rhodins (3-formyl-7,8-dihydroporphins). The characteristic of chlorins is the great relative intensity of the long wave peak.

		Soret band	IV		III		II		I
Chlorin	$\lambda_{max.}$ nm	388	493	484	540		588	610	638
	$\epsilon_{max.} \times 10^{-3}$	128	12·4	10·6	1·3		3·6	2·7	5·0 (in C_6H_6)
Octamethylchlorin	$\lambda_{max.}$ nm	390	488	496	522	548	594	618	646
	$\epsilon_{max.} \times 10^{-3}$	175	12·9	13·1	4·3	1·8	4·5	4·6	58·5 (in C_6H_6)

When a metal cation (e.g. Fe^{2+}) unites with a porphyrin two protons are lost from the two NH groups. This leaves the complex with a residual charge of zero

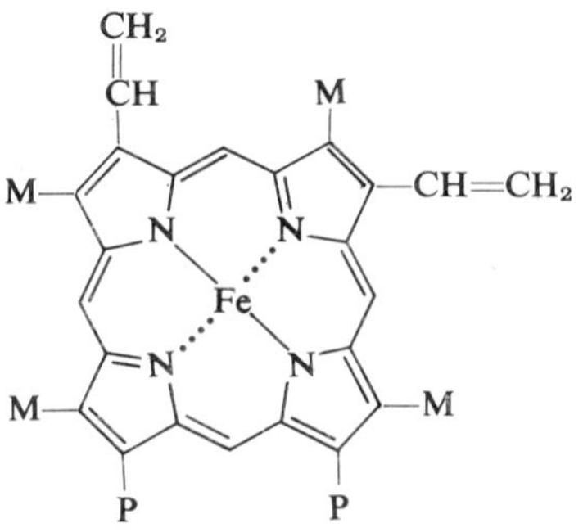

Protohaem
(haem or ferroprotoporphyrin)

Chlorocruorin
(has formyl instead of vinyl at position 2)

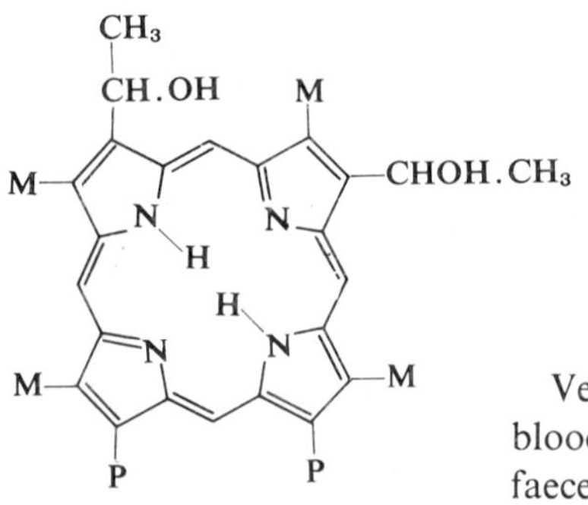

Haematoporphyrin
(haematohaem)

apart from that on any carboxylic acid side-chain. Protohaem or ferroprotoporphyrin is shown in the figure. With Fe^{3+} the complex as in protohaem in a chlorine atom is co-ordinated to the metal to form chlorohaemin or ferriprotoporphyrin chloride. In the presence of alkali the residual charge is neutralized by a hydroxyl ion yielding protohaematin (hydroxyferri-protoporphyrin).

Ferroporphyrins can add two extra ligands one on each side of the planar structure and the hexa-co-ordination complex assumes an octahedral structure seen for example in a pyridine haemochrome.

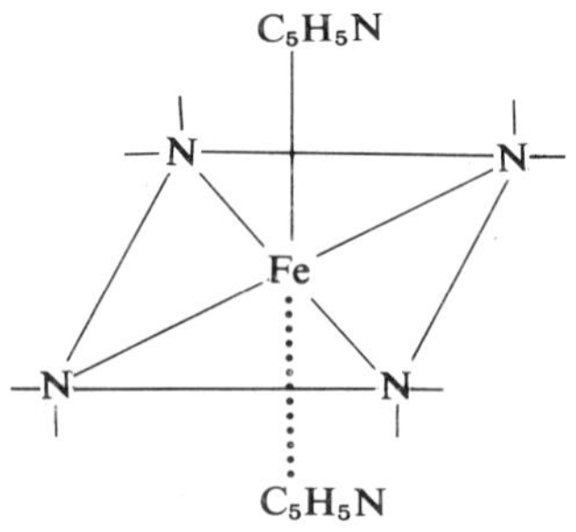

Very small amounts of free porphyrins occur in blood (coproporphyrins and protoporphyrins), in faeces (coproporphyrin) and in urine (uroporphyrin and coproporphyrin). The ingestion of blood and raw flesh by carnivorous animals is followed by degradation of haemoglobin and myoglobin etc., under the action of micro-organisms so that deuteroporphyrins and protoporphyrins appear on the faeces. Herbivorous animals consume large quantities of chlorophylls which are degraded to phylloerythrins.

Several different prosthetic groups have been identified as present in haemoproteins. The haemoglobins and myoglobins contain protohaem; this is also the prosthetic group of cytochromes b, a number of peroxidases and catalases and of erythrocruorin which replaces haemoglobin in many invertebrate species. Some polychaete worms have the pigment chlorocruorin instead of haemoglobin and the prosthetic group chlorocruorinhaem has at position 2 a formyl group instead of the vinyl group of protohaem.

Haem a

Haem a_2

Haem a is the prosthetic group of cytochromes a and a_3 of mammalian tissues and of some other cytochromes a. It is rather unstable, has a lipid side-chain and is difficult to separate from lipid congeners. Its structure was determined by Morrell *et al.* (1961), Lemberg & Legge (1949) and Morrison & Stotz (1961), and Marks *et al.* (1960). The bacterial cytochrome a_2 has for its prosthetic group an iron-chlorin (Barrett, 1956, 1959). The cytochromes c contain haem c, the structure of which was worked out by Theorell (1938) and Ehrenberg & Theorell (1955 *a*, *b*) taking advantage of earlier work by Hill and Keilin (1930).

Treatment of cytochrome c with sulphur dioxide liberates a water-soluble peptide-containing porphyrin. Classical degradations led to a structure shown below. Two cysteine moieties are linked by their S atoms to the α-carbon atom in the side-chains at positions 2 and 4 of haematoporphyrin. Cytochromes c from various species are very similar in respect of the aminoacids in a homologous region, e.g. Lys.Cys.Ala.Glu(NH_2). Cys.-His- recurs often but can be slightly modified.

Haemoglobins from different species have been isolated and crystallized and the absorption spectra of oxyhaemoglobin and reduced haemoglobin have been

Haem c

measured by very many workers (see Lemberg & Legge, 1949). Agreement on the wavelengths of maxima in the visible region for oxyhaemoglobin is good but less satisfactory for intensities of absorption. Three ultraviolet peaks occur, the Soret band maximum near 415 nm, a peak at *ca.* 343 nm associated with the presence of iron in the porphyrin moiety and a peak at 275 nm due to the aromatic aminoacid moieties in the globin. The absorption bands in the visible are well displayed in a 1 cm thickness of blood diluted 1/200 (v/v) using 0·1% sodium carbonate or 0·4% ammonia as solvent. Reduced haemoglobin shows two main maxima, one at 555 nm and the other (Soret band) at 430 nm. The 343 nm absorption of oxyhaemoglobin is replaced by an inflection while the 275 nm absorption is virtually unchanged (Fig. 10.2).

Figure 10.2 Absorption spectra of blood solution saturated with oxygen (solid curve) and the same solution reduced (dashed curve)

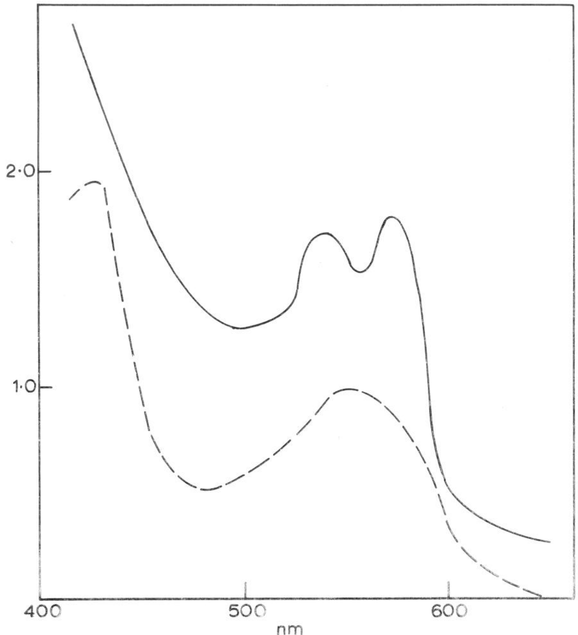

Table 10.2 Absorption peaks of haemoglobin

	$\lambda_{max.}$ nm	$\epsilon_{max.} \times 10^{-3}$	Infrared peaks	
			$\lambda_{max.}$ nm	$\epsilon_{max.}$
Haemoglobin (reduced)	555	13·25 (12·9–13·6)	760	390
	430	12·6 (11·8–13·4)	910	200
	350	30·0		
	275	31·6		
Oxyhaemoglobin α	516–578	15·65 (15·1–16·2)		
β	540–542	14·75 (14·2–15·3)	950	300
γ	418	126·7 (125–128·5)		
γ^{1}	345	28·4		
γ^{11}	275	35·5		
Carboxyhaemoglobin	568–572	14·35 (13·7–15·0)	weak end absorption	
	538–540	14·70 (14·1–15·3)		
	418	154		
	340	24·4		
	275	32		

Carboxyhaemoglobin shows five absorption peaks (see Table 10.2) and in the near infrared there is weak end-absorption from 700–1000 nm, whereas oxyhaemoglobin has stronger absorption with a flat peak at 925 nm and reduced haemoglobin shows a small maximum at 760 nm and a flat peak near 910 nm.

Many workers have made use of Hufner's points. These are the wavelengths of the second (β) peak in oxyhaemoglobin *ca.* 540 nm and the minimum near 560 nm (see Heilmeyer, 1943). The ratio of the extinctions at the two wavelengths is known as Hufner's Quotient and has a value near 1·63. [Today, it would be as well to ascertain the points and the ratio on the photoelectric spectrophotometer in use.] If in examining blood, the quotient falls outside the limits 1·58 and 1·68 the possibility of turbidity, instrumental error or decomposition should be investigated. For a 1 cm thickness of blood diluted 1/100 (v/v) the extinction $E = \log I_o/I$ at 540 nm multiplied by 0·00116 and that at 560 multiplied by 0·00189 should give the same concentration $\times 10^{-4}$ in g/100 cc blood approximately.

[Example, blood diluted 1/200 v/v: E 1 cm = 0·70 at 540·5 nm and 0·424 at 560 nm, $0{\cdot}7 \times 0{\cdot}00116 \times 200 \times 100 = 16{\cdot}24$ g/100 cc. $0{\cdot}424 \times 0{\cdot}00189 \times 200 \times 100 = 16{\cdot}03$ g/100 cc.]

In reduced haemoglobin the ratio E_{541}/E_{500} is about 0·866. Mixtures of haemoglobin and oxyhaemoglobin can readily be evaluated from observed quotients between this and 1·63.

Mixtures of carboxyhaemoglobin and oxyhaemoglobin can similarly be measured from the ratio E_{576}/E_{560}, which is near 1·72 for oxyhaemoglobin and 0·875 for carboxyhaemoglobin. With suitable calibration methaemoglobin (haem*i*globin) and oxyhaemoglobin mixtures can be analysed similarly.

Methaemoglobin (ferrihaemoglobin) is brown in solution and is obtained by the action of oxidizing agents effecting the change Fe(II) → Fe(III). It also occurs as an abnormal blood pigment. The following maxima are characteristic: $\lambda_{max.}$ nm 630, 500, 406, 275; $\epsilon_{max.} \times 10^{-3}$ 3·75, 9·5, 144 ($\pm$10), 32. The iron atom carries a positive charge which disappears in alkali when alkaline methaemoglobin (haemiglobin hydroxide) is formed: $\lambda_{max.}$ nm 577, 540, 411; $\epsilon_{max.} \times 10^{-3}$, 8·5, 9·7 80 ($\pm$10). Cyanmethaemoglobin (haemiglobin cyanide) can be obtained crystalline. Neutral cyanide added to methaemoglobin gives a bright red solution: $\lambda_{max.}$ nm 540, 414; $\epsilon_{max.} \times 10^{-3}$ 11·1, 98 ($\pm$6). The absorption shown by cyanmethaemoglobin lends itself to the determination of the sum of oxyhaemoglobin and methaemoglobin in mixtures and has the advantage of being unchanged over a rather wide pH range.

The red colour of meat is due to myoglobin, good preparations of which are conveniently obtained from horse muscle. It has been crystallized and its molecular weight is near 18 000. The prosthetic group is ferroprotoporphyrin IX. The absorption spectra exhibit peaks as follows:

Myoglobin (reduced)
$\lambda_{max.}$ nm 555 434; $\epsilon_{max.} \times 10^{-3}$ 13·3 113
Oxymyoglobin
$\lambda_{max.}$ nm 578 542 416; $\epsilon_{max.} \times 10^{-3}$ 13·3 13·2 108

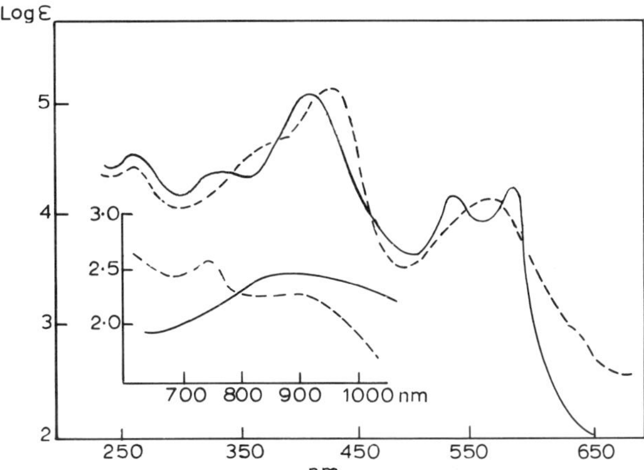

Figure 10.3 Absorption spectra of haemoglobin (dashed curve) and carboxyhaemoglobin (solid curve)

Myoglobin and haemoglobin are best distinguished via the carboxy compounds, as carboxymyoglobin shows a red shift of about 5 nm for the long wave band only.

When haemoglobin is treated with alkali (pH *ca.* 11), denaturation occurs to form haemochromogen (haemochrome): $\lambda_{max.}$ nm 558 (sharp), 528, 424; $\epsilon_{max.} \times 10^{-3}$ *ca.* 28, 12·8, 110. This forms a carbon monoxide derivative (CO-haemochrome): $\lambda_{max.}$ nm 571, 542; $\epsilon_{max.} \times 10^{-3}$ *ca.* 12, 12·6 or a pyridine haemochromogen in 20% pyridine $\lambda_{max.}$ nm 558, 528; $\epsilon_{max.} \times 10^{-3}$ *ca.* 28, 15·6. Also if haemoglobin is exposed to acid (pH 3–4) the protein–prosthetic group link is broken and the protein denatured. The colour becomes reddish brown and when oxygen is present the Fe (II) changes to Fe (III). The pigment so obtained is ferrihaemochromogen or 'acid haematin' and has a band in the red portion of the spectrum: $\lambda_{max.}$ 662; ϵ $6·7 \times 10^{-3}$ and a broader band near 505 nm together with a peak at 378 nm, ϵ $4·5 \times 10^{-3}$. If blood is diluted with 0·1 N HCl the haematin spectrum shows similarly $\lambda_{max.}$ nm 662, ~560, ~525, 376; $\epsilon_{max.} \times 10^{-3}$ 6·8, 11·7, 13·5, 46·0, 40·0.

The term haematin includes all the iron complexes of porphyrins and other tetrapyrroles. Haematin is a ferriporphyrin hydroxide, haemin is the corresponding chloride and haem or reduced haematin is the ferroporphyrin.

Haem	Haematin	Haemin
[N N Fe N N]°	[N OH N Fe N OH N]°	[N N Fe N N]$^+$ X^-

Haem shows a flat maximum near 572 nm, $\epsilon_{max.}$ $5·5 \times 10^{-3}$, the exact value of $\epsilon_{max.}$ apparently depending on pH. Haemin and haematin have similar spectra in acid. Acid haematin in glacial acetic acid shows $\lambda_{max.}$ nm 630–635, 540, 520, 400; $\epsilon_{max.} \times 10^{-3}$ 15·9, 31·6, 39·8, 141 (±10).

Freshly prepared haematin dissolved in alkali (10% w/v) shows a single peak at 580 nm, $\epsilon \times 10^{-3} = 10·5$. Crystallized haemin in 10% NaOH shows $\lambda_{max.}$ 610 nm, $\epsilon_{max.} \times 10^{-3}$ *ca.* 4·6, an inflection near 500 nm and a strong peak at 390 nm. At pH 11–13 akaline haematin has peaks at 492 and 385 nm, $\epsilon_{max.} \times 10^{-3}$ 6·55 and *ca.* 4·6 respectively. Purified haemin dissolved in blood serum and made alkaline showed $\lambda_{max.}$ 620 nm and peaks also occurred at 535 and 495 nm. Haematin dissolved in serum gave a similar curve. The 610 nm band is replaced by an inflection when the pH is between 9 and 11·5. Reduced haematin in 0·1 N NH_4OH shows a peak at 550 and an inflection at 575 nm.

10.1.1 Peroxidases and Catalases

The transfer of hydrogen to oxygen by an oxidase or an aerobic dehydrogenase could form water or hydrogen peroxide. The latter is toxic to tissues but peroxidases and catalases decompose the peroxide. Peroxidases occur in plant tissues only; their function is to combine with H_2O_2 so that the complex enables oxygen to accept hydrogen from various substrates. Catalases are widely distributed in plant and animal tissues. They specifically catalyse the breakdown of hydrogen peroxide into water and oxygen. The prosthetic group is identical with that of blood haemoglobin but the action of the iron atom is imperfectly understood. The enzyme is a highly effective catalyst ($10^5 - 10^7$ M^{-1} s^{-1} at 0°C, Saunders *et al.*, 1964). Catalases are crystallizable haemoproteins with four iron atoms per molecule, attached to porphyrin and protein. Protoporphyrin IX has been obtained from bacterial and other catalases but there are some catalases in which one of the four porphyrin groups has been converted to the bile pigment biliverdin.

The literature on catalases is very extensive and many methods of estimating enzyme activity have been described. One of the best is a version of the permanganate titration method (Bonnichsen *et al.*, 1947). These authors showed that the catalysed process has no induction period and is not a chain reaction.

A typical catalase forms three compounds with both

hydrogen peroxide and alkyl hydrogen peroxides, but in either case only one of the three is active:

	$\lambda_{max.}$ nm	$\epsilon_{max.} \times 10^{-3}$
Horse liver catalase	629	10·4
	536	13·5
	500	17
	405	?140–340
	266–280	very high
M. Lysodeikticus catalase	631·5	14
	545	17
	506	18

Compound I is green and has $\lambda_{max.}$ nm 655 and 405, the latter band having a very high ϵ value (295×10^{-3}). Compound II is red and has peaks at $\lambda_{max.}$ nm 568, 536 and 529. The Soret band at 410 nm is strong ($\epsilon = 260 \times 10^{-3}$). Compound II is only formed when there is excess of peroxide and its activity is 10^{-4} times that of Compound I (Chance, 1949). Compound III, formed in the presence of concentrated peroxide, was detected by Keilin & Hartree (1951). It has peaks at $\lambda_{max.}$ nm 585, 545 and 416 (ϵ values 52, 51 and 334×10^{-3} respectively). Catalase-CN^- (derived from horse liver) shows peaks at 553 and 425 nm ($\epsilon_{max.}$ 14·6 and $10{\cdot}9 \times 10^{-3}$) and an inflection near 580 nm (ϵ $1{\cdot}22 \times 10^{-3}$).

The spectrophotometric data on the kinetics of the catalase reaction (Chance, 1949) do not lend themselves to a brief summary. They imply, however, that catalase compound I collides with a second molecule of peroxide. The former contains two oxidizing equivalents and in the reaction with the latter a 'concerted' shift of 4 electrons is postulated whereby the free peroxide donates two electrons to the bound peroxide, leaving water and oxygen. There is no $Fe(II) \rightarrow Fe(III)$ change in the course of enzymatic activity.

At pH 7 free catalase has a hydroxyl group in the 'free' haem position. Numerous anions can here compete with hydroxyl to combine with the haem. The cyanide, azide and fluoride anions strongly inhibit catalase activity. Chance (1949) observed that free catalase and catalase–peroxide compound I had an isosbestic point at 435 nm. At that wavelength formation of the latter would not change the absorption. Under these circumstances the kinetics of the formation of the catalase–cyanide compound could be measured.

Horseradish root is the classical source of peroxidase but it can also be obtained from turnip or fig tree sap. A simple extract, concentrated by precipitating with ammonium sulphate and redissolving the precipitate in water, yields an enzyme preparation that catalyses the oxidation of many phenols and amines by hydrogen peroxide. One convenient test consists of the oxidation of pyrogallol to purpurogallin (Bach & Chodat, 1903; cf. also Keilin & Mann, 1937).

$$2\ \text{Pyrogallol} + 3H_2O_2 \xrightarrow{\text{Peroxidase}} \text{Purpurogallin} + 5H_2O + CO_2$$

The reagent consists of pyrogallol (5 g), H_2O_2 (50 mg) in 2 litres of water at pH 6. A measured volume is treated with peroxidase and the reaction is allowed to proceed for 5 minutes before being stopped by adding acid (10% sulphuric acid). The purpurogallin can be extracted with ether and estimated. The above method was worked out by Willstätter & Stoll (1917) but has been improved (Keilin & Hartree, 1951; Theorell & Maehly, 1950). The mesidine $\rightarrow$ 2,6-dimethylbenzoquinone-4-(2′,4′,6′-trimethyl)anil conversion has been much used (Saunders, 1941; Hughes *et al.*, 1954; Avi-Dor & Paul, 1953; Holmes-Siedle & Saunders, 1959).

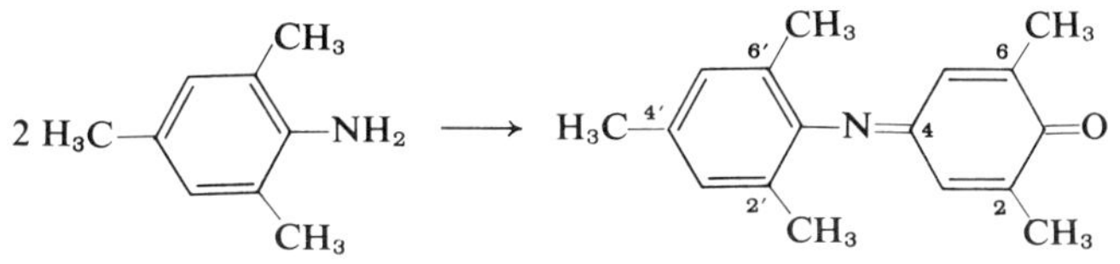

$$2\ H_3C\text{-}C_6H_2(CH_3)_2\text{-}NH_2 \longrightarrow H_3C\text{-}C_6H_2(CH_3)_2\text{-}N{=}C_6H_2(CH_3)_2{=}O$$

Mesidine 2,6-Dimethylbenzoquinone-4-(2′,4′,6′-Trimethyl) anil

or

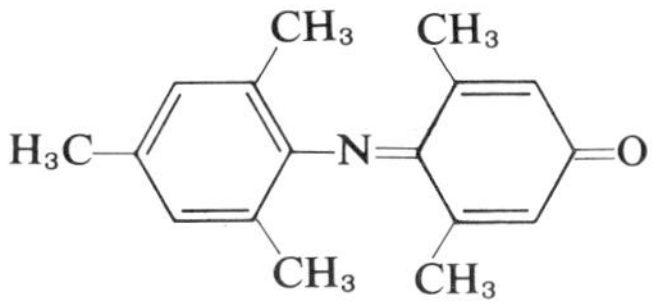

$$H_3C\text{-}C_6H_2(CH_3)_2\text{-}N{=}C_6H_2(CH_3)_2{=}O$$

In spite of some disadvantages mesidine is perhaps the most serviceable hydrogen donor for analytical studies.

Horseradish peroxidase (Keilin & Mann, 1937) can be split into a protein (molecular weight *ca.* 40 000) and protohaematin IX by hydrochloric acid and acetone at temperatures below 0°. The iron (Fe III) has six co-ordination links, four to porphyrin nitrogens, one to protein and the sixth is open to union with water, CN^- and other groups. Exchanges at the sixth co-ordination bond form an aspect of the mode of action. The ferriperoxidase is readily reduced by dithionite to the ferrous form; this can unite with carbon monoxide. Peroxidase (Fe III) reacts reversibly to give the products shown in Table 10.3 (see Saunders *et al.*, 1964).

Peroxidase, like catalase, forms three different complexes: Compound I (Theorell, 1942) is green and is formed as an unstable compound from equimolar amounts of peroxidase and H_2O_2. It shows peaks at 656–659 nm (ϵ *ca.* 3×10^{-3}) and 410 nm (ϵ *ca.* $4{\cdot}4\times10^{-3}$). Compound II is pink and is formed as a more stable complex between equimolar amounts of enzyme and substrate. It exhibits a broad inflection at 610—665 nm (ϵ 3×10^{-3}), a peak at 555 nm (ϵ $7{\cdot}4\times10^{-3}$) and 527–530 nm, ϵ $8{\cdot}1\times10^{-3}$ and a Soret band at 418 nm (ϵ $7{\cdot}7\times10^{-3}$). Compound III requires hydrogen peroxide to be in excess

$\lambda_{max.}$ nm	673	583	546	416
$\epsilon_{max.}\times10^{-3}$	2·8	8·7	10·0	9·3–10·6 (depending on pH)

The structures of the various complexes of peroxidase and hydrogen peroxide are not fully known, but it seems certain that higher states of oxidation than Fe (III) must be invoked (Polonovski *et al.*, 1938, 1939, 1941). George (1952, 1953, 1956) argues against peroxide as a component of compound I which probably contained iron in a state equivalent to Fe (V) or

Table 10.3 Horseradish peroxidase

Ferri-peroxidase						
—H_2O	$\lambda_{max.}$ nm	641·5	~540	497	403	
	$\epsilon_{max.}\times10^{-3}$	2·84	6·8	10·03	91	
—OH^-	$\lambda_{max.}$ nm	574·5	545		416	
	$\epsilon_{max.}\times10^{-3}$	6·86	8·64		88	
—CN^-	$\lambda_{max.}$ nm	~581·5	540		423	
	$\epsilon_{max.}\times10^{-3}$	8·5	11·0		93·9	
—F^-	$\lambda_{max.}$ nm	612	560	530	488	403
	$\epsilon_{max.}\times10^{-3}$	7·0	4·8	4·6	7·6	130
—N_3^-	$\lambda_{max.}$ nm	635	534	416		
	$\epsilon_{max.}\times10^{-3}$	1·7	8·3	11·5		
Ferro-peroxidase						
—H_2O	$\lambda_{max.}$ nm	~594·5	558	~515	440	
	$\epsilon_{max.}\times10^{-3}$	5·2	11·2	5·8	75	
—CN^-	$\lambda_{max.}$ nm	565·5	535	432·5		
	$\epsilon_{max.}\times10^{-3}$	18·5	13·0	135		
—CO	$\lambda_{max.}$ nm	573	542	423		
	$\epsilon_{max.}\times10^{-3}$	11·4	11·8	157		

(Saunders, 1952, 1957).

Fe^{3+}—O. The formation of compound II from compound I is a one-equivalent reduction accelerated by the presence of a hydrogen donor. The structure of compound III is not known but the effective oxidation state of the iron atom is +5 or +6, possibly Fe^{2+}—O_2.

Another sign of the complexity of the structural possibilities is seen in the effect of cold hydrochloric acid in acetone. Maehly (1952, 1954) observed that at pH < 2·5 a compound 'A' was formed which spontaneously underwent a change to compound 'B', then to 'C' with protohaemin peaks. This slowly changes to 'D'.

In horseradish peroxidase 'A' the Soret band is at 401·5 nm and

'A' → 'B'	(Soret peak 398·5 nm)
'B' → 'C'	(ferrioprotoporphyrin IX Cl_2, 620 and 500 nm) (Protohaemin I)
'C' → 'D'	(Protohaemin II)

'A' reacts with H_2O_2 but catalytic activity is lost when 'B' is formed, i.e. when the covalent link Fe–protein carboxyl is split, while the formation of 'C' is essentially the breaking of the link between one or both of the propionyl groups and the protein.

The porphyrin bands I–IV are probably due to $\pi-\pi^*$ transitions. Bands I and III are related because electron-attracting substituents introduced in the porphyrin ring cause them to increase in intensity whereas Bands II and IV are little influenced. The transitions point to a tautomeric balance on the pyrrole nitrogen atoms with their two free charges. When in acid solution protonation occurs or complexing occurs with Fe (II); the four bands are reduced to two (α- and β-bands) and from their responses to substitution the α-band belong to the porphyrin bands I and III and the β-band to bands II and IV. The Soret band is often designated 'γ'. It has very high ϵ values (> 100 000) and is almost certainly due to the conjugated double bond system with delocalized electrons largely unaffected by the metal atom in the middle of the ring system. Saunders *et al.* (1964) have used the designations 'D' and 'E' for two further bands in haemoproteins.

Band	Region, nm	$\epsilon_{max.}$
α	550–610	*ca.* 10 000+(covalent complexes) 5000 (ionic)
β	520–550	
D	590–650	Ionic Fe III haemoproteins, haems
E	480–510	

The D and E bands are probably due to charge-transfer from ligand to metal.

The changes in the Soret band peaks for horseradish peroxidase, the peroxidase–peroxide complex (compound I) and compounds II and III have been tentatively interpreted (Saunders *et al.*, 1964; Chance, 1954; Williams, 1958).

Further sources of peroxidases include milk, leucocytes and yeast. Agner (1941) isolated a green haemoprotein from pus using human empyema exudate and Schultz *et al.* (1957) obtained it (from rat chloroma) in a crystalline state. The designations verdoperoxidase and myeloperoxidase have been used (Agner, 1943; Lemberg & Legge, 1949). A small amount of copper (0·001%) persists in the best preparations. It is, like other peroxidases, a stable enzyme.

Verdoperoxidase (V) molecular weight 149 000. Two Fe atoms/molecule
(ϵ values per Fe atom)

V–H_2O	$\lambda_{max.}$ nm	690	625	570	500	430
(Ferri)	$\epsilon_{max.}\times10^{-3}$	10	15	25	20	160
V–H_2O	$\lambda_{max.}$ nm		637	590	475	430
(Ferro)	$\epsilon_{max.}\times10^{-3}$		40	24	148	80
V–H_2O_2 (Compound I)	$\lambda_{max.}$ nm	430				
(Compound II)	$\lambda_{max.}$ nm	625	458			

Lactoperoxidase (Elliott, 1932; Theorell and Åkesson, 1943) has a molecular weight near 92 700 (Fe 0·76%). Although it was crystallized 30 years ago, Morrison *et al.* (1957) found that two enzymatically active components coexist with a red protein, but it seems that one may be derived from the other.

Lactoperoxidase (L)

L–H_2O	$\lambda_{max.}$ nm	640	600	550	505	412
(Ferri)	$\epsilon_{max.}\times10^{-3}$	13·6	17·4	18·6	26·3	25·7
L–H_2O (Ferro)	$\lambda_{max.}$ nm	645	600	557–575		
L–H_2O_2						
(Compound I)	$\lambda_{max.}$ nm		416			
(Compound II)	$\lambda_{max.}$ nm	570	538	430		
(Compound III)	$\lambda_{max.}$ nm	617	524	440		

The precise nature of the prosthetic group in verdo- and lactoperoxidase has not been ascertained.

The advance of knowledge about iron-porphyrin enzymes during the past forty years is due to a host of investigators and the references at the end of this Chapter draw attention to only a few sources. The surveys made from time to time by Keilin, Chance, Theorell, George and Saunders afford some of the best portals of entry into this field.

10.1.2 Infrared Spectra of Porphyrins

Infrared measurements on porphin and porphyrins have been published by Falk & Willis (1951), Thomas & Martell (1956), Ueno & Martell (1956), Mason (1958) and Rimington *et al.* (1958). Potassium bromide disks seem to be the most convenient presentation and about 2 μg of material is needed. It is important to purify the specimens and to exclude water. Esters are often preferred to free porphyrins.

The assignments from 3400 to 1200 cm^{-1} present no special difficulties, thus the N—H stretching 3320 cm^{-1} vibration fits intramolecular hydrogen bonding and is absent from haems and other metal complexes. Carboxyl absorption (2700–3300 cm^{-1}) tends to mask the N—H band and this increases the advantage of using esters. Ester absorption at 1700–1720 cm^{-1} allows differentiation from COO^- at 1560 and 1400 cm^{-1}. When a formyl group is conjugated with a pyrrole ring, the carbonyl absorption is at about 1660 cm^{-1} and perhaps also at 1580 cm^{-1}. A band at 1375 cm^{-1} is ascribed to CH_3 and a 1350 cm^{-1} peak is assigned to pyrrolic C=N. The hydroxyethyl group is almost certainly responsible for a peak at 1070 cm^{-1}.

The absorption peaks in the region 600–1200 cm^{-1} are reproducible and empirically useful in confirming identifications, but the assignments are debatable (see Schwartz *et al.*, 1960 and for comparisons, Holman *et al.*, 1956, 1959).

10.1.3 Fluorescence Spectra

Many porphyrins are strongly fluorescent but metal complexes of iron and copper and conjugated proteins like haemoglobin and myoglobin are not. Moreover the fluorescence directly traceable to the tryptophan moieties of the proteins is quenched. Removal of iron allows porphyrin fluorescence to occur. An important early study was made by Dhéré (1937) and the topic has since been reviewed (Vannotti, 1954; Lemberg & Legge, 1949; Schwartz *et al.*, 1960).

The fluorescence spectra are measured in solvents such as pyridine, ethanol, dioxane or ammonium hydroxide and in general are characterized by four peaks in the visible region. In acid the presence of three peaks is more usual and 2 N hydrochloric acid has been widely used as solvent.

Clinical biochemistry requires the differential extraction of porphyrins from tissues prior to fluorometric assay. Wranne (1960), following Schwartz & Wikoff (1952), described well-tried procedures with full technical details (see also Schwartz *et al.*, 1960). Protoporphyrin in 2 N HCl is excited by radiations between 390 and 430 nm ($\lambda_{max.}$ 410 nm) with fluorescence at 550–700 nm ($\lambda_{max.}$ *ca.* 600 nm).

Under similar conditions coproporphyrin is excited maximally at 402 nm and shows a fluorescence peak at *ca.* 598 nm. Coproporphyrin occurs in urine and faeces. In fresh urine about half the 'total coproporphyrin' is present as a non-fluorescent precursor which is converted to ordinary fluorescent coproporphyrin on mild oxidation or on standing in air. The amounts excreted per day fall within the normal limits 75–300 μg, but in abnormal or diseased states considerable rises can occur.

In studying liver dysfunction or erythropoiesis it is useful to distinguish between urinary coproporphyrins I and III which can be separated by chromatography and subjected separately to fluorometric assay (Sweeney & Eales, 1964; see also Schwartz *et al.*, 1960).

The biosynthesis of porphyrins proceeds from δ-aminolaevulinic acid via porphobilinogen (PBG) to uroporphyrin III. Decarboxylation of four CH_2COOH units leads to coproporphyrin III and further decarboxylation and dehydrogenation of two CH_2CH_2-COOH groups leads to protoporphyrin IX.

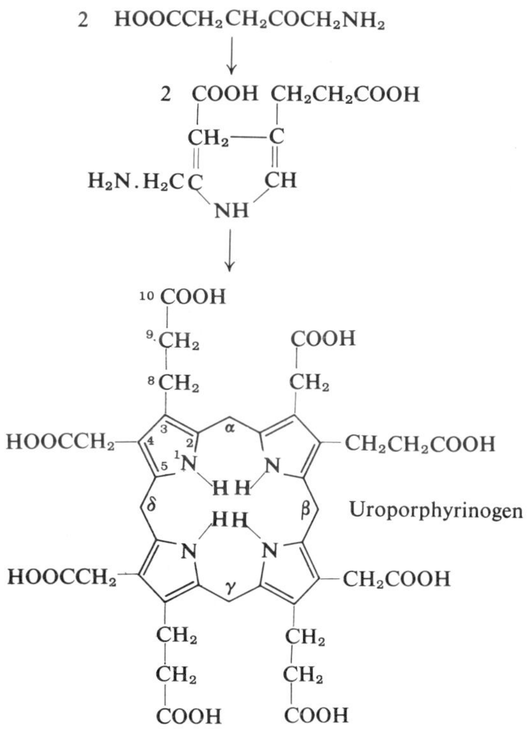

2 $HOOCCH_2CH_2COCH_2NH_2$

↓

2 [porphobilinogen: pyrrole ring bearing COOH–CH_2, CH_2CH_2COOH and $H_2N.H_2C$ substituents, NH]

↓

[Uroporphyrinogen structure: four pyrrole rings (positions 1–10, bridges α, β, γ, δ) with CH_2COOH and CH_2CH_2COOH side chains]

Uroporphyrinogen

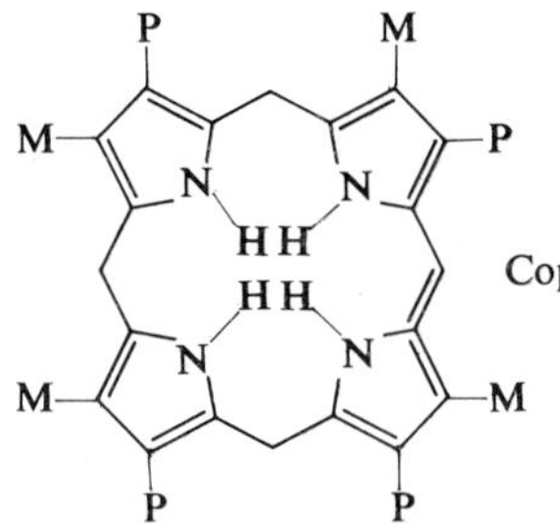

Coproporphyrinogen

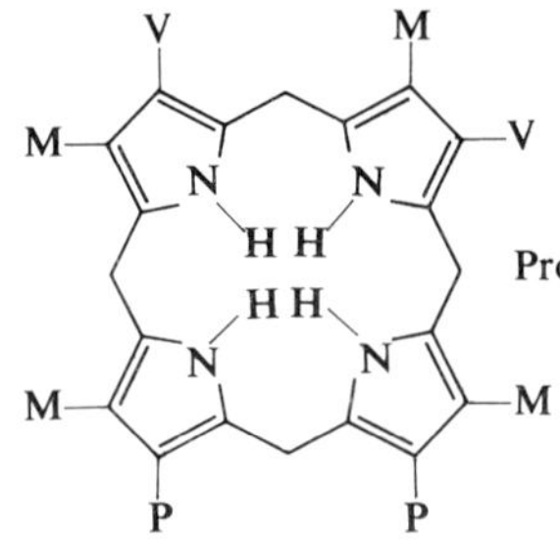

Protoporphyrinogen

Most of the free porphyrins obtained from tissue were actually present in reduced form as porphyrinogens (reduction at methine bridges and/or pyrrolic nitrogen). The fully reduced compounds do not fluoresce, neither does a partially reduced compound with $\lambda_{max.}$ 500 nm. Oxidation at acid pH by air or other means converts porphobilinogens into porphyrins.

Porphobilinogen in urine is detected by Ehrlich's reagent (*p*-dimethylaminobenzaldehyde, 0·7 g in 300 ml of a mixture of equal volumes of water and concentrated hydrochloric acid). Urine (*ca.* 10 ml) is mixed with an equal volume of reagent and two volumes of a saturated solution of sodium acetate in water are added. The product is red with $\lambda_{max.}$ 560–565 nm and a weaker peak at 520–525 nm. This reaction is not specific, but the PBG colour is developed almost fully without addition of sodium acetate, whereas interfering substances such as urobilinogen require the acetate supplement. Schwartz *et al.* (1960) discussed the methods of distinguishing between PBG and other Ehrlich-reagent reactors and also described a quantitative procedure culminating in spectrophotometry. Grinstein & Wintrobe (1948) determined protoporphyrin and coproporphyrin in red cells by methods requiring 5 ml as a minimum amount of erythrocytes (or 10–15 ml of blood containing anticoagulant). A rather complicated preparative sequence separates the two porphyrins, which are then determined either by absorption spectrophotometry or fluorescence. The method may be adapted to other tissues and to the measurement (if not identification) of liver pigments in fatal intermittent porphyria. Klüver (1955) used fluorescence to demonstrate the presence of traces of porphyrin in nervous tissue while Price & Schwartz (1956) elaborated the technique of fluorescence microscopy.

Faecal porphyrins include degradation products of chlorophyll and also deuteroporphyrin compounds of uncertain provenance; bile pigments can interfere but in man coproporphyrin (400–1100 μg/day) is the main endogenous porphyrin with very minor amounts of uroporphyrin (10–40 μg/day). The analytical problems are again dominated by the need for separations prior to spectroscopic determinations.

Erythrocyte porphyrin levels, together with abnormal urinary and faecal excretion of porphyrins, are of considerable diagnostic significance. Impairment of haem synthesis (whether it results from insufficient dietary intake of iron, or from lead poisoning or acute infective processes) causes a marked rise in red cell protoporphyrin. Increased erythropoiesis is accompanied by increases in urinary and faecal excretion of coproporphyrin. A rise is also characteristic of aplastic and pernicious anaemias. The diagnosis of lead poisoning can be confirmed by the observation that although an increase in urinary porphyrin excretion can be very large, faecal coproporphyrin remains normal. Impairment of both liver and kidney function can be examined in terms of relative urinary and faecal excretions. The different types of porphyria have inspired much investigation and the development of analytical procedures has been important (cf. Watson, 1959). See also Rimington & Sveinsson (1950), With (1955) and Haeger-Aronsen (1960).

10.1.4 Chlorophyll

The reduction of carbon dioxide in photosynthesis depends on a photosensitized decomposition of water to yield oxygen, the subsequent steps leading to carbohydrate being in the main enzymatically catalysed thermal processes. In photosynthetic bacteria the role of water is taken over by H_2, H_2S or H_2R where R is an organic radical. 'Chlorophyll' is the photosensitizing agent in the green plant and 'bacteriochlorophyll' functions in the photosynthetic bacteria. *Chlorophyll* is a designation for numerous related substances with somewhat different chemical structures and physical properties. It was right and proper for organic chemists to attach priority to establishing the structure

Table 10.4 Peak wavelengths of chlorophyll a forms (nm)

Designation	Ca 662	Ca 670	Ca 677	Ca 683	Ca 692	Ca 705
Average $\lambda_{max.}$	661·6	669·6	677·1	683·7	691·5	704·6
(S.D.)	0·09	0·05	0·09	0·16	0·39	0·90

(French *et al.*, 1971).

and the physical properties of individual substances. The degradation products and the biosynthetic intermediates were later to become equally important.

When, however, the photosynthetic processes were studied in depth it became clear (*a*) that the physiological units or chloroplasts were complex in organization and composition; besides the chlorophylls 'accessory' pigments (carotenoids or phycobilins in red and green algae) were present, (*b*) energy absorbed by different pigments proved capable of 'migration' so that most of it could be used in biosynthesis rather than emitted as fluorescence. The implication was clear that excited states had to be explored in detail.

There is now a plurality of chemically and physically distinguishable chlorophylls and a problem of excited states to which must be added the possibility of photochemical destruction. Thus in organic solvents exposure to light effects a (reversible) bleaching in the complete absence of oxygen. A reducing agent like ascorbic acid can reduce chlorophyll a dissolved in pyridine and exposed to light to a pink dihydro form:

$$\text{ascorbic acid} + \text{Chl a} \xrightarrow{h\nu} \text{Chl a 2H} + \text{dehydroascorbic acid}$$

Spectroscopic investigations with intact leaves or chloroplast preparations have revealed new complexities. Brown & French (1959) studied absorption spectra *in vivo* using an ingenious *differential* spectrophotometer which recorded the first derivative of absorbance (extinction) as a function of wavelength. The absorption spectra so obtained gave indications of masked maxima and by cooling to $-196°C$ more bands or inflections appeared. The observed direct absorption spectra were treated as summations of narrow symmetrical absorption bands, the absorbing entity in each case being a modified chlorophyll. At least four major forms of chlorophyll a co-exist in green leaves and algae. The peak wavelengths for the different components derived by curve analysis are strikingly constant as also are the half widths. Figure 10.4 and Table 10.4 illustrate the results.

When measurements were made at the temperature of liquid nitrogen on fractions from algal homogenates,

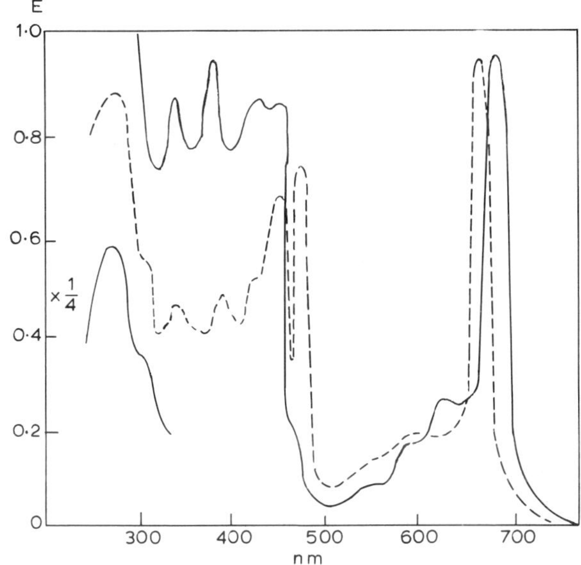

Figure 10.4 Absorption spectra of: solid curve, chlorophyll-protein from *Brassica nigra* (20°C). $\lambda_{max.}$ nm 276, ~290, 341, 436, ~600, 628, 673. Dashed curve, chlorophyll-protein from *Lepidium virginicum* (20°C). $\lambda_{max.}$ nm 265, ~291, 317, 339, 382, 420, 439, 617, 660

the spectra were very sharp and the analysis through summation of Gaussian component curves was fairly easy. The peaks and the relative intensities were used as 'input' in the curve-fitting process for direct spectra at room temperature. The results were consistent apart from a discrepancy over Ca 683.

The question must be faced whether the findings are to be regarded as unveiling fine structure or revealing plurality of chlorophylls. By measuring the area under the curve for each form, the distribution of the different component can be estimated (see Table 10.5).

Although the curve fitting as practised by French and his colleagues rests on assumptions about symmetry which may or may not be valid, it is difficult to resist the conclusions that two forms of chlorophyll b and at least four forms of chlorophyll a are very probable *in vivo* and that the quantitative evidence points to some novel type of heterogeneity rather than to unveiling masked fine structure.

The possible occurrence of chlorophyll–protein

Table 10.5 Apparent distribution of chlorophyll forms

	Chlorophyll b = 100		Chlorophyll a = 100					
(*a*) Spinach Fraction 1 23°C	Cb 640	Cb 650	Ca 662	Ca 670	Ca 677	Ca 683	Ca 692	Ca 700
	25	75	12	18	26	22	14	8
(*b*) Stichococcus								
−196	14	86	21	23	30	10	7	3
23	16	84	19	20	32	15	11	2

100 here = *ca.* 20 of total chlorophylls

complexes has been studied (Yakushiji *et al.*, 1963; Murata *et al.*, 1971; Murata & Murata, 1971). Leaves of *Brassica nigra* and *Lepidium virginicum* L. were used by Murata. The isolation procedure involved chromatography on DEAE-cellulose and gel filtration on Sephadex G-100. The molecular weights were 93 000 for the *Brassica* chlorophyll–protein and 40 000 for the *Lepidium* chlorophyll protein. Figure 10.4 shows the absorption spectra (see Murata & Fork, 1971).

Curve-fitting analysis based on measurements at −196°C indicated in the visible region four components having peaks at 662, 670, 676 and 684 nm with half widths 9–12 nm for the Brassica compound and 649, 658 and 668 nm in the *Lepidium* chlorophyll protein. Fluorescence spectra (excited at 437–440 nm) are shown in Fig. 10.5. The fluorescence was due to chlorophyll a even when the excitation was via the 496 nm absorption band in *Lepidium*, indicating efficient excitation transfer from chlorophyll a to chlorophyll b. The difficult problem in these spectra is the origin of the three ultraviolet bands at *ca.* 340, 382 and 437·5 nm (accepting that the 420 and 469 nm peaks are due to Soret bands). The three bands have separations of Δ cm^{-1} *ca.* 3290 and might well be due to a conjugated unsaturated system.

Figure 10.5 Fluorescence spectra of: Solid curve, chlorophyll-protein from *Brassica nigra* (20°C). Dashed curve, chlorophyll-protein from *Lepidium virginicum* (20°C) (from Murata & Murata, 1971)

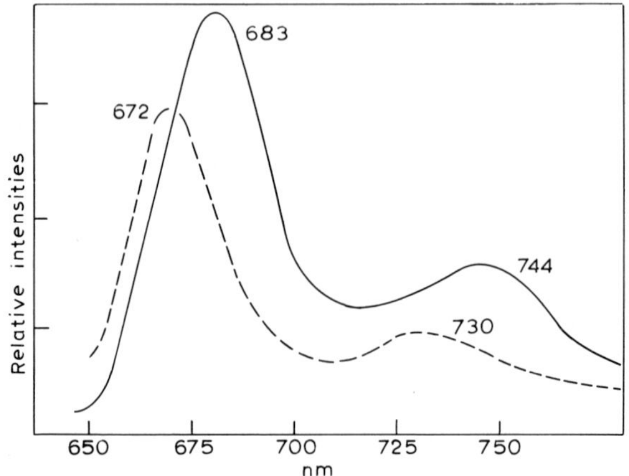

Fluorescence has proved a useful approach to numerous problems in photosynthesis. Although the various chlorophylls in the state of solution are strongly fluorescent, in the leaf or indeed in the isolated chloroplast fluorescence intensities are very low. This is because the absorbed energy is effectively transferred so as to effect primary photochemical change. Fluorescence then becomes a good weapon for elucidating interrupted sequential processes. When fluorescence occurs it provides an indication that absorbed radiation is being used inefficiently.

It is possible to measure the weak fluorescence (Litvin *et al.*, 1959, 1960) shown in leaves or cell suspensions at various temperatures and to confirm that protochlorophyll is a precursor of chlorophyll. The P_{660} form of phytochrome fluoresces maximally at 690 nm but the P_{730} form is non-fluorescent.

As already explained (p. 320) there are different forms of chlorophyll with distinctive absorption and fluorescence spectra. The fluorescence data in Table 10.7 are due to French *et al.* (1956). In *Porphyridium* absorption of light by phycoerythrin and phycocyanin is effective in phytosynthesis. *In vitro* light absorbed by these two pigments excites chlorophyll a fluorescence and in the intact organism the absorbed light is

Table 10.6 Absorption peaks of chlorophylls and related substances

Substance	Solvent	$\lambda_{max.}$ nm	α^x	Solvent	$\lambda_{max.}$ nm	
Chlorophyll a	Ether	410	85·2	Acetone		
		430	131·5		433	101·5
		533·5	4·22		536	4·78
		578	9·27		582	11·6
		615	16·3		618	19·6
		662	100·9		665	90·8
Chlorophyll b	Ether	430	62·7	Acetone	460	148
		455	174·8		(536)	6·37
		549	7·07		(558)	7·91
		595	12·7		600	14·3
		644	62·0		648·9	52·5
					(665)	10·8
Chlorophyll c	Ether	447	227	Acetone	442	—
		579·5	20·6		580	—
		628	220		628	—
Chlorophyll d	Ether	392	58·9			
		447	97·8			
		512	1·98			
		548·5	4·03			
		595	9·47			
		643	14·3			
		688	110·4			
Bacteriochlorophyll	Ether	358·5	80·5			
		391·5	52·8			
		(530)	3·0			
		577	22·9			
		(697)	10			
		773	100			
Phaeophytin a	Ether	408·5	132·0	Acetone	409	130·9
		471	5·1		472	6·3
		505	14·6		505	15·0
		534	12·6		536	13·12
		560	3·6		(558)	(5·23)
		609·5	9·8		610	11·9
		667	63·7		666·5	56·6
Phaeophytin b	Ether	412·5	83	Acetone	436	181
		434	216		527	149
		525·5	14·2		(558)	9·48
		599	9·5		600	10·7
		655	42·1		655	35·7
					(660)	10·4
Bacteriophaeophytin	Ether	357·5	127·9	Chloroform	363	111·9
		384	70·6		390	59·4
		525·5	31·9		533	29·5
		625	4·1		630	4·4
		680	12·0		687	13·0
		749	76·0		757·5	71·4

(French & Elliott, 1958; Jorgensen, 1962).

Table 10.7 Fluorescence of chlorophylls and phaeophytins

Designation of chlorophyll or phaeophytin	Solvent	Chlorophyll		Phaeophytin	
a	ether	668	724	673	721
	acetone	669	724		
b	ether	648	706	661	708
	acetone	652	706		
c	ether	629	682	648	717
d	ether	699	745	701	~740
Protochlorophyll	ether	631	683		
	acetone	626	680	648	713
Bacteriochlorophyll	ether	687	~735	761	
	also > 770				

Excitation by 405 or 436 nm, except for bacteriophaeophytin, where 525 nm was used.

effective in photosynthesis. This is important because Duysens (1951) and French & Young (1952) observed that the quantum efficiency for chlorophyll a fluorescence was greater with excitation at the absorption peak for phycoerythrin (548 nm) than with excitation at the chlorophyll a peak. The two photochemical systems SI and SII contain two distinct forms of chlorophyll a, namely a_1 and a_2 and the one linked to SII is highly fluorescent while the other is only weakly fluorescent.

SI Chl a_1 and carotenoids, absorption peaks at 680 and 440 nm $\lambda_{fl.}$ peak at 712 nm $-196°C$.

SII phycoerythrin phycocyanine, Chl a_2 absorption $\lambda_{max.}$ 544 nm $\lambda_{fl.}$ peak at 696 nm $-196°C$.

10.2 Bile pigments

The haemoglobin of erythrocytes at the end of their life is the principal source of bile pigments. When haemoglobin has been labelled by feeding ^{15}N-glycine, the incorporation of isotope into stercobilin, which can be obtained pure, reaches a peak 120 days later. There is also a subsidiary peak two to four days after feeding labelled glycine and this must be related to some aspect of the biosynthetic process, i.e. of haemopoiesis. The catabolic process normally accounts for 85% of the output but in several pathological states there is a large increase in bile pigment of haemopoietic origin. The catabolic processes occur in the reticulo-endothelial system and it has been estimated (Gray, 1953) that the breakdown of haemoglobin would account for about 300 mg/day of bile pigments.

The exact route whereby haem proteins are degraded *in vivo* to bile pigments is not certain (cf. Lemberg & Legge, 1949; Gray, 1960, 1962) although spectroscopic properties of intermediates have been noted. If haemoglobin were first to lose its globin moiety, and then its iron as ferric hydroxide, the residue would be protoporphyrin IX. Fission takes place by loss of a CH group, i.e. the bridge between the two rings containing vinyl groups. The dihydroxy compound biliverdin can equally well be regarded as a diketo compound. Reduction could then occur to bilirubin, two hydrogens being incorporated at the central methine group.

As will be seen later, bilirubin in blood plasma is bound to protein, mainly albumin. On reaching the liver, normal 'detoxication' processes result in formation of bilirubin diglucuronide (and perhaps some sulphate) and this is excreted into the bile. In the intestine considerable reduction takes place to form products like urobilinogen, stercobilinogen, stercobilin and related products. It will be noted that the coloured bile pigments (as distinct from the bilinogens which are colourless) exhibit sustained conjugation of double bonds and that this accounts for their chromophoric properties.

The nomenclature may be noted. A tetrapyrrole in which all four groups are linked by CH_2 bridges is called a bilane, while tetrapyrroles which have a central methine bridge (α-urobilin, urobilin IXa and stercobilin) are bilenes. Bilirubin and mesobilirubin which have the two outer bridges as methine groups are designated biladienes (ac). Mesobiliviolin is a biladiene (ab) and mesobilirhodin a biladiene (bc).

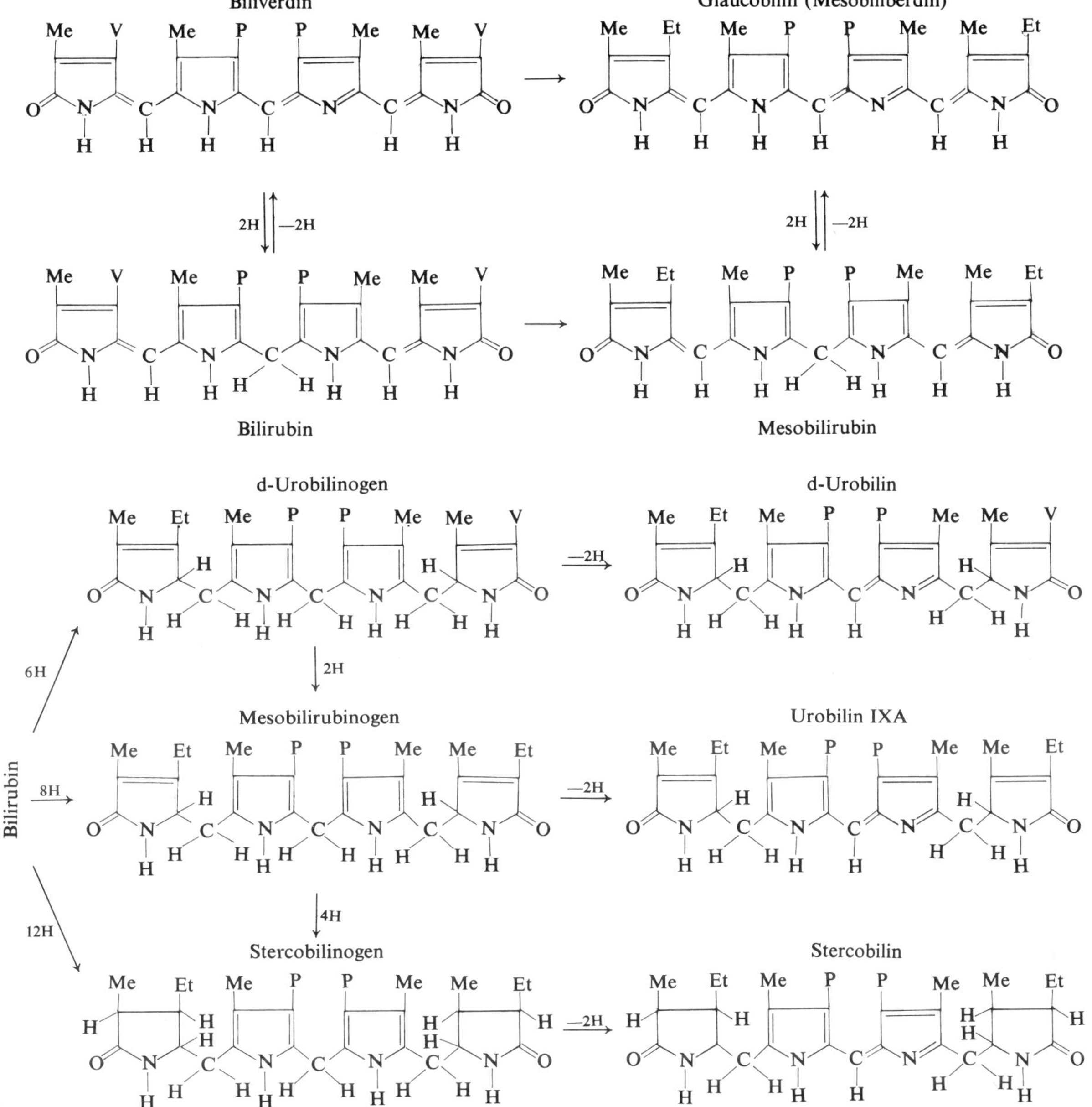

Biliverdin and glaucobilin are bilatrienes. The bilenes show $\lambda_{max.}$ 499 nm in chloroform and produce a zinc complex soluble in methanol with $\lambda_{max.}$ 510 nm; the chromophore has 5 conjugated double bonds and $\epsilon_{max.}$ is high (*ca.* 50 000). The biladienes bilirubin and mesobilirubin show $\lambda_{max.}$ at 454 and 426 nm with intensity of the same order (but variably recorded). The chromophore is

$$\mathrm{OC{-}(\overset{|}{C}{=}\overset{|}{C})_4{-}}$$

Mesobiliviolin and mesobilirhodin show (in 5% HCl) peaks at 598 and 560 nm consistent with a chomophore

$$\mathrm{OC{-}(\overset{|}{C}{=}\overset{|}{C})_7{-}}$$

The bilatrienes with conjugation maximal show peaks at 640–680 nm. As a first approximation the spectra conform with poly-ene and poly-en-one patterns.

The spectroscopy of bilirubin has been studied from different standpoints and the results are illuminating.

The purified substance is insoluble in water but in aqueous sodium hydroxide a dianion is formed with an absorption peak near 445 nm, $\epsilon_{max.}$ *ca.* 50 000. The absorption is independent of pH between 7·5 and 9·0. At pH 8·25 (25°, 0·1 **M** ionic strength) selective absorption varies with concentration in accordance with an equilibrium

$$B_2^{4-} \rightleftharpoons 2\ B^{2-}$$

between a monomer (B) and a dimer (B_2), the latter showing increased absorption between 475 and 600 nm. The ratio E_{520}/E_{430} is convenient for measuring the appearance of the dimer with 4 carboxyl ions. The dissociation constant is $K = 8{\cdot}5 \times 10^{-5}$ **M** (Brodersen, 1966).

Martin (1949) had earlier found that bilirubin (1 mg/100 ml in phosphate buffer pH 7·8 containing 20 mg/100 ml of ascorbic acid) showed $\lambda_{max.}$ 435 $E_{1cm}^{1\%}$ 800 ($\epsilon_{max.}$ 46 720). In the presence of excess of serum albumin $\lambda_{max.}$ was shifted to 460 nm without change in $\epsilon_{max.}$. There was stoichiometric interaction between bilirubin and serum albumin (and also α-globulin) but not between bilirubin and γ-globulin. Evidence was adduced that one molecule of serum albumin would bind three of bilirubin. It now seems, however, that there are two main sites each attaching one bilirubin molecule reversibly (Ostrow & Schmid, 1963; Lathe *et al.*, 1966).

Fog & Jellum (1963) and Fog & Bugge-Asperheim (1964) suggested, on the basis of infrared absorption, that hydrogen bonding occurs intramolecularly in bilirubin. Brodersen *et al.* (1967) confirmed this in a detailed investigation. Using HTO, it was found that four hydrogens in bilirubin exchanged with tritium (T), two in carboxyl groups and two attached to nitrogen in end rings, i.e. all four acidic protons. Exchange was also effected with deuterium and the infrared absorption of the freshly formed solid was obtained (KBr disks). The N—H stretching vibration at 3400 cm^{-1} (cyclic-γ-lactam) was not hydrogen-bonded (it would be expected at 3175 cm^{-1}). This peak lost half its intensity on deuteration and a new peak appeared at 2550 cm^{-1}. The shift was calculated as follows:

$$\frac{\nu_{N-D}}{\nu_{N-H}} = \left(\frac{M_N M_H / M_N + M_H}{M_N M_D / M_N + M_D}\right)^{1/2} = 2500\ cm^{-1}$$

(following Jones & Sandorfy, 1956).

In the solid state all four N—H bonds were unbonded. On the other hand, in chloroform solution the 3420 cm^{-1} peak was totally displaced on deuteration but only the N—H groups in the end rings were hydrogen bonded. In the solid state there was double hydrogen bonding between carboxyls whereas in solution the two central N—H bonds were hydrogen bonded.

Bilirubin glucuronide anion and bilirubin anion in aqueous solution show differences in electronic absorption spectra which are probably explicable in terms of intramolecular hydrogen bonding, and mixtures of bilirubin and its glucuronide do not give simply additive spectra, probably because the two solutes form a complex. Mertz & West (1956) avoided some of the difficulties by adding one volume of blood serum to 25 volumes of acetone whereby protein is thrown down and can be removed by centrifugation, leaving the bile pigment in solution. Unconjugated bilirubin is often the predominant bile pigment and it can be measured by direct spectrophotometry at 454 nm.

Overbeek *et al.* (1955) studied the kinetics of the formation of azobilirubin in a solvent consisting of ethanol, chloroform and water (6:3:1 v/v) containing hydrochloric acid (0·006 **M**). The constituents of the system are bilirubin, azobilirubin and hydroxypyrro-

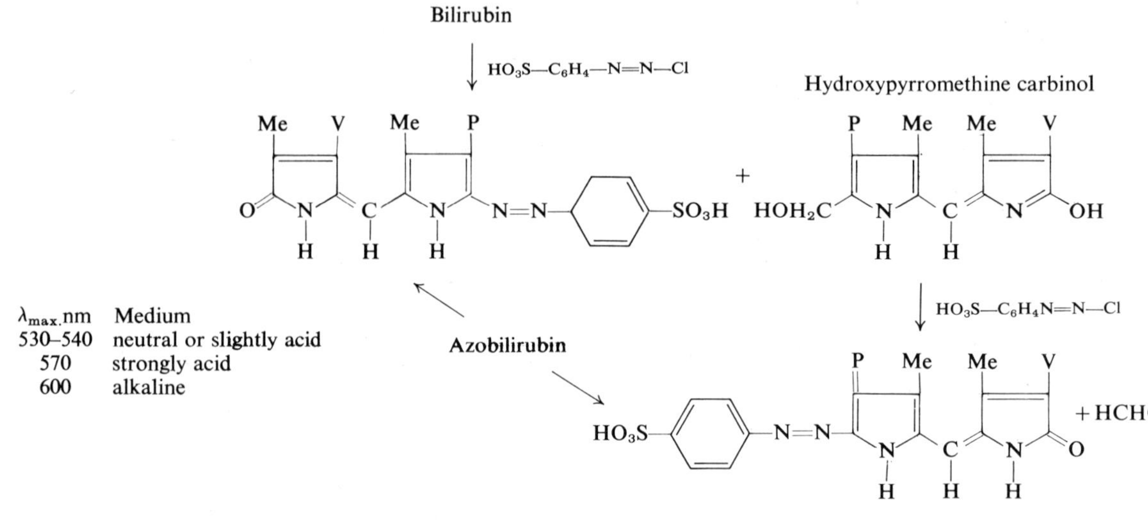

Bilirubin

$HO_3S—C_6H_4—N{=}N—Cl$

Hydroxypyrromethine carbinol

Me V Me P

P Me Me V

HOH_2C OH

$N{=}N$ SO_3H

H H H

H H

$HO_3S—C_6H_4N{=}N—Cl$

Azobilirubin

P Me Me V

HO_3S N=N O +HCHO

H H H

$\lambda_{max.}$nm	Medium
530–540	neutral or slightly acid
570	strongly acid
600	alkaline

methene carbinol (itself convertible to azobilirubin). The spectroscopic properties were:

	$\lambda_{max.}$	$\epsilon_{max.} \times 10^{-3}$ at 425	450	530 nm
Bilirubin (B)	454	45·8	55·2	0·5
Azobilirubin (A)	530	6·9	9·8	29·8
Hydroxypyrromethene-carbinol (H)	425	32·7	15·8	

The measurements at the three wavelengths suffice in theory to provide the concentrations of the three constituents:

$$C_A \times 10^5 = 0{\cdot}025E_{425} - 0{\cdot}05_1E_{454} + 3{\cdot}36_7E_{530}$$
$$C_B \times 10^5 = -1{\cdot}46_6E_{425} + 3{\cdot}03_4E_{454} - 0{\cdot}65_8E_{530}$$
$$C_H \times 10^5 = 5{\cdot}10_6E_{425} - 4{\cdot}23_8E_{454} + 0{\cdot}21_2E_{530}$$

The analyst should however be watchful for complicating factors.

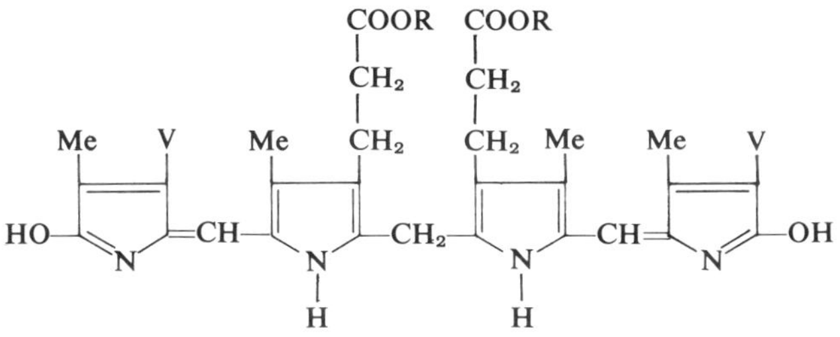

COOR COOR | CH2 CH2 | Me V Me CH2 CH2 Me Me V | HO—(N)=CH—(NH)—CH2—(NH)—CH=(N)—OH

Bilirubin R = H

R = glucuronide, Bilirubin diglucuronide

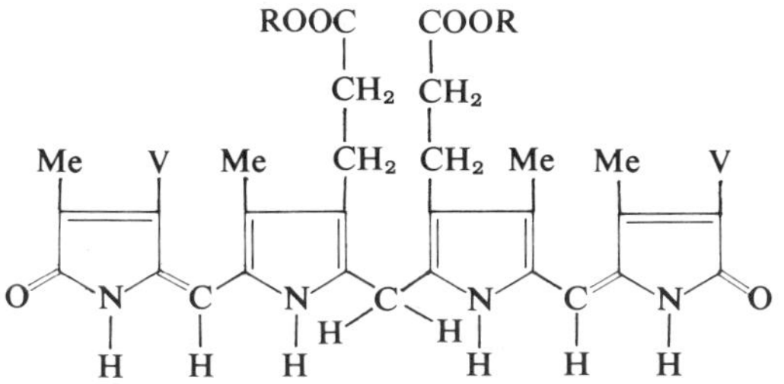

ROOC COOR | CH2 CH2 | Me V Me CH2 CH2 Me Me V | O=(NH)–C(H)–(NH)–C(H)–(NH)–C(H)–(NH)=O

Bilirubin forms red azo compounds by coupling with diazotized sulphanilic acid. Van den Bergh & Muller (1916) found that the bilirubin of bile and of sera in cholestatic jaundice reacts quickly (direct reaction) whereas the bilirubin of normal and some abnormal blood sera will only form a pigment in the presence of ethanol (the indirect reaction). Many years later it was shown that the directly reacting form of bilirubin is the diglucuronide (Billing *et al.*, 1957; Schmid, 1956; Talafant, 1956). The bilirubin undergoes cleavage at the central CH_2 with formation of formaldehyde, one molecule of azo-dye and one molecule of pyrromethene carbinol, which is itself then coupled to give a second molecule of azo-dye. In an acid medium the pigment exhibits $\lambda_{max.}$ 520 nm and in alkali a blue solution $\lambda_{axm.}$ 600 nm is formed. The glucuronide pigment is hydrolysed in alkali to give the unconjugated azo compound. Measurement of the blue colour 600 nm has advantages over readings at 520 nm.

The analytical problems of separating and determining bile pigments are dealt with in detail by Nosslin (1960) and Brodersen & Jacobsen (1969), and the clinical background is given by Gray (1953, 1961) and by Lathe & Billing (1958), Billing (1960) and Rimington & Sveinsson (1950).

10.2.1 Determination of δ-Aminolaevulinic Acid and Porphobilinogen in Urine*

Glycine, activated by pyridoxal phosphate and ferrous iron, forms a Schiff's base and reacts enzymatically with succinyl CoA to form α-amino-β-ketoadipic acid which in turn loses CO_2 to give δ-aminolaevulinic acid (ALA)

COOH | CH_2 | CH_2 | CO | $CH.NH_2$ | COOH —CO_2→ COOH | CH_2 | CH_2 | CO | $CH_2.NH_2$

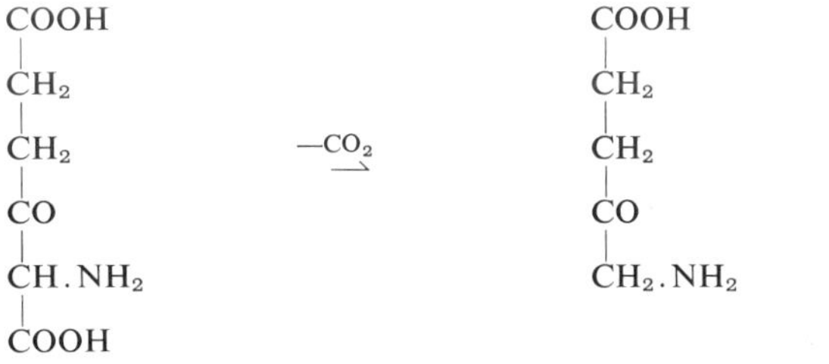

α-Amino-β-keto-adipic acid — β-Aminolaevulinic acid (ALA)

Two molecules of ALA undergo condensation to form one molecule of porphobilinogen (PBG) and water. Loss of NH_3 (4 molecules) and hydrogen (6

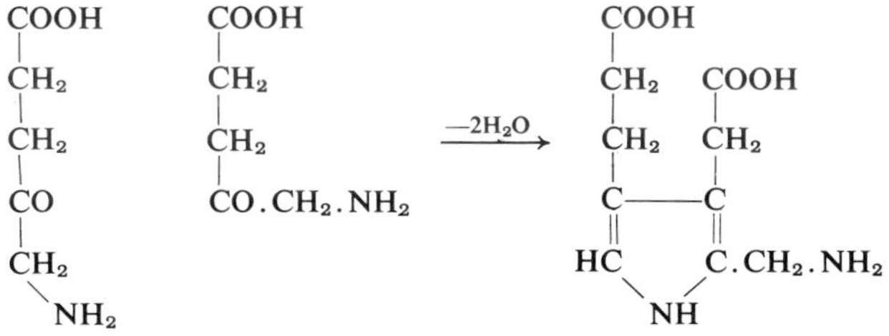

COOH | CH_2 | CH_2 | CO | CH_2 | NH_2 (ALA) + COOH | CH_2 | CH_2 | $CO.CH_2.NH_2$ (ALA) —$2H_2O$→ COOH | CH_2 | CH_2 | C=C(–CH_2–CH_2–COOH) | HC–NH–$C.CH_2.NH_2$ (PBG)

ALA ALA PBG

* With acknowledgements to Dr. W. H. Taylor of the Royal Liverpool United Hospital.

equivalents) leads to porphyrins. This can occur *in vitro* but *in vivo* it is catalysed by a dehydrase (ALAD). PBG is a precursor of porphyrins and on heating a slightly acid solution uroporphyrin (mainly Type III) is formed. Enzymatically 2 or 3 molecules of PBG are deaminated to give pyrrylmethane and ammonia. The deaminase also leads to uroporphyrinogen (UPG I) and this is isomerized enzymically to UPG III. Further enzymatic processes culminate in haem synthesis.

ALA is found in high concentration in the urine of patients with acute porphyria and is also present (in small amounts) in normal urine.

Lead in low concentration inhibits the synthesis of ALA from glycine *in vitro*, but erythrocytes from lead-poisoned rabbits synthesized protoporphyrin freely and Sano (1958) held that the first effect of lead poisoning on the biosynthesis of porphyrins is a large stimulus to ALA production from glycine. Lead inhibits *in vitro* the process ALA → PBG in haemolysed chicken erythrocytes and a similar inhibition occurs *in vivo* in lead-poisoned rabbits. There is increased urinary excretion of ALA in lead poisoning (rabbit, man).

The determination of ALA and PBG is therefore a matter of importance.

The analytical method is due to Mauzerall & Grannick (1956). Urine containing ALA and PBG is passed through a column of Dowex 2 resin in the acetate form. PBG is retained while urea and ALA are washed through. The washings are then passed through a column of Dowex 50 resin in the acid form and the urea is washed through. The ALA is eluted by means of sodium acetate solution and quantitatively converted to 3-acetyl-2-methyl-4-(3′) propionic acid by heating at pH 4·6 acetylacetone. The product, like

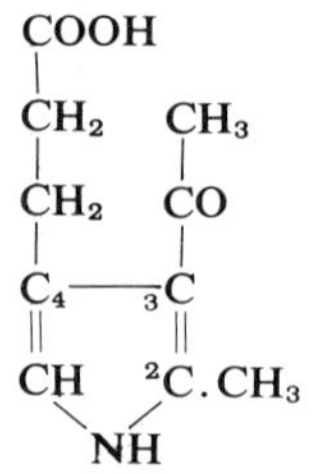

COOH
CH₂ CH₃
CH₂ CO
C₄ ³C
CH ²C.CH₃
NH

PBG, has a free α position and so forms a coloured complex with p-dimethylaminobenzaldehyde (Ehrlich reaction).

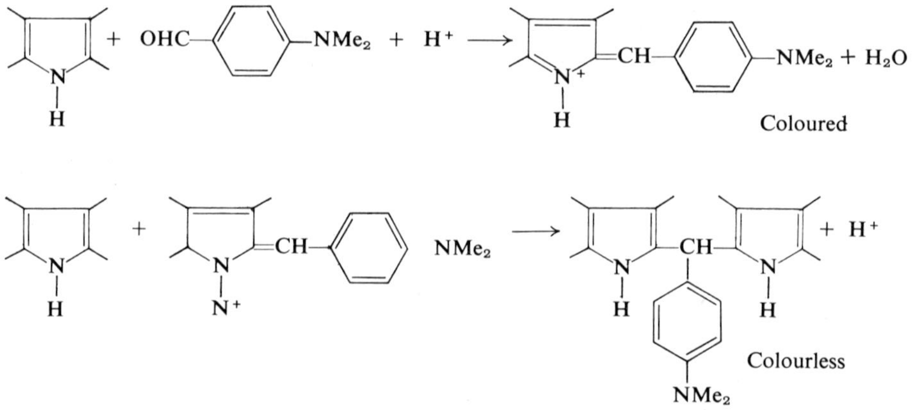

+ OHC— —NMe₂ + H⁺ → =CH— —NMe₂ + H₂O
N H   N⁺ H
Coloured

+ =CH— NMe₂ → —CH— + H⁺
N H   N N⁺   N H   N H   NMe₂
Colourless

The coloured reaction product is formed quickly but disappears slowly owing to the formation of a colourless dipyrrylphenylmethane (see Treibs & Herrmann, 1955). At low concentration of acid the coloured substance forms slowly and at high concentrations of acid both colour appearance and fading are speeded up. A modified Ehrlich reagent (II) (see below) allows the specification of experimental conditions in which the apparent molecular extinction coefficient of the Ehrlich coloured salt is about 6×10^4 as compared with 3×10^4 with the older reagent (I).

Reagents

1. Ehrlich reagent I. 2% (w/v) dimethylaminobenzaldehyde (DMAB) in 6 **N** HCl.
2. Ehrlich reagent II. DMAB (1 g) is dissolved in a mixture of glacial acetic acid (30 ml) and 70% perchloric acid (8 ml). When all the solid has dissolved the solution is made up to 50 ml with glacial acetic acid in a volumetric flask (perchloric acid is here 2 **N**; it must not exceed 4 **N** as spontaneous decomposition can then occur). The reagent should be freshly made each day.
3. Dowex-2XS 200–400 mesh is stirred with water and the supernatant decanted. The process is repeated to remove the finer particles. The resin is placed in a column and converted to the acetate form by washing with 3 **M** sodium acetate until the eluate is chloride-free ($AgNO_3$ test). The resin is then washed with water until the washings are neutral; it can then be stored in a closed vessel under twice its volume of water.
4. Dowex 50XS resin is washed similarly and converted to the sodium form by standing overnight under twice its volume of 2 **N** NaOH. The supernatant is decanted off and the resin stirred mechanically with two volumes of distilled water. The process is repeated until the washings are neutral. The resin is then reconverted to the acid form by successive treatments with 4 **N** HCl (1 vol.), 2 **N** HCl (6 vol.), 1 **N** HCl (2 vol.) and water. The resin may be stored under water (2 vol.). Both resins will keep for 3 months but should be washed with water before use to reduce possible contamination with bacteria.
5. Acetylacetone is redistilled.
6. Acetate buffer pH 4·6. Glacial acetic acid (57 ml) and $CH_3COONa.3H_2O$ (136 g) are made up to 1 l with distilled water.
7. Acetic acid 1 **N** and 0·2 **N**.
8. Sodium acetate 0·5 **M**.

The column for chromatography is 10 × 1 cm and the actual resin column after settling should be 2 cm ± 0·1 cm.

Porphobilinogen (PBG) is not readily available in a state of purity. Mauzerall & Granick (1956) found the apparent molecular extinction coefficient at 553 nm (Ehrlich reagent I) to be $3{\cdot}6 \times 10^4$ and with the Ehrlich reagent II $6{\cdot}1 \times 10^4$.

ALA can be purchased as hydrochloride. Colour tests on the product obtained by condensation with acetylacetone indicate $\epsilon_{max.}$ (553 nm) = $5{\cdot}52 \times 10^4$ using the Ehrlich reagent II.

Procedure

Urine (1 ml) is placed on a Dowex 2 column and runs through at about 6 drops/min. into a test tube. The column is washed with water (2 ml) twice. The combined eluate and washings are set aside (A), PBG is eluted from the column by 1 **N** acetic acid (2 ml) followed by 0·2 **N** acetic acid (2 ml). The combined eluates are made up to 20 ml in a volumetric flask with water. A 2 ml portion of the PBG solution is treated with Ehrlich reagent II (2 ml) and left to stand for 15 minutes when a cherry red colour is well developed. The optical density is then measured at 553 nm (1 cm cell) and a mixture of Ehrlich reagent II (2 ml) and distilled water (2 ml) is used as a blank.

The eluate A is transferred to a Dowex 50 column and the urea is eluted with distilled water (30 ml) (Ehrlich reagent I gives a yellow colour with urea). Then 0·5 **M** sodium acetate (3 ml) is added and this is followed by acetate buffer pH 4·6 (7 ml) to elute the ALA. This is collected in a 10 ml volumetric flask and acetylacetone (0·2 ml) is added and the solution made up to the mark with the pH 4·6 buffer. After thorough mixing the contents are transferred to a 15 ml stoppered test tube and heated on a boiling water bath for 10 minutes. A 2 ml portion of the cooled solution is mixed with Ehrlich reagent II (2 ml) and the coloured product measured as before. The blank is obtained by treating 0·5 **M** acetate exactly as the eluate from the Dowex column.

Calculation

$$\epsilon_{max.}\ (555\text{ nm}) = 61\,000 \text{ for PBG}$$

i.e.

$$\text{E for 226 g/l} = 61\,000$$

$$\text{E for 1 mg/100 ml} = \frac{610}{226}$$

The dilution of the original urine is finally 20-fold, hence

$$\text{mg of PBG/100 ml urine} = E_{555\text{ nm}} \times \frac{226}{610} \times 20$$

$$= E_{555\text{ nm}} \times 7{\cdot}4$$

For the coloured product from ALA, $\epsilon_{max.}$ (555 nm) = 55 200 (Ehrlich reagent II). Since the molecular weight of ALA is 131,

$$\text{E (m Mol/l)} = 55{\cdot}2$$

$$\text{E (1 mg/100 ml)} = 552/131$$

The urine used was diluted 20-fold so that

$$\text{mg of ALA/100 ml urine} = \frac{E \times 20 \times 131}{552} = 4{\cdot}74\ E$$

In normal urine the average concentration of ALA is 0·29 mg/100 ml (range 0·01–0·57) and for PBG 0·10 mg/100 ml (range 0–0·20) (see Haeger, 1960).

The method used for determining ALA in urine, depending on (*a*) the interaction of ALA and acetylacetone to form a pyrrole and (*b*) the interaction of the product and dimethylaminobenzaldehyde to give a coloured substance, is not very specific. Glucosamine and other substances also behave similarly.

It has been found that lead workers excrete an ALA-like substance in the urine and it is necessary to establish its precise nature. Haeger–Aronsen (1960) concentrated the ALA-like substances from (i) urine of workers exposed to lead, (ii) a lead-poisoned rabbit and (iii) authentic ALA. The absorption curves for the three 'Ehrlich' products (cherry red colour) were superposable (log E/λ) with a peak at 553 nm and a marked inflection near 520 nm.

The next step was the enzymatic dehydration of ALA and ALA-like substances. The enzyme δ-amino-laevulinic acid dehydrase, ALAD, is prepared from ox liver (Gibson *et al.*, 1955). It was found that the product of the enzymatic reaction with the human urine (lead worker) and the rabbit urine (lead poisoning) gave with DMAB a coloured solution with two absorption maxima 555 and 522 nm, respectively. PBG and DMAB gave exactly the same type of absorption curve. Uroporphyrin ($\lambda_{max.}$ 405 nm in 0·5 N HCl) was apparently obtained from the ALA-like substances. The spectroscopic (and other) evidence indicates that the substance in the urine after lead intoxication is indeed ALA.

Many workers in lead industries show a marked increase in ALA excretion but not of PBG. In lead-poisoned rabbits the urinary excretion of both substances rises. It is clear that determination of urinary ALA affords a very early and reliable indication of increased absorption of lead.

10.3 Cytochromes

Cytochromes are iron-containing haemoproteins which link anaerobic dehydrogenases with molecular oxygen. Enzyme systems in cells reduce the cytochromes by transfer of electrons and reciprocal systems allow reoxidation. In the presence of oxygen the major cytochrome component of mitochondria, cytochrome c, is reoxidized through the action of cytochrome oxidase and water is formed. The changes in each cytochrome may be monitored by means of absorption spectroscopy. Various inhibitors block the sequence of changes at identifiable stages.

The pioneer work in this field was carried out by MacMunn (1886) when he discovered his 'myohaematin'. Unfortunately the work was ill-received and fell into neglect; it was ahead of its time because too little was known to provide a favourable setting. Keilin (1925) revived interest in the topic and with great insight seized upon its potential significance. He recognized three haemoprotein components and designated them cytochromes a, b and c. In the visible region each showed well-defined selective absorption in the reduced state and the bands were designated α, β and γ. The wavelength of the long-wave (α) peak was greater than 585 nm for cytochrome a, around 557 nm for cytochrome b and near 550 nm for cytochrome c.

Since 1923 knowledge about cytochromes has steadily increased in amount and complexity as knowledge about mitochondria, microsomes and chloroplasts has grown and these substances now occupy central positions in biology. It is necessary, however, to begin with a simplified presentation and to proceed to a fuller account by stages. Table 10.8, due to Keilin &

Table 10.8 The cytochrome system in outline

Component	Absorption peaks nm	Notation	Properties
a	605 520 452	$a\alpha$ $a\beta$ $a\gamma$	Insoluble, thermolabile, does not combine with CO, KCN etc.; readily reduced
a_3	600 448	$a_3\alpha$ $a_3\beta$ $a_3\gamma$	Thermolabile
b (reduced)	564 530 432	$b\alpha$ $b\beta$ $b\gamma$	Occurs bound, thermolabile, autoxidizable (after reduction by ascorbic acid is reoxidized by air)
c	550 521 415	$c\alpha$ $c\beta$ $c\gamma$	Soluble at physiological pH values. Heat-stable. Detachable, not autooxidizable. Combines with NO but not CO
a_3^{+} + CO	590 432	a_3 $CO\alpha$ a_3 $CO\beta$ a_3 $CO\gamma$	
a_3^{+} + CN	590 450	a_3 $CN\alpha$ a_3 $CN\beta$ a_3 $CN\gamma$	

Keilin (1966), will serve as an introduction and below is illustrated a mitochondrial electron transfer sequence, albeit in an incomplete way.

Succinate		⟶				
↓	FAD	Fe(II) Cyt. b	Fe(III) Cyt. c	Fe(II) Cyt. a	Fe(III) Cyt. a_3	H_2O
Fumarate		⇅	⇅	⇅	⇅	
	$FAD.H_2$	Fe(III)	Fe(II)	Fe(III)	Fe(II)	$\frac{1}{2}O_2$

Cytochromes also occur in micro-organisms and Keilin & Keilin (1966) used the notation:

Cyt. a_1 $\lambda_{max.}$ 587–592 nm, a_2 628–632 nm
a $\lambda_{max.}$ 603–605 nm (598–600 nm?);
b $\lambda_{max.}$ 562–564 nm
c $\lambda_{max.}$ 550 nm, c_1 553 nm.

Baker's yeast, grown anaerobically, shows (nm) 585–590 ($a_1\alpha$), 556–566 (weak $b_1\alpha$), 521–532 (weak d?). Aerobically grown yeast shows four bands: a 603, b 564, c 550, d 522 nm, cytochrome b_2 (lactic dehydrogenase: Morton *et al.*, 1960); its cytochrome b_2 bands are at 557 and 530 nm and it has riboflavin phosphate as a second prosthetic group.

For many years the emphasis in cytochrome studies has been on the correlation of spectroscopic 'labels' with differing functions. This led to useful classifications but many problems had to be put aside. Kamen (1963) advanced a more sophisticated structural classification focused on bacterial cytochromes especially those of photosynthetic bacteria. c-Type cytochromes are often the most plentiful of the bacterial haemproteins. Although they resemble the mitochondrial cytochrome c in respect of absorption spectra, they cannot replace mammalian cytochrome c in the electron transport chain of a standard heart muscle preparation. None of the bacterial c-type cytochromes possess the acetylated *N*-terminal group which is found in the classical cytochrome c. Moreover they differ among themselves in respect of size, isoelectric point, redox potential, aminoacid compositions and peptide sequences. They function within chromatophores but do not have a terminal role in the reduction of molecular oxygen.

Vernon discovered the cytochromoids (Elsden *et al.*, 1953; Vernon & Kamen, 1954). These were present in trichloracetic acid extracts of *Rhodospirillum rubrum*. They may be defined as haem proteins with a haemoglobin-like absorption spectrum and a reactivity with ligands that do not react with cytochrome c (see Kamen, 1963). Cytochromoids I and III have been obtained from *R. rubrum* (I), from *Chromatium* (II) and from *Rhodopseudomonas palustrus*. It seems that I and II have a molecular weight about 28 000 and two haems, and in the case of II one haem has been shown to be attached by thio-ether bonds to two cysteinyl residues as in cytochrome c. The role of cytochromoids in early photochemical processes is discussed by Kamen (1961).

A cardinal fact is that the cytochrome system transfers electrons in steps, the sources being the substrates used in respiration and oxidative phosphorylation. The spectra are due to haemochromogens and the iron atoms can exist in oxidized (Fe III) and reduced (Fe II) states. Four types of haem are involved, protohaem, haem a, haem c and haem a_2. Yakushiji & Okunuki (1940) discovered an additional cytochrome which they designated c_1. It is now generally agreed that aerobic cells contain four main cytochromes each with its characteristic redox potential. Mammalian cytochrome c possesses thio-ether links with cysteine residues in the protein and this makes it difficult to release the haem. Treatment with silver sulphate, however, severs the links and haematohaem is set free and can be used to identify the cytochrome. Cytochromes of the a type unite with oxygen or carbon monoxide whereas those of the c type do not.

Studies carried out on a great range of living organisms have revealed a plurality of cytochromes b (b, b_1, b_2, b_3, b_4, b_5, b_6, b_7) and cytochromes c (c, c_1, c_2, c_3, c_4, c_5). Further reference will be made later to selected cytochromes and the relation between cytochromes a and a_3 will be discussed. The properties of so many bacterial cytochromes present persistent difficulties and complications.

At this stage it is convenient to consider cytochromes c. The classical cytochrome c from the heart muscle of horse or ox is a mitochondrial haem protein, the reduced form of which is oxidized by oxygen in the presence of the catalyst cytochrome oxidase. It has the following characteristics: isoelectric point pH 10·5, redox potential at physiological pH E_0^1 *ca.* +0·255 V; $\lambda_{max.}$ nm 550 (α), 520 (β), 415 (γ) in the ferro-form, and 530, 410, 358 and 275 nm in the oxidized form. The intensity of absorption at 530 nm ($A_\alpha \times 10^{-6}$ cm^2/mole) is near to 27·6. The iron content is probably 0·41% although 0·44–0·45% has been recorded. The molecular weight is near 12 350. The aminoacid sequence at the site of haem binding is: Lys.Cys X Gln-Cys-His where X is either alanine or serine. The haem prosthetic group is bound covalently by a thio-ether linkage as a result of condensation between the vinyl side-chains in positions 2 and 4 and cysteinyl residues of the single polypeptide chain of the protein. It has been shown that the link is to the α-carbon. There are 104 aminoacid residues in the molecule. Cytochrome c preparations from fishes, amphibia, birds, insects, reptiles, yeasts, moulds and fungi have been crystallized. In all cases the spectra in the ferro and ferri states are closely similar. Molecular weights (when known) are in the region 12 000–13 000 but small differences between species in the number of aminoacid residues and in sequencing have been noted.

Cytochromes c from algae are sometimes designated c-552 because the absorption maxima are slightly different in location, e.g. for *Euglena gracilis*:

cytochrome c-552 $\lambda_{max.}$
(Fe II) 552, 522, 416, 317 nm
(Fe III) 525, 410, 355–358, 275 nm

cytochrome c-556 $\lambda_{max.}$
(Fe II) 558, ~554, 525, 421, 317 nm
(Fe III) 530, 412, 365, 275 nm

Many algae contain a c-552 of molecular weight *ca.* 12 000 and E_0^1+0·300 to 0·38 V. In some cases the molecular weight is approximately doubled. Plants often show for their cytochrome c $\lambda_{max.}$ 552–553 nm. It is perhaps important that the nitrogen atoms of the porphyrin moiety and of two histidine residues from the polypeptide satisfy the ligands of the haem iron. The chromophore is buried in the peptide and this impedes direct action between Fe II and O_2. Acetylation of lysine residues modifies cytochrome c and effects a loss of electron transfer ring capacity.

The primary cytochrome c (Theorell & Åkesson, 1941; Tuppy & Dus, 1959; Margoliash *et al.*, 1961) shows however a remarkable degree of constancy. Differences are small, thus one lysyl residue is absent from tuna cytochrome c and although some additions or deletions occur in different species, the sequence His-Thr-Val-Glu following a cysteinyl residue is always present. Bacterial cytochromes c must however be treated as a distinct group.

The oxidized and reduced forms of mammalian cytochrome c are separable by chromatography (e.g. on Amberlite CG 50) and this suggests that a reversible change in conformation accompanies oxidation or reduction. Moreover, chromatography of horse-heart cytochrome c reveals several components derived from the ferro and ferri forms. The modified forms are apparently artefactual polymers and trichloracetic acid has been shown to hasten the process of polymerization. This work (Margoliash & Lustgarten, 1962) constitutes a warning against postulating new cytochromes without the fullest scrutiny.

The fact that cytochromes c are soluble, readily purified and of low molecular weight has all the same made them specially suitable for studying possible heterogeneity. Thus heart muscle cytochrome c exhibits fractions I–III on chromatography using the cation resin Amberlite. The minor component III consists of artefactual polymers recognized by spectroscopy (Schejter *et al.*, 1963). There seems, however, little doubt that native isocytochromes exist. Cytochrome c from (aerobic) baker's yeast consists of two monomers (Sels *et al.*, 1965; Taber & Sherman, 1964). The differences between iso-1- and iso-2-cytochrome c in respect of spectra, redox potentials and molecular weights are very small but there are definite differences

Table 10.9 Cytochrome c from ox heart or horse heart

Reduced		Oxidized	
$\lambda_{max.}$ nm	*$\epsilon_{max.} \times 10^{-3}$	$\lambda_{max.}$ nm	$\epsilon_{max.} \times 10^{-3}$
550	26·1 to 29·9	560	7·0
520	15·5 to 15·7	530–531	11·0 to 11·2
510			
415	128·5 to 133	409–410	105 to 109·5
317	34	361	19
272	32·5	275	23
$\lambda_{min.}$ nm			
537	7·0 to 7·4		
485	2·7		
Ferro minus ferri at 550 nm 18·5 to 21.			

* Different values from the literature for horse heart cytochrome c.

in aminoacid compositions and sequences. There may be other isocytochromes c in yeast. Bovine heart muscle cytochrome c has been subdivided into 4 forms, Cyt. I–IV, by disk electrophoresis. Flatmark (1966, 1967) obtained difference spectra (Cyt. III minus Cyt. I) in optical rotatory dispersion and in circular dichroic absorption spectra. The forms differ in conformation. All four forms are believed to exist *in vivo* but Cyt. I accounts for 90% and Cyt. II 9% of the total.

Cytochrome c_1 ($\lambda_{max.}$ 552 nm) (Yakushiji & Okunuki, 1940) was neglected until the observations were confirmed (Slater, 1949), but even then interpretation was difficult. Keilin & Hartree (1949), using low temperature spectroscopy, also observed a sharp band at 552 nm and postulated a cytochrome e but later (1955) became convinced that cytochrome e was the same as the Japanese workers' c_1. Cytochrome c_1 may be defined as a mitochondrial constituent obtained from animals, plants and yeasts (Okunuki, 1966). When purified by detergent treatment it exhibits absorption peaks in the reduced form at 553, 523 and 418 nm and at 523, 411 and 278 nm in the oxidized form. With an iron content of *ca.* 0·15% the minimum molecular weight is 36 700 but polymerization raises the operative molecular weight to *ca.* $1{\cdot}2 \times 10^6$. It combines neither with CO nor CN^-, which suggests that the prosthetic haem group is inaccessible and 'buried' in the structure of the apoprotein, as indeed is the case for cytochrome c. Cytochromes c and c_1 have the same haem moiety and both show haem side-chain-protein links. Cytochrome c_1, however, has little capacity to influence oxidase activity with cytochrome a unless cytochrome c is also present. However, it is functional in the succinate oxidase system or a NADH system. It must therefore precede cytochrome c in the electron transport sequence.

Material resembling mitochondrial cytochrome c is however present in rather plentiful amount in some strictly anaerobic bacteria such as the sulphate reducer *Desulphovibrio desulphuricans* (Postgate, 1954, 1956; Ishimoto *et al.*, 1954). Kamen & Vernon (1958) found that photosynthetic bacteria (including an anaerobic *Chlorobium* species) were rich in a c-type cytochrome.

These forms of cytochrome c also act as terminal substrates, e.g. in the reduction of SO_4^{2-}, NO_3^- or NO_2^- They also function in electron transport in chloroplasts and photosynthetic bacteria. It seems that various cytochromes exist to perform different but related functions and have developed differences to suit the biological roles (cf. Keilin & Hartree, 1955).

Four classes of soluble haem proteins have been recognized in photosynthetic bacteria:

1. c_2 type, molecular weight *ca.* 12 500–13 500, E_0^1 + 0·28 to 0·33 V, $\lambda_{max.}$ (Fe II) 550–552 nm. No reaction with CO. Sequencing reveals substantial homology with mammalian cytochrome c.
2. *Rhodospirillum rubrum* haem protein (RHP type) written c^1 if it contains one haem and cc^1 if it contains two haems. The Soret band is split into two (Fe II state) and there are two peaks in the α-region. The molecular weight is *ca.* 30 000 and there is a single peptide chain. In the Fe (III) state there is a peak at *ca.* 635 nm.
3. Small c-type is usually a minor constituent. There is a single haem ($\lambda_{max.}$ 555 nm) and the peptide contains 85–95 aminoacid residues.
4. A flavo-cytochrome c contains one molecule of a flavin (exact nature not established) in a protein of molecular weight given as either *ca.* 50 000 or *ca.* 72 000 ($\lambda_{max.}$ 552–553 nm) (cf. Kamen *et al.*, 1971).

Bartsch (1961, 1963) has carefully summarized the spectroscopic properties of purified cytochromes from photosynthetic bacteria (Table 10.10). In general it is desirable to use concentrations such that E = approximately 1·0 for the peak of the Soret bands and roughly three times greater concentrations are necessary for the longer wavelength peaks to be best recorded. Cytochromes of the c type are oxidized (ferricyanide) and passed through a Sephadex column to remove salts,

Table 10.10 Absorption maxima for c-type cytochromes of photosynthetic bacteria (pH 7; $\epsilon_M \times 10^{-3}$ litre/mol. cm.)

State	*R. rubrum* cyt. c_2 $\lambda_{max.}$ nm	ϵ_M	*Rps. palustris* cyt. c_2 $\lambda_{max.}$ nm	ϵ_M	*Chromatium* cyt. 553 $\lambda_{max.}$ nm	ϵ_M/3 haem
Oxidized	~560	6·6	~560	7·2	~562	18·9
	525–527	10·5	527	11·0	525	31·2
					480	31·5
	410–412·5	115	412	117	410	375
	357	30	—	—	—	—
	275	25	—	—	—	—
Reduced	551	27	552	26·7	553	75·6
	522	17	522	16·8	523	60
	416·5	132–143	417·5	136	416	435
	316	37	—	—	—	—
	272	34	—	—	—	—
CO ferro complex	—	—	—	—	555	62
	—	—	—	—	533	37
	—	—	—	—	525	45·6
	—	—	—	—	414	580

~ = shoulder or inflection.

two such treatments being needed to complete the oxidation. Cytochromoid c and *Chromatium* cytochrome 553 become fully oxidized in the course of purification. Sodium dithionite is used to reduce the cytochromes. Carbon monoxide complexes are prepared by equilibrating at 1 atmosphere pressure in an anaerobic cell. It is customary to plot (i) the absorption spectrum of the oxidized form and (ii) that of the reduced form and (iii) the difference spectrum (ferro-form minus ferri-form). Measurements are made at pH 7 in a potassium phosphate buffer.

Molecular extinction coefficients are calculated in two ways. In some cases the values are obtained from dry weights and molecular weights in the normal way. Values may however be calculated in terms of the haem concentration of each sample via measurements on the pyridine haemochromogen. (An aliquot is diluted in 0·1 **M** NaOH–25% pyridine (v/v) and reduced by dithionite, 1–2 mg per ml. After 5 minutes the absorption spectrum is measured against a solvent blank; $\lambda_{max.}$ occurs at $550 \pm 0{\cdot}5$ nm, indicating that the proteins contain cytochrome c-like haem groups.) Drabkin (1942) gave the intensity of absorption as $\alpha_M \times 10^{-3}$ litre/mol cm = 29.

Cytochrome c_2 of *Rhodospirillum rubrum* and *Rps. palustris* (molecular weight, *ca.* 12 000) has one haem bound to the peptide chain by two thio-ether bands (Paleus & Tuppy, 1959). It has a high redox potential and is probably analogous to cytochrome f of plants. *Chromatium* contains cytochrome 553 which has three cytochrome c type haem groups plus one flavin mononucleotide in a protein complex of molecular weight *ca.* 97 000. *Rps. spheroides* contains a similar cytochrome. It forms a CO-complex with one of the three haems. *Chromatium* also has the cytochromoid c already referred to (see Table 10.11).

Information concerning the b-group of cytochromes has also become very complex. The different members all have protohaem as the prosthetic group and the redox potentials are low (*ca.* 0 V). The α-band in the reduced state varies in location between 554 and 564 nm. Although the cytochromes b are autoxidizable the process occurs only slowly. Different organelles contain different members of the b family and sometimes, as in liver microsomes, two members co-exist (b and P_{450}) but act differently.

The classical cytochrome b of Keilin is present in a particulate preparation from bovine heart muscle. It is bound to protein in mitochondria and is also linked with cytochromes a and c_1. The effect of cholate or deoxycholate is to detach cytochromes and by regulating detergent concentrations, a b-c_1 complex can be obtained largely free from cytochrome a. Higher concentrations of cholate precipitate a cholate-b complex from which sodium dodecyl sulphate sets free the cytochrome b. After further purification it can be obtained as a monomeric form in a cationic detergent such as cetylmethylethylammonium bromide. In most detergents however, it exists as a polymer. Related cytochrome preparations have been obtained from blow-

Table 10.11 Absorption maxima of cytochromoid c from photosynthetic bacteria (pH 7, $\epsilon_M \times 10^{-3}$ litre/mol. cm.)

	R. rubrum		*Rps. palustris*		*Chromatium*	
State	$\lambda_{max.}$ nm	ϵ_M/2 haem	$\lambda_{max.}$ nm	ϵ_M/2 haem	$\lambda_{max.}$ nm	ϵ_M/2 haem
Oxidized	638	6	642	5·5	625	5·4
	497	22·3	500	20·8	495	20·4
	390	159–177	398	170	400	174–192
	282	44	283	42	280	63
Reduced	566	22	570	20	569	17
	550	23	552	20·8	547	19·6
	432	192	435	184	437	163
	423	196	426	192	426	188
CO ferro complex	564	24	570	22	565	23
	534	25·6	535	25·2	545	26·8
	416·5	480	418	524	418	564

fly larvae (*Musca domestica* L) and from *E. coli* (Goldberger *et al.*, 1961; Ohnishi & Okunuki, 1965). The ox heart muscle preparations are normally very highly polymerized and are autoxidizable, whereas the cytochromes b from *E. coli* and the larvae are neither polymerized nor autoxidizable.

The precise role of cytochrome b in the electron transfer system has not been established although Chance & Schoener (1966) and Green & Wharton (1963) have recorded many pertinent facts. The latter noted the evidence of four complexes in the mitochondrial system and pointed out that cytochrome b is functional in complex III, which lies between ubiquinone and cytochrome c in the sequence. Whether cytochrome b is functional in the complex between succinate and ubiquinone is debatable.

The recognition that the cytochromes b of liver microsomes were different from the main mitochondrial cytochrome b arose from the work of Yoshizawa (1951) and of Strittmatter & Ball (1952) supported by the studies of Pappenheimer & Williams (1954) on silkworm larvae. Cytochrome b_5 is a liver microsomal haemoprotein with $\lambda_{max.}$ *ca.* 557 nm with an inflection on the long wave side. It is reduced by NADP or NADPH. The preparation from calf liver showed α 557, β 526, γ 421 nm in the reduced state and in the oxidized state γ 412. The redox potential E_0^1 was +0·02 V. The molecular weight is 14 400 (Strittmatter, 1963, 1968) and the prosthetic group is protohaem. Despite its high concentration in microsomes a particular role for cytochrome b_5 remains uncertain.

An important cytochrome factor is known as P450

Table 10.12 Properties of cytochromes b

		Bovine heart muscle		Horsefly larvae		*E. coli*	
Source		$\lambda_{max.}$ nm	$\epsilon \times 10^{-3}$	$\lambda_{max.}$ nm	$\epsilon \times 10^{-3}$	$\lambda_{max.}$ nm	$\epsilon \times 10^{-3}$
(Fe II) (reduced form)							
	α	561·5–562·5	24·7	563	28·2	562	31·6
	β	531–532·5	—	530	—	531	—
	γ	428·5–429	—	428·5	—	427	—
(Fe III) (oxidized form)							
	γ	414·5	—	418·5	—	418	—
(reduced minus oxidized)		—	13·2–14·4	—	17·9	—	21·9
E_γ/E_α both in reduced state		—	6·0	—	6·5	—	5·7
Minimum mol. wt. $\times 10^{-3}$		21·3–28	—	23·4	—	—	12·0

and its properties will be discussed in outline. Klingenberg (1958) and Garfinkel (1958) noted that mammalian liver microsomes contained a pigment which when reduced formed a carbon monoxide complex with $\lambda_{max.}$ 450 nm. This cytochrome P450 complex is clearly revealed by difference spectrophotometry, using dithionite as reductant.

When microsomes are exposed anaerobically to deoxycholate or snake venom, a soluble derived pigment P420 Soret peak for the reduced —CO pigment at 420 nm results (Omura & Sato, 1962). Similar changes can be effected by solvents (Mason *et al.*, 1965) or by ureas, and partially reversed by glutathione or polyols (Ichikawa & Yamano, 1967, 1969), but as in the case for other solubilized cytochromes of the b type the biological activity tends to be lost.

Good preparations of cytochrome P450 can be obtained by using a bacterial proteinase (Nishibyashi & Sato, 1968). P450 may be described as a phospholipid–protohaem–sulphide–protein complex which plays a part in microsomal hydroxylating systems. In the reduced state it shows absorption peaks at 559, 530 and 427 nm and in the oxidized state at 414 nm. A similar steroid-hydroxylating system occurs in adrenal cortex mitochondria (Omura *et al.*, 1966).

The sequence of events in the microsomal system was seen (Omura *et al.*, 1965) as:

$NADPH_2$	Fp (oxidized)	Fe (II)	Cyt. P450 (oxidized)
		Protein	
NADP	Fp (reduced)	Fe (III)	Cyt. P450 (reduced)

but others (Holtzman *et al.*, 1968; Gigon *et al.*, 1968) envisage the final stages rather differently:

	RCH_3	$R.CH_2OH$
Fe(II)	Cyt. P450 (oxidized)	
Protein		
Fe(III)	Cyt. P450 RCH_3 (reduced)	O_2

Here a complex between the substrate and P450 is formed and reduction of this is regarded as the rate-limiting step.

The formation of a CO-P450 complex raises the question of possible photochemical effects. There is evidence (Cooper *et al.*, 1965) in respect of liver microsomes and for adrenal cortex microsomes (Estabrook *et al.*, 1963) that inhibition of hydroxylation brought about by carbon monoxide can be reversed by exposure to visible light (particularly that at 450 nm). Cytochrome P450 is clearly important in hydroxylation processes, but other factors may operate (Kato & Takanaka, 1968; Greim *et al.*, 1966, 1970).

When ethyl isocyanide is used instead of carbon monoxide for attachment to reduced P450 it appears that two pH-dependent interconvertible reduced forms can exist ($\lambda_{max.}$ 430 or 455 nm) (Imai & Sato, 1966, 1968), but Sladek & Mannering (1966) suggested that the two peaks in the isocyanide spectrum corresponded with distinct haemoproteins.

Pre-treatment of P450 preparations with 3-methylcholanthrene (Sladek & Mannering, 1966) induced formation of a distinct form of cytochrome-CO complex with $\lambda_{max.}$ 448 nm instead of 450 nm. This is designated P_1450 (see Schenkman *et al.*, 1969). Like P450, the new cytochrome P_1450 yields two pH-dependent interconvertible reduced forms. Preferential induction of P_1450 is opposed by ethionine and actinomycin D (Alvares *et al.*, 1967 *a*, 1968).

There is no doubt that P450 can activate oxygen so that drugs can be oxidized (Omura *et al.*, 1965; Cooper *et al.*, 1965). Interactions between cytochrome P450 and various organic substances (including some hormones) bring about changes in absorption spectra consistent with two rather different effects: (*a*) Type I, in

RCH_2OH	
	RCH_3
Cyt. P450–0	
	H_2O
O_2	

which $\lambda_{max.}$ is near 385 nm and $\lambda_{min.}$ near 420 nm and (*b*) Type II, with $\lambda_{max.}$ 430 nm and $\lambda_{min.}$ 394 nm (Remmer *et al.*, 1966; Schenkman *et al.*, 1967).

Accepting that microsomes are derived from the endoplasmic reticulum on homogenization of liver cells, it is pertinent to recall that many foreign organic chemicals become bound to the endoplasmic reticulum, the amounts so attached being stoichiometrically related to the spectroscopically ascertained P450 content. Carbon monoxide prevents the formation of the substrate-cytochrome P450 complex, despite the fact that different active sites are involved (Orrenius & Ernster, 1965).

The cytochrome P450-'drug' complex is reduced by $NADPH_2$ (Gigon *et al.*, 1968) but the rate depends on whether the complex belongs to Type I or Type II. Methylcholanthrene increases one form (455 nm) in the isocyanide complex but phenobarbital increases both the 430 and 455 nm forms.

Cytochrome P450 is thus a designation for a cytochrome system rather than for a single substance. The system is evidently complex and it is difficult at present to distinguish satisfactorily between a plurality of active sites and a plurality of haemoproteins. Both may be relevant.

Microsomal particles are rich in cytochrome b_5 and this hampers study of P450. The mammalian adrenal gland *mitochondria* are, however, rich in P450 (Harding *et al.*, 1964) unaccompanied by b_5. Sonication of the mitochondrial fractions releases particles containing P450 which can then be studied spectroscopically. The liver microsomes of drug-treated animals show induced increases in P450 but not of cytochrome b_5 (Ichikawa *et al.*, 1967; Kinoshita & Horie, 1967). MacLennan *et al.* (1967) obtained from cholate-treated ox liver microsomes P450 free from b_5. The spectroscopic properties are now given as shown in Table 10.13

Miyake *et al.* (1968) analysed a microsomal rabbit liver sub-particle that contained haemoprotein P450 and flavoprotein:

	μmoles/mg protein	Method
Total haem	1·7	(pyridine haemochrome)
Microsomal Fe	2·5	(e.s.r.)
Total Fe	3·4	
P450	2·5	(spectrophotometric)
Total flavin	0·2	
Cytochrome b_5	0·02	

The total lipids (604 μg/mg protein) were mainly phospholipid (269 μg/mg protein) and triglycerides (300 μg/mg protein).

Williams (1968) drew attention to the importance of high and low-spin states in this context.

Protohaem-proteins	(Soret)	$\lambda_{max.}$ nm		
Fe (II) high-spin	445	550	585	
Fe (II) low-spin	425	520	550	
Fe (III) high-spin	405	500	625	1000
Fe (III) low-spin	415	530	560	
Fe (II CO) low-spin	430	550	(580)	

It should be noted that when liver microsomes are

Table 10.13 Haemoprotein P450

	(i)		(ii)		(iii)	
Preparation	$\lambda_{max.}$ nm	ϵ_{mM} of haem	$\lambda_{max.}$ nm	ϵ_{mM} of haem	$\lambda_{max.}$ nm	ϵ_{mM} of haem
Oxidized	650	2·4	—	—	640	7·1
	570	7·3	567	18·5	566	18·0
	535	8·5	532	19	532	20·5
	415	108	415	138	414·5	136
	—	—	—	—	355	70
Reduced	555	11·7	573	21·5	541	23
	420	85	414	96	408	116
Reduced + CO	555	11·4	552	21	550	22·5
	449	114	450	131	446	130
	424	—	—	—	422	725

(i) P450 particles from rat or rabbit liver microsomes (Nishibayashi *et al.*, 1968).
(ii) P450 particles obtained without lipolysis or proteolysis from phenobarbital-induced rabbit liver (Miyake *et al.*, 1968).
(iii) Soluble haemoprotein P450 from *Rhizobium japonic* bacteroids (Appleby, 1968).

digested with steapsin, cytochrome b_5 is preferentially solubilized, leaving P450 largely in the bound form especially if glycerol is added to the mixture, since it tends to inhibit the P450 → P420 change. Appleby (1968) obtained from nitrogen-fixing *Rhizobium* bacteroids a *soluble* haemoprotein P450 (molecular weight *ca.* 50 000). The P450 → P420 change of microsomal haemoprotein catalysed by phospholipase or deoxycholate showed phospholipid to be part of the native P450. The *Rhizobium* P450 did not require phospholipid to retain its structure integrity. There is some evidence of two protohaems with an adjacent copper atom.

Three points need to be made: (1) P450 is important because of its role in the hydroxylation of steroids and of 'foreign' organic substances, (2) it is an unusual haemoprotein in respect of absorption spectra and electron paramagnetic resonance spectra, (3) P450 is imperfectly understood and it may be unsound to link the *Rhizobium* component to the mammalian one.

Cytochrome b_2 occurs in baker's yeast as a moiety within L-lactate-ferricytochrome c reductase, which contains equimolar amounts of flavin mononucleotide protohaem with a minimum molecular weight of 80 000 or *ca.* 180 000 by direct measurement. The absorption spectrum of the protohaem largely conceals the flavin contribution. The reduced enzyme when treated with methylene blue gave absorption peaks at 385 nm and 455 nm consistent with oxidation of reduced FMN but not of the protohaem. A cytochrome b_2 devoid of FMN was obtained from yeast (Yamashita & Okunuki, 1962) and crystallized after chromatography. This preparation (molecular weight 20 000) is probably a denaturation product of the native enzyme and its spectral absorption peaks are at 557, 528 and 422 nm.

Chloroplasts contain cytochrome b_{563} (also known as cytochrome b_6) and a cytochrome b_{559} from the wavelengths of the α-bands. Cytochrome b_{563} is believed to be associated with photosystem I in photosynthesis and to enter into the cyclic process from the primary electron acceptor back to the reaction centre chlorophyll P_{700} (Vernon *et al.*, 1969). Cytochrome b functions in photosynthesis in association with photosystem II but there are rival theories about its role (Fan & Cramer, 1970; Knaff & Arnon, 1969). The redox potentials E_0^1 for b_{563} and b_{559} as observed by different workers do not agree very well but there is evidence to suggest that a value near +0·35 V is reasonable for the latter whereas a negative value is characteristic of the former. Cytochrome f also occurs in chloroplasts (see Wilson & Dutton, 1971).

The classical demonstration of the properties of cytochrome oxidase (Warburg & Negelein, 1928, 1929) rested on the absorption spectrum of its complex with carbon monoxide. A paper by Kubowitz & Haas (1932) was of considerable theoretical and technical interest in determining the absorption spectrum of oxygen-transporting enzyme by measuring the photochemical breakdown of the CO-addition compound.

Various light sources were used to provide 30 practically monochromatic irradiations at wavelengths between the limits 250 and 672 nm. Experiments were made on *B. pasteurianum* (acetic acid bacteria) and *Torula utilis* (yeast) cells. Respiration was reduced to about 30 by carbon monoxide and restored to 60 by irradiation and from the intensities of monochromatic irradiation needed to effect the increase, the absorption spectrum was calculated. The acetic acid bacteria kept in darkness showed a halving of respiration with a CO/O_2 ratio of 1:2, while the yeast respiration fell to half only when the CO/O_2 ratio was 10:1.

If activity is partially inhibited by CO, let n be the residual activity and $1-n$ activity lost by CO poisoning; we can write $V=n/1-n$. The dark value can be written n_d and the 'light' value n_l so that the distribution changes from

$$V_d=(n_d)/(1-n_d) \quad \text{to} \quad V_e=(n_l)/(1-n_l)$$

and $$\Delta V=(n_l)/(1-n_l)-(n_d)/(1-n_d)$$

The light sensitivity L is given by:

$$L=\frac{(\Delta V/i)}{V}=\frac{[n_l/1-n_l]-[n_d/1-n_d]}{i(n_d/1-n_d)}$$

provided the ratio CO/O_2 is the same in the dark as in the light. Now put Nhν for one quantum/mole and z_d for the decomposition constant of the dark reaction for the CO-addition compound and β for the light absorption coefficient:

$$L=\frac{1}{\mathrm{Nh}\nu}\,\frac{\beta}{z_d}$$

and when two frequencies are used

$$\frac{\beta_1}{\beta_2}=\frac{L_1}{L_2}\cdot\frac{\lambda_2}{\lambda_1}$$

For the best results $V_l=2V_d$, i.e.

$$\frac{V}{V_d}=1$$

and $i = 1/L$. A reference line is chosen to compare β values and the H_g 436 nm blue line was chosen with 546 and 366 nm as subsidiary monochromatic radiations. All the data were expressed as

$$\frac{\beta\lambda_0}{\beta\lambda} = \frac{L\lambda}{L\lambda_0} \quad \frac{\lambda_0}{\lambda}$$

where $\lambda_0 = 436$ nm and the absorption curve was obtained by plotting $\beta\lambda/\beta 436$ nm against λ.

Keilin & Keilin (1966) replotted the data from several papers on the photochemical spectra and drew attention to a number of difficulties. They agreed that 'in the main the photochemical absorption spectrum represents the carbon monoxide compound of cytochrome oxidase', the variability of the spectrum reflected perhaps light absorption by other intracellular components and in particular the cytochromes closely linked to the oxidase. It is noteworthy that the protein band at 280 nm was consistently recorded, making it likely that the quanta absorbed by the tyrosine and tryptophan moieties are transmitted to the haem and utilized for the photochemical fission of the Fe–CO link just as readily as the quanta absorbed by the haem chromophore itself. Some of the difficulties have been removed by Chance and his colleagues (Chance *et al.*, 1953; Castor & Chance, 1955, 1959).

In the course of their studies on heart muscle preparations Keilin & Hartree (1939) distinguished between cytochrome a and cytochrome a_3. The crude preparations showed 5 bands in the visible region, a narrow band at 605 nm (cyt. a reduced), a band at 567 nm (fused α bands of b and c), 529 nm (β bands of b and c) and flavoprotein bands at 455 and 495 nm. The γ bands (Soret bands) occurred at 415, 432 and 448 nm ($c\gamma$, $b\gamma$ and $a\gamma$ respectively).

Respiratory inhibitors (CO, KCN, NaN_3, H_2S, NH_2OH, NaF, etc.) were studied using heart muscle preparations containing the cytochrome system reduced by sodium succinate:

(i) CO did not change $b\alpha$, $c\alpha$ and $(b+c)\beta$ but $a\alpha$ underwent a marked blue shift and became broader; absorption was stronger at 590 nm while $a\gamma$ (448 nm) disappeared and was replaced by a weak band at 452 nm.
(ii) Under strictly anaerobic conditions KCN had marked effects on $a\alpha$ and $a\gamma$ but under aerobic conditions only $a\gamma$ was affected.
(iii) Other inhibitors gave rather similar results.

The interpretation was advanced that there were two a-type cytochromes, the bands at 605 and 450 nm being made up of overlapping bands. The new cytochrome designated was a_3 (a_1 and a_2 having been used for bacterial cytochromes a). The band $a_3\alpha$ is weak compared with $a\alpha$ and is near 600 nm but $a_3\gamma$ (448 nm) is strong compared with $a\gamma$ (452 nm). The addition compound a_3CO shows the $a_3\alpha$ band displaced to 590 nm and the $a_3\gamma$ band is displaced to 432 nm (overlapping the b_γ band) and a weak $a\gamma$ band at 452 nm can then be seen.

Keilin identified a_3 with cytochrome oxidase because it combined reversibly with the known reversible inhibitors of respiration and its peaks with CO (a_3^{++} CO α at 590, a_3^{++} CO γ at 432 nm) agreed rather closely with the photochemical absorption spectra. That cytochrome oxidase contained copper seemed to Keilin possible but not certain. He also felt that the co-existence of a and a_3, the identity of their prosthetic groups and other resemblances did not preclude the possibility that they were 'very intimately connected if not inter-convertible'.

Okunuki & Yakashuji (1948) and Wainio (1955) argued against the existence of cytochrome a_3. As cytochrome a was purified it was found to contain equimolar amounts of copper and haem a. It was ultracentrifugally homogeneous although the polymer could be depolymerized to give products with enhanced oxidase activity. It was further argued that cytochrome a needs native cytochrome c for cytochrome oxidase activity. A tertiary complex O_2—cytochrome a—cytochrome c was postulated to accept electrons from donors via cytochrome c_1.

10.3.1 Cytochrome Oxidase

Important matters concerning this problem remain to be settled. Some of the evidence follows. It is convenient to accept the term cytochrome oxidase for the complex which acts as an oxidase towards cytochrome c. The complex contains cytochrome a, cytochrome a_3, functional copper and other constituents.

Cytochrome a, Takemori *et al.* (1961) (molecular weight 530 000) is a pentamer with a minimum molecular weight *ca.* 100 000 per haem residue. The copper and iron are in the ratio 1·05 (Cu per cytochrome a monomer *ca.* 1·18) so that probably there is one iron and one copper atom. Reduced cytochrome a [Fe(II)] shows $\lambda_{max.}$ 605 nm, $\epsilon_{max.}$ 20 640 (α) with $\lambda_{max.}$

444 nm(γ). Oxidized cytochrome a has peaks at 600 (α), 424 (γ) and 340 and 280 nm. The difference spectrum (for the reduced form) has $\epsilon_{605}-\epsilon_{630}=16\,500$. The copper is not readily set free; it withstands 100° for 30 minutes, and dialysis for 48 hours against copper chelating agents does not remove it. Trichloroacetic acid will however liberate it. The bound copper is divalent [Cu(II)] but is reduced to Cu(I), in the presence of substrates under anerobic conditions. The haem has the following side-chains:

position 2 —CH(OH).CH_2R'' position 4 —CH=CHR′
position 8 —CHO, where R′ may be H and R″ $C_{15}H_{31}$
[$C_{15}H_{31}$ = —CH_2CH_2—CH(Me).$CH_2CH_2CH_2$CH(Me).
$CH_2CH_2CHMe_2$]

Aldehyde reagents strongly inhibit cytochrome oxidase activity. Purified cytochrome is autoxidizable only in the presence of cytochrome c. A complex between equimolar amounts of a and c is formed and has a pH optimum of 7·0. With appropriate reductants the complex uses up much oxygen.

Cytochrome a does not react with carbon monoxide but cytochrome a_3 reacts readily. Carbon monoxide, like cyanide ion, sulphide, azide or hydroxylamine, inhibits the typical cytochrome oxidase obtained from heart muscle. Melnick's classical photochemical absorption spectrum ($\lambda_{max.}$ 589, 510, 450 nm) agreed with the spectrum of Warburg's respiratory ferment.

The absorption spectrum of a_3 shows:

$\lambda_{max.}$ nm	$\epsilon_{max.}$
590–593	135
432–433	156
283	87

The complexity of the problem is shown by the work of Caughey *et al.* (1968). The lack of chemical detail is such that the mode of action of cytochrome oxidase 'can hardly avoid oversimplification'. The smallest active unit contains two haems and two copper atoms (Orii & Okunuki, 1967). Caughey *et al.* (1967), using mild methods, isolated almost all the haem from ox heart muscle. This haem A differed from protohaem by having a formyl group in place of 8-methyl, an unsaturated alkyl chain of 16 or 17 carbon atoms in place of a vinyl group (possibly 2-vinyl) and a nitrogen-containing group (*ca.* $C_6H_{12}NO_4$). Chromatographic fractionation of haem a gave two major zones, a_α and a_β. The absorption spectra of the two haems revealed small but quite significant differences in band widths and relative intensities. The electronic spectra fit the idea that both haems have similar or even identical unsaturated groups conjugated to the ring but that there are some structural differences consistent with two unequal cytochrome a components.

Yonetani (1961) crystallized cytochrome oxidase from bovine heart muscle. It was apparently homogeneous (molecular weight *ca.* 500 000) and made up of 4–6 units of molecular weight 70 000–100 000 each containing 1 haem, 1 copper and 1 iron per mole (Takemori *et al.*, 1961). The photochemical action spectrum of the CO complex (Yonetani & Kidder, 1963) resembled those based on oxygen uptake (Castor & Chance, 1955) and was significantly different from those of reduced cytochrome oxidase in the presence of CO. The spectrum of cytochrome oxidase was a summation of two separate spectra, presumably of a and a_3. According to Yonetani (1961), in the reduced state the 605 nm maximum was due as to 72% and 28% to a and a_3 and the γ peak at 444 nm was due to 46% a and 54% a_3. Horie *et al.* (1968), in a reinvestigation, estimated that the a/a_3, contributions at 605 nm were 4/1.

Horie (1968) discussed the nature of a cytochrome 'a' preparation free of a_3. Morrison & Horie (1963) found that, in the presence of cyanide, borohydride selectively reduced the carbonyl group of haem a of a_3 to a secondary alcohol group. There were some disadvantages in using cyanide, and borohydride alone in higher concentration effected reduction but the concomitant rise in pH tended to alter cytochrome a. This was avoided by adjusting the pH to 6·9, adding ascorbic acid and ammonium sulphate. To prepare 'a' free from a_3, the a_3 carbonyl-reduced product (8 ml) was treated with $Na_2S_2O_4$ (8 mg) and after 10 minutes the mixture was diluted 5-fold and allowed to stand at 0°. Precipitation was effected by 35% saturation with ammonium sulphate.

The figures show that the preparation treated with borohydride is spectroscopically substantially cytochrome a free from a_3 (see also Kuboyama & King, 1964). The spectroscopic work of Horie is of considerable interest.

Morrison (1968) provided evidence that cytochrome oxidase gave two fractions (molecular weights 580 000 and 290 000 respectively). The smaller fraction consisted of two 120 000 units and one 60 000 unit, each unit having one haem and one copper. The haem was extracted with chloroform–pyridine and it was shown to be a dicarboxylic acid containing a formyl group, a hydroxyl group and a ketone group (i.e. 7 oxygens/Fe). No evidence was obtained for a hexosamine group.

Nicholls (1968), accepting that a and a_3 are 'spectroscopically distinct but associated chemically and physically', takes the view that a and a_3 provide an example of an allosteric system in which two haem a moieties are linked so that binding or redox changes at one haem affect the reactivity of the other haem. His arguments rest essentially on the interpretation both of spectra and of redox potentials. Reduced cytochrome a exhibits variable activity; two forms of reduced a may exist, one in the presence of reduced a_3 and the other in the presence of oxidized a_3. The former would be reactive, reducing cytochrome a_3 as soon as it has reacted with O_2 while the latter would be unreactive.

Mason *et al.* (1968), in answer to P. Nicholls, agreed that an alternative explanation of their results would be that on removal of copper, cytochrome oxidase can 'lose its capacity to be activated by Tween 80 in the polarographic assay', which would amount to saying that removal of copper could affect the capacity of the enzyme to change its conformation.

There is no doubt that in crystalline cytochrome oxidase copper is present in stoichiometric amount. Equally, it is clear that an infrared absorption peak occurs at *ca.* 830 nm in cytochrome oxidase.

A typical cytochrome oxidase preparation $(a+a_3)$ shows $\lambda_{max.}$ 600 nm and 424 nm. Okunuki (1966) found that in a cytochrome a preparation of good quality, oxidase activity was reduced by one half in the presence of a copper chelating agent. The copper was Cu(II) but during the transfer of reducing power to oxygen there was formation of Cu(I). The catalysis includes transfer of reducing equivalent from cytochrome c to cytochrome a to cytochrome a_3 to oxygen, with copper participating. In fact a distinction must be drawn between functioning copper and non-functional copper.

An elegant set of experiments has shown how part of the copper can be removed from the enzyme with concomitant loss of activity. Replacement of Cu(I) in a suitable form restores the activity.

Nair and Mason (1966, 1967) treated cytochrome oxidase under anaerobic conditions with bathocupreinedisulphonate (2,9-dimethyl-4,7-diphenyl-1,10-phenanthroline disulphonate) in the presence of reduced cytochrome c. About 45% of the copper present in the enzyme preparation was removed, leaving a copper-depleted and relatively inactive enzyme (*ca.* 15% of the original activity). The ratio of haem a to cytochrome a_3 (approximately 2) was unchanged. Copper was restored in the form of cuprous acetonitrile perchlorate $Cu(CH_3CN)_4ClO_4$, and the original cytochrome oxidase activity was more than fully regained.

Electron spin resonance (e.s.r.) spectroscopy showed that in cytochrome oxidase copper is bound by specific (functional) and non-specific (non-functional) systems of ligands. A unique e.s.r. signal accounted for 40% of the functional copper [Cu(I)] but the remainder of the functional copper was not detected by e.s.r. When the enzyme was denatured in the presence of a sulphydryl-binding agent the e.s.r. then corresponded accurately with total, chemically measurable copper. The functional copper Cu(I) and Cu(II) must be held by two ligand systems—a unique structural state. Nair and Mason consider it possible that copper converts cytochrome a_3 to cytochrome a 'either by direct interaction with the haem or by producing some specific conformational change'. The reconstitution reaction was inhibited by chloromercuriphenyl sulphonate, implicating —SH as a significant factor in the system. The evidence points strongly to cytochrome oxidase from bovine heart muscle being copper-dependent.

D. E. Griffiths had earlier (1961) stated a strong case for a role of copper in cytochrome oxidase. He pointed out how the Cu/protein ratio rose during purification but the equimolar relationship Cu/haem a persisted at all stages. Further, as the Cu/protein ratio rose, so did the specific activity of the enzyme. The copper is difficult to remove but if it is all taken away there is a complete loss of activity. Copper in cytochrome oxidase undergoes reversible oxidation–reduction and the well-known inhibitors (cyanide, azide) prevent reoxidation of reduced copper. Also electron paramagnetic resonance (e.p.r.) confirms some reduction of copper on cytochrome a by reduced cytochrome c and reoxidation of reduced Cu by oxygen. Two kinds of copper union have emerged, one reducible by substrate and not by borohydride (active Cu) and a second part readily reducible by borohydride but not by substrate. This is readily chelated and is largely 'inactive'. An important development was the observation that an infrared absorption band ($\lambda_{max.}$ 815–830 nm) in cytochrome oxidase could be attributed to copper. If the enzyme was dialysed against 0.01 **M** cyanide plus alkaline buffer of high ionic strength, there was a decline in specific activity parallel with a decline in the intensity of the 830 nm band which was thought to arise from a protein–copper link. There is a good proportionaility between e.p.r.-detectable Cu(II) and the intensity of the 830 nm peak. But not all workers are of the opinion that the 'undetectable' functional Cu(II) does not absorb at 830 nm (Wharton & Tzagoloff, 1963).

Gibson & Wharton (1968), using the 'flow flash' technique, confirmed the earlier work of Griffiths and Wharton that the 815–830 nm band is due to a copper chromophore (the band is also seen in plastocyanin, ascorbic and oxidase, *pseudomonas* blue protein and laccase) and that it is proportional to the 600 nm band. The extinction coefficient 815–830 nm, $\epsilon = 1400$ is quite high.

Van Gelder and Slater (1966) titrated cytochrome c oxidase with NADH and phenazine methosulphate and found that equimolar amounts of haem a and copper were simultaneously reduced. Functionally inactive copper (present in the enzyme or added as Cu^{2+}) was not reduced. Two kinds of functionally active copper were found in equal proportions. The cyanide-sensitive copper contributed one half of the difference spectrum (oxidized minus reduced) at 830 nm.

According to Beinert, the following depicts the likely interpretation:

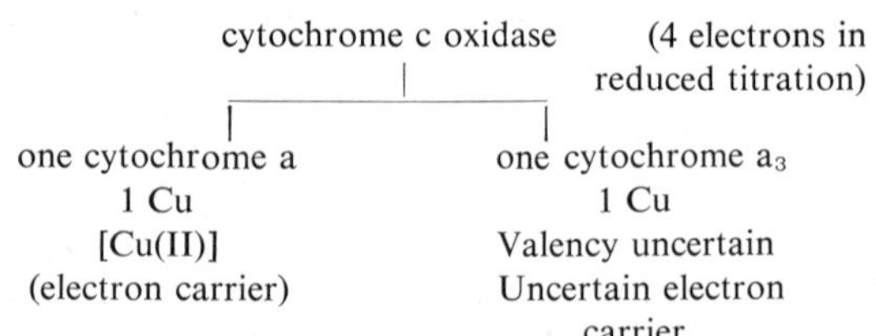

40% of Cu is Cu(II) by e.p.r. spectroscopy (? 50%)
e.p.r. detectable Cu(II) absorbs at 830 nm (rapidly reduced by cytochrome c)

Total Cu minus e.p.r.-detectable Cu may or may not be Cu(II) but must be able to take up or release one electron equivalent per Cu. On the other hand Tzagaloff & MacLennan (1967) held that the 830 nm absorbing entity was not part of cytochrome a_3 and concluded that the copper of cytochrome a was responsible for the near infrared band and for the e.p.r. effect, whereas the copper of cytochrome a_3 was responsible for neither.

It may be that cytochrome oxidase is a multiple protein sooner or later to be split into sub-units with copper proteins separable from haem proteins. Although the case for associating the 830 nm absorption band with the e.p.r.-detectable Cu(II) is strong, it does not compel assent and Chance seems still to doubt whether the correlation is a true one. The reaction of oxygen with the Cu(II) ion of cytochrome oxidase is only 1/10 as fast as the reaction of oxygen with the Fe(II) haematin of the oxidase.

Perhaps, however, the most disconcerting fact is the observation that the cytochrome oxidase prepared from *Pseudomonas aeruginosa* contains no copper. This material has most of the properties of a cytochrome oxidase: it contains two kinds of haem and exhibits spectra (in 0·1 **M** phosphate buffer pH 6) of the expected type:
FeIII 635, 525, 412, 280 nm, and after dithionite reduction, FeII 625, 554, 549, 523, ~460, 418 nm. Isoelectric point 5·8; molecular weight *ca.* 90 000, probably 2 Fe/mole.
This cytochrome oxidase reacts with the *Pseudomonas* blue protein (molecular weight 16 600, 1 Cu, $\lambda_{max.}$ 630 nm, $\varepsilon_{max.}$ 6950 (oxidized form)).

Any attempt to take stock of the 'cytochrome situation' must reflect the limitations and interests of the writer as well as his intentions.

Enormous progress has been made in collecting facts pertinent to the study of major topics such as oxidative phosphorylation, electron transport and photosynthesis. The factual material includes direct absorption spectra of the haemoproteins in oxidized and reduced states, circular dichroic spectra and electron spin resonance. Redox potentials, autooxidation and protein conformations affect the issues. Individual cytochromes have been isolated, purified and the prosthetic groups have been identified. Modes of linkage have been established, primary protein structures have been worked out and great progress made in sequencing. A vast range of living organisms has been studied and the distinction between procaryotic and eurkaryotic organisms has gained a new emphasis. The sequence studies have thrown much light on comparative biochemistry which may turn out to be important in the study of evolutionary change.

The progress has been great but it underlines the intractability of some central problems in biology. Students of oxidative phosphorylation and of the fundamental photosynthetic processes sometimes feel that there is almost too much information, but this is a passing phase and an intellectually satisfying synthesis is to be hoped for.

References

10.1, 10.2

Agner, K. (1941). *Acta physiol. Scand.*, **2**, Suppl. VIII.
Agner, K. (1943). *Adv. Enzymology*, **3**, 137.
Avi-Dor, Y. & Paul, K. G. (1953). *Acta. Chem. Scand.*, **7**, 44 (see also **8**, 649).
Bach, A. & Chodat, R. (1903). *Ber. dtsch chem. ges.*, **37**, 1342.
Barrett, J. (1956). *Biochem. J.*, **64**, 626.
Barrett, J. (1959). *Nature, Lond.*, **183**, 1185.
Billing, B. H. (1960). *Advances in Clinical Chemistry*, Vol. II. Academic Press, New York, London.
Billing, B. H. *et al.* (1957). *Biochem. J.*, **65**, 774.
Bonnichsen, R. K., Chance, B. & Theorell, H. (1947). *Acta. Chem. Scand.*, **1**, 685.
Brodersen, R. (1966). *Acta. Chem. Scand.*, **20**, 2895.
Brodersen, R., Flodgaard, H. & Krogh-Hansen, J. (1967). *Acta. Chem. Scand.*, **21**, 2284.
Brodersen, R. & Jacobsen, J. (1969). In *Methods of Biochemical Analysis*. Vol. 17, p. 31. Ed. D. Glick. Interscience Publishers, New York, London, Sydney and Toronto.
Brown, J. S. & French, C. S. (1959). *Plant Physiol.*, **34**, 305.
Chance, B. (1949). *J. Biol. Chem.*, **179**, 1331.
Chance, B. (1954). *Chem. Abs.*, **48**, 10839.
Chance, B. (1956). In *Currents in Biochemical Research*. Ed. D. E. Green. Interscience Publishers, New York.
Chance, B. & Fergusson, R. R. (1954). In *The Mechanisms of Enzyme Action*. Ed. W. D. McElroy and B. Glass. Johns Hopkins Press, Baltimore.
Chance, B. & Maehly, A. C. (1955). In *Methods in Enzymology*. Vol. 2. Ed. S. P. Colowick and N. O. Kaplin. Academic Press, New York.
Dhere, C. (1937). *La fluorescence en biochimie*. Les Presses Universitaires, Paris.
Duyssens, L. N. M. (1951). *Nature, Lond.*, **168**, 548.
Ehrenberg, A. & Theorell, H. (1955*a*). *Acta. chem. Scand.*, **9**, 1193. (1955*b*). *Nature, Lond.*, **176**, 158.
Elliott, K. A. C. (1932). *Biochem. J.*, **26**, 10.
Falk, J. E. (1936). In *Comprehensive Biochemistry*, Vol. 9, p. 3. Ed. M. Florkin and E. H. Stotz. Elsevier, Amsterdam.
Falk, J. E. & Willis, J. B. (1951). *Austral. J. Sci. Res.*, Ser. A4 to 5, 579.
Fischer, H. & Orth, H. (1937). In *Die Chimie des Pyrrols*, II, part 1. Akademische Verlagsgesellschaft, Leipzig.
Fischer, N. & Stern, A. (1940). *Ibid.*, II, part 2.
Fog, J. & Bugge-Asperheim, B. (1964). *Nature, Lond.*, **203**, 756.
Fog, J. & Jellum, P. (1963). *Nature, Lond.*, **198**, 88.
French, C. S., Brown, J. S. & Lawrence, M. C. (1971). In *Carnegie Institution Yearbook*, **70**, 487.
French, C. S. & Elliott, R. F. (1958). In *Carnegie Institution Yearbook*, **70**, 487.
French, C. S. *et al.* (1956). *Plant Physiol.*, **31**, 369.
R. L. (1956). *Plant Physiol.*, **31**, 369.
French, C. S. & Young, V. K. (1952). *J. Gen. Physiol.*, **35**, 873.
George, P. (1952). *Nature, Lond.*, **169**, 612. (1953). *Science, N.Y.*, **117**, 220.
George, P. (1956). In *Currents in Biochemical Research*, p. 338. Ed. D. E. Green. Interscience Publishers, New York.
Gibson, K. D. *et al.* (1955). *Biochem. J.*, **61**, 618.
Gray, C. H. (1953). *The Bile Pigments*. Methuen, London.
Gray, C. H. (1961). In *The Bile Pigments in Health and Disease*. C. C. Thomas, Springfield, Illinois.
Gray, C. H. (1963). In *Comprehensive Biochemistry*. Ed. M. Florkin and E. H. Stotz, Vol. 9, p. 98. Elsevier, Amsterdam.
Grinstein, M. & Wintrobe, C. J. J. (1948). *J. Biol. Chem.*, **157**, 671.
Haeger-Aronsen, B. (1960). *Scand. J. Clin. Lab. Invest.*, **12** Suppl. 47, 37.
Heilmeyer, L. (1943). In *Spectrophotometry in Medicine*. Trans. A. Jordan and T. L. Tipple. Hilger, London.
Hill, R. & Keilin, D. (1930). *Proc. Roy. Soc.*, **B107**, 286.
Holman, R. T. & Edmondson, P. R. (1956). *Anal. Chem.*, **28**, 1533.
Holman, R. T., Ener, S. & Edmondson, P. R. (1959). *Arch. Biochem. Biophys.*, **80**, 72.
Holmes-Siedle, A. G. & Saunders, B. C. (1959). *Chem. & Ind.*, 164.
Hughes, G. M. K. *et al.* (1954). *Chem. & Ind.*, 47
Jones, R. N. & Sandorfy, C. (1956). Chemical Applications of Spectroscopy, Ch. IV. In *Techniques of Organic Chemistry*. Vol. IX. Ed. W. West. Interscience Publishers, New York, London, Sydney and Toronto.
Jorgensen, C. K. (1962). *Absorption Spectra and Chemical Binding*. Pergamon Press, Oxford.
Keilin, D. & Hartree, E. F. (1951). *Biochem. J.*, **49**, 88.
Keilin, D. & Keilin, J. (1966). In *The History of Cell Respiration and Cytochrome*. Cambridge University Press.
Keilin, D. & Mann, T. (1937). *Proc. Roy. Soc.*, **B122**, 119.
Klüver, H. (1955). Biochem. Developing Nervous Systems, p. 137. In *Proc. ſst Intern. Neurochem.*
Lathe, G. H. & Billing, B. H. (1958). *Amer. J. Med.*, **24**, 111.
Lathe, G. H., Lord, R. & Tothill, C. (1966). In *Transport Function of Plasma Proteins*, p. 129. Ed. P. Desgrez & P. M. de Traverse. Elsevier, Amsterdam.
Lenberg, R. & Falk, J. E. (1951). *Biochem. J.*, **49**, 674.
Lemberg, R. & Legge, J. W. (1949). In *Haematin Compounds and Bile Pigments*. Interscience Publishers, New York.
Litvin, F. F. *et al.* (1959). *Dokl. Akad. Nauk. SSSR.*, **127**, 699; (1960). *Ibid.*, **135**, 287.
Maehly, A. C. (1949). *Z. Vitaminforsch. Hormon in Fermentforsch*, **3**, 115.
Maehly, A. C. (1952). *Biochem. Biophys. Acta.*, **8**, 1.

Maehly, A. C. (1953). *Arch. Biochim. Biophys.*, **44**, 430.
Maehly, A. C. (1955). *Ibid.*, **56**, 507.
Marks, G. S., Dougall, D. K., Bullock, E. & MacDonald, S. F. (1960). *J. Amer. Chem. Soc.*, **82**, 3183.
Martin, N. H. (1949). *J. Amer. Chem. Soc.*, **71**, 1230.
Mauzerell, D. & Granick, S. (1956). *J. Biol. Chem.*, **219**, 435.
Mason, S. F. (1958). *J. Chem. Soc.*, 976.
Mertz, J. E. & West, C. D. (1956). *Amer. J. Dis. Child*, **91**, 19.
Morell, D. B. *et al.* (1961). In *Haematin Enzymes.* Ed. J. E. Falk, R. K. Lemberg and R. K. Morton. Pergamon Press, Oxford, London.
Morrison, M., Hamilton, H. B. & Stotz, E. (1957). *J. Biol. Chem.*, **228**, 767.
Morrison, M. & Stotz, E. (1957). *J. Biol. Chem.*, **228**, 123.
Morrison, M. & Stotz, E. (1961). In *Haematin Enzymes*, p. 335. Ed. J. F. Falk *et al.* Pergamon, Oxford.
Murata, N. & Fork, D. C. (1971). *Carnegie Institution Year Book*, **70**, 468; also *Biochim. Biophys. Acta.*, **245**, 356.
Murata, T. & Murata, N. (1971). *Carnegie Institution Year Book*, **70**, 505.
Nosslin, B. (1960). *Scand. J. Clin. Lab. Invest.*, **12**, Suppl. 49.
Ostrow, J. D. & Schmid, R. (1963). *J. Clin. Invest.*, **42**, 1286.
Overbeek, J. T. G., Vink, C. L. & Deensta, H. (1955). *Rec. Trav. Chim.*, **74**, 85.
Phillips, J. N. (1963). In *Comprehensive Biochemistry*. Vol. 9, p. 35. Ed. M. Florkin and E. H. Stotz. Elsevier, Amsterdam.
Polonovski, M., Jayle, M. & Klotz, G. (1938). *Comptes rend. Soc. Biol.*, **128**, 1072; *Ibid.*, (1939). *Bull. Soc. Chim. Biol.*, **21**, 48.
Polonovski, M., Jayle, M. & Fraudet, G. (1941). *Comptes Rend. Acad. Sci., Paris* **213**, 740, 887.
Price, G. R. & Schwartz, S. (1956). In *Physical Techniques in Biological Research.* Vol. 3. Ed. G. Oster and A. W. Pollister. Academic Press, New York.
Rimington, C., Mason, S. F. & Kennard, O. (1958). *Spectrochem. Acta.*, **12**, 65.
Rimington, C. & Sveinsson, S. L. (1950). *Scand. J. Clin. Lab. Invest.*, **2**, 209.
Sano, S. (1955). *Acta Haemat. Jap.*, **18**, 625, 631.
Saunders, B. C. (1952). 2nd Internation. Congress Biochemistry, Paris.
Saunders, B. C. (1957). In *Peroxidase, Action and Use in Organic Synthesis.* Royal Institute Chemistry Lectures, 1957, No. 1.
Saunders, F., Dorfman, A. & Koser, S. (1941). *J. Biol. Chem.*, **138**, 64.
Saunders, B. C., Holmes-Siedle, A. G. & Stark, B. P. (1964). In *Peroxidase.* Butterworths, London.
Schmid, R. (1956). *Science, N.Y.*, **124**, 76.
Schultz, J., Gordon, A. & Shay, H. (1957). *J. Amer. Chem. Soc.*, **79**, 1632.
Schwartz, S., Berg, M. H., Bossenmaier, I. & Dinsmore, H. (1960). In *Methods of Biochemical Analysis.* Vol. 8, p. 221. Ed. D. Glick. Elsevier, Amsterdam.
Schwartz, S. & Wikoff, H. M. (1952). *J. Biol. Chem.*, **194**, 563.
Stern, A. & Wenderlein, H. (1934). *Z. Physikal Chem.*, **170A**, 337.
Stern, A. & Wenderlein, H. (1935). *Ibid.*, **174A**, 81.
Stern, A. & Wenderlein, H. (1936 *a*). *Ibid.*, **175A**, 405.
Stern, A. & Wenderlein, H. (1936 *b*). *Ibid.*, **176**, 81.
Sweeney, G. D. & Eales, L. (1964). *Scand. J. Clin. Lab. Invest.*, **16**, 250.
Talafant, E. (1956). *Nature, Lond.*, **178**, 312.
Theorell, H. (1938). *Biochem. Z.*, **298**, 242.
Theorell, H. (1942). *Enzymologia*, **10**, 250.
Theorell, H. (1947). *Adv. Enzymology*, **7**, 265.
Theorell, H. & Åkesson, Å. (1943). *Arkiv. Kemi Min. Geol.*, **17B**, No. 7.
Theorell, A. & Maehly, A. C. (1950). *Acta. Chem. Scand.*, **4**, 422.
Thomas, D. W. & Martell, A. E. (1956). *J. Amer. Chem. Soc.*, **78**, 1338.
Treibs, A. & Herrmann, E. (1955). *Z. physiol. Chem.*, **299**, 171.
Ueno, K. & Martell, A. E. (1956). *J. Phys. Chem.*, **60**, 934.
Van den Bergh, A. A. H. & Muller, P. (1916). *Biochem. Z.*, **77**, 90.
Vannotti, A. (1954). In *Porphyrins* (trans. C. Rimington). Adam Hilger, London.
Watson, C. J. (1959). Porphyrin metabolism, Chapter II in *Diseases of Metabolism.* Ed. G. C. Duncan. Saunders, Philadelphia.
Williams, R. J. P. (1956). *Chem. Rev.*, **56**, 299.
Williams, R. J. P. (1958). *Disc. Faraday Soc.*, **26**, 123.
Willstatter, R. & Stoll, A. (1917). *Leibigs Ann.*, **416**, 21.
With, T. K. (1955). *Scand. J. Clin. Lab. Invest.*, **7**, 193.
Wranne, L. (1960). *Acta Paediat.*, Suppl., **124**, 1.
Yakushiji, E., Uchino, K., Sugimura, Y., Shiraton, I. & Takamiya, F. (1963). *Biochem. Biophys. Acta*, **75**, 293.

10.3

Alvares, A. P. & Mannering, G. J. (1967). *Fed. Proc.*, **26**, 242.
Alvares, A. P., Schilling, G. R., Levin, W. & Kuntzman, R. (1967). *Biochem. Biophys. Res. Commun.*, **29**, 521; (1968). *J. Pharmacol. Exp. Therap.*, **163**, 417.
Appleby, C. A. (1968). In *Structure and Functions of Cytochromes*, p. 666. Eds. K. Okunuki, M. D. Kamen & I. Sekuzu, University of Tokyo Press.
Bartsch, R. G. (1961). *Fed. Proc.*, **20**, 43,
Bartsch, R. G. (1963). In *Bacterial Photosynthesis*, p. 475. Eds. H. Gest, A. San Pietro & K. P. Vernon. Antioch Press, Yellow Springs, Ohio.
Beinert, H., Griffiths, D. E., Wharton, D. C. & Sands, R. H. (1962). *J. Biol. Chem.*, **237**, 2337.
Beinert, H. & Palmer, C. (1965). *Adv. Enzymol.*, **27**, 105, and in *Oxidases and related systems.* Vol. 2. p. 567. Eds. T. E. King, G. S. Mason and M. Morrison, Wiley, New York.

Beinert, H., van Gelden, B. F. & Hansen, R. E. (1968). In *Structure and Function of Cytochromes*, p. 141. Eds. K. Okunuki, M. D. Kamen & I. Sekuzu. University of Tokyo Press.

Castor, L. N. & Chance, B. (1955). *J. Biol. Chem.*, **217**, 453.

Castor, L. N. & Chance, B. (1959). *J. Biol. Chem.*, **234**, 1587.

Caughey, W. S., Davies, J. L., Funksman, W. H. & McKay, S. (1968). In *Structure and Functions of Cytochromes*, p. 20. Eds. K. Okunuki, M. D. Kamen & I. Sekuzu. University of Tokyo Press.

Caughey, W. S., York, J. L. & Iber, P. K. (1967). In *Magnetic Resonances in Biological Systems*, p. 25. Eds. A. Ehrenberg, B. G. Malinström & T. Vänguard. Pergamon Press, Oxford.

Chance, B. & Schoener, B. (1966). *J. Biol. Chem.*, **241**, 4567, 4574, 4577.

Chance, B., Smith, L. & Caston, L. N. (1953). *Biochim. Biophys. Acta.*, **12**, 289.

Cooper, D. Y., Levine, S., Narasimhylu, S., Rosenthal, O. & Estabrook, R. W. (1965). *Science, N.Y.*, **147**, 400.

Drabkin, D. (1942). *J. Biol. Chem.*, **146**, 605.

Elsden, S. R., Kamen, M. D. & Vernon, L. P. (1953). *J. Amer. Chem. Soc.*, **75**, 6347.

Estabrook, R. W., Cooper, D. Y. & Rosenthal, O. (1963). *Biochem. Z.*, **338**, 741.

Fan, H. N. & Cramer, W. A. (1970). *Biochim. Biophys. Acta.*, **215**, 200.

Flatmark, T. (1966). *Acta. Chem. Scand.*, **20**, 1476.

Flatmark, T. (1967). *J. Biol. Chem.*, **242**, 2454.

Gallagher, C. H., Judah, J. H. & Rees, K. R. (1956). *Proc. Roy. Soc. (Lond.)*, **B145**, 134.

Garfinkel, D. (1958). *Arch. Biochem. Biophys.*, **75**, 493.

Gibson, Q. H. & Wharton, D. C. (1968). In *Structure and Functions of Cytochromes*, p. 5. Eds. K. Okunuki, M. D. Kamen & I. Sekuzu. University of Tokyo Press.

Gigon, P. L., Gram, T. E. & Gillette, J. R. (1968). *Biochem. Biophys. Res. Commun.*, **31**, 558.

Goldberger, R., Smith, A. L., Tisdale, H. D. & Bornstein, R. (1961). *J. Biol. Chem.*, **236**, 2788.

Green, D. E. & Wharton, D. C. (1963). *Biochem. Z.*, **338**, 335.

Greim, H. & Remmer, H. (1966). *Arch. Exp. Path. Pharmak.*, **25**, 16.

Greim, H., Schenkmann, J. B., Klotzbeucher, M. & Rembold, H. (1970). *Biochem. Biophys. Acta.*, **201**, 205.

Griffiths, D. E. & Wharton, D. C. (1961). *J. Biol. Chem.*, **236**, 1856 and 1857.

Harding, B. W., Wong, S. H. & Nelson, D. H. (1964). *Biochim. Biophys. Acta.*, **92**, 415.

Holtzman, J. L., Gram, T. E., Gigon, P. L. & Gillette, J. R. (1968). *Biochem. J.*, **110**, 407.

Horie, S. (1968). In *Structure and Function of Cytochromes*, p. 96. Eds. K. Okunuki, M. D. Kamen & I. Sekuzu. University of Tokyo Press.

Horie, S. & Morrison, M. (1963). *J. Biol. Chem.*, **238**, 1855 and 2859.

Ichikawa, Y. and Yamano, T. (1967). *Biochim. Biophys. Acta.*, **131**, 490. (1969). *Ibid.*, **147**, 518.

Imai, Y. & Sato, R. (1966). *Biochem. Biophys. Res. Commun.*, **22**, 620.

Imai, Y. & Sato, R. (1966) *Biochem. Biophys. Res. Commun.*, **23**, 5.

Imai, Y. & Sato, R. (1968). *J. Biochem.*, **63**, 270, 370, 380.

Ishimoto, M., Koyama, J., Omura, T. & Nagai, Y. (1954). *J. Biochem. Japan.*, **41**, 537.

Kamen, M. D. (1961). In *Biological Structure and Function of Cytochromes*, II, p. 277. Eds. T. W. Goodwin & O. Lindberg. Academic Press, London.

Kamen, M. D. (1963). In *Bacterial Photosynthesis*, p. 61. Eds. H. Gest, A. San Petro & L. P. Vernon. Antioch Press, Yellow Springs, Ohio, U.S.A.

Kamen, M. D., Dus, K. M., Flatmark, T. & Kerk, H. de (1971). In *Electron and Coupled Energy Transfer in Biological Systems*. Eds. T. E. King & M. Klingenberg. Dekker, New York.

Kamen, M. D. & Vernon, P. (1958). *Biochim. Biophys. Acta.*, **17**, 10.

Kato, R. & Takanaka, A. (1968). *J. Biochem.*, **63**, 406.

Keilin, D. (1925). *Proc. Roy. Soc.* B, **98**, 312.

Keilin, D. & Hartree, E. F. (1937). *Proc. Roy. Soc.*, B, **122**, 298.

Keilin, D. & Hartree, E. F. (1939). *Proc. Roy. Soc.* B, **127**, 1767.

Keilin, D. & Hartree, E. F. (1949). *Nature, Lond.*, **164**, 254.

Keilin, D. & Hartree, E. F. (1955). *Nature, Lond.*, **176**, 200.

Keilin, D. & Keilin, J. (1966). In *The History of Cell Respiration and Cytochrome*, p. 235. Cambridge University Press, London.

Keilin, D. & Keilin, J. (1966). In *Structure & Function of Cytochromes*. Eds. K. Okunuki, M. D. Kamen & I. Sekuzu. University of Tokyo Press.

Kinoshita, T. & Horie, S. (1967). *J. Biochem.*, **61**, 76.

Klingenberg, M. (1958). *Arch. Biochem. Biophys.*, **75**, 376.

Knaff, D. B. & Arnon, D. I. (1969). *Proc. Nat. Acad. Sci. U.S.*, **64**, 715.

Kubowitz, F. & Haas, E. (1932). *Biochem. Z.*, **255**, 247.

Kubayama, M. & King, T. E. (1964). *Biochim. Biophys. Acta.*, **92**, 618.

MacLennan, D. H., Tzagoloff, A. & McConnell, D. G. (1967). *Biochim. Biophys. Acta.*, **131**, 59.

MacMunn, C. A. (1886). *Phil. Trans. Roy. Soc.* B, **177**, 267.

MacMunn, C. A. (1886). *J. Physiol.*, **8**, 57.

Margoliash, E. & Lustgarten, J. (1962). *J. Biol. Chem.*, **237**, 3397.

Margoliash, E. & Schejter, A. (1966). *Adv. Protein Chem.*, **21**, 114.

Margoliash, E., Smith, E. L., Kreil, G. & Tuppy, H. (1961). *Nature, Lond.*, **192**, 1125.

Mason, H. S., North, J. C. & Vanneste, M. (1965). *Fed. Proc.*, **24**, 1172.

Mason, H. S., Nau, P. M. & Ganapathy, K. (1968). In *Structure and Function of Cytochromes*, p. 138. Eds. K. Okunuki, M. D. Kamen & I. Sekuzu. University of Tokyo Press.

Melnick, J. L. (1942). *J. Biol. Chem.*, **146**, 385.

Miyake, Y., Gayler, J. L. & Mason, H. S. (1968). In *Structure and Function of Cytochromes*, p. 656. Eds. K. Okunuki, M. D. Kamen & I. Sekuzu. University of Tokyo Press.

Morrison, M. (1968). In *Structure and Function of Cytochromes*, p. 89. Eds. K. Okunuki, M. D. Kamen & I. Sezuku. University of Tokyo Press.

Morrison, M. & Horie, S. (1963). *Biochem. Biophys. Res. Commun.*, **10**, 160.

Morrison, M., Horie, S. & Mason, H. S. (1963). *J. Biol. Chem.*, **238**, 2220.

Morton, R. K. (1961). *Nature, Lond.*, **192**, 727.

Morton, R. K., Armstrong, J. McD. & Appleby, C. A. (1961). In *Haematin Enzymes*. Vol. II, p. 501. Pergamon Press, London.

Morton, R. K. & Shepley, K. (1961). *Nature, Lond.*, **192**, 639.

Nair, P. M. & Mason, H. S. (1966). *Biochem. Biophys. Res. Commun.*, **23**, 12.

Nair, P. M. & Mason, H. S. (1967). *J. Biol. Chem.*, **242**, 1406.

Nicholls, P. (1968). In *Structure and Function of Cytochromes*, p. 76. Eds. K. Okunuki, M. D. Kamen & I. Sekuzu. University of Tokyo Press.

Nishibayashi, H., Omura, T., Sato, R. & Estabrook, R. W. (1968). In *Structure and Function of Cytochromes*, p. 658. Eds. K. Okunuki, M. D. Kamen & I. Sekuzu. University of Tokyo Press.

Nishibayashi, H. & Sato, R. (1968). *J. Biochem.*, **63**, 766.

Okunuki, K. (1966). In *Comprehensive Biochemistry*. Vol. 14, p. 232. Eds. M. Florkin & E. H. Stotz. Elsevier, Amsterdam.

Okunuki, K. & Yakashuji, E. (1948). *Proc. Jap. Acad.*, **24**, 12.

Omura, T. & Sato, R. (1962). *J. Biol. Chem.*, **237**, 1375.

Omura, T., Sato, R., Cooper, D. Y., Rosenthal, O. & Estabrook, R. W. (1965). *Fed. Proc.*, **24**, 1187.

Omura, T., Sanders, E., Estabrook, R. W., Cooper, D. Y. & Rosenthal, O. (1966). *Arch. Biochem. Biophys.*, **117**, 660.

Onishi, K. & Okunuki, K. (1965). *Biochim. Biophys. Acta.*, **99**, 574.

Orrenius, S. & Ernster, L. (1964). *Biochem. Biophys. Res. Commun.*, **10**, 60. (1967). *Biochem. J.*, **103**, 3P.

Orrenius, S. & Ernster, L. (1967). *Life Science*, **6**, 1473.

Orii, Y. & Okunuki, K. (1967). *J. Biochem.*, **61**, 388.

Pappenheimer, A. M. & Williams, C. M. (1954). *J. Biol. Chem.*, **209**, 915.

Paleus, S. & Tuppy, H. (1959). *Acta. Chem. Scand.*, **13**, 641.

Postgate, J. (1954). *Biochem. J.*, **56**, XI.

Postgate, J. (1956). *J. Gen. Microbiol.*, **15**, 156.

Remmer, H., Schenkman, J. B., Estabrook, R. W., Sasame, H., Gillette, J. R., Narasimhulu, S., Cooper, D. Y. & Rosenthal, O. (1966). *Molec. Pharmacol.*, **2**, 187.

Schejter, A., Glauser, S. C., George, P. & Margoliash, E. (1963). *Biochim. Biophys. Acta.*, **73**, 641.

Schenkman, J. B., Greim, H., Zange, M. & Remmer, H. (1969). *Biochim. Biophys. Acta.*, **171**, 23.

Schenkman, J. B., Remmer, H. & Estabrook, R. W. (1967). *Molec. Pharmacol.*, **3**, 113.

Secs, A. A., Fukuhara, H., Pere, G. & Sloniski, P. P. (1965). *Biochim. Biophys. Acta.*, **95**, 486.

Sladek, N. E. & Mannering, G. (1966). *Biochem. Biophys. Res. Commun.*, **24**, 668.

Slater, E. C. (1949). *Nature, Lond.*, **163**, 532.

Strittmatter, P. & Ball, E. G. (1952). *Proc. Natl. Acad. Sci. USA*, **38**, 19.

Strittmatter, P. (1963). In *The Enzymes*. Vol. 8, p. 113. Ed. P. D. Boyer, H. Lardy & K. Myrbach. Academic Press, New York.

Strittmatter, P. (1968). In *Biological Oxidations*, p. 171. Ed. T. P. Singer. Wiley (Interscience), New York.

Taber, H. W. & Sherman, F. (1964). *Ann. N.Y. Acad. Sci.*, **121**, 600.

Takemori, S. (1960). *J. Biochem. (Tokyo)*, **47**, 382.

Takemori, S., Sekuzu, I. & Okunuki, K. (1961). *Biochim. Biophys. Acta.*, **51**, 464.

Theorell, H. (1938). *Biochem. Z.*, **298**, 242.

Theorell, H. & Åkesson, Å. (1941). *J. Amer. Chem. Soc.*, **63**, 1804.

Tuppy, H. & Dus, K. (1959). *Monatsh. Chem.*, **89**, 407.

Tzagoloff, A. & Wharton, D. C. (1965). *J. Biol. Chem.*, **240**, 2628.

Ullrich, V., Cohen, B., Cooper, D. Y. & Estabrook, R. W. (1968). In *Structure and Functions of Cytochromes*, p. 649. Eds. K. Okunuki, M. D. Kamen & I. Sekuzu. University of Tokyo Press.

Van Gelder, B. G. & Slater, E. C. (1966). In *The Biochemistry of Copper*, p. 246. Eds. J. Peisach, P. Aisen & W. E. Blumberg. Academic Press, New York.

Vernon, L. P. & Kamen, M. D. (1954). *J. Biol. Chem.*, **211**, 643.

Wainio, W. W. (1955). *J. Biol. Chem.*, **112**, 723.

Warburg, O. & Negelein, E. (1928). *Biochem. Z.*, **193**, 339; **200**, 414; **202**, 202; and (1929). *Ibid.*, **214**, 26 and 64.

Wharton, D. C. & Tzagoloff, A. (1963). *Biochem. Biophys. Res. Commun.*, **13**, 121.

Williams, R. J. P. (1968). In *Structure and Function of Cytochromes*, p. 645. Eds. K. Okunuki, M. D. Kamen and I. Sekuzu. University of Tokyo Press.

Wilson, D. F. & Dutton, P. L. (1971). In *Electron and Coupled Energy Transfer in Biological Systems*. Eds. T. E. King & M. Klingenberg. Dekker, New York.

Yakushuji, E. & Okunuki, K. (1940). *Proc. Imp. Acad. Tokyo*, **16**, 229.

Yamashita, J. & Okunuki, K. (1963). *J. Biochem.*, **52**, 117.

Yonetani, T. (1961). *J. Biol. Chem.*, **236**, 1630.

Yonetani, T. & Kidder, G. W. (1964). *J. Biol. Chem.*, **238**, 386.

11 Copper Proteins

11.1 Metalloproteins

Absorption spectroscopy is potentially valuable in studying active sites of metallo-enzymes provided that highly purified materials become available in the necessary amounts. Vallee & Williams (1968 *a*, *b*) noted qualitative differences between the spectra of selected metalloproteins and the spectra of simpler 'model' compounds containing the metal in question. The absorption bands shown by many metallo-enzymes have yet to be explained satisfactorily. A steadily increasing number of enzymes are being seen to contain zinc as an essential component of the holo-enzyme; but zinc, however, is diamagnetic and its complexes are colourless so that this metal is defective as a 'probe'. On the other hand cobalt is paramagnetic and moreover gives rise to well-defined absorption spectra. These properties make it very serviceable in exploring sites of enzymic activity (Vallee *et al.*, 1969).

Some zinc-containing enzymes can have the metal replaced by cobalt and yet can retain enzymatic activity. Thus, the alkaline phosphatase of *Escherichia coli* contains four atoms of zinc per mole of protein (Simpson *et al.*, 1968). Treatment with 8-hydroxy-quinoline-5-sulphonic acid quickly removes exactly half the zinc and enzymatic activity falls almost to zero. The remaining two atoms of zinc per mole, which do not seem to be concerned in the catalysis, are removed very slowly. When zinc is returned to the apo-enzyme the results indicate again that only two out of four atoms are concerned in activity. In this alkaline phosphatase, cobalt can be added to the apoenzyme and the spectroscopic effects compared with acquired enzymatic activity. The uptake of two cobalt atoms results in a weak absorption spectrum resembling that of octahedral complex cobalt ions, e.g. $[Co(H_2O)_6]^{2+}$ has $\lambda_{max.}$ 510–515 nm, (ϵ *ca.* 5), $\lambda_{max.}$ 1200 nm (ϵ *ca.* 2) but no enzymatic activity ensues. Uptake of two additional cobalt atoms leads in the visible region to a complex absorption spectrum and restores enzymatic activity:

$\lambda_{max.}$ nm	510	555	~610	640
$\epsilon_{max.}$	280	350	210	250

This spectrum (Simpson & Vallee, 1968) bears some resemblance to those of cobalt halides in concentrated halogen acids, e.g. $CoCl_2$ in concentrated HCl (Brode & Morton, 1929) where the $CoCl_4^{2-}$ complex shows resolved absorption between 700 and 600 nm.

Cobalt can be introduced into other apoproteins and new absorption spectra appear, e.g.

(*a*) cobalt (II) carbonic anhydrase

$\lambda_{max.}$ nm	510	550	615	640	900	1250
$\epsilon_{max.}$	280	380	300	280	30	90

(Lindskog & Nyman, 1964; Lindskog & Ehrenberg, 1967).

(*b*) cobalt (II) carboxypeptidase

$\lambda_{max.}$ nm	~500	555	572	950
$\epsilon_{max.}$ *ca.*		150	150	25

(Latt & Vallee, 1969).

(*c*) cobalt (II) yeast alcohol dehydrogenase (containing both cobalt and zinc)

$\lambda_{max.}$ nm	620	657	710
$\epsilon_{max.}$ *ca.*	800	800	600

(Curdel & Iwatsubu, 1968).

The spectra suggest an unusual co-ordination environment for the metal at the active site. Findings about the circular dichroism of cobalt phosphatase bear the same interpretation.

Simpson & Vallee (1968) found that titration over the pH range 9 to 6 decreased the intensity of the absorption spectrum attributable to the two cobalt atoms associated with enzymatic activity, and over the same pH range the restoration of absorption bands and of catalytic activity to the apoenzyme ran in parallel.

Vallee's group extended the investigation to carboxypeptidases from pancreas (ox, pig and dog-fish) and other zinc-containing proteolytic enzymes. The cobalt carboxypeptidase at room temperature has poorly resolved peaks at 555 and 572 nm, and an infrared peak at 950 nm. At low temperature (4·2 K) in the presence of glycerol (*ca.* 45%) the two bands become well resolved ($\lambda_{max.}$ at 575 and 535 nm with inflections around 510 and 490 nm). Circular dichroic spectra for cobalt carboxypeptidase show a negative band near 538 nm and a shoulder near 500 nm, the long wave absorption band being optically inactive (cf. Latt *et al.*, 1969, 1970).

A neutral protease from *Bacillus subtilis* (molecular) weight 44 700, 1 Zn g atom/mole: Wagner & Prescott, 1966; Tsuru *et al.*, 1965; McConn *et al.*, 1964) can have the zinc replaced by cobalt (Vallee & Latt, 1970). Maxima appear at 525 and 555 nm with a shoulder at 475 nm. The intensities of absorption are much greater than for simple octahedral complexes (ϵ *ca.* 5) but lower than for cobalt carboxypeptidase. Circular dichroism is shown by the cobalt protease with peaks around 500 and 550 nm.

Vallee and Latt place cobalt alkaline phosphatase and cobalt carbonic anhydrase together in one group and neutral cobalt protease, cobalt carboxypeptidase, cobalt aldolase and cobalt enolase in another group.

Although no simple picture has yet emerged, it is evident that absorption spectra and circular dichroic spectra hold out promise in the study of active sites.

11.1.1. Absorption Spectra of Copper Compounds

The absorptive properties of many copper compounds depend primarily on the electronic configuration of copper: $1s^2\ 2s^2\ 2p^6\ 3s^2\ 3p^6\ 3d^{10}\ 4s$. Cuprous copper Cu(I) has a filled $3d$ shell and is diamagnetic. Cupric copper Cu(II) has a $3d^9\ 4s^2$ configuration and is paramagnetic because of the one unpaired electron.

The absorption spectra of the simpler Cu(I) compounds, e.g.

Cu(I) Cl^- in solution	$\lambda_{max.}$ nm 272	$\epsilon_{max.}$ 4300
Cu(I) Br^-	$\lambda_{max.}$ nm 274	$\epsilon_{max.}$ 8500

indicate that there is no visible colour and even the complexes Cu $(dip)_2^+$ and Cu $(phen)_2^+$ $\lambda_{max.}$ 435 nm show no blue colour.

The cupric ion Cu^{2+} in aqueous solution has six water molecules $Cu(H_2O)_6^{2+}$ and $\lambda_{max.}$ 793·5 nm is quite weak (low $\epsilon_{max.}$). The complexes $CuCl_4^{2-}$ and $CuBr_4^{2-}$ exhibit charge transfer bands in the visible region and more intense bands in the ultraviolet.

	$\lambda_{max.}$ nm	$\epsilon_{max.}$			$\lambda_{max.}$ nm	$\epsilon_{max.}$
$CuCl_4^{2-}$	1050	600	C.T.			
	377	1700	$\pi \rightarrow 3d$	$Cu(NH_3)_6^{2+}$	660	20
	274	4000	$\sigma \rightarrow 3d$		660	40
$CuBr_4^{2-}$	585	1000	$\pi \rightarrow 3d$		750	25
	512	2100	$\pi \rightarrow 3d$	$Cu(H_2O)_2^{2+}$	700	5
	338	4200	$\sigma \rightarrow 3d$		790	10
	266	2300	$\sigma \rightarrow 3d$		800	5

These spectra are not immediately useful in studying copper proteins.

In general copper complexes with planar (square)

and regular or distorted octahedral structures have quite low molecular extinction coefficients:

octahedral $\epsilon \simeq 1$–150; planar $\epsilon \simeq 5$–250

This is largely because the presence of a centre of symmetry makes certain transitions forbidden. There may however be transfer of charge (from ligand to metal or vice versa) in which the transitions will show coupling of electronic and vibrational motions.

A complex Cu(II) bistetraglycine shows $\lambda_{max.}$ 580 nm, $\epsilon_{max.}$ 95, but many blue copper proteins are at least ten times more strongly absorbing than this. Some synthetic blue copper compounds (e.g. copper chelates of oxalyldihydrazides of aldehydes such as acetaldehyde or of cyclohexanone) may have ϵ values near 600 nm up to 14 000. The copper is at the centre of a tetrahedron. The absorption band is regarded as due to charge transfer from π orbitals of the ligand to the central *d* orbitals or possibly from σ orbitals of the ligand to the *d* orbitals. Addition of Cu^{2+} to many proteins at pH 10 or above produces the well-known purple colour of the biuret reaction and it is possible that this affords the most direct approach to the catalytic copper proteins.

From dialysis studies, bovine serum albumin (BSA) binds a first Cu(II) ion more strongly than subsequent ones (Klotz & Curme, 1948). The binding site includes the terminal amino position because (Peters, 1960) one Cu(II) ion blocks the reaction in BSA of the terminal aspartate residue with DFNB (3,4-dinitrofluorobenzene). Chelation of the first Cu(II) by 3 or 4 ligands is likely (Breslow, 1964). Mild peptic digestion of BSA gives a peptide (1–24) having the terminal sequence Asp–Thr of BSA. This Asp fragment (1–24) adds one Cu(II) ion in a manner involving Asp–Thr–His (Peters & Hawn, 1967).

BSA in presence of a first Cu(II) ion shows at pH 5·5, $\lambda_{max.}$ 525–530 nm, at 5·3, $\lambda_{max.}$ 525–530 and 650–700 nm and at 4·9, $\lambda_{max.}$ 650–700 becomes stronger than 525–530 nm. The spectrum resembles that of copper triglycyl glycine (Breslow, 1964).

The fragment 1–24 shows $\lambda_{max.}$ 525–530 nm at pH 5·5, $\epsilon_{max.}$ being 30% higher than in BSA. When the pH is $<5·5$, the 650–700 nm maximum appears and becomes stronger as the pH falls. The effect of a second Cu(II) ion is revealed by difference spectroscopy. For BSA the difference spectrum indicates $\lambda_{max.}$ 650 nm and for fragment 1–24, $\lambda_{max.}$ is 600 nm.

A synthetic glycylglycylhistidine gives for the Cu(II) derivative $\lambda_{max.}$ 525 nm, $\epsilon_{max.}=90$ (see Bradshaw, Shearer & Gurd, 1968). The peptide Asp–Thr–His–Lys complex with Cu(II) shows $\lambda_{max.}$525 nm, $\epsilon_{max.}=$ 90 at pH 6. A careful comparison showed that the tetrapeptide, the 1–24 fragment and BSA all show $\lambda_{max.}$at 530 nm and $\epsilon_{max.}$ *ca.* 93–98. An investigation by Sarkar & Wigfield (1967) on copper chelates showed that the above values for $\epsilon_{max.}$ were relatively high.

Spectroscopic characterization of copper (II) complexes

	Alanine		Histidine		Histamine		Imidazole	
pH	$\lambda_{max.}$ nm	$\epsilon_{max.}$	$\lambda_{max.}$ nm	$\epsilon_{max.}$	$\lambda_{max.}$ nm	$\epsilon_{max.}$	$\lambda_{max.}$ nm	$\epsilon_{max.}$
5·5	650	38	625	66	655	37	680	24·5
6·5	630	44·5	640	84·2	610	65·5	635	34
8·0	625	51·5	642	92	600	80	620	40
10·0	620	48·6	645	90·5	600	80·5	—	—

Molar ratio Cu(II) to ligand 1:2, except for imidazole, where it is 1/10.
(0·2M amino acid, pH adjusted by NaOH or HCl Cu(II) acetate added.)

Mixed complexes were studied in a further paper by Sarkar *et al.* (1968). It was concluded that in BSA (human) with about 565 aminoacid residues there were about 17 disulphide bridges. The first Cu(II) binding involved an N-terminal asparagyl residue and a ternary complex with histidine and threonine.

Many years ago Kober & Haw (1916), reviewing experience with cupric peptide or protein complexes in solution, found that the peptides fell into three classes:

(i) blue $\lambda_{max.}$ 630 nm—two N ligands
(ii) purple 540 nm—three N ligands
(iii) red 505 nm—four N ligands

The spectral shifts towards shorter wavelengths correspond to enhancement of the ligand field as donor oxygen atoms of carbonyl groups (or water molecules) are replaced by donor nitrogen atoms of peptide groups. The pH effect is explained by the progressive replacement of oxygen ligands by peptide N atoms which lose protons. The ligand field of the four ligands closest to the copper atom determines the position of $\lambda_{max.}$. This is thought to correspond to a *d–d* transition in the Cu^{2+} spectrum.

It seems probable that the absorption spectra of hepatocuprein, erythrocuprein, cerebrocuprein and of a peptide derived from neonatal hepatic mitochondro-cuprein are to be accounted for in terms of co-ordination of copper as seen in the simpler peptides. Extinction coefficients are low and the positions of absorption maxima fit this concept.

Difficulties however remain. In copper proteins each copper atom may be held by 3, 4, 5 or 6 ligands from the protein. These are not easily identifiable. Even the exact nature of the links holding copper in the glycylglycine derivative is uncertain. Electron paramagnetic resonance spectra on 1 Cu/1 glycylglycine indicate different complexes at pH 7 and pH 10 (Gould & Mason, 1966). Another fact must be faced. Tyrosinases, which indubitably contain copper [and monovalent copper Cu(I)] exhibit no colour. Moreover adrenal dopamine β-hydrolase also contains copper and is colourless; in this case the copper is unambiguously Cu(II). Furthermore haemocyanins in the reduced (or deoxygenated) form are likewise colourless, despite the presence of Cu(I). The copper in Cu(I) complexes is sometimes firmly held and it is puzzling to find that the absorption spectra in the visible and ultraviolet for apohaemocyanin and deoxygenated haemocyanin do not differ significantly. The *absence* of selective absorption is the problem here. On the other hand oxyhaemocyanin, concerned essentially with oxygen uptake and transport, shows two absorption bands at 585 nm and 346 nm, the former fairly strong (ϵ 750) and the latter quite strong (ϵ 3100). There is as yet no fully satisfactory explanation of this spectrum which is referred to later.

There are many copper enzymes with characteristic 'copper blue' colours. Broadly, for Cu(II) complexes this means an absorption peak around 597–610 nm with $\epsilon_{max.} \nless 770$ and $\ngtr 5000$ with laccase $\lambda_{max.}$ 610 nm, $\epsilon_{max.}$ 1300 as illustrative. The position on the wavelength scale is not inconsistent with what is known about copper peptide compounds. The main difficulty to be disposed of is the tenfold or greater enhancement of the intensity of absorption per copper atom. Several of the copper enzymes exhibit more than one absorption peak, sometimes referred to as fine structure. It might be thought that vibrational fine structure is involved but the plurality of absorption peaks makes this idea somewhat unhelpful.

The case of caeruloplasmin is interesting. This serum protein has 8 Cu atoms [4 Cu(I) and 4 Cu(II)] in a molecular weight of 160 000, $\lambda_{max.}$ is at 605 nm, $\epsilon_{max.}$ *ca.* 1200 at pH 7 with inflections at 330, 450 and 750 nm; $\lambda_{max.}$ is given also as 625 nm and at pH 4 the only absorption band is at 400 nm (G. Curzon, quoted by G. Morpurgo and R. J. P. Williams in Ghiretti's *Haemocyanin*). Various added ligands changed the spectra:

	$\lambda_{max.}$ nm	
Azide	350–375,	420
Cyanate	380	
Cyanide	no bands	300–900
Thiocyanide	375	420
Reduced caeruloplasmin	no bands	300–900

It seems here that when Cu^{2+} enters into combination with these 'inhibitors' a charge-transfer band arises (*ca.* 390 nm).

The strong blue band of caeruloplasmin is attributed to the reducing nature of *all* the ligands of the protein. The analogy with the biuret colours can be sustained and the intensity of the band *ca.* 600 nm will vary as the symmetry and as the ligand atoms other than $>N^-$ are changed. The blue colour arises if the symmetry is unusually low for copper [Cu(II)] and one or more of the ligands is strongly reducing. Caeruloplasmin is unusual in that half its copper is Cu(I). It is not an oxygen carrier but its oxidase properties towards numerous substrates are well verified. It is sometimes called ferroxidase (Osaki, McDermott and Frieden in Ghiretti's *Haemocyanin*) because of its action on Fe(II). An abbreviated picture is:

H_2O O_2

caeruloplasmin ⇄ reduced caeruloplasmin

Fe(II) ⇌ Fe(III)

R RH_2

transferrin

Presumably there is a change in the Cu(II)/Cu(I) ratio as the caeruloplasmin functions, but it seems unlikely that all the Cu(II) is converted to Cu(I).

Vallee & Williams (1968) propose that the copper atoms created in the protein an entatic state (Gr. *entasis*, stretched or under tension), a catalytically poised state *intrinsic to the active site*. The metal and its ligands are jointly responsible for generating the entatic state. This is an interesting concept and it would be easy to imagine the joint action of Cu(I) at one site and Cu(II) at another generating a special situation. It is perhaps less easy to visualize why 8 copper atoms should be needed in the native protein which is fully effective as a bio-catalyst.

In the case of the tyrosinases, the product from Neurospora has 1 Cu(I) per molecule (molecular weight 33 000) while the tyrosinase of mushrooms has 4 Cu(I) atoms in a molecule of molecular weight 120 000. This is a case of a colourless copper enzyme in which any difference in specificity must be ascribed to different 'entatic' states.

The case of a ascorbate oxidase with two Cu(I) and six Cu(II) atoms in a protein weight of 146 000 needs further study.

[Does the Cu(II) responsible for the 606 nm band, (intensity $\epsilon_{max.}$ 770 calculated for 8 Cu or 6 Cu) have a dynamic or a static role? If in oxido-reduction processes only the Cu(I) $\rightarrow$ Cu(II) is functional, does the pre-existing Cu(II) have an entatic role? Hitherto might it not have been said that the six Cu(II) atoms were largely irrelevant—or is the 2:6 ratio largely an accident with the six Cu(II) atoms reducible?]

We are left with the idea that in many of the Cu(II) copper proteins the absorption spectrum is roughly explicable for $\lambda_{max.}$ and that diminishing symmetry in the (octahedral) arrangement around the copper atom is the most plausible explanation of the tenfold (or more) increase in $\epsilon_{max.}$ compared with Cu(II) peptides.

From the point of view of absorption spectra two further problems arise. One is the origin of the 830 nm band in cytochrome oxidase and the other is the absorption spectrum shown by oxyhaemocyanin.

The fact that the 830 nm peak is also seen in ascorbate oxidase, in laccase, in Pseudomonas blue protein and plastocyanin adds strength to the evidence that it is due to a *copper* protein. It is moreover proportional in intensity to the intensity of the peak *ca.* 600 nm. The molecular extinction coefficient ($\epsilon = 1400$) is quite reasonably high. It is associated with Cu(II) although only half the Cu(II) of cytochrome oxidase appears to be reducible and both halves contribute to the 830 nm absorption. It would be most interesting to know whether Cu(II) peptides have a band at 830 nm. It is however certain that with an intensity of $\epsilon_{max.}$ *ca.* 1400 the original difficulty of assimilating copper protein spectra ($\lambda_{max.}$ 600 nm, $\epsilon_{max.} = 1000$) with copper peptide spectra remains. The parallelism 830/600 nm in terms of intensity means that if high absorption at 600 nm is ascribable to molecular dissymmetry, high absorption at 830 nm must be attributed to the *same* dissymmetry.

The simplest view (and perhaps at this stage the reasonable view) is that all active Cu(II) proteins have a tripartite spectrum in the visible region (*ca.* 450–460, 594–625, 725–850 nm for $\lambda_{max.}$) and that the differences in position are due to rather small differences in binding and relative intensities. On this basis there would be no *special* problem of the origin of the 830 nm peak; there would however remain the general problem of the absorption of copper proteins in the visible and near infrared. Whether there is in cytochrome oxidase a separate copper protein or a special site of attachment on the haemoprotein is in the present context not very relevant. The haemocyanins present yet another problem. The principal absorption band (at 347 nm) has a high intensity and is pH-dependent and like the 585 nm band it only appears in the *oxy*haemocyanins.

Morpurgo & Williams (1968) see the 585 nm band as a rather low intensity version of the normal Cu(II) protein visible absorption, with similar resolution indicated from circular dichroism. Whether the Cu(I) of reduced haemocyanin yields any Cu(II) on oxygenation is debatable. The second band (at 345 nm) is very like the charge transfer band from O_2^{2-} to Cu(II)" and reference is made to solid cupric peroxides with $\lambda_{max.}$ 388 nm, $\epsilon_{max.} = 2000$. They ask: 'is oxyhaemocyanin really then a Copper (II) system $Cu(I)O_2Cu(I) \rightleftharpoons Cu(II)O_2^{2-}Cu(II)$ or is there a *thermal* equilibrium?'. In order that oxygen should be lost reversibly, one place in the co-ordination 'environment' of copper must be vacant in deoxygenated haemocyanin. Only further work can resolve the difficulties.

11.2 Haemocyanins

Haemocyanins are copper-containing proteins concerned in the transport of oxygen in various classes of invertebrates, Gastropoda, Cephalopoda, Crustacea and Xiphosura. They occur in the blood and their most characteristic property is shown by the cuttlefish (*Sepia officinalis*) where the haemocyanin changes from colourless to blue as it is oxygenated in passing through the gills. The following species have been used in studies on haemocyanin:

Octopus vulgaris, *O. dolfleini* (giant octopus), *Loligo pealii* (squid), *Maja squinado* (spider crab), *Carcinus moenas*, *Cancer majester*, *C. borealis*, *Helix pomatia* (edible snail), *Busycon canaliculatum* (whelk), *Limulus polyphemus* (*horseshoe crab*), *Homerus americanus* (lobster).

Haemocyanin is always found in the haemolymph, never in blood cells or tissues (Redfield, 1934, 1950). Rich sources are octopus haemolymph, *ca.* 9%, Limulus 4·6–9·8%, spider crab 2·2%. In most cases it is the predominating solute and generally accounts for over 90% of the haemolymph protein. In some species the haemolymph will clot, but in molluscs it does not do so and low-speed centrifugation suffices to remove any cells. Salts are eliminated by dialysis against water. Some haemocyanins crystallize from purified solutions (Hentze, 1901). The crystals vary in composition and in form with the species. The copper content is about 0·25% for molluscan haemocyanin and near 0·18% for the haemocyanins of arthropods. The aminoacid made-up of the protein of *Helix pomatia* (Claesson, Moring, 1956) is rather different from that of *Limulus polyphemus* (Barron & Johnson, quoted by Ghiretti, 1962) but broadly the haemocyanins are similar in make-up.

Oxygenated haemocyanins are blue in colour. The absorption spectra show bands with peaks at 550–560, 345 and 278 nm, the last being due (at least in part) to the aromatic aminoacids present in the protein. The visible absorption (550–560 nm) and the 345 nm band are both due to a copper complex and vanish on deoxygenation or removal of copper with cyanide (Kubowitz, 1938). Addition of cuprous copper can restore both absorption bands. Haemocyanins from different species exhibit some variability in the maxima over the ranges 338–346 nm and 556–575 nm, perhaps because the sites at which copper is held are not identical in the different apoproteins.

Molecular weights have been estimated from osmotic pressure, light scattering and sedimentation measurements. Although there are very clear differences between species the haemocyanins are all large molecules with the molecular weight ranging between 400 000 and 8 900 000. The minimum unit corresponding with one gram atom of copper in molluscan haemocyanin is about 25 000 and for haemocyanin of arthropods 37 000. These values must be doubled since two copper atoms are needed to bind one molecule of oxygen. The larger haemocyanins contain some 400 copper atoms per molecule. The big protein is important physiologically. This type of oxygen carrier is not like the erythrocyte enclosed in a membrane and it is retained in the haemolymph by virtue of its great molecular size. The giant molecules consist at pH 6 of squat cylinders of length *ca.* 335 Å and diameter *ca.* 300 Å. Purified haemocyanin undergoes reversible disaggregation depending on concentration, pH and ionic environment (Svedberg & Pedersen, 1940). Dissociation is least at pH values near the isoelectric point, and a number of haemocyanins then consist of a single component. *Limulus* haemocyanin has 4 constituents in the pH range 5·2–10·5 (Svedberg, 1937). In general, unitary haemocyanins tend to be divided into half molecules and eighth or tenth molecules at pH values above or below the stability range on either side of the isoelectric point. *Helix pomatia* haemocyanin splits into two equal parts and into approximately eighth parts in the presence of a variety of added solutes (electrolytes and non-electrolytes) (Brohult & Claesson, 1939; Brohult, 1940).

It has however been found that disaggregation into halves (e.g. by addition of sodium chloride) does not exceed 75% of the total *Helix* haemocyanin. By means of ultracentrifugation the 'undissociable' material could be separated from the halved molecules (Lontie, 1957). This work led to the idea that the *Helix* haemocyanin consisted of α-haemocyanin (dissociated into halves by sodium chloride) and β-haemocyanin, resistant to dissociation (Lontie, 1957). The two components can be separated and β-haemocyanin crystallizes. Lontie's group (Lontie, 1958; Hairwegh *et al.*, 1961; Lontie *et al.*, 1962; Lontie & Witters, 1966) have found that both α- and β-haemocyanin from *Helix* have high molecular weights (8 690 000 and 9 030 000 respectively). For the α-form the stability range for the whole molecule is from pH 5 to pH 7. Dissociation into half molecules is maximal at pH 4·2 and 7·8, while at pH 3·9 and 7·9 disaggregation into tenths reaches 50%. For the β-form the stability range for the whole molecule is wider (pH 4·2–7·5); there is no splitting into halves but division into tenths is virtually complete at pH 3·8 and pH 8.

The measurement of absorption spectra for very large molecules is influenced by light scattering and minor differences recorded between α- and β-forms may not be very significant. Spectra are usually measured at an alkaline pH (8·2) at which dissociation into tenths is complete. The spectra for α-haemocyanin are variable but the molecular extinction coefficients at the three absorption peaks 278, 346, 580 nm of β-haemocyanin are more reproducible. Changes occur however on storage, extinctions at 346 and 580 nm decreasing substantially. This is due to the appearance of Cu^{2+} and is accompanied by loss of oxygen capacity and a decrease in free sulphydryl groups. Cysteine, hydrogen peroxide and hydroxylamine will all cause the disappearance of Cu^{2+} and oxygen capacity returns with the restoration of most of the original absorbance at 580 and 346 nm. Ageing thus amounts in the main to oxidation of Cu^{+} at the expense of the 'transport' oxygen. Copper can be set free from haemocyanin by acid (Hentze, 1901) and becomes removable by dialysis at pH 4. Many chelating agents fail to detach the copper of intact haemocyanins at pH 7 or a little above. Thiocyanate and thiourea diminish the 'copper bands' by oxygen expulsion rather than liberation of copper. The bound copper does not exchange with radioactive Cu^{2+} (Rombauts & Lontie, 1960).

The best method of removing copper and obtaining undenatured apohaemocyanin is to treat haemocyanin with cyanide at pH 8·2 in the presence of Cu^{2+} which broadens the region of stability of the intact protein (Felsenfeld, 1954; Lontie & Witters, 1966). The absorption spectrum of the colourless copper-free protein reveals only the aromatic aminoacid residues. Haemocyanin (*Helix pomatia*) can be reconstituted by the addition of monovalent copper stabilized by acetonitrile, both the absorption spectrum (580 and 346 nm bands) and the capacity to bind oxygen being restored. Octopus haemocyanin can similarly be deprived of its copper and regenerated by means of a Cu^{+} complex (Ghiretti, 1962). The removal of copper from octopus haemocyanin sets free only one sulphydryl group for each four copper atoms (Ghiretti-Magaldi & Nuzzolo, 1965). The degree to which haemocyanin is regenerated indicates that removal of copper does not significantly affect the protein structure. Cohen & Van Holde (1964) obtained haemocyanin from the squid (*Loligo pealii*) and dialysed it against 0·01 **M** potassium cyanide (in the presence of 0·01 **M** $MgCl_2$ and phosphate buffer pH 6·6) until the protein no longer showed the 345 nm absorption band. The haemocyanin itself showed at pH 8·2 $E^{1\%}_{1cm}$ 4·25 at 345 nm and $E^{1\%}_{1cm}$ 0·25 at 585 nm. The apohaemocyanin showed $E^{1\%}_{1cm}$ 0·6 and 0·1 at the two wavelengths. The small absorption of the apoprotein was probably due to light scattering. On the basis of a copper content of 0·26% the corrected absorption of oxyhaemocyanin shows $\lambda_{max.}$ 345 nm and 580 nm ($\epsilon_{max.}$ 8900 and 370). The optical rotatory dispersions of apohaemocyanin and deoxygenated haemocyanin differed from that of the oxygenated (blue) haemocyanin only in respect of the negative Cotton effects associated with the absorption bands at 345 and 580 nm. This indicates that neither deoxygenation nor removal of copper significantly changes the conformation of the protein structure. Van Bruggen *et al.* (1962) confirmed this view on the basis of electron micrographs of *Helix* haemocyanin and apohaemocyanin.

Recent studies on molluscan haemocyanin (Nardi *et al.*, 1962; Ghiretti-Magaldi *et al.*, 1965) indicate that the aminoacids present in the specimens from different species are reasonably the same. If the results are expressed in terms of a unit containing two copper atoms (molecular weight 50 800) aspartic and glutamic acids account for 21–24% (*ca.* 94 residues out of *ca.* 430). Leucine was consistently high (*ca.* 38 residues) followed by alanine, phenylalanine, valine, glycine, proline, serine, threonine, histidine and lysine which for five molluscan haemocyanins each accounted for very approximately 1 in 18 of the residues. Methionine, cystine (half) and methionine were consistently less prominent, while arginine corresponded with roughly one residue in 24. Imidazole residues (Lontie, 1958), the SH groups of cysteine and the guanidinyl groups of arginine and the amino groups of lysine are possible for the site of copper attachment (Ghiretti, 1966). Numerous other modes of attachment are possible and aminoacid sequences must clearly be established before much progress can be made. Preliminary studies (Gruber, 1968) on trypsin digestion indicate that more than 45 peptides are readily formed. It is perhaps important that only one terminal aminoacid, namely arginine, was found. Alkali, as has been noted, disaggregates *Helix* haemocyanin: the smallest biological sub-unit (molecular weight 500 000) capable of oxygenation to give a blue colour is a succinylated material.

The two bands (580 nm and 346 nm) of haemocyanin are due to the same absorbing entity. Hairwegh & Lontie (1960) made this clear when they found for *Helix* pomatia haemocyanin that as the product 'aged' under normal storage conditions the two peaks decreased in intensity in a strictly parallel manner. There was a corresponding but slower behaviour for β-haemocyanin and in haemocyanin from *Limulus polyphemus*. The decrease in absorption was proportional to the diminution in oxygen capacity.

The haemocyanin of the whelk (*Murex trunculus*) was studied by Wood & Bannister (1967), who found that ethyl isocyanide removed oxygen atoms (but not copper) even more efficiently than thiourea or thiocyanate (Rombauts & Lontie, 1960 *a*, *b*). The reaction, involving disappearance of the 580 and 346 absorption bands, goes nearly to completion in 1 minute at room temperature. It seems that ethyl isocyanide exchanges for oxygen.

The binding of copper in haemocyanin was studied by Thomson, Hines & Mason (1959), starting from an observation by Klotz, Urquhart & Fiess (1952) that a bovine serum albumin–copper complex with $\lambda_{max.}$ 375 nm had the copper attached through a sulphydryl group. In haemocyanin one SH group was revealed for 4·2 copper atoms displaced. Since oxygen combines with (deoxygenated) haemocyanin to the extent of one O_2 per two Cu atoms, the active centre must contain two Cu so that only half the pairs would seem to involve a sulphydryl link. Unfortunately the removal of copper by cyanide itself introduces complications and the results are not easy to interpret. Ghiretti-Magaldi & Nuzzolo (1965) found that after removal of copper from the haemocyanin of *Eriphia spinofrons* no new thiol group was revealed above the one SH/mole protein. *Octopus* haemocyanin however showed one SH per 4 atoms copper revealed after removal of copper.

Photo-oxidation of haemocyanin in the presence of methylene blue (Wood & Bannister, 1967) strengthens Lontie's suggestion (Lontie, 1958) that the copper is bound through imidazole groups. It is known that this type of aerobic photo-oxidation of proteins leads *inter alia* to destruction of histidyl residues. Photo-oxidation of *Murex* haemocyanin liberated most of the copper, with loss of selective absorption at 346 nm and of oxygen-binding capacity. Histidine residues are thus very probably concerned in the copper binding. Further experiments (Wood, Salisbury & Bannister, 1968) using an improved method of determining freed —SH groups showed that *Murex* apohaemocyanin contained no more than 0·1 mole thiol/per copper atom removed.

Electron paramagnetic resonance spectra have been tried in the hope of settling whether oxygenation of haemocyanin means Cu(I) to Cu(II). The EPR test is based on the paramagnetism of Cu(II) and the diamagnetic nature of Cu(I) and it might seem that the method should permit proof that copper can have a direct role in electron transport catalysed by copper proteins. Disappearance of the e.p.r. signal for Cu^{2+} upon 'reduction' does not afford unambiguous evidence that the valency of copper has been changed. False negatives for Cu(II) may appear when the Cu(II) is involved in spin coupling with oxygen or another ligand or in the formation of a diamagnetic dimer or a covalent Cu–Cu bond (Nakamura & Mason, 1960). In some copper proteins the valency of copper is indeterminate and perhaps indeterminable.

This does not mean that e.p.r. measurements are to be discounted here; indeed positive signals are valuable. Negative results however need to be backed up by chemical determinations using selective reagents for Cu(I) like biquinoline or neocupreine. The important question for haemocyanin is whether in the coloured oxyhaemocyanin the Cu remains Cu(I) or becomes Cu(II) or has an indeterminable valency.

Very full studies of the physical properties of haemocyanins have been made by Van Holde & Cohen (1964). In a later paper Van Holde (1967) discussed circular dichroism and absorption spectra for the haemocyanins of *Octopus vulgaris* and *Loligo pealei* (cf. also H. S. Mason, 1964, in *Oxygen in the Animal Organism*, Ed. F. Dickens and E. Neil, N.Y., p. 49). It was found that the 346 nm band corresponded to a fairly strong negative dichroic band simple in structure. In the visible region, oxygenated haemocyanin shows two dichroic bands opposite in sign and located near 440 and 570 nm with probably a third band near 700 nm. The middle band had perhaps 1·5 times the intensity of the flanking bands. Van Holde summarized the work (see Table 11.1). He notes that three copper proteins have ultraviolet bands (330, 332 and 347 nm) and that in several instances the visible absorption

Table 11.1 Some properties of copper proteins

	Cu/mole	Cu^{2+}/mole max.	e.p.r.	Absorption spectra I $\lambda_{max.}$ nm	I $\epsilon_{max.}$	II $\lambda_{max.}$ nm	II $\epsilon_{max.}$	III $\lambda_{max.}$ nm	III $\epsilon_{max.}$	IV $\lambda_{max.}$ nm	IV $\epsilon_{max.}$
Haemocyanin	many pairs	0	–	—	—	—	—	—	—	—	—
Oxy-haemocyanin	many pairs	?	–	347	8900	440	500	570	500	700	< 500
Caeruloplasmin	8	4	+	332	500	459	140	610	1300	794	260
Laccase (fungal)	4	2	+	330	1000	450 ?	970	610	4000*	850	< 1000
Plastocyanin	2	2	+	—	—	460	590	597	4900	770	1700
R. vernicifera (blue protein)	1	1	+	—	—	450	970	608	4030	850	700
Pseudomonas aeruginosa blue	—	—	—	—	—	500	weak	625	~3000	725	weak
P. dentrificans blue	—	—	—	—	—	450	weak	594	~3000	800	1000
Erythrocuprein 'inert' Cu proteins	—	—	—	—	—	—	—	655	*ca.* 280	—	—
Cerebrocuprein 'inert' Cu proteins	—	—	—	—	—	—	—	655	*ca.* 400	—	—

*(Cf. Vallee & Williams, 1968).

can be resolved into three components. He concludes that in haemocyanin the copper is at least partially in the Cu(II) form and he does not find the absence of an e.p.r. signal decisive against this view (cf. Frieden *et al.*, 1965). The 346 nm band may be due to a $Cu^{+} \rightarrow Cu^{2+}$ charge-transfer or to charge-transfer between copper and oxygen. In any case the very high absorbance remains difficult to explain.

11.2.1. A Note on Copper Valency: Indeterminate or Indeterminable?

In the formation of a copper complex by co-ordination

$$R_1R_2R_3N + Cu^{2+} \rightleftharpoons R_1R_2R_3N-Cu^{2+}$$

there is (partial) donation of the electron pair of the base to the vacant orbital of acid (charge transfer). In forming $Cu(NH_3)_4^{2+}$ the lone pair electrons of ammonia (the ligand) are transferred to the central copper atom. The net charge on the ion will be less than 2 and perhaps less than 1. The ion will have its charge neutralized but it need not be reduced; it will still be divalent Cu(II) formally.

In copper proteins Cu(II) there will be electron donation from co-ordinating groups of proteins (perhaps also water of hydration). Substrate molecules have labile π electrons and if the copper site of the blue protein is tetrahedral, a π electron from the substrate will merge with the central *d* orbital of copper. This delocalization does not mean that Cu(II) has been reduced to Cu(I) until the co-ordination bond is broken and the substrate leaves a π electron at the Cu site.

Deoxygenated haemocyanin contains Cu(I) ions which do not generate an e.s.r. signal or show any visible region absorption. The electron configuration of oxyhaemocyanin is regarded as consistent with

$$2Cu(I) + O_2 \rightleftharpoons Cu(I) \rightarrow O_2 \leftarrow Cu(I)$$

The blue colour of oxyhaemocyanin arises from the charge transfer induced by the bridged O_2 and not as the result of charge transfer between the metal and O_2 which has been assumed to cause the blue colour. In haemocyanin the copper has found its niche in the protein and seems to be firmly bound. But (reduced) haemocyanin is virtually colourless. Molecular oxygen converts haemocyanin to oxyhaemocyanin.

Entry of molecular oxygen can mean (i) that oxygen changes (potentiates) the copper absorption spectrum, (ii) that cuprous copper (Cu I) has an effect on the properties of oxygen, (iii) that both phenomena occur.

Griffith (1956) suggests that the most easily donated electrons of the oxygen molecule will be the π electrons so that each Cu will accept electrons in a σ orbital from a filled oxygen π orbital (*cf.* Mandeles *et al.*, 1954). The back donation will be from a $d\pi$ orbital of the copper to a higher half filled orbital of the oxygen. Both orbitals are at 90° so the two Cu(I) ions are oriented at 90° to the oxygen molecule (Leirulov & Meister, 1954). The oxygen must be in an exposed position if two Cu(I) atoms are each within a niche (see Fig. 11.1).

If the blue colour of oxyhaemocyanin is a puzzle so is blue colour of liquid oxygen (−183°) and blue colour of long layers of water (Bader & Ogryzio, 1964; Ogryzio, 1965). Gaseous oxygen shows energy levels 1269 mm, ^{1}Ac, 761·9 nm $'\Sigma g^r$ and 699 very weak, whereas liquid oxygen displays 699 (very weak), 634 nm (main) with 575, 530, 480 as vibrational sub-

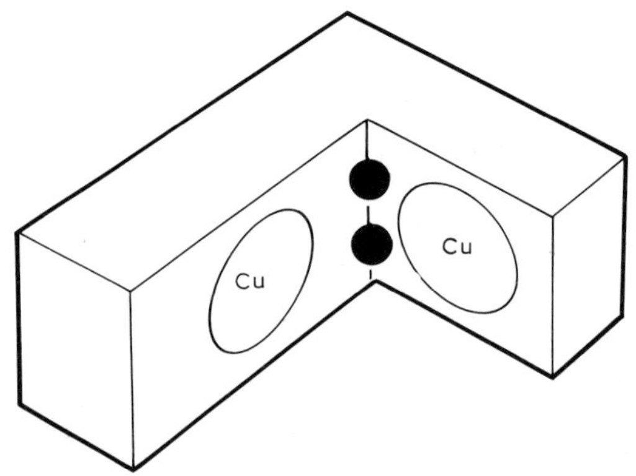

Figure 11.1

levels. Ellis & Knesser (1933) held that the liquid oxygen bands occurred when a single photon simultaneously elevated two electrons on two different molecules to excited states. Twice the energy needed to excite the 'Ac state' (1269 nm) is possessed by a photon at 634 nm.

Bibliography and References

(A) *The Biochemistry of Copper.* (1966). Eds. J. Peisach, P. Aisen & W. E. Blumberg. Academic Press, New York, London.

(B) *Physiology and Biochemistry of the Haemocyanins.* (1968). Ed. F. Ghiretti. Academic Press, New York, London.

(C) *Oxidases and Related Oxidase Systems.* (1965) Eds. T. E. King, H. S. Mason & M. Morrison. John Wiley, New York.

(D) *Oxygenases.* (1962). Ed. O. Hayaishi. Academic Press, New York, London.

Bader, L. W. & Ogryzlo, E. A. (1964). *Disc. Faraday Soc.*, **37**, 46.

Bradshaw, R. A., Shearer, W. T. S. & Gurd, F. N. R. (1968). *J. Biol. Chem.*, **243**, 3817.

Breslow, E. (1964). *J. Biol. Chem.*, **239**, 3252.

Brode, W. R. & Morton, R. A. (1938). *Proc. Roy. Soc.*, A, **120**, 21.

Brohult, S. (1940). *Nova Acta Regiae Soc. Sci. Upsaliensus*, **12**, 1–69.

Brohult, S. & Claesson, S. (1939). *Nature, Lond.*, **144**, 111.

Claesson, I. Möring (1956). *Arkiv Kemi*, 10.

Cohen, L. B. & Van Holde, R. E. (1964). *Biochemistry*, **3**, 1809.

Curdel, A. & Iwatsubu, M. (1968). *FEBS Lett.*, **1**, 133.

Curzon, G. In (B). p. 121.

Curzon, G. & Cummings, J. N. (1966). in (A).

Curzon, G. & Vallet, L. (1960). *Biochem. J.*, **74**, 279.

Ellis, J. W. & Knesser, H. O. (1933). *Z. Physik.*, **80**, 583.

Felsenfeld, G. (1954). *J. Cell. Comp. Physiol.*, **42**, 33.

Frieden, E., Osaki, S. & Kobayashi, H. (1965). *J. Gen. Physiol.*, **49**, 213.

Ghiretti, F. (1962). in (D).

Ghiretti-Magaldi, A. & Nuzzolo, C. (1965). *Comp. Biochem. Physiol.*, **15**, 249.

Gould, D. C. & Ehrenberg, A. (1968). in (B).

Gould, D. C. & Mason, H. S. (1966). in (A).

Griffith, J. S. (1956). *Proc. Roy. Soc.* A, **235**, 23.

Gruber, M. (1968). in (B).

Hairwegh, K. & Lontie, R. (1960). *Nature, Lond.*, **185**, 854.

Hentze, M. (1901). *Z. Physiol. Chem.*, **33**, 370.

Klotz, I. M. & Curme, H. G. (1948). *J. Amer. Chem. Soc.*, **70**, 930.

Klotz, I. M. & Klotz, T. A. (1955). *Science, N.Y.*, **121**, 477.

Klotz, I. M., Urquhart, J. M. & Fiess, H. A. (1957). *Arch. Biochem. Biophys.*, **68**, 784.

Kober, A. & Haw, A. B. (1916). *J. Amer. Chem. Soc.*, **38**, 457.

Kubowitz, F. (1937). *Biochem. Z.*, **292**, 221; (1938). *Ibid.*, **296**, 443; (1938). *Ibid.*, **299**, 32.

Latt, S. A., Holmquist, B. & Vallee, B. L. (1969). *Biochem. Biophys. Res. Commun.*, **137**, 333.

Latt, S. A. & Vallee, B. L. (1969). *Fed. Proc.*, **28**, 691.

Latt, S. A., Auld, D. S. & Vallee, B. L. (1970). *Proc. Nat. Acad. Sci., USA*, **67**, 1383.

Leirulov, L. & Meister, A. (1954). *J. Biol. Chem.*, **209**, 265.

Lindskog, S. & Ehrenberg, A. (1967). *J. Molec. Biol.*, **24**, 133.

Lindskog, S. & Nyman, P. O. (1964). *Biochem. Biophys. Acta.*, **85**, 462.

Lontie, R. (1957). In (A) and (B).

Lontie, R. (1958). *Clinica Chimica Acta.*, **3**, 68–71.

Lontie, R., Brauns, G., Cooreman, H. & Vandof, A. (1962). *Arch. Biochem. Biophys.*, Suppl. 1, 295–300.

Lontie, R. & Witters, R. (1966). in (A).

McConn, J. D., Tsuru, D. & Yasanobu, K. T. (1964). *J. Biol. Chem.*, **240**, 2415.

McGarvey, B. R. (1966). *Transition Metal Chem.*, **3**, 180; (1967). *J. Phys. Chem.*, **71**, 51.

Mandeles, S., Koppelman, R. & Hanke, M. E. (1954). *J. Biol. Chem.*, **209**, 327.

Mason, H. S. (1963). *Biochem. Biophys. Res. Commun.*, **10**, 11.

Mason, H. S. (1966). in (A).

Morpurgo, G. & Williams, R. J. P. (1968). in (B).

Nair, P. M. & Mason, H. S. (1967). *J. Biol. Chem.*, **242**, 1406.

Nakamura, T. & Mason, H. S. (1960). *Biochem. Biophys. Res. Commun.*, **3**, 297.

Nardi, G., Ghiretti, A., Caserta, G., Zito, R. & Ghiretti, F. (1962). *Bull. Soc. Ital. Biol. Sper.*, **38**, 1839, 1851, 1855.

Ogryzlo, E. A. (1965). *J. Chem. Educ.*, **42**, 647.

Osaki, S., McDermott, J. A. & Frieden, E. (1964). *J. Biol. Chem.*, **237**, PC.364; (also in (A) and (B)).

Peters, T. (1960). *Biochem. Biophys. Acta.*, **39**, 546.

Peters, T. & Blumenstock, F. A. (1967). *J. Biol. Chem.*, **242**, 1574.

Peters, T. & Hawn, C. (1967). *J. Biol. Chem.*, **242**, 1506.

Redfield, A. C. (1934). *Biol. Rev.*, **9**, 176; (1950). In *Copper Metabolism. A Symposium*, p. 174. Ed. W. D. McElroy and B. Glass. Johns Hopkins Press, Baltimore; (1968). In (A).

Rombauts, W. & Lontie, R. (1960). *Arch. intern. Physiol. Biochem.*, **67**, 514; also (1960), *ibid.*, **68**, 230 and 695.

Sarkar, B., Bersohn, M. & Wigfield, Y. (1968). *Can. J. Biochem.*, **46**, 595.

Sarkar, B. Wigfield, Y. (1967). *Canad. J. Biochem.*, **46**, 601.

Simpson, R. T., Riordan, J. F. & Vallee, B. L. (1933). *Biochemistry*, **2**, 616.

Simpson, R. T. & Vallee, B. L. (1966). *Biochemistry*, **5**, 1760; (1968). *Fed. Proc.*, **27**, 291.

Svedberg, T. (1937). *Chem. Rev.*, **20**, 81. (See also (1937). *Nature, Lond.*, **139**, 1051.)

Svedberg, T. & Pedersen, K. G. (1940). *The Ultracentrifuge*. Oxford University Press.

Thompson, L. C. G., Hines, M. & Mason, H. S. (1959). *Arch. Biochem. Biophys.*, **83**, 88.

Tsuru, D., McConn, J. D. & Kasunobu, K. T. (1965). *J. Biol. Chem.*, **240**, 2415.

Vallee, B. L. (1969). *Ann. N.Y. Acad. Sci.*, **158**, 377.

Vallee, B. L. & Williams, R. L. P. (1968). *Chem. in Brit.*, **4**, 397; *also* (1968). *Proc. Nat. Acad. Sci. U.S.A.*, **59**, 498.

Vallee, B. L. & Latt, S. A. (1970). In *Structure–formation relationships of proteolytic enzymes*. Ed. P. Desneulle, H. Neurath & M. Ottesen. Munksgaard, Copenhagen.

Van Bruggen, E. F. J., Wiebenza, E. H. & Gruber, M. (1962). *J. Molec. Biol.*, **4**, 1; *also* (1962) *J. Molec. Biol.*, **7**, 249.

Van Holde, K. E. (1967). *Biochemistry*, **6**, 93.

Van Holde, K. E. & Cohen, L. B. (1964). *Analyt. Chem.*, **27**, 1195; *also* (1964). *Biochemistry*, **3**, 1803, 1809.

Wagner, F. W. & Prescott, J. M. (1966). *Comp. Biochem. Physiol.*, **17**, 191; *Fed. Proc.*, **25**, 590.

Williams, R. J. P. (1953). *Biol. Res.*, **28**, 331.

Wood, E. J. & Bannister, W. H. (1967). *Biochem. J.*, **104**, 42P; *also* (1967). *Nature, Lond.*, **215**, 1091.

Wood, E. J., Salisbury, C. M. & Bannister, W. H. (1968), *Biochem. J.*, **108**, 26–27P.

12 Steroids and Related Substances

12.1 Ultraviolet Absorption

Vast developments have taken place in the past half century in the chemistry and biochemistry of the sterols. Some of the work is important in relation to nutrition, e.g. that on provitamins and vitamins D. There is also much that is highly relevant to endocrinology. Synthetic and modified sterols are increasingly available and have found many uses. Some of these topics are discussed elsewhere in the present volume but here our concern is with the relations between chemical constitution and selective absorption (see Dorfman, 1953; Morton, 1962; Rao, 1967; Gillam & Stern, 1970; Scott, 1964, deals comprehensively with u.v. absorption data and colour lists for steroids).

Steroids with a single double bond (and unconjugated dienes) show $\lambda_{max.}$ near 190–200 nm. Modern photoelectric spectrophotometers (especially when new) are capable of accurate measurements at 180–210 nm. When the chromophore is X—CH=CHY, as in Δ^2-, Δ^6- and Δ^{11}-steroids, ϵ values at 210 nm are near 1000; substitution to give X—CH=C< as in Δ^4-, Δ^5-, Δ^7-, $\Delta^{9(11)}$- and Δ^{14}-steroids causes ϵ to rise to 2000–4500 at 210 nm and the chromophore >C=C< which occurs in $\Delta^{8(9)}$-, $\Delta^{8(14)}$- and $\Delta^{9(10)}$-steroids shows ϵ 210 nm $\not<$ 4500 (Bladon *et al.*, 1952). The 'weighting' caused by substituents also involves a bathochromic shift.

Alkyl substituted acyclic conjugated dienes also exhibit displacements. Thus, butadiene has $\lambda_{max.}$ 217 nm, $\epsilon_{max.}$ 22 400 in ethanol and the introduction of one or two methyl groups displaces $\lambda_{max.}$ to 222–223 nm, while further substitution moves it to 226 nm without much variation in $\epsilon_{max.}$, culminating in $\lambda_{max.}$ 231 nm for $Me_2C{=}CH{-}CH{=}CH_2$ and 242 nm

Table 12.1 Absorption spectra of some steroids with unconjugated double bonds

(In cyclohexane except where indicated)

	$\lambda_{max.}$ nm	$\epsilon \times 10^{-3}$	
Δ^4-Cholestene	193	10	
Δ^4-Cholestene 3β-ol	199·5	8·7	
Δ^4-Cholestene 3β-ol acetate	196	13·5	
Δ^5-Cholestene	189	8·9	
Cholesterol (Δ^5-cholesten 3β-ol)	189·5	9·9	
Cholesteryl acetate	189	10·0	
Δ^7-Cholestene	180	10·3	
	190·5	8·9	(ethanol)
Δ^2-Cholestene	180	6·6	
Δ^2-Cholestene 3β-ol acetate	182	8·74	
Δ^7-Cholestene	205	5·37	
	179·5	10·5	
Δ^7-Ergostenylacetate	203·5	5·57	
$\Delta^{8(14)}$-Ergostenylacetate	206	12·0	
Stigmasterol	188	21·7	

for $Me_2C{=}CH{-}CH{=}CMe_2$. Substitution at carbons 2 and 3, e.g. $Me_2C{=}CMe{-}CMe{=}CMe_2$ greatly reduces $\epsilon_{max.}$ to 4200 and no actual peak is seen (Forbes *et al.*, 1964). The simpler conjugated dienes are planar (s-*trans* conformation). In 1,1,3-trimethylbutadiene $\lambda_{max.}$ is at 231 nm, but $\epsilon_{max.}$ is only 10 000, a decrease attributed to non-planarity. Further substitution enhances the non-planarity.

Cyclic dienes are illustrated below:

	CH_2 / CH_2	CMe_2 / CMe_2	
λ_{max}, nm	220	260	259
ϵ_{max}, $\times 10^{-3}$	10	0·75	10

Cyclic dienes include *cis* and *trans* forms as well as homoannular and heteroannular structures. Cisoid dienes ($\epsilon_{max.}$ 5–15 × 10^3) absorb less strongly than transoid dienes ($\epsilon_{max.}$ 12–28 × 10^3). The latter display peaks at longer wavelengths than the former (cf. Fieser & Fieser, 1959; Scott, 1964). The effect of ring size on monocyclic diene absorption was discussed by Braude (1954). The smaller rings can only have the *cis-cis* configuration.

The predictability of many conjugated diene spectra has been examined by Booker *et al.* (1940) and by Woodward (1941, 1942). The steroidal dienes illustrate the position very well. With an exocyclic double bond at carbon 3 or 7 absorption at 233–239 nm is strong. In cholesta-3,5-diene (and the related androst-3,5-diene-17-one and pregna-3,5-20-triene), three peaks are seen at 228, 235 and ~243 nm, respectively. Cholest-4,6-diene has $\lambda_{max.}$ 238 nm, while the related 3-acetoxy compound shows fine structure and androst-4,6-diene-3β,17β-diol has very similar absorption. When the two conjugated double bonds are in the same ring highly characteristic spectra occur. In particular the $\Delta^{5,7}$ chromophore which occurs in ergosterol, 7-dehydrocholesterol, 7-dehydrostigmasterol and 7-dehydrositosterol shows reproducible resolution but some variations in $\epsilon_{max.}$ have been recorded for various compounds containing the chromophore.

Conjugated trienes can be compared with decatrienoic acid $CH_3CH_2(CH{=}CH)_3CH_2COOH$, which has $\lambda_{max.}$ 265 nm and $\epsilon_{max.}$ 43 000. Calciferol shows $\lambda_{max.}$ 265 nm, $\epsilon_{max.}$ 18 200, tachysterol 281 nm, ϵ_{max} 24 600 and previtamin D_2 has $\lambda_{max.}$ 262 nm, ϵ_{max} 9000 (see p. 398). In the cyclic conjugated trienes the diene absorption of for example the $\Delta^{5,7}$ chromophore is displaced and in most cases good resolution persists. The $\Delta^{5,8(14),9(11)}$ triene may well have two partials acting in the main independently. Extinction coefficients are however not quantitatively predictable; e.g. cholest-3,5,7-triene and ergost-3,5,7,22-tetraene have the same chromophore but the intensities for the former are only about 81% of those for the latter.

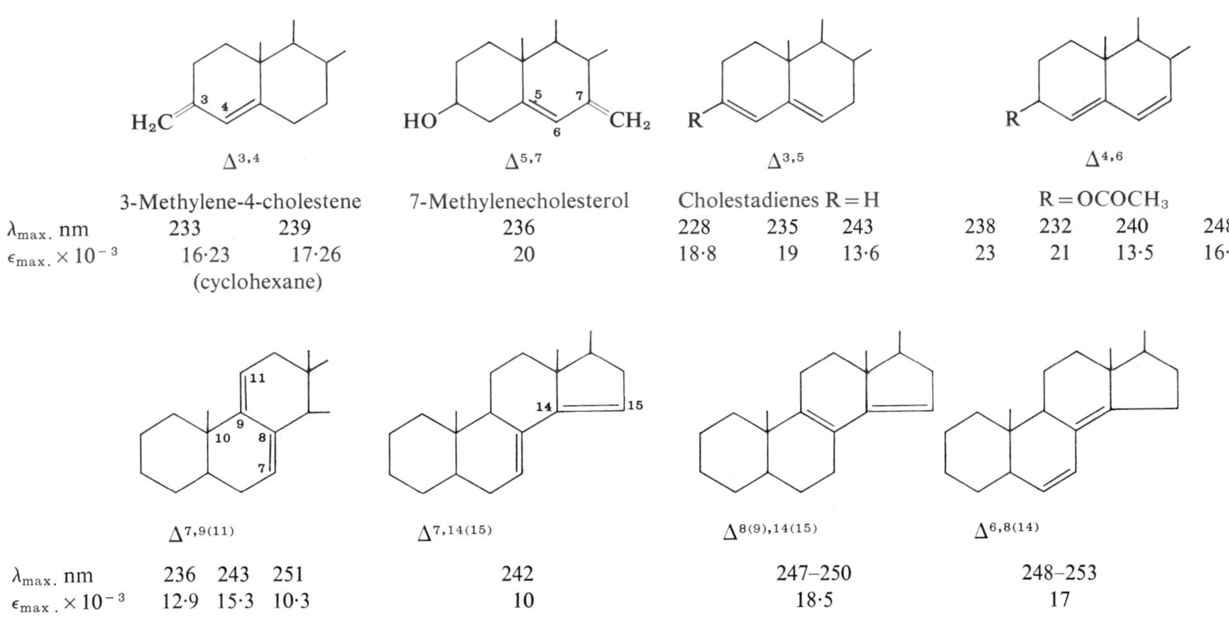

	$\Delta^{2,4}$	$\Delta^{5,7}$	$\Delta^{6,8(9)}$	$\Delta^{4,6,8(14)}$
$\lambda_{max.}$ nm	266 275 287	262 271 282 293	275	283
$\epsilon_{max.} \times 10^{-3}$	6·3	7·7 11·4 11·9 6·9	5·3	33·0
			(*iso*dehydrocholesterol)	

	$\Delta^{2,4,6}$-Triene	$\Delta^{3,5,7}$-Triene	$\Delta^{5,7,9(11)}$	$\Delta^{5,8(14),9(11)}$	$\Delta^{5,7,14}$
$\lambda_{max.}$ nm	296 306 320	302 315 331	312 324 339	243 285	319
$\epsilon_{max.} \times 10^{-3}$	14·5	16·8 19·8 13·4	10·4 11·8 7·4	12·0 9·1	16·2

C_9H_{17}

AcO

Ergost-5,7,9(11),22-tetra-en-3-one
3-Enolacetate

$\lambda_{max.}$ nm	339	356	375
$\epsilon_{max.} \times 10^{-3}$	15	17·4	13

C_8H_{17}

HO

Cholest-4,6,8(9),11-tetra-en-3β-ol
355
13·5

Very many examples of unsaturated steroid ketones are known and the absorption spectra have been recorded for the following enone chromophores:

	$\lambda_{max.}$ nm	$\epsilon_{max.} \times 10^{-3}$	$\epsilon_{max.} \times 10^{-1}$
Δ^{7}-3-one	229–232	9·78–12·6	
Δ^{4}-3-one	239–244	10·5–22	
	295–312		6–14·6
Δ^{4}-6-one	244	6·3	
	300–320		8·5–12
Δ^{5}-4-one	241	3·2–7·2	
Δ^{5}-7-one	238–240	8·9–15·1	
Δ^{7}-6-one	252	13·4	
	323–333		10–16
$\Delta^{8(9)}$-7-one	253	8·3–13·2	
$\Delta^{8(9)}$-11-one	254	8·7–9·6	
$\Delta^{8(14)}$-7-one	263	7·8–10·8	
$\Delta^{8(14)}$-15-one	259	12·7–13·3	
$\Delta^{9(11)}$-12-one	238–243	8·3–15·8	
	311–322		7·1–12

The spectra conform on the whole reasonably well with the empirical rules (p. 29) but it is possible to overstress predictability, particularly in respect of intensities of absorption.

Saturated ketones exhibit one broad band in the region 270–300 nm, $\epsilon_{max.}$ *ca.* 75–100. A bathochromic shift occurs with cyclohexanone at 292 nm and with cyclopentanone to 299 nm. Saturated steroid ketones need to be highly purified if the measured spectra are to be trusted. Authentic values of $\lambda_{max.}$ will fall between 282 and 297 nm. In diketones which do not enolize the two carbonyls function additively.

Steroid dienones clearly show the effect of a second conjugated double bond:

Cholesterol derivative:

	$\lambda_{max.}$ nm	$\epsilon_{max.} \times 10^{-3}$
-3,5-diene-7-one	277	24·4
-4,6-diene-3-one	284	26·3
-3,5-diene-2-one	290	12·6
-2,4-diene-6-one	314	7·26
-1,4-diene-3-one	245	14·5
-7,9-diene-12-one	239	3·7
	292	13·1

The 5-en-7-one chromophore (239 nm, ϵ *ca.* 12 000) and the 3,5-diene chromophore (234–239 nm, ϵ *ca.* 19 000) may be compared with the related dienone (277 nm, $\epsilon_{max.}$ 24 000). A marked red shift occurs with some increase in $\epsilon_{max.}$ Much the same obtains for the comparison between the 4,6-diene-3-one and the simpler related diene and enone. The 2,4-diene has $\lambda_{max.}$ 275 nm but in the 2,4-diene-6-one the peak is shifted to 314 nm. On the other hand the second double bond in the 1,4-diene-3-one makes little difference to the spectra of the related enones. In the case of the 7,9-diene-12-one a red shift to 292 nm occurs.

Steroid ene-diones occur and some of the chromophores are noted below. Dorfman (1953) and later Scott (1964) have dealt in detail with the effects of different groupings adjacent to enone chromophore and Scott has neatly worked out the effects of enolization on the spectra of conjugated steroid ketones.

A good example is shown in the case of the $\Delta^{4,7,9}$-3-one which can form an enolacetate with four conjugated double bonds. The trienones further illustrate the effects of sustained conjugation:

	$\lambda_{max.}$ nm	$\epsilon_{max.} \times 10^{-3}$
$\Delta^{1,4,6}$-3-one	223	13·5
	256	11·9
	298	15·3
$\Delta^{1,3,5}$-7-one	230	18·6
	278	3·7
	348	11·0
$\Delta^{4,6,8(9)}$-3-one	244	17·8
	288	2·2
	388	12·3
$\Delta^{4,6,8(14)}$-3-one	348	26·5

The element of predictability in these spectra is very well surveyed by Scott (1964) and Jaffé & Orchin (1962) summarize the empirical rules and quote many convincing examples.

The conjugated enones and dienones are of great practical importance as will be seen from the substances shown below.

Testosterone in CH_3OH

$\lambda_{max.}$ nm	241	308–310	$\lambda_{min.}$ nm	287
$\epsilon_{max.} \times 10^{-2}$	165	0·82		0·68

Testosterone propionate

$\lambda_{max.}$ nm	241·5	310	$\lambda_{min.}$ nm	287
$\epsilon_{max.} \times 10^{-2}$	170	0·73		58
		(in $CHCl_3$)		
$\lambda_{max.}$ nm	246	320	$\lambda_{min.}$ nm	286
$\epsilon_{max.} \times 10^{-2}$	148·4	0·61		0·34

17α-Methyltestosterone in CH_3OH | Ethisterone in CH_3OH

	17α-Methyltestosterone			Ethisterone	
$\lambda_{max.}$ nm	241·5	~286	~308	241	~300
$\epsilon_{max.} \times 10^{-2}$	161	2·8	1·4	165	1·0
			(in $CHCl_3$)		(in $CHCl_3$)
$\lambda_{max.}$ nm	245	285	325	246	flat 280–325
$\epsilon_{max.} \times 10^{-2}$	153·6	2·6	0·75	160	0·8

Norethisterone (in CH_3OH)
(17α-Ethynyl-19-nortestosterone)

$\lambda_{max.}$ nm	241	310	$\lambda_{min.}$	286 nm
$\epsilon_{max.} \times 10^{-2}$	182	0·93		0·63

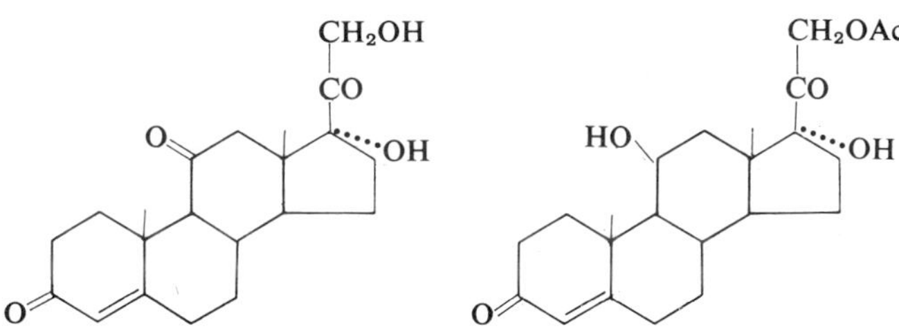

D-Cortisone (in CH_3OH) | D-Hydrocortisone (cortisol) acetate

	D-Cortisone		D-Hydrocortisone acetate		
$\lambda_{max.}$ nm	238	~285	242	300	
$\epsilon_{max.} \times 10^{-2}$	161	3·72	166	1·61	(in CH_3OH)

D-Cortisone acetate essentially similar

β-Sitosterone
241
164 (in CH_3OH)

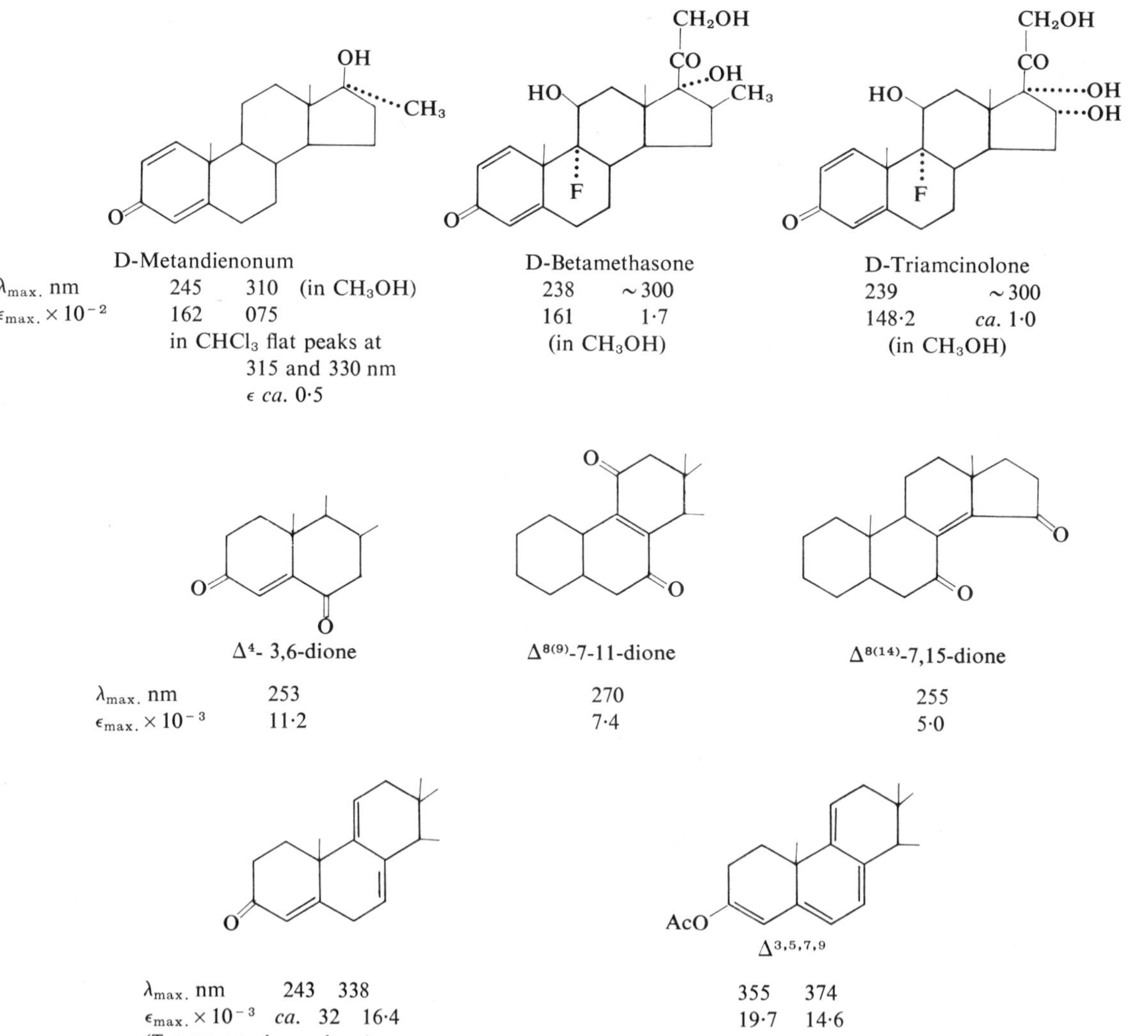

Aromatic chromophores are seen in neoergosterol and 22-dihydroneoergostatriene and the related oestra-5,7,9-triene-3-ol-17-one, both of which show absorption spectra resembling those of tetraalkylbenzenes. An extra double bond in an adjacent ring increases the intensity of absorption but the differences between compounds containing this 'styrene' chromophore invite further study.

In β-dl-equilane and 15-dehydroequilin-7-one the naphthalene chromophore is clearly evident and the $\alpha\beta$-unsaturated ketone chromophore is isolated and masked. In 14-dehydroequilenene there is a considerable increase in absorption on the long-wave side of 300 nm which obscures the region where the naphthalene chromophore has narrow bands of rather low intensity.

The absorption spectra of oestrone, oestradiol and oestriol show the phenol chromophore with $\lambda_{max.}$ 280 nm, $\epsilon_{max.}$ 2300 as would be expected for a dialkyl phenol. Equilin has an additional double bond which is unconjugated and the spectrum resembles that of oestrone. The effect of a conjugated double bond is shown opposite.

It is always possible that ultraviolet absorption curves for dilute solutions in polar solvents are less informative than spectra in non-polar solvents or spectra at reduced temperatures. Another possible method of refinement is to study first derivative spectra.

Olson & Alway (1960), attracted by French's (1957) use of derivative spectra, devised an attachment to the Cary Model II recording spectrophotometer which permitted measurement of the curve dA/dX

C$_9$H$_{17}$

HO

Neoergosterol

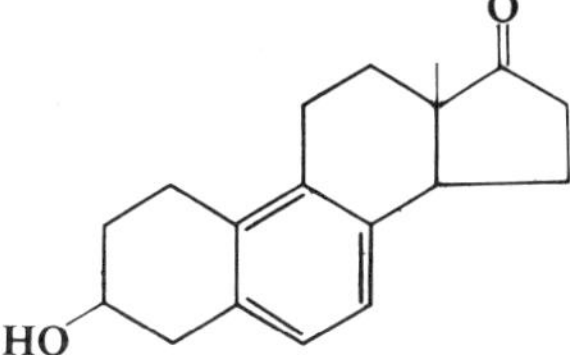

Dihydroneoergostatriene

O

HO

$\Delta^{5,7,9}$-Oestratriene-3-ol-17-one

	Neoergosterol	Dihydroneoergostatriene			$\Delta^{5,7,9}$-Oestratriene-3-ol-17-one	
$\lambda_{max.}$ nm	268	224	269	277	270	278
$\epsilon_{max.} \times 10^{-2}$	5·0	100	4	2·5	3·45	2·4

C$_9$H$_{17}$

C$_9$H$_{17}$

$\lambda_{max.}$ nm	224	231	268	308	314	323	270
$\epsilon_{max.} \times 10^{-3}$	26	6·3	6·3	very low			11·5

O

β-dl equilane

15-Dehydroequilene-17-one

14-Dehydroequilenene

	β-dl equilane			15-Dehydroequilene-17-one			14-Dehydroequilenene			
$\lambda_{max.}$ nm	231	282	322	230	270	321	255	285	293	305
$\epsilon_{max.} \times 10^{-3}$	93·3	5·7	0·91	77·6	6·8	0·66	45·7	12·6	15·5	12·9

where $X = k\lambda$. This is not quite the same as the first derivative spectrum $dA/d\lambda$ and the heights in the recorded spectrum are influenced both by concentration and rate of wavelength scan. It does however provide a good indication of the location of 'hidden' sub-maxima in poorly resolved direct absorption curves. The molecular extinction coefficients at individual wavelengths are, however, measured using the direct absorption curves.

Olson & Alway (1960) studied testosterone and related compounds. Figure 12.1 shows how (in a solvent not stated) there is one maximum for testosterone and several inflections. Each maximum on the first derivative curve corresponds with the point of steepest slope on a section of the direct absorption

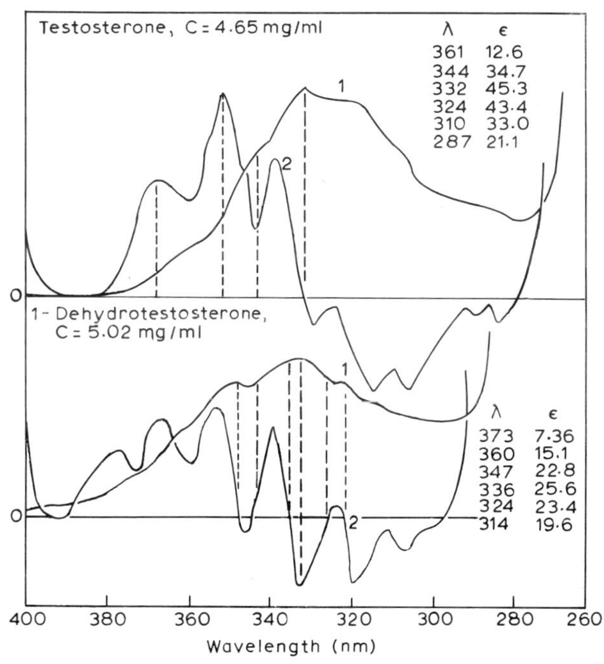

Figure 12.1 Relationship between the ultraviolet spectrum (1) and the derivative ultraviolet spectrum (2) for testosterone and 1-dehydrotestosterone (from Olson & Alway, 1960)

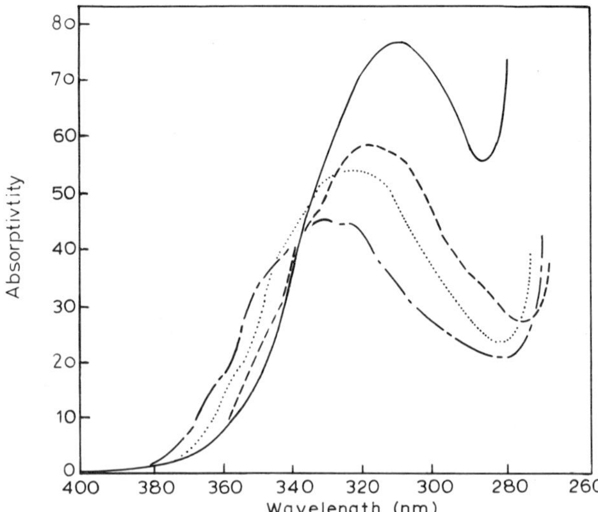

Figure 12.2 Effect of solvent upon the ultraviolet spectrum of testosterone: ———, ethyl alcohol; – – – – –, methylene chloride; ……., acetonitrile; — · — · — · —, diethylene glycol dimethyl ether (from Olson & Alway, 1960)

Figure 12.3 Effect of solvent upon the derivative ultraviolet spectrum of testosterone: (1) Diethylene glycol dimethyl ether; (2) dioxane; (3) acetonitrile; (4) methylene chloride; (5) chloroform (0·7% ethyl alcohol); (6) 95% ethyl alcohol; (7) methanol (from Olson & Alway, 1960)

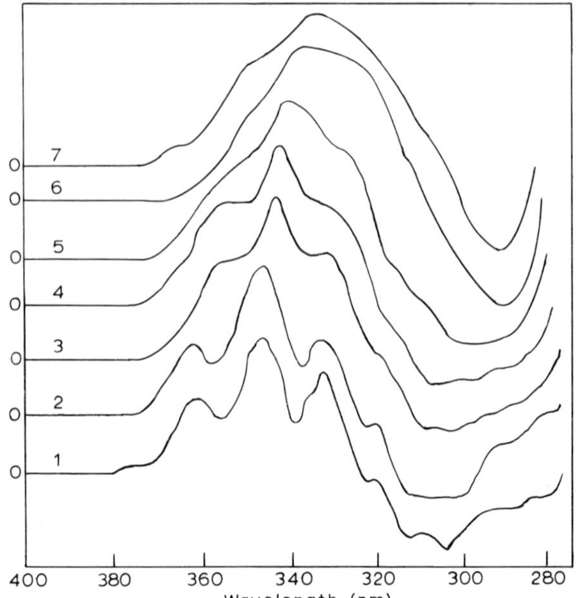

while the derivative minima and zero values represent the true positions of the inflections and maxima on the direct curve. The testosterone absorption is measured in the region 260–380 nm ($n \rightarrow \pi$ low intensity transition). Figure 12.1 also shows the direct absorption curve for 1-dehydrotestosterone where the indications of resolution are more definite and the derivative spectrum adds definition to some of the less evident inflections.

The authors went on to measure the direct and derivative spectra of testosterone in different solvents. Figures 12.2 and 12.3 illustrate the results. The more polar solvents displace $\lambda_{max.}$ in the direction of shorter wavelengths and signs of fine structure are best seen in the less polar solvents. In the direct spectra there is a near approach to one or more isosbestic points and this would suggest that intermolecular complexes or solvent–solute complexes deserve to be considered. The fine structure in a well-resolved spectrum should show a constant frequency difference (Δ cm^{-1}) which is not particularly clear from the results on testosterone or 1-dehydrotestosterone. (The results of low temperature spectra or vapour absorption spectra would be relevant in a fuller analysis.) The fact that the precise significance of the measurements of derivative spectra is uncertain does not, however, prevent them from being empirically useful.

Olson & Alway went on to determine the spectra of progesterone, its 11α- and 11β-hydroxy relatives and the 17α-hydroxyprogesterone (Fig. 12.4). The derivative spectra exhibit very similar structure but there are distinct differences in intensities of absorption, particularly for the 17α-hydroxy compound. Hydrogen bonding between the 20-carbonyl and the 17-hydroxy groups is thought to be responsible for an approximate doubling of ϵ values. The derivative spectra of 6α- and 6β-methyltestosterones are distinctly different, the 6β compound closely resembling the parent substance (Fig. 12.5). Similarly the derivative spectrum of progesterone resembles that of its 6β- rather than its 6α-methyl analogue. The 6α-methyl compounds show a sharp maximum near 300 nm in the derivative spectra which is diagnostic, although its origin is none too certain.

Derivative spectra can be of considerable service, but other methods of tackling the problems raised should not be neglected.

The absorption spectra of so many steroids are dominated by the Δ^4-3-one chromophore that direct ultraviolet absorption cannot be specific. Even with

Figure 12.4 Effect of 11- and 17-hydroxylation on the derivative ultraviolet spectrum of progesterone (from Olson & Alway, 1960)

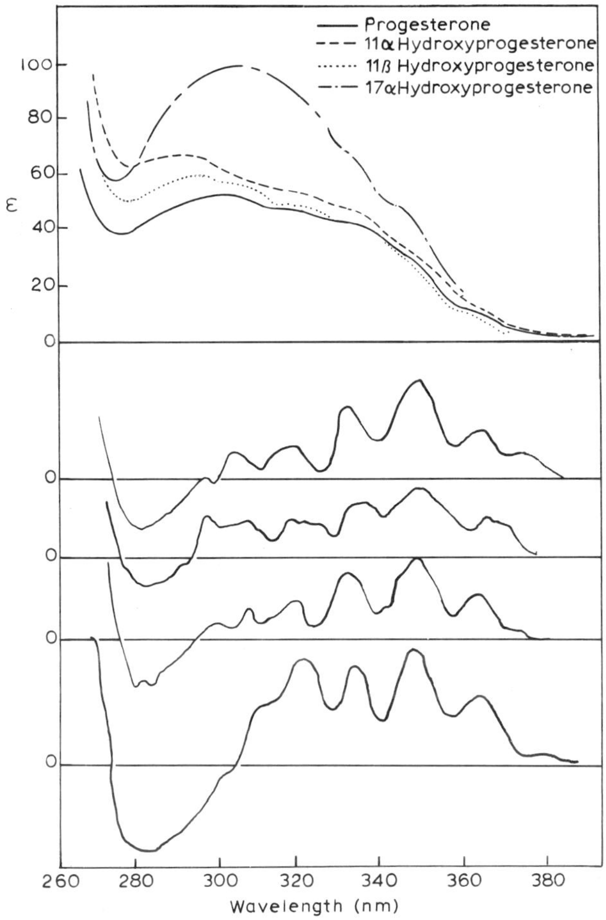

the rather similar but less common $\Delta^{1,4}$-diene-3-one chromophore a similar difficulty arises. Methods have however been found to circumvent the low specificity.

Görög & Szepesi (1972) improved on an oxidation of 21-aminocorticosteroids with mercuric chloride, the calomel formed being used as a measure of the reaction (Görög *et al.*, 1969). Instead they carried out the oxidation with a mercury (II)–EDTA complex and condensed the resulting 20-keto-21-aldehyde with 4,5-dimethyl-*o*-phenylenediamine to form a quinoxaline derivative. This compound was then determined spectrophotometrically at 351 nm in a strongly acid medium, the molecular extinction coefficient of 12 500 being high enough to make the method quite sensitive.

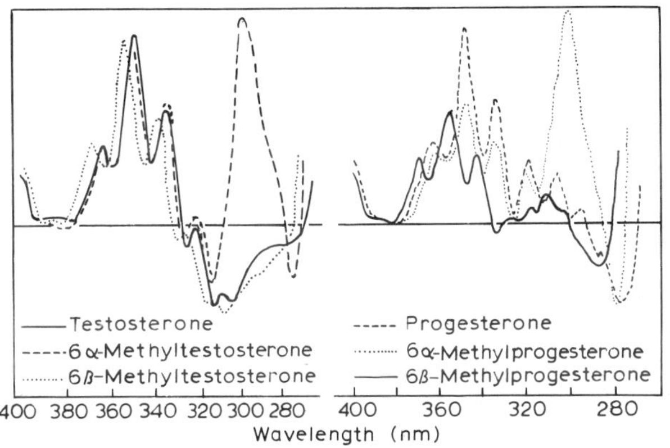

Figure 12.5 Effect of 6α- and 6β-methyl substitution on the derivative ultraviolet spectra of testosterone and progesterone (from Olson & Alway, 1960)

$$Na_3HY + HgCl_2 \rightarrow NaH(HgY) + 2NaCl$$

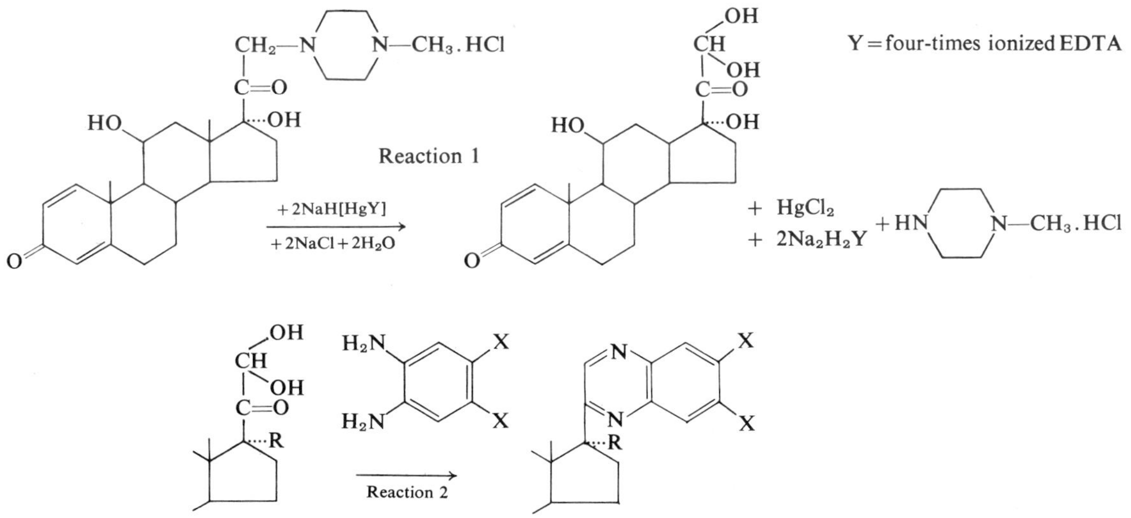

Reaction 1

$+2NaH[HgY]$ → $+2NaCl+2H_2O$

$+ HgCl_2 + 2Na_2H_2Y + HN\langle\rangle N{-}CH_3.HCl$

Y = four-times ionized EDTA

Reaction 2

(X = H or CH_3 and R = H or OH)

21-Hydroxycorticosteroids (including prednisolone) do not react with the mercury(II)–EDTA complex and this gives the method practical specificity.

Görög & Szepesi (1972) gave details for determining depersolone in solutions dispensed for injection, or as drops for nose, ear or eye, or in ointment.

The spectrophotometric method can be varied by using the readily accessible compound *o*-phenylenediamine which gives a quinoxaline derivative with $\lambda_{max.}$ 331 nm and $\epsilon_{max.}$ 19 000. It is necessary to make the measurements in a strongly acid medium so that the quinoxaline derivatives are fully protonized (Görög & Szepesi, 1972).

21-Hydroxycorticosteroids and 21-aminocorticosteroids can be oxidized to 20-keto-21-aldehydes with copper (II) acetate, and prednisolone free from aminocorticosteroids can be determined by this procedure.

Görög & Tuba (1972) devised a specific method for the determination of 21-halo-corticosteroid derivatives and the sulphonic acid esters of 21-hydroxycorticosteroids. Its basis is the formation of 21-quaternary pyridinium salts and their measurement in alkaline solution as yellow 'quaternary pyridinium enolates'. The position of $\lambda_{max.}$ lies between 413 and 417 nm for different corticosteroid derivatives and $\epsilon_{max.}$ is between 20 240 and 21 900 with a coefficient variation of 0·7 per cent. The method permits the determination of the above compounds when they are present as corticosteroid contaminants at the level of 0·1%.

A sample of the material to be tested (0·5–1 mg) is dissolved in pyridine (10 ml) and 5 **N** hydrochloric acid (0·1 ml) is added. The mixture is kept in a loosely stoppered flask for 1 hour in a boiling water bath. The solution is cooled, transferred to a 100 ml calibrated flask with ethanol, 1 ml of tetramethylammonium chloride is added and the solution is made up to volume with ethanol.

Table 12.2 Absorption peaks of quaternary pyridinium enolates

	$\lambda_{max.}$ nm	$\epsilon_{max.} \times 10^{-3}$
Prednisolone-21-mesylate	414	21·7
Prednisolone-21-tosylate	414	21·9
21-Deoxy-21-bromoprednisolone	414	21·05
21-Chloroprogesterone	417	20·8
11β-Hydroxy-21-chloro-progesterone	413	20·5
Prednisolone*	414	18·75
Prednisolone†	392	11·62

* In 0·01 N tetramethylammonium hydroxide in ethanol containing 10% v/v pyridine.

† In 0·01 N $(CH_3)_4NOH$ in water containing 10% pyridine.

CH_2X — $C{=}O$ — R; Pyridine, reaction 1 → CH_2—$\overset{+}{N}$(pyridine) X^- — $C{=}O$ — R

CH_2—N(pyridine) — $C{=}O$ — OH (steroid) + $CH_3C_6H_4SO_3^-$ ⇌ (reaction 2, OH^-) HC—N(pyridine) = $C{-}O^-$ — OH (steroid) + $CH_3C_6H_4SO_3H$

12.2 Infrared Absorption

As has been seen, numerous important steroids show very similar ultraviolet absorption spectra and many ingenious ways of evading the consequences of this lack of specificity have been tried. Infrared absorption spectra have however proved complementary to ultraviolet absorption in the sense that one is serviceable when the other is not. There is a very rich literature for infrared spectra of steroids (Dobriner *et al.*, 1953; Roberts *et al.*, 1958; Jones & Dobriner, 1949; Rosenkratz, 1955; Jones & Sandorfy, 1956; Fieser & Fieser, 1959; Rao, 1963).

The steroid ring system is planar and groups attached to the nucleus which are above the plane of the ring and on the same side as the angular methyl group at C10 are designated β and the convention is to draw the bonds with a heavy continuous line. Groups below the plane of the ring system are designated α and the bonds are drawn with broken lines. The allo or 5α-steroids show *trans* relationships at C5–C10–C9–C8–C14–C13. In the 5β or normal series C5 and C10 substituents are *cis* and C10–9–8–14–13 *trans*.

Cyclohexane can exist in two forms free from angle strain, the chair and boat conformations, the former being the more stable. Of the two substituents at each CH_2 one is parallel to the threefold axis of symmetry of the ring (axial a) and the other is approximately in the plane of the ring (equatorial e). The details of steroid nomenclature are, however, readily available and it is assumed that there is no need to repeat them here.

Table 12.3 Ethylenic double bonds in steroids

Position of C=C	C=C stretching cm^{-1} ($CHCl_3$)	C=C—H stretching cm^{-1} (CCl_4)	C—H bending cm^{-1} (CS_2)
Δ^1	1664	3021–2995	754–700
Δ^2	1655	3034	774–664
Δ^3	1647	3015	773–671
Δ^4	1657	3040	810
Δ^5	1672–1667	3030	803–799, 814–812, 840–830
Δ^6	1639–1633	3017–3000	772–704
Δ^7	1666–1664	3040–3013	847–827
$\Delta^{9(11)}$	1648–1643	3042	852–827
Δ^{11}	1628–1620	3050–3033	839–703
Δ^{14}	1648–1646	3055	807–797
Δ^{16}	1630–1621		
Δ^{22}	1666–1604		974–970

After Cole (1956).

Table 12.4 Absorption peaks for unsaturated steroid ketones

	C=O stretch cm^{-1}	C=C stretch cm^{-1}
Δ^1-3-one	1684–1680	1609–1604
Δ^4-3-one	1681–1677	1615–1609
Δ^5-7-one	1682–1674	—
Δ^8-11-one	1660	—
Δ^{15}-17-one	1716	1587
Δ^{16}-20-one	1670–1666	1592–1588
$\Delta^{1,4}$-diene-3-one	1671–1663	1621, 1606–1603
$\Delta^{4,6}$-diene-3-one	1669–1666	1619–1616, 1587
$\Delta^{3,5}$-diene-7-one	1663	1627, 1598
Δ^4-en-3,6-dione	1692	—

Infrared spectra provide substantial guidance in the location of unsaturation, as will be seen. The evidence of Tables 12.3 and 12.4 has proved useful.

Saturated ketones with carbonyl groups at positions 1–12, (5α and 5β) show in CCl_4 or CS_2 the C=O stretch at 1719–1704 cm^{-1} mainly near 1710 cm^{-1}. If the carbonyl is in the cyclopentane ring the peak is at 1749–1738 cm^{-1} (C=O at C17 *ca.* 1743 cm^{-1}).

Hydroxysteroids show free hydroxyl stretching vibrations near 3630 cm^{-1} (in CCl_4) but the band has no positional specificity. The C–O vibration is stereochemically diagnostic.

A/B ring fusion	Configuration and conformation*	Principal C—O† cm^{-1}
trans	3β (e)	1040–1035
trans	3α (a)	1004–999
cis	3α (e)	1038–1035
cis	3β (a)	1033–1030
C=C	3β (e)	1056–1047
C=C	3α (a)	1034

* (e) = equatorial, (a) = axial. † In CS_2 solution.

In 3-hydroxysteroids the C–O band between 1050 and 1000 cm^{-1} has $\epsilon_{max.}$ 1660–220, while the bands in C17-β(e)-hydroxysteroids (1012–1008 cm^{-1}) and 20α (1100–1090 cm^{-1}) or 20β-hydroxysteroids (1016–1008 cm^{-1}) have lower molecular extinction coefficients

($\epsilon_{max.}$ 60–120). If there are two or more hydroxyls in the steroid molecule ϵ values tend to be additive.

Overlapping of absorption bands in the region 3000–2850 cm^{-1} reduces the diagnostic value of C—H stretching vibration of methyl and methylene groups. The C—H bonds attached to unsaturated groups have higher stretching frequencies, e.g. Δ^5, (C=CH) *ca.* 3300 cm^{-1}. The CH_3 and CH_2 bending vibrations are however much more serviceable. Thus ring system CH_2 groups exhibit (in CCl_4) a peak at 1450 (1464–1446) cm^{-1} while CH_2 groups adjacent to a double bond have 1438 (1445–1432) cm^{-1} and side-chain CH_2 groups have $\nu_{max.}$ at 1468 (1470–1461) cm^{-1}. A methylene group adjacent to a carbonyl in a 5-membered ring D displays a peak at 1408 (1411–1404) cm^{-1}. Compounds with a carbonyl at position 3 or 7 are readily distinguishable in that the methylene at C2 or C4 in a 3-ketone gives rise to a well-marked peak at about 1420 cm^{-1} (e.g. cholestan-3-one) whereas in cholestan-7-one there is a weak maximum at 1433 cm^{-1}. Deuterium exchange at the relevant methylene groups causes these bands to disappear.

12.3 Steroidal Sapogenins

Steroidal sapogenins, as would be expected from their structures, possess little interest in respect of ultraviolet absorption but the spectra in sulphuric acid are more definite (Walens *et al.*, 1954). In the infrared, attention has been devoted particularly to the region 3100–2750 cm^{-1} (Smith *et al.*, 1959). The following assignments were made (cm^{-1}): =CH— 3035, 3000–2960; asymmetric—CH_3 2950; asymmetric

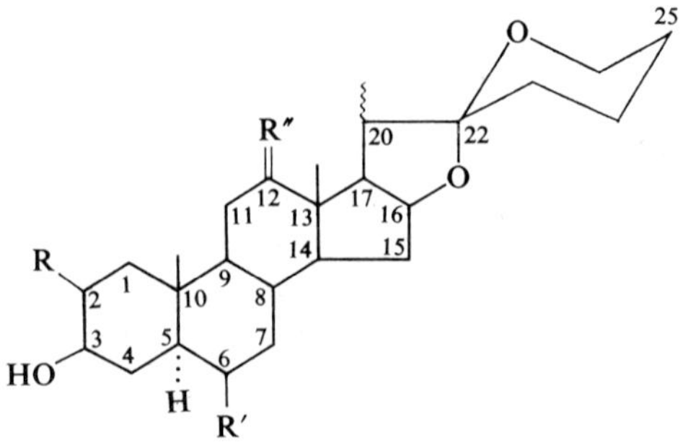

20α or β. A/B *trans* 25D 3β-OH

R=R′=H R″=2H	Tigogenin
R=R′=H R″=O	Hecogenin
R=α-OH R′=H R″=H	Gitogenin
R=α-OH R′=H R″=O²	Manogenin
R=H R′=α-OH R″=H_2	Chlorogenin
R=H R″=2H; 20α, 25L, 3β-OH	Yamogenin
R=H R″=2H; 25D, 3β-OH	Diosgenin

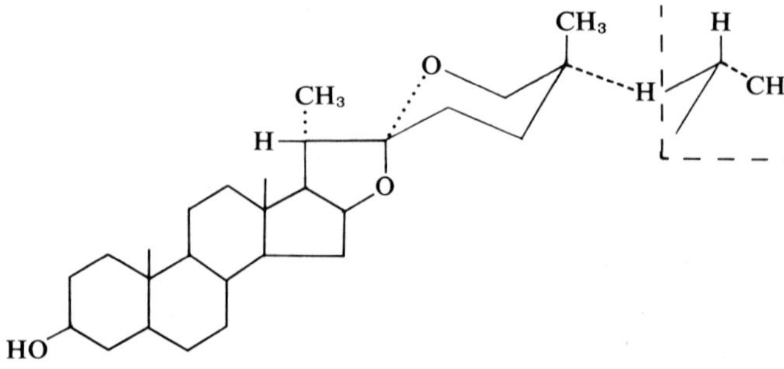

20α. A/B *cis* 25L 3β-OH=Sarsopogenin

20α. A/B *cis* 25D 3β-OH=Smilagenin

—CH_2 2930, 2920, 2910; symmetric —CH_3 2870; symmetric —CH_2— 2860, 2850, 2830. Nuclear magnetic resonance spectra have been applied with success to the problems of conformation at C20, C22 and C25 (Rosen *et al.*, 1969). The available information on hydroxy- and polyhydroxysapogenenins has been assembled by Yamaguchi (1970). The spectra of luvigenin and meteogenin which have an aromatic ring A show peaks at 263–265 and 270–271, $\epsilon_{max.}$ *ca.* 200–260.

12.4 Cardenolides

Cardenolides are characterized by an additional (lactone) ring at position 17. Hydroxyls can occur at position 3, 12, 14 and glycosides are often found at position 3. A very large number of natural compounds are covered by the generalization that the principal absorption peaks occur at 216–218 nm, $\epsilon_{max.}$ *ca.* 14 500 and 300 nm (sometimes partially resolved), $\epsilon_{max.}$ 30–60. The $\alpha\beta$-unsaturation of the lactone determines the absorption spectrum, the saturated steroid nucleus and the glycosidic moieties contributing negligibly to the absorption. If as in cyanogenin there is another carbonyl chromophore, the effect is largely additive $\lambda_{max.}$ 216 nm, $\epsilon_{max.}$ 15 500; 308 nm, 28·2; and for calotropagenin $\lambda_{max.}$ 218 nm, $\epsilon_{max.}$ 17 400; 309 nm, 24 (Reichstein, 1950*a*, *b*, 1957). An additional conjugated double bond as in anhydrogitoxenin causes new absorption to be seen near 270 nm, $\epsilon_{max.}$ *ca.* 16 600 with shoulders at 215 and 230 nm (Murphy, 1957).

The poor specificity of ultraviolet absorption in the cardenolides make the spectra in concentrated sulphuric acid of more than usual interest:

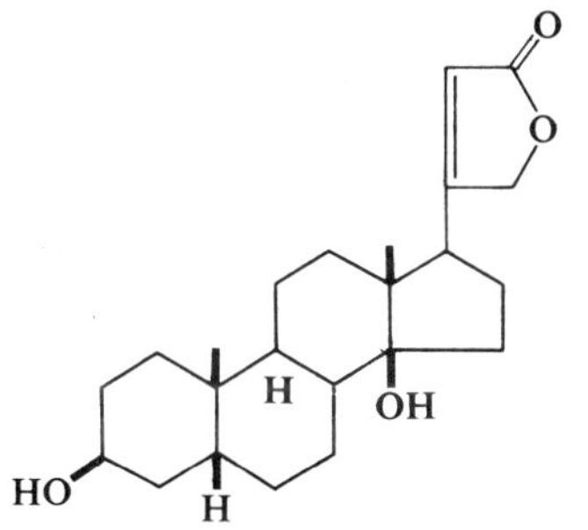

	Digitoxigenin	Gitoxigenin	Digoxigenin
$\lambda_{max.}$ nm	235	235	235
	350	320	320
	415	490	320
		530	

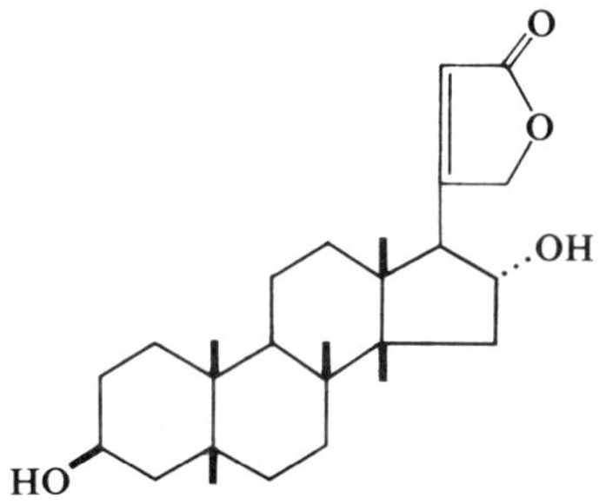

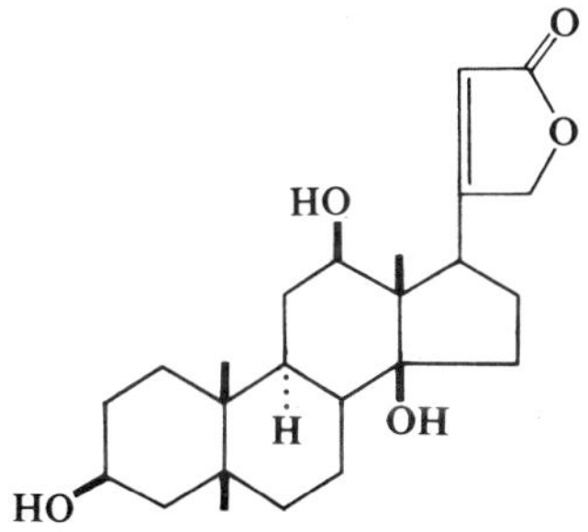

Brown *et al.* (1960) were able to distinguish quite sharply between three different types of cardenolides.

A great deal has been done on the infrared spectra of cardenolides. The principal vibrations are readily assigned. The C═O of the butanoid ring has ν cm^{-1} *ca.* 1739 and the C═C stretching vibration is at 1621,

1618 cm^{-1}. A good collection of data has been made by Yamaguchi (1970).

The bufadienolides have a somewhat different substituent at C–17, e.g. as in bufalin or gamabufotalin. The characteristic spectrum has $\lambda_{max.}$ 300 nm, $\epsilon_{max.}$ 5500–5600 with a minimum near 250 nm, $\epsilon_{min.}$ at 500–600 (ethanol). Several naturally occurring compounds also have a formyl group at C–10 (instead of CH_3) and there are often glucosidic residues at C–3 (e.g. bovoside is a 3-thevetosyl derivative). Scillirosodin also has a peak at *ca.* 225 nm, ϵ *ca.* 17 800, but there is a secondary chromophore. In fact for a whole range of bufadienolides the absorption peak at 300 nm is dominant and $\epsilon_{max.}$ is constant within quite narrow limits (Katz *et al.*, 1953–1958; Meyer *et al.*, 1960; Renz *et al.*, 1959; Lichti & Wartburg, 1960). In the infrared the α-pyrone is characterized by peaks at 1538 and 830 cm^{-1}.

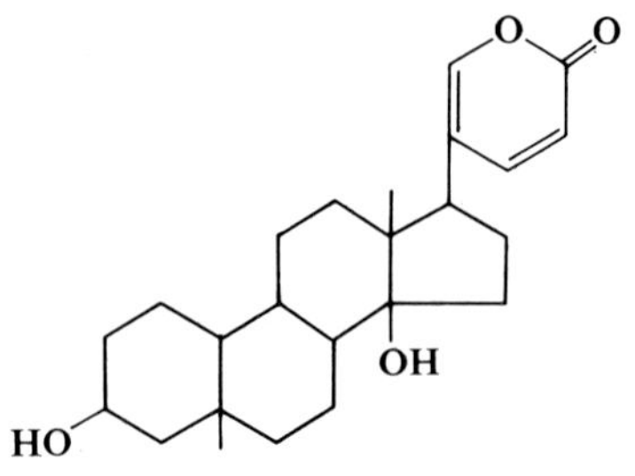

Bufalin

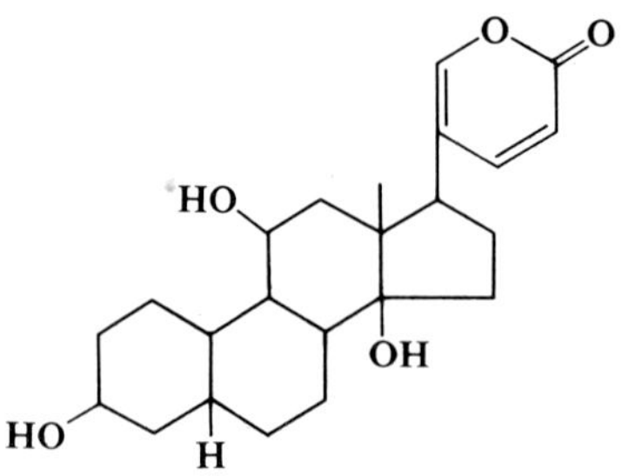

Gamabufatalin

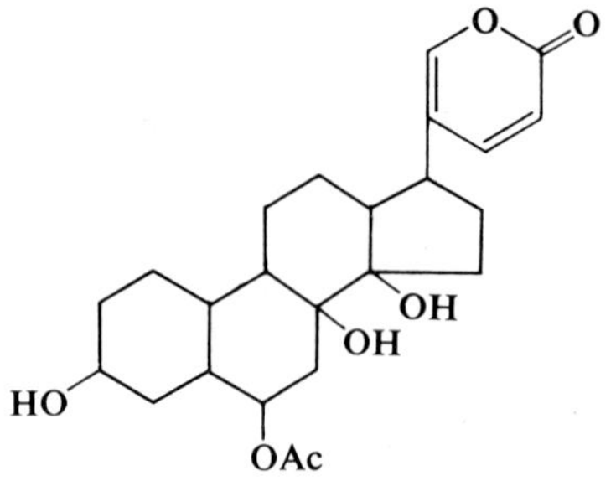

Scillirosidin

12.5 Mass Spectroscopy in Steroid Analysis

This is potentially of great value (Budzickiewicz & Djerassi, 1962; Bergstrom *et al.*, 1960, 1961; Beynon, 1960). The method was rather slow to gain full recognition in this area owing to technical difficulties (cf. Peterson, 1962 *a*) which have since been overcome. The possibilities of the method are well illustrated by Peterson (1962 *b*) in a study of 30 steroids, all substituted pregnanes or pregnenes. Most of the compounds were homogeneous by gas chromatography and thin layer chromatography. The amount used for mass spectrophotometry was *ca.* 1 mg. The inlet system was maintained at 225°C. In general it was easy to ascertain the molecular weight unambiguously but in some cases, e.g. Δ^5-3-acetates, a molecule of acetic acid was lost so readily that the parent peak was not recorded. 3-Acetates without the Δ^5-double bond also eliminate acetic acid but not to the complete exclusion of the present peak. Three examples from Peterson's paper are given opposite.

Hydroxyl groups at positions 3, 5 or 17 cause elimination of water and 3,5-dihydroxycompounds may lose two molecules of water with exclusion of the parent peak. The results on 30 related compounds provide a sound base on which analysts can build.

Three cholestanone mass spectra studied by Djerassi *et al.* (1962) illustrate the 'cracking' pattern. (See p. 372).

It may be concluded that the combination of chromatographic methods of separation and mass spectroscopy constitutes a very powerful means of studying steroids and steroid mixtures.

Djerassi *et al.* (1959) have determined the mass spectra of various steroids with one or two aromatic rings.

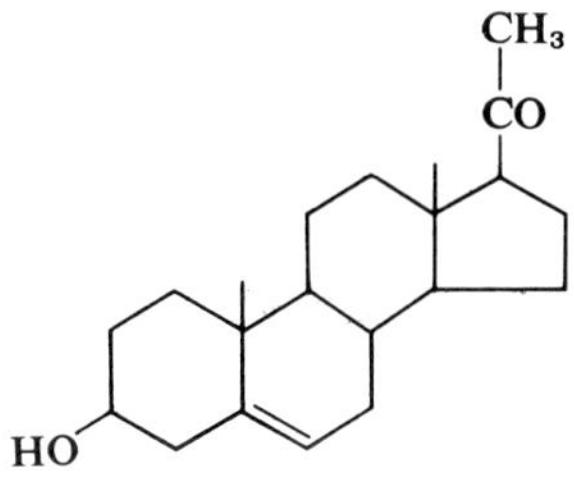

Pregnenolone

	Molecular weight	Relative intensity
M^+	316	100
	317	23
Minus CH_3	301	20·7
Minus H_2O	298	70
Minus CH_3 & H_2O	283	54·8
Minus $2H_2O$	280	0·7
Minus $—COCH_3$	273	5·5
Minus $COCH_3$ and H_2O	255	23·0
Minus ring D+H (CHO)+H	231	43·4
Minus ring D+H (H_2O)	213	27·7

Other values relative intensities >30
161, 159, 145, 133, 121, 120, 119, 107, 105
95, 93, 91, 85, 81, 77, 67, 57, 55, 54, 53
(After L. Peterson, 1962)

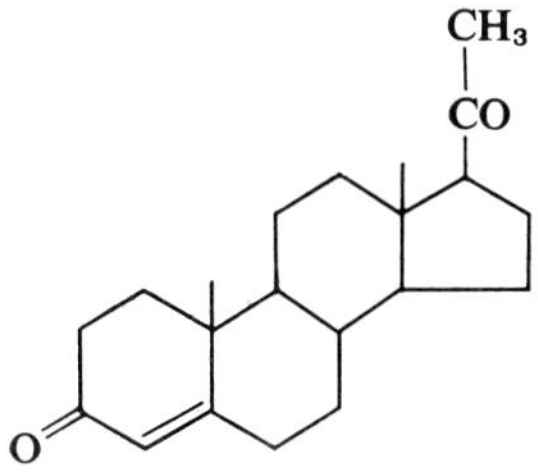

Progesterone

Molecular weight	Relative intensity
314	90·3
315	22
299	10·5
296	5·5
281	3·0
—	—
271	8·0
272	72·5
273	15
253	4·3
244	27
229	57·0
211	5·0

199, 133, 124(100), 107, 105
95, 93, 91, 79, 77, 67, 55, 53

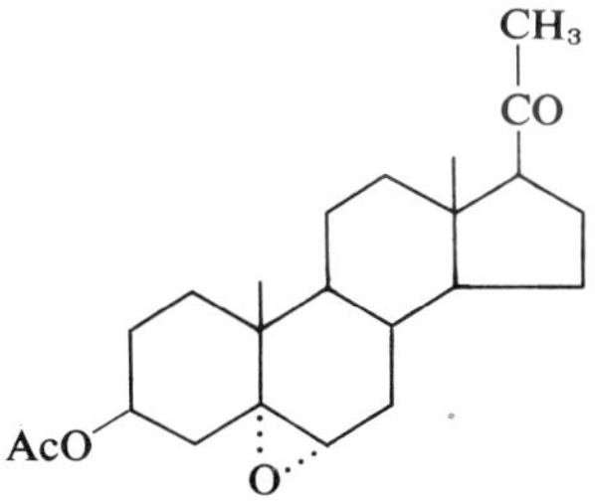

5α,6α-Epoxyallopregnane-3β-ol-20-one-3-acetate

Molecular weight	Relative intensity	
374	4·1	
375	0·9	
359	0·5	
356	2·6	
—	—	
331	0·9	
314	100	(374—CH_3COOH)
315	24·8	
299	29·5	(374—CH_3COOH—CH_3)
297	33·6	(374—CH_3 COOH—H_2O)
—	—	
282	16·5	(374—CH_3COOH—CH_3—H_2O)
271	9·4	(374—CH_3COOH—CO.CH_3)
253	—	(374—CH_3COOH—H_2O—$COCH_3$)

133, 120, 107, 105, 95
91, 81, 79, 77, 67, 55, 53

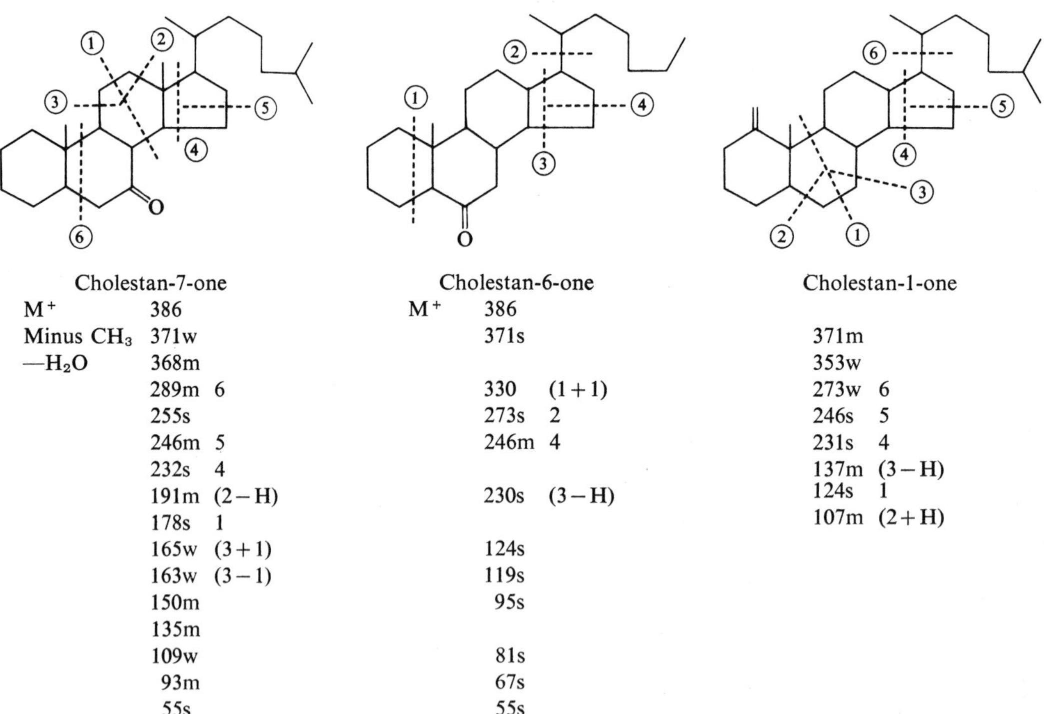

Cholestan-7-one

M^+	386	
Minus CH_3	371w	
$-H_2O$	368m	
	289m	6
	255s	
	246m	5
	232s	4
	191m	(2−H)
	178s	1
	165w	(3+1)
	163w	(3−1)
	150m	
	135m	
	109w	
	93m	
	55s	

Cholestan-6-one

M^+	386	
	371s	
	330	(1+1)
	273s	2
	246m	4
	230s	(3−H)
	124s	
	119s	
	95s	
	81s	
	67s	
	55s	

Cholestan-1-one

371m	
353w	
273w	6
246s	5
231s	4
137m	(3−H)
124s	1
107m	(2+H)

(After C. Djerassi *et al.*, 1962)

12.6 Oestrogens

The oestrogenic substances oestradiol, oestrone, oestriol, equilin and equilenin are all phenols. This permits them to be extracted from organic solvents into aqueous alkali and so to be separated from nearly all steroids. As free phenols they give rise in the ultraviolet to fluorescence which is intense enough to permit detection at very low concentrations (Duggan *et al.*, 1957). Both β-oestradiol and oestrone in 90% ethanol have an excitation peak at 285 nm with a fluorescence maximum at 325–330 nm, the intensity for oestradiol being about 8 times that shown by oestrone. Equilin excited at 290 nm gives a fluorescence maximum near 345 nm at about 1/10 the intensity of the β-oestradiol fluorescence. Equilenin with its tripartite absorption spectrum (250, 290, 340 nm) shows a fluorescence peak at 370 nm ten times as intense as the emission for β-oestradiol.

Oestrogens treated with strong sulphuric or phosphoric acid develop a new absorption spectrum usually with a maximum near 430–460 nm. Equilin and equilenin show peaks at 470 and 480 nm respectively. The products are powerfully fluorescent, $\lambda_{max.}$ *ca.* 500 nm (Bates, 1954). A careful investigation (Bauld, 1956, 1960) showed that oestrone or oestradiol dissolved in 75% sulphuric acid gave rise to a broad absorption band extending from 400 to 500 nm with its peak near 455 nm. The fluorescence extended from 450 to 600 nm with a maximum at 479–484 nm. Because the absorption and fluorescence spectra overlap it is necessary to excite the fluorescence by means of radiation on the short-wave side outside the fluorescence spectrum and for this the mercury line 436 nm is very suitable. Because of the risk of error due to inner filter effects, in a new series of experiments the relationship between concentration and fluorescence intensity needs to be verified. Nakao & Aizawa (1956) have described a chromatographic method for separating oestradiol, oestrone and oestriol before carrying out fluorimetric determinations in 75% sulphuric acid (cf. Udenfriend, 1962).

The Kober reaction has undergone many modifications. It consists of heating the oestrogen in the presence of a phenol and a strong aqueous sulphuric acid solution. The phenol diminishes the fluorescence and increases the intensity of an unknown chromophore, giving a pink colour $\lambda_{max.}$ 515–520 nm. To obtain good results it is necessary to separate oestrone, oestradiol and oestriol usually by one form or another of chromatography. The preferred phenol is hydroquinone and the optimum concentrations of acid are 60% v/v for oestradiol, 66% for oestrone and 76% for oestriol. When the reaction has occurred the mixture is in each case diluted to about 55% sulphuric acid, and the pink solution is measured at 515–520 nm. The value of $\epsilon_{max.}$ is between 25 000 and 50 000 depending on the oestrogen and the details of the procedure. The very high value for $\epsilon_{max.}$ greatly increases the sensitivity, but even so there is a considerable risk of significant irrelevant absorption appearing at 520 nm in fractions from urine. Calibration with pure oestrogens and strictly standardized experimental conditions are advisable and in some cases correction for irrelevant absorption will increase the reproducibility and reliability of the results.

Table 12.5 Absorption spectra of oestrogens

Substance	$\lambda_{max.}$ nm	$\epsilon_{max.}$	$\lambda_{max.}$ nm
71-α-Oestradiol	280	2 000	
17-β-Oestradiol	780	2 000	300 in alkali
Oestrone	280	2 300	300 in alkali
Oestrone-3-acetate	260	1 040	
Equilin	280	2 000	
Estriol	280	2 300	300 in alkali
6-Dehydro-oestradiol	262	10 000	
	302	2 950	
6-Dehydro-oestrone	220	30 900	
	262	8 910	
	304	2 750	
β-Oestradiol-6-one	256	8 000	
	326	3 000	
Equilenin	230	—	
	270	7 080	
	282	7 420	
	292	5 500	
	328	3 800	
	340	4 790	

Oakey *et al.* (1967) determined oestrogens in pregnancy urine by the Kober method and gave full details of the procedure. They measured absorbances at 472, 514 and 556 nm and arrived at a corrected $E_{514\,nm}$ from the equation $E_{514}^{corr.} = 2E_{514} - (E_{472} + E_{556})$. These authors also measured the normal range (calculated as oestriol) week by week for the last 12 weeks.

(See also Katzman & Elliot, 1963; Brown, 1955; Bauld, 1966; Bauld & Greenway, 1957; Nocke, 1961).

12.7 Adrenal Cortical Steroids

More than 40 steroids have been isolated from adrenal glands and their structures determined. Replacement therapy in adrenalectomized animals and in patients suffering from Addison's disease opened up vast regions of research and numerous fruitful bio-assay procedures were devised. Several groups in Europe and the U.S.A. succeeded in purifying and characterizing many steroid components.

It will be possible in the present work to take no more than a narrow look at adrenal cortical steroids and the main emphasis must be on the spectroscopic aspects.

Of the 40 or so steroids, seven are of special interest in that bio-assays show that they are hormones capable of influencing survival, and the metabolism of carbohydrate, of protein or of minerals (sodium

Figure 12.6 Effect of carbonyl groups in adrenal cortical steroids on their infrared absorption spectra. Ester CO, 1758 cm^{-1}; $C_{20}CO$, 1730 cm^{-1}; $C_{11}CO$, 1712 cm^{-1}; C_3CO ($\alpha\beta$-unsaturated), 1678 cm^{-1} (from Jones, 1958)

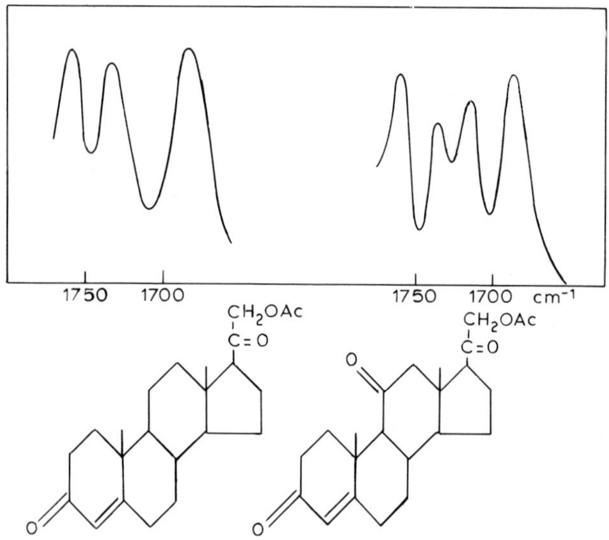

and potassium ions). It is convenient to distinguish between cortisone and hydrocortisone, which are powerful as glucocorticords, and aldosterone and desoxycorticosterone, specially potent as mineralocorticoids. The ultraviolet absorption is dominated by the intense band near 240 nm, $\epsilon_{max.}$ order 20 000 which is characteristic of $\Delta^{4,5}$-3-ones. All the 'active' adrenal steroids have this $\alpha\beta$-unsaturated ketonic absorption which tends to mask that of carbonyl groups at carbons 11 or 20.

Infrared absorption spectra for carbonyl compounds have been of great service in this field. In steroids an unconjugated carbonyl group at carbon 3 gives rise to a peak at 1716 cm^{-1}, at carbon 17 at 1745 cm^{-1} and if both are present peaks are seen at 1719 and 1745 cm^{-1}, the intensities being additive. A carbonyl at position 20 gives rise to a band near 1710 cm^{-1}. 11-Keto and 12-keto groups are rather similar with bands at 1710–1712 cm^{-1}. $\alpha\beta$-Unsaturation (Δ^4-3-ones) show 1677 cm^{-1}. The acetate band is around 1740 cm^{-1}.

The constitutions shown (right) illustrate how different carbonyl groups in adrenal cortical steroids register independently. This is of obvious diagnostic value.

	Carbonyls at positions	Ethylenic bond at	Hydroxyls at
Desoxycorticosterone	3, 20	4(5)	21
17-α-Hydroxydesoxy-corticosterone	3, 20	4(5)	17, 21
Corticosterone	3, 20	4	11, 21
11-Dehydrocortico-sterone	3, 11, 20	4	21
Cortisol	3, 20	4	11, 17, 21
Cortisone	3, 11, 20	4	17, 21
Aldosterone	3, 20 13(CHO*)	4	11, 21

*can form hemiacetal

12.8 Determination of Urinary Steroids

Allen (1950), recognizing the usefulness of some colour reactions for the determination of urinary steroids, made a careful study of methods of eliminating absorption due to 'irrelevant' impurities. He was particularly concerned with Kober's (1931) phenolsulphonic acid test for urinary oestrogens and Zimmermann's *m*-dinitrobenzene test for urinary 17-ketosteroids.

Allen *et al.* (1950) developed a colour test for dehydroisoandrosterone and related steroids for the diagnosis of adrenocortical tumours. The urine is hydrolysed and extracted with butyl alcohol (or ether) and a portion (usually equivalent to $\ngtr 1/200$ of a 24 hour specimen of urine) of the dry residue is treated with 2 ml of a reagent consisting of concentrated H_2SO_4 (4 vols) and 90% ethyl alcohol (1 vol) followed by heating at 55°C for 12 minutes. At that point 95% ethyl alcohol (3 ml) is added and a blue colour results ($\lambda_{max.}$ 600 nm). A purple colour tended to be produced in cases of proved adreno-

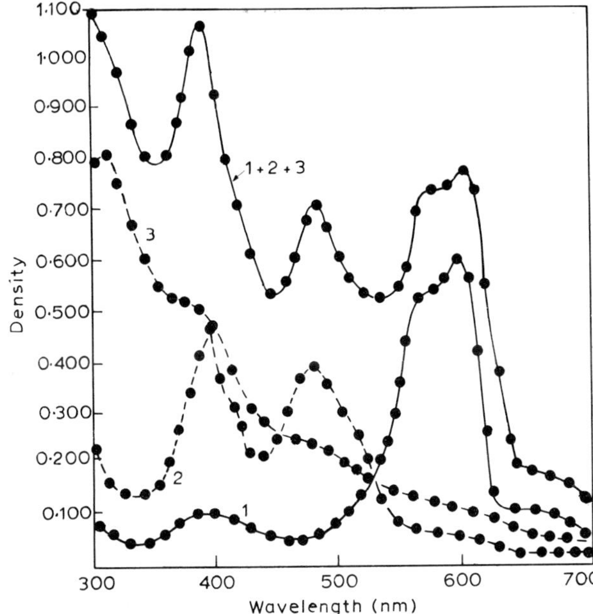

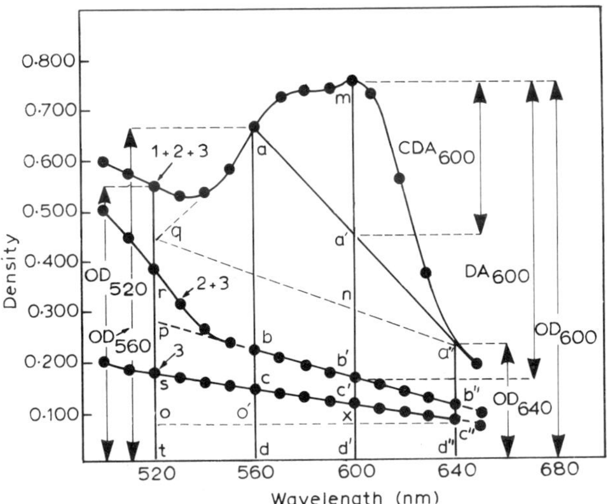

Figure 12.7 (*a*) Colour test on : (1) dehydroisoandrosterone, 0·05 mg; (2) pregnanediol, 0·5 mg; (3) urinary extract containing no dehydroisoandrosterone; (1)+(2)+(3) composite curve, the sum of (1), (2) and (3)
(*b*) Enlarged portion of (*a*); solid line is the summation curve. OD_{600}=gross optical density at 600 nm (0·75). (For pure dehydroisoandrosterone, OD_{600}=0·592, OD_{640}=0·125, OD_{650}=0·445 from figure)

$OD_{640}=0{\cdot}225$ $\quad$ $OD_{560}=0{\cdot}662$

$0{\cdot}592-\dfrac{0{\cdot}445+0{\cdot}125}{2}$ $\quad$ $0{\cdot}750-\dfrac{0{\cdot}662+0{\cdot}225}{2}$

$0{\cdot}592-0{\cdot}285=0{\cdot}307$ $\quad$ $0{\cdot}750-0{\cdot}443=0{\cdot}307$

OD corr.=0·307+0·285 (from Allen, 1950)

cortical tumours and a brown colour was characteristic of female pseudohermaphrodism.

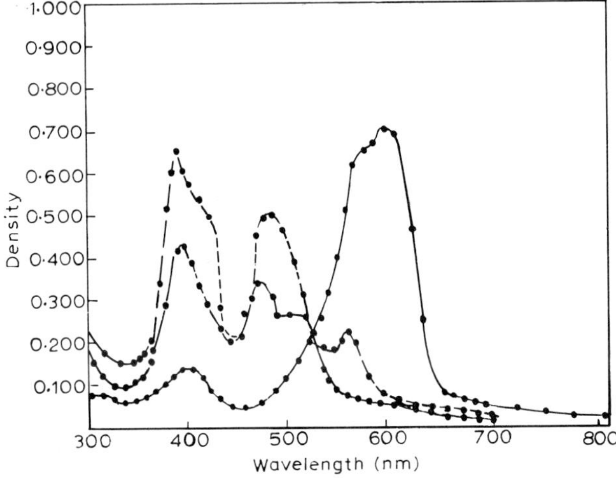

Figure 12.8 Colour test results for: ———, dehydroisoandrosterone (0·05 mg); — — —, androsterone (0·25 mg); - - - - -, pregnanediol (0·50 mg). 5 ml solution, 1 cm cell (from Allen *et al.*, 1950)

The above procedure was applied to pure substances with the results shown in Figs. 12.7, 12.8 and 12.9 and Table 12.6. Saturated substances like pregnanediol, pregnanolone, androstane diol and androsterone produced negligible colour.

Figure 12.9 Colour test results for: ———, desoxycorticosterone (0·05) mg; — — —, oestrone (0·2 mg); - - - - -, testosterone (0·025 mg). 5 ml solution, 1 cm cell (from Allen *et al.*, 1950)

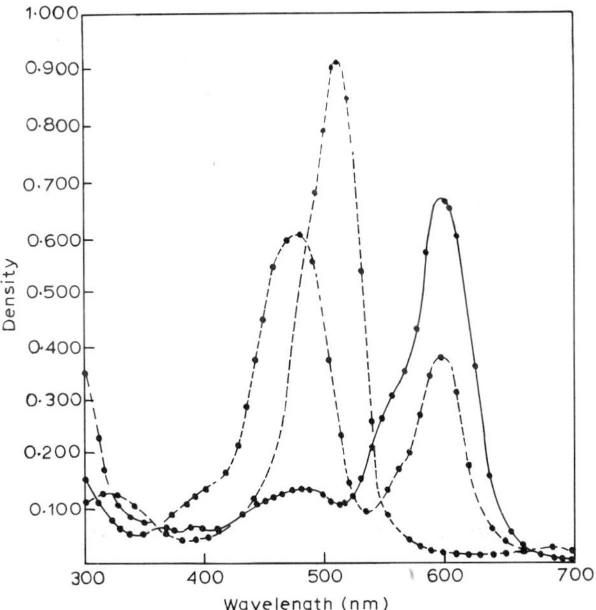

Table 12.6 Intensities of absorption produced by various steroids in Allen's modification of the sulphuric acid colour test (Calculated for 100 μg each, 5 ml solution, 1 cm cell)

Compound	E_{480mm}	E_{600mm}
αβ-Unsaturated ketones (3 keto-$\Delta^{4,5}$)		
Testosterone	2·42	1·50
Desoxycorticosterone	0·27	1·336
Androstenedione	0·652	0·46
Progesterone	0·046	0·033
17-Hydroxy-11-dehydrocorticosterone	0·110	0·015
3-Hydroxy compounds		
Dehydroisoandrosterone	0·169	1·230
Androstenediol	0·512	0·800
Pregnenolone	0·188	0·068

Allen (1950) found that a composite curve obtained by the summation of curves for dehydroisoandrosterone, pregnanediol (in large amounts) and a control urine extract resulted in virtually linear absorption by constituents other than dehydroisoandrosterone in the region 560–640 nm. Applying the Morton–Stubbs (1946) procedure it was easy to correct the absorption.

In a general form

$$E_{adj.\lambda_{max.}} = E_{obs.\lambda_{max.}} - \frac{E_a + E_b}{2}$$

where E_a and E_b are observed intensities of absorption at a and b nm, a and b being equidistant from $\lambda_{max.}$ (oestriol $\lambda_{max.}$ 450 nm, dehydroisoandrosterone 600 nm). $E_{adj.}$ (adjusted) varies as the concentration but represents only part of the absorption. From it, however, $E_{corr.}$ can be calculated easily (see Fig. 12.7 *b*).

12.8.1 Pregnanediol in Urine

The following method (Klopper *et al.*, 1955) is highly serviceable (private report by W. H. Taylor). Urine (1/20th of a 24 hour specimen) is placed in a flask (500 ml, RB) and diluted to 150 ml with distilled water. Toluene (50 ml) is added and the mixture is refluxed (glass beads). Concentrated HCl is then added (15 ml) down the condenser and boiling is continued for 10 min. After cooling quickly the contents are transferred to a separating funnel (500 ml). The aqueous layer is removed and re-extracted with toluene (50 ml) with vigorous shaking for several minutes. The combined toluene extracts are thoroughly shaken with 25 ml alkaline sodium chloride solution (25% NaCl in **N** NaOH). The aqueous layer and any curdy precipitate are discarded. The toluene layer is washed several times with alkaline permanganate (fresh saturated solution of $KMnO_4$ in **N** NaOH, i.e. *ca.* 4%) and after drying (Na_2SO_4) the toluene solution is filtered and the sulphate washed with fresh toluene. The volume is then reduced to about 10 ml.

This is then chromatographed in alumina (deactivated to Brockman 2–3). 'Impurities' are removed by 0·8% ethanol in benzene (25 ml) and the pregnanediol is eluted by 3% ethanol in benzene (25 ml). The benzene is removed in a rotary evaporator. While still warm add benzene (2 ml) and acetyl chloride (2 ml) and leave for 1 hour at room temperature to complete the acetylation of the pregnanediol.

The acetate is washed into a separating funnel with light petroleum and after washing (water 30 ml 8% $NaHCO_3$ 25 ml, water 2×25 ml) the petrol solution is rechromatographed on alumina. When the petrol has run through pregnanediol, acetate is eluted by benzene (25 ml). The solvent is evaporated and dried over calcium chloride for at least 1 hour.

To the residue in a stoppered tube anhydrous Na_2SO_3 is added (10 mg) and then concentrated H_2SO_4 (10 ml). The stopper is inserted and the solution is well shaken and left at 25° for 17 hours overnight. The intensity of absorption is measured at 430 nm.

See also §26.5 for pregnanetriol determination.

12.9 Limonoids

Since the basic structure of limonin was established (Arigoni *et al.*, 1960) immense progress has been made in the study of limonoids and quassinoids, both from the standpoints of constitution and biogenesis (Conolly *et al.*, 1970) and spectroscopic methods have contributed largely to the advances. Ultraviolet absorption data have played a useful part (cf. Vitaglioni & Comin, 1972). Some examples are given below with representative structures and chromophores. Attention is drawn to the diosphenol chromophore and to the structure of cedrelone obtained from the heartwood of *Cedrela toona* (Gopinath *et al.*, 1961; Grant *et al.*, 1963; Hodges *et al.*, 1963; Kubota *et al.*, 1960; Bevan *et al.*, 1963; Barton *et al.*, 1961).

Nuclear magnetic resonance spectra have been much used in structural studies (Dreyer, 1965; Powell, 1966; Ohochuku & Taylor, 1969). Mass spectra have played a large part (Fourrey *et al.*, 1968) and optical rotatory dispersion and circular dichroism have also proved serviceable (Dreyer, 1968).

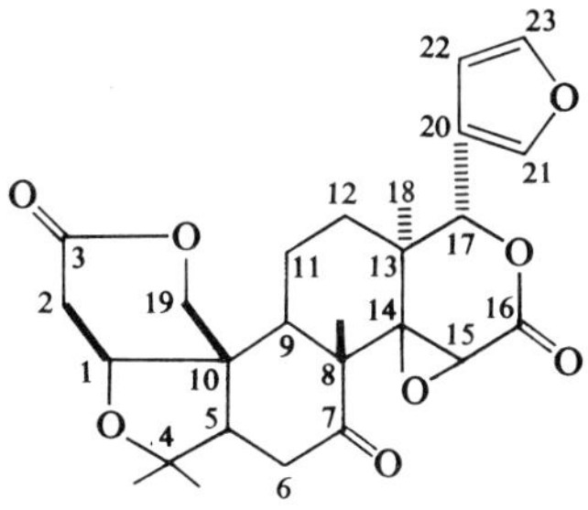

Limonin

$\lambda_{max.}$ nm	207	290
$\epsilon_{max.} \times 10^{-3}$	6·3	0·028

cm^{-1} 1760 (δ-lactone), 1710 (C=O), 1605, 1542, 1514 876 (furan)

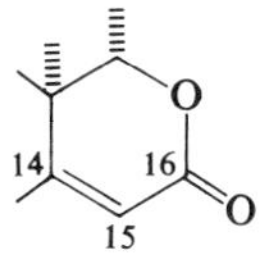

Deoxyliminin

$\lambda_{max.}$ nm	214
$\epsilon_{max.} \times 10^{-3}$	17·0
cm^{-1}	1747, 1714, 1699

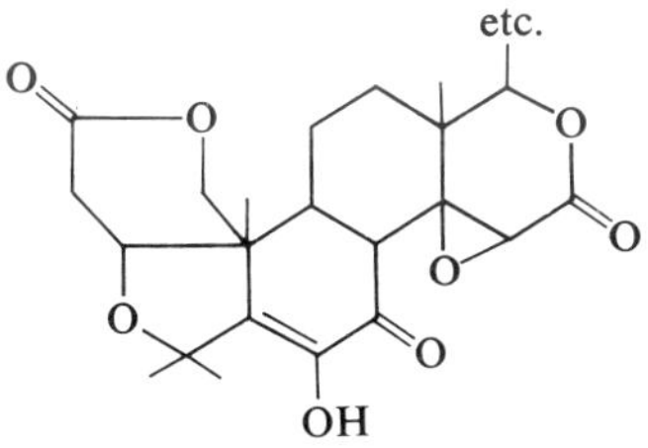

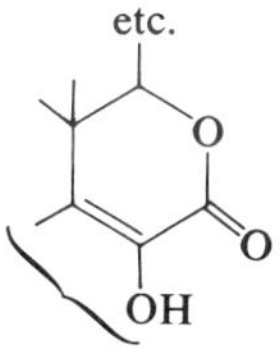

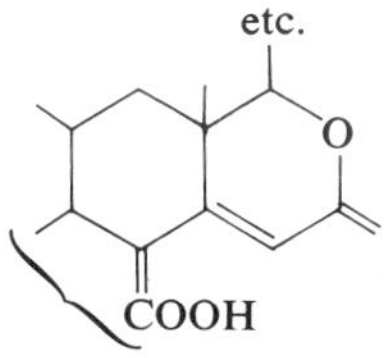

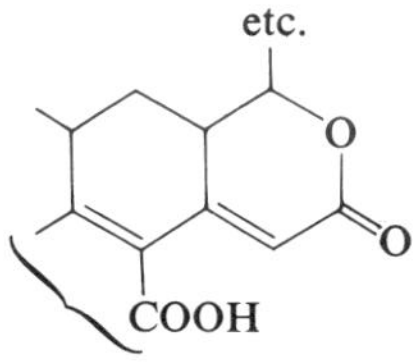

	Diosphenols		α-Oxo-lactone		*Cis*oid diene	*Trans*oid diene	
$\lambda_{max.}$ nm	278	338	256	290	255	230	284
$\epsilon_{max.} \times 10^{-3}$	11·0	6·3	5·3	4·6	7·8	6·3	16·4
	(EtOH)	(EtOH+ NaOH)	(EtOH)	(EtOH+ NaOH)			
	(acetate $\lambda_{max.}$ 245 nm)		(acetate $\lambda_{max.}$ 218 nm)				

n.m.r. (J)

	a	b	c	d	
CH_3—4 d	8·94	8·86	8·88	8·87	J=6·5
CH_3—13	8·85*	8·82	8·12s	8·10s	J=7·0*
CH_3—8 s	8·45	8·88*	8·79	8·79	
CH_3—10 s	8·45	8·34*	8·50	8·42	
CH_3O—2 s	6·36	6·42	6·41	6·29	
H—3 d	4·54	4·59	4·68	4·59	J=2·5
H—7 m	5·64	5·66	5·74	5·64	

s=singlet
d=doublet
m=multiplet

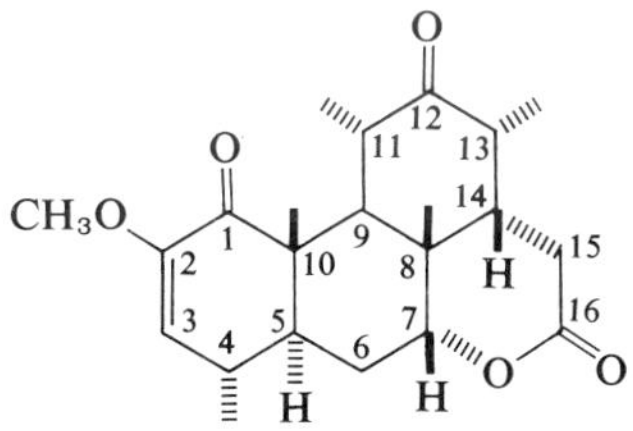

Paraine (*a*) —CrO_3→ Isoparaine (*b*)

	Paraine (*a*)	Isoparaine (*b*)
$\lambda_{max.}$ nm	270	261
$\epsilon_{max.} \times 10^{-3}$	4·68	6·03
	(EtOH)	(EtOH)

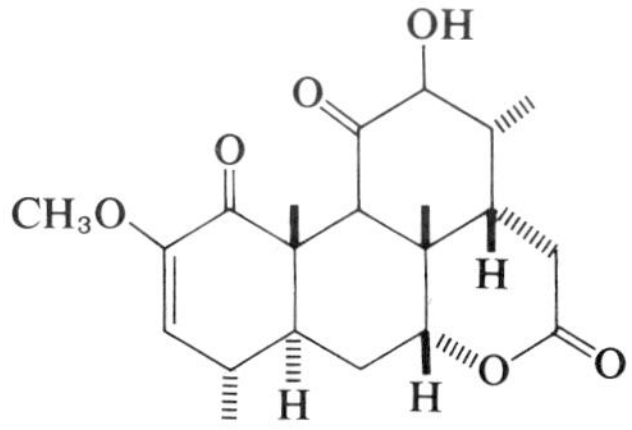

	Infrared peaks cm^{-1} (Nujol)				
	a	b	c	d	
H-bonded OH	3400	3400	3300		broad
δ-lactone	~1735	1730	1720	1740	
ring-ketone	1720				
αβ-enone	1680	1690	1690 1670	1690	
	1640	1630	1655 1630	1640	

Data from: Vitagliono & Comin (1972). Casinovi *et al.* (1969)

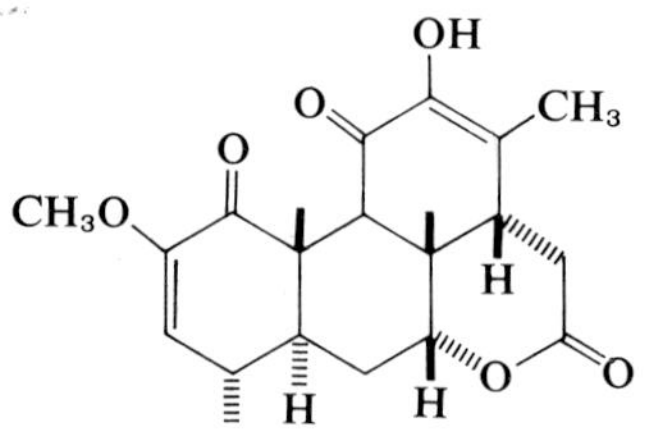

12-Norquassin (*c*)

$\lambda_{max.}$ nm	268	265	317
$\epsilon_{max.} \times 10^{-3}$	7·08	6·46	4·47
	(EtOH)	(EtOH+NaOH)	

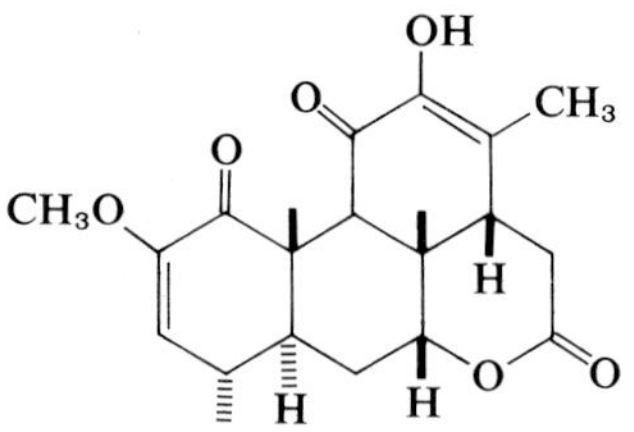

Quassin (*d*)
255
13·50
(EtOH)

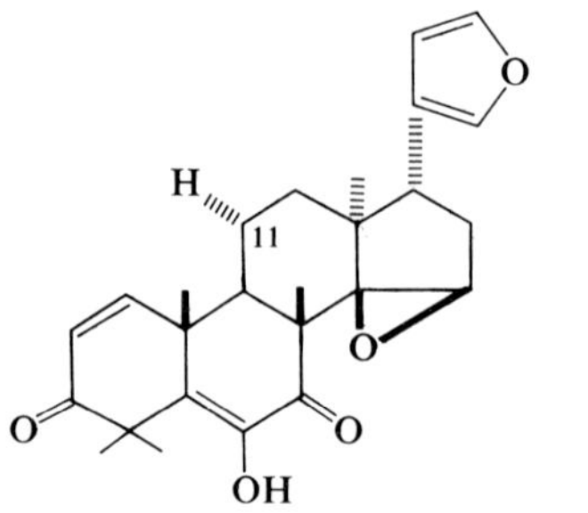

— OAc at position II
= Anthothecol

Cedrelone

$\lambda_{max.}$ nm	217	279	327
$\epsilon_{max.} \times 10^{-3}$	11·8	9·1	5·53
cm^{-1}	3425 (OH), 1678, 1685 (αβ unsatd. C=O) 1505, 878 (furan ring)		

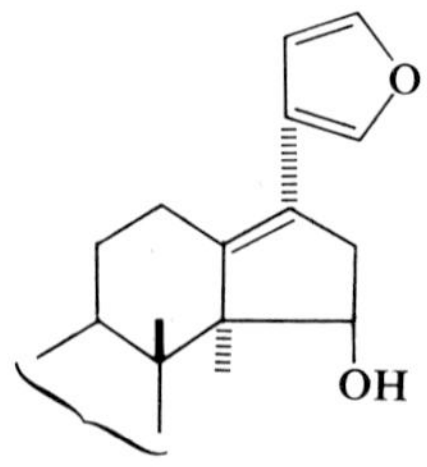

Isocedrolone
242
3-Vinyfuran
system

Acetate (anthothecol)

$\lambda_{max.}$ nm	219	281	224	326
$\epsilon_{max.} \times 10^{-3}$	130	11·6	8·32	4·6
	(EtOH)		(EtOH+NaOH)	
cm^{-1}	3400, 1740, 1670, 1655, 1620, 1504, 1235, 877			

12.10 Ultraviolet Absorption of Bile Acid Degradation Products: Colour Reactions of Bile Acids

Care must be taken in extending any procedure outside the properly tested and calibrated conditions.

Strong acids, e.g. sulphuric acid (and less often phosphoric and hydrochloric acids), form ultraviolet-absorbing derivatives from bile acids. The spectra vary with the bile acid, the concentration of sulphuric acid, the time and the temperature of heating. Steroids, many of which produce ultraviolet-absorbing products under similar conditions, must first be removed. Cholic acid and its conjugated acid show $\lambda_{max.}$ 389 nm (an intense band) while dihydroxy acids show a weaker band with its peak at 310–315 nm. Monohydroxycholanic acids display an even weaker absorption band ($\lambda_{max.}$ 315 nm).

If however 65% sulphuric acid is used very different spectra appear (Kier, 1952):

	$\lambda_{max.}$ nm
Conjugated and free cholic acid	320
Free dihydroxy acids (60° 10–30 min.)	380–385
(Mosbach *et al.*, 1954)	

It must be said that both the $\epsilon_{max.}$ values at 380–385 nm and the time taken to attain them differ for the various dihydroxy acids, and the free and conjugated dihydroxy acids behave differently. Conjugated deoxycholic acids (60° for 1 hour) result in $\lambda_{max.}$ 389 nm, while the conjugated chenodeoxycholic acids show $\lambda_{max.}$ 305 nm weakly. Nevertheless, the procedure can give trustworthy results.

Methods of this kind have been used successfully (Matschiner *et al.*, 1957; Sjövall, 1959, 1964; Eriksson & Sjövall, 1955) and there is scope for ingenuity, but every problem needs to be treated as an investigation.

A reagent consisting of concentrated sulphuric acid and acetic acid (9:1) was used by Minebeck (1938) and later by Singer & Fitschen (1961) and Wysocki *et al.* (1955). It has the advantage of giving very similar spectra for free and conjugated bile acids, and unlike the 65% sulphuric acid reagent it permits determination of total di- and trihydroxycholanic acids. Unfortunately the molecular extinction coefficients for different free and conjugated dihydroxy acids are not identical, a fact which imposes rather severe limitations on the method.

Fluorescence spectra of bile acids in strong sulphuric acid have been described and a near approach to a reliable method has been made by Levin *et al.* (1961). There are, however, again many complications especially if impurities are present.

Various empirical procedures result in reproducible appearance of coloured substances with well-defined absorption bands:

(a) purified conjugated or free cholic acid (a known amount evaporated to dryness) is treated with 16 **N** sulphuric acid to which a 0·5% aqueous solution of freshly distilled furfuraldehyde has been added (6 ml acid, 2 ml aldehyde solution). After heating at 65° for 13 minutes, acetic acid (5 ml) is added and the mixture well stirred before measuring the intensity of absorption at 620 nm of the blue solution,

(b) chenodeoxycholic acid (free) is treated with fresh ethyl acetate–concentrated sulphuric acid 15:1 (3 ml) followed by acetic anhydride (2 ml). The well-stirred mixture is left for 12–15 minutes and the intensity of absorption is measured at 615 nm (this is a Liebermann–Burchard type reaction and cholesterol must not be a contaminant of the bile acid).

Provided that the preliminary separations (whether by gas–liquid chromatography or adsorption chromatography (alumina) or thin layer chromatography) have been successful the spectroscopic approach is promising but it affords no safe escape from the need to effect separation.

12.11 The Liebermann-Burchard Reaction

This well-tried analytical procedure (Liebermann, 1885; Burchard, 1890; Grigaut, 1910) has gained new power by the application of modern spectroscopic methods to it.

A suitable sterol is dissolved in chloroform or glacial acetic acid and treated with acetic anhydride containing concentrated sulphuric acid (5% v/v). A blue-green colour appears, usually slowly, and slowly disappears. Absorption peaks occur at 405 and 625 nm, but develop at different rates. The 405 nm absorption is much more sensitive to light than the 625 nm absorption.

The reaction was at first treated as specific for sterols having a hydroxyl at position 3 and a double bond at position 5 (although its empirical value was recognized before the sterol ring system had been established). The reaction has proved very useful for determining cholesterol and cholesteryl esters (Abell *et al.*, 1952; Anderson & Keys, 1956). Boyd (1962) critically reviewed the use of the 625 nm absorption in the determination of cholesterol for clinical purposes.

Truter & Bush (1962) surveyed the problem of determining the steroid constituents of wool wax. By measuring absorption at four carefully chosen wavelengths twice, once at 20 minutes and once at 40 minutes after mixing the Liebermann–Burchard solution under strictly standardized conditions, they were able to determine four major components with considerable accuracy. With the aid of chromatography, the procedure was extended to a 7-component mixture with reasonable success.

Moore & Baumann (1952) using a modified Liebermann–Burchard procedure found that Δ^7-sterols reacted quickly and produced maximum colour in the test after 1·5–2 minutes, at which time practically no colour had arisen from Δ^5-sterols. The distinction between fast-reacting and slow-reacting sterols has been put to effective use by Glover (1962). 7-Dehydrocholesterol is determined by (corrected) ultraviolet absorption and the result, deducted from the total estimated fast-reacting sterol, gives the lathosterol content. See also p. 238.

References

Abell, L. L. *et al.* (1952). *J. Biol. Chem.*, **195**, 357.

Allen, W. M. (1950). *J. Clin. Endocrin.*, **10**, 71.

Allen, W. M., Hayward, S. J. & Pinto, A. (1950). *J. Clin. Endocrin.*, **10**, 54.

Anderson, J. T. & Keys, A. (1956). *Clin. Chem.*, **2**, 145.

Arigoni, D., Barton, D. H. R., Corey, E. J., Jeger, O., Caglioti, L., Sukh Dev, Ferrini, P. G., Glazier, E. R., Melera, A., Pradhan, S. K., Schaffner, K., Sternhell, S., Templeton, J. F. & Tobinga, S. (1960). *Experientia*, **16**, 41.

Barton, D. H. R., Pradhan, S. K., Sternhell, S. & Templeton, J. F. (1961). *J. Chem. Soc.*, **255**.

Bates, R. W. (1954). *Progress in Hormone Research*, **9**, 95.

Bauld, W. S. (1956). *Biochem. J.*, **63**, 488.

Bauld, W. S. & Greenway, R. M. (1957). In *Methods of Biochemical Analysis*, **9**, 337. Ed. D. Glick. Interscience, New York.

Bauld, W. S., Givner, M. L., Engel, L. L. & Goldzieher, J. W. (1960). *Canad. J. Biochem. & Physiol.*, **38**, 213.

Bergstrom, S. (1961). *Svensk. Kem. Tidskr.*, **73**, 566.

Bergstrom, S., Ryhage, R. & Stenhagen, E. (1960). *J. Lipid Res.*, **1**, 361.

Bevan, C. W. L. *et al.* (1963). *J. Chem. Soc.*, 983.

Beynon, O. H. (1960). In *Mass Spectrometry and its application to Organic Chemistry.* Elsevier, Amsterdam.

Bladon, P., Henbest, H. B. & Wood, G. W. (1952). *J. Chem. Soc.*, 2537.

Booker, H., Evans, L. K. & Gillam, A. E. (1940). *J. Chem. Soc.*, 1453.

Bourquin, J. P. *et al.* (1959). *Helv. Chim. Acta.*, **42**, 259.

Boyd, G. S. (1962). In *Determination of Sterols.* Soc. Anal. Chem., London.

Braude, E. A. (1954). *Chem. & Ind.*, 1557.

Brown, B. T. *et al.* (1960). *J. Amer. Pharm. Assoc.*, **49**, 777.

Brown, J. B. (1955). *Biochem. J.*, **60**, 185.

Budjikiewicz, H. & Djerassi, C. (1962). *J. Amer. Chem. Soc.*, **84**, 1430.

Burchard, H. (1890). *Chem. Centr.*, **61**, 25.

Casinovi, C. G., Grandolini, G., Marini-Bettolo, C. B. & Bellavita, V. (1969). *Ann. Chem.*, **59**, 230.

Cole, A. R. H. (1956). *Fortsch. Chem. Org. Naturstoffe*, **13**, 27.

Connolly, J. F., Overton, K. H. & Polonsky, J. (1970). In *Progress in Phytochemistry.* Vol. 2. Ed. L. Reinhold and Y. Liwschitz. Interscience Publishers, London, New York.

Djerassi, C. *et al.* (1959). *J. Amer. Chem. Soc.*, **84**, 4544.

Dobriner, K., Katzenellenbogen, E. & Jones, R. N. (1953). In *Infrared Absorption Spectra of Steroids—an Atlas.* Interscience Publishers, London, New York.

Dorfman, L. (1953). *Chem. Rev.*, **53**, 47.

Dreyer, D. L. (1965). *Tetrahedron Lett.*, **21**, 75.

Dreyer, D. L. (1968). *Tetrahedron Lett.*, **24**, 3273.

Duggan, D. E., Bowman, R. L., Brodie, B. B. & Udenfriend, S. (1957). *Arch. Biochem. Biophys.*, **68**, 1.

Eriksson, S. & Sjövall, J. (1955). *Arkiv. Kemi.*, **8**, 303.

Fieser, L. F. & Fieser, M. (1959). In *Steroids.* Reinhold, New York.

Forbes, W. F., Shilton, R. & Balsubramanian, A. (1964). *J. Org. Chem.*, **29**, 3527.

Fourrey, J. L., Das, B. C. & Polonsky, J. (1968). *Org. Mass Spectrometry*, 819.

French, C. S. (1957). *Proc. Instr. Soc. Am.*, **8**, 83.

French, C. S., Church, A. B. & Epply, R. W. (1954). *Carnegie Institute, Washington, Year Book*, **53**, 182.

Gillam, A. E. & Stern, E. S. (1970). In *Electronic Absorption Spectroscopy in Organic Chemistry*. 3rd Edition. Ed. E. S. Stern and E. J. Timmons. Arnold, London.

Glover, J. (1962). In *Determination of Sterols*. Soc. Anal. Chem., London.

Gopinath, K. W., Gofindachari, T. R., Parthasarthy, P. C., Viswananthan, K., Arigoni, D. & Wildman, W. C. (1961). *Proc. Chem. Soc.*, 446.

Görög, S. & Szepesi, G. (1972). *Analyt. Chem.*, **44**, 1079; see also (1972). *Analyst*, **97**, 519.

Görög, S., Tuba, Z. & Egyed, I. (1969). *Analyst*, **94**, 1044.

Görög, S. & Tuba, Z. (1972). *Analyst*, **97**, 519, 523.

Grant, I. J., Hamilton, J. A., Hamor, T. A., Robertson, J. M. & Sim, G. A. (1963). *J. Chem. Soc.*, 2506.

Grigaut, A. (1910). *C. R. Soc. Biol.*, **68**, 791, 827.

Hodges, R., McGeachin, S. G. & Raphael, R. A. (1963). *J. Chem. Soc.*, 2515.

Jaffe, H. H. & Orchin, M. (1962). In *Theory and Applications of Ultraviolet Spectroscopy*. Chapter 10. John Wiley, New York and London.

Jones, R. N. & Dobriner, K. (1949). In *Vitamins & Hormones*, Vol. 7, p. 293.

Jones, R. N. & Sandorfy, C. (1956). In *Techniques of Organic Chemistry*. Vol. 9. Ed. A. Weissberger. *Chemical Applications of Spectroscopy*. Ed. W. West. Interscience Publishers, New York.

Jones, R. N. & Roberts, G. (1958). *J. Amer. Chem. Soc.*, **80**, 6121.

Katz, A. *et al.* (1953–1958). *Helv. Chim. Acta.*, **36**, 1344, 1417; **37**, 451, 833; **41**, 1399.

Katzman, P. A. & Elliott, W. H. (1963). In *Comprehensive Biochemistry*. Vol. 10, p. 47. Ed. M. Florkin and E. H. Stotz. Elsevier, Amsterdam, New York.

Kier, L. C. (1952). *J. Lab. Clin. Med.*, **40**, 762.

Klopper, W. J., Michie, E. A. & Brown, J. B. (1955). *J. Endocrinol.*, **12**, 209.

Kober, S. (1931). *Biochem. Z.*, **239**, 209.

Kubota, T., Kamikawa, T., Tokoroyama, T. & Matsuura, T. (1960). *Tetrahedron Lett.*, **8**, 1.

Levin, S. J., Irvin, J. L. & Johnston, C. G. (1961). *Anal. Chem.*, **33**, 856.

Lichti, H. & Wartburg, Avon. (1960). *Helv. Chim. Acta.*, **43**, 1666.

Liebermann, C. (1885). *Ber. dtsch Chem. Ges.*, **18**, 1803.

Matschiner, J. T., Mahowald, T. A., Elliott, W. H., Doisy, E. A. Jr., Hsie, S. L. & Doisy, E. A. (1957). *J. Biol. Chem.*, **225**, 771.

Meyer, K. *et al.* (1960). *Helv. Chim. Acta.*, **43**, 1955, 1960.

Minibeck, H. (1938). *Biochem. Z.*, **297**, 29.

Moore, P. R. & Baumann, C. A. (1952). *J. Biol. Chem.*, **195**, 613.

Morton, R. A. (1962). In *Comprehensive Biochemistry*. Vol. 3, p. 89. Ed. M. Florkin and E. H. Stotz. Elsevier, Amsterdam, New York.

Mosbach, E. H., Kalinsky, H. J., Halpern, E. & Kendall, F. E. (1954). *Arch. Biochem. Biophys.*, **51**, 402.

Murphy, J. E. (1957). *J. Amer. Pharm. Assoc. Sci. Ed.*, **46**, 170.

Nakao, T. & Aizawa, Y. (1956). *Endocrinol. Japan*, **3**, 92.

Nocke, W. (1961). *Biochem. Z.*, **78**, 593.

Oakey, R. E., Bradshaw, L. R. A., Eccles, S. S., Stitch, S. R. & Heys, R. F. (1967). *Clin. Chem. Acta.*, **15**, 35.

Ohuchuku, N. S. & Taylor, D. A. H. (1969). *J. Chem. Soc.* (C), 864.

Olson, E. C. & Alway, C. D. (1960). *Anal. Chem.*, **32**, 370.

Peterson, L. (1962 *a*). *Anal. Chem.*, **34**, 1851; (1962 *b*). *Anal. Chem.*, **34**, 1781.

Powell, J. W. (1966). *J. Chem. Soc.*, 1794.

Rao, C. N. R. (1963). In *Chemical Applications of Infrared Spectroscopy*. Academic Press, New York, London; (1967). In *Ultraviolet and Visible Spectroscopy, Chemical Applications*. 2nd Edition. Butterworths, London.

Reichstein, T. (1950 *a*). *Helv. Chim. Acta.*, **33**, 1993; *b* (see also **33**, 1013).

Reichstein, T. (1957). *Helv. Chim. Acta.*, **40**, 284.

Renz, J. *et al.* (1959). *Helv. Chim. Acta*, **42**, 1620.

Roberts, G., Gallagher, B. & Jones, R. N. (1958). In *Infrared Absorption Spectra of Steroids—an Atlas*. Interscience Publishers, New York, Vol. II.

Rosen, W. E., Ziegler, J. B., Shabica, A. C. & Shoolery, J. N. (1969). *J. Amer. Chem. Soc.*, **81**, 1687.

Rosenkrantz, H. (1955). In *Method of Biochemical Analysis*. Vol. II. Ed. D. Glick. Interscience Publishers, New York.

Scott, A. I. (1964). In *Interpretation of Ultraviolet Spectra of Natural Products*. Pergamon Press, Oxford.

Singer, E. J. & Fitschen, W. H. (1961). *Anal. Biochem.*, **2**, 292.

Sjövall, J. (1959). *Clin. Chim. Acta.*, **4**, 652.

Sjövall, J. (1964). In *Methods of Biochemical Analysis*. Vol. 12, p. 97. Ed. D. Glick. Interscience Publishers.

Smith, A. M. & Eddy, C. R. (1959). *Anal. Chem.*, **31**, 1539.

Truter, E. V. & Bush, B. (1962). In *Determination of Sterols*. Soc. Anal. Chem., London.

Udenfriend, S. (1962). In *Fluorescence Assay in Biology and Medicine*. Academic Press, New York, London.

Vitaglioni, J. C. & Comin, J. (1972). *Phytochem.*, **11**, 807.

Walens, H. A., Turner, A. & Wall, M. E. (1954). *Anal. Chem.*, **26**, 325.

Woodward, R. B. (1941). *J. Amer. Chem. Soc.*, **63**, 1123; (1942). *Ibid.*, **64**, 72, 76.

Wysocki, A. P., Portman, O. W. & Mann, G. V. (1955). *Arch. Biochem. Biophys.*, **55**, 213.

Yamaguchi, K. (1970). In *Spectral Data of Natural Products*. Elsevier, Amsterdam.

Zimmermann, W. (1935). *J. Physiol.*, **233**, 257; (1944). *Vitamins & Hormones*, **5**, 1.